T H I R D   E D I T I O N

A SECOND COURSE IN BUSINESS STATISTICS: REGRESSION ANALYSIS

THIRD EDITION

A SECOND COURSE IN BUSINESS STATISTICS: REGRESSION ANALYSIS

WILLIAM MENDENHALL
University of Florida

TERRY SINCICH
University of South Florida

DELLEN PUBLISHING COMPANY
San Francisco

COLLIER MACMILLAN PUBLISHERS
London

divisions of Macmillan, Inc.

On the cover: The cover, *Gardena*, was executed by Los Angeles artist Peter Alexander in 1988. The acrylic work measures 30″ × 33″. Alexander's work may be seen at the James Corcoran Gallery in Los Angeles and the Charles Cowles Gallery in New York. His work is also in the permanent collections of the Los Angeles County Museum, the Museum of Modern Art in New York City, and the Metropolitan Museum.

Permissions: Dellen Publishing Company
400 Pacific Avenue
San Francisco, California 94133

Orders: Dellen Publishing Company
c/o Macmillan Publishing Company
Front and Brown Streets
Riverside, New Jersey 08075

Collier Macmillan Canada, Inc.

LIBRARY OF CONGRESS CATALOGING IN PUBLICATION DATA

Mendenhall, William.
    A second course in business statistics: regression analysis
  William Mendenhall, Terry Sincich.—3rd ed.
    Includes index.
    1. Commercial statistics. 2. Statistics. 3. Regression analysis.
I. Sincich, Terry. II. Title.
HF1017.M46 1989   88-18893
519.5—dc19   CIP

Printing: 2 3 4 5 6 7 8 9   Year: 9 0 1 2

ISBN 0-02-380510-2

C O N T E N T S

This book is designed for two types of business statistics courses. The early chapters, combined with a selection of the case study chapters, are designed for use in the second half of a two-semester (or two-quarter) introductory statistics sequence for undergraduate business majors. Or, the book can be used for a course in applied regression analysis for MBA or Ph.D. students in business administration.

At first glance, these two uses for the book may seem inconsistent. How could a text be appropriate for both undergraduate and graduate students? The answer lies in the content. In contrast to a course in statistical theory, the level of mathematical knowledge required for an applied regression analysis course is minimal. Consequently, the difficulty encountered in learning the mechanics is much the same for both undergraduate and graduate students. The challenge is in the application—diagnosing practical problems, deciding on the appropriate linear model for a given situation, and knowing which inferential technique will answer a manager's practical questions. This takes *experience*, and it explains why a student can take an undergraduate course in applied regression analysis and still benefit from covering the same ground in a graduate course.

It is difficult to identify the amount of material that should be included in the second semester of a two-semester sequence in introductory business statistics. Chapter 1 should be included to identify the course goals, and a few lectures should be devoted to Chapter 2 to make certain that all students possess a common background knowledge of the basic concepts covered in a first-semester (first-quarter) course, Chapter 3 (Simple Linear Regression), Chapter 4 (Multiple Regression), Chapter 5 (Some Problems Encountered in Performing a Multiple Regression Analysis), Chapter 6 (Residual Analysis), and optional Chapter 7 (Model-Building) provide the core for an applied regression analysis course. These chapters could be supplemented by the addition of Chapter 8 (The Analysis of Variance for Designed Experiments), Chapter 9 (Time Series Modeling and Forecasting), and some of the case study chapters.

In our opinion, the quality of an applied course is not measured by the number of topics covered or the amount of material memorized by the students. The measure is how well they can apply the techniques covered in the course to the solution of real business problems. Consequently, we advocate moving on to new topics only after the students have demonstrated ability (through testing) to apply the techniques under discussion. In-class consulting sessions, where a case study is presented and the students have the opportunity to diagnose the problem and recommend an appropriate method of analysis, are very helpful in teaching applied regression analysis. This approach is particularly useful in helping students master the difficult topic of model selection and model-building (Chapters 4–7) and relating questions about the model to the real-world questions of a business manager. The case study chapters (Chapters 11–16) illustrate the type of material that might be used for this purpose.

A course in applied regression analysis for graduate students would start in the same manner as the undergraduate course, but would move more rapidly over the review material and would more than likely be supplemented by Appendix A (The Mechanics of a Multiple Regression Analysis), Appendix C (How to Perform Regression Analysis Using Four Computer Program Packages: SAS, SPSS^x, Minitab, and BMDP), Chapter 10 (Special Topics in Regression), and other chapters selected by the instructor. As in the undergraduate course, we recommend the use of case studies and in-class consulting sessions to help students develop an ability to formulate appropriate statistical models and to interpret the results of their analyses.

Although the scope and coverage remain the same, the third edition contains several substantial changes, additions, and enhancements:

1. **Chapter 4: Log model interpretations** We have added practical interpretations of the β coefficients and the standard deviation s of the log model presented in Chapter 4 (Multiple Regression), Section 4.8.
2. **Chapter 6: Partial residuals** A discussion of partial residuals and partial residual plots is now included in Chapter 6 (Residual Analysis), Section 6.2.
3. **Chapter 8: More emphasis on data collection methods for designed experiments** Several users of the first edition were disappointed to find that the chapter on Methods for Collecting Data (Chapter 8 in the first edition) was deleted from the second edition of the text. This material has been reincorporated into Chapter 8 (The Analysis of Variance for Designed Experiments) of the third edition. New examples and exercises have been added which emphasize the data collection phase of designed experiments, including completely randomized designs, randomized block designs, and factorial experiments.
4. **Chapter 8: More multiple comparisons procedures** In addition to Tukey's Studentized range procedure, Chapter 8 now includes a new section on other multiple comparisons procedures for means in an analysis of variance (Section 8.9)—the Scheffé and Bonferroni techniques.
5. **Case Study 11: Residential property sale price data updated** The data set for the case study on predicting sale prices of residential properties (Chapter 11) has been updated to reflect current economic trends.
6. **Case Study 15: Residual analysis of bidding data** The case study on modeling low bid price (Chapter 15) has been expanded to include a residual analysis of the regression model (Section 15.5).
7. **Appendix C: Computer commands for both mainframe and PC versions** The commands for conducting a regression analysis using any one of four popular statistical computer program packages (SAS, SPSS^x, Minitab, or BMDP), provided in Appendix C, have been updated to reflect the new PC versions of the software.

8. **More exercises with real data**   Many new "real-life" applied exercises and examples have been added throughout the text. Most of these are extracted from business and economic journals.

9. **Updated computer printouts**   All computer printouts that appear throughout the text reflect the most recent version of the statistical software package (SAS, SPSS[x], Minitab, or BMDP).

The main features of this text are:

1. **Readability**   We have purposely tried to make this a teaching (rather than a reference) text. Concepts are explained in a logical intuitive manner using worked examples.

2. **Emphasis on model-building**   The formulation of an appropriate statistical model is fundamental to any regression analysis. This topic is treated in Chapters 4–7 and is emphasized throughout the text.

3. **Emphasis on developing skill to use regression analysis**   In addition to teaching the basic concepts and methodology of regression analysis, this text stresses its use, as a tool, in solving business problems. Consequently, a major objective of the text is to develop a skill in applying regression analysis to appropriate real-life situations.

4. **Numerous realistic examples and exercises**   The text contains many worked examples that illustrate important aspects of model construction, data analysis, and the interpretation of results. Numerous exercises are located at the ends of key sections and at the ends of chapters.

5. **Case study chapters**   The text contains six case study chapters, each of which addresses a real-life business problem. The student can see how regression analysis was used to answer the practical questions posed by the problem, proceeding with the formulation of appropriate statistical models to the analysis and interpretation of sample data.

6. **Data sets**   The text contains four complete data sets that are associated with the case studies (Chapters 11–16). These can be used by instructors and students to practice model-building and data analyses.

7. **Computer orientation**   Instructions on how to use any of four popular statistical computer program packages, SAS, SPSS[x], Minitab, and BMDP, are provided in Appendix C. The printouts of the respective packages are presented and discussed throughout the text.

The text is also accompanied by the following supplementary material:

1. **Solutions manual**   A student's exercise solutions manual presents the solutions to selected exercises (approximately half of the exercises contained in the text).

2. **Data sets available on diskette or tape**   The four data sets in Appendices E–H are available on either an IBM PC diskette (for PC computing) or a 7-track, nonlabeled magnetic computer tape (for mainframe computing).

We want to thank the many people who contributed time, advice, and other assistance to this project. We owe particular thanks to the many reviewers who provided suggestions and advice at the onset of the project and for the succeeding editions:

Mohammed Askalani, Mankato State University (Minnesota)
Ken Boehm, Pacific Telesis (California)
James Daly, California State Polytechnic Institute at San Luis Obispo
Robert Elrod, Georgia State University
James Ford, University of Delaware
Carol Ghomi, University of Houston
James Holstein, University of Missouri at Columbia
Steve Hora, Texas Technological University
Ann Kittler, Ryerson College (Toronto)
James T. McClave, University of Florida
John Monahan, North Carolina State University
P. V. Rao, University of Florida
Tom Rothrock, Info Tech, Inc.
Ray Twery, University of North Carolina at Charlotte
Joseph Van Matre, University of Alabama at Birmingham
William Weida, United States Air Force Academy
Dean Wichern, Texas A & M University
James Willis, Louisiana State University

For the third edition, Susan Reiland deserves special recognition for her excellent line-by-line review of the manuscript and for managing the production of the text. We are particularly grateful to Herman Kelting, Jim McClave, and Ronald Ward (all of the University of Florida), Tom Rothrock (Info Tech, Inc.), Ed Crapo (Alachua County Property Appraiser), and Mike Jacob (Florida Power Corporation), who provided the data sets and background information used in the case studies (Chapters 11–16). And, finally, we give special thanks to Carol Springer for a superb job of proofreading and transforming our notes into typed copy.

OBJECTIVE

To explain what regression analysis is and to describe its applications

CONTENTS

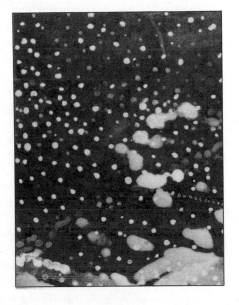

WHAT IS REGRESSION ANALYSIS?

WHAT IS STATISTICS?

Before we attempt to describe the meaning of regression analysis, we will review some of the basic ideas of statistics. Then we will see how regression analysis, which is a particular type of statistical methodology, can be useful in making business decisions.

A recent survey of 500 purchasing agents provides a simple example of an application of statistics in business. The survey was conducted to sample the opinions of industrial purchasing agents concerning their intentions for the coming year. Did they intend to increase or decrease expenditures and by what percentage? Using this information, an economist can estimate the percentage increase (or decrease) for all industrial concerns and can use this estimate in other calculations concerning the future direction of the economy. From this simple example we can identify the elements of a statistical problem and the contribution of statistics in business decision-making.

Statistics is concerned with the solution of a specific type of inferential problem: How can you infer the nature of a **population**, a large set of measurements, based on information contained in a **sample** from the population? And then, having made the inference, how can you assess its reliability?

For our example, an economist would like to know the percentage increase (or decrease) associated with every industrial purchasing agent in the United States. However, obtaining all the data (the population) would be not only costly and time-consuming, but also most likely impossible to accomplish. Consequently, the economist has resorted to selecting a **random sample** of 500 responses from the totality of responses associated with all industrial purchasing agents in the United States. Using the information in the **sample measurements**, the economist can **estimate** the **mean** percentage increase in purchases, a **measure of variation** in the percentage increases, and other numerical characteristics of the population.

How can statistical methods help the economist or any other business person who wants to make inferences based on sample data? First, statistical methods can help determine how to use the sample data to make inferences about the population from which the sample was selected. Second, the reliability of the inferences can be described. For example, if you estimate the mean percentage increase in purchasing budgets to be 3.9%, statistical methodology might tell you that with a high probability, your sample should produce an estimate that is within 0.7% of the true population percent increase for *all* purchasing agents. Third, statistical methodology can tell you how to design your sample survey (how to select your sample from the population). The reliability of your inferences will usually depend on the number of observations in the sample and on how the data were collected. Improper sampling procedures can result in worthless inferences and a loss in time and money. Consequently, knowing how to design the sampling procedure is all-important.

Most applications of statistics in business and economics are much more complicated than the example of the survey of purchasing agents. Many use sample data to investigate the relationships among a group of variables, ultimately to create a model of some response (profit, productivity, price, and so forth) that can be used to predict its value in the future. The applications of these techniques in business and economics will be the subject of this text.

| | | | | | | | | | | | |
S E C T I O N 1.2

MODELING A RESPONSE

The objective of many statistical investigations is to estimate or make a decision about a population mean. Or, if the population represents a collection of measurements on a variable (often called a **response variable**) y, we might want to predict the value of y at some future time when a single observation is to be selected from the population.

For example, suppose a tax assessor wants to adjust the assessed values on the residential properties in a medium-sized city. The assessor could select a random sample of residential sales during the past year, note the percentage increase y in each over the current assessed value, and then use these percentage increases to estimate the mean percentage increase $E(y)$ of all residential properties in the city.* The assessor could then adjust the assessed value of each residential property in the city by this percentage.

Adjusting the assessed value of every residential property in the city by the mean percentage increase in property value is tantamount to using the mean percentage increase as a **model** for the true percentage increase in the value of each residential property. We know that the actual percentage increase for a particular property will depend on location, square footage of heated space, size of lot, and many other factors. Consequently, the real percentage increases in value for all residential properties may have the distribution shown in Figure 1.1. Thus, the assessor is modeling the percentage increase in value y for a particular property by stating that y is equal to the mean increase $E(y)$ plus or minus some random amount, which is unknown to the assessor; that is,

$$y = E(y) + \text{Random error}$$

Since the assessor does not know the value of the random error for a particular property, a reasonable strategy would be to increase the assessed value by the estimate of the mean amount $E(y)$.

FIGURE 1.1
Distribution of Percentage Increase in Residential Property Values

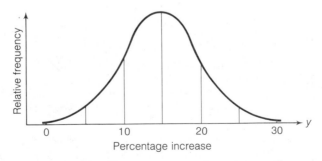

*The symbol $E(y)$ will be used to denote the mean or **expected value** of a response variable y.

Suppose the assessor estimates the mean increase in residential property values to be 10% and, for the sake of argument, let us assume this is a very accurate estimate of $E(y)$. Further, suppose you own a house in the city and you are advised by an experienced appraiser that an increase of 10% in the assessed value of your property (and subsequent property taxes) is too high. How can you justify the disagreement between the appraiser and the tax assessor? Both possess models that predict property values, but they are led to different conclusions.

If you place yourself in the position of the appraiser, you will be able to answer this question. The appraiser has actually seen the house and is therefore using a more sophisticated model than is the assessor. In particular, the appraiser is taking into consideration a number of variables that are highly related to the value of a residential property (such as location, square footage, size of lot, and so forth), and is using these variables to obtain a more accurate model for the percentage increase in residential property values. This model might be based purely on experience and exist only in the appraiser's mind. Or, the appraiser might have obtained a mathematical model (an equation) relating percentage price increase to square footage, location, and so forth, by sampling individual residential sales and measuring the percentage increase, square footage, location, and so forth, of each. The process of finding the mathematical model that best fits the data is part of the process known as a **regression analysis**.

For example, suppose the appraiser decided to relate percentage price increase y to a single variable x, defined as the square footage of heated space in the residence. The appraiser might select a random sample of residential sales, record y and x for each sale, and then plot them on a graph as shown in Figure 1.2. Finding the equation of the smooth curve that best fits the data points is part of a regression analysis. Once obtained, this equation (a graph of which is superimposed on the data points in Figure 1.2) provides a model for estimating the mean percentage increase for houses with any specific amount of heated floor space. The appraiser can use the model to predict the percentage price increase for an individual residential property that has a known amount of floor space. As you can see from Figure 1.2, the model would also predict with some error (most of the points do not lie exactly on the curve), but the error of prediction will be much less than the error obtained using the model represented in Figure 1.1. As shown in Figure 1.1, a good estimate of the percentage increase for a

FIGURE 1.2

Relating Percentage Price Increase in Residential Property to Heated Floor Space

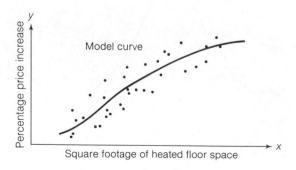

residential property would be a value near the center of the distribution, say, the mean. Since this prediction does not take square footage of floor space into account, the error of prediction will be larger than the error of prediction for the model of Figure 1.2. Consequently, we would state that the model utilizing the information provided by the square footage variable is superior to the model represented in Figure 1.1.

| | | | | | | | | | | | | |

SECTION 1.3

WHAT IS REGRESSION ANALYSIS?

Regression analysis is a branch of statistical methodology concerned with relating a response y to a set of independent, or predictor, variables $x_1, x_2, \ldots, x_k$. The goal is to build a good model—a prediction equation relating y to the independent variables—that will enable us to predict y for given values of $x_1, x_2, \ldots, x_k$, and to do so with a small error of prediction. When using the model to predict y for a particular set of values of $x_1, x_2, \ldots, x_k$, we will want a measure of the reliability of our prediction. That is, we will want to know how large the error of prediction might be. All these elements are parts of a regression analysis, and the resulting prediction equation is often called a **regression model**.

For example, a property appraiser might like to relate percentage price increase y to the two independent variables x_1, square footage of heated space, and x_2, lot size. This model could be represented by a **response surface** (see Figure 1.3) that traces the mean percentage price increase $E(y)$ for various combinations of x_1 and x_2. To predict the percentage price increase y for a given residential property with $x_1 = 2,000$ square feet of heated space and lot size $x_2 = 0.7$ acre, you would locate the point $x_1 = 2,000$, $x_2 = 0.7$ on the x_1, x_2-plane (see Figure 1.3). The height of the surface above that point gives the mean percentage increase in price $E(y)$, and this is a reasonable value to use to predict the percentage price increase for a property with $x_1 = 2,000$ and $x_2 = 0.7$.

FIGURE 1.3

Mean Percentage Price Increase as a Function of Heated Square Footage, x_1, and Lot Size, x_2

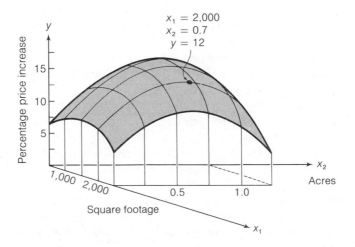

The response surface is a convenient method for modeling a response y that is a function of two **quantitative**, or **numerical**, independent variables, x_1 and x_2. The mathematical equivalent of the response surface shown in Figure 1.3 might be given by the expression

$$E(y) = \beta_0 + \beta_1 x_1 + \beta_2 x_2 + \beta_3 x_1 x_2 + \beta_4 x_1^2 + \beta_5 x_2^2$$

where $E(y)$ is the mean percentage price increase for a set of values x_1 and x_2, and $\beta_0, \beta_1, \ldots, \beta_5$ are constants with values that would have to be estimated from the sample data. Note that finding a response surface that approximates the mean value of y is analogous to a geographer's contour mapping problem of modeling the surface of the earth. For the geographer, x_1 and x_2 identify a particular point on the earth's surface and y gives the elevation of ground level at that point.

In practice, the appraiser would construct a model that takes into account other quantitative variables, as well as **qualitative** (nonnumerical) independent variables such as location and type of construction. In the following chapters we will show how to construct a model relating a response to both quantitative and qualitative independent variables, and we will fit the model to a set of sample data using a regression analysis.

The preceding description of regression analysis is oversimplified, but it provides a preliminary view of the methodology that is the subject of this text. In addition to predicting y for specific of values $x_1, x_2, \ldots, x_k$, a regression model can also be used to estimate the mean value of y for given values of $x_1, x_2, \ldots, x_k$ and to answer other questions concerning the relationship between y and one or more of the independent variables. The practical values attached to these inferences will be illustrated by examples in the following chapters.

SECTION 1.4

THE VALUE OF REGRESSION ANALYSES

Regression analysis of data is a very powerful statistical tool. It provides a technique for building a statistical predictor of a response and enables you to place a bound (an approximate upper limit) on your error of prediction. For example, suppose you manage a construction company and you would like to predict the profit y per construction job as a function of a set of independent variables $x_1, x_2, \ldots, x_k$. If you could find the right combination of independent variables and could postulate a reasonable mathematical equation to relate y to these variables, you could possibly deduce which of the independent variables were causally related to profit per job and then control these variables in order to achieve a higher company profit. In addition, you could use the forecasts in corporate planning. The following examples illustrate a few of the many useful applications of regression analysis in business and economics.

EXAMPLE 1.1

Real estate: As suggested in Section 1.3, regression models for predicting property values are used by tax assessors, appraisers, and real estate investment managers to forecast (or predict) the value of a parcel of real estate.   ∎

EXAMPLE 1.2

Management: In order to reward their executives appropriately, many large corporations receive advice from consulting firms as to the amount of compensation that each executive should receive. To provide this advice, the consulting firm collects information on the compensation y received by a large number of corporate executives. For each of these executives, the firm records the values of many independent variables, some of which are the following:

1. Experience (in years)
2. Education (in years)
3. Number of employees supervised
4. Corporate assets (dollars)
5. Age of the executive
6. Whether the executive is on the company's board of directors
7. Whether the executive has international responsibility
8. Company profits

The consulting firm then uses a regression analysis to build a good prediction equation for y, an executive's annual compensation, as a function of the independent variables listed above. If it is successful, the company can sell its services to both participating and nonparticipating corporations, providing them with reasonable compensation projections for their executives.   ■

EXAMPLE 1.3

Finance: Some commodity traders have successfully used regression analysis to forecast the price y of a commodity. For example, we might expect the supply of wheat, and hence its price y, to be directly related to (1) the amount of wheat planted in the United States, the Soviet Union, Europe, and so forth; (2) the weather conditions in various locations; and (3) the projected sizes of the crops of competing grains. Using the recorded prices of wheat at a specific time over a period of a year and corresponding information on the independent variables listed above, the commodity trader could use a regression analysis to build a model relating wheat price y to the independent variables. The model could then be used to predict the price of wheat at the time of year used in the data collection, based on the present values of the independent variables. If the general economic conditions remained stable and representative of the conditions prevalent over the period during which the sample data were observed, the regression analysis would provide a measure of the prediction error that would apply to predictions of future wheat prices.   ■

S E C T I O N 1.5

SUMMARY

Better business management requires a better understanding of the phenomena that affect the variables that measure a business's health—namely, profit, sales, inventory, various measures of product demand, and others. To achieve this understanding, we seek the assistance of mathematical models that relate the mean value of a response (such as profit) to various variables such as the advertising budget, size of inventory, and so forth. Since we know that even a perfect mathematical description of this relationship will still predict the response with error, we include a random component in the model to account for the many other variables that have been purposely or inadvertently excluded.

The mathematical relationship that we have described forms a model for the relative frequency distribution of the population of response measurements that would be generated when the process is in a specific state. For example, a model might represent a relative frequency distribution of the population of monthly profits that a business might generate, now and in the immediate future, when the business is operating with a $1,000,000 inventory and a $500,000 annual advertising budget. Estimating the unknown parameters for this population, i.e., the unknown parameters in the model, and using the model to make predictions with known reliability, is the objective of a **regression analysis**.

As a prerequisite for studying this topic, we assume that you have had an introductory course in statistics. However, since these courses vary somewhat in content, we will present an optional review of some of the important concepts and methods in Chapter 2. Other important basic concepts will be reviewed in the text as the need arises.

C H A P T E R 2

OBJECTIVE

To review the basic concepts of statistics that are an essential prerequisite to a study of regression analysis

CONTENTS

A REVIEW OF BASIC CONCEPTS (OPTIONAL)

SECTION 2.1

INTRODUCTION

Although this text assumes a prerequisite introductory course in statistics, courses vary somewhat in content and the manner in which they present statistical concepts. To be certain that we are starting with a common background, we will use this chapter to review some basic definitions and concepts. Coverage is optional.

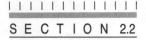

SECTION 2.2

POPULATIONS, SAMPLES, AND RANDOM SAMPLING

Business phenomena are most often characterized by sets of numerical measurements which exist in fact or are part of an ongoing operation and hence are conceptual. These sets of data are usually large, and they are called **populations**. Some examples of business phenomena and the corresponding populations and types are shown in Table 2.1.

TABLE 2.1
Some Typical Populations

PHENOMENON	POPULATION	TYPE
Price of new residential construction this year	Set of prices of all new residential properties sold this year	Existing
Profit per job in a construction company	Set of profits for all jobs performed recently or to be performed in the near future	Part existing, part conceptual
Number of daily customers at a bank	Set of numbers of daily customers over the recent past and future	Part existing, part conceptual

---

DEFINITION 2.1

A **population** is a set of measurements that describes some phenomenon.

Many populations are too large to measure (because of time and cost); others cannot be measured because they are conceptual. Thus, we are often required to select a subset of values from a population and to make inferences about the population based on information contained in a **sample**. This is the major objective of modern statistics.

DEFINITION 2.2

A **sample** is a subset of measurements selected from a population.

Probability theory is employed in inferring the nature of a population from information contained in a sample. We observe the sample data and then consider the likelihood of observing these particular measurements for populations possessing various characteristics. Generally speaking, we infer that the sample was selected from the population most likely to have produced the observed sample. For example, if you toss a coin ten times and observe ten heads, you should infer either that the coin that generated the sample was biased in favor of heads or that something had gone wrong with your sampling (coin-tossing) procedure. Since the probability of observing a particular sample depends on how the sample was selected, the sampling procedure plays an important role in statistical inference.

The most common type of sampling procedure is one that gives every different sample of fixed size in the population an equal probability of selection. This is called a **random sample**.

DEFINITION 2.3

If n elements are selected from a population in such a way that every sample combination of n elements in the population has an equal probability of being selected, the n elements are said to be a **random sample**.

SECTION 2.3

DESCRIBING SETS OF DATA

The word *inference* implies description. For example, to infer the nature of a company, we would describe its product, annual sales volume, number of employees, annual profit, and so forth. Similarly, to infer the nature of a set of measurements, such as a population or a sample, we need to be able to describe the set.

A useful graphical method for describing data is provided by a **relative frequency distribution**. This type of graph shows the proportions of the total set of measurements that fall in various intervals on the scale of measurement. For example, Figure 2.1 shows the prices of new residential properties sold in a particular region in 1987. The area over a particular interval under a relative frequency distribution curve is proportional to the fraction of the total number of measurements that fall in that interval. In Figure 2.1, the fraction of the total number of residential house prices that falls between $70,000 and $80,000 is proportional to the shaded area. **If we take the total area under the distribution curve as equal to 1, then the shaded area is equal to the fraction of house prices that fall between $70,000 and $80,000.**

FIGURE 2.1
Relative Frequency Distribution: Prices of New Residential Properties

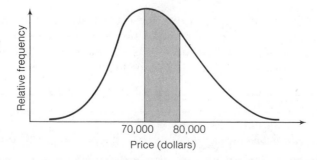

The variable measured in generating a population, denoted by the symbol y, is called a **random variable**. Observing a single value of y is equivalent to selecting a single measurement from the population. The probability that it will assume a value in an interval, say a to b, is given by its relative frequency or **probability distribution**. The total area under a probability distribution curve is always assumed to equal 1. Hence, the probability that a measurement on y will fall in the interval between a and b is equal to the shaded area shown in Figure 2.2.

FIGURE 2.2

Probability Distribution for a Random Variable

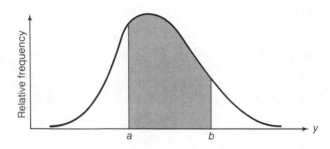

Numerical descriptive measures provide a second method for describing a set of measurements. These measures, which locate the center of the data set and its spread, actually enable you to construct an approximate mental image of the relative frequency distribution of the data set.

One of the most common measures of central tendency is the **mean**, or arithmetic average, of a data set. Thus, if we denote the sample measurements on a random variable y by the symbols $y_1, y_2, y_3, \ldots$, the sample mean is defined as follows:

DEFINITION 2.4

The **mean** of a sample of n measurements $y_1, y_2, \ldots , y_n$ is

$$\bar{y} = \frac{\sum\limits_{i=1}^{n} y_i}{n}$$

The mean of a population, or equivalently, the expected value of y, $E(y)$, is usually unknown in a practical situation (we will want to infer its value based on the sample data). Most texts use the symbol μ to denote the mean of a population. Thus, we will use the following notation:

NOTATION

Sample mean: $\bar{y}$

Population mean: $E(y) = \mu$

The spread or variation of a data set is measured by its **range**, its **variance**, or its **standard deviation**.

DEFINITION 2.5

The **range** of a sample of n measurements $y_1, y_2, \ldots, y_n$ is the difference between the largest and smallest measurements in the sample.

EXAMPLE 2.1

If a sample consists of measurements 3, 1, 0, 4, 7, find the sample mean and the sample range.

SOLUTION

The sample mean and range are

$$\bar{y} = \frac{\sum_{i=1}^{n} y_i}{n} = \frac{15}{5} = 3$$

Range $= 7 - 0 = 7$ ∎

The variance of a set of measurements is defined to be the average of the *squares of the deviations* of the measurements about their mean. Thus, the population variance, which is usually unknown in a practical situation, would be the mean or expected value of $(y - \mu)^2$, or $E[(y - \mu)^2]$. We use the symbol σ^2 to represent the variance of a population:

$$E[(y - \mu)^2] = \sigma^2$$

The quantity usually termed the **sample variance** is defined as follows:

DEFINITION 2.6

The **variance** of a sample of n measurements $y_1, y_2, \ldots, y_n$ is defined to be

$$s^2 = \frac{\sum_{i=1}^{n} (y_i - \bar{y})^2}{n - 1} = \frac{\sum_{i=1}^{n} y_i^2 - \left(\sum_{i=1}^{n} y_i\right)^2 / n}{n - 1}$$

Note that the sum of squares of deviations in the sample variance is divided by $(n - 1)$, rather than n. Division by n produces estimates that tend to underestimate σ^2. Division by $(n - 1)$ corrects this problem.

EXAMPLE 2.2

Calculate the sample variance for the sample 3, 1, 0, 4, 7.

SOLUTION

We first calculate

$$\sum_{i=1}^{n} (y_i - \bar{y})^2 = \sum_{i=1}^{n} y_i^2 - \frac{\left(\sum_{i=1}^{n} y_i\right)^2}{n} = 75 - \frac{(15)^2}{5} = 30$$

Then

$$s^2 = \frac{\sum_{i=1}^{n}(y_i - \bar{y})^2}{n - 1} = \frac{30}{4} = 7.5$$ ∎

The concept of a variance is important in theoretical statistics, but its square root, called a **standard deviation**, is the quantity most often used to describe data variation.

DEFINITION 2.7

The **standard deviation** of a set of measurements is equal to the square root of their variance. Thus, the standard deviations of a sample and a population are

Sample standard deviation: s

Population standard deviation: σ

The standard deviation of a set of data takes on meaning in light of a theorem (Tchebysheff's theorem) and a rule of thumb.* Basically, they give us the following guidelines:

GUIDELINES FOR INTERPRETING A STANDARD DEVIATION

1. For *any* data set (population or sample), at least $\frac{3}{4}$ of the measurements will lie within 2 standard deviations of their mean.
2. For *most* data sets of moderate size (say 25 or more measurements), approximately 95% of the measurements will lie within 2 standard deviations of their mean.

EXAMPLE 2.3

You are informed that for a large data set, the mean and standard deviation are

$$\bar{y} = 20 \qquad s = 3$$

Construct a mental image of the relative frequency distribution for the data.

SOLUTION

Although we do not know the exact shape of the distribution, we can imagine a distribution that would possibly be similar to that shown in Figure 2.3. The distribution would center about $\bar{y} = 20$, and most of the distribution would fall within $2s = 2(3) = 6$ of the mean.

*For a more complete discussion and a statement of Tchebysheff's theorem, see the references listed at the end of this chapter.

FIGURE 2.3

A Graphical Visualization of a Distribution with $\bar{y} = 20$, $s = 3$

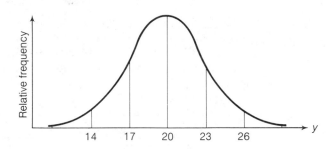

■

Numerical descriptive measures calculated from sample data are called **statistics**. Numerical descriptive measures of the population are called **parameters**. In a practical situation, we will not know the population relative frequency distribution (or equivalently, the population distribution for y). We will usually assume that it has unknown numerical descriptive measures, such as its mean μ and standard deviation σ, and by inferring (using **sample statistics**) the values of these parameters, we infer the nature of the population relative frequency distribution. Sometimes we will assume that we know the shape of the population relative frequency distribution and use this information to help us make our inferences. When we do this, we are postulating a model for the population relative frequency distribution, and we must keep in mind that the validity of the inference may depend on how well our model fits reality.

DEFINITION 2.8

Numerical descriptive measures of a population are called **parameters**.

DEFINITION 2.9

A **sample statistic** is a quantity calculated from the observations in a sample.

EXERCISES 2.1–2.8

2.1 Compute $\bar{y}$, s^2, and s for each of the following data sets:
a. 1, 5, 0, 2, 5, 7, 1   **b.** 1, 2, 0, 0, 5, 4
c. 10, 8, 12, 2   **d.** 3, 4, 10, 2

2.2 Compute $\bar{y}$, s^2, and s for each of the following data sets:
a. 1, 1, 20, 20, 8   **b.** 2, 100, 104, 2
c. −1, −3, −2, 0, −3, −3   **d.** $\frac{1}{5}, \frac{1}{5}, \frac{1}{5}, \frac{2}{5}, 0.2, \frac{4}{5}$

2.3 Travelers who have no intention of showing up often fail to cancel their hotel reservations in a timely manner. These travelers are known, in the parlance of the hospitality trade, as "no-shows." To protect against no-shows and late cancellations, hotels invari-ably overbook rooms. A recent study examined the problems of overbooking rooms in the hotel industry. The following data, extracted from the study, represent the daily numbers of late cancellations and/or no-shows for a random sample of 30 days at a 500-room hotel.

LATE CANCELLATIONS AND/OR NO-SHOWS									
18	16	16	16	14	18	16	18	14	19
15	19	9	20	10	10	12	14	18	12
14	14	17	12	18	13	15	13	15	19

Source: Toh, R. S. "An inventory depletion overbooking model for the hotel industry," *Journal of Travel Research*, Vol. 23, No. 4, Spring 1985 p. 27.

a. Compute $\bar{y}$, s^2, and s for the data set.
b. What percentage of the measurements would you expect to find in the interval $\bar{y} \pm 2s$?
c. Count the number of measurements that actually fall within the interval of part **b**, and express the interval count as a percentage of the total number of measurements. Compare this result with the answer to part **b**.
d. Based on these results, how many rooms should be overbooked each day?

2.4 Each week, the syndicated "Money Funds Report" appears in daily newspapers across the country. The report lists average maturity and 7-day yields for over 200 taxable money funds that are available to individual investors. Shown here are the average maturity (in days) for the week ending May 7, 1987, for each of 50 money funds with assets of $100 million or more.

57	60	77	24	51	61	79	35	26	6
37	12	13	20	41	60	63	54	67	47
65	67	38	30	23	52	73	43	43	51
58	2	12	28	35	63	44	65	1	24
29	8	19	52	48	76	31	24	30	54

a. Compute $\bar{y}$, s^2, and s for the data set.
b. What percentage of the measurements would you expect to find in the interval $\bar{y} \pm 2s$?
c. Count the number of measurements that actually fall within the interval of part **b**, and express the interval count as a percentage of the total number of measurements. Compare this result with the answer to part **b**.

2.5 Many businesses now consider personal computers (PCs) crucial to their corporate strategy. Travelers Insurance Company, for example, has set a goal of a PC on every employee's desk by 1990 (*Wall Street Journal*, Sept. 16, 1985). The accompanying table lists the companies that had installed the most PCs for the workplace in 1985.

COMPANY	INSTALLED PCs	COMPANY	INSTALLED PCs
General Motors	31,000	Intel	4,000
General Electric	18,000	Security Pacific	3,000
Westinghouse	12,000	Lockheed	3,000
Citicorp	10,000	Allied Products	2,500
Du Pont	10,000	Chevron	2,400
Ford Motor Co.	9,000	Union Carbide	2,200
Pacific Bell	7,000	Mobil	2,000
Chase Manhattan Bank	6,700	PG&E	2,000
Exxon	6,000	Chemical Bank	1,700
Peat, Marwick, Mitchell	5,600	Time	1,700
United Technologies	5,000	J. P. Morgan & Co.	1,600
McDonnell Douglas	5,000	American Can	1,500
Hughes Aircraft	5,000	Wells Fargo	1,400
Aetna Life & Casualty	5,000	Upjohn	1,400
TRW	5,000	Northwestern Mutual	1,400
Boeing	5,000	LTV Steel	1,200
Sperry	4,700	Pillsbury	1,000
General Dynamics	4,700	General Mills	1,000
Travelers	4,500	Nabisco	1,000
Merrill Lynch	4,400	Federal Express	1,000
Dun & Bradstreet	4,300	R. J. Reynolds	1,000
Touche Ross & Co.	4,100	Bechtel	1,000
3M	4,000	Teledyne	1,000

Source: Infoworld, 1985.

a. Compute $\bar{y}$, s^2, and s for this data set.

b. What percentage of the measurements would you expect to find in the interval $\bar{y} \pm 2s$?

c. Count the number of measurements that actually fall within the interval of part **b**, and express the interval count as a percentage of the total number of measurements. Compare this result with the answer to part **b**.

d. Suppose the data point for General Motors was omitted from the analysis. Would you expect $\bar{y}$ to increase or decrease? s?

e. Calculate $\bar{y}$ and s for the data set with General Motors excluded. Compare these results with your answer to part **d**.

f. Answer parts **b** and **c** using the recalculated $\bar{y}$ and s. This problem illustrates the dramatic effect a single observation can have on the analysis.

2.6 The Federal Trade Commission (FTC) annually ranks varieties of domestic cigarettes according to their tar, nicotine, and carbon monoxide contents. The U.S. surgeon general considers each of these three substances hazardous to a smoker's health. Each year, the FTC tests and ranks approximately 200 brands of cigarettes. The table on page 18 contains the tar, nicotine, and carbon monoxide contents (in milligrams) for a sample of 25 (filter) brands in a recent year.

BRAND	TAR	NICOTINE	CARBON MONOXIDE
Alpine	14.1	.86	13.6
Benson & Hedges	16.0	1.06	16.6
Bull Durham	29.8	2.03	23.5
Camel Lights	8.0	.67	10.2
Carlton	4.1	.40	5.4
Chesterfield	15.0	1.04	15.0
Golden Lights	8.8	.76	9.0
Kent	12.4	.95	12.3
Kool	16.6	1.12	16.3
L&M	14.9	1.02	15.4
Lark Lights	13.7	1.01	13.0
Marlboro	15.1	.90	14.4
Merit	7.8	.57	10.0
Multifilter	11.4	.78	10.2
Newport Lights	9.0	.74	9.5
Now	1.0	.13	1.5
Old Gold	17.0	1.26	18.5
Pall Mall Lights	12.8	1.08	12.6
Raleigh	15.8	.96	17.5
Salem Ultra	4.5	.42	4.9
Tareyton	14.5	1.01	15.9
True	7.3	.61	8.5
Viceroy Rich Lights	8.6	.69	10.6
Virginia Slims	15.2	1.02	13.9
Winston Lights	12.0	.82	14.9

Source: Federal Trade Commission.

a. Calculate $\bar{y}$, s^2, and s for the sample of 25 tar contents.

b. What percentage of the measurements would you expect to find in the interval $\bar{y} \pm 2s$?

c. Count the number of measurements that actually fall within the interval of part b, and express the interval count as a percentage of the total number of measurements. Compare this result with the answer to part b.

d. Repeat parts a–c for the sample of nicotine contents.

e. Repeat parts a–c for the sample of carbon monoxide contents.

2.7 Given a data set whose largest value is 900 and smallest value is 50, what would you estimate the standard deviation to be? Explain the logic behind the procedure you used to estimate the standard deviation.

2.8 It is well known that worker absenteeism is costly and leads to decreased production efficiency at most firms. A study was recently conducted to investigate employee absenteeism at a medium-sized assembly and packaging plant in the United States. Workers in each of three departments—packaging, assembly, and maintenance—were monitored over a 2-year period and the number of days of unanticipated absences due to sickness was recorded each week. The accompanying table gives the mean and standard deviation of number of days absent per 100 employees for each department.

	PACKAGING DEPARTMENT	ASSEMBLY DEPARTMENT	MAINTENANCE DEPARTMENT
Mean number of days per 1,000 employees	39.24	17.38	14.56
Standard deviation	9.88	5.16	6.88

Source: Moch, M. K. and Fitzgibbons, D. E. "The relationship between absenteeism and production efficiency: An empirical assessment," *Journal of Occupational Psychology*, Vol. 58, 1985, pp. 39–47.

a. Use the information in the table to sketch your mental images of the three relative frequency distributions. Construct them on the same graph so that you can see how they appear relative to each other.
b. Estimate the proportion of weeks in which between 19.48 and 59.00 days of unanticipated absences per 100 workers occur in the packaging department.
c. In a typical week, how many days of unanticipated absences per 100 workers would you expect to occur in the assembly department?
d. Repeat part **c** for the maintenance department.

| | | | | | | | | | | | |

SECTION 2.4

THE NORMAL PROBABILITY DISTRIBUTION

One of the most commonly used models for a population relative frequency distribution is the **normal probability distribution** shown in Figure 2.1. The normal distribution is symmetric about its mean μ, and its spread is determined by the value of its standard deviation σ. Three normal curves with different means and standard deviations are shown in Figure 2.5.

FIGURE 2.4
A Normal Probability Distribution

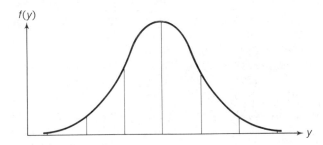

FIGURE 2.5
Several Normal Distributions with Different Means and Standard Deviations

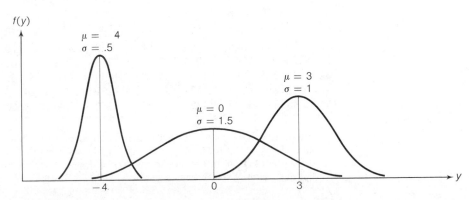

Computing the area over an interval under the normal probability distribution can be a difficult task.* Consequently, we will use the computed areas listed in Table 1 of Appendix D. A partial reproduction of this table is shown in Table 2.2. As you can see from the normal curve above the table, the entries give areas under the normal curve between the mean of the distribution and a standardized distance,

$$z = \frac{y - \mu}{\sigma}$$

to the right of the mean. Note that z is the number of standard deviations σ between μ and y. The distribution of z, which has mean $\mu = 0$ and standard deviation $\sigma = 1$, is called a **standard normal distribution**.

TABLE 2.2 Reproduction of Part of Table 1 of Appendix D

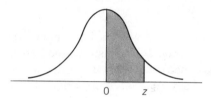

z	.00	.01	.02	.03	.04	.05	.06	.07	.08	.09
.0	.0000	.0040	.0080	.0120	.0160	.0199	.0239	.0279	.0319	.0359
.1	.0398	.0438	.0478	.0517	.0557	.0596	.0636	.0675	.0714	.0753
.2	.0793	.0832	.0871	.0910	.0948	.0987	.1026	.1064	.1103	.1141
.3	.1179	.1217	.1255	.1293	.1331	.1368	.1406	.1443	.1480	.1517
.4	.1554	.1591	.1628	.1664	.1700	.1736	.1772	.1808	.1844	.1879
.5	.1915	.1950	.1985	.2019	.2054	.2088	.2123	.2157	.2190	.2224
.6	.2257	.2291	.2324	.2357	.2389	.2422	.2454	.2486	.2517	.2549
.7	.2580	.2611	.2642	.2673	.2704	.2734	.2764	.2794	.2823	.2852
.8	.2881	.2910	.2939	.2967	.2995	.3023	.3051	.3078	.3106	.3133
.9	.3159	.3186	.3212	.3238	.3264	.3289	.3315	.3340	.3365	.3389
1.0	.3413	.3438	.3461	.3485	.3508	.3531	.3554	.3577	.3599	.3621
1.1	.3643	.3665	.3686	.3708	.3729	.3749	.3770	.3790	.3810	.3830
1.2	.3849	.3869	.3888	.3907	.3925	.3944	.3962	.3980	.3997	.4015
1.3	.4032	.4049	.4066	.4082	.4099	.4115	.4131	.4147	.4162	.4177
1.4	.4192	.4207	.4222	.4236	.4251	.4265	.4279	.4292	.4306	.4319
1.5	.4332	.4345	.4357	.4370	.4382	.4394	.4406	.4418	.4429	.4441

*Students with knowledge of calculus should note that the probability that y assumes a value in the interval $a < y < b$ is $P(a < y < b) = \int_a^b f(y)\, dy$, assuming the integral exists. The value of this definite integral can be obtained to any desired degree of accuracy by approximation procedures. For this reason, it is tabulated for the user.

EXAMPLE 2.4

Suppose y is a normal random variable with $\mu = 50$ and $\sigma = 15$. Find $P(30 < y < 70)$, the probability that y will fall within the interval $30 < y < 70$.

SOLUTION

Refer to Figure 2.6. Note that $y = 30$ and $y = 70$ lie the same distance from the mean $\mu = 50$, with $y = 30$ below the mean and $y = 70$ above it. Then, because the normal curve is symmetric about the mean, the probability A_1 that y falls between $y = 30$ and $\mu = 50$ is equal to the probability A_2 that y falls between $\mu = 50$ and $y = 70$. The z-score corresponding to $y = 70$ is

$$z = \frac{y - \mu}{\sigma} - \frac{70 - 50}{15} = 1.33$$

Therefore, the area between the mean $\mu = 50$ and the point $y = 70$ is given in Table 1 (and Table 2.2) at the intersection of the row corresponding to $z = 1.3$ and the column corresponding to .03. This area (probability) is $A_2 = .4082$. Since $A_1 = A_2$, A_1 also equals .4082, and it follows that the probability that y falls in the interval $30 < y < 70$ is $P(30 < y < 70) = 2(.4082) = .8164$. The z-scores corresponding to $y = 30$ ($z = -1.33$) and $y = 70$ ($z = 1.33$) are shown in Figure 2.7.

FIGURE 2.6
Normal Probability
Distribution:
$\mu = 50$, $\sigma = 15$

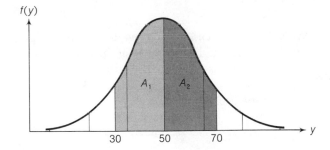

FIGURE 2.7
A Distribution of z-Scores
(a Standard Normal
Distribution)

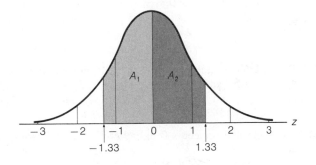

EXAMPLE 2.5

Use Table 1 of Appendix D to determine the area to the right of the z-score 1.64 for the standard normal distribution. That is, find $P(z \geq 1.64)$.

SOLUTION

The probability that a normal random variable will fall more than 1.64 standard deviations to the right of its mean is indicated in Figure 2.8. Because the normal distribution is symmetric, half of the total probability (.5) lies to the right of the mean and half to the left. Therefore, the desired probability is

$$P(z \geq 1.64) = .5 - A$$

where A is the area between $\mu = 0$ and $z = 1.64$, as shown in the figure. Referring to Table 1, we find that the area A corresponding to $z = 1.64$ is .4495. So

$$P(z \geq 1.64) = .5 - A = .5 - .4495 = .0505$$

FIGURE 2.8
Standard Normal
Distribution:
$\mu = 0, \sigma = 1$

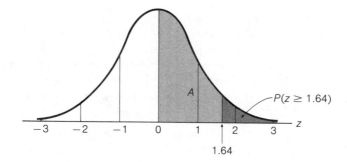

We will not be making extensive use of the table of areas under the normal curve, but you should know some of the common tabulated areas. In particular, you should note that the area between $z = -2.0$ and $z = 2.0$, which gives the probability that y falls in the interval $\mu - 2\sigma < y < \mu + 2\sigma$, is .9544 and agrees with guideline 2 of Section 2.3.

EXERCISES 2.9–2.20

2.9 Use Table 1 of Appendix D to calculate the area under the standard normal probability distribution between the following pairs of z-values:
    a. $z = 0$ and $z = 2$ **b.** $z = 0$ and $z = 1.5$
    c. $z = 0$ and $z = 3$ **d.** $z = 0$ and $z = .5$

2.10 Repeat Exercise 2.9 for each of the following pairs of z-values:
    a. $z = -1.5$ and $z = 1$ **b.** $z = -.5$ and $z = .75$
    c. $z = -2$ and $z = -1$ **d.** $z = -3$ and $z = 1.5$
    e. $z = .5$ and $z = 3.5$ **f.** $z = -.5$ and $z = 2$

2.11 Use Table 1 of Appendix D to find each of the following:
    a. $P(-1 \leq z \leq 1)$ **b.** $P(-1.96 \leq z \leq 1.96)$
    c. $P(-1.645 \leq z \leq 1.645)$ **d.** $P(-3 \leq z \leq 3)$

2.12 Find a value of z, call it z_0, such that
    a. $P(z \geq z_0) = .05$ **b.** $P(z \geq z_0) = .025$
    c. $P(z \leq z_0) = .025$ **d.** $P(z \geq z_0) = .0228$

2.13 Find a value of z, call it z_0, such that
  a. $P(z \leq z_0) = .0013$ **b.** $P(-z_0 \leq z \leq z_0) = .95$
  c. $P(-z_0 \leq z \leq z_0) = .90$ **d.** $P(-z_0 \leq z \leq z_0) = .6826$
  e. $P(-z_0 \leq z \leq 0) = .0596$

2.14 Given that the random variable y has a normal probability distribution with mean 100 and variance 64, draw a sketch (i.e, graph) of the frequency function of y. Locate μ and the interval $\mu \pm 2\sigma$ on the graph. Find the following probabilities:
  a. $P(\mu - 2\sigma \leq y \leq \mu + 2\sigma)$ **b.** $P(y \geq 108)$
  c. $P(y \leq 92)$ **d.** $P(92 \leq y \leq 116)$
  e. $P(92 \leq y \leq 96)$ **f.** $P(76 \leq y \leq 124)$

2.15 Use Table 1 of Appendix D to calculate the area under the standard normal distribution between the following pairs of z-values:
  a. -1.96 and 1.96 **b.** -1.645 and 1.645 **c.** -3 and 3
  d. -3 and 2 **e.** 1 and 3 **f.** -1.5 and 3

2.16 How reliable are security analysts' forecasts of corporate earnings growth? According to David Dreman of *Forbes* magazine, "Astrology might be better" (*Forbes*, Mar. 26, 1984). Dreman reported on a study of annual earnings estimates made by institutional brokerage analysts covering the more widely followed companies between 1977 and 1981. The study revealed that "the average annual error by the analysts was a staggering 31.3% over the 5-year period." Suppose the annual forecast error by security analysts is normally distributed with a mean of 31.3% and a standard deviation of 10%. Suppose you obtain a security analyst's forecast of annual earnings for a particular corporation.
  a. What is the probability that the forecast error will be between 20% and 25%?
  b. What is the probability that the forecast error will be greater than 50%?

2.17 A company that sells annuities must base the annual payout on the probability distribution on the length of life of the participants in the plan. Suppose the probability distribution of the lifetimes of the participants in the plan is approximately a normal distribution with $\mu = 68$ years and $\sigma = 3.5$ years.
  a. What proportion of the plan participants would receive payments beyond age 70?
  b. What proportion of the plan participants would receive payments beyond age 75?

2.18 The U.S. Department of Agriculture (USDA) has recently patented a process that uses a bacterium for removing bitterness from citrus juices (*Chemical Engineering*, Feb. 3, 1986). In theory, almost all the bitterness could be removed by the process, but for practical purposes the USDA aims at 50% overall removal. Suppose a USDA spokesman claims that the percentage of bitterness removed from an 8-ounce glass of freshly squeezed citrus juice is normally distributed with mean 50.1% and standard deviation 10.4%. To test this claim, the bitterness removal process is applied to a randomly selected 8-ounce glass of citrus juice. Find the probability that the process removes less than 33.7%. Based on this probability, what can you infer about the spokesman's claim?

2.19 Value Line is an advisory service that provides investors with a forecast of the movement of the stock market. Each week Value Line selects one stock which it believes has the highest probability of moving upward in the coming months. In the early 1980's, Value Line's stock selection strategy was highly successful. However, one year its selections performed below par. According to an issue of Value Line's *Selection &*

Opinion, which lists the performance of its stock selections to date, the mean percentage change in stock price since the date of recommendation of the 52 stocks was −19.9 and the standard deviation was 18.6. Assume that the percentage change in stock price for this period can be modeled by the normal probability distribution.

a. What is the probability that the percentage change of a stock recommended by Value Line during this period is greater than 0? (This is the probability that the stock will "gain" or move upward.)

b. What is the probability that the percentage change of a stock recommended by Value Line during this period is less than −50?

c. Determine the percentage change below which only 5% of the stocks recommended by Value Line fell during the period.

2.20 The inherent risk involved with marketing a new product is a key consideration for market researchers. Various statistical models have been developed to aid speculators and businesses in making such decisions. One area of marketing research that uses decision models is called *break-even analysis*. The underlying assumption of break-even analysis is that the demand for a product is normally distributed, with known mean and standard deviation. Of interest is the relationship between actual demand D and the break-even point BE, where BE is defined as the number of units of the product the company must sell in order to "break even" on the investment. Shih (1981) developed two practical decision criteria using break-even analysis.

> *Decision Rule A:*   Market the new product if the chance is better than 50% that demand D will exceed the break-even point BE, i.e., if $P(D \geq BE) > .5$.

> *Decision Rule B:*   For a specified level of risk p, where $0 \leq p \leq 1$, market the new product if $P(D \geq BE) > p$.

Note that Decision Rule A is a special case of Decision Rule B, with specified level of risk $p = .5$. Suppose a company will utilize break-even analysis to decide whether to market a new type of ceiling fan. From past experience, the company knows that the number of ceiling fans of this type sold per year follows a normal distribution with $\mu = 4,000$ and $\sigma = 500$. Marketing researchers have also determined that the company needs to sell 3,500 units in order to break even for the year.

a. According to Decision Rule A, should the company market the new ceiling fans?

b. Use your knowledge of the normal distribution to show that if $P(D \geq BE) > .5$, then it must be true that $\mu \geq BE$.

c. Suppose that the minimum level of risk the company is willing to tolerate is $p = .8$. Use Decision Rule B to arrive at a decision.

d. Refer to part c. Find the value of BE such that the probability of at least breaking even is equal to the specified level of risk, $p = .8$.

| | | | | | | | | | | | |

S E C T I O N   2.5

SAMPLING DISTRIBUTIONS AND THE CENTRAL LIMIT THEOREM

Since we will use sample statistics to make inferences about population parameters, it is natural that we would want to know something about the reliability of the resulting inferences. For example, if we use a statistic to estimate the value of a population mean μ, we will want to know how close to μ our estimate is likely to fall. To answer this question, we need to know the probability distribution of the statistic.

The probability distribution for a statistic based on a random sample of n measurements could be generated in the following way. For purposes of illustration, we will suppose we are sampling from a population with $\mu = 10$ and $\sigma = 5$, the sample statistic is $\bar{y}$, and the sample size is $n = 25$. Draw a single random sample of 25 measurements from the population and suppose that $\bar{y} = 9.8$. Return the measurements to the population and try again. That is, draw another random sample of $n = 25$ measurements and see what you obtain for an outcome. Now, perhaps, $\bar{y} = 11.4$. Replace these measurements, draw another sample of $n = 25$ measurements, calculate $\bar{y}$, and so on. If this sampling process were repeated over and over again an infinitely large number of times, you would generate an infinitely large number of values of $\bar{y}$ that could be arranged in a relative frequency distribution. This distribution, which would appear as shown in Figure 2.9, is the probability distribution (or **sampling distribution**, as it is commonly called) of the statistic $\bar{y}$.

FIGURE 2.9

Sampling Distribution for $\bar{y}$ Based on a Sample of $n = 25$ Measurements

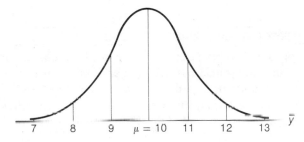

DEFINITION 2.10

The **sampling distribution** of a sample statistic calculated from a sample of n measurements is the probability distribution of the statistic.

In actual practice, the sampling distribution of a statistic is obtained mathematically or by simulating the sampling on a computer using the procedure described previously.

If $\bar{y}$ has been calculated from a sample of $n = 25$ measurements selected from a population with mean $\mu = 10$ and standard deviation $\sigma = 5$, the sampling distribution shown in Figure 2.9 provides all the information you may wish to know about its behavior. For example, the probability that you will draw a sample of 25 measurements and obtain a value of $\bar{y}$ in the interval $9 \leq \bar{y} \leq 10$, will be the area under the sampling distribution over that interval.

Generally speaking, if we use a statistic to make an inference about a population parameter, we want its sampling distribution to center about the parameter (as is the case in Figure 2.9) and the standard deviation of the sampling distribution, called the **standard error of estimate**, to be as small as possible.

Two theorems provide information on the sampling distribution of a sample mean.

THEOREM 2.1

If $y_1, y_2, \ldots, y_n$ represent a random sample of n measurements from a large (or infinite) population with mean μ and standard deviation σ, then, regardless of the form of the population relative frequency distribution, the mean and standard error of estimate of the sampling distribution of $\bar{y}$ will be

Mean: $E(\bar{y}) = \mu$

Standard error of estimate: $\sigma_{\bar{y}} = \dfrac{\sigma}{\sqrt{n}}$

THEOREM 2.2   THE CENTRAL LIMIT THEOREM

For large sample sizes, the mean $\bar{y}$ of a sample from a population with mean μ and standard deviation σ has a sampling distribution that is approximately normal, **regardless of the probability distribution of the sampled population**. The larger the sample size, the better will be the normal approximation to the sampling distribution of $\bar{y}$.

Theorems 2.1 and 2.2 together imply that for sufficiently large samples, the sampling distribution for the sample mean $\bar{y}$ will be approximately normal with mean μ and standard error $\sigma_{\bar{y}} = \sigma/\sqrt{n}$. The parameters μ and σ are the mean and standard deviation of the sampled population.

To illustrate Theorems 2.1 and 2.2, we randomly selected 1,000 random samples of $n = 11$ measurements from a population with a **uniform relative frequency distribution** with $\mu = .5$ and $\sigma = .29$, as shown in Figure 2.10. The sample mean $\bar{y}$ was calculated for each of the 1,000 samples. The resulting relative frequency distribution, along with a superimposed normal curve, is shown in Figure 2.11. You can see that the relative frequency distribution of 1,000 sample means (which approximates the sampling distribution of $\bar{y}$) is approximately normal, and that, in fact, the mean of the approximating normal distribution is

$\mu = .5$

The standard error is

$$\sigma_{\bar{y}} = \frac{\sigma}{\sqrt{n}} = \frac{.29}{\sqrt{11}} = .09$$

FIGURE 2.10
Relative Frequency
Distribution of Sampled
Population

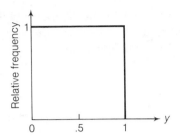

FIGURE 2.11 Relative Frequency Distribution for $\bar{y}$ in 1,000 Samples from a Uniform Distribution with Normal Frequency Function Superimposed

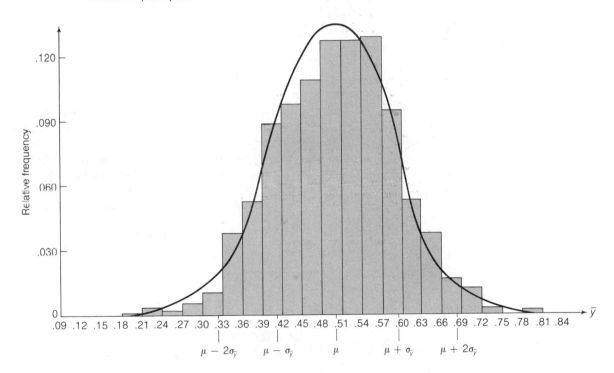

Note that for this example, the sample size $n = 11$ is relatively small and the normal curve approximation to the sampling distribution is quite good. This will not always be true, but it does illustrate the power of the Central Limit Theorem.

The Central Limit Theorem can also be used to justify the fact that the *sum* of the sample measurements possesses a sampling distribution that is approximately normal for large sample sizes. In fact, since many statistics are obtained by summing or averaging random quantities, the Central Limit Theorem helps to explain why many statistics have mound-shaped (or approximately normal) sampling distributions.

As we proceed, we will encounter many different sample statistics, and we will need to know their sampling distributions in order to evaluate their reliabilities for making inferences. These sampling distributions will be described as the need arises.

| | | | | | | | | | | | |

S E C T I O N 2.6

ESTIMATING A POPULATION MEAN μ

We can make an inference about a population parameter in two ways:

1. Estimate its value
2. Make a decision about its value (i.e., test a hypothesis about its value)

In this section we will illustrate the concepts involved in estimation, using the estimation of a population mean as an example. Tests of hypotheses will be discussed in Section 2.7.

To estimate a population parameter, we choose a sample statistic with a sampling distribution that centers about the parameter and has a small standard error. If the mean of the sampling distribution of a statistic equals the parameter we are estimating, we say that the statistic is an **unbiased estimator** of the parameter. If not, we say that it is **biased**.

In Section 2.5, we noted that the sampling distribution of the sample mean is approximately normally distributed for moderate to large sample sizes and that it possesses a mean μ and standard error $\sigma/\sqrt{n}$. Therefore, as shown in Figure 2.12, $\bar{y}$ is an unbiased estimator of the population mean μ, and the probability that $\bar{y}$ will fall within $1.96\sigma_{\bar{y}} = 1.96\sigma/\sqrt{n}$ of the true value of μ is approximately .95.

FIGURE 2.12
Sampling Distribution of $\bar{y}$

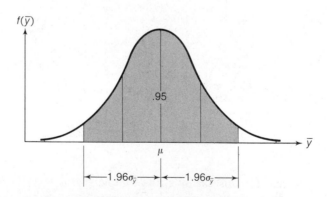

Since $\bar{y}$ will fall within $1.96\sigma_{\bar{y}}$ of μ approximately 95% of the time, it follows that the interval

$$\bar{y} - 1.96\sigma_{\bar{y}} \quad \text{to} \quad \bar{y} + 1.96\sigma_{\bar{y}}$$

will enclose μ approximately 95% of the time in repeated sampling. This interval is called a 95% **confidence interval**, and .95 is called the **confidence coefficient**.

Notice that μ is fixed and that the confidence interval changes from sample to sample. The probability that a confidence interval calculated using the formula

$$\bar{y} \pm 1.96\sigma_{\bar{y}}$$

will enclose μ is approximately .95. Thus, the confidence coefficient measures the confidence that we can place in a particular confidence interval.

Confidence intervals can be constructed using any desired confidence coefficient. For example, if we define $z_{\alpha/2}$ to be the value of a standard normal variable that places the area $\alpha/2$ in the right-hand tail of the z-distribution (see Figure 2.13), then a $100(1 - \alpha)\%$ confidence interval for μ is given in the box:

LARGE-SAMPLE $100(1 - \alpha)\%$ CONFIDENCE INTERVAL FOR μ

$$\bar{y} \pm z_{\alpha/2}\sigma_{\bar{y}}$$

where $z_{\alpha/2}$ is the z-value with an area $\alpha/2$ to its right (see Figure 2.13) and $\sigma_{\bar{y}} = \sigma/\sqrt{n}$. The parameter σ is the standard deviation of the sampled population and n is the sample size. If σ is unknown, its value may be approximated by the sample standard deviation. The sample size n must be at least 30.

FIGURE 2.13

Locating $z_{\alpha/2}$ on the Standard Normal Curve

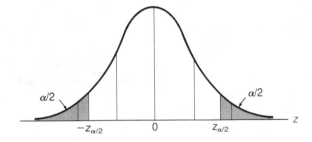

The confidence interval shown in the box is called a large-sample confidence interval because the sample size must be large enough to ensure approximate normality for the sampling distribution of $\bar{y}$. Also, and even more important, you will rarely, if ever, know the value of σ, so its value must be estimated using the sample standard deviation s. This approximation for σ will be adequate only when $n \geq 30$.

Typical confidence coefficients and corresponding values of $z_{\alpha/2}$ are shown in Table 2.3 on page 30.

TABLE 2.3
Commonly Used Values
of $z_{\alpha/2}$

CONFIDENCE COEFFICIENT $(1 - \alpha)$	α	$\alpha/2$	$z_{\alpha/2}$
.90	.10	.05	1.645
.95	.05	.025	1.96
.99	.01	.005	2.576

EXAMPLE 2.6

Unoccupied seats on flights cause airlines to lose revenue. Suppose a large airline wants to estimate the mean number of unoccupied seats per flight over the past year. To accomplish this, the records of 225 flights are randomly selected from the files, and the number of unoccupied seats is noted for each of the sampled flights. The sample mean and standard deviation are

$$\bar{y} = 11.6 \text{ seats} \qquad s = 4.1 \text{ seats}$$

Estimate μ, the mean number of unoccupied seats per flight during the past year, using a 90% confidence interval.

SOLUTION

The general form of the 90% confidence interval for a population mean is

$$\bar{y} \pm z_{\alpha/2}\sigma_{\bar{y}} = \bar{y} \pm z_{.05}\sigma_{\bar{y}}$$
$$= \bar{y} \pm 1.645\left(\frac{\sigma}{\sqrt{n}}\right)$$

For the 225 records sampled, we have

$$11.6 \pm 1.645\left(\frac{\sigma}{\sqrt{225}}\right)$$

Since we do not know the value of σ (the standard deviation of the number of unoccupied seats per flight for all flights during the year), we use our best approximation, the sample standard deviation s. Then the 90% confidence interval is, approximately,

$$11.6 \pm 1.645\left(\frac{4.1}{\sqrt{225}}\right) = 11.6 \pm .45$$

or, from 11.15 to 12.05. That is, the airline can be 90% confident that the mean number of unoccupied seats per flight was between 11.15 and 12.05 during the sampled year. ∎

The large-sample method for making inferences about a population mean μ assumes either that σ is known or that the sample size is large enough ($n \geq 30$) so that the sample standard deviation s can be used as a good approximation to σ. The technique for finding a $100(1 - \alpha)\%$ confidence interval for a population mean μ for small sample sizes requires that the sampled population have a normal probability distribution. The formula, which is similar to the one for a large-sample confidence interval for μ, is

$$\bar{y} \pm t_{\alpha/2}\left(\frac{s}{\sqrt{n}}\right)$$

The quantity $t_{\alpha/2}$ is directly analogous to the standard normal value $z_{\alpha/2}$ used in finding a large-sample confidence interval for μ except that it is an upper-tail t-value obtained from a Student's t-distribution. Thus, $t_{\alpha/2}$ is an upper-tail t-value such that an area $\alpha/2$ lies to its right.

Like the standardized normal (z) distribution, a Student's t-distribution is symmetric about the value $t = 0$, but it is more variable than a z-distribution. The variability depends on the number of **degrees of freedom, df**, which in turn depends on the number of measurements available for estimating σ^2. The smaller the number of degrees of freedom, the greater will be the spread of the t-distribution. For this application of a Student's t-distribution, df $= n - 1$. As the sample size increases (and df increases), the Student's t-distribution looks more and more like a z-distribution, and for $n \geq 30$, the two distributions will be nearly identical. A Student's t-distribution based on df $= 4$ and a standard normal distribution are shown in Figure 2.14. Note the corresponding values of $z_{.025}$ and $t_{.025}$.

FIGURE 2.14

The $t_{.025}$ Value in a t-Distribution with 4 df and the Corresponding $z_{.025}$ Value

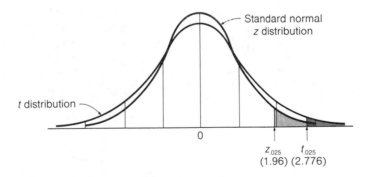

The upper-tail values of the Student's t-distribution are given in Table 2 of Appendix D. An abbreviated version of the t-table is presented in Table 2.4 on page 32. To find the upper-tail t-value based on 4 df that places .025 in the upper tail of the t-distribution, we look in the row of the table corresponding to df $= 4$ and the column corresponding to $t_{.025}$. The t-value is read as 2.776 and is shown in Figure 2.14.

The process of finding a small-sample confidence interval for μ is given in the next box.

> ### SMALL-SAMPLE CONFIDENCE INTERVAL FOR μ
>
> $$\bar{y} \pm t_{\alpha/2}\left(\frac{s}{\sqrt{n}}\right)$$
>
> where $t_{\alpha/2}$ is a t-value based on $(n - 1)$ degrees of freedom, such that the probability that $t > t_{\alpha/2}$ is $\alpha/2$.
>
> *Assumption:*   The relative frequency distribution of the sampled population is approximately normal.

TABLE 2.4

Reproduction of a Portion of Table 2 of Appendix D

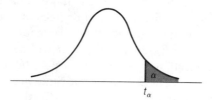

DEGREES OF FREEDOM	$t_{.100}$	$t_{.050}$	$t_{.025}$	$t_{.010}$	$t_{.005}$
1	3.078	6.314	12.706	31.821	63.657
2	1.886	2.920	4.303	6.965	9.925
3	1.638	2.353	3.182	4.541	5.841
4	1.533	2.132	2.776	3.747	4.604
5	1.476	2.015	2.571	3.365	4.032
6	1.440	1.943	2.447	3.143	3.707
7	1.415	1.895	2.365	2.998	3.499
8	1.397	1.860	2.306	2.896	3.355
9	1.383	1.833	2.262	2.821	3.250
10	1.372	1.812	2.228	2.764	3.169
11	1.363	1.796	2.201	2.718	3.106
12	1.356	1.782	2.179	2.681	3.055
13	1.350	1.771	2.160	2.650	3.012
14	1.345	1.761	2.145	2.624	2.977
15	1.341	1.753	2.131	2.602	2.947

EXAMPLE 2.7

When food prices began their rapid increase in the early 1970's, some of the major television networks began periodically to purchase a grocery basket full of food at supermarkets around the country. They always bought the same items at each store so they could compare food prices. Today, such comparisons continue to be made periodically throughout the United States. Suppose you want to estimate the mean price for a grocery basket in a specific geographical region of the country. You purchase the specified items at a random sample of 20 supermarkets in the region. The mean and standard deviation of the costs at the 20 supermarkets are

$$\bar{y} = \$56.84 \qquad s = \$2.63$$

Form a 95% confidence interval for the mean cost μ of a grocery basket for this region.

SOLUTION

If we assume that the distribution of costs for the grocery basket at all supermarkets in the region is approximately normal, we can use the t-statistic to form the confidence interval. For a confidence coefficient of .95, we need the tabulated value of t with df $= n - 1 = 19$:

$$t_{\alpha/2} = t_{.025} = 2.093$$

Then the confidence interval is

$$\bar{y} \pm t_{.025}\left(\frac{s}{\sqrt{n}}\right) = 56.84 \pm 2.093\left(\frac{2.63}{\sqrt{20}}\right)$$

$$= 56.84 \pm 1.23 \quad \text{or} \quad (55.61, 58.07)$$

Thus, we are reasonably confident that the interval from \$55.61 to \$58.07 contains the true mean cost μ of the grocery basket. This is because, if we were to employ our interval estimator on repeated occasions, 95% of the intervals constructed would contain μ. ∎

EXAMPLE 2.8

Refer to Example 2.7. Suppose you want to estimate the mean cost of a grocery basket correct to within \$1.00 with probability approximately equal to .95. How many supermarkets would you have to include in your sample?

SOLUTION

We will interpret the phrase, "correct to within \$1.00 . . . equal to .95" to mean that we want half the width of the confidence interval for μ to equal \$1.00. That is, we want

$$t_{.025}\left(\frac{s}{\sqrt{n}}\right) = 1.00$$

In order to be able to use this equation to find n, we need approximate values for $t_{.025}$ and s. Since we know from Example 2.7 that the confidence interval was larger than desired for $n = 20$, it is clear that our sample size must be larger than 20. Consequently, $t_{.025}$ will be very close to 2 and this value will provide a good approximation to $t_{.025}$. A good measure of the data variation is given by the standard deviation computed in Example 2.7. We substitute $t_{.025} \approx 2$ and $s \approx 2.63$ into the equation and solve for n:

$$t_{.025}\left(\frac{s}{\sqrt{n}}\right) = 1.00$$

$$2\left(\frac{2.63}{\sqrt{n}}\right) = 1.00$$

$$\sqrt{n} = 5.26$$

$$n = 27.7 \quad \text{or} \quad n = 28$$

Remember that this sample size is an approximate solution because we approximated the value of $t_{.025}$ and the value of s that might be computed from the prospective data. Nevertheless, $n = 28$ will be reasonably close to the sample size needed to estimate the mean grocery basket cost correct to within $1.00. ∎

EXERCISES 2.21–2.33

2.21 The table contains 50 random samples of random digits, $y = 0, 1, 2, 3, \ldots, 9$, where the probabilities corresponding to the values of y are given by the formula $p(y) = \frac{1}{10}$. Each sample contains $n = 6$ measurements.

SAMPLE	SAMPLE	SAMPLE	SAMPLE
8, 1, 8, 0, 6, 6	7, 6, 7, 0, 4, 3	4, 4, 5, 2, 6, 6	0, 8, 4, 7, 6, 9
7, 2, 1, 7, 2, 9	1, 0, 5, 9, 9, 6	2, 9, 3, 7, 1, 3	5, 6, 9, 4, 4, 2
7, 4, 5, 7, 7, 1	2, 4, 4, 7, 5, 6	5, 1, 9, 6, 9, 2	4, 2, 3, 7, 6, 3
8, 3, 6, 1, 8, 1	4, 6, 6, 5, 5, 6	8, 5, 1, 2, 3, 4	1, 2, 0, 6, 3, 3
0, 9, 8, 6, 2, 9	1, 5, 0, 6, 6, 5	2, 4, 5, 3, 4, 8	1, 1, 9, 0, 3, 2
0, 6, 8, 8, 3, 5	3, 3, 0, 4, 9, 6	1, 5, 6, 7, 8, 2	7, 8, 9, 2, 7, 0
7, 9, 5, 7, 7, 9	9, 3, 0, 7, 4, 1	3, 3, 8, 6, 0, 1	1, 1, 5, 0, 5, 1
7, 7, 6, 4, 4, 7	5, 3, 6, 4, 2, 0	3, 1, 4, 4, 9, 0	7, 7, 8, 7, 7, 6
1, 6, 5, 6, 4, 2	7, 1, 5, 0, 5, 8	9, 7, 7, 9, 8, 1	4, 9, 3, 7, 3, 9
9, 8, 6, 8, 6, 0	4, 4, 6, 2, 6, 2	6, 9, 2, 9, 8, 7	5, 5, 1, 1, 4, 0
3, 1, 6, 0, 0, 9	3, 1, 8, 8, 2, 1	6, 6, 8, 9, 6, 0	4, 2, 5, 7, 7, 9
0, 6, 8, 5, 2, 8	8, 9, 0, 6, 1, 7	3, 3, 4, 6, 7, 0	8, 3, 0, 6, 9, 7
8, 2, 4, 9, 4, 6	1, 3, 7, 3, 4, 3		

a. Use the 300 random digits to construct a relative frequency distribution for the data. This relative frequency distribution should approximate $p(y)$.

b. Calculate the mean of the 300 digits. This will give an accurate estimate of μ (the mean of the population) and should be very near to $E(y)$, which is 4.5.

c. Calculate s^2 for the 300 digits. This should be close to the variance of y, $\sigma^2 = 8.25$.

d. Calculate $\bar{y}$ for each of the 50 samples. Construct a relative frequency distribution for the sample means to see how close they lie to the mean of $\mu = 4.5$. Calculate the mean and standard deviation of the 50 means.

2.22 Refer to Exercise 2.21. To see the effect of sample size on the standard deviation of the sampling distribution of a statistic, combine pairs of samples (moving down the columns of the table) to obtain 25 samples of $n = 12$ measurements. Calculate the mean for each sample.

a. Construct a relative frequency distribution for the 25 means. Compare this with the distribution prepared for Exercise 2.21, which is based on samples of $n = 6$ digits.

b. Calculate the mean and standard deviation of the 25 means. Compare the standard deviation of this sampling distribution with the standard deviation of the sampling distribution in Exercise 2.21. What relationship would you expect to exist between the two standard deviations?

2.23 As part of a study to determine the relationship between the length of time patients wait in the physician's office and certain demand and cost factors, Sloan and Lorant (1977) obtained data on the typical patient waiting times for 4,500 physicians in the five largest specialties—general practice, general surgery, internal medicine, obstetrics/gynecology, and pediatrics. They reported a mean waiting time of 24.7 minutes and a standard deviation of 19.3 minutes. One important aspect in the study of waiting times is the effect of policy changes on the waiting time distribution. Suppose that 4,500 typical waiting times were collected over a period of time at a health clinic and that the clinic wants to change its operating procedure in order to increase the number of patients treated by a physician in a given period of time. To determine what effect the change in operating procedure will have on the mean patient waiting time, the clinic places the new procedure in operation and then collects a sample of $n = 100$ waiting times. The sample yielded the following results:

$$\bar{y} = 19.8 \text{ minutes} \qquad s = 20.6 \text{ minutes}$$

a. Assuming that the distribution of patient waiting times has not changed (i.e., $\mu = 24.7$ and $\sigma = 19.3$), describe the sampling distribution of $\bar{y}$, the mean of the sample of 100 patient waiting times.

b. Find $P(\bar{y} \leq 19.8)$.

c. Based on the probability in part b, what can you infer about the waiting time distribution under the new operating procedure? Explain.

2.24 By definition, an *entrepreneur* is "one who undertakes to start and conduct an enterprise or business, assuming full control and risks" (Funk and Wagnall's *Standard Dictionary*). Thus, a distinguishing characteristic of entrepreneurs is their propensity for taking risks. R. H. Brockhaus (1980) used a choice dilemma questionnaire (CDQ) to measure the risk-taking propensities of successful entrepreneurs. He found that entrepreneurs had a mean CDQ score of 71 and a standard deviation of 12. (Lower scores are associated with a greater propensity for taking risks.) In a random sample of $n = 50$ entrepreneurs, let $\bar{y}$ be the sample mean CDQ score.

a. Describe the sampling distribution of $\bar{y}$.

b. Find $P(69 \leq \bar{y} \leq 72)$.

c. Find $P(\bar{y} \leq 67)$.

d. Would you expect to observe a sample mean CDQ score of 67 or lower? Explain.

2.25 Explain what is meant by the statement, "We are 95% confident that our interval estimate contains μ."

2.26 To estimate the average rate of fuel consumption of their snowmobiles, a manufacturer tested 36 of them and obtained the following results: $\bar{y} = 20.3$ miles per gallon, $s = 2$ miles per gallon. Estimate the average rate of fuel consumption of all snowmobiles produced, using a 90% confidence interval.

2.27 Refer to Exercise 2.26. In order to continue marketing their snowmobiles, the manufacturer must produce snowmobiles that do not exceed the 78 decibel noise limit recently set by the government. To determine the average decibel level of the snowmobiles it currently has in stock, 49 are randomly selected and run at full throttle, and their noise levels are measured. The sample statistics, expressed in decibels, are $\bar{y} = 75$ and $s = 14$. Estimate the average decibel level of the snowmobiles the manufacturer currently has in stock, using a 99% confidence interval.

2.28 CPA firms generally charge $25 to $35 per hour for their auditing services. Because external audits can become quite expensive, many companies are creating or augmenting internal audit departments to lower auditing costs. Wanda A. Wallace (1984), CPA and Professor at Southern Methodist University, conducted a study of the audit departments of 32 diverse companies. Her main interest was to determine the effect that internal audit departments had on external audit fees. The mean external audit fee paid by the 32 companies in the study period was $779,030, and the standard deviation was $1,083,162. Calculate a 95% confidence interval for the mean external audit fee paid by all companies in the study period.

2.29 In what ways are the distributions of the z-statistic and t-statistic alike? How do they differ?

2.30 Let t_0 be a particular value of t. Use Table 2 of Appendix D to find t_0 values such that the following statements are true:
a. $P(t \geq t_0) = .025$ where $n = 10$ **b.** $P(t \geq t_0) = .01$ where $n = 5$
c. $P(t \leq t_0) = .005$ where $n = 20$ **d.** $P(t \leq t_0) = .05$ where $n = 12$

2.31 Each year, thousands of manufacturer sales promotions are conducted by North American package good companies, yet promotion managers are frequently dissatisfied with their results. An exploratory study was conducted by K. G. Hardy (1986) to examine the objectives and impact of such sales promotions. A sample of Canadian package goods companies provided information on examples of past sales promotions, including trade promotions. For the 21 "successful" trade promotions (where "success" is determined by the company managers) identified in the sample, the mean incremental profit was $53,000 and the standard deviation was $95,000.

a. Find a 90% confidence interval for the true mean incremental profit of "successful" trade promotions.
b. State the assumptions required for this confidence interval to be valid.
c. Give two ways in which you could reduce the width of the interval in part **a**. Which would you recommend?

2.32 Many North American cities have built or are considering building light rail transit (LRT) systems as an alternative to the heavy rail transit systems which use large passenger trains and subways. LRT systems are similar to the early 1900's streetcar, but are typically longer, quieter, faster, and more comfortable. In one study, the characteristics of LRT operations were examined in ten cities which have built or are planning to build LRT systems (*Journal of the American Planning Association*, Spring 1984). One characteristic of importance to urban planners is the farebox recovery rate, computed by dividing passenger revenues by operating costs. The sample of ten cities had a mean farebox recovery rate of .604 and a standard deviation of .163.

a. Estimate the true mean farebox recovery rate for LRT systems in North American cities, using a 95% confidence interval.

b. What assumption is required for the confidence interval procedure of part **a** to be valid?

2.33 During the budget preparation process at a hospital, forecasts of inpatient utilization (measured in patient days) for the coming fiscal year play a pivotal role. Planners at Sisters of St. Joseph of Peace Health and Hospital Services, Bellevue, Washington, have developed a budget early warning technique (BEWT) for forecasting future fiscal year utilizations (MacStravic, January/February 1986). The method involves collecting past data on monthly utilization and calculating, for each month, the ratio of monthly utilization to the utilization for the entire fiscal year preceding the month. Confidence intervals established on this utilization ratio enable planners to establish budget forecasts. At one of the smaller (50 beds) hospitals, the April utilization ratios for each of the past 10 years were calculated and found to have a sample mean of .0817 and sample standard deviation of .0069. Establish a 99% confidence interval for μ, the true mean April utilization ratio. Interpret the interval.

SECTION 2.7

TESTING A HYPOTHESIS ABOUT A POPULATION MEAN

The procedure involved in testing a hypothesis about a population parameter can be illustrated with the procedure for a test concerning a population mean μ.

A statistical test of a hypothesis is composed of four parts:

1. A **null hypothesis**, denoted by the symbol H_0, which is the hypothesis that we postulate is true
2. An **alternative** (or **research**) **hypothesis**, denoted by the symbol H_a, which is counter to the null hypothesis and is what we want to support
3. A **test statistic**, calculated from the sample data, that functions as a decision-maker
4. A **rejection region**—if the value of a test statistic, calculated from a particular sample, falls in the rejection region, we reject the null hypothesis and accept the alternative hypothesis.

The test statistic for testing the null hypothesis that a population mean μ equals some specific value, say μ_0, is the sample mean $\bar{y}$ or the standardized normal variable

$$z = \frac{\bar{y} - \mu_0}{\sigma_{\bar{y}}} \qquad \text{where} \quad \sigma_{\bar{y}} = \frac{\sigma}{\sqrt{n}}$$

The logic employed in deciding whether sample data *disagree* with this hypothesis can be seen by viewing the sampling distribution of $\bar{y}$ shown in Figure 2.15 (page 38). If the population mean μ is in fact equal to μ_0 (i.e., if the null hypothesis is true), then the mean $\bar{y}$ calculated from a sample should fall, with high probability, within $2\sigma_{\bar{y}}$ of μ_0. If $\bar{y}$ falls too far away from μ_0, or if the standardized distance

$$z = \frac{\bar{y} - \mu_0}{\sigma_{\bar{y}}}$$

is too large, we conclude that the data disagree with our hypothesis, and we reject the null hypothesis.

FIGURE 2.15

The Sampling Distribution of $\bar{y}$ for $\mu = \mu_0$

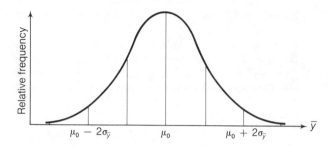

If we want to detect the alternative hypothesis that $\mu > \mu_0$, we locate the boundary of the rejection region in the upper tail of the z-distribution, as shown in Figure 2.16(a), at the point z_α. Similarly, to detect $\mu < \mu_0$, we place the rejection region in the lower tail of the z-distribution, as shown in Figure 2.16(b). These are called **one-tailed statistical tests**. To detect *either* $\mu > \mu_0$ or $\mu < \mu_0$— that is, $\mu \neq \mu_0$—we split α equally between the two tails of the z-distribution and reject the null hypothesis if $z > z_{\alpha/2}$ or $z < -z_{\alpha/2}$, as shown in Figure 2.16(c). This is called a **two-tailed statistical test**.

The z-test, which is summarized in the next box, is called a *large-sample test* because we will rarely know σ and hence will need a sample size that is large enough so that the sample standard deviation s will provide a good approximation to σ. Normally, we recommend that the sample size be $n \geq 30$.

FIGURE 2.16   Location of the Rejection Region for Various Alternative Hypotheses

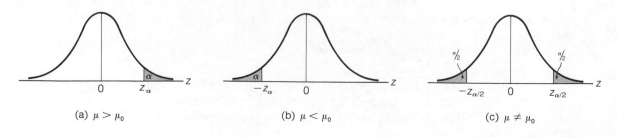

LARGE-SAMPLE ($n \geq 30$) TEST OF HYPOTHESIS ABOUT μ

ONE-TAILED TEST

H_0: $\mu = \mu_0$

H_a: $\mu < \mu_0$
     (or H_a: $\mu > \mu_0$)

Test statistic: $z = \dfrac{\bar{y} - \mu_0}{\sigma_{\bar{y}}}$

Rejection region: $z < -z_\alpha$
          (or $z > z_\alpha$)

where z_α is chosen so that
$P(z > z_\alpha) = \alpha$

TWO-TAILED TEST

H_0: $\mu = \mu_0$

H_a: $\mu \neq \mu_0$

Test statistic: $z = \dfrac{\bar{y} - \mu_0}{\sigma_{\bar{y}}}$

Rejection region: $z < -z_{\alpha/2}$
          or $z > z_{\alpha/2}$

where $z_{\alpha/2}$ is chosen so that
$P(z > z_{\alpha/2}) = \alpha/2$

We illustrate with an example.

EXAMPLE 2.9

Building specifications in a certain city require that the average breaking strength of residential sewer pipe be more than 2,400 pounds per foot of length (i.e., per lineal foot). A sampling of the strengths of 70 sections of pipe produced by a manufacturer yielded a sample mean and standard deviation of

$$y = 2,430 \qquad s - 190$$

Do these statistics, calculated from the sample data, present sufficient evidence to indicate that the manufacturer's pipe meets the city's specifications? Use $\alpha = .05$.

SOLUTION

Since we wish to determine whether $\mu > 2,400$, the elements of the test are

$$H_0: \quad \mu = 2,400 \qquad H_a: \quad \mu > 2,400$$

Test statistic: $z = \dfrac{\bar{y} - 2,400}{\sigma_{\bar{y}}} = \dfrac{\bar{y} - 2,400}{\sigma/\sqrt{n}} \approx \dfrac{\bar{y} - 2,400}{s/\sqrt{n}}$

Rejection region: $z > 1.645$ for $\alpha = .05$

Substituting the sample statistics into the test statistic, we have

$$z \approx \frac{\bar{y} - 2,400}{s/\sqrt{n}} = \frac{2,430 - 2,400}{190/\sqrt{70}} = 1.32$$

The sample mean of 2,430, although greater than 2,400, is only $1.32\sigma_{\bar{y}}$ above that value. Therefore, the sample does not provide sufficient evidence to conclude that the sewer pipe meets the city's strength specifications. ∎

The reliability of a statistical test is measured by the probability of making an incorrect decision. For example, the probability of rejecting the null hypothesis and accepting the alternative hypothesis when the null hypothesis is true (called a **Type I error**) is α, the tail probability used in locating the rejection region. A second type of error could be made if we accepted the null hypothesis when, in fact, the alternative hypothesis is true (a **Type II error**). Thus, you never "accept" the null hypothesis unless you know the probability of making a Type II error. Since this probability (denoted by the symbol β) is often unknown, it is a common practice to defer judgment if a test statistic falls in the nonrejection region.

A small-sample test of the null hypothesis $\mu = \mu_0$ using a Student's t-statistic is based on the assumption that the sample was randomly selected from a population with a normal relative frequency distribution. The test is conducted in exactly the same manner as the large-sample z-test except that we use

$$t = \frac{\bar{y} - \mu_0}{s/\sqrt{n}}$$

as the test statistic and we locate the rejection region in the tail(s) of a Student's t-distribution with df $= n - 1$. We summarize the technique for conducting a small-sample test of hypothesis about a population mean in the box.

SMALL-SAMPLE TEST OF HYPOTHESIS ABOUT μ

ONE-TAILED TEST	TWO-TAILED TEST
H_0: $\mu = \mu_0$	H_0: $\mu = \mu_0$
H_a: $\mu < \mu_0$	H_a: $\mu \neq \mu_0$
(or $\mu > \mu_0$)	
Test statistic: $t = \dfrac{\bar{y} - \mu_0}{s/\sqrt{n}}$	Test statistic: $t = \dfrac{\bar{y} - \mu_0}{s/\sqrt{n}}$
Rejection region: $t < -t_\alpha$	Rejection region: $t < -t_{\alpha/2}$
(or $t > t_\alpha$)	or $t > t_{\alpha/2}$
where t_α is based on $(n - 1)$ df	where $t_{\alpha/2}$ is based on $(n - 1)$ df

Assumption: The population from which the sample is drawn is approximately normal.

EXAMPLE 2.10

A major car manufacturer wants to test a new engine to see whether it meets new air pollution standards. The mean emission μ for all engines of this type must be less than 20 parts per million of carbon. Ten engines are manufactured for testing purposes, and the mean and standard deviation of the emissions for this sample of engines are determined to be

$\bar{y} = 17.1$ parts per million $s = 3.0$ parts per million

Do the data supply sufficient evidence to allow the manufacturer to conclude that this type of engine meets the pollution standard? Assume that the manufacturer is willing to risk a Type I error with probability equal to $\alpha = .01$.

SOLUTION

The manufacturer wants to establish the research hypothesis that the mean emission level μ for all engines of this type is less than 20 parts per million. The elements of this small-sample one-tailed test are

$$H_0: \quad \mu = 20 \qquad H_a: \quad \mu < 20$$

Test statistic: $\quad t = \dfrac{\bar{y} - 20}{s/\sqrt{n}}$

Assumption: The relative frequency distribution of the population of emission levels for all engines of this type is approximately normal.

Rejection region: For $\alpha = .01$ and df $= n - 1 = 9$, the one-tailed rejection region (see Figure 2.17) is $t < -t_{.01} = -2.821$

We now calculate the test statistic:

$$t = \frac{\bar{y} - 20}{s/\sqrt{n}} = \frac{17.1 - 20}{3.0/\sqrt{10}} = -3.06$$

Since the calculated t falls in the rejection region (see Figure 2.17), the manufacturer concludes that $\mu < 20$ parts per million, and the new engine type meets the pollution standard. Are you satisfied with the reliability associated with this inference? The probability is only $\alpha = .01$ that the test would support the research hypothesis when in fact it was false.

FIGURE 2.17
Rejection Region for
Example 2.10

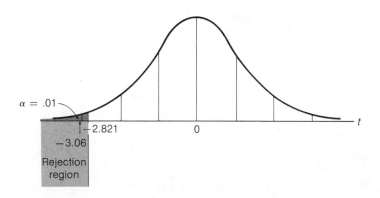

To conclude, we mention a second method that is useful in reporting the results of a statistical test. Some data analyzers indicate the degree to which the test statistic contradicts the null hypothesis (and hence supports the alternative hypothesis). This quantity, called the **level of significance**, or **p-value**, of the test, is the probability of observing a value of the test statistic at least as contradictory to the null hypothesis as the observed value of the test statistic, *assuming the null hypothesis is true.* For example, suppose you conducted a large-sample z-test to detect the alternative hypothesis $\mu > 100$, and the computed value of the test statistic was 2.12. Then the level of significance for this one-tailed test is the probability of observing a z-value larger than 2.12 if $\mu = 100$, or

$$P(z > 2.12) = .0170$$

If the test were two-tailed (i.e., the alternative hypothesis were $\mu \neq 100$), then values more contradictory to the null hypothesis $\mu = 100$ would be z-values greater than 2.12 or less than -2.12. The level of significance for this test would be $2(.0170) = .0340$.

EXERCISES 2.34–2.47

2.34 What is the difference between a research hypothesis and a null hypothesis?

2.35 Define each of the following:
a. Type I error   **b.** Type II error   **c.** α **d.** β

2.36 In hypothesis testing, who or what determines the size of the rejection region?

2.37 If you test a hypothesis and reject the null hypothesis in favor of your research hypothesis, does your test prove that the research hypothesis is correct? Explain.

2.38 When do you risk making a Type I error? A Type II error?

2.39 For each of the following rejection regions, sketch the sampling distribution for z and indicate the location of the rejection region:
a. $z > 1.96$ **b.** $z > 1.645$ **c.** $z > 2.576$ **d.** $z < -1.29$
e. $z < -1.645$ or $z > 1.645$ **f.** $z < -2.576$ or $z > 2.576$

2.40 If the rejection region is defined as in Exercise 2.39, what is the probability that a Type I error will be made in each case?

2.41 According to the May 1985 Current Population Survey, American workers receive an average of 9 hours of overtime pay per month (*Monthly Labor Review*, November 1986). Suppose the personnel manager at a large manufacturing plant suspects that the mean number of overtime pay hours worked per month by plant workers exceeds the national average. To check this hypothesis, the manager obtains the payroll records of a random sample of 50 workers and records the number of overtime pay hours worked by each in the last month. The following statistics were obtained:

$$\bar{y} = 12.3 \text{ hours} \qquad s = 4.1 \text{ hours}$$

Do these statistics support the manager's hypothesis? Test using $\alpha = .05$.

2.42 The "major bottom indicator" of the New York Stock Exchange (NYSE) was coined by Gerald Appel (1984) to describe a stock market recovery signal. Appel noticed that when there is a large negative difference (-4.0 points or more) between the weekly close of the NYSE index and the 10-week moving average, the market is almost always deeply oversold and ready for an upward turn. Appel recommends buying stock 3 weeks after the date of the largest negative divergence. *Barron's* reported the percentage gains for stock held 26 weeks after the "buy" signal for a sample of eight dates. These values are reproduced in the accompanying table. If the major bottom indicator is, in fact, a trustworthy indicator of upward turns in the market in the long run, then the true mean percentage gain for the 26-week holding period will be greater than 0%. Test this hypothesis and state any assumptions necessary for the procedure to be valid. Let $\alpha = .05$.

DATE OF BUY SIGNAL	PERCENTAGE GAIN IN NYSE INDEX 26 WEEKS AFTER BUY
June 12, 1970	21.1
December 21, 1973	−7.7
October 4, 1974	30.6
November 17, 1978	7.1
November 16, 1979	3.6
April 11, 1980	32.7
October 16, 1981	−2.7
July 9, 1982	33.8

2.43 A machine is set to produce nails with a mean length of 1 inch. Nails that are too long or too short do not meet the customer's specifications and must be rejected. To avoid producing too many rejects, the nails produced by the machine are sampled from time to time and tested as a check to see whether the machine is still operating properly, i.e., producing nails with a mean length of 1 inch. Suppose 50 nails have been sampled, and $\bar{y} = 1.02$ inches and $s = .04$ inch. Does the sample evidence indicate that the machine is producing nails with a mean not equal to 1 inch, i.e., is the production process out of control? Let $\alpha = .01$.

2.44 Most supermarket chains give each store manager a detailed plan showing exactly where each product belongs on each shelf. Since more products are picked up from shelves at eye level than from any others, the most profitable and fastest-moving items are placed at eye level to make them easy for shoppers to reach. Traditionally, the eye-level shelf has been slightly under 5 feet from the floor—just the right height for the average female shopper, 5-feet 4-inches tall. But nowadays, more men are shopping than ever before. Since the average male shopper is 5-feet 10-inches tall, how will this affect the eye-level shelf? To investigate this, a random sample of 50 supermarkets from across the country were selected and the height of the eye-level shelf (i.e., the shelf with the most popular products) was recorded for each. The results were

$\bar{y} = 62$ inches $s = 5$ inches

a. Is there evidence that the average height of the eye-level shelf at supermarkets is now higher than 60 inches from the floor (i.e., higher than the traditional height of 5 feet)?

b. Calculate the level of significance (p-value) of the test.

2.45 In any bottling process, a manufacturer will lose money if the bottles contain either more or less than is claimed on the label. Accordingly, bottlers pay close attention to the amount of their product being dispensed by bottle-filling machines. Suppose a quality control inspector for a catsup company is interested in testing whether the mean number of ounces of catsup per family-size bottle is 20 ounces. The inspector samples nine bottles, measures the weight of their contents, and finds that $\bar{y} = 19.7$ ounces and $s = .3$ ounce. Does the sample evidence indicate that the catsup-dispensing machine is in need of adjustment? Test using $\alpha = .05$. What assumptions are required in order for this test to be valid?

2.46 How do the makers of Kleenex know exactly how many tissues to put in a box? According to the *Wall Street Journal* (Sept. 21, 1984), the marketing experts at Kimberly Clark Corporation have "little doubt that the company should put 60 tissues in each pack." The researchers determined that 60 is "the average number of times people blow their nose during a cold" by asking hundreds of customers to count and record their Kleenex use. Suppose a random sample of 250 Kleenex customers yielded the following summary statistics on the number of times they blew their noses when they had a cold:

$$\bar{y} = 57 \qquad s = 26$$

Is this sufficient evidence to dispute the researchers' claim? Test at $\alpha = .05$.

2.47 An experiment was conducted in England to compare methods of estimating the diet metabolizable energy content of commercial cat foods (*Feline Practice*, February 1986). Method A assumes a digestibility coefficient for crude protein of .91 (called the Atwater factor for crude protein). To determine the validity of this method, the researchers monitored the diets of 28 adult domestic short-haired cats. The cats were fed a diet of commercial canned cat food over a 3-week period. At the end of the trial, the apparent digestibility coefficient was determined for each cat, with these results:

$$\bar{y} = .81 \qquad s = .042$$

Test the hypothesis that the mean digestibility coefficient for crude protein in cats is less than .91, the value assumed by Method A. Use $\alpha = .01$.

S E C T I O N   2.8

ESTIMATION AND A TEST OF HYPOTHESIS CONCERNING THE DIFFERENCE BETWEEN TWO POPULATION MEANS

The reasoning employed in constructing a confidence interval and performing a statistical test for comparing two population means is identical to that discussed in Sections 2.6 and 2.7. The procedures are based on the assumption that we have selected *independent* random samples from the two populations. The parameters and sample sizes for the two populations, the sample means, and the sample variances are shown in Table 2.5. The objective of the sampling is to make an inference about the difference $(\mu_1 - \mu_2)$ between the two population means.

Because the sampling distribution of the difference between the sample means $(\bar{y}_1 - \bar{y}_2)$ is approximately normal for large samples, the large-sample techniques are based on the standardized normal z-statistic. Since the variances of the populations, σ_1^2 and σ_2^2, will rarely be known, we will estimate their values using s_1^2 and s_2^2.

TABLE 2.5

	POPULATION	
	1	2
Sample size	n_1	n_2
Population mean	μ_1	μ_2
Population variance	σ_1^2	σ_2^2
Sample mean	$\bar{y}_1$	$\bar{y}_2$
Sample variance	s_1^2	s_2^2

To employ these large-sample techniques, we recommend that both sample sizes be at least 30. The large-sample confidence interval and test are summarized in the boxes.

LARGE-SAMPLE CONFIDENCE INTERVAL FOR $(\mu_1 - \mu_2)$

$$(\bar{y}_1 - \bar{y}_2) \pm z_{\alpha/2}\sigma_{(\bar{y}_1 - \bar{y}_2)}* = (\bar{y}_1 - \bar{y}_2) \pm z_{\alpha/2}\sqrt{\frac{\sigma_1^2}{n_1} + \frac{\sigma_2^2}{n_2}}$$

Assumptions: The two samples are randomly and independently selected from the two populations. The sample sizes, n_1 and n_2, are large enough so that $\bar{y}_1$ and $\bar{y}_2$ each have approximately normal sampling distributions and so that s_1^2 and s_2^2 provide good approximations to σ_1^2 and σ_2^2. This will be true if $n_1 \geq 30$ and $n_2 \geq 30$.

LARGE-SAMPLE TEST OF HYPOTHESIS ABOUT $(\mu_1 - \mu_2)$

ONE-TAILED TEST

H_0: $(\mu_1 - \mu_2) = D_0$
H_a: $(\mu_1 - \mu_2) < D_0$
     [or H_a: $(\mu_1 - \mu_2) > D_0$]

TWO-TAILED TEST

H_0: $(\mu_1 - \mu_2) = D_0$
H_a: $(\mu_1 - \mu_2) \neq D_0$

where D_0 = Hypothesized difference between the means (this is often 0)

Test statistic: $z = \dfrac{(\bar{y}_1 - \bar{y}_2) - D_0}{\sigma_{(\bar{y}_1 - \bar{y}_2)}}$ *Test statistic:* $z = \dfrac{(\bar{y}_1 - \bar{y}_2) - D_0}{\sigma_{(\bar{y}_1 - \bar{y}_2)}}$

where $\sigma_{(\bar{y}_1 - \bar{y}_2)} = \sqrt{\dfrac{\sigma_1^2}{n_1} + \dfrac{\sigma_2^2}{n_2}}$

Rejection region: $z < -z_\alpha$
     [or $z > z_\alpha$ when
     H_a: $(\mu_1 - \mu_2) > D_0$]

Rejection: $z < -z_{\alpha/2}$
     or $z > z_{\alpha/2}$

Assumptions: Same as for the large-sample confidence interval above.

---

*The symbol $\sigma_{(\bar{y}_1 - \bar{y}_2)}$ is used to denote the standard error of the distribution of $(\bar{y}_1 - \bar{y}_2)$.

EXAMPLE 2.11

The management of a restaurant wants to determine whether a new advertising campaign has increased its mean daily income (gross). The daily income for 50 randomly selected business days prior to the campaign's beginning was recorded. After conducting the advertising campaign and allowing a 20-day period for the advertising to take effect, the restaurant management recorded the income for another 30 randomly selected business days. These two samples will allow management to make an inference about the effect of the advertising campaign on the restaurant's daily income. A summary of the results of the two samples is shown in the table. Do these samples provide sufficient evidence for the management to conclude that the mean income has been increased by the advertising campaign? Test using $\alpha = .05$.

BEFORE CAMPAIGN	AFTER CAMPAIGN
$n_1 = 50$	$n_2 = 30$
$\bar{y}_1 = \$1,255$	$\bar{y}_2 = \$1,330$
$s_1 = \$215$	$s_2 = \$238$

SOLUTION

We can best answer this question by performing a test of hypothesis. Defining μ_1 as the mean daily income before the campaign and μ_2 as the mean daily income after the campaign, we will attempt to support the research (alternative) hypothesis that $\mu_1 < \mu_2$ [i.e., that $(\mu_1 - \mu_2) < 0$]. Thus, we will test the null hypothesis $(\mu_1 - \mu_2) = 0$, rejecting this hypothesis if $(\bar{y}_1 - \bar{y}_2)$ equals a large negative value. The elements of the test are as follows:

H_0: $(\mu_1 - \mu_2) = 0$ (i.e., $D_0 = 0$)

H_a: $(\mu_1 - \mu_2) < 0$ (i.e., $\mu_1 < \mu_2$)

Test statistic: $z = \dfrac{(\bar{y}_1 - \bar{y}_2) - D_0}{\sigma_{(\bar{y}_1 - \bar{y}_2)}} = \dfrac{(\bar{y}_1 - \bar{y}_2) - 0}{\sigma_{(\bar{y}_1 - \bar{y}_2)}}$

Rejection region: $z < -z_\alpha = -1.645$ (see Figure 2.18)

FIGURE 2.18
Rejection Region for Advertising Campaign of Example 2.11

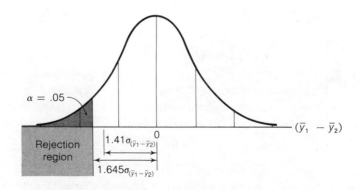

$\alpha = .05$

Rejection region

$1.41\sigma_{(\bar{y}_1 - \bar{y}_2)}$

$1.645\sigma_{(\bar{y}_1 - \bar{y}_2)}$

0

$(\bar{y}_1 - \bar{y}_2)$

We now calculate

$$z = \frac{(\bar{y}_1 - \bar{y}_2)}{\sigma_{(\bar{y}_1 - \bar{y}_2)}} = \frac{(1{,}255 - 1{,}330)}{\sqrt{\dfrac{\sigma_1^2}{n_1} + \dfrac{\sigma_2^2}{n_2}}}$$

$$\approx \frac{-75}{\sqrt{\dfrac{s_1^2}{n_1} + \dfrac{s_2^2}{n_2}}} = \frac{-75}{\sqrt{\dfrac{(215)^2}{50} + \dfrac{(238)^2}{30}}} = \frac{-75}{53.03} = -1.41$$

As you can see in Figure 2.18, the calculated z-value does not fall in the rejection region. The samples do not provide sufficient evidence, at $\alpha = .05$, for the restaurant management to conclude that the advertising campaign has increased the mean daily income. ∎

EXAMPLE 2.12

Find a 95% confidence interval for the difference in mean daily incomes before and after the advertising campaign of Example 2.11, and discuss the implications of the confidence interval.

SOLUTION

The 95% confidence interval for $(\mu_1 - \mu_2)$ is

$$(\bar{y}_1 - \bar{y}_2) \pm z_{\alpha/2} \sqrt{\frac{\sigma_1^2}{n_1} + \frac{\sigma_2^2}{n_2}}$$

Once again, we will substitute s_1^2 and s_2^2 for σ_1^2 and σ_2^2, because these quantities will provide good approximations to σ_1^2 and σ_2^2 for samples as large as $n_1 = 50$ and $n_2 = 30$. Then, the 95% confidence interval for $(\mu_1 - \mu_2)$ is

$$(1{,}255 - 1{,}330) \pm 1.96 \sqrt{\frac{(215)^2}{50} + \frac{(238)^2}{30}} = -75 \pm 103.95$$

Thus, we estimate the difference in mean daily income to fall in the interval −$178.95 to $28.95. In other words, we estimate that μ_2, the mean daily income *after* the advertising campaign, could be larger than μ_1, the mean daily income *before* the campaign, by as much as $178.95 per day or it could be less than μ_1 by as much as $28.95 per day.

Now what should the restaurant management do? You can see that the sample sizes collected in the experiment were not large enough to detect a difference in $(\mu_1 - \mu_2)$. To be able to detect a difference (if in fact a difference exists), management will have to repeat the experiment and increase the sample sizes. This will reduce the width of the confidence interval for $(\mu_1 - \mu_2)$. The restaurant management's best estimate of $(\mu_1 - \mu_2)$ is the point estimate $(\bar{y}_1 - \bar{y}_2) = -\$75$. Thus, management must decide whether the cost of conducting the advertising campaign is outweighed by a possible gain in mean daily income estimated at $75 (but which might be as large as $178.95 or could be as low as −$28.95). Based on this analysis, management will decide whether to continue the experiment or reject the new advertising program as a poor investment. ∎

The small-sample statistical techniques are based on the assumptions that both populations have normal probability distributions and that the variation within the two populations is of the same magnitude, i.e., $\sigma_1^2 = \sigma_2^2$. When these assumptions are approximately satisfied, we can employ a Student's t-statistic to find a confidence interval and test a hypothesis concerning $(\mu_1 - \mu_2)$. The techniques are summarized in the accompanying boxes.

SMALL-SAMPLE CONFIDENCE INTERVAL FOR $(\mu_1 - \mu_2)$ (INDEPENDENT SAMPLES)

$$(\bar{y}_1 - \bar{y}_2) \pm t_{\alpha/2} \sqrt{s_p^2\left(\frac{1}{n_1} + \frac{1}{n_2}\right)}$$

where

$$s_p^2 = \frac{(n_1 - 1)s_1^2 + (n_2 - 1)s_2^2}{n_1 + n_2 - 2}$$

is a "pooled" estimate of the common population variance and $t_{\alpha/2}$ is based on $(n_1 + n_2 - 2)$ df.

Assumptions: 1. Both sampled populations have relative frequency distributions that are approximately normal.
2. The population variances are equal.
3. The samples are randomly and independently selected from the populations.

SMALL-SAMPLE TEST OF HYPOTHESIS ABOUT $(\mu_1 - \mu_2)$ (INDEPENDENT SAMPLES)

ONE-TAILED TEST

H_0: $(\mu_1 - \mu_2) = D_0$
H_a: $(\mu_1 - \mu_2) < D_0$
      [or H_a: $(\mu_1 - \mu_2) > D_0$]

Test statistic: $t = \dfrac{(\bar{y}_1 - \bar{y}_2) - D_0}{\sqrt{s_p^2\left(\frac{1}{n_1} + \frac{1}{n_2}\right)}}$

Rejection region: $t < -t_\alpha$
      [or $t > t_\alpha$ when
   H_a: $(\mu_1 - \mu_2) > D_0$]

where t_α is based on $(n_1 + n_2 - 2)$ df

TWO-TAILED TEST

H_0: $(\mu_1 - \mu_2) = D_0$
H_a: $(\mu_1 - \mu_2) \neq D_0$

Test statistic: $t = \dfrac{(\bar{y}_1 - \bar{y}_2) - D_0}{\sqrt{s_p^2\left(\frac{1}{n_1} + \frac{1}{n_2}\right)}}$

Rejection region: $t < -t_{\alpha/2}$
      or $t > t_{\alpha/2}$

Assumptions: Same as for the small-sample confidence interval for $(\mu_1 - \mu_2)$ in the previous box.

EXAMPLE 2.13

Suppose you want to estimate the difference in annual operating costs for automobiles with rotary engines and those with standard engines. You find eight owners of cars with rotary engines and twelve owners of cars with standard engines, who have purchased their cars within the last 2 years and are willing to participate in the experiment. Each of the 20 owners keeps accurate records of the amount spent on operating his or her car (including gasoline, oil, repairs, etc.) for a 12-month period. All costs are recorded on a per-1000-mile basis to adjust for differences in mileage driven during the 12-month period. The results are summarized in the table.

ROTARY	STANDARD
$n_1 = 8$	$n_2 = 12$
$\bar{y}_1 = \$56.96$	$\bar{y}_2 = \$52.73$
$s_1 = \$4.85$	$s_2 = \$6.35$

Estimate the true difference $(\mu_1 - \mu_2)$ between the mean operating cost per 1,000 miles for cars with rotary and standard engines. Use a confidence coefficient of .90.

SOLUTION

The objective of this experiment is to obtain a 90% confidence interval for $(\mu_1 - \mu_2)$. To use the small-sample confidence interval for $(\mu_1 - \mu_2)$, you will have to make the following assumptions:

1. The operating cost per 1,000 miles is normally distributed for cars with rotary engines and for cars with standard engines. Since these costs are actually averages (because we observe them on a per-1,000-mile basis), the Central Limit Theorem lends credence to this assumption.
2. The variance in cost is the same for the two types of cars. Under these circumstances, we might expect the variation in costs from automobile to automobile to be about the same for both types of engines.
3. The samples are randomly and independently selected from the two populations. We have randomly chosen 20 different owners for the two samples in such a way that the cost measurement for one owner is not dependent on the cost measurement for any other owner. Therefore, this assumption would be valid.

The first step in performing the test is to calculate the pooled estimate of variance:

$$s_p^2 = \frac{(n_1 - 1)s_1^2 + (n_2 - 1)s_2^2}{n_1 + n_2 - 2}$$

$$= \frac{(8 - 1)(4.85)^2 + (12 - 1)(6.35)^2}{8 + 12 - 2}$$

$$= 33.7892$$

Then, the 90% confidence interval for $(\mu_1 - \mu_2)$, the difference in mean operating costs for the two types of automobiles, is

$$(\bar{y}_1 - \bar{y}_2) \pm t_{\alpha/2} \sqrt{s_P^2 \left(\frac{1}{n_1} + \frac{1}{n_2} \right)} = (56.96 - 52.73) \pm t_{.05} \sqrt{33.7892 \left(\frac{1}{8} + \frac{1}{12} \right)}$$

$$= 4.23 \pm 1.734(2.653) = 4.23 \pm 4.60$$

or $(-.37, 8.83)$, where $t_{\alpha/2}$ is based on $(n_1 + n_2 - 2)$ df. Therefore, we estimate the difference in mean operating costs between cars with rotary engines and those with standard engines to fall in the interval from $-\$.37$ to \$8.83. In other words, we estimate the mean operating costs for rotary engines to be anywhere from \$.37 less than to \$8.83 more than the operating costs per 1,000 miles for standard engines. Although the sample means seem to suggest that standard cars are cheaper to operate, there is insufficient evidence to indicate that $(\mu_1 - \mu_2)$ differs from 0 because the interval includes 0 as a possible value for $(\mu_1 - \mu_2)$. To show a difference in mean operating costs (if it exists), you will have to increase the sample sizes and, thereby, narrow the width of the confidence interval for $(\mu_1 - \mu_2)$.   ■

EXAMPLE 2.14

Refer to Example 2.13. Suppose you wanted to estimate the difference in mean operating costs $(\mu_1 - \mu_2)$ correct to within \$3.00 with probability near .90. How many automobiles of each type must be selected in order to obtain this estimate?

SOLUTION

This problem is similar to Example 2.8. We want to estimate $(\mu_1 - \mu_2)$ with a confidence interval that has a half-width less than or equal to \$3.00. That is, we want

$$t_{.05} \sqrt{s_P^2 \left(\frac{1}{n_1} + \frac{1}{n_2} \right)} = 3.00$$

We will use equal numbers for each type of automobile, i.e., we will let $n_1 = n_2$. Since the degrees of freedom for $t_{.05}$ depends on n (and hence is unknown) and since the value of s_P^2 depends on the data we collect, we will approximate both these values. You can see from the tabulated values of $t_{.05}$ (Table 2 of Appendix D) that $t_{.05}$ approaches $z_{.05} = 1.645$ as n gets close to 30. Consequently, $t_{.05}$ might be 1.64. The best approximation to s_P^2 would be the value computed from the samples in Example 2.13, $s_P^2 = 33.7892$. Substituting these values into the equation and solving for $n_1 = n_2 = n$, we obtain

$$t_{.05} \sqrt{s_P^2 \left(\frac{1}{n_1} + \frac{1}{n_2} \right)} = 3.00$$

$$1.64 \sqrt{33.7892 \left(\frac{1}{n} + \frac{1}{n} \right)} = 3.00$$

$$\sqrt{n} = 4.49$$

$$n = 20.2$$

Therefore, we will need approximately 20 automobiles of each type to estimate the difference in mean costs correct to within $3.00 with probability near .90. This is an approximate solution for n_1 and n_2, but it will be fairly close to the exact solution. Since the solution to the equation is slightly larger than 20, you may wish to include 21 automobiles in each sample to be fairly certain that you have enough information about $(\mu_1 - \mu_2)$.   ■

EXAMPLE 2.15

Suppose that instead of estimation, we had wanted to test the null hypothesis that the mean operating costs in Example 2.13 were equal. Conduct the test against the alternative hypothesis $(\mu_1 - \mu_2) \neq 0$. Test using $\alpha = .10$.

SOLUTION

The elements of the test are

$$H_0: \ (\mu_1 - \mu_2) = 0 \quad (\text{i.e., } D_0 = 0) \qquad H_a: \ (\mu_1 - \mu_2) \neq 0$$

$$\text{Test statistic:} \quad t = \frac{(\bar{y}_1 - \bar{y}_2) - D_0}{\sqrt{s_p^2\left(\dfrac{1}{n_1} + \dfrac{1}{n_2}\right)}}$$

$$= \frac{(56.96 - 52.73) - 0}{\sqrt{33.7892\left(\dfrac{1}{8} + \dfrac{1}{12}\right)}}$$

$$= 1.59$$

Rejection region: Since s_p^2 is based on $(n_1 + n_2 - 2) = 8 + 12 - 2 = 18$ df, we will reject the null hypothesis and conclude that the mean operating costs differ if $t > 1.734$ or $t < -1.734$. The rejection region will appear as shown in Figure 2.19.

FIGURE 2.19
Rejection Region for the Test of Example 2.15

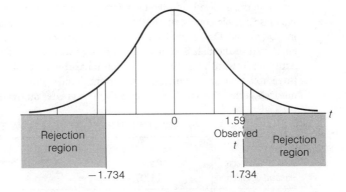

Conclusion: Since the observed value of $t = 1.59$ does not fall in the rejection region, there is insufficient evidence to conclude that the mean operating costs differ for the two engines. Note that this conclusion agrees with the conclusion of Example 2.13.   ■

The two-sample t-statistic is a powerful tool for comparing population means when the necessary assumptions are satisfied. It has also been shown to retain its usefulness when the sampled populations are only approximately normally distributed. And, when the sample sizes are equal, the assumption of equal population variances can be relaxed. That is, when $n_1 = n_2$, σ_1^2 and σ_2^2 can be quite different and the test statistic will still have (approximately) a Student's t-distribution.

EXERCISES 2.48–2.58

2.48 Describe the sampling distribution of $(\bar{y}_1 - \bar{y}_2)$.

2.49 A large supermarket chain is interested in determining whether a difference exists between the mean shelf-life (in days) of brand S bread and brand H bread. Random samples of 50 freshly baked loaves of each brand were tested, with the results shown in the table.

BRAND S	BRAND H
$\bar{y}_1 = 4.1$	$\bar{y}_2 = 5.2$
$s_1 = 1.2$	$s_2 = 1.4$

a. Is there sufficient evidence to conclude that a difference does exist between the mean shelf-lives of brand S and brand H? Test using $\alpha = .05$.

b. Construct a 90% confidence interval for $(\mu_1 - \mu_2)$. Give an interpretation of your confidence interval.

2.50 Are Japanese managers and their workers more motivated than their American counterparts? Recent claims in the literature tout the superiority of the Japanese-inspired "Theory Z" management for business productivity. The "Theory Z" philosophy emphasizes trusting, intimate, and subtle relationships between management and workers; directness and confrontation are avoided. To investigate this phenomenon, researchers surveyed middle-aged Japanese and American business managers. The Japanese sample consisted of 100 managers selected at a 2-day management seminar in Tokyo and Osaka, while the American sample was made up of 211 managers employed in the Bell System. Each manager was administered the Sarnoff Survey of Attitudes Toward Life (SSATL), which measures motivation for upward mobility in three areas: advancement, forward striving, and money. The SSATL scores are summarized in the table (higher scores indicate a greater motivation for upward mobility).

	AMERICAN MANAGERS	JAPANESE MANAGERS
Sample size	211	100
Mean SSATL score	65.75	79.83
Standard deviation	11.07	6.41

Source: Howard, A., Shudo, K., and Umeshima, M. "Motivation and values among Japanese and American managers," *Personnel Psychology*, Winter 1983, Vol. 36, No. 4, pp. 883–898.

a. Do the data provide sufficient evidence to indicate that Japanese managers, on average, are more motivated for upward mobility than American managers? Test using $\alpha = .05$.

b. Find a 95% confidence interval for the difference in mean SSATL scores for American and Japanese managers. Interpret the interval.

2.51 According to many experts, the single most important goal of any corporation should be maximization of shareholder wealth (measured as stock-price maximization). But do the stocks of companies that adopt such a long-term financial strategy perform better than other stocks? To answer this question, Pantalone and Welch gathered stock price and corporate strategy information on a sample of 111 companies listed in the *Standard & Poor's 500.* Of these, 23 were identified as shareholder-wealth maximizers and 88 as having alternative corporate goals based on a survey questionnaire sent to chief executives. The accompanying table gives summary information on the return on equity (measured in percent) earned by the two groups of companies over a 5-year period (1977–1981). Is there sufficient evidence to indicate that companies that aim to maximize shareholder wealth have a higher mean return on equity than companies with alternative corporate goals? Test using $\alpha = .05$.

	MAXIMIZE SHAREHOLDER WEALTH	ALTERNATIVE CORPORATE GOALS
Number of companies	23	88
Mean return on equity	16.55	14.80
Standard deviation	7.61	5.17

Source: Pantalone, C. and Welch, J. B. "The usefulness of public information about corporate goals," *Quarterly Journal of Business & Economics,* Vol. 25, No. 4, Autumn 1986, pp. 29–39 (Table II).

2.52 The label *Machiavellian* was derived from the 16th-century Florentine writer Niccolo Machiavelli, who described ways of manipulating others to accomplish one's objective. Critics often accuse marketers of being manipulative and unethical, or Machiavellian in nature. Researchers at Texas Tech University explored the question of whether "marketers are more Machiavellian than others." The Machiavellian scores (measured by the Mach IV Scale) for a sample of marketing professionals were recorded and compared to the Machiavellian scores for other groups of people, including a sample of college students in an earlier study. The results are summarized in the accompanying table (higher scores are associated with Machiavellian attitudes).

	MARKETING PROFESSIONALS	COLLEGE STUDENTS
Sample size	1,076	1,782
Mean score	85.7	90.7
Standard deviation	13.2	14.3

Source: Hunt, S. D. and Chonko, L. B. "Marketing and Machiavellianism," *Journal of Marketing,* Summer 1984, Vol. 48, No. 3, pp. 30–42. Reprinted by permission of the American Marketing Association.

a. Construct a 99% confidence interval for the difference in mean Machiavellian scores between marketing professionals and college students.
b. Based on the interval constructed in part a, comment on the statement that "marketers are more Machiavellian than college students."

2.53 Two manufacturers of corrugated fiberboard each claim that the strength of their product tests at more than 360 pounds per square inch, on the average. As a result of consumer complaints, a consumer products testing firm believes that firm A's product is stronger than firm B's. To test its belief, 100 fiberboards were chosen randomly from firm A's inventory and 100 were chosen from firm B's inventory. The results of tests run on the samples are summarized in the accompanying table.

A	B
$\bar{y}_1 = 365$	$\bar{y}_2 = 352$
$s_1 = 23$	$s_2 = 41$

a. Does the sample information support the consumer products testing firm's belief? Test using $\alpha = .05$.

b. What assumptions did you make in conducting the test in part **a**? Do you think such assumptions could comfortably be made in practice? Why or why not?

c. Does the sample information support firm A's claim that the mean strength of its corrugated fiberboard is more than 360 pounds per square inch? Test using $\alpha = .10$.

2.54 In order to use the t-statistic to test for differences in the means of two populations, what assumptions must be made about the two sampled populations? What assumptions must be made about the two samples?

2.55 A key aspect of organizational buying is negotiation. S. W. Clopton, a professor in the Marketing Department at Virginia Polytechnic Institute, investigated several issues pertaining to buyer–seller negotiations. One aspect of the analysis involved a comparison of two types of bargaining strategies: competitive bargaining and coordinative bargaining. A *competitive* strategy is characterized by inflexible behavior aimed at forcing concessions, while a *coordinative* strategy involves a problem-solving orientation to negotiations with a high degree of trust and cooperation. A sample of organizational buyers was recruited to participate in a particular negotiation experiment. In one negotiation setting, where the maximum profit was fixed, eight buyers used the competitive bargaining strategy and eight buyers used the coordinative bargaining strategy. Summary statistics on individual savings for the two types of buyers are provided in the table. Is there evidence of a difference in mean buyer savings for the two types of bargaining strategies? Test using $\alpha = .05$.

	COMPETITIVE BARGAINING	COORDINATIVE BARGAINING
Sample size	8	8
Mean savings	$1,706.25	$2,106.25
Standard deviation	$532.81	$359.99

Source: Clopton, S. W. "Seller and buying firm factors affecting industrial buyer's negotiation behavior and outcome," *Journal of Marketing Research*, February 1984, Vol. 21, No. 1, pp. 39–53. Reprinted by permission of the American Marketing Association.

2.56 Suppose you are the personnel manager for a company and you suspect a difference in the mean length of work time lost due to sickness for two types of employees: those who work at night versus those who work during the day. In particular, you suspect that the mean time lost for the night shift exceeds the mean time lost for the day shift. To check your theory, you randomly sample the records for ten employees for each shift category and record the number of days lost due to sickness within the past year. The data are shown in the accompanying table.

NIGHT SHIFT, 1		DAY SHIFT, 2	
21	2	13	18
10	19	5	17
14	6	16	3
33	4	0	24
7	12	7	1
$\bar{y}_1 = 12.8$		$\bar{y}_2 = 10.4$	
$\sum_{i=1}^{n} y_i^2 = 2{,}436$		$\sum_{i=1}^{n} y_i^2 = 1{,}698$	

a. Calculate s_1^2 and s_2^2.

b. Show that the pooled estimate of the common population standard deviation, σ, is 8.86. Look at the range of the observations within each of the two samples. Does it appear that the estimate, 8.86, is a reasonable value for σ?

c. If μ_1 and μ_2 represent the mean number of days per year lost due to sickness for the night and day shifts, respectively, test the null hypothesis $H_0: \mu_1 = \mu_2$ against the alternative $H_a: \mu_1 > \mu_2$. Use $\alpha = .05$. Do the data provide sufficient evidence to indicate that $\mu_1 > \mu_2$?

d. What assumptions must be satisfied so that the t-test from part c is valid?

2.57 An experiment was conducted to determine if individuals could be taught how to make decisions rationally and if such training would improve the quality of their career decisions. A questionnaire was administered to a sample of 147 community college students in California who were enrolled in career guidance classes in order to assess the decision-making style of each. Of these, 69 were classified as rational decision-makers, a style characterized by a systematic and logical approach to decision-making. Several weeks later, these 69 subjects were randomly divided into two groups. The experimental group (34 students) received instruction in rational decision-making while the control group (35 students) did not. At the end of the instruction period, all subjects completed a multiple-choice test designed to assess the extent to which an individual knows how to apply rational principles in job-decision situations. The results are summarized in the accompanying table. (Higher scores indicate higher adherence to the rational style of decision-making.)

	EXPERIMENTAL GROUP	CONTROL GROUP
Sample size	34	35
Mean	50.26	47.34
Standard deviation	6.67	11.52

Source: Krumboltz, J. D., Kinnier, R. T., Rude, S. S., Sherba, D. S., and Hamel, D. A. "Teaching a rational approach to career decision making: Who benefits most?" *Journal of Vocational Behavior*, Vol. 29, No. 8, August 1986, pp. 1–6 (Table 1).

a. Do the samples provide sufficient evidence to conclude that students who receive training in rational decision-making score higher, on average, on the multiple-choice exam than students who receive no training? Test using $\alpha = .05$.

b. Construct a 90% confidence interval for the difference between the mean test scores of the two groups of students. Interpret the interval. Is the result consistent with your answer in part **a**?

2.58 Suppose your plant purifies its waste and discharges the water into a local river. An EPA inspector has collected water specimens of the discharge at your plant and also water specimens in the river upstream from your plant. Each water specimen is divided into five parts, the bacteria count is read on each, and the mean count for each specimen is reported. The average bacteria count readings for each of six specimens are reported in the table for the two locations.

PLANT DISCHARGE	UPSTREAM
30.1	29.7
36.2	30.3
33.4	26.4
28.2	27.3
29.8	31.7
34.9	32.3

a. Why would the bacteria count readings shown in the table tend to be approximately normally distributed?

b. Do the data provide sufficient evidence to indicate that the mean bacteria count for the discharge at your plant exceeds the mean of the count upstream? Use $\alpha = .05$.

SECTION 2.9

COMPARING TWO POPULATION VARIANCES: INDEPENDENT RANDOM SAMPLES

Suppose you want to use the two-sample t-statistic to compare the mean productivity of two paper mills. However, you are concerned that the assumption of equal variances of the productivity for the two plants may be unrealistic. It would be helpful to have a statistical procedure to check the validity of this assumption.

The common statistical procedure for comparing population variances σ_1^2 and σ_2^2 is to make an inference about the ratio, σ_1^2/σ_2^2, using the ratio of the sample variances, s_1^2/s_2^2. Thus, we will attempt to support the research hypothesis that the ratio σ_1^2/σ_2^2 differs from 1 (i.e., the variances are unequal) by testing the null hypothesis that the ratio equals 1 (i.e., the variances are equal).

$$H_0: \quad \frac{\sigma_1^2}{\sigma_2^2} = 1 \quad (\sigma_1^2 = \sigma_2^2) \qquad H_a: \quad \frac{\sigma_1^2}{\sigma_2^2} \neq 1 \quad (\sigma_1^2 \neq \sigma_2^2)$$

We will use the test statistic

$$F = \frac{s_1^2}{s_2^2}$$

To establish a rejection region for the test statistic, we need to know how s_1^2/s_2^2 is distributed in repeated sampling. That is, we need to know the sampling distribution of s_1^2/s_2^2. As you will subsequently see, the sampling distribution of s_1^2/s_2^2 depends on two of the assumptions already required for the t-test, namely:

1. The two sampled populations are normally distributed.
2. The samples are randomly and independently selected from their respective populations.

When these assumptions are satisfied and when the null hypothesis is true (i.e., $\sigma_1^2 = \sigma_2^2$), the sampling distribution of s_1^2/s_2^2 is an **F-distribution** with $(n_1 - 1)$ df and $(n_2 - 1)$ df, respectively. The shape of the F-distribution depends on the degrees of freedom associated with s_1^2 and s_2^2, i.e., $(n_1 - 1)$ and $(n_2 - 1)$. An F-distribution with 7 and 9 df is shown in Figure 2.20. As you can see, the distribution is skewed to the right.

FIGURE 2.20
An F-Distribution with 7 and 9 df

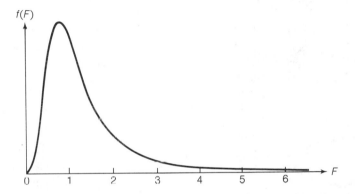

When the population variances are unequal, we expect the ratio of the sample variances, $F = s_1^2/s_2^2$, to be either very large or very small. Therefore, we will need to find F-values corresponding to the tail areas of the F-distribution in order to establish the rejection region for our test of hypothesis. The upper-tail F-values can be found in Tables 3, 4, 5, and 6 of Appendix D. Table 4 is partially reproduced in Table 2.6 (page 58). It gives F-values that correspond to $\alpha = .05$ upper-tail areas for different degrees of freedom. The columns of the tables correspond to various degrees of freedom for the numerator sample variance s_1^2, while the rows correspond to the degrees of freedom for the denominator sample variance s_2^2. Thus, if the numerator degrees of freedom is 7 and the denominator degrees of freedom is 9, we look in the seventh column and ninth row to find $F_{.05} = 3.29$. As shown in Figure 2.21, $\alpha = .05$ is the tail area to the right of 3.29 in the F-distribution with 7 and 9 df. That is, if $\sigma_1^2 = \sigma_2^2$, the probability that the F-statistic will exceed 3.29 is $\alpha = .05$.

TABLE 2.6   Reproduction of Part of Table 4 of Appendix D: $\alpha = .05$

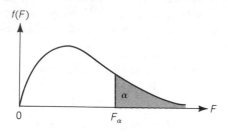

ν_2 \ ν_1	1	2	3	4	5	6	7	8	9
1	161.4	199.5	215.7	224.6	230.2	234.0	236.8	238.9	240.5
2	18.51	19.00	19.16	19.25	19.30	19.33	19.35	19.37	19.38
3	10.13	9.55	9.28	9.12	9.01	8.94	8.89	8.85	8.81
4	7.71	6.94	6.59	6.39	6.26	6.16	6.09	6.04	6.00
5	6.61	5.79	5.41	5.19	5.05	4.95	4.88	4.82	4.77
6	5.99	5.14	4.76	4.53	4.39	4.28	4.21	4.15	4.10
7	5.59	4.74	4.35	4.12	3.97	3.87	3.79	3.73	3.68
8	5.32	4.46	4.07	3.84	3.69	3.58	3.50	3.44	3.39
9	5.12	4.26	3.86	3.63	3.48	3.37	3.29	3.23	3.18
10	4.96	4.10	3.71	3.48	3.33	3.22	3.14	3.07	3.02
11	4.84	3.98	3.59	3.36	3.20	3.09	3.01	2.95	2.90
12	4.75	3.89	3.49	3.25	3.11	3.00	2.91	2.85	2.80
13	4.67	3.81	3.41	3.18	3.03	2.92	2.83	2.77	2.71
14	4.60	3.74	3.34	3.11	2.96	2.85	2.76	2.70	2.65

NUMERATOR DEGREES OF FREEDOM (ν_1); DENOMINATOR DEGREES OF FREEDOM (ν_2)

FIGURE 2.21
An F-Distribution for 7 and 9 df: $\alpha = .05$

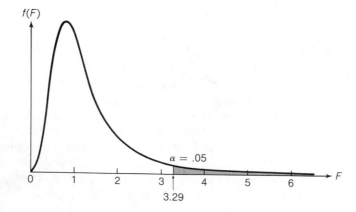

Suppose we want to compare the variability in production for two paper mills and we have obtained the following results:

SAMPLE 1	SAMPLE 2
$n_1 = 13$ days	$n_2 = 18$ days
$\bar{y}_1 = 26.3$ production units	$\bar{y}_2 = 19.7$ production units
$s_1 = 8.2$ production units	$s_2 = 4.7$ production units

To form the rejection region for a two-tailed F-test we want to make certain that the upper tail is used, because only the upper-tail values of F are shown in Tables 3, 4, 5, and 6. To accomplish this, **we will always place the larger sample variance in the numerator of the F-test**. This has the effect of doubling the tabulated value for α, since we double the probability that the F-ratio will fall in the upper tail by always placing the larger sample variance in the numerator. In effect, we make the test two-tailed by putting the larger variance in the numerator rather than establishing rejection regions in both tails.

Thus, for our production example, we have a numerator s_1^2 with df $= n_1 - 1 = 12$ and a denominator s_2^2 with df $= n_2 - 1 = 17$. Therefore, the test statistic will be

$$F = \frac{\text{Larger sample variance}}{\text{Smaller sample variance}} = \frac{s_1^2}{s_2^2}$$

and we will reject H_0: $\sigma_1^2 = \sigma_2^2$ for $\alpha = .10$ if the calculated value of F exceeds the tabulated value:

$$F_{.05} = 2.38 \quad \text{(see Figure 2.22)}$$

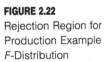

FIGURE 2.22
Rejection Region for
Production Example
F-Distribution

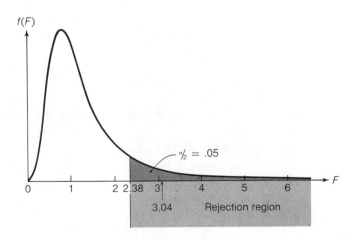

Now, what do the data tell us? We calculate

$$F = \frac{s_1^2}{s_2^2} = \frac{(8.2)^2}{(4.7)^2} = 3.04$$

and compare it to the rejection region shown in Figure 2.22. Since the calculated F-value, 3.04, falls in the rejection region, the data provide sufficient evidence to indicate that the population variances differ. Consequently, we would be reluctant to use the two-sample t-statistic to compare the population means, since the assumption of equal population variances is apparently untrue.

What would you have concluded if the value of F calculated from the samples had not fallen in the rejection region? Would you conclude that the null hypothesis of equal variances is true? No, because then you risk the possibility of a Type II error (accepting H_0 when H_a is true) without knowing the probability of this error (the probability of accepting H_0: $\sigma_1^2 = \sigma_2^2$ when in fact it is false). Since we will not consider the calculation of β for specific alternatives in this text, when the F-statistic does not fall in the rejection region, we simply conclude that **insufficient sample evidence exists to refute the null hypothesis that** $\sigma_1^2 = \sigma_2^2$.

The F-test for equal population variances is summarized in the next box.

F-TEST FOR EQUAL POPULATION VARIANCES

ONE-TAILED TEST	**TWO-TAILED TEST**
H_0: $\dfrac{\sigma_1^2}{\sigma_2^2} = 1$ (or $\sigma_1^2 = \sigma_2^2$)	H_0: $\dfrac{\sigma_1^2}{\sigma_2^2} = 1$ (or $\sigma_1^2 = \sigma_2^2$)
H_a: $\dfrac{\sigma_1^2}{\sigma_2^2} > 1$ (or $\sigma_1^2 > \sigma_2^2$)	H_a: $\dfrac{\sigma_1^2}{\sigma_2^2} \neq 1$ (or $\sigma_1^2 \neq \sigma_2^2$)
Test statistic: $F = \dfrac{s_1^2}{s_2^2}$	*Test statistic:* $F = \dfrac{\text{Larger sample variance}}{\text{Smaller sample variance}} = \dfrac{s_1^2}{s_2^2}$ where the samples are numbered so that $s_1^2 > s_2^2$
Rejection region: $F > F_\alpha$ where the degrees of freedom for F are:	*Rejection region:* $F > F_{\alpha/2}$ where the degrees of freedom for F are:
$n_1 - 1 =$ df numerator sample variance	$n_1 - 1 =$ df larger (numerator) sample variance
$n_2 - 1 =$ df denominator sample variance	$n_2 - 1 =$ df smaller (denominator) sample variance

Assumptions: 1. Both sampled populations are normally distributed.
2. The samples are random and independent.

EXAMPLE 2.16

Refer to Example 2.13, in which we used the two-sample t-statistic to compare the mean operating cost of cars with rotary and standard engines. Perform the F-test to check the assumption that the population variances are equal. Use $\alpha = .10$.

SOLUTION

The elements of the test are as follows:

$$H_0: \quad \frac{\sigma_1^2}{\sigma_2^2} = 1 \qquad H_a: \quad \frac{\sigma_1^2}{\sigma_2^2} \neq 1$$

$$\text{Test statistic:} \quad F = \frac{\text{Larger sample variance}}{\text{Smaller sample variance}} = \frac{s_1^2}{s_2^2}$$

To find the rejection region, we proceed as follows: The numerator degrees of freedom are df $= n_2 - 1 = 12 - 1 = 11$ and the denominator degrees of freedom are $n_1 - 1 = 8 - 1 = 7$. From Table 4 of Appendix D, we find

$$F_{.05} \approx 3.60$$

(Since no table value is given for a numerator df $= 11$, we average the entries at 10 and 12 to obtain $F_{.05} \approx 3.60$.)
   We now calculate

$$F = \frac{s_2^2}{s_1^2} = \frac{(6.35)^2}{(4.85)^2} = 1.71$$

This F-value is not in the rejection region. Therefore, there is insufficient evidence at the $\alpha = .10$ level to refute the assumption of equal population variances. ∎

The following example shows that the F-statistic is sometimes used to compare population variances in their own right, rather than just to check the validity of an assumption.

EXAMPLE 2.17

Suppose an investor wants to compare the risks associated with two different stocks, where the risk of a given stock is measured by the variation in daily price changes. Suppose we obtain random samples of 25 daily price changes for stock 1 and 25 for stock 2. The sample results are summarized in the table.

STOCK 1	STOCK 2
$n_1 = 25$	$n_2 = 25$
$\bar{y}_1 = .250$	$\bar{y}_2 = .125$
$s_1 = .76$	$s_2 = .46$

Compare the risks associated with the two stocks by testing the null hypothesis that the variances of the price changes for the stocks are equal. Use $\alpha = .10$.

SOLUTION

$$H_0: \quad \frac{\sigma_1^2}{\sigma_2^2} = 1 \qquad H_a: \quad \frac{\sigma_1^2}{\sigma_2^2} \neq 1$$

Test statistic: $\quad F = \dfrac{\text{Larger sample variance}}{\text{Smaller sample variance}} = \dfrac{s_1^2}{s_2^2}$

Rejection region: $\quad F > F_{.05}$

where F is based on $n_1 - 1 = 24$ df and $n_2 - 1 = 24$ df

Assumptions: 1. The changes in daily stocks have relative frequency distributions that are approximately normal.

2. The stock samples are randomly and independently selected from a set of daily stock reports.

We calculate

$$F = \frac{s_1^2}{s_2^2} = \frac{(.76)^2}{(.46)^2} = 2.73$$

The calculated F exceeds the rejection value of 1.98. Therefore, we conclude that the variances of daily price changes *differ* for the two stocks. In fact, it appears that the risk, as measured by the variance of daily price changes, is greater for stock 1 than for stock 2. How much reliability can we place in this inference? Only one time in ten (since $\alpha = .10$), on the average, would this statistical test lead us to conclude erroneously that σ_1^2 and σ_2^2 were different when in fact they were equal. ∎

The previous examples demonstrate how to conduct a two-tailed F-test when the alternative hypothesis is $H_a: \sigma_1^2 \neq \sigma_2^2$. One-tailed tests for determining whether one population variance is larger than another population variance (i.e., $H_a: \sigma_1^2 > \sigma_2^2$) are conducted similarly. However, the α-value no longer needs to be doubled since the area of rejection lies only in the upper (or lower) tail area of the F-distribution. The procedure for conducting an upper-tailed F-test is outlined in the previous box. Whenever you conduct a one-tailed F-test, be sure to write H_a in the form of an upper-tailed test. This can be accomplished by numbering the populations so that the variance hypothesized to be larger in H_a is associated with population 1 and the hypothesized smaller variance is associated with population 2.

EXERCISES 2.59–2.64

2.59 Under what conditions does the sampling distribution of s_1^2/s_2^2 have an F-distribution?

2.60 Use Tables 3, 4, 5, and 6 of Appendix D to find F_α for α, numerator df, and denominator df equal to:
a. .05, 8, 7   **b.** .01, 15, 20   **c.** .025, 12, 5
d. .01, 5, 25   **e.** .10, 5, 10   **f.** .05, 20, 9

2.61 Refer to the English study on estimating the metabolizable energy (ME) content of commercial cat foods in Exercise 2.47. In addition to the sample of 28 cats that were fed canned food, a second (independent) sample of 29 cats were fed a commercial brand of dry cat food. The ME content for each cat was measured (in kilocalories per gram) using Method A, with the results shown in the table.

	CANNED FOOD	DRY FOOD
Sample size	28	29
Mean ME content	.96	3.70
Standard deviation	.26	.48

Source: Kendall, P. T., Burger, I. N., and Smith, P. M. "Methods of estimation of the metabolizable energy content of cat foods," *Feline Practice*, Vol. 15, No. 2, February 1986, pp. 38–44.

Conduct a test to determine whether the variation in ME content of cats fed canned food differs from the variation in ME content of cats fed dry food. Use $\alpha = .10$.

2.62 U.S. corporations, especially automakers, have felt the effects of increased competition from foreign corporations. According to the *Wall Street Journal* (Feb. 9, 1984), many U.S. corporations are countering foreign competition by merging with other large firms to "lead to more efficient, more effective, and more competitive combinations." One of the major concerns of these types of mergers, however, is the effect on the variability of returns to shareholders after stock acquisition takes place. Consider a U.S. tool company which has recently merged with another U.S. corporation in order to combat foreign competition. Samples of monthly returns, before and after the merger, yielded the summary statistics provided in the table. Is there evidence of a difference in the variation of the monthly returns before and after the merger? Test using $\alpha = .02$.

	BEFORE MERGER	AFTER MERGER
Months sampled	10	6
Mean return	8.5%	7.1%
Standard deviation	4.3%	4.8%

2.63 Suppose your firm has been experimenting with two different physical arrangements of its assembly line. It has been determined that both arrangements yield approximately the same average number of finished units per day. To obtain an arrangement that produces greater process control, you suggest that the arrangement with the smaller variance in the number of finished units produced per day be permanently adopted. Two independent random samples yield the results listed in the accompanying table. Do the samples provide sufficient evidence to conclude at $\alpha = .10$ that the variances of the two arrangements differ? If so, which arrangement would you choose? If not, what would you suggest the firm do?

ASSEMBLY LINE 1	ASSEMBLY LINE 2
$n_1 = 21$ days	$n_2 = 21$ days
$s_1^2 = 1,432$	$s_2^2 = 3,761$

2.64 The quality control department of a paper company measures the brightness (a measure of reflectance) of finished paper on a periodic basis throughout the day. Two instruments that are available to measure the paper specimens are subject to error, but they can be adjusted so that the mean readings for a control paper specimen are the same for both instruments. Suppose you are concerned about the precision of the two instruments, namely that the variation in readings from instrument 2 exceeds that for instrument 1. To check this theory, five measurements of a single paper sample are made on both instruments. The data are shown in the table. Do the data provide sufficient evidence to indicate that instrument 2 is less precise than instrument 1? Test using $\alpha = .05$.

INSTRUMENT 1	INSTRUMENT 2
29	26
28	34
30	30
28	32
30	28

SECTION 2.10

SUMMARY

The preceding sections summarize many of the basic concepts and methods presented in an introductory statistics course. We presented the concepts of a **population**, **random sampling**, and the ultimate objective of most statistical investigations, **making an inference about a population based on information contained in a sample**. Because inference implies description, we first considered two methods for describing a set of data—a graphical method using a **relative frequency distribution** and **numerical descriptive methods** that provide measures of centrality and variability for a data set.

Proceeding to inference, we noted that quantities computed from sample data—**statistics**—are used to estimate population numerical descriptive measures—**parameters**—and to make decisions about their values. In order to evaluate the properties of these statistics, we need to know the probabilities that they will assume specific sets of values in repeated sampling, i.e., we need to know their **probability sampling distributions**. If we know the sampling distribution for a statistic, we can make probabilistic statements that measure the **reliability** of the statistic when it is used as an estimator or as the basis of a decision.

Finally, we summarized the basic concepts involved in **interval estimation** and **tests of hypotheses**. In particular, we presented **large- and small-sample confidence intervals** and **statistical tests for making inferences about a single population mean** and for **comparing two population means or variances based on independent random sampling**.

To aid in the formulation of confidence intervals and test statistics, we present two summary tables. Table 2.7 contains a list of parameters and their corresponding estimators and standard errors. Once you have identified the parameter of interest in Table 2.7, confidence intervals and test statistics are formed as shown in Table 2.8.

TABLE 2.7 Some Population Parameters and Corresponding Estimators and Standard Errors

PARAMETER (θ)	ESTIMATOR ($\hat{\theta}$)	STANDARD ERROR ($\sigma_{\hat{\theta}}$)	ESTIMATE OF STANDARD ERROR ($s_{\hat{\theta}}$)
μ Mean (average)	$\bar{y}$	$\dfrac{\sigma}{\sqrt{n}}$	$\dfrac{s}{\sqrt{n}}$
$\mu_1 - \mu_2$ Difference between means (averages)	$\bar{y}_1 - \bar{y}_2$	$\sqrt{\dfrac{\sigma_1^2}{n_1} + \dfrac{\sigma_2^2}{n_2}}$	$\sqrt{\dfrac{s_1^2}{n_1} + \dfrac{s_2^2}{n_2}}$, $n_1 \geq 30,\ n_2 \geq 30$ $\sqrt{s_p^2\left(\dfrac{1}{n_1} + \dfrac{1}{n_2}\right)}$, either $n_1 < 30$ or $n_2 < 30$ where $s_p^2 = \dfrac{(n_1 - 1)s_1^2 + (n_2 - 1)s_2^2}{n_1 + n_2 - 2}$
$\dfrac{\sigma_1^2}{\sigma_2^2}$ Ratio of variances	$\dfrac{s_1^2}{s_2^2}$	(not necessary)	(not necessary)

TABLE 2.8
Formulation of Confidence
Intervals for a Population
Parameter θ and Test
Statistics for H_0: $\theta = \theta_0$,
where $\theta = \mu$ or $(\mu_1 - \mu_2)$

SAMPLE SIZE	CONFIDENCE INTERVAL	TEST STATISTIC
Large	$\hat{\theta} \pm z_{\alpha/2} s_{\hat{\theta}}$	$z = \dfrac{\hat{\theta} - \theta_0}{s_{\hat{\theta}}}$
Small	$\hat{\theta} + t_{\alpha/2} s_{\hat{\theta}}$	$t = \dfrac{\hat{\theta} - \theta_0}{s_{\hat{\theta}}}$

Note: The test statistic for testing H_0: $\sigma_1^2/\sigma_2^2 = 1$ is
$F = s_1^2/s_2^2$ (see the box on page 60).

| | | | | | | | | | | | | | |

SUPPLEMENTARY EXERCISES 2.65–2.94

2.65 For each of the following data sets, compute $\bar{y}$, s, and s^2.
a. 11, 2, 2, 1, 9   **b.** 22, 9, 21, 15
c. 1, 0, 1, 10, 11, 11, 0   **d.** 4, 4, 4, 4

2.66 Tchebysheff's theorem states that at least $1 - (1/K^2)$ of a set of measurements will
lie within K standard deviations of the mean of the data set. Use Tchebysheff's theorem
to find the fraction of a set of measurements that will lie within:
a. 2 standard deviations of the mean ($K = 2$)
b. 3 standard deviations of the mean
c. 1.5 standard deviations of the mean

2.67 Use Table 1 of Appendix D to calculate the area under the standard normal probability
distribution between the following pairs of z-values:
a. $z = -1$ and $z = 1$ **b.** $z = -2$ and $z = 2$
c. $z = -.5$ and $z = .5$ **d.** $z = -3$ and $z = 3$

2.68 Use Table 1 of Appendix D to find each of the following:
a. $P(z \geq 2)$ **b.** $P(z \leq -2)$ **c.** $P(z \geq -1.96)$
d. $P(z \geq 0)$ **e.** $P(z \leq -.5)$ **f.** $P(z \leq -1.96)$

2.69 Suppose the random variable y has mean $\mu = 30$ and standard deviation $\sigma = 5$. How many standard deviations away from the mean of y are each of the following y-values?

   a. $y = 10$ **b.** $y = 32.5$ **c.** $y = 30$ **d.** $y = 60$

2.70 The variation in the rates of return on a bond is often used to measure the level of risk associated with buying the bond. The greater the variation, the higher the level of risk. Writing in his monthly "Your Finances" column for the *American Bar Association Journal*, Tom Potter presented the results of a simulation to compare the performance of zero-coupon bonds (those for which the interest coupons have been removed and which therefore pay no interest) to bonds with coupon payments attached. The rates of return on zero-coupon bonds are determined solely by the changes in bond prices, while the return on coupon-attached bonds depends also on the market interest rate. The simulation study yielded means and standard deviations of the rates of return for both zero-coupon and coupon-attached bonds at a fixed interest rate. The results for a market interest rate of 14% are shown in the table.

	ZERO-COUPON BONDS	COUPON-ATTACHED BONDS
Mean	12.48	12.48
Standard deviation	20.80	14.56

Source: Potter, T. "Your finances," *American Bar Association Journal*, August 1984, Vol. 70, No. 8, pp. 160–161.

   a. How many standard deviations away from the mean is a zero-coupon bond with a rate of return of -20%?

   b. How many standard deviations away from the mean is a coupon-attached bond with a rate of return of -20%?

   c. From your knowledge of the standard deviation, if a bond has a rate of return of -20%, is it more likely to be a zero-coupon bond or a coupon-attached bond? Explain.

2.71 Find a value of z, call it z_0, such that

   a. $P(z \geq z_0) = .5$ **b.** $P(z \leq z_0) = .5199$

   c. $P(z \geq z_0) = .3300$ **d.** $P(z_0 \leq z \leq .59) = .5845$

2.72 The random variable y has a normal distribution with $\mu = 80$ and $\sigma = 10$. Find the following probabilities:

   a. $P(y \leq 75)$ **b.** $P(y \geq 90)$ **c.** $P(60 \leq y \leq 70)$

   d. $P(y \geq 75)$ **e.** $P(y = 75)$ **f.** $P(y \leq 105)$

2.73 A firm believes the internal rate of return for its proposed investment can best be described by a normal distribution with mean 15% and standard deviation 3%. What is the probability that the internal rate of return for the investment will be:

   a. Greater than 20% or less than 10%?

   b. At least 6%?

   c. More than 16.5%?

2.74 Labor productivity, as measured by output per hour, rose 4.4% in manufacturing in the United States in 1985. The accompanying table gives the 1985 percent changes in manufacturing productivity for 12 countries, including the United States.

COUNTRY	1985 PERCENT CHANGE IN MANUFACTURING PRODUCTIVITY
United States	4.4
Canada	3.2
Japan	5.0
France	3.3
Germany	5.6
Italy	3.1
United Kingdom	3.4
Belgium	4.6
Denmark	.7
Netherlands	3.1
Norway	.9
Sweden	2.7

Source: Neef, A. "International trends in productivity and unit labor costs in manufacturing," *Monthly Labor Review*, Vol. 109, No. 12, December 1986, p. 13.

a. Compute $\bar{y}$, s^2, and s for the data set.

b. What percentage of the measurements would you expect to find in the interval $\bar{y} \pm 2s$?

c. Count the number of measurements that actually fall within the interval of part b, and express the interval count as a percentage of the total number of measurements. Compare this result with the answer to part b.

2.75 The distribution of the demand (in number of items per unit time) for a product can often be approximated by a normal probability distribution. For example, a bakery has found that the demand for its standard loaf of white bread is 7,200 loaves per day and the standard deviation of the demand is 300 loaves. Based on cost considerations, the company has decided that its best strategy is to produce a sufficient number of loaves so that it will fully supply demand on 94% of all days.

a. How many loaves of bread should the company produce on a given day?

b. Based on the production in part a, what percentage of days will the company be left with more than 500 loaves of unsold bread?

2.76 Due to a variety of factors, such as free agency, cable and pay television, inflation, and spendthrift owners, the average salary of a major-league baseball player has escalated from $19,000 in 1967 to $412,600 in 1987 (*New York Times*, April 11, 1987). Assume that the 1987 salaries of major-league baseball players are approximately normally distributed with a mean of $412,600 and a standard deviation of $200,000.

a. Find the probability that a randomly selected player will earn over $1,000,000 in 1987.

b. Calculate the interval $\mu \pm 3\sigma$ and locate it on a sketch of the normal distribution. Would you agree that your best guess at the minimum 1987 salary is $\mu - 3\sigma$ and the maximum 1987 salary is $\mu + 3\sigma$? Explain.

c. In 1987, the minimum salary of major-league baseball players was $62,500 (normally paid to rookies) while the maximum salary was $2,412,500 (paid to superstar Jim Rice of the Boston Red Sox). Do these figures agree with your answer to part b? If not, give a reason why.

2.77 Each year, Mattel, Inc., a manufacturer of dolls and toys, conducts marketing forecasts to estimate consumer demand for its new products. Recently, Mattel introduced new "He-Man" dolls (e.g., Masters of the Universe) in an attempt to appeal to young boys. After a period of time, Mattel's public relations director announced that sales for "He-Man" dolls were 150% over their expected sales (*Wall Street Journal*, Feb. 24, 1984). Let y be the percentage by which actual sales of a new Mattel product exceed its expected sales. (A negative percentage implies actual sales fell below expected sales.) Assume that past experience shows that y has a normal distribution with mean 6% and standard deviation 50%.

a. Find the probability that actual sales of "He-Man" dolls would exceed expected sales by more than 150%.

b. Find the probability that actual sales of "He-Man" dolls would be less than expected sales.

2.78 When long-term resource allocation decisions (e.g., building a domed stadium or constructing a major highway system) meet with unexpected major setbacks, the decision-maker must decide whether to abandon the project before irretrievable costs are incurred or to continue on in the face of probable losses. Surprisingly, researchers have found that decision-makers all too often persist, or even allocate more resources, in these situations. In an effort to explain this pattern of resource allocation behavior, a study was conducted at the University of Arizona (Northcraft and Neale, 1986). Twenty undergraduate business school students were asked to make resource allocation decisions for fictional Sunburst Investments, which had undertaken construction of an office park and tennis club. For one portion of the study, each student was asked to rate the options of finishing and not finishing the project on a 7-point scale (1 = very negative, 7 = very positive). The differences between the ratings of the two options (finishing option rating minus not finishing option rating) were calculated, and the mean and standard deviation of the 20 sample differences obtained were

$$\bar{y} = -2.05 \qquad s = 1.85$$

a. Conduct a test to determine whether subjects rate the finishing option more negative, on average, than the option of not finishing the project (i.e., test the hypothesis that the mean difference in ratings is less than 0). Use $\alpha = .10$.

b. What assumptions are required for the test in part **a** to be valid?

2.79 Suppose you want to estimate the mean percentage gain in per-share value for growth-type mutual funds over a specific 2-year period. Ten mutual funds are randomly selected from the population of all the commonly listed funds. The percentage gain figures are shown here.

12.1	−3.7	7.6	6.8	−2.3
4.6	8.4	18.1	9.2	3.0

a. Find a 90% confidence interval for the mean percentage gain for the population of funds.

b. State the assumptions required in order for this confidence interval to be valid.

2.80 Suppose a sample of $n = 50$ items is drawn from a population of manufactured products and the weight y of each item is recorded. Prior experience has shown that the weight has a probability distribution with $\mu = 6$ ounces and $\sigma = 2.5$ ounces.

Then $\bar{y}$, the sample mean, will be approximately normally distributed (because of the Central Limit Theorem).

a. Calculate $\mu_{\bar{y}}$ and $\sigma_{\bar{y}}$.

b. What is the probability that the manufacturer's sample has a mean weight between 5.75 and 6.25 ounces?

c. What is the probability that the manufacturer's sample has a mean weight less than 5.5 ounces?

d. How would the sampling distribution of $\bar{y}$ change if the sample size n were increased from 50 to 100?

2.81 What are the top corporate executives being paid? In order to answer this question, *Business Week* magazine conducts a survey of corporate executives each year. *Business Week*'s 1985 survey of executives at 258 companies revealed a 9% increase in salaries and bonuses from 1984. The top 25 corporate executives and their 1985 total cash compensations (salary and bonus) are shown in the table. Assume that these represent a sample of the highest-paid corporate executives in the United States.

CORPORATE EXECUTIVE (COMPANY)	TOTAL 1985 CASH COMPENSATION (in thousands)
Victor Posner (DWG)	$12,739
Lee A. Iacocca (Chrysler)	11,426
T. Boone Pickens, Jr. (Mesa Petroleum)	8,431
Drew Lewis (Warner Amex)	6,000
Robert L. Mitchell (Celanese)	4,756
Sidney J. Sheinberg (MCA)	4,484
Robert Anderson (Rockwell)	3,636
Clifton C. Garvin, Jr. (Exxon)	3,561
David S. Lewis (General Dynamics)	3,351
John H. Gutfreund (Phibro-Salomon)	3,206
George B. Beitzel (IBM)	3,017
Roy A. Anderson (Lockheed)	2,976
Joseph B. Flavin (Singer)	2,972
Peter A. Cohen (Shearson Lehman Bros.)	2,911
Michael D. Rose (Holiday)	2,909
Frank Price (MCA)	2,894
Richard J. Schmeelk (Phibro-Salomon)	2,839
Steven J. Ross (Warner Communications)	2,800
Gerald Greenwald (Chrysler)	2,780
David R. Banks (Beverly Enterprises)	2,778
Richard M. Furland (Squibb)	2,760
Henry Kaufman (Phibro-Salomon)	2,760
Harold K. Sperlich (Chrysler)	2,760
Spencer Scott (Citadel Holding)	2,647
Peter J. Buchanan (First Boston)	2,533

Source: "Executive pay: How the boss did in '85," *Business Week*, May 5, 1986, p. 49.

a. Calculate the mean and standard deviation of the sample of 1985 total cash compensations.

b. Describe the sampling distribution of the sample mean $\bar{y}$.

c. What is the probability that $\bar{y}$ will fall within 2 standard deviations of μ, the true mean compensation paid to the top corporate executives in 1985?

2.82 Let t_0 be a particular value of t. Use Table 2 of Appendix D to find the values such that the following statements are true:

a. $P(t \le t_0) = .10$ when $n = 23$
b. $P(t \ge t_0) = .005$ when $n = 3$
c. $P(t \le -t_0 \quad \text{or} \quad t \ge t_0) = .05$ when $n = 7$
d. $P(t \le -t_0 \quad \text{or} \quad t \ge t_0) = .01$ when $n = 24$

2.83 If rejection of the null hypothesis for a particular test would cause your firm to go out of business, would you want α to be small or large? Explain.

2.84 A manufacturing company is interested in determining whether there is a significant difference between the average number of units produced per day by two different machine operators. A random sample of ten daily outputs was selected from the outputs over the past year for each operator. The data on number of items produced per day are shown in the table.

OPERATOR 1	OPERATOR 2
$n_1 = 10$	$n_2 = 10$
$\bar{y}_1 = 35$	$\bar{y}_2 = 31$
$s_1^2 = 17.2$	$s_2^2 = 19.1$

a. Do the samples provide sufficient evidence at the .05 significance level to conclude that a difference does exist between the mean daily outputs of the machine operators?
b. Find a 90% confidence interval for $(\mu_1 - \mu_2)$. Explain clearly the meaning of your confidence interval.

2.85 In order to conduct the hypothesis test required in Exercise 2.84, you had to assume that $\sigma_1^2 = \sigma_2^2$. Perform an F-test with $\alpha = .10$ to check that assumption. Explain the significance of your result.

2.86 A company is interested in estimating the mean number of days of sick leave, μ, taken by all its employees. The firm's statistician selects at random 100 personnel files and notes the number of sick days taken by each employee. The following sample statistics are computed:

$\bar{y} = 12.2$ days $s = 10$ days

a. Estimate μ using a 90% confidence interval.
b. How many personnel files would the statistician have to select in order to estimate μ to within 2 days with 99% confidence?
c. Do the data support the research hypothesis that μ, the mean number of sick days taken by the employees, is greater than 10.9 days? Test using $\alpha = .05$.

2.87 C. S. Patterson (Concordia University, Montreal) conducted an investigation of the financing policies and practices of large regulated public utilities (*Financial Management*, Summer 1984). One goal of Patterson's research was to determine whether public utilities operate at a debt level that maximizes shareholder wealth. A sample of 47 publicly held electric utilities identified in *Moody's Manual* as having revenues of $300 million or more took part in the study. The actual 1982 debt ratios (defined as long-term debt divided by total capital) of the companies were recorded, with the following results:

$$\bar{y} = .485 \qquad s = .029$$

Prior to giving their actual 1982 debt ratios, the companies estimated the mean debt ratio at which they should operate in order to maximize shareholder wealth as .459. Is there sufficient evidence to indicate that the actual mean 1982 debt ratio of public utilities differs from the optimum value of .459? Test using $\alpha = .10$.

2.88 The EPA sets a limit of 5 parts per million on PCB (a dangerous substance) in water. A major manufacturing firm producing PCB for electrical insulation discharges small amounts from the plant. The company management, attempting to control the PCB in its discharge, has given instructions to halt production if the mean amount of PCB in the effluent exceeds 3 parts per million. A random sampling of 50 water specimens produced the following statistics:

$$\bar{y} = 3.1 \text{ parts per million} \qquad s = .5 \text{ part per million}$$

a. Do these statistics provide sufficient evidence to halt the production process? Use $\alpha = .01$.
b. If you were the plant manager, would you want to use a large or a small value for α for the test in part a?

2.89 A company purchases large quantities of naphtha in 50-gallon drums. Because the purchases are ongoing, small shortages in the drums can represent a sizable loss to the company. The weights of the drums vary slightly from drum to drum, so the weight of the naphtha is determined by removing it from the drums and measuring it. Suppose the company samples the contents of 20 drums, measures the naphtha in each, and calculates $\bar{y} = 49.70$ gallons and $s = .32$ gallon. Do the sample statistics provide sufficient evidence to indicate that the mean fill per 50-gallon drum is less than 50 gallons? Use $\alpha = .10$. What assumptions are required in order for this test to be valid?

2.90 Scientists have labeled benzene, a chemical solvent commonly used to synthesize plastics, as a possible cancer-causing agent. Studies have shown that people who work with benzene more than 5 years have 20 times the incidence of leukemia than the general population. As a result, the federal government lowered the maximum allowable level of benzene in the workplace from 10 parts per million (ppm) to 1 ppm (*Florida Times-Union*, Apr. 2, 1984). Suppose a steel manufacturing plant, which exposes its workers to benzene daily, is under investigation by the Occupational Health and Safety Commission. Twenty air samples, collected over a 1-month period and examined for benzene content, yielded the following summary statistics:

$$\bar{y} = 2.1 \text{ ppm} \qquad s = 1.7 \text{ ppm}$$

Is the steel manufacturing plant in violation of the new government standards? Test the hypothesis that the mean level of benzene at the steel manufacturing plant is greater than 1 ppm (use $\alpha = .05$). What assumption is required for the hypothesis test to be valid?

2.91 Studies have shown that in a nonbusiness (e.g., academic) setting, those who tend to have job mobility are predominantly better performers. In order to examine the performance turnover relationship in a business setting, G. F. Dreher (University of Kansas) examined the personnel records of a large national oil company. Dreher's sample consisted of 174 employees who were classified as "stayers" (those who stayed with the company from 1964 through 1979) and 355 former employees who were classified as "leavers" (those who left the company at varying points during the 15-year period). The company's annual performance appraisals corresponding to the initial years of service were used to form an initial performance rating for each employee. Summary statistics on initial performance for the two groups of employees are provided in the table. Is there evidence of a difference between the mean initial performance ratings of "stayers" and "leavers"? Test using $\alpha = .01$.

STAYERS	LEAVERS
$n_1 = 174$	$n_2 = 355$
$\bar{y}_1 = 3.51$	$\bar{y}_2 = 3.24$
$s_1 = .51$	$s_2 = .52$

Source: Dreher, G. H. "The role of performance in the turnover process," *Academy of Management Journal*, March 1982, Vol. 25, No. 1, pp. 137–147.

2.92 Refer to Exercise 2.91. Dreher also measured the rates of career advancement (number of promotions per year) for each employee. The data are summarized in the table. Suppose we want to compare the variation in rates of career advancement for the two groups. Do the data provide sufficient evidence to indicate that the variation in rates of career advancement differs for "leavers" and "stayers"? Test using $\alpha = .02$.

STAYERS	LEAVERS
$n_1 = 174$	$n_2 = 355$
$\bar{y}_1 = .43$	$\bar{y}_2 = .31$
$s_1 = .20$	$s_2 = .31$

2.93 Suppose you have been offered similar jobs in two different locales. To help in deciding which job to accept, you would like to compare the cost of living in the two cities. One of your primary concerns is the cost of housing, so you obtain a copy of a newspaper from each locale and begin to study the housing prices in the classified advertisements. One convenient method for getting a general idea of prices is to compute the prices on a per-square-foot basis. This is done by dividing the price of the house by the heated area (in square feet) of the house. Random samples of 63 advertisements in locale 1 and 78 in locale 2 produce the results listed in the table. Is there evidence that the mean housing price per square foot differs in the two locales? Test using $\alpha = .10$.

LOCALE 1	LOCALE 2
$\bar{y}_1 = \$23.40$ per square foot	$\bar{y}_2 = \$25.20$ per square foot
$s_1 = \$2.50$ per square foot	$s_2 = \$2.80$ per square foot

2.94 Random samples of residential sale prices were recorded for eight properties for each of two neighborhoods. The data, extracted from Appendix E (see Case Study 11), are shown in the accompanying table.

NEIGHBORHOOD	
A	B
$ 85,000	$172,500
112,000	69,500
80,000	102,000
117,000	365,000
72,500	87,500
102,000	60,000
93,500	95,000
140,000	85,000

a. Do the data provide sufficient evidence to indicate a difference in the variability of the sale prices between the two neighborhoods? Test using $\alpha = .10$.

b. What assumptions are necessary in order for the test in part a to be valid?

c. Do the data provide sufficient evidence to indicate a difference in mean sale prices between the two neighborhoods? Test using $\alpha = .05$.

d. What assumptions are necessary in order for the test in part c to be valid?

e. Find a 90% confidence interval for the mean sale price of residential properties in neighborhood A.

REFERENCES

Appel, G. "Have we hit bottom? A trusty indicator signals a possible turn in the market," *Barron's*, February 27, 1984, p. 24.

Brockhaus, R. H. "Risk-taking propensity of entrepreneurs," *Academy of Management Journal*, September 1980, Vol. 23, No. 3, pp. 509–520.

Dreman, D. "Astrology might be better," *Forbes*, March 26, 1984, p. 242.

Hamburg, M. *Statistical Analysis for Decision Making*, 2nd ed. New York: Harcourt Brace Jovanovich, 1977.

Hardy, K. G. "Key success factors for manufacturer's sales promotions in package goods," *Journal of Marketing*, Vol. 50, No. 7, July 1986, p. 16.

Kendall, P. T., Burger, I. N., and Smith, P. M. "Methods of estimation of the metabolizable energy content of cat foods," *Feline Practice*, Vol. 15, No. 2, February 1986, pp. 38–44.

MacStravic, R. S. "An early warning technique," *Hospital & Health Services Administration*, January/February 1986, pp. 86–98.

Mendenhall, W. and Reinmuth, J. *Statistics for Management and Economics*, 4th ed. Boston: Duxbury Press, 1982.

Neter, J., Wasserman, W., and Whitmore, G. A. *Fundamental Statistics for Business and Economics*, 4th ed. Boston: Allyn and Bacon, 1973.

Northcraft, G. B. and Neale, M. A. "Opportunity costs and the framing of resource allocation decisions," *Organizational Behavior and Human Decision Processes*, Vol. 37, No. 6, June 1986, pp. 348–356.

Patterson, C. S. "The financing objectives of large U.S. electrical utilities," *Financial Management*, Summer 1984.

Shih, W. "A general decision model for cost-volume-profit analysis under uncertainty: A reply," *The Accounting Review*, Vol. 56, No. 2, 1981, pp. 404–408.

Sincich, T. *Business Statistics by Example*, 2nd ed. San Francisco: Dellen, 1986.

Sloan, F. A. and Lorant, J. H. "The role of patient waiting time: Evidence from physicians' practices," *Journal of Business*, October 1977, Vol. 50, pp. 486–507.

Wallace, W. A. "Internal auditors can cut outside CPA costs," *Harvard Business Review*, March–April 1984, Vol. 62, No. 2, pp. 16–20.

OBJECTIVE

To present the basic concepts of regression analysis based on a simple linear relation between a response y and a single predictor variable x

CONTENTS

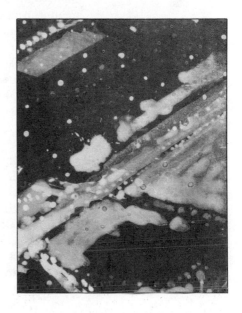

SIMPLE LINEAR REGRESSION

INTRODUCTION

As noted in Chapter 1, much business research is devoted to the topic of **modeling**, i.e., trying to describe how variables are related. For example, an econometrician might be interested in modeling the relationship between the level of consumption expenditure and the level of disposable personal income. An advertising agency might want to know the relationship between a firm's sales revenue and the amount spent on advertising. And an investment firm may be interested in relating the performance of the stock market to the current discount rate of the Federal Reserve Board.

The simplest graphical model for relating a response variable y to a single independent variable x is a straight line. In this chapter we will discuss **simple linear (straight-line) models** and will show how to fit them to a set of data points using the **method of least squares**. We will then show how to judge whether a relationship exists between y and x, and how to use the model either to estimate $E(y)$, the mean value of y, or to predict a future value of y for a given value of x. The totality of these methods is called a **simple linear regression analysis**.

Most models for business response variables are much more complicated than implied by a straight-line relationship. Nevertheless, the methods of this chapter are very useful, and they set the stage for the formulation and fitting of more complex models in succeeding chapters. Thus, this chapter will provide an intuitive justification for the techniques employed in a regression analysis, and it will identify most of the types of inferences that we will want to make using a **multiple regression analysis** later in this book.

PROBABILISTIC MODELS

An important consideration in merchandising a product is the amount of money spent on advertising. Suppose you want to model the monthly sales revenue of an appliance store as a function of the monthly advertising expenditure. The first question to be answered is this: Do you think an exact relationship exists between these two variables? That is, can the exact value of sales revenue be predicted if the advertising expenditure is specified? We think you will agree that this is not possible for several reasons. Sales depend on many variables other than advertising expenditure—for example, time of year, state of the general economy, inventory, and price structure. However, even if many variables are included in the model (the topic of Chapter 4), it is still unlikely that we can predict the monthly sales *exactly*. There will almost certainly be some variation in sales due strictly to **random phenomena** that cannot be modeled or explained. We will refer to all unexplained variations in sales—caused by important but unincluded variables or by unexplainable random phenomena—as **random error**.

If we construct a model that hypothesizes an exact relationship between variables, it is called a **deterministic model**. For example, if we believe that monthly sales revenue y will be exactly ten times the monthly advertising expenditure x, we write

$$y = 10x$$

This represents a **deterministic** relationship between the variables y and x.

On the other hand, if we believe that the model should be constructed to allow for random error, then we hypothesize a **probabilistic model**. This includes both a deterministic component and a random error component. For example, if we hypothesize that the sales y is related to advertising x by

$$y = 10x + \text{Random error}$$

we are hypothesizing a **probabilistic** relationship between y and x. Note that the deterministic component of this probabilistic model is $10x$.

GENERAL FORM OF PROBABILISTIC MODELS

$y = $ Deterministic component + Random error

where y is the variable to be predicted

As you will subsequently see, the random error will play an important role in testing hypotheses or finding confidence intervals for the deterministic portion of the model and will enable us to estimate the magnitude of the error of prediction when the model is used to predict some value of y to be observed in the future.

We begin with the simplest of probabilistic models—a **first-order linear model*** that graphs as a straight line. The elements of the straight-line model are summarized in the box.

A FIRST-ORDER (STRAIGHT-LINE) MODEL

$$y = \beta_0 + \beta_1 x + \varepsilon$$

where

$y = $ **Dependent** variable (variable to be modeled—sometimes called the **response** variable)

$x = $ **Independent**[†] variable (variable used as a **predictor** of y)

ε (epsilon) $= $ Random error component

β_0 (beta zero) $= $ y-intercept of the line, i.e., point at which the line intercepts or cuts through the y-axis (see Figure 3.1 on page 78)

β_1 (beta one) $= $ Slope of the line, i.e., amount of increase (or decrease) in the deterministic component of y for every 1-unit increase in x (see Figure 3.1)

---

*A general definition of the expression *first-order* is given in Section 7.3.

†The word *independent* should not be interpreted in a probabilistic sense. The phrase *independent variable* is used in regression analysis to refer to a predictor variable for the response y.

FIGURE 3.1

The Straight-Line Model

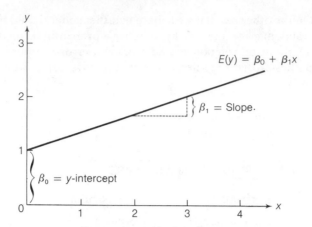

In Section 3.4 we make the standard assumption that the average of the random errors is zero, i.e., $E(\varepsilon) = 0$. Then the deterministic component of the straight-line probabilistic model represents the line of means $E(y) = \beta_0 + \beta_1 x$. Note that we use Greek symbols β_0 and β_1 to represent the y-intercept and slope of the line. They are population parameters with numerical values that will be known only if we have access to the entire population of (x, y) measurements.

It is helpful to think of regression modeling as a five-step procedure:

STEP 1 Hypothesize the deterministic component of the probabilistic model.

STEP 2 Use the sample data to estimate unknown parameters in the model.

STEP 3 Specify the probability distribution of the random error term, and estimate any unknown parameters of this distribution.

STEP 4 Statistically check the usefulness of the model.

STEP 5 When satisfied that the model is useful, use it for prediction, estimation, and so on.

In this chapter we will skip step 1 and deal only with the straight-line model. Chapters 4 and 7 will discuss how to build more complex models.

EXERCISES 3.1–3.5

3.1 In each case graph the line that passes through the points.
  a. (0, 2) and (2, 6)   **b.** (0, 4) and (2, 6)
  c. (0, − 2) and (−1, −6)   **d.** (0, −4) and (3, −7)

3.2 The equation for a straight line (deterministic) is

$$y = \beta_0 + \beta_1 x$$

If the line passes through the point (0, 1), then $x = 0, y = 1$ must satisfy the equation. That is,

$$1 = \beta_0 + \beta_1(0)$$

Similarly, if the line passes through the point (2, 3), then $x = 2$, $y = 3$ must satisfy the equation:

$$3 = \beta_0 + \beta_1(2)$$

Use these two equations to solve for β_0 and β_1, and find the equation of the line that passes through the points (0, 1) and (2, 3).

3.3 Find the equations of the lines passing through the four sets of points given in Exercise 3.1.

3.4 Plot the following lines:
  a. $y = 3 + 2x$ **b.** $y = 1 + x$ **c.** $y = -2 + 3x$
  d. $y = 5x$ **e.** $y = 4 - 2x$

3.5 Give the slope and y-intercept for each of the lines defined in Exercise 3.4.

SECTION 3.3

FITTING THE MODEL: THE METHOD OF LEAST SQUARES

Suppose an appliance store conducts a 5-month experiment to determine the effect of advertising on sales revenue. The results are shown in Table 3.1. (The number of measurements is small and the measurements themselves are unrealistically simple to avoid arithmetic confusion in this initial example.) The straight-line model is hypothesized to relate sales revenue y to advertising expenditure x. That is,

$$y = \beta_0 + \beta_1 x + \varepsilon$$

The question is this. How can we best use the information in the sample of five observations in Table 3.1 to estimate the unknown y-intercept β_0 and slope β_1?

To gain some information on the approximate values of these parameters, it is helpful to plot the sample data. Such a plot, called a **scattergram**, locates each of the five data points on a graph, as in Figure 3.2 (page 80). Note that the scattergram suggests a general tendency for y to increase as x increases. If you place a ruler on the scattergram, you will see that a line may be drawn through three of the five points, as shown in Figure 3.3. To obtain the equation of this visually fitted line, notice that the line intersects the y-axis at $y = -1$, so the y-intercept is -1. Also, y increases exactly 1 unit for every 1-unit increase in x, indicating that the slope is $+1$. Therefore, the equation is

$$\bar{y} = -1 + 1(x) = -1 + x$$

where $\bar{y}$ is used to denote the predictor of y based on the visually fitted model.

TABLE 3.1

MONTH	ADVERTISING EXPENDITURE x, hundreds of dollars	SALES REVENUE y, thousands of dollars
1	1	1
2	2	1
3	3	2
4	4	2
5	5	4

FIGURE 3.2
Scattergram for Data
in Table 3.1

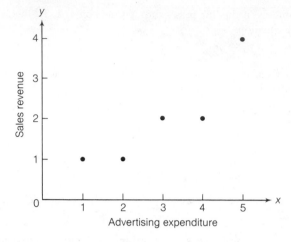

FIGURE 3.3
Visual Straight-Line Fit to
the Data in Table 3.1

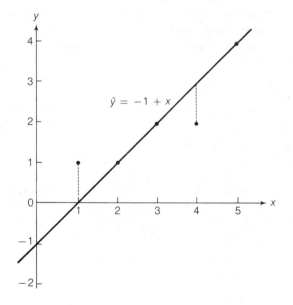

$\hat{y} = -1 + x$

One way to decide quantitatively how well a straight line fits a set of data is to determine the extent to which the data points deviate from the line. For example, to evaluate the visually fitted model in Figure 3.3, we calculate the magnitude of the **deviations**, i.e., the differences between the observed and the predicted values of y. These deviations, or **errors**, are the vertical distances between observed and predicted values of y (see Figure 3.3). The observed and predicted values of y, their differences, and their squared differences are shown in Table 3.2. Note that the **sum of the errors** equals 0 and the **sum of squares of the errors (SSE)**, which gives greater emphasis to large deviations of the points from the line, is equal to 2.

TABLE 3.2
Comparing Observed and
Predicted Values for the
Visual Model

x	y	$\hat{y} = -1 + x$	$(y - \hat{y})$		$(y - \hat{y})^2$
1	1	0	$(1 - 0) =$	1	1
2	1	1	$(1 - 1) =$	0	0
3	2	2	$(2 - 2) =$	0	0
4	2	3	$(2 - 3) = -1$		1
5	4	4	$(4 - 4) =$	0	0
			Sum of errors =	0	Sum of squared errors (SSE) = 2

You can see by shifting the ruler around the graph that is possible to find many lines for which the sum of the errors is equal to 0, but it can be shown that there is one (and only one) line for which the *SSE is a minimum*. This line is called the **least squares line**, the **regression line**, or **least squares prediction equation**.

To find the least squares line for a set of data, assume that we have a sample of n data points which can be identified by corresponding values of x and y, say $(x_1, y_1), (x_2, y_2), \ldots, (x_n, y_n)$. For example, the $n = 5$ data points shown in Table 3.2 are (1, 1), (2, 1), (3, 2), (4, 2), and (5, 4). The straight-line model for the response y in terms of x is

$$y = \beta_0 + \beta_1 x + \varepsilon$$

The line of means is

$$E(y) = \beta_0 + \beta_1 x$$

and the fitted line, which we hope to find, is represented as

$$\hat{y} = \hat{\beta}_0 + \hat{\beta}_1 x$$

The "hats" can be read as "estimator of." Thus, $\hat{y}$ is an estimator of the mean value of y, $E(y)$, and a predictor of some future value of y; and $\hat{\beta}_0$ and $\hat{\beta}_1$ are estimators of β_0 and β_1, respectively.

For a given data point, say the point (x_i, y_i), the observed value of y is y_i and the predicted value of y would be obtained by substituting x_i into the prediction equation:

$$\hat{y}_i = \hat{\beta}_0 + \hat{\beta}_1 x_i$$

The deviation of the ith value of y from its predicted value, called the **ith residual**, is

$$(y_i - \hat{y}_i) = [y_i - (\hat{\beta}_0 + \hat{\beta}_1 x_i)]$$

Then the sum of squares of the deviations of the y-values about their predicted values (i.e., the **sum of squares of residuals**) for all of the n data points is

$$SSE = \sum_{i=1}^{n} [y_i - (\hat{\beta}_0 + \hat{\beta}_1 x_i)]^2$$

The quantities $\hat{\beta}_0$ and $\hat{\beta}_1$ that make the SSE a minimum are called the **least squares estimates** of the population parameters β_0 and β_1, and the prediction equation $\hat{y} = \hat{\beta}_0 + \hat{\beta}_1 x$ is called the **least squares line**.

DEFINITION 3.1

The **least squares line** is one that has a smaller SSE than any other straight-line model.

The values of $\hat{\beta}_0$ and $\hat{\beta}_1$ that minimize the SSE are given (proof omitted) by the formulas in the box.*

FORMULAS FOR THE LEAST SQUARES ESTIMATES

$$Slope: \quad \hat{\beta}_1 = \frac{SS_{xy}}{SS_{xx}}$$

$$y\text{-}intercept: \quad \hat{\beta}_0 = \bar{y} - \hat{\beta}_1 \bar{x}$$

where

$$SS_{xy} = \sum_{i=1}^{n} x_i y_i - \frac{\left(\sum_{i=1}^{n} x_i\right)\left(\sum_{i=1}^{n} y_i\right)}{n}$$

$$SS_{xx} = \sum_{i=1}^{n} x_i^2 - \frac{\left(\sum_{i=1}^{n} x_i\right)^2}{n}$$

n = Sample size

Preliminary computations for finding the least squares line for the advertising–sales example are contained in Table 3.3. We can now calculate†

$$SS_{xy} = \sum x_i y_i - \frac{\left(\sum x_i\right)\left(\sum y_i\right)}{5} = 37 - \frac{(15)(10)}{5}$$

$$= 37 - 30 = 7$$

$$SS_{xx} = \sum x_i^2 - \frac{\left(\sum x_i\right)^2}{5} = 55 - \frac{(15)^2}{5}$$

$$= 55 - 45 = 10$$

---

*Students who are familiar with calculus should note that the values of β_0 and β_1 that minimize SSE $= \sum_{i=1}^{n} (y_i - \hat{y}_i)^2$ are obtained by setting the two partial derivatives $\partial SSE/\partial \beta_0$ and $\partial SSE/\partial \beta_1$ equal to 0. The solutions to these two equations yield the formulas shown in the box. Furthermore, we denote the *sample* solutions to the equations by $\hat{\beta}_0$ and $\hat{\beta}_1$, where the "ˆ" (hat) denotes that these are sample estimates of the true population intercept β_0 and slope β_1.

†Since summations will be used extensively from this point on, we will omit the limits on Σ when the summation includes all the measurements in the sample, i.e., when the summation is $\Sigma_{i=1}^{n}$, we will write Σ.

TABLE 3.3
Preliminary Computations for the Advertising–Sales Example

x_i	y_i	x_i^2	x_iy_i
1	1	1	1
2	1	4	2
3	2	9	6
4	2	16	8
5	4	25	20
Totals $\Sigma x_i = 15$	$\Sigma y_i = 10$	$\Sigma x_i^2 = 55$	$\Sigma x_iy_i = 37$

Then, the slope of the least squares line is

$$\hat{\beta}_1 = \frac{SS_{xy}}{SS_{xx}} = \frac{7}{10} = .7$$

and the y-intercept is

$$\hat{\beta}_0 = \bar{y} - \hat{\beta}_1\bar{x} = \frac{\sum y_i}{5} - \hat{\beta}_1\frac{\left(\sum x_i\right)}{5}$$

$$= \frac{10}{5} - (.7)\frac{(15)}{5}$$

$$= 2 - (.7)(3) = 2 - 2.1 = -.1$$

The least squares line is then

$$\hat{y} = \hat{\beta}_0 + \hat{\beta}_1 x = -.1 + .7x$$

The graph of this line is shown in Figure 3.4.

FIGURE 3.4
The Line $\hat{y} = -.1 + .7x$ Fit to the Data

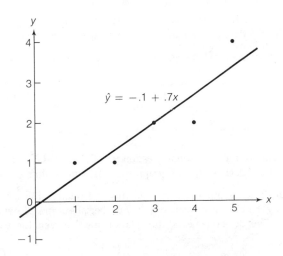

TABLE 3.4
Comparing Observed and
Predicted Values for the
Least Squares Model

x	y	$\hat{y} = -.1 + .7x$	$(y - \hat{y})$	$(y - \hat{y})^2$
1	1	.6	$(1 - .6) = \ \ .4$	.16
2	1	1.3	$(1 - 1.3) = -.3$	.09
3	2	2.0	$(2 - 2.0) = \ \ 0$	.00
4	2	2.7	$(2 - 2.7) = -.7$	.49
5	4	3.4	$(4 - 3.4) = \ \ .6$	.36
			Sum of errors = $\ \ 0$	SSE = 1.10

The observed and predicted values of y, the deviations of the y-values about their predicted values, and the squares of these deviations are shown in Table 3.4. Note that the sum of squares of the deviations, SSE, is 1.10, and (as we would expect) this is less than the SSE = 2.0 obtained in Table 3.2 for the visually fitted line.

To summarize, we have defined the best-fitting straight line to be the one that satisfies the least squares criterion; that is, the sum of the squared errors will be smaller than for any other straight-line model. This line is called the **least squares line**, and its equation is called the **least squares prediction equation**.

EXERCISES 3.6–3.11

3.6 Use the method of least squares to fit a straight line to these six data points:

x	1	2	3	4	5	6
y	1	2	2	3	5	5

a. What are the least squares estimates of β_0 and β_1?
b. Plot the data points and graph the least squares line on the scattergram.

3.7 Use the method of least squares to fit a straight line to these five data points:

x	−2	−1	0	1	2
y	4	3	3	1	−1

a. What are the least squares estimates of β_0 and β_1?
b. Plot the data points and graph the least squares line on the scattergram.

3.8 A car dealer is interested in modeling the relationship between the number of cars sold by the firm each week and the number of salespeople who work on the showroom floor. The dealer believes the relationship between the two variables can best be described by a straight line. The sample data shown in the table were supplied by the car dealer.

WEEK OF	NUMBER OF CARS SOLD y	NUMBER OF SALESPEOPLE ON DUTY x
January 30	20	6
June 3	6	2
March 2	10	4
October 26	18	6
February 7	11	3

a. Construct a scattergram for the data.
b. Assuming the relationship between the variables is best described by a straight line, use the method of least squares to estimate the y-intercept and slope of the line.
c. Plot the least squares line on your scattergram.
d. According to your least squares line, approximately how many cars should the dealer expect to sell in a week if five salespeople are kept on the showroom floor each day? [*Note:* A measure of the reliability of these predictions will be discussed in Section 3.9.]

3.9 Find the least squares line to describe the relationship between a company's sales and the total sales for a particular industry. The data are shown in the table. Plot the data points and graph the least squares line as a check on your calculations.

YEAR	COMPANY SALES y, millions of dollars	INDUSTRY SALES x, millions of dollars
1982	.5	10
1983	1.0	12
1984	1.0	13
1985	1.4	15
1986	1.3	14
1987	1.6	15

3.10 In order for a company to maintain a competitive edge in the marketplace, spending on research and development (R&D) is essential. To determine the optimum level for R&D spending and its effect on a company's value, a simple linear regression analysis was performed. Data collected for the largest R&D spenders (based on 1981–1982 averages) were used to fit the straight-line model

$$y = \beta_0 + \beta_1 x + \varepsilon$$

where

y = Price/earnings (P/E) ratio

x = R&D expenditures/sales (R/S) ratio

The data for 20 of the companies used in the study are provided in the table.

COMPANY	P/E RATIO y	R/S RATIO x	COMPANY	P/E RATIO y	R/S RATIO x
1	5.6	.003	11	8.4	.058
2	7.2	.004	12	11.1	.058
3	8.1	.009	13	11.1	.067
4	9.9	.021	14	13.2	.080
5	6.0	.023	15	13.4	.080
6	8.2	.030	16	11.5	.083
7	6.3	.035	17	9.8	.091
8	10.0	.037	18	16.1	.092
9	8.5	.044	19	7.0	.064
10	13.2	.051	20	5.9	.028

Source: Wallin, C. C. and Gilman J. J. "Determining the optimum level for R&D spending," *Research Management*, Vol. 14, No. 5, September/October 1986, pp. 19–24 (adapted from Figure 1, p. 20).

a. Construct a scattergram for the data.

b. Find the least squares prediction equation.

c. Plot the least squares line on your scattergram.

d. Use the least squares line to predict the P/E ratio for a company with an R/S ratio of .070. [*Note:*   We will find a measure of reliability for this prediction in Section 3.9.]

3.11 An appliance company is interested in relating the sales rate of 17-inch color television sets to the price per set. To do this, the company randomly selected 15 weeks in the past year and recorded the number of sets sold and the price at which the sets were being sold during that week. The data are shown in the table.

WEEK	NUMBER OF SETS SOLD y	PRICE x, dollars	WEEK	NUMBER OF SETS SOLD y	PRICE x, dollars
1	55	350	9	20	400
2	54	360	10	45	340
3	25	385	11	50	350
4	18	400	12	35	335
5	51	370	13	30	330
6	20	390	14	30	325
7	45	375	15	53	365
8	19	390			

a. Find the least squares line relating y to x.

b. Plot the data and graph the least squares line as a check on your calculations.

S E C T I O N 3.4

MODEL ASSUMPTIONS

In the advertising–sales example presented in Section 3.3, we assumed that the probabilistic model relating the firm's sales revenue y to advertising dollars x is

$$y = \beta_0 + \beta_1 x + \varepsilon$$

Recall that the least squares estimate of the deterministic component of the model $\beta_0 + \beta_1 x$ is

$$\hat{y} = \hat{\beta}_0 + \hat{\beta}_1 x = -.1 + .7x$$

Now we turn our attention to the random component ε of the probabilistic model and its relation to the errors of estimating β_0 and β_1. In particular, we will see how the probability distribution of ε determines how well the model describes the true relationship between the dependent variable y and the independent variable x.

We will make four basic assumptions about the general form of the probability distribution of ε:

ASSUMPTION 1 The mean of the probability distribution of ε is 0. That is, the average of the errors over an infinitely long series of experiments is 0 for each setting of the independent variable x. This assumption implies that the mean value of y, $E(y)$, for a given value of x is $E(y) = \beta_0 + \beta_1 x$.

ASSUMPTION 2 The variance of the probability distribution of ε is constant for all settings of the independent variable x. For our straight-line model, this assumption means that the variance of ε is equal to a constant, say σ^2, for all values of x.

ASSUMPTION 3 The probability distribution of ε is normal.

ASSUMPTION 4 The errors associated with any two different observations are independent. That is, the error associated with one value of y has no effect on the errors associated with other y-values.

The implications of the first three assumptions can be seen in Figure 3.5 (page 88), which shows distributions of errors for three particular values of x, namely x_1, x_2, and x_3. Note that the relative frequency distributions of the errors are normal, with a mean of 0, and a constant variance σ^2 (all the distributions shown have the same amount of spread or variability). A point which lies on the straight line shown in Figure 3.5 represents the mean value of y for a given value of x.

We will denote this mean value as $E(y)$. Then, the line of means is given by the equation

$$E(y) = \beta_0 + \beta_1 x$$

FIGURE 3.5
The Probability
Distribution of ε

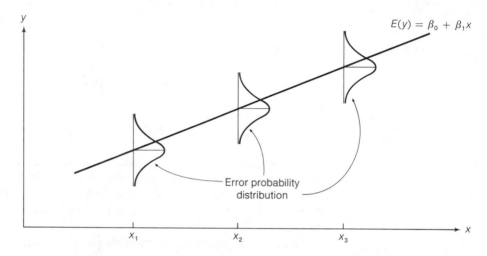

Various techniques exist for checking the validity of these assumptions, and there are remedies to be applied when the assumptions appear to be invalid. Most of these are beyond the scope of this text, but we will mention a few in Chapter 6. In actual practice, the assumptions need not hold exactly in order for least squares estimators and test statistics (to be described subsequently) to possess the measures of reliability that we would expect from a regression analysis. The assumptions will be satisfied adequately for many applications encountered in business.

S E C T I O N   3.5

AN ESTIMATOR OF σ^2

It seems reasonable to assume that the greater the variability of the random error ε (which is measured by its variance σ^2), the greater will be the errors in the estimation of the model parameters β_0 and β_1, and in the error of prediction when $\hat{y}$ is used to predict y for some value of x. Consequently, you should not be surprised, as we proceed through this chapter, to find that σ^2 appears in the formulas for all confidence intervals and test statistics that will be used.

In most practical situations, σ^2 will be unknown, and we must use the data to estimate its value. The best (proof omitted) estimate of σ^2 is s^2, which is obtained by dividing the sum of squares of residuals,

$$\text{SSE} = \sum (y_i - \hat{y}_i)^2$$

by the number of degrees of freedom associated with this quantity. We use 2 df to estimate the y-intercept and slope in the straight-line model, leaving $(n - 2)$ df for the error variance estimation (see the formulas in the box).

ESTIMATION OF σ^2

$$s^2 = \frac{\text{SSE}}{\text{Degrees of freedom for error}} = \frac{\text{SSE}}{n-2}$$

where

$$\text{SSE} = \sum (y_i - \hat{y}_i)^2$$

$$= \text{SS}_{yy} - \hat{\beta}_1 \text{SS}_{xy} \quad \text{(calculation formula)}$$

$$\text{SS}_{yy} = \sum (y_i - \bar{y})^2 = \sum y_i^2 - \frac{\left(\sum y_i\right)^2}{n}$$

Warning: When performing these calculations, you may be tempted to round the calculated values of SS_{yy}, $\hat{\beta}_1$, and SS_{xy}. Be certain to carry at least six significant figures for each of these quantities to avoid substantial errors in the calculation of the SSE.

In the advertising–sales example, we previously calculated SSE = 1.10 for the least squares line $\hat{y} = -.1 + .7x$. Recalling that there were $n = 5$ data points, we have $n - 2 = 5 - 2 = 3$ df for estimating σ^2. Thus,

$$s^2 = \frac{\text{SSE}}{n-2} = \frac{1.10}{3} = .367$$

is the estimated variance, and

$$s = \sqrt{.367} = .61$$

is the estimated standard deviation of ε.

You may be able to obtain an intuitive feeling for s by recalling the interpretation given to a standard deviation in Chapter 2 and remembering that the least squares line estimates the mean value of y for a given value of x. Since s measures the spread of the distribution of y-values about the least squares line, we should not be surprised to find that most of the observations lie within $2s$ or $2(.61) = 1.22$ of the least squares line. For this simple example (only five data points), all five data points fall within $2s$ of the least squares line. In Section 3.9, we will use s to evaluate the error of prediction when the least squares line is used to predict a value of y to be observed for a given value of x.

INTERPRETATION OF s, THE ESTIMATED STANDARD DEVIATION OF ε

We expect most of the observed y-values to lie within $2s$ of their respective least squares predicted values, $\hat{y}$.

EXERCISES 3.12–3.16

3.12 Suppose you fit a least squares line to nine data points and calculate SSE = .219. Find s^2, the estimator of σ^2, the variance of the random error term ε.

3.13 Calculate SSE and s^2 for the least squares line plotted in:
a. Exercise 3.6   **b.** Exercise 3.7   **c.** Exercise 3.8
d. Exercise 3.9   **e.** Exercise 3.10   **f.** Exercise 3.11

3.14 An electronics dealer believes that there is a positive linear relationship (i.e., $\beta_1 > 0$) between the number of hours of quadraphonic programming on a city's FM stations and sales of quadraphonic systems. Records for the dealer's sales during the last 6 months and the amount of quadraphonic programming for the corresponding months are given in the table.

MONTH	AVERAGE AMOUNT OF QUADRAPHONIC PROGRAMMING x, hours	NUMBER OF QUADRAPHONIC SYSTEMS SOLD y
1	33.6	7
2	36.3	10
3	38.7	13
4	36.6	11
5	39.0	14
6	38.4	18

a. Fit a least squares line to the data.
b. Plot the data and graph the least squares line as a check on your calculations.
c. Calculate SSE and s^2.
d. Calculate s and interpret its value.

3.15 The Consumer Attitude Survey, performed by the Bureau of Economic and Business Research, University of Florida, is conducted using random-digit telephone dialings to Florida households. The reliability of a telephone survey such as this depends on the refusal rate—that is, the percentage of dialed households that refuse to take part in the study. One factor thought to be related to refusal rate is personal income. The accompanying table gives the refusal rate y and personal income per capita x for 12 randomly selected Florida counties from the 1983 survey.

COUNTY	REFUSAL RATE y	PER CAPITA INCOME x
1	.296	$ 7,737
2	.498	12,330
3	.386	12,058
4	.327	9,927
5	.500	6,904
6	.333	9,463
7	.429	11,466
8	.422	10,000
9	.441	10,052
10	.191	8,636
11	.526	7,445
12	.405	9,059

Source: Bureau of Economic and Business Research, University of Florida.

a. Construct a scattergram for the data.
b. Find the least squares prediction equation.
c. Graph the least squares line on the scattergram.
d. Use the least squares prediction equation to predict the refusal rate for a Florida county with a per capita income of $8,000. [*Note:*   We will find a measure of the reliability of this prediction in Section 3.9.]
e. Calculate SSE and s^2.
f. Calculate s and interpret its value.

3.16 A company keeps extensive records on its new salespeople on the premise that sales should increase with experience. A random sample of seven new salespeople produced the data on experience and sales shown in the table.

MONTHS ON JOB x	MONTHLY SALES y, thousands of dollars
2	2.4
4	7.0
8	11.3
12	15.0
1	.8
5	3.7
9	12.0

a. Fit a least squares line to the data.
b. Plot the data and graph the least squares line.
c. Predict the sales that a new salesperson would be expected to generate after 6 months on the job. After 9 months.
d. Calculate SSE and s^2.
e. Calculate s and interpret its value.

S E C T I O N   3.6

ASSESSING THE UTILITY OF THE MODEL: MAKING INFERENCES ABOUT THE SLOPE β_1

Refer to the advertising–sales data of Table 3.1 and suppose that the appliance store's sales revenue is *completely unrelated* to the advertising expenditure. What could be said about the values of β_0 and β_1 in the hypothesized probabilistic model

$$y = \beta_0 + \beta_1 x + \varepsilon$$

if x contributes no information for the prediction of y? The implication is that the mean of y, i.e., the deterministic part of the model $E(y) = \beta_0 + \beta_1 x$, does not change as x changes. Regardless of the value of x, you always predict the same value of y. In the straight-line model, this means that the true slope, β_1, is equal to 0. Therefore, to test the null hypothesis that x contributes no information for the prediction of y against the alternative hypothesis that these variables are linearly related with a slope differing from 0, we test

$$H_0: \quad \beta_1 = 0 \qquad H_a: \quad \beta_1 \neq 0$$

If the data support the alternative hypothesis, we will conclude that x does contribute information for the prediction of y using the straight-line model [although the true relationship between $E(y)$ and x could be more complex than a straight line]. Thus, to some extent, this is a test of the utility of the hypothesized model.

The appropriate test statistic is found by considering the sampling distribution of $\hat{\beta}_1$, the least squares estimator of the slope β_1.

SAMPLING DISTRIBUTION OF $\hat{\beta}_1$

If we make the four assumptions about ε (see Section 3.4), then the sampling distribution of $\hat{\beta}_1$, the least squares estimator of the slope, will be a normal distribution with mean β_1 (the true slope) and standard deviation

$$\sigma_{\hat{\beta}_1} = \frac{\sigma}{\sqrt{SS_{xx}}} \quad \text{(See Figure 3.6.)}$$

FIGURE 3.6
Sampling Distribution of $\hat{\beta}_1$

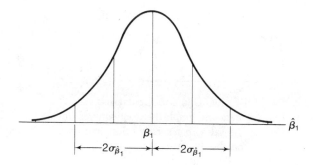

Since σ will usually be unknown, the appropriate test statistic will generally be a Student's t-statistic formed as follows:

$$t = \frac{\hat{\beta}_1 - \text{Hypothesized value of } \beta_1}{s_{\hat{\beta}_1}} \quad \text{where} \quad s_{\hat{\beta}_1} = \frac{s}{\sqrt{SS_{xx}}}$$

$$= \frac{\hat{\beta}_1 - 0}{s/\sqrt{SS_{xx}}}$$

Note that we have substituted the estimator s for σ, and then formed $s_{\hat{\beta}_1}$ by dividing s by $\sqrt{SS_{xx}}$. The number of degrees of freedom associated with this t-statistic is the same as the number of degrees of freedom associated with s. Recall that this will be $(n - 2)$ df when the hypothesized model is a straight line (see Section 3.5).

The test of the utility of the model is summarized in the next box.

For the advertising–sales example, we will choose $\alpha = .05$ and, since $n = 5$, df $= (n - 2) = 5 - 2 = 3$. Then the rejection region for the two-tailed test is

$$t < -t_{.025} = -3.182 \quad \text{or} \quad t > t_{.025} = 3.182$$

A TEST OF MODEL UTILITY

ONE-TAILED TEST	**TWO-TAILED TEST**

H_0: $\beta_1 = 0$ H_0: $\beta_1 = 0$

H_a: $\beta_1 < 0$ H_a: $\beta_1 \neq 0$

     (or H_a: $\beta_1 > 0$)

Test statistic: $t = \dfrac{\hat{\beta}_1}{s_{\hat{\beta}_1}} = \dfrac{\hat{\beta}_1}{s/\sqrt{SS_{xx}}}$ Test statistic: $t = \dfrac{\hat{\beta}_1}{s_{\hat{\beta}_1}} = \dfrac{\hat{\beta}_1}{s/\sqrt{SS_{xx}}}$

Rejection region: $t < -t_\alpha$ Rejection region: $t < -t_{\alpha/2}$ or

            (or $t > t_\alpha$) $t > t_{\alpha/2}$

where t_α is based on $(n - 2)$ df   where $t_{\alpha/2}$ is based on $(n - 2)$ df

Assumptions: The four assumptions about ε listed in Section 3.4.

We previously calculated $\hat{\beta}_1 = .7$, $s = .61$, and $SS_{xx} = 10$. Thus,

$$t = \frac{\hat{\beta}_1}{s/\sqrt{SS_{xx}}} = \frac{.7}{.61/\sqrt{10}} = \frac{.7}{.19} = 3.7$$

Since this calculated t-value falls in the upper tail rejection region (see Figure 3.7), we reject the null hypothesis and conclude that the slope β_1 is not 0. The sample evidence indicates that x contributes information for the prediction of y using a linear model for the relationship between sales revenue and advertising.

FIGURE 3.7

Rejection Region and Calculated t-Value for Testing Whether the Slope $\beta_1 = 0$

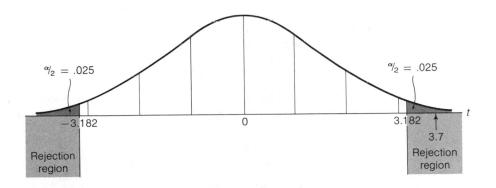

What conclusion can be drawn if the calculated t-value does not fall in the rejection region? We know from previous discussions of the philosophy of hypothesis testing that such a t-value does *not* lead us to accept the null hypothesis. That is, we do not conclude that $\beta_1 = 0$. Additional data might indicate that β_1 differs from 0, or a more complex relationship may exist between x and y, requiring the fitting of a model other than the straight-line model. We will discuss several such models in Chapter 4.

Another way to make inferences about the slope β_1 is to estimate it using a confidence interval. This interval is formed as shown in the box.

A 100(1 − α)% CONFIDENCE INTERVAL FOR THE SLOPE β_1

$$\hat{\beta}_1 \pm t_{\alpha/2}s_{\hat{\beta}_1} \qquad \text{where} \quad s_{\hat{\beta}_1} = \frac{s}{\sqrt{SS_{xx}}}$$

and $t_{\alpha/2}$ is based on $(n-2)$ df

For the advertising–sales example, a 95% confidence interval for the slope β_1 is

$$\hat{\beta}_1 \pm t_{.025}s_{\hat{\beta}_1} = .7 \pm 3.182\left(\frac{s}{\sqrt{SS_{xx}}}\right)$$

$$= .7 \pm 3.182\left(\frac{.61}{\sqrt{10}}\right) = .7 \pm .61$$

Thus, we estimate that the interval from .09 to 1.31 includes the slope parameter β_1.

Since all the values in this interval are positive, it appears that β_1 is positive and that the mean of y, $E(y)$, increases as x increases. However, the rather large width of the confidence interval reflects the small number of data points (and, consequently, a lack of information) in the experiment. We would expect a narrower interval if the sample size were increased.

EXERCISES 3.17–3.25

3.17 Do the data provide sufficient evidence to indicate that β_1 differs from 0 for the least squares analyses in the following exercises? Use $\alpha = .05$.
a. Exercise 3.6   **b.** Exercise 3.7   **c.** Exercise 3.8
d. Exercise 3.9   **e.** Exercise 3.10   **f.** Exercise 3.11

3.18 Do the data in Exercise 3.14 provide sufficient evidence to indicate that sales, y, tend to increase as the number of hours of programming, x, increases (i.e., that $\beta_1 > 0$)? Test using $\alpha = .10$.

3.19 Do the data in Exercise 3.15 provide sufficient evidence to indicate that refusal rate y is linearly related to per capita income x? Test using $\alpha = .01$.

3.20 Do the data in Exercise 3.16 support the theory that sales increase as experience of a salesperson increases? Test using $\alpha = .05$.

3.21 Refer to the shareholder-wealth maximization study in Exercise 2.51. The researchers conducted an analysis of return on equity to compare companies that aim at maximizing shareholder wealth to those with alternative corporate goals (*Quarterly Journal of Business & Economics*, Autumn 1986). Monthly holding-period returns were averaged for the stocks of shareholder wealth-maximizer companies and for companies holding alternative goals, and the difference between the averages (called monthly portfolio return differences) were calculated for each of $n = 72$ months.

The return differences y, were then regressed on monthly *Standard & Poor's 500 Composite Stock Index*, x, using the straight-line model

$$y = \beta_0 + \beta_1 x + \varepsilon$$

The regression results are summarized here:

$$\hat{y} = .00008 - .01748x \qquad SSE = .322 \qquad SS_{xx} = 2.914$$

a. Conduct a test to determine whether the monthly portfolio return differences are linearly related to the monthly *Standard & Poor's 500* Stock Index. Use $\alpha = .10$.
b. Construct a 90% confidence interval for the slope of the straight-line model.
c. Interpret the confidence interval in part **b**, and explain what it tells you about the relationship between monthly portfolio return differences and the monthly *Standard & Poor's 500* Stock Index.

3.22 Some economists fear that the current unemployment compensation system in the United States distorts the number of layoffs in a downturn of the business cycle. The hypothesis is that the unemployment compensation subsidy causes firms to lay off more people than they would if they knew those laid off would receive no outside subsidy. The accompanying table gives the unemployment compensation subsidy rate x (as a percentage of total revenues) and the layoff rate y (number of workers per 1,000) for 11 industries.

INDUSTRY	SUBSIDY RATE x	LAYOFF RATE y
Apparel	57%	12.54
Chemicals	32	1.70
Construction	31	7.10
Electrical machinery	29	8.38
Fabricated metals	27	11.72
Food	36	5.10
Machinery	32	4.44
Misc. manufacturing	61	9.82
Primary metals	23	7.34
Retail	27	1.98
Wholesale trade	33	1.86

Source: Tropel, R. H. "On layoffs and unemployment insurance," *American Economic Review*, 1983, Vol. 83, pp. 541–559.

a. Fit the model $E(y) = \beta_0 + \beta_1 x$.
b. Do the data provide sufficient evidence to indicate that the unemployment compensation subsidy rate x contributes information for the prediction of the layoff rate y? Test using $\alpha = .05$.
c. Find a 95% confidence interval for the mean increase in layoff rate y for each percentage-point increase in subsidy rate x. [*Hint:* Find a 95% confidence interval for β_1.]

3.23 Buyers are often influenced by bulk advertising of a particular product. For example, suppose you have a product that sells for 25¢. If it is advertised at 2/50¢, 3/75¢, or

4/$1, some people may think they are getting a bargain. To test this theory, a store manager advertised an item for equal periods of time at five different bulk rates and observed the volume sold, as listed in the table. Do the data provide sufficient evidence to indicate that sales increase as the number in the bulk increases?

ADVERTISED NUMBER IN BULK SALE x	VOLUME SOLD y
1	27
2	36
3	34
4	63
5	52

3.24 Starting salary is considered to be an important indicator of success in the job market by many graduating MBAs. The factors that influence starting salary are many and varied, but one variable intrinsically interesting to MBA students is Graduate Management Aptitude Test (GMAT) score. The starting salaries and GMAT scores for a sample of ten MBAs who graduated from the University of Florida's MBA program in 1987 are given in the table.

STARTING SALARY y	GMAT x	STARTING SALARY y	GMAT x
$40,000	510	$28,000	520
33,000	510	34,000	560
40,000	550	30,000	530
35,000	600	32,500	530
28,000	600	31,000	590

Source: Graduate College of Business Administration, University of Florida.

a. Find the least squares prediction equation for the model

$$y = \beta_0 + \beta_1 x + \varepsilon$$

Plot the data on a scattergram and graph the least squares line.

b. Find SSE and s^2 for the data.

c. Is there sufficient evidence to indicate that GMAT score x contributes information for the prediction of starting salary y? Test using $\alpha = .10$.

d. Construct a 95% confidence interval for the true slope β_1. Interpret the interval.

3.25 A large car-rental agency sells its cars after using them for a year. Among the records kept for each car are mileage and maintenance costs for the year. To evaluate the performance of a particular car model in terms of maintenance costs, the agency wants to use a 95% confidence interval to estimate the mean increase in maintenance costs for each additional 1,000 miles driven. Assume the relationship between maintenance cost and miles driven is linear. Use the data in the accompanying table to accomplish the objective of the rental agency.

CAR	MILES DRIVEN x, thousands	MAINTENANCE COST y, dollars
1	54	326
2	27	159
3	29	202
4	32	200
5	28	181
6	36	217

SECTION 3.7

THE COEFFICIENT OF CORRELATION

The claim is often made that the crime rate and the unemployment rate are "highly correlated." Another popular belief is that the Gross National Product (GNP) and the rate of inflation are "correlated." Some people even believe that the Dow Jones Industrial Average and the lengths of fashionable skirts are "correlated." Thus, the term *correlation* implies a relationship between two variables.

The **Pearson product moment correlation coefficient** r, defined in the box, provides a quantitative measure of the strength of the linear relationship between x and y, just as does the least squares slope $\hat{\beta}_1$. However, unlike the slope, the correlation coefficient r is *scaleless*. The value of r is always between -1 and $+1$, regardless of the units of measurement used for the variables x and y.

DEFINITION 3.2

The **Pearson product moment coefficient of correlation** r is a measure of the strength of the *linear* relationship between two variables x and y. It is computed (for a sample of n measurements on x and y) as follows:

$$r = \frac{SS_{xy}}{\sqrt{SS_{xx}SS_{yy}}}$$

Note that r is computed using the same quantities used in fitting the least squares line. Since both r and $\hat{\beta}_1$ provide information about the utility of the model, it is not surprising that there is a similarity in their computational formulas. In particular, note that SS_{xy} appears in the numerators of both expressions and, since both denominators are always positive, r and $\hat{\beta}_1$ will always be of the same sign (either both positive or both negative). A value of r near or equal to 0 implies little or no linear relationship between y and x. In contrast, the closer r is to 1 or -1, the stronger the linear relationship between y and x. And, if $r = 1$ or $r = -1$, all the points fall exactly on the least squares line. Positive values of r imply that y increases as x increases; negative values imply that y decreases as x increases. Each of these situations is portrayed in Figure 3.8 (page 98).

FIGURE 3.8
Values of *r* and Their
Implications

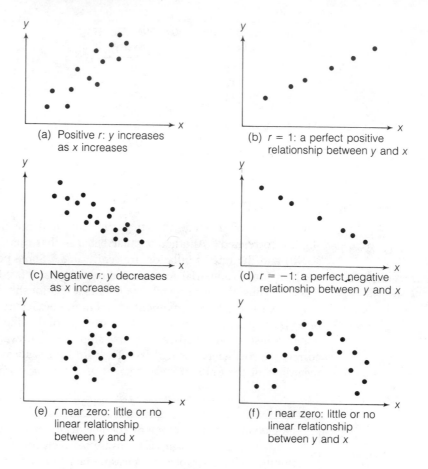

(a) Positive *r*: *y* increases
as *x* increases

(b) *r* = 1: a perfect positive
relationship between *y* and *x*

(c) Negative *r*: *y* decreases
as *x* increases

(d) *r* = −1: a perfect negative
relationship between *y* and *x*

(e) *r* near zero: little or no
linear relationship
between *y* and *x*

(f) *r* near zero: little or no
linear relationship
between *y* and *x*

EXAMPLE 3.1

A firm wants to know the correlation between the size of its sales force and its
yearly sales revenue. The records for the past 10 years are examined, and the
results listed in Table 3.5 are obtained. Calculate the coefficient of correlation *r*
for the data.

TABLE 3.5

YEAR	NUMBER OF SALESPEOPLE x	SALES y, hundred thousand dollars
1978	15	1.35
1979	18	1.63
1980	24	2.33
1981	22	2.41
1982	25	2.63
1983	29	2.93
1984	30	3.41
1985	32	3.26
1986	35	3.63
1987	38	4.15

SOLUTION

We need to calculate SS_{xy}, SS_{xx}, and SS_{yy}:

$$SS_{xy} = \sum x_i y_i - \frac{\left(\sum x_i\right)\left(\sum y_i\right)}{10} = 800.62 - \frac{(268)(27.73)}{10} = 57.456$$

$$SS_{xx} = \sum x_i^2 - \frac{\left(\sum x_i\right)^2}{10} = 7,668 - \frac{(268)^2}{10} = 485.6$$

$$SS_{yy} = \sum y_i^2 - \frac{\left(\sum y_i\right)^2}{10} = 83.8733 - \frac{(27.73)^2}{10} = 6.97801$$

Then, the coefficient of correlation is

$$r = \frac{SS_{xy}}{\sqrt{SS_{xx}SS_{yy}}} = \frac{57.456}{\sqrt{(485.6)(6.97801)}} = \frac{57.456}{58.211} = .99$$

Thus, the size of the sales force and sales revenue are very highly correlated—at least over the past 10 years. The implication is that a strong positive linear relationship exists between these variables (see Figure 3.9). We must be careful, however, not to jump to any unwarranted conclusions. For instance, the firm may be tempted to conclude that the best thing it can do to increase sales is to hire a large number of new salespeople. The implication of such a conclusion is that there is a *causal* relationship between the two variables. However, **high correlation does not imply causality.** The fact is, many things have probably contributed both to the increase in the size of the sales force and to the increase in sales revenue. The firm's expertise has undoubtedly grown, the economy has inflated (so that 1987 dollars are not worth as much as 1978 dollars), and perhaps the scope of products and services sold by the firm has widened. We must be careful not to infer a causal relationship on the basis of high sample correlation. The only safe conclusion when a high correlation is observed in the sample data is that a linear trend may exist between x and y.

FIGURE 3.9

Scattergram for Example 3.1

■

WARNING

High correlation does *not* imply causality. If a large positive or negative value of the sample correlation coefficient r is observed, it is incorrect to conclude that a change in x causes a change in y. The only valid conclusion is that a linear trend *may* exist between x and y.

Keep in mind that the correlation coefficient r measures the correlation between x-values and y-values in the sample, and that a similar linear coefficient of correlation exists for the population from which the data points were selected. The **population correlation coefficient** is denoted by the symbol ρ (rho). As you might expect, ρ is estimated by the corresponding sample statistic, r. Or, rather than estimating ρ, we might want to test

$$H_0: \quad \rho = 0 \quad \text{against} \quad H_a: \quad \rho \neq 0$$

That is, we might want to test the hypothesis that x contributes no information for the prediction of y, using the straight-line model against the alternative that the two variables are at least linearly related. However, we have already performed this identical test in Section 3.6 when we tested $H_0: \beta_1 = 0$ against $H_a: \beta_1 \neq 0$. It can be shown that the null hypothesis $H_0: \rho = 0$ is equivalent to the hypothesis $H_0: \beta_1 = 0$. When we tested the null hypothesis $H_0: \beta_1 = 0$ in connection with the advertising–sales example, the data led to a rejection of the null hypothesis for $\alpha = .05$. This implies that the null hypothesis of a zero linear correlation between the two variables (advertising and sales) can also be rejected at $\alpha = .05$. The only real difference between the least squares slope $\hat{\beta}_1$ and the coefficient of correlation r is the measurement scale. Therefore, the information they provide about the utility of the least squares model is to some extent redundant. Furthermore, the slope $\hat{\beta}_1$ gives us additional information on the amount of increase

---

TEST OF HYPOTHESIS FOR LINEAR CORRELATION

ONE-TAILED TEST	**TWO-TAILED TEST**
$H_0: \quad \rho = 0$	$H_0: \quad \rho = 0$
$H_a: \quad \rho > 0$	$H_a: \quad \rho \neq 0$
$\qquad$ (or $H_a: \quad \rho < 0$)	
Test statistic: $\quad r$	*Test statistic:* $\quad r$
Rejection region: $\quad r > r_\alpha$	*Rejection region:* $\quad r > r_{\alpha/2}$
(or $r < -r_\alpha$)	or $r < -r_{\alpha/2}$

where the distribution of the sample correlation coefficient r depends on the sample size n, and r_α and $r_{\alpha/2}$ are the critical values obtained from Table 8 of Appendix D, such that

$$P(r > r_\alpha) = \alpha \quad \text{and} \quad P(r > r_{\alpha/2}) = \alpha/2$$

(or decrease) in y for every 1-unit increase in x. For this reason, we recommend using the slope to make inferences about the existence of a positive or negative linear relationship between two variables.

For those who prefer to test for a linear relationship between two variables using the coefficient of correlation r, we outline the procedure in the preceding box.

SECTION 3.8

THE COEFFICIENT OF DETERMINATION

Another way to measure the contribution of x in predicting y is to consider how much the errors of prediction of y were reduced by using the information provided by x.

To illustrate, suppose a sample of data produces the scattergram shown in Figure 3.10(a). If we assume that x contributes no information for the prediction of y, the best prediction for a value of y is the sample mean $\bar{y}$, which graphs as the horizontal line shown in Figure 3.10(b). The vertical line segments in Figure 3.10(b) are the deviations of the points about the mean $\bar{y}$. Note that the sum of squares of deviations for the model $\hat{y} = \bar{y}$ is

$$SS_{yy} = \sum (y_i - \bar{y})^2$$

Now suppose you fit a least squares line to the same set of data and locate the deviations of the points about the line as shown in Figure 3.10(c). Compare the deviations about the prediction lines in parts (b) and (c) of Figure 3.10. You can see that:

1. If x contributes little or no information for the prediction of y, the sums of squares of deviations for the two lines,

$$SS_{yy} = \sum (y_i - \bar{y})^2 \quad \text{and} \quad SSE = \sum (y_i - \hat{y}_i)^2$$

will be nearly equal.

2. If x does contribute information for the prediction of y, then SSE will be smaller than SS_{yy}. In fact, if all the points fall on the least squares line, then SSE = 0.

FIGURE 3.10 A Comparison of the Sum of Squares of Deviations for Two Models

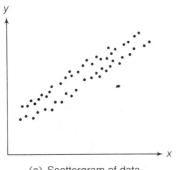

(a) Scattergram of data

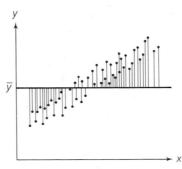

(b) Assumption: x contributes no information for predicting y
$\hat{y} = \bar{y}$

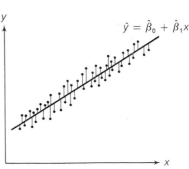

(c) Assumption: x contributes information for predicting y
$\hat{y} = \hat{\beta}_0 + \hat{\beta}_1 x$

A convenient way of measuring how well the least squares equation $\hat{y} = \hat{\beta}_0 + \hat{\beta}_1 x$ performs as a predictor of y is to compute the reduction in the sum of squares of deviations that can be attributed to x, expressed as a proportion of SS_{yy}. This quantity, called the **coefficient of determination**, is

$$\frac{SS_{yy} - SSE}{SS_{yy}}$$

In simple linear regression, it can be shown that this quantity is equal to the square of the simple linear coefficient of correlation r.

DEFINITION 3.3

The **coefficient of determination** is

$$r^2 = \frac{SS_{yy} - SSE}{SS_{yy}} = 1 - \frac{SSE}{SS_{yy}}$$

It represents the proportion of the sum of squares of deviations of the y-values about their mean that can be attributed to a linear relationship between y and x. (In simple linear regression, it may also be computed as the square of the coefficient of correlation r.)

Note that r^2 is always between 0 and 1, because r is between -1 and $+1$. Thus, an r^2 of .60 means that the sum of squares of deviations of the y-values about their predicted values has been reduced 60% by the use of $\hat{y}$, instead of $\bar{y}$, to predict y.

EXAMPLE 3.2

Calculate the coefficient of determination for the advertising–sales example. The data are repeated in Table 3.6.

TABLE 3.6

ADVERTISING EXPENDITURE x, hundreds of dollars	SALES REVENUE y, thousands of dollars
1	1
2	1
3	2
4	2
5	4

SOLUTION

We first calculate

$$SS_{yy} = \sum y_i^2 - \frac{\left(\sum y_i\right)^2}{5} = 26 - \frac{(10)^2}{5}$$

$$= 26 - 20 = 6$$

From previous calculations,

$$\text{SSE} = \sum (y_i - \hat{y}_i)^2 = 1.10$$

Then, the coefficient of determination is given by

$$r^2 = \frac{\text{SS}_{yy} - \text{SSE}}{\text{SS}_{yy}} = \frac{6.0 - 1.1}{6.0} = \frac{4.9}{6.0}$$

$$= .82$$

By using the advertising expenditure x to predict y with the least squares line

$$\hat{y} = -.1 + .7x$$

the total sum of squares of deviations of the five sample y-values about their predicted values has been reduced 82%. ∎

CASE STUDY 3.1

As evidenced by the cost overruns of public building projects, the initial estimate of the ultimate cost of a structure is often rather poor. These estimates usually rely on a precise definition of the proposed building in terms of working drawings and specifications. However, cost estimators do not take random error into account, so that no measure of reliability is possible for their deterministic estimates. Crandall and Cedercreutz (1976) propose the use of a probabilistic model to make cost estimates. They use regression models to relate cost to independent variables such as volume, amount of glass, floor area, and so forth. Crandall and Cedercreutz's rationale for choosing this approach is that "one of the principal merits of the least squares regression model, for the purpose of preliminary cost estimating, is the method of dealing with anticipated error." They go on to point out that when random error is anticipated, "statistical methods, such as regression analysis, attack the problem head on."

Crandall and Cedercreutz initially focused on the cost of mechanical work (heating, ventilating, and plumbing), since this part of the total cost is generally difficult to predict. Conventional cost estimates rely heavily on the amount of ductwork and piping used in construction, but this information is not precisely known until too late to be of use to the cost estimator. One of several models discussed was a simple linear model relating mechanical cost to floor area. Based on the data associated with 26 factory and warehouse buildings, the least squares prediction equation given in Figure 3.11 (page 104) was found. It was concluded that floor area and mechanical cost are linearly related, since the t-statistic (for testing H_0: $\beta_1 = 0$) was found to equal 3.61, which is significant with an α as small as .002. Thus, floor area should be useful when predicting the mechanical cost of a factory or warehouse. In addition, the regression model enables the reliability of the predicted cost to be assessed.

The value of the coefficient of determination r^2 was found to be .35. This tells us that only 35% of the variation among mechanical costs is accounted for by the differences in floor areas. Since there is only one independent variable in the model, this relatively small value of r^2 should not be too surprising. If other

FIGURE 3.11

Simple Linear Model
Relating Cost to Floor
Area

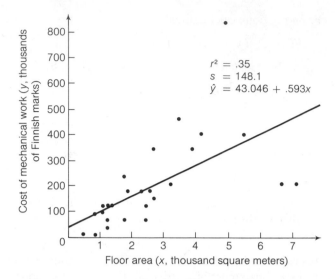

$r^2 = .35$
$s = 148.1$
$\hat{y} = 43.046 + .593x$

variables related to mechanical cost were included in the model, they would probably account for a significant portion of the remaining 65% of the variation in mechanical cost not explained by floor area. In the next chapter we discuss this important aspect of relating a response to more than one independent variable.

EXERCISES 3.26–3.33

3.26 Find the correlation coefficient and the coefficient of determination for the sample data listed in the table, and interpret your results.

YEAR	NUMBER OF 18-HOLE AND LARGER GOLF COURSES IN THE UNITED STATES	U.S. DIVORCE RATE PER 1,000 POPULATION
1960	2,725	2.2
1965	3,769	2.5
1970	4,845	3.5
1975	6,282	4.8
1980	6,856	5.2
1985	7,230	4.9

Source: United States Bureau of the Census, *Statistical Abstracts of the United States 1976–1987.*

3.27 To what extent is the age of members of a board of directors related to a firm's financial performance? In order to explore this question, three Pennsylvania State University professors collected data on the composition of the boards of directors and financial performance of 399 firms listed in the *Fortune 500* (*Quarterly Journal of Business and Economics*, Autumn 1984). The correlation coefficient between the average age of corporate boards of directors and the ratio of operating earnings to assets (a measure of financial performance) was found to be $r = -.1153$.

a. Interpret the value of r.

b. Is there evidence to indicate that average age of corporate boards of directors is linearly related to ratio of operating earnings to assets? Test using $\alpha = .01$.

3.28 Find the correlation coefficient and the coefficient of determination for the sample data listed in the table, and interpret your results.

YEAR	GROSS NATIONAL PRODUCT Billions of 1972 dollars	NEW HOUSING STARTS Thousands
1960	736.8	1,296
1965	925.9	1,510
1970	1,075.3	1,469
1975	1,191.7	1,171
1980	2,631.7	1,313
1985	3,998.5	1,745

Source: United States Bureau of the Census, *Statistical Abstracts of the United States 1976–1987.*

3.29 In 1984, federal government outlays for elementary, secondary, and vocational education were cut, yet Scholastic Aptitude Test (SAT) scores increased. Has such a relationship existed in the past? According to *Fortune* (October 29, 1984), "for the decade ending in 1984 . . . federal spending on education is strongly and negatively correlated with both verbal and math SAT scores." The correlation coefficient between verbal scores and federal spending on education for the past $n - 10$ years is $r - -.92$, while the correlation between math scores and federal spending is $r = -.71$.

a. Do the data support *Fortune*'s claim that verbal SAT scores and federal spending on education are "strongly and negatively correlated"? Test using $\alpha = .05$.

b. Do the data support *Fortune*'s claim that math SAT scores and federal spending on education are "strongly and negatively correlated"? Test using $\alpha = .05$.

c. Calculate the coefficient of determination for a straight-line model relating verbal SAT scores to federal spending. Interpret this value.

d. Calculate the coefficient of determination for a straight-line model relating math SAT scores to federal spending. Interpret this value.

3.30 Data on monthly sales y, price per unit during the month x_1, and amount spent on advertising x_2, for a product are shown in the table for a 5-month period. Based on this sample, which variable—price or advertising expenditure—appears to provide more information about sales? Explain.

MONTH	TOTAL MONTHLY SALES y, thousands of dollars	PRICE PER UNIT x_1, dollars	AMOUNT SPENT ON ALL FORMS OF ADVERTISING x_2, hundreds of dollars
June	40	.85	6.0
July	50	.76	5.0
August	55	.75	8.0
September	30	1.00	7.5
October	45	.80	5.5

3.31 Best and Kahle (1985) investigated several factors thought to influence market share for capital-equipment businesses. One variable considered was product quality, measured as the difference between the percentage of sales derived from products superior to competition and the percentage of sales inferior to competition. Based on data collected for 333 capital-equipment businesses, the correlation between market share and product quality was found to be $r = .373$.

a. Is there sufficient evidence to indicate that product quality and market share for capital equipment businesses are positively correlated? Test using $\alpha = .01$.

b. Calculate r^2. Interpret its value.

3.32 The accompanying table shows a portion of the experimental data obtained in a study of the radial tension strength of concrete pipe. The concrete pipe used for the experiment had an inside diameter of 84 inches and a wall thickness of approximately 8.75 inches. In addition, it was reinforced with cold drawn wire. The variable y is the load (in pounds per foot) until the first crack in a pipe specimen was observed. The variable x is the age of the specimen (in days) at the time of the test.

y	x	y	x
11,450	20	10,540	25
10,420	20	9,470	31
11,142	20	9,190	31
10,840	25	9,540	31
11,170	25		

Source: Heger, F. J. and McGrath, T. J. "Radial tension strength of pipe and other curved flexural members," *Journal of the American Concrete Institute*, Vol. 80, No. 1, 1983, pp. 33–39.

a. Construct a scattergram for the data. After examining the scattergram, do you think that x and y are correlated? If correlation is present, is it positive or negative?

b. Find the correlation coefficient r and interpret its value.

c. Do the data provide sufficient evidence to indicate that x and y are linearly correlated? Test using $\alpha = .05$.

d. Find the coefficient of determination r^2 and interpret its value.

3.33 A major portion of the effort expended in developing commercial computer software is associated with program testing. A study was undertaken to assess the potential usefulness of various product- and process-related variables in identifying error-prone software (*IEEE Transactions on Software Engineering*, April 1985). A straight-line model relating the number y of module defects to the number x of unique operands in the module was fit to the data collected for a sample of software modules. The coefficient of determination for this analysis was $r^2 = .74$.

a. Interpret the value of r^2.

b. Based on this value, would you infer that the straight-line model is a useful predictor of number y of module defects? Explain.

**USING THE MODEL
FOR ESTIMATION
AND PREDICTION**

If we are satisfied that a useful model has been found to describe the relationship between sales revenue and advertising, we are ready to accomplish the original objectives for building the model: using it to estimate or to predict sales on the basis of advertising dollars spent.

The most common uses of a probabilistic model can be divided into two categories. **The first is the use of the model for estimating the mean value of y, $E(y)$, for a specific value of x.** For our example, we may want to estimate the mean sales revenue for *all* months during which $400 ($x = 4$) is spent on advertising. **The second use of the model entails predicting a particular y-value for a given x.** That is, if we decide to spend $400 next month, we want to predict the firm's sales revenue for that month.

In the case of estimating a mean value of y, we are attempting to estimate the mean result of a very large number of experiments at the given x-value. In the second case, we are trying to predict the outcome of a single experiment at the given x-value. In which of these model uses do you expect to have more success, i.e., which value—the mean or individual value of y—can we estimate (or predict) with more accuracy?

Before answering this question, we first consider the problem of choosing an estimator (or predictor) of the mean (or individual) y-value. We will use the least squares model

$$\hat{y} = \hat{\beta}_0 + \hat{\beta}_1 x$$

both to estimate the mean value of y and to predict a particular value of y for a given value of x. For our example, we found

$$\hat{y} = -.1 + .7x$$

so that the estimated mean value of sales revenue for all months when $x = 4$ (advertising = $400) is

$$\hat{y} = -.1 + .7(4) = 2.7$$

or $2,700 (the units of y are thousands of dollars). The identical value is used to predict the y-value when $x = 4$. That is, both the estimated mean value and the predicted value of y equal $\hat{y} = 2.7$ when $x = 4$, as shown in Figure 3.12.

FIGURE 3.12
Estimated Mean Value
and Predicted Individual
Value of Sales Revenue y
for $x = 4$

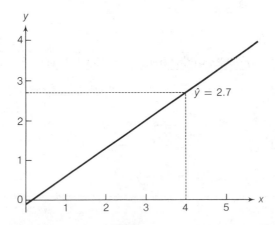

The difference in these two model uses lies in the relative accuracy of the estimate and the prediction. These accuracies are best measured by the repeated sampling errors of the least squares line when it is used as an estimator and as a predictor, respectively. These errors are given in the box.

SAMPLING ERRORS FOR THE ESTIMATOR OF THE MEAN OF y AND THE PREDICTOR OF AN INDIVIDUAL y FOR $x = x_p$

1. The standard deviation of the sampling distribution of the estimator $\hat{y}$ of the mean value of y at a particular value of x, say x_p, is

$$\sigma_{\hat{y}} = \sigma \sqrt{\frac{1}{n} + \frac{(x_p - \bar{x})^2}{SS_{xx}}}$$

where σ is the standard deviation of the random error ε.

2. The standard deviation of the prediction error for the predictor $\hat{y}$ of an individual y-value for $x = x_p$ is

$$\sigma_{(y-\hat{y})} = \sigma \sqrt{1 + \frac{1}{n} + \frac{(x_p - \bar{x})^2}{SS_{xx}}}$$

where σ is the standard deviation of the random error ε.

The true value of σ will rarely be known. Thus, we estimate σ by s and calculate the estimation and prediction intervals as shown in the next two boxes.

A $100(1 - \alpha)$% CONFIDENCE INTERVAL FOR THE MEAN VALUE OF y FOR $x = x_p$

$$\hat{y} \pm t_{\alpha/2}(\text{Estimated standard deviation of } \hat{y})$$

or

$$\hat{y} \pm t_{\alpha/2}s \sqrt{\frac{1}{n} + \frac{(x_p - \bar{x})^2}{SS_{xx}}}$$

where $t_{\alpha/2}$ is based on $(n - 2)$df

A $100(1 - \alpha)$% PREDICTION INTERVAL FOR AN INDIVIDUAL y FOR $x = x_p$

$$\hat{y} \pm t_{\alpha/2}[\text{Estimated standard deviation of } (y - \hat{y})]$$

or

$$\hat{y} \pm t_{\alpha/2}s \sqrt{1 + \frac{1}{n} + \frac{(x_p - \bar{x})^2}{SS_{xx}}}$$

where $t_{\alpha/2}$ is based on $(n - 2)$ df

EXAMPLE 3.3

Find a 95% confidence interval for mean monthly sales when the appliance store spends $400 on advertising.

SOLUTION

For a $400 advertising expenditure, $x_p = 4$ and, since $n = 5$, df $= n - 2 = 3$. Then the confidence interval for the mean value of y is

$$\hat{y} \pm t_{\alpha/2}s \sqrt{\frac{1}{n} + \frac{(x_p - \bar{x})^2}{SS_{xx}}}$$

or

$$\hat{y} \pm t_{.025}s \sqrt{\frac{1}{5} + \frac{(4 - \bar{x})^2}{SS_{xx}}}$$

Recall that $\hat{y} = 2.7$, $s = .61$, $\bar{x} = 3$, and $SS_{xx} = 10$. From Table 2 of Appendix D, $t_{.025} = 3.182$. Thus, we have

$$2.7 \pm (3.182)(.61) \sqrt{\frac{1}{5} + \frac{(4 - 3)^2}{10}} = 2.7 \pm (3.182)(.61)(.55)$$

$$= 2.7 \pm 1.1$$

We estimate that the interval from $1,600 to $3,800 encloses the mean sales revenue when the store spends $400 a month on advertising. Note that we used a small amount of data for purposes of illustration in fitting the least squares line and that the width of the interval could be decreased by using a larger number of data points. ∎

EXAMPLE 3.4

Predict the monthly sales for next month if a $400 expenditure is to be made on advertising. Use a 95% prediction interval.

SOLUTION

To predict the sales for a particular month for which $x_p = 4$, we calculate the 95% prediction interval as

$$\hat{y} \pm t_{\alpha/2}s \sqrt{1 + \frac{1}{n} + \frac{(x_p - \bar{x})^2}{SS_{xx}}} = 2.7 \pm (3.182)(.61) \sqrt{1 + \frac{1}{5} + \frac{(4 - 3)^2}{10}}$$

$$= 2.7 \pm (3.182)(.61)(1.14) = 2.7 \pm 2.2$$

Therefore, we predict that the sales next month will fall in the interval from $500 to $4,900. As in the case of the confidence interval for the mean value of y, the prediction interval for y is quite large. This is because we have chosen a simple example (only five data points) to fit the least squares line. The width of the prediction interval could be reduced by using a larger number of data points. ∎

A comparison of the confidence interval for the mean value of y and the prediction interval for some future value of y for a $400 advertising expenditure ($x = 4$) is illustrated in Figure 3.13 (page 110). It is important to note that the

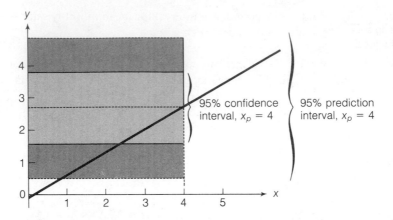

prediction interval for an individual value of y will always be wider than the confidence interval for a mean value of y. You can see this by examining the formulas for the two intervals, and you can see it in Figure 3.13.

The error in estimating the mean value of y, $E(y)$, for a given value of x, say x_p, is the distance between the least squares line and the true line of means, $E(y) = \beta_0 + \beta_1 x$. This error, $[\hat{y} - E(y)]$, is shown in Figure 3.14. In contrast, the error $(y_p - \hat{y})$ in predicting some future value of y is the sum of two errors—the error of estimating the mean of y, $E(y)$, shown in Figure 3.14, plus the random error that is a component of the value of y to be predicted (see Figure 3.15). Consequently, the error of predicting a particular value of y will always be larger than the error of estimating the mean value of y for a particular value of x. Note from their formulas that both the error of estimation and the error of prediction take their smallest values when $x_p = \bar{x}$. The farther x lies from $\bar{x}$, the larger will be the errors of estimation and prediction. You can see why this is true by noting the deviations for different values of x between the line of means $E(y) = \beta_0 + \beta_1 x$ and the predicted line of means $\hat{y} = \hat{\beta}_0 + \hat{\beta}_1 x$ shown in Figure 3.15. The deviation is larger at the extremities of the interval where the largest and smallest values of x in the data set occur.

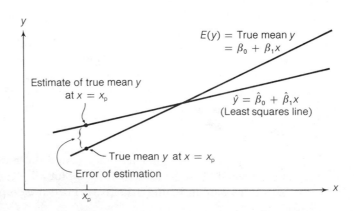

FIGURE 3.15
Error of Predicting a
Future Value of y for a
Given Value of x

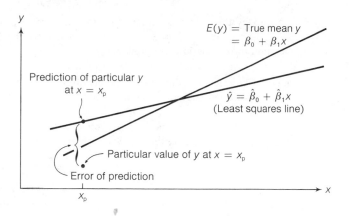

EXERCISES 3.34–3.41

3.34 A simple linear regression analysis for $n = 20$ data points produced the following results:

$$\hat{y} = 2.1 + 3.4x \qquad SS_{xx} = 4.77$$

$$\bar{x} = 2.5 \qquad SS_{yy} = 59.21$$

$$\bar{y} = 10.6 \qquad SS_{xy} = 16.22$$

a. Find SSE and s^2.
b. Find a 95% confidence interval for $E(y)$ when $x = 2.5$. Interpret this interval.
c. Find a 95% confidence interval for $E(y)$ when $x = 2.0$. Interpret this interval.
d. Find a 95% confidence interval for $E(y)$ when $x = 3.0$. Interpret this interval.
e. Examine the widths of the confidence intervals obtained in parts **b**, **c**, and **d**. What happens to the width of the confidence interval for $E(y)$ as the value of x moves away from the value of $\bar{x}$?
f. Find a 95% prediction interval for a value of y to be observed in the future when $x = 3.0$. Interpret its value.

3.35 Refer to Exercise 3.9. Find a 95% confidence interval for the mean yearly company sales when industry sales are $11 million.

3.36 Refer to Exercise 3.10. The ratio of research and development expenditures to sales (R/S) ratio for a company is .070.
a. Find a 90% prediction interval for the company's price/earnings (P/E) ratio.
b. Find a 90% confidence interval for the average (mean) P/E ratio of all firms with an R/S ratio of .070.
c. Compare and comment on the sizes of the intervals in parts **a** and **b**.
d. Could you reduce the size of either or both intervals by increasing your sample size? Explain.

3.37 Explain why for a particular x-value, the prediction interval for an individual y-value will always be wider than the confidence interval for a mean value of y.

3.38 Explain why the confidence interval for the mean value of y for a particular x value, say x_p, gets wider the farther x_p is from $\bar{x}$. What are the implications of this phenomenon for estimation and prediction?

3.39 Refer to Exercise 3.15. Find a 95% prediction interval for the refusal rate of a Florida county with a per capita income of $8,000.

3.40 Refer to Exercise 3.25. Find a 90% confidence interval for the mean maintenance cost per rental car during the first year if it is driven 35,000 miles.

3.41 Refer to the study of the radial tension strength of concrete pipe in Exercise 3.32. The data are reproduced in the table. The variable y is the load (in pounds per foot) until the first crack in a pipe was observed, and the variable x is the age of the specimen (in days) at the time of the test.

y	x	y	x
11,450	20	10,540	25
10,420	20	9,470	31
11,142	20	9,190	31
10,840	25	9,540	31
11,170	25		

Source: Heger, F. J. and McGrath, T. J. "Radial tension strength of pipe and other curved flexural members," *Journal of the American Concrete Institute*, Vol. 80, No. 1, 1983, pp. 33–39.

a. Find the least squares prediction equation relating load y and age x.
b. Test the hypothesis $H_0: \beta_1 = 0$ (at $\alpha = .05$), and show that the result agrees with your answer to Exercise 3.32, part **c**.
c. Find a 95% prediction interval for the crack load of a 35-day-old concrete specimen.
d. Why might the prediction interval in part **c** be less reliable than expected? Explain.

SECTION 3.10

SIMPLE LINEAR REGRESSION: AN EXAMPLE

In the previous sections we have presented the basic elements necessary to fit and use a straight-line regression model. In this section we will assemble these elements by applying them in an example.

Suppose a fire insurance company wants to relate the amount of fire damage in major residential fires to the distance between the residence and the nearest fire station. The study is to be conducted in a large suburb of a major city; a sample of 15 recent fires in this suburb is selected. The amount of damage y and the distance x between the fire and the nearest fire station are recorded for each fire. The results are given in Table 3.7.

TABLE 3.7
Fire Damage Data

DISTANCE FROM FIRE STATION x, miles	FIRE DAMAGE y, thousands of dollars
3.4	26.2
1.8	17.8
4.6	31.3
2.3	23.1
3.1	27.5
5.5	36.0
.7	14.1
3.0	22.3
2.6	19.6
4.3	31.3
2.1	24.0
1.1	17.3
6.1	43.2
4.8	36.4
3.8	26.1

STEP 1 First, we hypothesize a model to relate fire damage y to the distance x from the nearest fire station. We will hypothesize a straight-line probabilistic model:

$$y = \beta_0 + \beta_1 x + \varepsilon$$

STEP 2 Next, we use the data to estimate the unknown parameters in the deterministic component of the hypothesized model. We make some preliminary calculations:

$$SS_{xx} = \sum x_i^2 - \frac{\left(\sum x_i\right)^2}{15} = 196.16 - \frac{(49.2)^2}{15}$$
$$= 196.160 - 161.376 = 34.784$$

$$SS_{yy} = \sum y_i^2 - \frac{\left(\sum y_i\right)^2}{15} = 11{,}376.48 - \frac{(396.2)^2}{15}$$
$$= 11{,}376.480 - 10{,}464.962667 = 911.517333$$

$$SS_{xy} = \sum x_i y_i - \frac{\left(\sum x_i\right)\left(\sum y_i\right)}{15} = 1{,}470.65 - \frac{(49.2)(396.2)}{15}$$
$$= 1{,}470.650 - 1{,}299.536 = 171.114$$

Then the least squares estimates of the slope β_1 and intercept β_0 are

$$\hat{\beta}_1 = \frac{SS_{xy}}{SS_{xx}} = \frac{171.114}{34.784} = 4.9193307$$

$$\hat{\beta}_0 = \bar{y} - \hat{\beta}_1 \bar{x} = \frac{396.2}{15} - 4.9193307\left(\frac{49.2}{15}\right)$$

$$= 26.413333 - (4.9193307)(3.28)$$
$$= 10.2777929$$

And the least squares equation is

$$\hat{y} = 10.278 + 4.919x*$$

This prediction equation is graphed in Figure 3.16, along with a plot of the data points.

FIGURE 3.16

Least Squares Model for the Fire Damage Data

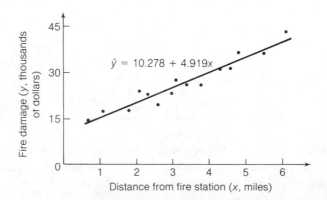

STEP 3 Now, we specify the probability distribution of the random error component ε. The assumptions about the distribution will be identical to those listed in Section 3.4. Although we know that these assumptions are not completely satisfied (they rarely are for any practical problem), we are willing to assume they are approximately satisfied for this example. We have to estimate the variance σ^2 of ε, so we calculate

$$\text{SSE} = \sum (y_i - \hat{y}_i)^2 = \text{SS}_{yy} - \hat{\beta}_1 \text{SS}_{xy}$$

where the last expression represents a shortcut formula for the SSE. Thus,

$$\text{SSE} = 911.517333 - (4.9193307)(171.114)$$
$$= 911.517333 - 841.766358 = 69.750972$$

To estimate σ^2, we divide the SSE by the degrees of freedom available for error, $n - 2$. Thus,

$$s^2 = \frac{\text{SSE}}{n-2} = \frac{69.75092}{15-2} = 5.3655$$
$$s = \sqrt{5.3655} = 2.3163$$

---

*The values for SS_{xy}, $\bar{x}$, and SS_{xx} are exact for this example, while values for $\bar{y}$ and SS_{yy} are reported to six decimal places. In general, at least six-place decimal accuracy should be retained for all calculations. Otherwise, the calculated values of $\hat{\beta}_0$, SSE, and s^2 may be substantially in error.

STEP 4 We can now check the utility of the hypothesized model, that is, whether x really contributes information for the prediction of y using the straight-line model. First, test the null hypothesis that the slope β_1 is 0, i.e., that there is no linear relationship between the fire damage and the distance from the nearest fire station. We test:

$$H_0: \quad \beta_1 = 0 \qquad H_a: \quad \beta_1 \neq 0$$

Test statistic: $\quad t = \dfrac{\hat{\beta}_1 - 0}{s_{\hat{\beta}_1}} = \dfrac{\hat{\beta}_1}{s/\sqrt{SS_{xx}}}$

Assumptions: Those made about ε in Section 3.4.

For $\alpha = .05$, we will reject H_0 if

$$t > t_{\alpha/2} \quad \text{or} \quad t < -t_{\alpha/2}$$

where for $n = 15$, df $= n - 2 = 15 - 2 = 13$ and $t_{.025} = 2.160$. We then calculate the t-statistic:

$$t = \frac{\hat{\beta}_1}{s_{\hat{\beta}_1}} = \frac{\hat{\beta}_1}{s/\sqrt{SS_{xx}}} = \frac{4.9193307}{2.3163/\sqrt{34.784}} = \frac{4.9193307}{.392748}$$

$$= 12.5$$

This large t-value leaves little doubt that distance between the fire and the fire station contributes information for the prediction of fire damage and that fire damage increases as the distance increases.

We gain additional information about the relationship by forming a confidence interval for the slope β_1. A 95% confidence interval is

$$\hat{\beta}_1 \pm t_{.025}s_{\hat{\beta}_1} = 4.919 \pm (2.160)(.392748)$$

$$= 4.919 \pm .848 \quad \text{or} \quad (4.071, 5.767)$$

We estimate that the interval from \$4,071 to \$5,767 encloses the mean increase (β_1) in fire damage per additional mile distance from the fire station.

Another measure of the utility of the model is the coefficient of correlation r:

$$r = \frac{SS_{xy}}{\sqrt{SS_{xx}SS_{yy}}} = \frac{171.114}{\sqrt{(34.784)(911.517333)}}$$

$$= \frac{171.114}{178.062} = .96$$

The high correlation confirms our conclusion that β_1 differs from 0; it appears that fire damage and distance from the fire station are highly correlated.

The coefficient of determination is

$$r^2 = (.96)^2 = .92$$

which implies that 92% of the sum of squares of deviations of the y-values about $\bar{y}$ is explained by the distance x between the fire and the fire station. All signs point to a strong linear relationship between x and y.

STEP 5 We are now prepared to use the least squares model. Suppose the insurance company wants to predict the fire damage if a major residential fire were to occur 3.5 miles from the nearest fire station, i.e., $x_p = 3.5$. The predicted value is

$$\hat{y} = \hat{\beta}_0 + \hat{\beta}_1 x_p = 10.278 + (4.919)(3.5)$$

$$= 10.278 + 17.216 = 27.5$$

(we round to the nearest tenth to be consistent with the units of the original data in Table 3.7). If we want a 95% prediction interval, we calculate

$$\hat{y} \pm t_{.025}s\sqrt{1 + \frac{1}{n} + \frac{(x_p - \bar{x})^2}{SS_{xx}}} = 27.5 \pm (2.16)(2.3163)\sqrt{1 + \frac{1}{15} + \frac{(3.5 - 3.28)^2}{34.784}}$$

$$= 27.5 \pm (2.16)(2.3163)\sqrt{1.0681}$$

$$= 27.5 \pm 5.2 \quad \text{or} \quad (22.3, 32.7)$$

The model yields a 95% prediction interval, for fire damage in a major residential fire 3.5 miles from the nearest station, of $22,300 to $32,700.

One caution before closing: We would not use this prediction model to make predictions for homes less than .7 mile or more than 6.1 miles from the nearest fire station. A look at the data in Table 3.7 reveals that all the x-values fall between .7 and 6.1. It is dangerous to use the model to make predictions outside the region in which the sample data fall. A straight line might not provide a good model for the relationship between the mean value of y and the value of x when stretched over a wider range of x-values.

WARNING

Using the least squares prediction equation to estimate the mean value of y or to predict a particular value of y for values of x that fall *outside the range* of the values of x contained in your sample data may lead to errors of estimation or prediction that are much larger than expected. Although the least squares model may provide a very good fit to the data over the range of x-values contained in the sample, it could give a poor representation of the true model for values of x outside this region.

S E C T I O N   3.11

A COMPUTER PRINTOUT FOR SIMPLE LINEAR REGRESSION

The computations required for a simple linear regression analysis, even if performed with the aid of a pocket or desk calculator, can become tedious and cumbersome, especially if the number of measurements is large. Many institutions have installed statistical computer program packages which fit a straight-line regression model by the method of least squares. These computer packages enable the user to greatly decrease the burden of calculation. In this section, we locate, discuss, and interpret the elements of a simple linear regression on a computer printout from one of the more popular statistical computer program packages, called the SAS® System.* At your institution, other computer packages (such as BMDP, Minitab, and SPSSˣ) may be installed. These packages also have simple linear regression procedures available and their respective printouts are similar to that of the SAS System. (Computer printouts for BMDP, Minitab, and SPSSˣ are shown and discussed in Chapter 4.) Regardless of the package(s) to which you have access, you should be able to understand the regression results after reading the discussion that follows.

The SAS printout for the simple linear regression analysis of the data relating fire damage to distance from fire station in Table 3.7 is shown in Figure 3.17.[†]

FIGURE 3.17 SAS Printout for Fire Damage Model

ANALYSIS OF VARIANCE

SOURCE	DF	SUM OF SQUARES	MEAN SQUARE	F VALUE	PROB>F
MODEL	1	841.76636	841.76636	156.886	0.0001
ERROR	13	69.75097535	5.36545964		
C TOTAL	14	911.51733			

ROOT MSE	2.316346	R-SQUARE	0.9235	
DEP MEAN	26.41333	ADJ R-SQ	0.9176	
C.V.	8.769609			

PARAMETER ESTIMATES

VARIABLE	DF	PARAMETER ESTIMATE	STANDARD ERROR	T FOR H0: PARAMETER=0	PROB > |T|
INTERCEP	1	10.27792855	1.42027781	7.237	0.0001
X	1	4.91933073	0.39274775	12.525	0.0001

OBS	ID	ACTUAL	PREDICT VALUE	STD ERR PREDICT	LOWER95% PREDICT	UPPER95% PREDICT	RESIDUAL
1	3.4	26.2000	27.0037	0.5999	21.8344	32.1729	-0.8037
2	1.8	17.8000	19.1327	0.8340	13.8141	24.4514	-1.3327
3	4.6	31.3000	32.9068	0.7915	27.6186	38.1951	-1.6068
4	2.3	23.1000	21.5924	0.7112	16.3577	26.8271	1.5076
5	3.1	27.5000	25.5279	0.6022	20.3573	30.6984	1.9721
6	5.5	36.0000	37.3342	1.0573	31.8334	42.8351	-1.3342
7	0.7	14.1000	13.7215	1.1766	8.1087	19.3342	0.3785
8	3	22.3000	25.0359	0.6081	19.8622	30.2097	-2.7359
9	2.6	19.6000	23.0682	0.6550	17.8678	28.2686	-3.4682
10	4.3	31.3000	31.4311	0.7198	26.1908	36.6713	-0.1311
11	2.1	24.0000	20.6085	0.7566	15.3442	25.8729	3.3915
12	1.1	17.3000	15.6892	1.0444	10.1999	21.1785	1.6108
13	6.1	43.2000	40.2858	1.2587	34.5906	45.9811	2.9142
14	4.8	36.4000	33.8907	0.8450	28.5640	39.2175	2.5093
15	3.8	26.1000	28.9714	0.6320	23.7843	34.1585	-2.8714
16	3.5		27.4956	0.6043	22.3239	32.6672	

SUM OF RESIDUALS	7.57172E-14
SUM OF SQUARED RESIDUALS	69.75098
PREDICTED RESID SS (PRESS)	93.21169

The least squares estimates of the y-intercept and slope are found (shaded) under the column labeled PARAMETER ESTIMATE (in the middle portion of the printout) in the rows labeled INTERCEP and X, respectively. Note that the estimate of the y-intercept, $\hat{\beta}_0 = 10.27792855$, and the estimate of the slope,

*SAS is the registered trademark of SAS Institute, Inc., Cary, North Carolina, U.S.A.

[†]See Appendix C for the SAS commands used to generate the printout.

$\hat{\beta}_1 = 4.91933073$, given in the printout agree with our calculations made by hand in Section 3.10.

The values of SSE, s^2, and s are given (shaded) in the upper middle portion of the printout. The SSE is found by locating the entry under the column heading SUM OF SQUARES in the row labeled ERROR: SSE = 69.75097535. The estimate of σ^2 is given as MEAN SQUARE for ERROR and is located to the immediate right of SSE. This value is $s^2 = 5.36545964$. The estimate of σ, given next to the heading ROOT MSE, is $s = 2.316346$. Except for rounding errors, our hand-computed values of s^2 and s agree with the figures given in the printout.

The coefficient of determination is given (shaded) next to the heading R-SQUARE in the upper portion of the printout. This value is $r^2 = .9235$. Due to roundoff, our hand-calculated value of r^2 is accurate to only two decimal places. Note that the coefficient of correlation r is not given on the SAS printout.

The value of the test statistic for testing $H_0: \beta_1 = 0$ against $H_a: \beta_1 \neq 0$ is given (shaded) under the column heading T FOR H0: PARAMETER = 0 in the middle portion of the printout in the row corresponding to the VARIABLE X. The value shown here is $t = 12.525$, which agrees with our computed test statistic. To determine whether this value falls within the rejection region, check the shaded quantity to the right under PROB > |T|. This quantity is the **observed significance level** or **p-value** of the test. Generally, if the p-value is smaller than .05, there is evidence to reject the null hypothesis that the slope is 0 in favor of the alternative hypothesis that the slope differs from 0. In this example, a significance level of .0001 indicates that distance from fire station x and fire damage y are linearly related (a result consistent with the test conducted in Section 3.10). The p-value for a one-tailed test of $H_0: \beta_1 = 0$ is found by taking half the PROB > |T| value reported in SAS.

Prediction intervals for individual y-values are reported in the lower portion of the SAS printout. To find a 95% prediction interval for fire damage y when the distance from the fire station is $x = 3.5$ miles, first locate the value 3.5 (shaded) in the column labeled ID. [The ID (identification) variable represents the independent variable x.] The lower and upper endpoints of the prediction interval are found (shaded) in the ID = 3.5 row under the columns labeled LOWER 95% PREDICT and UPPER 95% PREDICT, respectively. These values, 22.3239 and 32.6672, agree (except for rounding) with our hand-calculated values in Section 3.10.

In Chapter 4 we will discuss the interpretation of those portions of the SAS printout that were not mentioned here. However, the important elements of a simple linear regression analysis have been located, and you should be able to use this discussion as a guide to interpreting simple linear regression computer printouts. Details on how to perform a simple linear regression analysis on the computer using any one of the four program packages, SAS, SPSS^x, Minitab, or BMDP, are provided in Appendix C.

EXERCISES 3.42–3.45

3.42 A portion of the SAS printout for the model $E(y) = \beta_0 + \beta_1 x$ fit to $n = 10$ data points is shown here.

SAS Printout for Exercise 3.42

SOURCE	DF	SUM OF SQUARES	MEAN SQUARE	F VALUE	PROB>F
MODEL	1	127.6935	127.6935	62.26	0.0001
ERROR	8	16.4065	2.0508		
TOTAL	9	144.1000		R-SQUARE	ROOT MSE
				.8861	1.4321

VARIABLE	PARAMETER ESTIMATE	STANDARD ERROR	T FOR H0: PARAMETER = 0	PROB>|T|
INTERCEPT	55.2942	7.4529	7.42	0.0001
X	-.7716	.0978	-7.89	0.0001

ID	ACTUAL	PREDICT VALUE	RESIDUAL	LOWER 95% PREDICT	UPPER 95% PREDICT
58	10	10.5436	.5436	6.8647	14.225

a. Give the least squares prediction equation.

b. Locate the values of SSE, s^2, s, and r^2 on the printout.

c. What is the value of the test statistic for testing model adequacy?

d. Locate and interpret the p-value of the test for model adequacy.

e. Give the lower and upper endpoints for a 95% prediction interval for y when $x = 58$. Interpret the interval.

3.43 The Federal Trade Commission (FTC) annually ranks American cigarette brands according to carbon monoxide and nicotine content (measured in milligrams). In an

SAS Printout for Exercise 3.43

ANALYSIS OF VARIANCE

SOURCE	DF	SUM OF SQUARES	MEAN SQUARE	F VALUE	PROB>F
MODEL	1	283.93484	283.93484	325.699	0.0001
ERROR	8	6.97415579	0.87176947		
C TOTAL	9	290.90900			

ROOT MSE	0.933686	R-SQUARE	0.9760	
DEP MEAN	9.79	ADJ R-SQ	0.9730	
C.V.	9.53714			

PARAMETER ESTIMATES

VARIABLE	DF	PARAMETER ESTIMATE	STANDARD ERROR	T FOR H0: PARAMETER=0	PROB > |T|
INTERCEP	1	-0.53490343	0.64380408	-0.831	0.4302
X	1	15.52617057	0.86031171	18.047	0.0001

OBS	ID	ACTUAL	PREDICT VALUE	STD ERR PREDICT	LOWER95% PREDICT	UPPER95% PREDICT	RESIDUAL
1	0.86	13.6000	12.8176	0.3396	10.5265	15.1087	0.7824
2	0.17	1.2000	2.1045	0.5182	-0.3579	4.5670	-0.9045
3	0.72	9.8000	10.6439	0.2990	8.3831	12.9048	-0.8439
4	0.1	1.2000	1.0177	0.5687	-1.5034	3.5388	0.1823
5	0.4	5.4000	5.6756	0.3730	3.3570	7.9942	-0.2756
6	0.93	12.7000	13.9044	0.3730	11.5858	16.2230	-1.2044
7	0.67	10.5000	9.8676	0.2953	7.6094	12.1258	0.6324
8	1.02	15.4000	15.3018	0.4248	12.9363	17.6673	0.0982
9	0.57	10.0000	8.3150	0.3064	6.0490	10.5811	1.6850
10	1.21	18.1000	18.2518	0.5541	15.7481	20.7555	-0.1518

SUM OF RESIDUALS	-6.66134E-16
SUM OF SQUARED RESIDUALS	6.974156
PREDICTED RESID SS (PRESS)	9.781261

attempt to determine the relationship between carbon monoxide content (y) and nicotine content (x), the FTC fit the straight-line model $y = \beta_0 + \beta_1 x + \varepsilon$ to data collected for a sample of ten brands of cigarettes. The SAS printout is reproduced on page 119.

a. Give the least squares prediction equation.
b. Locate the values of SSE, s^2, s, and r^2 on the printout.
c. What is the value of the test statistic for testing model adequacy?
d. Locate and interpret the p-value of the test for model adequacy.
e. Give the lower and upper endpoints for a 95% prediction interval for y when $x = .4$ milligram. Interpret the interval.

3.44 In retailing ready-to-wear fashion, price concession—in the form of a markdown— is the traditional vehicle for selling slow-moving merchandise. In order to investigate the effect of a markdown on the rate of sale of slow-moving items, P. G. Carlson (Emory University) studied the markdown policy at Rich's department store in Atlanta. Nine styles of budget junior dresses were selected for analysis. For each style, the initial rate of sale, x, and the postmarkdown rate of sale, y, were recorded. (The initial rate of sale is the average rate of sale from the day the style was put on the floor to the time of the first markdown, while the postmarkdown rate of sale is the average rate of sale from the day of the first markdown to the time of either the second markdown or the end of the fashion life.) The nine data points were subjected to a simple linear regression analysis. The SAS printout appears here.

SAS Printout for Exercise 3.44

SOURCE	DF	SUM OF SQUARES	MEAN SQUARE	F VALUE	PROB > F
MODEL	1	.1603	.1603	14.71	.01
ERROR	7	.0764	.0109		
TOTAL	8	.2367		R-SQUARE	ROOT MSE
				.6772	.1044

VARIABLE	PARAMETER ESTIMATE	STANDARD ERROR	T FOR H0: PARAMETER = 0	PROB > !T!
INTERCEPT	-.0292	.0902	-.32	.5000
X	2.2985	.5999	3.83	.0200

Source: Carlson, P. G. "Fashion retailing: The sensitivity of rate of sale to markdown," *Journal of Retailing*, Spring 1983, Vol. 59, No. 1, pp. 67–76.

a. Write the least squares prediction equation.
b. Locate SSE on the printout and interpret its value.
c. Locate s on the printout and interpret its value.
d. Is there sufficient evidence to indicate that postmarkdown rate of sale y and initial rate of sale x are linearly related? Test at $\alpha = .05$.

3.45 The SAS printout for the straight-line model relating sale price y of a residential property in neighborhood D and total appraised value of x of the property (both measured in thousands of dollars), for a sample of 20 properties selected from Appendix E, is provided here.
a. Give the least squares prediction equation.
b. Locate the values of SSE, s^2, s, and r^2 on the printout.
c. What is the value of the test statistic for testing model adequacy?

SAS Printout for Exercise 3.45

ANALYSIS OF VARIANCE

SOURCE	DF	SUM OF SQUARES	MEAN SQUARE	F VALUE	PROB>F
MODEL	1	58.31306210	58.31306210	3.729	0.0694
ERROR	18	281.48694	15.63816322		
C TOTAL	19	339.80000			

ROOT MSE	3.954512	R-SQUARE	0.1716	
DEP MEAN	41.1	ADJ R-SQ	0.1256	
C.V.	9.621683			

PARAMETER ESTIMATES

VARIABLE	DF	PARAMETER ESTIMATE	STANDARD ERROR	T FOR H0: PARAMETER=0	PROB > |T|
INTERCEP	1	30.11691649	5.75599468	5.232	0.0001
X	1	0.31605996	0.16367390	1.931	0.0694

OBS	ID	ACTUAL	PREDICT VALUE	STD ERR PREDICT	LOWER95% PREDICT	UPPER95% PREDICT	RESIDUAL
1	31	44.0000	39.9148	1.0764	31.3044	48.5251	4.0852
2	31	45.0000	39.9148	1.0764	31.3044	48.5251	5.0852
3	32	44.0000	40.2308	0.9922	31.6652	48.7964	3.7692
4	33	40.0000	40.5469	0.9295	32.0124	49.0814	-0.5469
5	35	36.0000	41.1790	0.8852	32.6653	49.6927	-5.1790
6	40	45.0000	42.7593	1.2330	34.0568	51.4619	2.2407
7	29	40.0000	39.2827	1.2914	30.5428	48.0225	0.7173
8	40	40.0000	42.7593	1.2330	34.0568	51.4619	-2.7593
9	41	39.0000	43.0754	1.3522	34.2951	51.8557	-4.0754
10	36	45.0000	41.4951	0.9076	32.9710	50.0192	3.5049
11	27	43.0000	38.6505	1.5463	29.7299	47.5711	4.3495
12	31	33.0000	39.9148	1.0764	31.3044	48.5251	-6.9148
13	39	40.0000	42.4433	1.1251	33.8055	51.0810	-2.4433
14	43	42.0000	43.7075	1.6141	34.7340	52.6810	-1.7075
15	28	36.0000	38.9666	1.4151	30.1426	47.7906	-2.9666
16	36	37.0000	41.4951	0.9076	32.9710	50.0192	-4.4951
17	33	41.0000	40.5469	0.9295	32.0124	49.0814	0.4531
18	32	42.0000	40.2308	0.9922	31.6652	48.7964	1.7692
19	30	38.0000	39.5987	1.1774	30.9302	48.2672	-1.5987
20	48	52.0000	45.2878	2.3420	35.6320	54.9436	6.7122

SUM OF RESIDUALS	8.88178E-14
SUM OF SQUARED RESIDUALS	281.4869
PREDICTED RESID SS (PRESS)	388.2499

d. Locate and interpret the *p*-value of the test for model adequacy.

e. Give the lower and upper endpoints for a 95% prediction interval for *y* when $x = 40$. Interpret the interval.

| | | | | | | | |

S E C T I O N 3.12

REGRESSION THROUGH THE ORIGIN (OPTIONAL)

In practice, occasionally it is known in advance that the true line of means $E(y)$ passes through the point ($x = 0$, $y = 0$), called the **origin**. For example, a chain of convenience stores may be interested in modeling sales *y* of a new diet soft drink as a linear function of amount *x* of the new product in stock for a sample of stores. Or, a graphic designer may be interested in the linear relationship between the number *x* of company logos produced in a month and the total monthly logo production cost *y* for the past 12 months. In both cases, it is known that the regression line must pass through the origin. The convenience store chain knows that if one of its stores chooses not to stock the new diet soft drink, it will have zero sales of the new product. Likewise, if the graphic designer fails to produce any company logos during a month, total production costs will be 0.

FORMULAS FOR REGRESSION THROUGH THE ORIGIN

$$y = \beta_1 x + \varepsilon$$

Least squares slope:

$$\hat{\beta}_1 = \frac{\sum x_i y_i}{\sum x_i^2}$$

Estimate of σ^2:

$$s^2 = \frac{\text{SSE}}{n - 1}, \quad \text{where SSE} = \sum y_i^2 - \hat{\beta}_1 \sum x_i y_i$$

Estimate of $\sigma_{\hat{\beta}_1}$:

$$s_{\hat{\beta}_1} = s \Big/ \sqrt{\sum x_i^2}$$

Estimate of $\sigma_{\hat{y}}$ for estimating $E(y)$ when $x = x_p$:

$$s_{\hat{y}} = s \left(\frac{x_p}{\sqrt{\sum x_i^2}} \right)$$

Estimate of $\sigma_{(y - \hat{y})}$ for predicting y when $x = x_p$:

$$s_{(y - \hat{y})} = s \sqrt{1 + \frac{x_p^2}{\sum x_i^2}}$$

For situations in which we know that the regression line passes through the origin, the y-intercept is $\beta_0 = 0$ and the probabilistic straight-line model takes the form

$$y = \beta_1 x + \varepsilon$$

When the regression line passes through the origin the formula for the least squares estimate of the slope β_1 differs from the formula given in Section 3.3. Several other formulas required to perform the regression analysis also are different. These new computing formulas are provided in the preceding box.

Note that the denominator of s^2 is $n - 1$, not $n - 2$ as in the previous sections. This is because we need to estimate only a single parameter β_1 rather than both β_0 and β_1. Consequently, we have one additional degree of freedom for estimating σ^2, the variance of ε. Tests and confidence intervals for β_1 are carried out exactly as outlined in the previous sections, except that the t-distribution is based on $(n - 1)$ df. The test statistic and confidence intervals are given in the next box.

TESTS AND CONFIDENCE INTERVALS FOR REGRESSION THROUGH THE ORIGIN

Test statistic for H_0: $\beta_1 = 0$:

$$t = \frac{\hat{\beta}_1 - 0}{s_{\hat{\beta}_1}} = \frac{\hat{\beta}_1}{s \big/ \sqrt{\sum x_i^2}}$$

$100(1 - \alpha)\%$ confidence interval for β_1:

$$\hat{\beta}_1 \pm t_{\alpha/2} s_{\hat{\beta}_1} = \hat{\beta}_1 \pm t_{\alpha/2} s \big/ \sqrt{\sum x_i^2}$$

$100(1 - \alpha)\%$ confidence interval for $E(y)$:

$$\hat{y} \pm t_{\alpha/2} s_{\hat{y}} = \hat{y} \pm t_{\alpha/2} s \left(\frac{x_p}{\sqrt{\sum x_i^2}} \right)$$

$100(1 - \alpha)\%$ prediction interval for y:

$$\hat{y} \pm t_{\alpha/2} s_{(y-\hat{y})} = \hat{y} \pm t_{\alpha/2} s \sqrt{1 + \frac{x_p^2}{\sum x_i^2}}$$

where the distribution of t is based on $(n - 1)$ df

We illustrate the procedure for conducting simple linear regression through the origin in the following example.

EXAMPLE 3.5

Suppose we want to fit a straight-line model relating the number of items produced by a particular manufacturing process, x, and the total variable cost, y, involved in the production. For this process, total variable cost will be 0 when total output is 0. Table 3.8 gives the total output and total variable cost for a sample of $n = 7$ production days. Use the data to fit a straight-line regression model through the origin and calculate SSE.

TABLE 3.8
Sample Data for
Example 3.5

TOTAL OUTPUT x	TOTAL VARIABLE COST y, dollars
10	10
15	12
20	20
20	21
25	22
30	20
30	19

SOLUTION

The model we want to fit is $y = \beta_1 x + \varepsilon$. Preliminary calculations for estimating β_1 and calculating SSE are contained in Table 3.9 (page 124).

TABLE 3.9
Preliminary Calculations for Example 3.5

x_i	y_i	x_i^2	x_iy_i	y_i^2
10	10	100	100	100
15	12	225	180	144
20	20	400	400	400
20	21	400	420	441
25	22	625	550	484
30	20	900	600	400
30	19	900	570	361
Totals		$\Sigma x_i^2 = 3{,}550$	$\Sigma x_iy_i = 2{,}820$	$\Sigma y_i^2 = 2{,}330$

The estimate of the slope is

$$\hat{\beta}_1 = \frac{\Sigma x_iy_i}{\Sigma x_i^2} = \frac{2{,}820}{3{,}550} = .7943662$$

and the least squares line is

$$\hat{y} = .794x$$

The value of SSE for the line is

$$SSE = \Sigma y_i^2 - \hat{\beta}_1 \Sigma x_iy_i$$
$$= 2{,}330 - (.7943662)(2{,}820) = 89.8873$$

The graph of the least squares line with the observations is shown in Figure 3.18.

FIGURE 3.18
The Line $\hat{y} = .794x$ Fit to the Data in Table 3.8

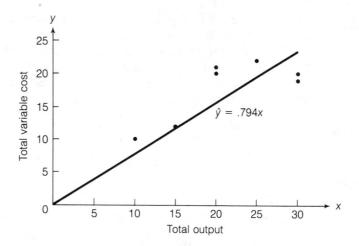

EXAMPLE 3.6

Refer to Example 3.5. Conduct the appropriate test for model adequacy. If the model is deemed adequate, predict the total variable cost y for a day in which $x = 23$ items are produced with a 95% prediction interval.

SOLUTION

The appropriate test for model adequacy is

$$H_0: \quad \beta_1 = 0 \qquad H_a: \quad \beta_1 > 0$$

(We choose to do an upper-tailed test since it is reasonable to assume that if a linear relationship exists between total output x and total variable cost y, it is a positive one.)

In order to calculate the test statistic, we first compute s, where

$$s = \sqrt{\frac{\text{SSE}}{n-1}} = \sqrt{\frac{89.8873}{6}} = 3.87056$$

Then the test statistic is

$$t = \frac{\hat{\beta}_1}{s \big/ \sqrt{\sum x_i^2}} = \frac{.7943662}{3.87056 / \sqrt{3,550}} = 12.23$$

For $\alpha = .05$, we will reject the null hypothesis if $t > t_{.05}$, where the distribution of t is based on $(n-1) = 6$ df. From Table 2 of Appendix D, $t_{.05} = 1.943$. Thus, we will reject H_0 if

$$t > 1.943$$

Since the calculated value of t, 12.23, exceeds the critical value, there is sufficient evidence (at $\alpha = .05$) to conclude that the model is adequate for predicting total variable cost y.

To calculate a 95% prediction interval for total variable cost y when 23 units are produced, we first substitute $x = 23$ into the least squares prediction equation $\hat{y} = .794x$:

$$\hat{y} = .794(23) = 18.26$$

From Table 2 of Appendix D, $t_{.025} = 2.447$ (based on 6 df). Then, our 95% prediction interval is

$$\hat{y} \pm t_{.025} s \sqrt{1 + \frac{x_p^2}{\sum x_i^2}}$$

$$= 18.26 \pm 2.447(3.87056) \sqrt{1 + \frac{(23)^2}{3,550}}$$

$$= 18.26 \pm 2.447(3.87056)(1.07192)$$

$$= 18.26 \pm 10.15 \quad \text{or} \quad (8.11, 28.41)$$

We predict with 95% confidence that the total variable cost will range between \$8.11 and \$28.41 on a day in which 23 units are produced. ∎

Warning: There are several situations where it is dangerous to fit the model $E(y) = \beta_1 x$. If you are not certain that the regression line passes through the origin, it is a safe practice to fit the more general model $E(y) = \beta_0 + \beta_1 x$. If the line of means does, in fact, pass through the origin, the estimate of β_0 will differ from the true value $\beta_0 = 0$ by only a small amount. For all practical purposes, the least squares prediction equations will be the same.

On the other hand, you may know that the regression passes through the origin (see Example 3.5), but are uncertain about whether the true relationship between y and x is linear or curvilinear. In fact, most theoretical relationships are *curvilinear*. Yet, we often fit a linear model to the data in such situations because we believe that a straight line will make a good approximation to the mean response $E(y)$ over the region of interest. The problem is that this straight line is not likely to pass through the origin (see Figure 3.19). By forcing the regression line through the origin, we may not obtain a very good approximation to $E(y)$. For these reasons, regression through the origin should be used with extreme caution.

FIGURE 3.19

Using a Straight Line to Approximate a Curvilinear Relationship When the True Relationship Passes Through the Origin

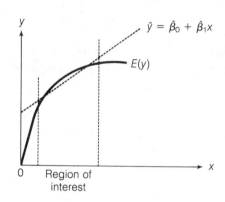

EXERCISES 3.46–3.51

3.46 Consider the eight data points shown in the table.

x	−4	−2	0	2	4	6	8	10
y	−12	−7	0	6	14	21	24	31

a. Fit a straight-line model through the origin, i.e., fit $E(y) = \beta_1 x$.
b. Calculate SSE, s^2, and s.
c. Do the data provide sufficient evidence to indicate that x and y are positively linearly related?
d. Construct a 95% confidence interval for β_1.
e. Construct a 95% confidence interval for $E(y)$ when $x = 7$.
f. Construct a 95% prediction interval for y when $x = 7$.

3.47 Consider the five data points shown in the table.

x	0	1	2	3	4
y	0	−8	−20	−30	−35

a. Fit a straight-line model through the origin, i.e., fit $E(y) = \beta_1 x$.
b. Calculate SSE, s^2, and s.
c. Do the data provide sufficient evidence to indicate that x and y are negatively linearly related?
d. Construct a 95% confidence interval for β_1.
e. Construct a 95% confidence interval for $E(y)$ when $x = 1$.
f. Construct a 95% prediction interval for y when $x = 1$.

3.48 Consider the ten data points shown in the table.

x	30	50	70	90	100	120	140	160	180	200
y	4	10	15	21	21	22	29	34	39	41

a. Fit a straight-line model through the origin, i.e., fit $E(y) = \beta_1 x$.
b. Calculate SSE, s^2, and s.
c. Do the data provide sufficient evidence to indicate that x and y are positively linearly related?
d. Construct a 95% confidence interval for β_1.
e. Construct a 95% confidence interval for $E(y)$ when $x = 125$.
f. Construct a 95% prediction interval for y when $x = 125$.

3.49 A pharmaceutical company has developed a new drug designed to reduce a smoker's reliance on tobacco. Since certain dosages of the drug may reduce one's pulse rate to dangerously low levels, the product-testing division of the pharmaceutical company wants to model the relationship between decrease in pulse rate, y (beats/minute), and dosage, x (cubic centimeters). Different dosages of the drug were administered to eight randomly selected patients, and 30 minutes later the decrease in each patient's pulse rate was recorded. The results are given in the accompanying table. Initially, the company considered the model $y = \beta_1 x + \varepsilon$, since in theory, a patient who receives a dosage of $x = 0$ should show no decrease in pulse rate ($y = 0$).

PATIENT	DOSAGE x, cubic centimeters	DECREASE IN PULSE RATE y, beats/minute
1	2.0	12
2	4.0	20
3	1.5	6
4	1.0	3
5	3.0	16
6	3.5	20
7	2.5	13
8	3.0	18

a. Fit a straight-line model that passes through the origin.
b. Is there evidence of a linear relationship between drug dosage and decrease in pulse rate? Test at $\alpha = .10$.
c. Find a 99% prediction interval for the decrease in pulse rate corresponding to a dosage of 3.5 cubic centimeters.

3.50 Consider the relationship between the total weight of a shipment of 50-pound bags of flour, y, and the number of bags in the shipment, x. Since a shipment containing $x = 0$ bags (i.e., no shipment at all) has a total weight of $y = 0$, a straight-line model of the relationship between x and y should pass through the point $x = 0$, $y = 0$. Hence, the appropriate model might be

$$y = \beta_1 x + \varepsilon$$

From the records of past flour shipments, 15 shipments were randomly chosen and the data in the table recorded.

WEIGHT OF SHIPMENT	NUMBER OF 50-POUND BAGS IN SHIPMENT	WEIGHT OF SHIPMENT	NUMBER OF 50-POUND BAGS IN SHIPMENT
5,050	100	7,162	150
10,249	205	24,000	500
20,000	450	4,900	100
7,420	150	14,501	300
24,685	500	28,000	600
10,206	200	17,002	400
7,325	150	16,100	400
4,958	100		

a. Find the least squares line for the given data under the assumption that $\beta_0 = 0$. Plot the least squares line on a scattergram of the data.

b. Find the least squares line for the given data using the model

$$y = \beta_0 + \beta_1 x = \varepsilon$$

(i.e., do not restrict β_0 to equal 0). Plot this line on the same scatterplot you constructed in part **a**.

c. Refer to part **b**. Why might $\hat{\beta}_0$ be different from 0 even though the true value of β_0 is known to be 0?

d. The estimated standard error of $\hat{\beta}_0$ is equal to

$$s\sqrt{\frac{1}{n} + \frac{\bar{x}^2}{SS_{xx}}}$$

Use the t-statistic,

$$t = \frac{\hat{\beta}_0 - 0}{s\sqrt{\frac{1}{n} + \frac{\bar{x}^2}{SS_{xx}}}}$$

to test the null hypothesis H_0: $\beta_0 = 0$ against the alternative H_a: $\beta_0 \neq 0$. Use $\alpha = .10$. Should you include β_0 in your model?

3.51 In order to satisfy the Public Service Commission's energy conservation requirements, an electric utility company must develop a reliable model for projecting the number of residential electric customers in its service area. The first step is to study the effect of changing population on the number of electric customers. The information shown in the accompanying table was obtained for the service area from 1978 to 1987.

Since a service area with 0 population obviously would have 0 residential electric customers, one could argue that regression through the origin is appropriate.

a. Fit the model $y = \beta_1 x + \varepsilon$ to the data.

YEAR	POPULATION IN SERVICE AREA x, hundreds	RESIDENTIAL ELECTRIC CUSTOMERS IN SERVICE AREA y
1978	262	14,041
1979	319	16,953
1980	361	18,984
1981	381	19,870
1982	405	20,953
1983	439	22,538
1984	472	23,985
1985	508	25,641
1986	547	27,365
1987	592	29,967

b. Is there evidence that x contributes information for the prediction of y? Test using $\alpha = .01$.

c. Now fit the more general model $y = \beta_0 + \beta_1 x + \varepsilon$ to the data. Is there evidence (at $\alpha = .01$) that x contributes information for the prediction of y?

d. Which model would you recommend?

SECTION 3.13

A SUMMARY OF THE STEPS TO FOLLOW IN A SIMPLE LINEAR REGRESSION ANALYSIS

We have introduced an extremely useful tool in this chapter—the method of least squares for fitting a prediction equation to a set of data. This procedure, along with associated statistical tests and estimations, is called a **regression analysis**. In five steps we showed how to use sample data to build a model relating a dependent variable y to a single independent variable x.

STEPS TO FOLLOW IN A SIMPLE LINEAR REGRESSION ANALYSIS

1. The first step is to hypothesize a **probabilistic model**. In this chapter, we confined our attention to the **first-order (straight-line) model**,

$$y = \beta_0 + \beta_1 x + \varepsilon$$

2. The second step is to use the method of least squares to estimate the unknown parameters in the **deterministic component**, $\beta_0 + \beta_1 x$. The least squares estimates yield a model $\hat{y} = \hat{\beta}_0 + \hat{\beta}_1 x$ with a **sum of squared errors (SSE)** that is smaller than the SSE for any other straight-line model.

3. The third step is to specify the probability distribution of the **random error component ε**.

4. The fourth step is to assess the utility of the hypothesized model. Included here are making inferences about the **slope β_1**, calculating the **coefficient of correlation r**, and calculating the **coefficient of determination r^2**.

5. Finally, if we are satisfied with the model, we are prepared to use it. We used the model to **estimate the mean y-value, $E(y)$**, for a given x-value and to **predict an individual y-value** for a specific value of x.

The concepts introduced in this chapter will be developed more fully in Chapter 4.

3.52 Consider the data on 1986 U.S. production and price of eggs provided in the table.

MONTH	UNITED STATES FARM PRODUCTION OF EGGS x, millions of cases	WHOLESALE PRICE PER DOZEN y, cents
January	16.3	70.6
February	14.7	65.7
March	16.4	76.9
April	15.7	62.6
May	16.1	62.0
June	15.6	57.3
July	15.8	69.4
August	15.9	70.0
September	15.4	69.4
October	16.1	66.3
November	15.9	74.1
December	13.8	72.8

Source: Survey of Current Business, Vol. 67, No. 3, March 1987. U.S. Department of Commerce, Bureau of Economic Analysis.

a. Construct a scattergram of the data.
b. Find the least squares line for the data and plot it on your scattergram.
c. Define β_1 in the context of this problem.
d. Test the hypothesis that the number of cases of eggs produced per month contributes no information for the prediction of the monthly wholesale price per dozen when a linear model is used (use $\alpha = .05$). Draw the appropriate conclusions.
e. Find a 90% confidence interval for β_1. Interpret your result.
f. Find the coefficient of correlation for the data.
g. Find the coefficient of determination for the linear model you constructed in part b. Interpret your result.
h. In January 1987, 13.8 million cases of eggs were produced on the farm. Find a 90% prediction interval for the January 1987 wholesale price of eggs per dozen.
i. Explain why the prediction interval in part h is so wide. How could you reduce the width of the interval?

3.53 Refer to the experiment on estimating the diet metabolizable energy (ME) content of commerical cat foods in Exercise 2.47. In one phase of the study, simple linear regression analysis was used to relate the estimated ME content, x (in kilocalories per gram), to the actual ME content y for cats fed a diet of canned and dry foods (*Feline Practice*, February 1986). The results for the two groups of cats are reported in the table.

DIET	$\hat{\beta}_0$	$\hat{\beta}_1$	$s_{\hat{\beta}_0}$	$s_{\hat{\beta}_1}$	r	s
Canned foods ($n = 28$)	.02	.96	.04	.04	.97	.05
Dry foods ($n = 29$)	.47	.84	.26	.09	.88	.15

a. Is there sufficient evidence to indicate that estimated ME content x is a useful predictor of actual ME content y for cats fed a diet of canned foods? Test using $\alpha = .05$.

b. Is there sufficient evidence to indicate that estimated ME content x is a useful predictor of actual ME content y for cats fed a diet of dry foods? Test using $\alpha = .05$.

c. Calculate the coefficient of determination r^2 for both regressions. Interpret these values.

d. Interpret the values of s for both regressions. What is the practical implication of this result?

3.54 Consider the data on air and rail passengers provided in the accompanying table.

YEAR	PASSENGERS CARRIED BY SCHEDULED AIR CARRIERS IN THE UNITED STATES y, millions	PASSENGERS CARRIED BY RAILROADS IN THE UNITED STATES x, millions
1950	19	488
1955	12	433
1960	62	327
1965	103	306
1970	169	289
1975	205	275
1980	297	350
1985	380	82

Source: 1986 Statistical Abstract of the United States, U.S. Department of Commerce, Bureau of the Census.

a. Find the correlation coefficient and coefficient of determination for the data and interpret your result.

b. Do the data provide sufficient evidence to indicate correlation between x and y?

3.55 In the business world, *Machiavellian* is a term often used to describe one who employs aggressive, manipulative, exploiting, and devious moves in order to achieve personal and corporate objectives. Hunt and Chonko (*Journal of Marketing*, Summer 1984) investigated Machiavellian tactics in marketing. One question concerned the relationship between age and Machiavellianism. Do young marketers tend to be more Machiavellian than older marketers? A sample of 1,076 members of the American Marketing Association were administered a questionnaire which measured tendency toward Machiavellianism. (The higher the score, the greater the tendency toward Machiavellianism.) The correlation coefficient between age and Machiavellian score was found to be $r = -.20$.

a. Is there evidence of a negative linear relationship between age of marketers and Machiavellian score? Test using $\alpha = .01$.

b. Calculate the coefficient of determination r^2 for the linear relationship between age of marketers and Machiavellian score. Interpret this value.

3.56 Many successful businesses have adopted high market share as a strategic objective. However, Michael Hergert, writing in *Business Economics* (October 1984), warns that "generalizations about the attractiveness of seeking high market share may prove misleading to strategic planners." Hergert investigated the relationship between 1980 market share and profitability (measured as return on assets) for a sample of approximately 5,400 businesses registered with the Securities and Exchange Commission. A simple linear regression analysis relating return on assets y and market share x yielded the following results:

$$\hat{y} = .103 + .128x$$
$$r^2 = .001 \qquad t = 2.51$$

a. Is there evidence to indicate that return on assets y and market share x are positively linearly related? Test using $\alpha = .05$.

b. Interpret the value of r^2.

c. Based on your answers to parts **a** and **b**, would you recommend a corporate strategy based on the premise that "a bigger market share yields a higher profit"?

3.57 A large supermarket chain has its own store brand for many grocery items. They tend to be priced lower than other brands. For a particular item, the chain wants to study the effect of varying the price for the major competing brand on the sales of the store brand item, while the prices for the store brand and all other brands are held fixed. The experiment is conducted at one of the chain's stores over a 7-week period and the results are shown in the table.

WEEK	MAJOR COMPETITOR'S PRICE x, cents	STORE BRAND SALES y
1	37	122
2	32	107
3	29	99
4	35	110
5	33	113
6	31	104
7	35	116

a. Find the least squares line relating store brand sales y to major competitor's price x.

b. Plot the data and graph the line as a check on your calculations.

c. Does x contribute information for the prediction of y?

d. Calculate r and r^2 and interpret their values.

e. Find a 90% confidence interval for mean store brand sales when the competitor's price is 33¢.

f. Suppose you were to set the competitor's price at 33¢. Find a 90% prediction interval for next week's sales.

3.58 Most investment firms provide estimates of systematic risks of securities, called *betas*. A stock's beta measures the relationship between the stock's rate of return and the average rate of return for the market as a whole. The term *beta* derives its name from the beta coefficient for the slope in simple linear regression, where the dependent variable is the stock's rate of return y and the independent variable is the market rate of return x. Stocks with beta-values (i.e., slopes) greater than 1 are considered "aggressive" securities since their rates of return are expected to move (upward or downward) faster than the market as a whole. In contrast, stocks with beta-values less than 1 are called "defensive" securities since their rates of return move much slower than the market. A stock with a beta-value near 1 is called a "neutral" security for its rate of return mirrors the market. The data in the table are monthly rates of return (in percent) for a particular stock and the market as a whole for seven randomly selected months.

MONTH	STOCK RATE OF RETURN y	MARKET RATE OF RETURN x
1	12.0	7.2
2	−1.3	.0
3	2.5	2.1
4	18.6	11.9
5	9.0	5.3
6	−3.8	−1.2
7	−10.0	−4.7

a. Fit the straight-line model $E(y) = \beta_0 + \beta_1 x$.
b. Find r^2 and s and interpret their values.
c. Is there sufficient evidence to indicate that market rate of return x is a useful predictor of stock rate of return y? Test using $\alpha = .01$.
d. Construct a 99% confidence interval for β_1, the mean increase in stock rate of return for every 1% increase in market rate of return.
e. Based on your answer to part d, would you classify this stock as aggressive, defensive, or neutral?

3.59 Refer to Exercise 3.58. Does a stock's beta-value depend on the length of the horizon over which the rates of return are calculated? Since some brokerage firms base their beta-values on monthly data and others on annual data, the question is an important one for investors. Haim Levy (University of Florida) investigated the relationship between length of horizon (in months) and average beta-value for each of the three types of stocks. Varying the length of horizon from 1 to 30 months, Levy calculated rates of return for 144 stocks over the years 1946–1975. The stocks were divided into 38 aggressive, 38 defensive, and 68 neutral stocks based on their beta-values. The table on page 134 gives the average beta-value for different horizons for each of the stock types.
a. Find the least squares simple linear regression equation relating length of horizon x to average beta-value y for (i) aggressive stocks, (ii) defensive stocks, and (iii) neutral stocks.

LENGTH OF HORIZON		BETA-VALUES	
	Aggressive Stocks	Defensive Stocks	Neutral Stocks
1	1.37	.50	.98
3	1.42	.44	.95
6	1.53	.41	.94
9	1.69	.39	1.00
12	1.83	.40	.98
15	1.67	.38	1.00
18	1.78	.39	1.02
24	1.86	.35	1.14
30	1.83	.33	1.22

Source: Levy, H. "Measuring risk and performance over alternative investment horizons," *Financial Analysts Journal*, March–April 1984, pp. 61–68.

b. For each type of stock, test the hypothesis that length of horizon x is a useful linear predictor of average beta-value y. Test using $\alpha = .05$.

c. For each type of stock, construct a 95% confidence interval for the slope of the line. Which stocks have beta-values that increase linearly as length of horizon increases? Which stocks have beta-values that decrease linearly as length of horizon increases?

3.60 Paper-and-pencil honesty tests of employees are common among retail stores, financial institutions, and warehouse operations where employees have access to cash and merchandise. These tests are less costly than polygraphs and can be used in states where pre-employment polygraph examinations are illegal. P. R. Sackett and M. M. Harris (University of Illinois) reviewed a number of studies that examined the validity of paper-and-pencil honesty tests (*Personnel Psychology*, Summer 1984). In one study, $n = 80$ applicants for retail management positions were given both an honesty test and a polygraph examination. The correlation coefficient between the scores of the two tests was $r = .48$. Another, independent study of $n = 17$ warehouse employees showed a correlation coefficient of $r = -.41$ between honesty test scores and dollar amount of money and merchandise stolen.

a. Is there sufficient evidence to indicate that paper-and-pencil honesty test scores and polygraph examination scores of retail management applicants are positively correlated? Test using $\alpha = .05$.

b. Calculate and interpret the coefficient of determination r^2 for the linear relationship between honesty test scores and polygraph examination scores.

c. Is there sufficient evidence to indicate that paper-and-pencil honesty test scores of warehouse employees are correlated with dollar amount of theft? Test using $\alpha = .05$.

d. Calculate and interpret the coefficient of determination r^2 for the linear relationship between honesty test scores and dollar amount of theft.

3.61 "In the analysis of urban transportation systems it is important to be able to estimate expected travel time between locations." Cook and Russell collected data in the city of Tulsa on the urban travel times and distances between locations for two types of vehicles, large hoist compactor trucks and passenger cars. A simple linear regression analysis was conducted for both sets of data (y = urban travel time in minutes,

x = distance between locations in miles) with the results given in the accompanying table.

PASSENGER CARS	TRUCKS
$\hat{y} = 2.50 + 1.93x$	$\hat{y} = 1.85 + 3.86x$
$r^2 = .676$; p-value $< .05$	$r^2 = .758$; p-value $< .01$

Source: Cook, T. M. and Russell, R. A. "Estimating urban travel times: A comparative study," *Transportation Research*, June 1980, 14A, pp. 173–175.

a. Is there sufficient evidence to indicate that distance between locations is linearly related to urban travel time for passenger cars? Test at $\alpha = .05$.

b. Is there sufficient evidence to indicate that distance between locations is linearly related to urban travel time for trucks? Test at $\alpha = .01$.

c. Interpret the values of r^2 for the two prediction equations.

d. Estimate the mean urban travel time for all passenger cars traveling a distance of 3 miles on Tulsa's highways.

e. Predict the urban travel time for a particular truck traveling a distance of 5 miles on Tulsa's highways.

f. Explain how we could attach a measure of reliability to the inferences derived in parts **d** and **e**.

3.62 In placing a weekly order, a concessionaire who provides services at a baseball stadium must know what size crowd is expected during the coming week in order to know how much food and other supplies to order. Since advanced ticket sales give an indication of expected attendance, food needs might be predicted on the basis of the advanced sales.

HOT DOGS PURCHASED DURING WEEK y, thousands	ADVANCED TICKET SALES FOR WEEK x, thousands
39.1	54.0
35.9	48.1
20.8	28.8
42.4	62.4
46.0	64.4
40.7	59.5
29.9	42.3

a. Use the data in the table from 7 previous weeks of home games to develop a simple linear model for hot dogs purchased as a function of advanced ticket sales.

b. Plot the data and graph the line as a check on your calculations.

c. Do the data provide sufficient information to indicate that advanced ticket sales provide information for the prediction of hot dog demand?

d. Calculate r^2 and interpret its value.

e. Find a 90% confidence interval for the mean number of hot dogs purchased when the advance ticket sales equal 50,000.

f. If the advance ticket sales this week equal 55,000, find a 90% prediction interval for the number of hot dogs that will be purchased this week at the game.

3.63 A certain manufacturer evaluates the sales potential for a product in a new marketing area by selecting several stores within the area to sell the product on a trial basis for a 1-month period. The sales figures for the trial period are then used to project sales for the entire area. [*Note:* The same number of trial stores are used each time.]

TOTAL SALES DURING TRIAL PERIOD x, hundreds of dollars	TOTAL SALES DURING FIRST MONTH FOR ENTIRE AREA y, hundreds of dollars
16.8	48.2
14.0	46.8
18.3	54.3
22.1	59.7
14.9	48.3
23.2	67.5

a. Use the data in the table to develop a simple linear model for predicting first-month sales for the entire area based on sales during the trial period.
b. Plot the data and graph the line as a check on your calculations.
c. Do the data provide sufficient evidence to indicate that total sales during the trial period contribute information for predicting total sales during the first month?
d. Use a 90% prediction interval to predict total sales during the first month for the entire area if the trial sales equal $2,000.

3.64 The management of a manufacturing firm is considering the possibility of setting up its own market research department rather than continuing to use the services of a market research firm. The management wants to know what salary should be paid to a market researcher, based on years of experience. An independent consultant checks with several other firms in the area and obtains the information shown in the table on market researchers.

ANNUAL SALARY y, thousands of dollars	EXPERIENCE x, years
16.3	2
16.2	1.5
25.0	11
29.1	15
25.4	9
21.9	6

a. Fit a least squares line to the data.
b. Plot the data and graph the line as a check on your calculations.
c. Calculate r and r^2 Explain how these values measure the utility of the model.
d. Estimate the mean annual salary of a market researcher with 8 years of experience. Use a 90% confidence interval.
e. Predict the salary of a market researcher with 7 years of experience using a 90% prediction interval.

3.65 Although the income tax system is structured so that people with higher incomes should pay taxes at a higher percentage of their incomes, there are many loopholes and tax shelters available for individuals with higher incomes. A sample of seven individual tax returns gave the data listed in the table on income and percentage of total income paid in taxes last year.

INDIVIDUAL	GROSS INCOME x, thousands of dollars	TAXES PAID y, percentage of total income
1	35.8	16.7
2	80.2	21.4
3	14.9	15.2
4	7.3	10.1
5	9.1	12.2
6	150.7	19.6
7	25.9	17.3

a. Fit a least squares line to the data.
b. Plot the data and graph the line as a check on your calculations.
c. Calculate r and r^2 and interpret their values.
d. Find a 90% confidence interval for the mean percentage of total income paid in taxes for individuals with gross incomes of $70,000.

3.66 A tire company sells five different kinds of automobile tires. The more expensive tires are the best in terms of toughness, durability, and mileage guarantee, but they are not in great demand because of their high prices. The data in the table represent the total number of sales for a 1-year period for tires sold at five different prices.

TIRE COST x, dollars	NUMBER SOLD y, units of 100
20	13
35	57
45	85
60	43
70	17

a. Fit a least squares line to the data.
b. Plot the data and graph the line as a check on your calculations.
c. Test $H_0: \beta_1 = 0$ using a two-tailed test at $\alpha = .05$. Does nonrejection of H_0 imply no relationship between tire price and sales volume?

3.67 How much influence does the board of directors have on the length of tenure of the chief executive of a large corporation? Donald L. Helmich* examined the relationship between the size of the board (i.e., the number of members on the board) and the tenure lengths of newly appointed corporate presidents for a sample of 54 petro-chemical companies. From each corporation the chief executive office tenures during

*Helmich, D. L. "Board size variation and rates of succession in the corporate presidency," *Journal of Business Research*, March 1980, Vol. 8, pp. 51–63. Reprinted by permission. Copyright 1980 by Elsevier North Holland, Inc.

the period 1947–1977 and the number of directors serving on the board during each year of the 30-year period were obtained. Using this information, Helmich measured the following variables for each of the 54 companies:

y = Average succession rate at the corporation (a measure of the rate of replacement of the chief executive)

x = Variation coefficient for corporation board size (a measure of the variability of board size adjusted for annual and long-term linear trends due to organizational growth or decline)

Helmich subjected the 54 (x, y) data points to a simple linear regression analysis. The slope of the least squares line was found to be $\hat{\beta}_1 = .254$, while a test of the null hypothesis H_0: $\beta_1 = 0$ resulted in a p-value less than .05. Based on this result, Helmich concluded that "board size variation had a significant positive influence upon the rate of succession in the office of chief executive over the 30-year interval of study for the total sample of 54 companies. That is, the higher the variation in board size, the more rapid the change in leadership in the chief executive office." Do you agree with this assessment?

3.68 Refer to Exercise 3.67. Knowledge of the (apparent) positive linear relationship between variation in board size and rate of succession becomes important during the process of succession planning by directors in the corporate structure. However, what may be even more important is knowing whether this positive relationship is exhibited by profitable as well as unprofitable companies. To examine this phenomenon, Helmich categorized each of the sampled companies as unsuccessful or successful, based on profit performance. The linear regression was then performed independently on each of the two groups ($n = 27$ observations in each group) of data. The results are summarized in the table.

UNSUCCESSFUL COMPANIES	SUCCESSFUL COMPANIES
$\hat{\beta}_1 = .603$	$\hat{\beta}_1 = -.230$
p-value < .05	p-value < .05

a. Is there evidence of a linear relationship between rate of succession and variation in board size for unsuccessful companies? If so, does the relationship appear to be positive, as suggested by the previous results?

b. Answer part **a** for the group of successful companies.

*3.69 If you have studied calculus, use the information provided in the footnote on page 82 to derive the formulas for the least squares estimators $\hat{\beta}_0$ and $\hat{\beta}_1$.

3.70 Use Table 7 in Appendix D to select a random sample of $n = 20$ residential properties from the 2,251 properties in Appendix F. Record the sale price y and the total appraised property value (land plus improvements) x for each property.

a. Fit a simple linear regression model to the data.

b. Do the data provide sufficient evidence to indicate that the total appraised value contributes information for the prediction of sale price?

*Exercises marked with an asterisk are optional exercises.

c. Calculate the coefficient of determination and interpret its value.

d. Use the prediction equation to find 90% prediction interval for the sale price of a property that has a total appraised value of $70,000.

ON YOUR OWN . . .

The Gross National Product (GNP) is one of the nation's best-known economic indicators. Many economists have developed models to forecast future values of the GNP. There are surely a large number of variables that should be included if an accurate prediction is to be made. For the moment, however, consider the simple case of choosing one important variable to include in a simple straight-line model for GNP.

First, list three independent variables, x_1, x_2, and x_3, that you think might be (individually) strongly related to the GNP. Next, obtain ten yearly values (preferably the last ten) of the three independent variables and the GNP.*

a. Use the least squares formulas given in this chapter to fit three straight-line models—one for each independent variable—for predicting the GNP.

b. Interpret the sign of the estimated slope coefficient $\hat{\beta}_1$ in each case, and test the utility of the model by testing $H_0: \beta_1 = 0$ against $H_a: \beta_1 \neq 0$.

c. Calculate the coefficient of determination r^2 for each model. Which of the independent variables predicts the GNP best over the 10 sample years when a straight-line model is used? Is this variable necessarily best in general (i.e., for all years)? Explain.

REFERENCES

Best, R. J. and Kable, J. R. "Managing a capital equipment business's market share," *Industrial Marketing Management*, Vol. 14, 1985, pp. 159–164.

Chou, Ya-lun *Statistical Analysis with Business and Economic Applications*, 2nd ed. New York: Holt, Rinehart, and Winston, 1975. Chapter 17.

Cochran, P. L., Wartick, S. L., and Wood, R. A. "The average age of boards and financial performance, revisited," *Quarterly Journal of Business and Economics*, Vol. 23, No. 4, Autumn 1984, pp. 57–63.

Crandall, J. S. and Cedercreutz, M. "Preliminary cost estimates for mechanical work," *Building Systems Design*, October–November 1976, Vol. 73, pp. 35–51.

Draper, N. and Smith, H. *Applied Regression Analysis*. 2nd ed. New York: Wiley, 1981. Chapter 1.

Hergert, M. "Market share and profitability: Is bigger really better?" *Business Economics*, October 1984, pp. 45–48.

Hunt, S. D. and Chonko, L. B. "Marketing and Machiavellianism," *Journal of Marketing*, Summer 1984, Vol. 48, No. 3, pp. 30–42.

Miller, R. B. and Wichern, D. W. *Intermediate Business Statistics: Analysis of Variance, Regression, and Time Series*. New York: Holt, Rinehart, and Winston, 1977. Chapter 5.

Neter, J., Wasserman, W., and Kutner, M. H. *Applied Linear Statistical Models*, Homewood, Ill.: Richard D. Irwin, 1985. Chapters 2–3.

Sackett, P. R. and Harris, M. M. "Honesty testing for personnel selection: A review and critique," *Personnel Psychology*, Summer 1984, Vol. 37, No. 2, pp. 221–243.

*The assumption that the random errors are independent is debatable for time series data. For purposes of illustration, we will assume they are approximately independent. The problem of dependent errors is discussed in Chapter 6 and 9.

OBJECTIVE

To extend the methods of Chapter 3; to develop a procedure for predicting a response y based on the values of two or more independent variables; to illustrate the types of practical inferences that can be drawn from this type of analysis

CONTENTS

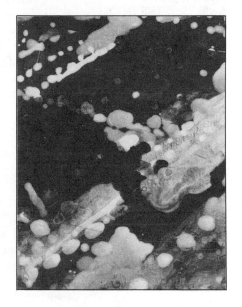

MULTIPLE REGRESSION

SECTION 4.1

A MULTIPLE REGRESSION ANALYSIS: THE MODEL AND THE PROCEDURE

Most practical applications of regression analysis utilize models that are more complex than the first-order (straight-line) model. For example, a realistic probabilistic model for monthly sales revenue would include more than just the advertising expenditure discussed in Chapter 3 in order to provide a good predictive model for sales. Factors such as season, inventory on hand, sales force, and productivity are a few of the many variables that might influence sales. Thus, we would want to incorporate these and other potentially important independent variables into the model if we need to make accurate predictions.

Probabilistic models that include terms involving x^2, x^3 (or higher-order terms), or more than one independent variable are called **multiple regression models**, or **linear statistical models**. The general form of these models is

$$y = \beta_0 + \beta_1 x_1 + \beta_2 x_2 + \cdots + \beta_k x_k + \varepsilon$$

The dependent variable y is now written as a function of k independent variables, $x_1, x_2, \ldots, x_k$. The random error term is added to make the model probabilistic rather than deterministic. The value of the coefficient β_i determines the contribution of the independent variable x_i, given that the other $(k-1)$ independent variables are held constant, and β_0 is the y-intercept. The coefficients β_0, β_1, $\ldots$, β_k will usually be unknown, since they represent population parameters.

At first glance it might appear that the regression model shown above would not allow for anything other than straight-line relationships between y and the independent variables, but this is not true. Actually, $x_1, x_2, \ldots, x_k$ can be functions of variables as long as the functions do not contain unknown parameters. For example, the dollar sales y in new housing in a region could be a function of the independent variables

x_1 = Mortgage interest rate

x_2 = (Mortgage interest rate)^2 = x_1^2

x_3 = Unemployment rate in the region

and so on. You could even insert a cyclical term (if it would be useful) of the form $x_4 = \sin t$, where t is a time variable. The multiple regression model is quite versatile and can be made to model many different types of response variables.

The same steps we followed in developing a straight-line model are applicable to the multiple regression model:

STEP 1   First, hypothesize the form of the model. This involves the choice of the independent variables to be included in the model.

STEP 2   Next, estimate the unknown parameters $\beta_0, \beta_1, \ldots, \beta_k$.

STEP 3   Then, specify the probability distribution of the random error component ε and estimate its variance, σ^2.

STEP 4   The fourth step is to check the utility of the model.

STEP 5 Finally, use the fitted model to estimate the mean value of y or to predict a particular value of y for given values of the independent variables.

The intial step—hypothesizing the form of the model—is the subject of optional Chapter 7. In this chapter we will assume that the form of the model is known, and we will discuss steps 2–5 for a given model.

CASE STUDY 4.1

Towers, Perrin, Forster & Crosby (TPF&C), an international management consulting firm, has developed a unique and interesting application of multiple regression analysis. Many firms are interested in evaluating their management salary structure, and TPF&C uses multiple regression models to accomplish this salary evaluation. The Compensation Management Service, as TPF&C calls it, measures both the internal and external consistency of a company's pay policies to determine whether they reflect the management's intent.

The dependent variable y used to measure executive compensation is annual salary. The independent variables used to explain salary structure include the executive's age, education, rank, and bonus eligibility; number of employees under the executive's direct supervision; as well as variables that describe the company for which the executive works, such as annual sales, profit, and total assets.

The initial step in developing models for executive compensation is to obtain a sample of executives from various client firms, which TPF&C calls the Compensation Data Bank. The data for these executives are used to estimate the model coefficients (the β parameters), and these estimates are then substituted into the linear model to form a prediction equation. To predict a particular executive's compensation, TPF&C substitutes into the prediction equation the values of the independent variables that pertain to the executive (the executive's age, rank, and so on). This application of multiple regression analysis will be developed more fully in Section 4.8.

| | | | | | | | | | | | | |

S E C T I O N   4.2

MODEL ASSUMPTIONS

We noted in Section 4.1 that the multiple regression model is of the form

$$y = \beta_0 + \beta_1 x_1 + \beta_2 x_2 + \cdots + \beta_k x_k + \varepsilon$$

where y is the response variable that you want to predict; $\beta_0, \beta_1, \ldots, \beta_k$ are parameters with unknown values; $x_1, x_2, \ldots, x_k$ are independent information-contributing variables that are measured without error; and ε is a random error component. Since $\beta_0, \beta_1, \ldots, \beta_k$ and $x_1, x_2, \ldots, x_k$ are nonrandom, the quantity

$$\beta_0 + \beta_1 x_1 + \beta_2 x_2 + \cdots + \beta_k x_k$$

represents the deterministic portion of the model. Therefore, y is composed of two components—one fixed and one random—and, consequently, y is a random variable.

$$\overbrace{y = \beta_0 + \beta_1 x_1 + \beta_2 x_2 + \cdots + \beta_k x_k +}^{\substack{\text{Deterministic} \\ \text{portion of model}}} \quad \overbrace{\varepsilon}^{\substack{\text{Random} \\ \text{error}}}$$

We will assume (as in Chapter 3) that the random error can be positive or negative and that for any setting of the x-values, $x_1, x_2, \ldots, x_k$, ε has a normal probability distribution with mean equal to 0 and variance equal to σ^2. Further, we assume that the random errors associated with any (and every) pair of y-values are probabilistically independent. That is, the error ε associated with any one y-value is independent of the error associated with any other y-value. These asssumptions are summarized in the accompanying box.

ASSUMPTIONS ABOUT THE RANDOM ERROR ε

1. For any given set of values of $x_1, x_2, \ldots, x_k$, ε has a normal probability distribution with mean equal to 0 and variance equal to σ^2.
2. The random errors are independent (in a probabilistic sense).

The assumptions that we have described for a multiple regression model imply that the mean value $E(y)$ for a given set of values of $x_1, x_2, \ldots, x_k$ is equal to

$$E(y) = \beta_0 + \beta_1 x_1 + \beta_2 x_2 + \cdots + \beta_k x_k$$

Models of this type are called **linear** statistical models because $E(y)$ is a *linear function* of the unknown parameters $\beta_0, \beta_1, \ldots, \beta_k$.

All the estimation and statistical test procedures described in this chapter depend on the data satisfying the assumptions described in this section. Since we will rarely, if ever, know for certain whether the assumptions are actually satisfied in practice, we will want to know how well a regression analysis works, and how much faith we can place in our inferences when certain assumptions are not satisfied. We will have more to say on this topic in Chapters 5 and 6. First, we need to discuss the methods of a regression analysis more thoroughly and show how they are used in a practical situation.

S E C T I O N 4.3

FITTING THE MODEL: THE METHOD OF LEAST SQUARES

The method of fitting multiple regression models is identical to that of fitting the first-order (straight-line) model—namely, the method of least squares. That is, we choose the estimated model

$$\hat{y} = \hat{\beta}_0 + \hat{\beta}_1 x_1 + \cdots + \hat{\beta}_k x_k$$

that minimizes

$$\text{SSE} = \sum (y_i - \hat{y}_i)^2$$

As in the case of the straight-line model, the sample estimates $\hat{\beta}_0, \hat{\beta}_1, \ldots, \hat{\beta}_k$ will be obtained as solutions to a set of simultaneous linear equations.*

The primary difference between fitting the simple and multiple regression models is computational difficulty. The $(k + 1)$ simultaneous linear equations that must be solved to find the $(k + 1)$ estimated coefficients $\hat{\beta}_0, \hat{\beta}_1, \ldots, \hat{\beta}_k$ are often difficult (sometimes physically impossible) to solve with a pocket or desk calculator. Consequently, we resort to the use of computers. As with the straight-line model, many computer packages have been developed to fit a multiple regression model by the method of least squares. We will present output from several of the more popular computer packages, commencing, in our first example, with the computer output for the SAS System. Since the SAS System regression output is similar to that of most other package regression programs, you should have little trouble interpreting regression output from other packages as you encounter them in future examples and exercises or at your computer center.

To illustrate, suppose we theorize that monthly electrical usage y in all-electric homes is related to the size x of the home by the model

$$y = \beta_0 + \beta_1 x + \beta_2 x^2 + \varepsilon$$

To estimate the unknown parameters β_0, β_1, and β_2, values of y and x were collected for each of ten homes during a particular month. The data are shown in Table 4.1.

TABLE 4.1

SIZE OF HOME x, square feet	MONTHLY USAGE y, kilowatt-hours
1,290	1,182
1,350	1,172
1,470	1,264
1,600	1,493
1,710	1,571
1,840	1,711
1,980	1,804
2,230	1,840
2,400	1,956
2,930	1,954

Notice that we include a term involving x^2 in the model because we expect curvature in the graph of the response model relating x to y. The term involving x^2 is called a **second-order**, or **quadratic**, term. Figure 4.1 (page 146) illustrates that the electrical usage appears to increase in a curvilinear manner with the size

---

*Students who are familiar with calculus should note that $\hat{\beta}_0, \hat{\beta}_1, \ldots, \hat{\beta}_k$ are the solutions to the set of equations $\partial SSE/\partial \beta_0 = 0$, $\partial SSE/\partial \beta_1 = 0, \ldots, \partial SSE/\partial \beta_k = 0$. The solution, given in matrix notation, is presented in Appendix A.

FIGURE 4.1

Scattergram of the Home
Size–Electrical Usage
Data

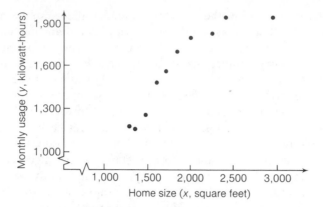

of the home. This provides some support for the inclusion of the second-order term x^2 in the model.

Part of the output from the SAS System multiple regression routine for the data in Table 4.1 is reproduced in Figure 4.2. The least squares estimates of the β parameters appear in the column labeled PARAMETER ESTIMATE. You can see that $\hat{\beta}_0 = -1{,}216.1$, $\hat{\beta}_1 = 2.3989$, and $\hat{\beta}_2 = -.00045$. Therefore, the equation that minimizes the SSE for the data is

$$\hat{y} = -1{,}216.1 + 2.3989x - .00045x^2$$

The minimum value of SSE, 15,332.6, also appears in the printout. [*Note:* Much detail on the printout has not yet been discussed. We will continue throughout this chapter to shade the aspects of the printout that are under discussion.]

FIGURE 4.2   Portion of the SAS Printout for the Home Size–Electrical Usage Data

ANALYSIS OF VARIANCE

SOURCE	DF	SUM OF SQUARES	MEAN SQUARE	F VALUE	PROB>F
MODEL	2	831069.55	415534.77	189.710	0.0001
ERROR	7	15332.55363	2190.36480		
C TOTAL	9	846402.10			

ROOT MSE	46.80133	R-SQUARE	0.9819	
DEP MEAN	1594.7	ADJ R-SQ	0.9767	
C.V.	2.934805			

PARAMETER ESTIMATES

VARIABLE	DF	PARAMETER ESTIMATE	STANDARD ERROR	T FOR H0: PARAMETER=0	PROB > |T|
INTERCEP	1	-1216.14389	242.80637	-5.009	0.0016
X	1	2.39893018	0.24583560	9.758	0.0001
XX	1	-0.000450040	0.000059077	-7.618	0.0001

Note that the graph of the multiple regression model (Figure 4.3, a response curve) provides a good fit to the data of Table 4.1. Furthermore, the small value of $\hat{\beta}_2$ does *not* imply that the curvature is insignificant, since the numerical value of $\hat{\beta}_2$ is dependent on the scale of the measurements. We will test the contribution of the second-order coefficient β_2 in Section 4.5.

FIGURE 4.3
Least Squares Model for
the Home Size–Electrical
Usage Data

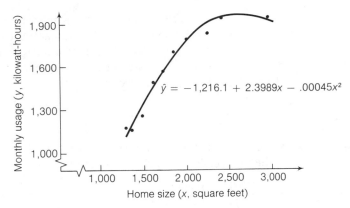

$$\hat{y} = -1,216.1 + 2.3989x - .00045x^2$$

The ultimate goal of this multiple regression analysis is to use the fitted model to predict electrical usage y for a home of a specific size (area) x. And, of course, we will want to give a prediction interval for y so that we will know how much faith we can place in the prediction. That is, if the prediction model is used to predict electrical usage y for a given size of home x, what will be the error of prediction? To answer this question, we need to estimate σ^2, the variance of ε.

SECTION 4.4

ESTIMATION OF σ^2, THE VARIANCE OF ε

You will recall that σ^2 is the variance of the random error ε. If $\sigma^2 = 0$, all the random errors will equal 0 and the prediction equation $\hat{y}$ will be identical to $E(y)$, i.e., $E(y)$ will be estimated without error. In contrast, a large value of σ^2 implies large (absolute) values of ε and larger deviations between the prediction equation $\hat{y}$ and the mean value $E(y)$. Consequently, the larger the value of σ^2, the greater will be the error in estimating the model parameters $\beta_0, \beta_1, \ldots, \beta_k$ and the error in predicting a value of y for a specific set of values of $x_1, x_2, \ldots, x_k$. Thus, σ^2 plays a major role in making inferences about $\beta_0, \beta_1, \ldots, \beta_k$, in estimating $E(y)$, and in predicting y for specific values of $x_1, x_2, \ldots, x_k$.

Since the variance σ^2 of the random error ε will rarely be known, we must use the results of the regression analysis to estimate its value. You will recall that σ^2 is the variance of the probability distribution of the random error ε for a given set of values for $x_1, x_2, \ldots, x_k$, and hence that it is the mean value of the squares of the deviations of the y-values (for given values of $x_1, x_2, \ldots, x_k$) about the mean value $E(y)$.* Since the predicted value $\hat{y}$ estimates $E(y)$ for each of the data points, it seems natural to use

$$\text{SSE} = \sum (y_i - \hat{y}_i)^2$$

to construct an estimator of σ^2.

For example, in the second-order model describing electrical usage as a function of home size, we found that SSE = 15,332.6. We now want to use this quantity

*Remember, we stated in Section 4.2 that $y = E(y) + \varepsilon$. Therefore, ε is equal to the deviation $y - E(y)$. Also, by definition, the variance of a random variable is the expected value of the square of the deviation of the random variable from its mean. According to our model, $E(\varepsilon) = 0$. Therefore, $\sigma^2 = E(\varepsilon^2)$.

to estimate the variance of ε. Recall that the estimator for the straight-line model was $s^2 = SSE/(n-2)$ and note that the denominator is (n − number of estimated β parameters), which is ($n-2$) in the first-order (straight-line) model. Since we must estimate one more parameter, β_2, for the second-order model $y = \beta_0 + \beta_1 x + \beta_2 x^2 + \varepsilon$, the estimator of σ^2 is

$$s^2 = \frac{SSE}{n-3}$$

That is, the denominator becomes ($n-3$) because there are now three β parameters in the model. The numerical estimate for this example is

$$s^2 = \frac{SSE}{10-3} = \frac{15{,}332.6}{7} = 2{,}190.36$$

In many computer printouts and textbooks, s^2 is called the **mean square for error (MSE)**. This estimate of σ^2 is shown in the column titled MEAN SQUARE in the SAS printout in Figure 4.2.

For the general multiple regression model

$$y = \beta_0 + \beta_1 x_1 + \beta_2 x_2 + \cdots + \beta_k x_k + \varepsilon$$

we must estimate the ($k+1$) parameters $\beta_0, \beta_1, \beta_2, \ldots, \beta_k$. Thus, the estimator of σ^2 is SSE divided by the quantity (n − Number of estimated β parameters).

ESTIMATOR OF σ^2 FOR MULTIPLE REGRESSION MODEL WITH k INDEPENDENT VARIABLES

$$MSE = \frac{SSE}{n - \text{Number of estimated } \beta \text{ parameters}}$$

$$= \frac{SSE}{n-(k+1)}$$

We will use the estimator of σ^2 both to check the utility of the model (Sections 4.5 and 4.6) and to provide a measure of the reliability of predictions and estimates when the model is used for those purposes (Section 4.7). Thus, you can see that the estimation of σ^2 plays an important part in the development of a regression model.

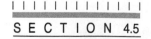

SECTION 4.5

ESTIMATING AND TESTING HYPOTHESES ABOUT THE β PARAMETERS

Sometimes the individual β parameters in a model have particular practical significance and we want to estimate their values or test hypotheses about them. For example, if electrical usage y is related to home size x by the straight-line relationship

$$y = \beta_0 + \beta_1 x + \varepsilon$$

then β_1 has a very practical interpretation. That is, you saw in Chapter 3 that β_1 is the mean increase in kilowatt-hours of electrical usage y for a square foot increase in home size x.

As proposed in the preceding sections, suppose the electrical usage y is related to home size x by the quadratic model

$$y = \beta_0 + \beta_1 x + \beta_2 x^2 + \varepsilon$$

Then the mean value of y for a given value of x is

$$E(y) = \beta_0 + \beta_1 x + \beta_2 x^2$$

What is the practical interpretation of β_2? As noted earlier, the parameter β_2 measures the curvature of the response curve shown in Figure 4.3. If $\beta_2 > 0$, the slope of the curve will increase as x increases [see Figure 4.4(a)]. If $\beta_2 < 0$, the slope of the curve will decrease as x increases, as shown in Figure 4.4(b).

FIGURE 4.4

The Interpretation of β_2 for a Second-Order Model

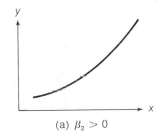
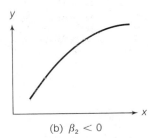

(a) $\beta_2 > 0$ (b) $\beta_2 < 0$

Intuitively, we would expect the electrical usage y to rise almost proportional to home size x. Then, eventually, as the size of the home increases, the increase in electrical usage for a 1-unit increase in home size might begin to decrease. Thus, a forecaster of electrical usage would want to determine whether this type of curvature actually was present in the response curve, or, equivalently, the forecaster would want to test the null hypothesis

$H_0: \quad \beta_2 = 0 \quad$ (No curvature in the response curve)

against the alternative hypothesis

$H_a: \quad \beta_2 < 0 \quad$ (Downward curvature in the response curve)

A test of this hypothesis can be performed using a Student's t-test.

The t-test utilizes a test statistic that is analogous to that used to make inferences about the slope of the straight-line model (Section 3.6). The t-statistic is formed by dividing the sample estimate $\hat{\beta}_2$ of the population coefficient β_2 by the estimated standard deviation of the sampling distribution of $\hat{\beta}_2$:

$$\text{Test statistic:} \quad t = \frac{\hat{\beta}_2}{s_{\hat{\beta}_2}}$$

We use the symbol $s_{\hat{\beta}_2}$ to represent the estimated standard deviation of $\hat{\beta}_2$. The formula for computing $s_{\hat{\beta}_2}$ (presented in Appendix A) is very complex, but its computation is performed automatically as part of most standard multiple regression computer analyses. Thus, most computer packages list the estimated standard deviation $s_{\hat{\beta}_i}$ for each estimated model coefficient $\hat{\beta}_i$. In addition, they usually give the calculated t-values for testing $H_0: \beta_i = 0$ for each coefficient in the model.

The rejection region for the test is found in exactly the same way as the rejection regions for the t-tests in Chapters 2 and 3. That is, we consult Table 2 of Appendix D to obtain an upper-tail value of t. This is a value t_α such that $P(t > t_\alpha) = \alpha$. We can then use this value to construct rejection regions for either one- or two-tailed tests. To illustrate, in the electrical usage example, the error degrees of freedom is $(n - 3) = 7$, the denominator of the estimate of σ^2. Then the rejection region (shown in Figure 4.5) for a one-tailed test with $\alpha = .05$ is

$$\text{Rejection region:} \quad t < -t_\alpha; \quad \alpha = .05, \quad df = 7$$
$$t < -1.895$$

FIGURE 4.5
Rejection Region for Test of $H_0: \beta_2 = 0$

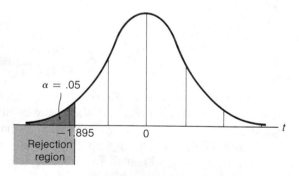

In Figure 4.6 we again show a portion of the SAS printout for the electrical usage example. The following quantities are shaded:

1. The estimated coefficients, $\hat{\beta}_0$, $\hat{\beta}_1$, and $\hat{\beta}_2$
2. The sum of squared errors (SSE)
3. The mean square for error (MSE, estimate of σ^2, the variance of ε)

The estimated standard deviations for the model coefficients appear under the column labeled STANDARD ERROR. The t-statistics for testing the null

FIGURE 4.6 SAS Printout for Electrical Usage Example

ANALYSIS OF VARIANCE

SOURCE	DF	SUM OF SQUARES	MEAN SQUARE	F VALUE	PROB>F
MODEL	2	831069.55	415534.77	189.710	0.0001
ERROR	7	15332.55363	2190.36480		
C TOTAL	9	846402.10			

ROOT MSE		46.80133	R-SQUARE	0.9819	
DEP MEAN		1594.7	ADJ R-SQ	0.9767	
C.V.		2.934805			

PARAMETER ESTIMATES

VARIABLE	DF	PARAMETER ESTIMATE	STANDARD ERROR	T FOR H0: PARAMETER=0	PROB > |T|
INTERCEP	1	-1216.14389	242.80637	-5.009	0.0016
X	1	2.39893018	0.24583560	9.758	0.0001
XX	1	-0.000450040	0.000059077	-7.618	0.0001

hypotheses that the coefficients $\beta_0, \beta_1, \ldots, \beta_k$ individually equal 0 appear under the column headed T FOR H0: PARAMETER = 0. The t-value corresponding to the test of the null hypothesis H_0: $\beta_2 = 0$ is the last one in the column, i.e., $t = -7.62$. Since this value falls in the rejection region, i.e., it is less than -1.895, we conclude that the second-order term $\beta_2 x^2$ makes an important contribution to the prediction model of electrical usage.

The SAS printout shown in Figure 4.6 also lists the two-tailed observed significance levels (i.e., p-values) for each t-value. These values appear under the column headed PROB > $|T|$. The observed significance level .0001 corresponds to the quadratic term, and this implies that we would reject H_0: $\beta_2 = 0$ in favor of H_a: $\beta_2 \neq 0$ at any α level larger than .0001. Since our alternative was one-sided, H_a: $\beta_2 < 0$, the observed significance level is half that given in the printout, i.e., p-value = $\frac{1}{2}(.0001) = .00005$. Thus, there is very strong evidence that the mean electrical usage increases more slowly per square foot for large houses than for small houses.

We can also form a 95% confidence interval for the parameter β_2 as follows:

$$\hat{\beta}_2 \pm t_{\alpha/2}s_{\hat{\beta}_2} = -.000450 \pm (2.365)(.0000591)$$

or $(-.000590, -.000310)$. Note that the t-value 2.365 corresponds to $\alpha/2 = .025$ and $(n - 3) = 7$ df. This interval constitutes a 95% confidence interval for β_2, the rate of change in curvature in mean electrical usage as home size is increased. Note that all values in the interval are negative, reconfirming the conclusion of our test.

Testing a hypothesis about a single β parameter that appears in any multiple regression model is accomplished in exactly the same manner as described for the second-order electrical usage model. The form of the t-test is shown in the box on page 152.

TEST OF AN INDIVIDUAL PARAMETER COEFFICIENT IN THE MULTIPLE REGRESSION MODEL

$$y = \beta_0 + \beta_1 x_1 + \beta_2 x_2 + \cdots + \beta_k x_k + \varepsilon$$

ONE-TAILED TEST   **TWO-TAILED TEST**

H_0: $\beta_i = 0$ H_0: $\beta_i = 0$

H_a: $\beta_i > 0$ H_a: $\beta_i \neq 0$

  (or $\beta_i < 0$)

Test statistic:* $t = \dfrac{\hat{\beta}_i}{s_{\hat{\beta}_i}}$ Test statistic:* $t = \dfrac{\hat{\beta}_i}{s_{\hat{\beta}_i}}$

Rejection region: $t > t_\alpha$ Rejection region: $t > t_{\alpha/2}$

  (or $t < -t_\alpha$)   or $t < -t_{\alpha/2}$

where

  n = Number of observations
  k = Number of independent variables in the model

and $t_{\alpha/2}$ is based on $[n - (k + 1)]$ df

Assumptions: See Section 4.2 for the assumptions about the probability
distribution of the random error component ε.

EXAMPLE 4.1

A collector of antique grandfather clocks believes that the price received for the clocks at an antique auction increases with the age of the clocks and with the number of bidders. Thus, the following model is hypothesized:

$$y = \beta_0 + \beta_1 x_1 + \beta_2 x_2 + \varepsilon$$

where

  y = Auction price
  x_1 = Age of clock (years)
  x_2 = Number of bidders

A sample of 32 auction prices of grandfather clocks, along with their age and the number of bidders, is given in Table 4.2. The model $y = \beta_0 + \beta_1 x_1 + \beta_2 x_2 + \varepsilon$ is fit to the data, and a portion of the SAS printout is shown in Figure 4.7. Test the hypothesis that the auction price increases as the number of bidders increases (and age is held constant), i.e., $\beta_2 > 0$. Use $\alpha = .05$.

---

*To test the null hypothesis that a parameter β_i equals some value other than 0, say H_0: $\beta_i = \beta_{i0}$, use the test statistic $t = (\hat{\beta}_i - \beta_{i0})/s_{\hat{\beta}_i}$. All other aspects of the test will be as described in the box.

TABLE 4.2
Auction Price Data

AGE x_1	NUMBER OF BIDDERS x_2	AUCTION PRICE y	AGE x_1	NUMBER OF BIDDERS x_2	AUCTION PRICE y
127	13	1,235	170	14	2,131
115	12	1,080	182	8	1,550
127	7	845	162	11	1,884
150	9	1,522	184	10	2,041
156	6	1,047	143	6	854
182	11	1,979	159	9	1,483
156	12	1,822	108	14	1,055
132	10	1,253	175	8	1,545
137	9	1,297	108	6	729
113	9	946	179	9	1,792
137	15	1,713	111	15	1,175
117	11	1,024	187	8	1,593
137	8	1,147	111	7	785
153	6	1,092	115	7	744
117	13	1,152	194	5	1,356
126	10	1,336	168	7	1,262

FIGURE 4.7 Portion of the SAS Printout for Example 4.1

```
                        ANALYSIS OF VARIANCE

                          SUM OF         MEAN
        SOURCE     DF    SQUARES        SQUARE       F VALUE     PROB>F

        MODEL       2   4277159.70    2138579.85    120.651     0.0001
        ERROR      29    514034.52   17725.32812
        C TOTAL    31   4791194.22

              ROOT MSE     133.1365    R-SQUARE      0.8927
              DEP MEAN     1327.156    ADJ R-SQ      0.8853
              C.V.          10.03171

                        PARAMETER ESTIMATES

                        PARAMETER      STANDARD    T FOR H0:
        VARIABLE   DF    ESTIMATE        ERROR   PARAMETER=0   PROB > |T|

        INTERCEP    1  -1336.72205    173.35613     -7.711       0.0001
        X1          1    12.73619884    0.90238049   14.114       0.0001
        X2          1    85.81513260    8.70575681    9.857       0.0001
```

SOLUTION

The hypothesis of interest concerns the parameter β_2. Specifically,

H_0: $\beta_2 = 0$ H_a: $\beta_2 > 0$

Test statistic: $t = \dfrac{\hat{\beta}_2}{s_{\hat{\beta}_2}}$

Rejection region: For $\alpha = .05$, $t > t_{.05}$
where df $= n - (k + 1) = 32 - 3 = 29$
or $t > 1.699$ (see Figure 4.8 on page 154)

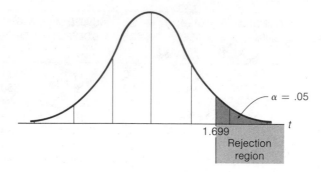

The calculated t-value, $t = 9.86$, is indicated in Figure 4.7. This value exceeds 1.699 and therefore falls in the rejection region. Thus, the collector can conclude that the mean auction price of the clocks increases as the number of bidders increases, when age is held constant. We could have arrived at the same conclusion by observing that the one-tailed observed significance level for the test is p-value $= .0001/2 = .00005$. Any value of α larger than .00005 will lead us to reject H_0.

Note that the values $\hat{\beta}_1 = 12.74$ and $\hat{\beta}_2 = 85.82$ (shaded in Figure 4.7) are easily interpreted. We estimate that the mean auction price increases $12.74 per year of age of the clock when the number of bidders is held constant, and the mean price increases by $85.82 per additional bidder, for clocks of a fixed age.

Be careful not to try to interpret the estimated intercept $\hat{\beta}_0 = -1,336.72$ in the same way as we interpreted $\hat{\beta}_1$ and $\hat{\beta}_2$. You might think that this implies a negative price for clocks 0 years of age with 0 bidders. However, these zeros are meaningless numbers in this example, since the ages range from 108 to 194 and the number of bidders ranges from 5 to 15. Keep in mind that we are modeling y within the range of values observed for the predictor variables and that interpretations of the models for values of the independent variables outside their sampled ranges can be very misleading.

Some computer programs use an F-test to test hypotheses concerning the individual β parameters. If you conduct a two-tailed t-test and reject the hypothesis if $t > t_{\alpha/2}$ or $t < -t_{\alpha/2}$, the corresponding F-test will imply rejection if the computed value of F (which is equal to the square of the computed t-statistic) is larger than F_α, because the square of the Student's t with ν (Greek nu) degrees of freedom is equal to an F-statistic with 1 df in the numerator and ν df in the denominator. Thus, $t_{\alpha/2}^2 = F_\alpha$, where t is based on ν df and F possesses 1 numerator and ν denominator degrees of freedom, respectively. As an example, when we tested the curvature parameter β_2 in the second-order model relating electrical usage to home size, the computed t-value was -7.62 (see Figure 4.6). The equivalent F-statistic yields

$$F = t^2 = (-7.62)^2 = 58.06$$

Suppose we wanted to conduct a two-tailed statistical test, i.e., $H_0: \beta_2 = 0$ and $H_a: \beta_2 \neq 0$. The upper-tail rejection region for a two-tailed test with $\alpha = .05$ is

$$F > F_{.05} \quad \text{where } F_{.05} \text{ is based on } \nu_1 = 1 \text{ df and } \nu_2 = 7 \text{ df}$$

or

$$F > 5.59$$

Note that the F-value, 5.59, is equal to the square of 2.365, the value of t that corresponds to $t_{.025}$ with 7 df. In other words, you can conduct a two-tailed test of the null hypothesis $H_0: \beta_i = 0$, using either a two-tailed t-test or a one-tailed F-test. If you want to conduct a one-tailed test to detect $H_a: \beta_i > 0$ (or $H_a: \beta_i < 0$), the F-test will not suffice. You will have to conduct the test using a t-statistic.

EXERCISES 4.1–4.5

4.1 How is the number of degrees of freedom available for estimating σ^2, the variance of ε, related to the number of independent variables in a regression model?

4.2 An employer believes that factory workers who are with the company longer tend to invest more in a company investment program per year than workers with less time with the company. The following model is believed to be adequate in modeling the relationship between annual amount invested y and years working for the company x:

$$y = \beta_0 + \beta_1 x + \beta_2 x^2 + \varepsilon$$

The employer checks the records for a sample of 50 factory employees for a previous year, and fits the above model to get $\hat{\beta}_2 = .0015$ and $s_{\hat{\beta}_2} = .000712$. The basic shape of a second-order model depends on whether $\beta_2 < 0$ or $\beta_2 > 0$. Test to determine whether the employer can conclude that $\beta_2 > 0$. Use $\alpha = .05$.

4.3 Real estate appraisers rely heavily on multiple regression analysis in their evaluation of property. Typically, the sale price y of a home is modeled as a function of several home-related variables (e.g., home size, home condition, location, and so forth). For example, an article in *The Real Estate Appraiser and Analyst* (Spring 1986) considered the following regression model:

$$y = \beta_0 + \beta_1 x_1 + \beta_2 x_2 + \varepsilon$$

where

$x_1 =$ home size (in square feet)
$x_2 =$ home condition rating (1 = poor, . . . , 10 = excellent)

Data collected for $n = 10$ recent home sales were used in the analysis. The data are reproduced in the table, and a SAS printout is provided.

SALE PRICE y, $ thousands	HOME SIZE x_1, hundreds of sq. ft.	CONDITION RATING x_2, 1 to 10
60.0	23	5
32.7	11	2
57.7	20	9
45.5	17	3
47.0	15	8
55.3	21	4
64.5	24	7
42.6	13	6
54.5	19	7
57.5	25	2

Source: Andrews, R. L. and Ferguson, J. T. "Integrating judgment with a regression appraisal," *The Real Estate Appraiser and Analyst*, Vol. 52, No. 2, Spring 1986 (Table I).

a. Plot sale price y against home size x_1. Do you detect a linear relationship between the two variables? Is it positive or negative?

b. Plot sale price y against home condition x_2. Do you detect a linear relationship between the two variables? Is it positive or negative?

c. Is there evidence to indicate that sale price and home size are linearly related? Test using $\alpha = .01$.

d. Calculate a 99% prediction interval for β_2. Interpret the result.

SAS Printout for Exercise 4.3

ANALYSIS OF VARIANCE

SOURCE	DF	SUM OF SQUARES	MEAN SQUARE	F VALUE	PROB>F
MODEL	2	819.32795	409.66398	350.866	0.0001
ERROR	7	8.17304573	1.16757796		
C TOTAL	9	827.50100			

ROOT MSE	1.080545	R-SQUARE	0.9901	
DEP MEAN	51.73	ADJ R-SQ	0.9873	
C.V.	2.088817			

PARAMETER ESTIMATES

VARIABLE	DF	PARAMETER ESTIMATE	STANDARD ERROR	T FOR H0: PARAMETER=0	PROB > |T|
INTERCEP	1	9.78227061	1.63048067	6.000	0.0005
X1	1	1.87093528	0.07617357	24.561	0.0001
X2	1	1.27814078	0.14440032	8.851	0.0001

4.4 Refer to the *Feline Practice* (February 1986) study on estimating the diet metabolizable energy (ME) content of commercial cat foods in Exercises 2.47 and 3.53. Three factors thought to influence ME content (y) of dry cat food are the crude protein (x_1), acid ether abstract (x_2), and nitrogen-free extract (x_3) content of the food. Data collected for 28 cats fed a diet of dry food were used to fit the multiple regression model

$$y = \beta_0 + \beta_1 x_1 + \beta_2 x_2 + \beta_3 x_3 + \varepsilon$$

with the following results:

$$\hat{y} = 2.44 + .45x_1 + 3.43x_2 + .10x_3$$

$$s_{\hat{\beta}_1} = .38 \qquad s_{\hat{\beta}_2} = .73 \qquad s_{\hat{\beta}_3} = .36$$

a. Is there sufficient evidence to indicate that crude protein x_1 is positively related to ME content y? Test using $\alpha = .05$.

b. Find a 95% confidence interval for β_3. Interpret your result.

c. The F-statistic for testing $H_0: \beta_2 = 0$ was found to be $F = 22.08$. Is there sufficient evidence to indicate that ME content y is linearly related to acid ether abstract x_2? Test using $\alpha = .05$.

d. Refer to part c. Calculate the t-statistic for testing $H_0: \beta_2 = 0$ and show that $F = t^2$.

4.5 In the mid 1800's, the U.S. census inquired about the real property and personal wealth of individual households. Using census information from 1860 and 1870, J. R. Kearl and C. L. Pope (Brigham Young University) examined the mobility of Utah households as measured by their wealth holdings (*The Review of Economics and Statistics*, May 1984). Holding occupation, time of entry into the economy, nativity, sex, place of residence, and internal migration constant, Kearl and Pope fit the quadratic model $E(y) = \beta_0 + \beta_1 x + \beta_2 x^2$, where y is personal wealth (in dollars) of a Utah household and x is age (in years) of the head of household. The results of the regression are summarized as follows:

$$\hat{y} - 52.39 + 74.21x - .71x^2$$

$$s_{\hat{\beta}_1} = 5.38 \qquad s_{\hat{\beta}_2} = 4.73 \qquad n > 20{,}000$$

a. Graph the least squares prediction equation.

b. Is there evidence of a quadratic relationship in the wealth–age relationship for Utah households during 1860–1870? Test using $\alpha = .10$.

S E C T I O N 4.6

CHECKING THE UTILITY OF A MODEL: R^2 AND THE ANALYSIS OF VARIANCE F-TEST

Conducting t-tests on each β parameter in a model is not a good way to determine whether a model is contributing information for the prediction of y. If we were to conduct a series of t-tests to determine whether the independent variables are contributing to the predictive relationship, we would be very likely to make one or more errors in deciding which terms to retain in the model and which to exclude. For example, suppose all the β parameters (except β_0) are equal to 0. Although the probability of concluding that any single β parameter differs from 0 is only α, the probability of rejecting *at least one* of a set of t-tests when H_0 is true is much higher. You can see why this is true by considering the following analogy: The probability of observing a head on a single toss of a coin is only .5, but the probability of observing *at least one* head in five tosses of a coin is .97. Thus, in multiple regression models for which a large number of independent variables are being considered, conducting a series of t-tests may cause the experimenter to include a large number of insignificant variables and exclude some useful ones. If we want to test the utility of a multiple regression model, we will need a global test (one that encompasses all the β parameters). We would also like to find some statistical quantity that measures how well the model fits the data.

We begin with the easier problem—finding a measure of how well a linear model fits a set of data. For this we use the multiple regression equivalent of r^2, the coefficient of determination for the straight-line model (Chapter 3). We define the **sample multiple coefficient of determination R^2** as

$$R^2 = 1 - \frac{\sum (y_i - \hat{y}_i)^2}{\sum (y_i - \bar{y})^2} = 1 - \frac{\text{SSE}}{\text{SS}_{yy}}$$

where $\hat{y}_i$ is the predicted value of y_i for the model. Just as for the simple linear model, R^2 is a sample statistic that represents the fraction of the sample variation of the y-values (measured by SS_{yy}) that is attributable to the regression model. Thus, $R^2 = 0$ implies a complete lack of fit of the model to the data, and $R^2 = 1$ implies a perfect fit, with the model passing through every data point. In general, the closer the value of R^2 is to 1, the better the model fits the data.

DEFINITION 4.1

The **multiple coefficient of determination, R^2**, is defined as

$$R^2 = 1 - \frac{\text{SSE}}{\text{SS}_{yy}}$$

where $\text{SSE} = \sum (y_i - \hat{y}_i)^2$, $\text{SS}_{yy} = \sum (y_i - \bar{y})^2$, and $\hat{y}_i$ is the predicted value of y_i for the multiple regression model.

To illustrate, the value $R^2 = .9819$ for the electrical usage example is shaded in Figure 4.9. This very high value of R^2 implies that 98.2% of the sample variation is attributable to, or explained by, the independent variable (home size) x. Thus, R^2 is a sample statistic that tells how well the model fits the data, and thereby represents a measure of the utility of the entire model.

The fact that R^2 is a sample statistic implies that it can be used to make inferences about the utility of the entire for predicting y-values for specific settings of the independent variables. In particular, for the electrical usage data, the test

H_0: $\beta_1 = \beta_2 = 0$

H_a: At least one of the parameters β_1 and β_2 is nonzero

would formally test the global utility of the model. The test statistic used to test this null hypothesis is

FIGURE 4.9 Portion of the SAS Printout for Electrical Usage Example

```
                          ANALYSIS OF VARIANCE

                       SUM OF          MEAN
    SOURCE      DF     SQUARES        SQUARE       F VALUE      PROB>F

    MODEL       2     831069.55      415534.77     189.710      0.0001
    ERROR       7    15332.55363    2190.36480
    C TOTAL     9    846402.10

              ROOT MSE       46.80133     R-SQUARE      0.9819
              DEP MEAN       1594.7       ADJ R-SQ      0.9767
              C.V.            2.934805

                          PARAMETER ESTIMATES

                       PARAMETER      STANDARD      T FOR H0:
    VARIABLE   DF      ESTIMATE        ERROR       PARAMETER=0    PROB > |T|

    INTERCEP   1     -1216.14389    242.80637       -5.009        0.0016
    X          1        2.39893018    0.24583560     9.758        0.0001
    XX         1       -0.000450040   0.000059077   -7.618        0.0001
```

Test statistic: $F = \dfrac{\text{Mean square for model}}{\text{Mean square for error}}$

$$= \frac{\text{SS(model)}/k}{\text{SSE}/[n - (k + 1)]}$$

where n is the number of data points, k is the number of parameters in the model (not including β_0), and SS(Model) = SS(Total) − SSE. When H_0 is true, this F-test statistic will have an F probability distribution with k df in the numerator and $[n - (k + 1)]$ df in the denominator. The upper-tail values of the F-distribution are given in Tables 3, 4, 5, and 6 of Appendix D.

It can be shown (proof omitted) that an equivalent form of this test statistic is

$$F = \frac{R^2/k}{(1 - R^2)/[n - (k + 1)]}$$

Therefore, the F-test statistic becomes large as the coefficient of determination R^2 becomes large. To determine how large F must be before we can conclude at a given value of α that the model is useful for predicting y, we set up the rejection region as follows:

Rejection region: $F > F_\alpha$ where $\nu_1 = k$ df and $\nu_2 = n - (k + 1)$ df

For the electrical usage example, $n = 10$, $k = 2$, $n - (k + 1) = 7$, and $\alpha = .05$. Consequently, we will reject $H_0: \beta_1 = \beta_2 = 0$ if

$F > F_{.05}$ where $\nu_1 = 2$ and $\nu_2 = 7$

or

$F > 4.74$ (see Figure 4.10 on page 160)

FIGURE 4.10

Rejection Region for the
F-Statistic with $\nu_1 = 2$,
$\nu_2 = 7$, and $\alpha = .05$

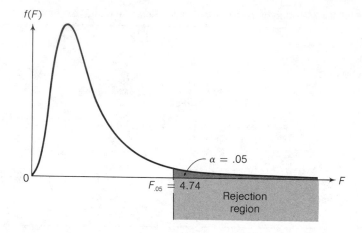

From the computer printout shown in Figure 4.9, we find that the computed F (shaded in the upper right-hand corner of the printout) is 189.71. Since this value greatly exceeds the tabulated value of 4.74, we conclude that at least one of the model coefficients β_1 and β_2 is nonzero. Therefore, this global F-test indicates that the second-order model $y = \beta_0 + \beta_1 x + \beta_2 x^2 + \varepsilon$ is useful for predicting electrical usage.

TEST OF THE OVERALL UTILITY OF A MULTIPLE REGRESSION MODEL: THE ANALYSIS OF VARIANCE F-TEST

H_0: $\beta_1 = \beta_2 = \cdots = \beta_k = 0$

H_a: At least one of the parameters, $\beta_1, \beta_2, \ldots, \beta_k$ differs from 0

Rejection region: $F > F_\alpha$

where the distribution of F depends on k numerator df and $n - (k + 1)$ denominator df

Test statistic:

$$F = \frac{\text{Mean square for model}}{\text{Mean square for error}} = \frac{\text{SS(model)}/k}{\text{SSE}/[n - (k + 1)]}$$

$$= \frac{R^2/k}{(1 - R^2)/[n - (k + 1)]}$$

where

n = Number of observations

k = Number of parameters in the model (excluding β_0)

R^2 = Multiple coefficient of determination

Values of F_α for α = .10, .05, .025, and .01 are given in Tables 3, 4, 5, and 6 of Appendix D.

We could arrive at the same decision by checking the observed significance level (p-value) of the F-test, given as PROB > F in the SAS printout. This value (shaded in Figure 4.9) indicates that we will reject H_0 for any α greater than $p = .0001$.

EXAMPLE 4.2

Refer to Example 4.1, in which an antique collector modeled the auction price y of grandfather clocks as a function of the age of the clock, x_1, and the number of bidders, x_2. The hypothesized model was

$$y = \beta_0 + \beta_1 x_1 + \beta_2 x_2 + \varepsilon$$

A sample of 32 observations was obtained, with the results summarized in the SAS printout repeated in Figure 4.11. Discuss the coefficient of determination R^2 for this example and then conduct the global F-test of model utility using $\alpha = .05$.

SOLUTION

The R^2 value is .89 (see Figure 4.11). This implies that 89% of the variation of the y-values (the auction prices) about their mean can be explained by the least squares model. We now test:

H_0: $\beta_1 = \beta_2 = 0$ [*Note*: $k = 2$]

H_a: At least one of the two model coefficients is nonzero

Test statistic: $F = \dfrac{\text{Mean square for model}}{\text{Mean square for error}} = \dfrac{\text{SS(Model)}/k}{\text{SSE}/[n - (k + 1)]}$

Rejection region: $F > F_\alpha$ where $\nu_1 = k$ and $\nu_2 = n - (k + 1)$

FIGURE 4.11 Portion of the SAS printout for Example 4.2

ANALYSIS OF VARIANCE

SOURCE	DF	SUM OF SQUARES	MEAN SQUARE	F VALUE	PROB>F
MODEL	2	4277159.70	2138579.85	120.651	0.0001
ERROR	29	514034.52	17725.32812		
C TOTAL	31	4791194.22			

ROOT MSE	133.1365	R-SQUARE	0.8927	
DEP MEAN	1327.156	ADJ R-SQ	0.8853	
C.V.	10.03171			

PARAMETER ESTIMATES

VARIABLE	DF	PARAMETER ESTIMATE	STANDARD ERROR	T FOR H0: PARAMETER=0	PROB > |T|
INTERCEP	1	-1336.72205	173.35613	-7.711	0.0001
X1	1	12.73619884	0.90238049	14.114	0.0001
X2	1	85.81513260	8.70575681	9.857	0.0001

For this example, $n = 32$, $k = 2$, and $n - (k + 1) = 32 - 3 = 29$. Then, for $\alpha = .05$, we will reject H_0: $\beta_1 = \beta_2 = 0$ if $F > F_{.05}$, i.e., if $F > 3.33$ (obtained from Table 4 of Appendix D). The computed value of the F-test statistic is 120.65 (see Figure 4.11). Since this value of F falls in the rejection region ($F = 120.65$ greatly exceeds $F_{.05} = 3.33$ and $\alpha = .05$ greatly exceeds $p = .0001$), the data provide strong evidence that at least one of the model coefficients is nonzero. The model appears to be useful for predicting auction prices. ■

Can we be sure that the best prediction model has been found if the global F-test indicates that a model is useful? Unfortunately, we cannot. There is no way of knowing whether the addition of other independent variables will further improve the utility of the model, as the following example indicates.

EXAMPLE 4.3

Refer to Examples 4.1 and 4.2. Suppose the collector, having observed many auctions, believes that the *rate of increase* of the auction price with age will be driven upward by a large number of bidders. Thus, instead of a relationship like that shown in Figure 4.12(a), in which the rate of increase in price with age is the same for any number of bidders, the collector believes the relationship is like that shown in Figure 4.12(b). Note that as the number of bidders increases from 5 to 15, the slope of the price versus age line increases. When the slope of the relationship between y and one independent variable (x_1) depends on the value of a second independent variable (x_2), as is the case here, we say that x_1 and x_2 **interact**.* A model that accounts for this type of interaction is written

$$y = \beta_0 + \beta_1 x_1 + \beta_2 x_2 + \beta_3 x_1 x_2 + \varepsilon$$

Note that the increase in the mean price $E(y)$ for each 1-year increase in age x_1 is no longer given by the constant β_1, but is now $\beta_1 + \beta_3 x_2$. That is, the amount $E(y)$ increases for each 1-unit increase in x_1 is *dependent on the number of bidders* x_2. Thus, the two variables x_1 and x_2 interact to affect y.

FIGURE 4.12
Examples of No Interaction and Interaction Models

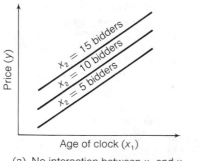

(a) No interaction between x_1 and x_2

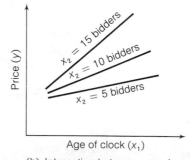

(b) Interaction between x_1 and x_2

The 32 data points listed in Table 4.2 were used to fit the first-order model with interaction. A portion of the SAS printout is shown in Figure 4.13.

Test the hypothesis that the price–age slope increases as the number of bidders increases, i.e., that age and number of bidders, x_2, interact positively.

*A more complete discussion of interaction is given in Section 4.10 and optional Chapter 7.

FIGURE 4.13 Portion of the SAS Printout for the Model with Interaction, Example 4.3

ANALYSIS OF VARIANCE

SOURCE	DF	SUM OF SQUARES	MEAN SQUARE	F VALUE	PROB>F
MODEL	3	4572547.99	1524182.66	195.188	0.0001
ERROR	28	218646.23	7808.79398		
C TOTAL	31	4791194.22			

ROOT MSE	88.36738	R-SQUARE	0.9544
DEP MEAN	1327.156	ADJ R-SQ	0.9495
C.V.	6.658401		

PARAMETER ESTIMATES

VARIABLE	DF	PARAMETER ESTIMATE	STANDARD ERROR	T FOR H0: PARAMETER=0	PROB > |T|
INTERCEP	1	322.75435	293.32515	1.100	0.2806
X1	1	0.87328775	2.01965115	0.432	0.6688
X2	1	-93.40991991	29.70767946	-3.144	0.0039
X1X2	1	1.29789828	0.21102602	6.150	0.0001

SOLUTION The model is

$$y = \beta_0 + \beta_1 x_1 + \beta_2 x_2 + \beta_3 x_1 x_2 + \varepsilon$$

and the hypothesis of interest to the collector concerns the parameter β_3. Specifically,

$$H_0: \quad \beta_3 = 0 \qquad H_a: \quad \beta_3 > 0$$

Test statistic: $t = \dfrac{\hat{\beta}_3}{s_{\hat{\beta}_3}}$

Rejection region: For $\alpha = .05$, $t > t_{.05}$
where df $= n - (k + 1)$

In this example, $n = 32$, $k = 3$, df $= n - (k + 1) = 32 - 4 = 28$, and thus, $t_{.05} = 1.701$.

The t-value corresponding to $\hat{\beta}_3$ is shaded in Figure 4.13. The value, $t = 6.15$, exceeds 1.701 and therefore falls in the rejection region. Thus, the collector can conclude that the rate of change of the mean price of the clocks with age increases as the number of bidders increases, i.e., x_1 and x_2 interact positively. Thus, it appears that the interaction term should be included in the model. ∎

One note of caution: Although the coefficient of x_2 is negative ($\hat{\beta}_2 = -93.41$) in Example 4.3, this does *not* imply that auction price decreases as the number of bidders increases. Since interaction is present, the rate of change (slope) of mean auction price with the number of bidders *depends on* x_1, the age of the clock. Thus, for example, the estimated rate of change of y with x_2 for a 150-year-old clock is

Estimated x_2 slope: $\hat{\beta}_2 + \hat{\beta}_3 x_1 = -93.41 + 1.30(150)$

$$= 101.59$$

In other words, we estimate that the auction price of a 150-year-old clock will *increase* by $101.59 for every additional bidder. Although this rate of increase will vary as x_1 is changed, it will remain positive for the range of values of x_1 included in the sample. Extreme care is needed in interpreting the signs and sizes of coefficients in a multiple regression model.

To summarize the discussion in this section, the value of R^2 is an indicator of how well the prediction equation fits the data. More importantly, it can be used (in the F-statistic) to determine whether the data provide sufficient evidence to indicate that the model contributes information for the prediction of y. Intuitive evaluations of the contribution of the model based on the computed value of R^2 must be examined with care. The value of R^2 will increase as more and more variables are added to the model. Consequently, you could force R^2 to take a value very close to 1 even though the model contributes no information for the prediction of y. In fact, R^2 will equal 1 when the number of terms in the model equals the number of data points. Therefore, **you should not rely solely on the value of R^2 to tell you whether the model is useful for predicting y. Use the analysis of variance F-test.**

EXERCISES 4.6–4.11

4.6 In hopes of increasing the company's share of the fine food market, researchers for a meat-processing firm that prepares meats for exclusive restaurants are working to improve the quality of its hickory-smoked hams. One of their studies concerns the effect of time spent in the smokehouse on the flavor of the ham. Hams that were in the smokehouse for varying amounts of time were each subjected to a taste test by a panel of ten food experts. The following model was thought to be appropriate by the researchers:

$$y = \beta_0 + \beta_1 t + \beta_2 t^2 + \varepsilon$$

where

y = Mean of the taste scores for the ten experts

t = Time in the smokehouse (hours)

Assume the least squares model estimated using a sample of 20 hams is

$$\hat{y} = 20.3 + 5.2t - .0025t^2$$

and that $s_{\hat{\beta}_2} = .0011$. The coefficient of determination is $R^2 = .79$.
a. Is there evidence to indicate that the overall model is useful? Test at $\alpha = .05$.
b. Is there evidence to indicate that the second-order term is important in this model? Test at $\alpha = .05$.

4.7 Because the coefficient of determination R^2 always increases when a new independent variable is added to the model, it is tempting to include many variables in a model to force R^2 to be near 1. However, doing so reduces the degrees of freedom available for estimating σ^2, which adversely affects our ability to make reliable inferences. As an example, suppose you want to use 18 economic indicators to predict next year's GNP. You fit the model

$$y = \beta_0 + \beta_1 x_1 + \beta_2 x_2 + \cdots + \beta_{17} x_{17} + \beta_{18} x_{18} + \varepsilon$$

where y = GNP and $x_1, x_2, \ldots, x_{18}$ are indicators. Only 20 years of data ($n = 20$) are used to fit the model, and you obtain $R^2 = .95$. Test to see whether this impressive looking R^2 is large enough for you to infer that this model is useful, i.e., that at least one term in the model is important for predicting GNP. Use $\alpha = .05$.

4.8 In Exercise 3.56 we gave the results of M. Hergert's simple linear regression analysis relating 1980 return on assets y to market share x for a sample of 5,400 businesses (*Business Economics,* October 1984). The main objective of the analysis was to investigate the conventional wisdom in business strategy that "a better market share yields a higher profit." The extremely small value of R^2 leads Hergert to suggest that "corporate strategy should be based on this premise only with great caution." In addition to the straight-line model, Hergert fit the quadratic model $E(y) = \beta_0 + \beta_1 x + \beta_2 x^2$. The results of the multiple regression are shown here:

$$\hat{y} = .093 + .441x - .409x^2 \qquad s_{\hat{\beta}_2} = .171 \qquad R^2 = .002$$

a. Graph the least squares prediction equation. (Let market share x range from 0% to 60%.)
b. Intepret the value of R^2.
c. Is there sufficient evidence to indicate that market share x is a useful predictor of return on assets y? Test using $\alpha = .05$.
d. As a result of the multiple regression analysis, Hergert concludes that "profits rise with size (of market share) up to some intermediate level and taper off thereafter." Do you agree with this statement? [*Hint:* Test for downward curvature using $\alpha - .05$.]

4.9 A study was conducted at Union Carbide to identify the optimal catalyst preparation conditions in the conversion of monoethanolamine (MEA) to ethylenediamine (EDA), a substance used commercially in soaps.* For each of ten selected catalysts, the following experimental variables were measured:

y = Rate of conversion of MEA to EDA

x_1 = Atom ratio of metal used in the experiment

x_2 = Reduction temperature

$x_3 = \begin{cases} 1 & \text{if high acidity support used} \\ 0 & \text{if low acidity support used} \end{cases}$

The data for the $n = 10$ experiments were used to fit the multiple regression model $E(y) = \beta_0 + \beta_1 x_1 + \beta_2 x_2 + \beta_3 x_3$. The results are summarized here:

$$\hat{y} = 40.2 - .808x_1 - 6.38x_2 - 4.45x_3 \qquad R^2 = .899$$
$$s_{\hat{\beta}_1} = .231 \qquad s_{\hat{\beta}_2} = 1.93 \qquad s_{\hat{\beta}_3} = .99$$

a. Is there sufficient evidence to indicate that the model is useful for predicting rate of conversion y? Test using $\alpha = .01$.
b. Conduct a test to determine whether atom ratio x_1 is a useful predictor of rate of conversion y. Use $\alpha = .05$.
c. Construct a 95% confidence interval for β_2. Interpret the interval.

*Hansen, J. L. and Best, D. C. "How to pick a winner." Paper presented at Joint Statistical Meetings, American Statistical Association and Biometric Society, August 1986, Chicago, Illinois.

4.10 Stock market analysts are continually searching for reliable predictors of stock price. Consider the problem of modeling the price per share, y, of electric utility stocks. Two variables which are thought to influence stock price are return on average equity, x_1, and annual rate of dividend, x_2. The stock prices, returns on equity, and dividend rates on a randomly selected day for a sample of 12 nuclear and 16 nonnuclear electric utility stocks are shown in the table. The interaction model

$$E(y) = \beta_0 + \beta_1 x_1 + \beta_2 x_2 + \beta_3 x_1 x_2$$

was fit to the data on each type of stock (nuclear and nonnuclear). The SAS printouts are provided here.

NUCLEAR STOCKS			NONNUCLEAR STOCKS		
y	x_1	x_2	y	x_1	x_2
21	15.1	2.36	25	15.2	2.60
31	15.0	3.00	20	13.9	2.14
26	11.2	3.00	15	15.8	1.52
11	12.1	1.96	34	12.8	3.12
24	16.3	3.00	20	6.9	2.48
8	11.9	1.40	33	14.6	3.08
18	14.9	1.80	28	15.4	2.92
23	11.8	2.56	30	17.3	2.76
13	13.4	2.06	23	13.7	2.36
14	16.2	1.94	24	12.7	2.36
35	17.1	2.96	25	15.3	2.56
13	13.3	2.20	26	15.2	2.80
			26	12.0	2.72
			20	15.3	1.92
			20	13.7	1.92
			13	13.3	1.60

Source: United Business Investment Report.

SAS Printout for Exercise 4.10: Nuclear Stocks

TYPE=NUC

ANALYSIS OF VARIANCE

SOURCE	DF	SUM OF SQUARES	MEAN SQUARE	F VALUE	PROB>F
MODEL	3	640.93485	213.64495	13.217	0.0018
ERROR	8	129.31515	16.16439437		
C TOTAL	11	770.25000			

ROOT MSE	4.020497	R-SQUARE	0.8321	
DEP MEAN	19.75	ADJ R-SQ	0.7692	
C.V.	20.35695			

PARAMETER ESTIMATES

VARIABLE	DF	PARAMETER ESTIMATE	STANDARD ERROR	T FOR H0: PARAMETER=0	PROB > |T|
INTERCEP	1	-17.55634046	40.05327713	-0.438	0.6727
X1	1	0.51898803	2.93598713	0.177	0.8641
X2	1	10.88942280	15.57141358	0.699	0.5042
X1X2	1	0.13221521	1.12491646	0.118	0.9093

SAS Printout for Exercise 4.10: Nonnuclear Stocks

TYPE=NO

ANALYSIS OF VARIANCE

SOURCE	DF	SUM OF SQUARES	MEAN SQUARE	F VALUE	PROB>F
MODEL	3	478.30855	159.43618	60.851	0.0001
ERROR	12	31.44145069	2.62012089		
C TOTAL	15	509.75000			

ROOT MSE	1.618679	R-SQUARE	0.9383
DEP MEAN	23.875	ADJ R-SQ	0.9229
C.V.	6.779806		

PARAMETER ESTIMATES

VARIABLE	DF	PARAMETER ESTIMATE	STANDARD ERROR	T FOR H0: PARAMETER=0	PROB > |T|
INTERCEP	1	-44.68177311	25.23972659	-1.770	0.1021
X1	1	2.87957851	1.74113100	1.654	0.1241
X2	1	25.06218058	10.02876655	2.499	0.0280
X1X2	1	-0.95900631	0.69103996	-1.388	0.1904

a. Write the least squares prediction equations for the two types of electric utility stock.

b. Is the model useful for predicting price of nuclear stocks? Nonnuclear stocks? Test using $\alpha = .05$.

c. Is there evidence of interaction between return on equity and dividend rate for the nuclear stock model? The nonnuclear stock model? Perform each test using $\alpha = .05$.

4.11 A utility company in a major city gave the average utility bills (listed in the table) for a standard-size home during the last year.

MONTH	AVERAGE MONTHLY TEMPERATURE x, °F	AVERAGE UTILITY BILL y, dollars
January	38	99
February	45	91
March	49	78
April	57	61
May	69	55
June	78	63
July	84	80
August	89	95
September	79	65
October	64	56
November	54	74
December	41	93

a. Plot the points on a scattergram.

b. Use the methods of Chapter 3 to fit the model

$$y = \beta_0 + \beta_1 x + \varepsilon$$

What would you conclude about the utility of this model?

c. Hypothesize another model that might better describe the relationship between the average utility bill and average temperature. If you have access to a computer package, fit the model and test its utility.

SECTION 4.7

USING THE MODEL FOR ESTIMATION AND PREDICTION

In Section 3.9 we discussed the use of the least squares line for estimating the mean value of y, $E(y)$, for some value of x, say $x = x_p$. We also showed how to use the same fitted model to predict, when $x = x_p$, some value of y to be observed in the future. Recall that the least squares line yielded the same value for both the estimate of $E(y)$ and the prediction of some future value of y. That is, both are the result of substituting x_p into the prediction equation $\hat{y} = \hat{\beta}_0 + \hat{\beta}_1 x$ and calculating $\hat{y}$. There the equivalence ends. The confidence interval for the mean $E(y)$ was narrower than the prediction interval for y, because of the additional uncertainty attributable to the random error ε when predicting some future value of y.

These same concepts carry over to the multiple regression model. For example, suppose we want to estimate the mean electrical usage for a given home size, say $x_p = 1,500$ square feet. Assuming the quadratic model represents the true relationship between electrical usage and home size, we want to estimate

$$E(y) = \beta_0 + \beta_1 x_p + \beta_2 x_p^2$$
$$= \beta_0 + \beta_1(1,500) + \beta_2(1,500)^2$$

Substituting into the least squares prediction equation yields the estimate of $E(y)$:

$$\hat{y} = \hat{\beta}_0 + \hat{\beta}_1(1,500) + \hat{\beta}_2(1,500)^2$$
$$= -1,216.144 - 2.3989(1,500) - .00045004(1,500)^2$$
$$= 1,369.7$$

To form a confidence interval for the mean, we need to know the standard deviation of the sampling distribution for the estimator $\hat{y}$. For multiple regression models, the form of this standard deviation is rather complex. However, some regression packages allow us to obtain the confidence intervals for mean values of y at any given setting of the independent variables. A portion of the SAS output for the electrical usage example is shown in Figure 4.14. The mean value and corresponding 95% confidence interval for $x_p = 1,500$ are shown in the columns labeled PREDICT VALUE, LOWER 95% MEAN, AND UPPER 95% MEAN. Note that

$$\hat{y} = 1,369.7$$

which agrees with our earlier calculation. The 95% confidence interval for the true mean of y is shown to be 1,325.0 to 1,414.3 (see Figure 4.15).

FIGURE 4.14
SAS Printout for
Estimated Mean Value
and Corresponding
Confidence Interval for
$x_p = 1,500$

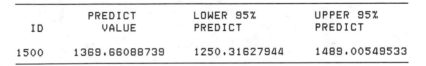

ID	PREDICT VALUE	LOWER 95% MEAN	UPPER 95% MEAN
1500	1369.66088739	1324.98831001	1414.33346477

FIGURE 4.15
Confidence Interval for
Mean Electrical Usage
When $x_p = 1,500$

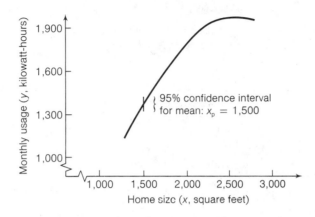

If we were interested in predicting the electrical usage for a particular 1,500-square-foot home, $\hat{y} = 1,369.7$ would be used as the predicted value. However, the prediction interval for a particular value of y will be wider than the confidence interval for the mean value. This is reflected by the printout shown in Figure 4.16, which gives the predicted value of y and corresponding 95% prediction interval for the predicted value. The predicted value for $x_p - 1,500$ is 1,369.7, and the prediction interval extends from 1,250.3 to 1,489.0. This interval is shown in Figure 4.17.

FIGURE 4.16
SAS Printout for Predicted
Value and Corresponding
Prediction Interval When
$x_p = 1,500$

ID	PREDICT VALUE	LOWER 95% PREDICT	UPPER 95% PREDICT
1500	1369.66088739	1250.31627944	1489.00549533

FIGURE 4.17
Prediction Interval for
Electrical Usage When
$x_p = 1,500$

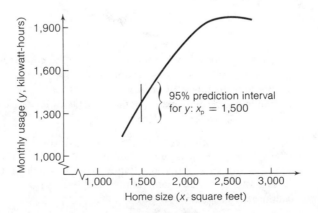

Unfortunately, not all computer packages have the capability to produce confidence intervals for means and prediction intervals for particular y-values. This is a rather serious oversight, since the estimation of mean values and the prediction of particular values represent the culmination of our model building efforts: using the model to make inferences about the dependent variable y.

| | | | | | | | | | | | |

EXERCISES 4.12–4.13

4.12 Refer to the sale price model in Exercise 4.3. The SAS printout with 95% confidence intervals for $E(y)$ is reproduced here.

SAS Printout for Exercise 4.12

ANALYSIS OF VARIANCE

SOURCE	DF	SUM OF SQUARES	MEAN SQUARE	F VALUE	PROB>F
MODEL	2	819.32795	409.66398	350.866	0.0001
ERROR	7	8.17304573	1.16757796		
C TOTAL	9	827.50100			

ROOT MSE	1.080545	R-SQUARE	0.9901
DEP MEAN	51.73	ADJ R-SQ	0.9873
C.V.	2.088817		

PARAMETER ESTIMATES

VARIABLE	DF	PARAMETER ESTIMATE	STANDARD ERROR	T FOR H0: PARAMETER=0	PROB > |T|
INTERCEP	1	9.78227061	1.63048067	6.000	0.0005
X1	1	1.87093528	0.07617357	24.561	0.0001
X2	1	1.27814078	0.14440032	8.851	0.0001

OBS	ID	ACTUAL	PREDICT VALUE	STD ERR PREDICT	LOWER95% MEAN	UPPER95% MEAN	RESIDUAL
1	23	60.0000	59.2045	0.4714	58.0899	60.3191	0.7955
2	11	32.7000	32.9188	0.8200	30.9799	34.8578	-0.2188
3	20	57.7000	58.7042	0.6375	57.1969	60.2116	-1.0042
4	17	45.5000	45.4226	0.4919	44.2595	46.5857	0.0774
5	15	47.0000	48.0714	0.6019	46.6481	49.4948	-1.0714
6	21	55.3000	54.1845	0.4276	53.1735	55.1955	1.1155
7	24	64.5000	63.6317	0.5705	62.2826	64.9808	0.8683
8	13	42.6000	41.7733	0.5710	40.4231	43.1235	0.8267
9	19	54.5000	54.2770	0.4206	53.2824	55.2717	0.2230
10	25	57.5000	59.1119	0.7657	57.3013	60.9226	-1.6119

SUM OF RESIDUALS	6.75016E-14
SUM OF SQUARED RESIDUALS	8.173046
PREDICTED RESID SS (PRESS)	21.09523

a. Locate the lower and upper limits for a 95% confidence interval for $E(y)$ when $x_1 = 17$ and $x_2 = 3$ (observation #4).

b. Interpret the interval in part a.

c. Would you recommend using the model to estimate the mean sale price for homes with 3,000 square feet? Explain.

4.13 Refer to the stock price model in Exercise 4.10. The SAS printout with 95% prediction intervals for price per share y of nuclear stocks is given here.

a. Locate the lower and upper limits for a 95% prediction interval for y when $x_1 = 13.3$ and $x_2 = 2.20$ (observation #12).

b. Interpret the interval in part a.

c. Would you recommend using the model to predict price per share of a nuclear stock with a dividend rate of 1.10? Explain.

SAS Printout for Exercise 4.13

ANALYSIS OF VARIANCE

SOURCE	DF	SUM OF SQUARES	MEAN SQUARE	F VALUE	PROB>F
MODEL	3	640.93485	213.64495	13.217	0.0018
ERROR	8	129.31515	16.16439437		
C TOTAL	11	770.25000			

ROOT MSE	4.020497	R-SQUARE	0.8321	
DEP MEAN	19.75	ADJ R-SQ	0.7692	
C.V.	20.35695			

PARAMETER ESTIMATES

VARIABLE	DF	PARAMETER ESTIMATE	STANDARD ERROR	T FOR H0: PARAMETER=0	PROB > |T|
INTERCEP	1	-17.55634046	40.05327713	-0.438	0.6727
X1	1	0.51898803	2.93598713	0.177	0.8641
X2	1	10.88942280	15.57141358	0.699	0.5042
X1X2	1	0.13221521	1.12491646	0.118	0.9093

OBS	ID	ACTUAL	PREDICT VALUE	STD ERR PREDICT	LOWER95% PREDICT	UPPER95% PREDICT	RESIDUAL
1	15.1	21.0000	20.6910	1.4471	10.8374	30.5447	0.3090
2	15	31.0000	28.8464	1.8979	18.5940	39.0989	2.1536
3	11.2	26.0000	25.3670	3.2339	13.4687	37.2654	0.6330
4	12.1	11.0000	13.2023	1.9178	2.9301	23.4745	-2.2023
5	16.3	24.0000	30.0368	2.3630	19.2826	40.7909	-6.0368
6	11.9	8.0000	6.0675	3.2551	-5.8616	17.9966	1.9325
7	14.9	18.0000	13.3236	2.2922	2.6513	23.9959	4.6764
8	11.8	23.0000	20.4386	1.9934	10.0902	30.7870	2.5614
9	13.4	13.0000	15.4800	1.3474	5.7018	25.2581	-2.4800
10	16.2	14.0000	16.1320	2.9501	4.6325	27.6315	-2.1320
11	17.1	35.0000	30.2433	2.7022	19.0724	41.4141	4.7567
12	13.3	13.0000	17.1715	1.2640	7.4528	26.8903	-4.1715

SUM OF RESIDUALS -8.19345E-14
SUM OF SQUARED RESIDUALS 129.3152
PREDICTED RESID SS (PRESS) 321.047

| | | | | | | | | | | |

SECTION 4.8

MULTIPLE REGRESSION: AN EXAMPLE

Let us return to the executive compensation example introduced in Case Study 4.1. Recall that the management consultant firm of Towers, Perrin, Forster & Crosby (TPF&C) uses a multiple regression model to project executive salaries. Suppose the list of independent variables given in Table 4.3 is to be used to build a model for the salaries of corporate executives.

TABLE 4.3

List of Independent Variables for Executive Compensation Example

INDEPENDENT VARIABLE	DESCRIPTION
x_1	Years of experience
x_2	Years of education
x_3	1 if male; 0 if female
x_4	Number of employees supervised
x_5	Corporate assets (millions of dollars)
x_6	x_1^2
x_7	$x_3 x_4$

STEP 1 The first step is to hypothesize a model relating executive salary to the independent variables listed in the table. TPF&C have found that executive compensation models that use the logarithm of salary as the dependent variable provide better predictive models than those using the salary as the dependent variable. This is probably because salaries tend to be incremented in *percentages* rather than dollar values. When a dependent variable undergoes percentage changes as the independent variables are varied, the logarithm of the dependent variable will be more suitable as a dependent variable. (We discuss transformations on the dependent variable in more detail in Chapter 6.) The model we propose is

$$y = \beta_0 + \beta_1 x_1 + \beta_2 x_2 + \beta_3 x_3 + \beta_4 x_4 + \beta_5 x_5 + \beta_6 x_6 + \beta_7 x_7 + \varepsilon$$

where y = log(Executive salary), $x_6 = x_1^2$ (second-order term in years of experience), and $x_7 = x_3 x_4$ (cross product or interaction term between sex and number of employees supervised). The variable x_3 is a **dummy variable**; it is used to describe an independent variable that is not measured on a numerical scale, but instead is **qualitative (categorical)** in nature. Sex is such a variable, since its values, male and female, are categories rather than numbers. Thus, we assign the value $x_3 = 1$ if the executive is male, $x_3 = 0$ if the executive is female. (For more detail on the use and interpretation of dummy variables, see Section 4.10 and optional Chapter 7.) The interaction term $x_3 x_4$ accounts for the fact that the relationship between the number of employees supervised, x_4, and corporate salary is dependent on sex, x_3. For example, as the number of supervised employees increases, a woman's salary (with all other factors being equal) might rise more rapidly than a man's. This concept (interaction) is also explained in more detail in Section 4.10 and optional Chapter 7.

STEP 2 Now, we estimate the model coefficients $\beta_0, \beta_1, \ldots, \beta_7$. Suppose a sample of 100 executives is selected, and the variables y and $x_1, x_2, \ldots, x_7$ are recorded (or, in the case of x_6 and x_7, calculated). The sample is then used as input for a SAS regression routine; the output is shown in Figure 4.18. The least squares model is

$$\hat{y} = 8.88 + .045x_1 + .033x_2 + .119x_3 + .00033x_4 + .0020x_5 - .00072x_6 + .00031x_7$$

Because we are using the logarithm of salary as the dependent variable, the β estimates have slightly different interpretations than previously discussed. In general, a parameter β in a log model represents the percentage increase (or decrease) in the dependent variable for a 1-unit increase in the corresponding independent variable. The percentage change is calculated by taking the antilogarithm of the β estimate and subtracting 1, i.e., $e^{\beta} - 1$.* For example, the percentage change in executive compensation associated with a 1-unit (i.e., 1-year) increase in years of education x_2 is $(e^{\beta_2} - 1) = (e^{.033} - 1) = .034$. Thus, when all other independent variables are held constant, we estimate executive salary to increase 3.4% for each additional year of education.

FIGURE 4.18 Portion of the SAS Printout for Executive Compensation Example

SOURCE	DF	SUM OF SQUARES	MEAN SQUARE	F VALUE	PROB > F
MODEL	7	27.06425564	3.85632223	1819.30	0.0001
ERROR	92	0.19551523	0.00212517		
TOTAL	99	27.25977087		R-SQUARE	ROOT MSE
				0.992828	0.0460995

VARIABLE	DF	PARAMETER ESTIMATE	STANDARD ERROR	T FOR H0: PARAMETER = 0	PROB > !T!
INTERCEPT	1	8.87878688	0.04612667	192.49	0.0001
X1 (EXPERIENCE)	1	0.04460301	0.00166257	26.83	0.0001
X2 (EDUCATION)	1	0.03326230	0.00270306	12.31	0.0001
X3 (SEX)	1	0.11892473	0.01724977	6.89	0.0001
X4 (EMPLOYEES SUPERVISED)	1	0.00033216	0.00001664	19.97	0.0001
X5 (ASSETS)	1	0.00201021	0.00002744	73.25	0.0001
X6 (= X1*X1)	1	-0.00071702	0.00004746	-15.11	0.0001
X7 (= X3*X4)	1	0.00031244	0.00001933	16.16	0.0001

STEP 3 The next step is to specify the probability distribution of ε, the random error component. We assume that ε is normally distributed, with a mean of 0 and a constant variance σ^2. Furthermore, we assume that the errors are independent. The estimate of the variance σ^2 is given in the SAS printout as

$$s^2 = \text{MSE} = \frac{\text{SSE}}{n - (k + 1)} = \frac{\text{SSE}}{100 - (7 + 1)} = .0021$$

and the estimate of the standard deviation σ, also given on the SAS printout as ROOT MSE, is $s = \sqrt{s^2} = .046$. Our interpretation is that most of the observed y-values (logarithms of salaries) lie within $2s = 2(.046) = .092$ of their least squares predicted values, $\hat{y}$. A more practical interpretation (in terms of salaries) is obtained, however, if we take the antilog of this value and subtract 1, similar to the manipulation in step 2. That is, we expect most of the observed executive salaries to lie within $e^{2s} - 1 = e^{.092} - 1 = .096$, or 9.6%, of their respective least squares predicted values.

STEP 4 We now want to see how well the model predicts salaries. First, note that $R^2 = .993$. This implies that 99.3% of the variation in y (the logarithm of salaries) for these 100 sampled executives is accounted for by the model. The utility of the model can be tested:

*The result is derived by expressing the percentage change in salary y, as $(y_1 - y_0)/y_0$, where y_1 = the value of y when, say $x = 1$, and y_0 = the value of y when $x = 0$. Now let $y^* = \log(y)$ and assume the log model is $y^* = \beta_0 + \beta_1 x$. Then

$$y = e^{y^*} = e^{\beta_0} e^{\beta_1 x} = \begin{cases} e^{\beta_0} & \text{when } x = 0 \\ e^{\beta_0} e^{\beta_1} & \text{when } x = 1 \end{cases}$$

Substituting, we have

$$\frac{y_1 - y_0}{y_0} = \frac{e^{\beta_0} e^{\beta_1} - e^{\beta_0}}{e^{\beta_0}} = e^{\beta_1} - 1$$

H_0: $\beta_1 = \beta_2 = \cdots = \beta_7 = 0$

H_a: At least one of the model coefficients is nonzero

$$\text{Test statistic:} \quad F = \frac{\text{Mean square for model}}{\text{MSE}} = \frac{\text{SS(Model)}/k}{\text{SSE}/[n - (k + 1)]}$$

Rejection region: For $\alpha = .05$, $F > F_{.05}$

where $\nu_1 = k = 7$ and $\nu_2 = n - (k + 1) = 92$

where from Table 4 of Appendix D, $F_{.05} \approx 2.1$. The test statistic is given on the SAS printout. Since $F = 1,819.3$ exceeds the tabulated value of F, we conclude that the model does contribute information for predicting executive salaries. It appears that at least one of the β parameters in the model differs from 0. Note that the observed significance level of the F-test, $p = .0001$, confirms this result.

We may be particularly interested in whether the data provide evidence that the mean salaries of executives increase as the asset value of the company increases, when all other variables (experience, education, etc.) are held constant. Putting it another way, we may want to know whether the data provide sufficient evidence to show that $\beta_5 > 0$. We use the following test:

H_0: $\beta_5 = 0$ H_a: $\beta_5 > 0$

$$\text{Test statistic:} \quad t = \frac{\hat{\beta}_5}{s_{\hat{\beta}_5}}$$

For $\alpha = .05$, $n = 100$, $k = 7$ and $df = n - (k + 1) = 92$, we will reject H_0 if $t > t_{.05}$, where (because the degrees of freedom of t are so large) $t_{.05} \approx z_{.05} = 1.645$. Thus, we reject H_0 if

$t > 1.645$

The t-value is indicated in Figure 4.18. The value corresponding to the independent variable x_5 is 73.25. Since this value exceeds 1.645, we find evidence that the mean salaries of executives do increase as the company assets increase, when all other variables are held constant. The observed significance level of the test, $p = .0001/2 = .00005$, confirms this result.

RECOMMENDATION FOR CHECKING THE UTILITY OF THE MODEL

1. First conduct the global F-test, H_0: $\beta_1 = \beta_2 = \cdots = \beta_k = 0$. If the model is deemed adequate (i.e., if you reject H_0) then proceed to step 2.

2. Conduct t-tests on those individual β parameters that you are particularly interested in. However, it is a safe practice to limit the number of β's that are tested. Conducting a series of t-tests leads to a high overall Type I error rate α.

STEP 5   The culmination of the modeling effort is to use the model for estimation and/or prediction. Suppose a firm is trying to determine fair compensation for an executive with the characteristics shown in Table 4.4. The least squares model can be used to obtain a predicted value for the logarithm of salary. That is,

$$\hat{y} = \hat{\beta}_0 + \hat{\beta}_1(12) + \hat{\beta}_2(16) + \hat{\beta}_3(0) + \hat{\beta}_4(400) + \hat{\beta}_5(160.1) + \hat{\beta}_6(144) + \hat{\beta}_7(0)$$

TABLE 4.4

Values of Independent Variables for a Particular Executive

$x_1 = 12$ years of experience
$x_2 = 16$ years of education
$x_3 = 0$ (female)
$x_4 = 400$ employees supervised
$x_5 = \$160.1$ million (the firm's asset value)
$x_6 = x_1^2 = 144$
$x_7 = x_3 x_4 = 0$

This predicted value is given in Figure 4.19, a partial reproduction of the SAS regression printout for this problem: $\hat{y} = 10.298$. The 95% prediction interval is also given: from 10.203 to 10.392. To predict the salary of an executive with these characteristics we take the antilog of these values. That is, the predicted salary is $e^{10.298} = \$29,700$ (rounded to the nearest hundred) and the 95% prediction interval is from $e^{10.203}$ to $e^{10.392}$ (or from \$27,000 to \$32,600). Thus, an executive with the characteristics in Table 4.4 should be paid between \$27,000 and \$32,600 to be consistent with the sample data.

FIGURE 4.19 Portion of the SAS Printout for Executive Compensation Problem

X1	X2	X3	X4	X5	X6	X7	PREDICT VALUE	LOWER 95% PREDICT	UPPER 95% PREDICT
12	16	0	400	160.1	144	0	10.29766682	10.20298295	10.39235070

EXERCISES 4.14–4.17

4.14 Before accepting a job, a computer at a major university estimates the cost of running the job in order to see if the user's account contains enough money to cover the cost. As part of the job submission, the user must specify estimated values for two variables—central processing unit (CPU) time and lines printed. While the CPU time required and the lines printed do not account for the complete cost of the run, it is thought that knowledge of their values should allow a good prediction of job cost. The following model is proposed to explain the relationship of CPU time and lines printed to job cost:

$$E(y) = \beta_0 + \beta_1 x_1 + \beta_2 x_2 + \beta_3 x_1 x_2$$

where

$y = $ Job cost

$x_1 = $ Lines printed

$x_2 = $ CPU time (tenths of a second)

Records from 20 previous runs were used to fit this model.

Portion of the SAS Printout for Exercise 4.14

SOURCE	DF	SUM OF SQUARES	MEAN SQUARE	F VALUE	PROB > F
MODEL	3	43.25090461	14.41696820	84.96	0.0001
ERROR	16	2.71515039	0.16969690		
CORRECTED TOTAL	19	45.96605500		R-SQUARE	ROOT MSE
				0.940931	0.41194283

VARIABLE	PARAMETER ESTIMATE	STANDARD ERROR	T FOR H0: PARAMETER = 0	PROB > !T!
INTERCEPT	0.04564705	0.21082636	0.22	0.8313
X1	0.00078505	0.00013537	5.80	0.0001
X2	0.23737262	0.03163301	7.50	0.0001
X1X2	-0.00003809	0.00001273	-2.99	0.0086

X1	X2	PREDICT VALUE	LOWER 95% MEAN	UPPER 95% MEAN
2000	42	8.38574865	7.32284845	9.44864885

    a. Identify the least squares model that was fit to the data.
    b. What are the values of SSE and s^2 (estimate of σ^2) for the data?
    c. What do we mean by the statement: This value of SSE (see part **b**) is minimum?

4.15 Refer to Exercise 4.14 and the portion of the SAS printout shown there.
    a. Is there evidence that the model is useful (as a whole) for predicting job cost? Test at $\alpha = .05$.
    b. Is there evidence that the variables x_1 and x_2 interact to affect y? Test at $\alpha = .01$.
    c. What assumptions are necessary for the validity of the tests conducted in parts **a** and **b**?
    d. Would you recommend conducting t-tests on β_1 and β_2? Explain.

4.16 Refer to Exercise 4.14 and the portion of the SAS printout shown in that exercise. Use a 95% confidence interval to estimate the mean cost of computer jobs that require 4.2 seconds of CPU time and print 2,000 lines. Interpret the interval.

4.17 As a result of the U.S. surgeon general's warnings about the health hazards of smoking, Congress banned television and radio advertising of cigarettes in January 1971. The banning of pro-smoking messages, however, also led to the virtual elimination of anti-smoking messages. In theory, if these anti-smoking commercials are more effective than pro-smoking commercials, the net effect of the Congressional ban will be to increase the consumption of cigarettes and therefore benefit the tobacco industry. To test this hypothesis, researchers at the University of Houston built a cigarette demand model based on data collected for 46 states over the 18-year period from 1963 to 1980 (*The Review of Economics and Statistics*, February 1986). For each state-year, the following independent variables were recorded:

    x_1 = Natural log of price of a carton of cigarettes

    x_2 = Natural log of minimum price of a carton of cigarettes in any neighboring state (This variable was included to measure the effect of "bootlegging" cigarettes in nearby states with lower tax rates.)

    x_3 = Natural log of real disposable income per capita

    x_4 = Per capita index of expenditures for cigarette advertising on television and radio (This value is 0 for the years 1971–1980, when the ban was in effect.)

The dependent variable of interest is y, the natural log of per capita consumption of cigarettes by persons of smoking age (14 years and older). The multiple regression model

$$E(y) = \beta_0 + \beta_1 x_1 + \beta_2 x_2 + \beta_3 x_3 + \beta_4 x_4$$

was fit to the $n = 828$ observations (48 states $\times$ 18 years) with the following results:

$$R^2 = .95 \qquad s = .047$$

a. Test the hypothesis that the model is useful for predicting y. Use $\alpha = .05$.
b. Interpret the value of s.
c. Give the null and alternative hypotheses appropriate for testing whether a decrease in per capita cigarette advertising expenditures is accompanied by an increase in per capita consumption of cigarettes over the period from 1963–1980.
d. The value of $\hat{\beta}_4$ was determined to be .033. Interpret this value.
e. Does the value $\hat{\beta}_4 = .033$ support the alternative hypothesis in part c? Explain.

| | | | | | | | | | | | | |
SECTION 4.9

OTHER COMPUTER PRINTOUTS

We have highlighted the key elements of a multiple regression analysis as they appear in the SAS System printout. However, there are a number of different statistical program packages; some of the most popular, in addition to the SAS System, are BMDP, Minitab, and SPSS^x (see the references at the end of Appendix C). Some can be used on all large computers produced by a specific manufacturer. Consequently, you may have access to one or more of these packages at your computer center.

The multiple regression computer programs for these packages may differ in what they are programmed to do, how they do it, and the appearance of their computer printouts, but all of them print the basic outputs needed for a regression analysis. For example, some will compute confidence intervals $E(y)$ and prediction intervals for y. Others will not. Some test the null hypotheses that the individual β parameters equal 0 using Student's t-tests, while others use F tests.* But all give the least squares estimates, the values of SSE, s^2, and other pertinent information.

To illustrate, the Minitab, SAS, SPSS^x, and BMDP regression analysis computer printouts for Example 4.3 are shown in Figure 4.20 (pages 178–179). For that example, we fit the model

$$y = \beta_0 + \beta_1 x_1 + \beta_2 x_2 + \beta_3 x_1 x_2 + \varepsilon$$

to $n = 32$ data points. The variables in the model were

$y = $ Auction price
$x_1 = $ Age of clock (years)
$x_2 = $ Number of bidders

*A two-tailed Student's t-test based on ν df is equivalent to an F-test where the F-statistic has 1 df in the numerator and ν df in the denominator. See Section 4.5.

FIGURE 4.20 Computer Printouts for Example 4.3

(a) Minitab Regression Printout

The regression equation is
$Y = 323 + 0.87 X1 - 93.4 X2 + 1.30 X1X2$

Predictor	Coef	Stdev	t-ratio
Constant	322.8	293.3	1.10
X1	0.873	2.020	0.43
X2	-93.41	29.71	-3.14
X1X2	1.2979	0.2110	6.15

s = 88.37   R-sq = 95.4%   R-sq(adj) = 94.9%

Analysis of Variance

SOURCE	DF	SS	MS
Regression	3	4572548	1524183
Error	28	218646	7809
Total	31	4791194	

SOURCE	DF	SEQ SS
X1	1	2554859
X2	1	1722301
X1X2	1	295388

(b) SAS Regression Printout

ANALYSIS OF VARIANCE

SOURCE	DF	SUM OF SQUARES	MEAN SQUARE	F VALUE	PROB>F
MODEL	3	4572547.99	1524182.66	195.188	0.0001
ERROR	28	218646.23	7808.79398		
C TOTAL	31	4791194.22			

ROOT MSE	88.36738	R-SQUARE	0.9544
DEP MEAN	1327.156	ADJ R-SQ	0.9495
C.V.	6.658401		

PARAMETER ESTIMATES

VARIABLE	DF	PARAMETER ESTIMATE	STANDARD ERROR	T FOR H0: PARAMETER=0	PROB > |T|
INTERCEP	1	322.75435	293.32515	1.100	0.2806
X1	1	0.87328775	2.01965115	0.432	0.6688
X2	1	-93.40991991	29.70767946	-3.144	0.0039
X1X2	1	1.29789828	0.21102602	6.150	0.0001

FIGURE 4.20 (Continued)

(c) SPSSx Regression Printout

```
LISTWISE DELETION OF MISSING DATA

EQUATION NUMBER 1   DEPENDENT VARIABLE..   Y

BEGINNING BLOCK NUMBER  1.  METHOD:  ENTER       X1      X2      X1X2

VARIABLE(S) ENTERED ON STEP NUMBER     1..   X1X2
                                       2..   X1
                                       3..   X2

MULTIPLE R          .97692           ANALYSIS OF VARIANCE
R SQUARE            .95436                          DF      SUM OF SQUARES      MEAN SQUARE
ADJUSTED R SQUARE   .94948           REGRESSION      3       4572547.98718    1524182.66239
STANDARD ERROR    88.36738           RESIDUAL       28        218646.23157       7808.79398

                                     F =  195.18797      SIGNIF F =  .0000

------------ VARIABLES IN THE EQUATION ------------

VARIABLE        B           SE B         BETA          T       SIG T

X1X2          1.297898     .211026     1.370324      6.150     .0000
X1             .873288    2.019651      .060855       .432     .6688
X2          -93.409920   29.707679     -.674705     -3.144     .0039
(CONSTANT)  322.754353  293.325147                   1.100     .2806

END BLOCK NUMBER   1   ALL REQUESTED VARIABLES ENTERED.
```

(d) BMDP Regression Printout

```
DEPENDENT VARIABLE. . . . . . . . . . . . . .   3 Y
TOLERANCE . . . . . . . . . . . . . . . . . .   0.0010
ALL DATA CONSIDERED AS A SINGLE GROUP

MULTIPLE R          0.9769            STD. ERROR OF EST.     88.3671
MULTIPLE R-SQUARE   0.9544

ANALYSIS OF VARIANCE
             SUM OF SQUARES     DF     MEAN SQUARE      F RATIO     P(TAIL)
REGRESSION   4572538.0000        3    1524179.0000     195.189      0.0000
RESIDUAL      218644.8125       28       7808.7422

                            STD.       STD. REG
VARIABLE     COEFFICIENT     ERROR      COEFF        T     P(2 TAIL)  TOLERANCE

INTERCEPT     322.75806     2.0196
X1     1        0.8733     29.7071      0.06       0.43     0.67       0.0823
X2     2      -93.4103     29.7071     -0.67      -3.14     0.00       0.0554
X1X2   4        1.2979      0.2110      1.37       6.15     0.00       0.0328

NUMBER OF INTEGER WORDS OF STORAGE USED IN PRECEDING PROBLEM     708
CPU TIME USED    0.097 SECONDS
```

INDEPENDENT
VARIABLES

Notice that the Minitab printout in Figure 4.20 gives the prediction equation at the top of the printout. The independent variables shown in the prediction equation and in the Predictor column are x_1, x_2, and x_1x_2. The product, x_1x_2, must be computed before the fitting begins. The inclusion of the squares or cross products of independent variables is treated in the same manner in the SAS, SPSS^x, and BMDP programs shown in Figure 4.20—that is, the values of the cross products and squares of the independent variables must be calculated prior to fitting the model.* The independent variables are listed in the SAS, SPSS^x, and BMDP columns titled VARIABLE.

PARAMETER ESTIMATES

The estimates of the regression coefficients appear opposite the identifying variable in the Minitab column titled Coef, in the BMDP column titled COEFFICIENT, in the SAS column titled PARAMETER ESTIMATE, and in the SPSS^x column titled B. Compare the estimates given in these four columns. Note that the Minitab printout gives the estimates with a lesser degree of accuracy (fewer decimal places) than SAS, SPSS^x, and BMDP. (Ignore the columns titled BETA in the SPSS^x printout and STD. REG COEFF in the BMDP printout. These are standardized estimates and will not be discussed in this text.)†

STANDARD ERRORS
OF ESTIMATES

The standard errors of the estimates are given in the Minitab column titled Stdev, in the SAS column titled STANDARD ERROR, in the SPSS^x column titled SE B, and in the BMDP column titled STD. ERROR.

t-TESTS FOR INDIVIDUAL
β's

The values of the test statistics for testing H_0: $\beta_i = 0$, where $i = 1, 2, 3$, are shown in the Minitab column titled t-ratio, in the SAS column titled T FOR H0: PARAMETER = 0, and in the SPSS^x and BMDP columns titled T. Note that the computed *t*-values shown in the various printouts are identical (except for the number of decimal places) and that Minitab does not give the significance level of the test. Consequently, to draw conclusions from the Minitab printout, you must compare the computed values of *t* with the critical values given in a *t*-table (Table 2 of Appendix D). In contrast, SAS gives the observed significance levels for the *t*-tests in the column titled PROB > |T|, SPSS^x in the column titled SIG T, and BMDP in the column titled P(2 TAIL). Note that these observed significance levels have been computed assuming that the tests are two-tailed. The observed significance levels for one-tailed tests would equal one-half of these values.

SSE AND s^2

The Minitab printout gives the value of SSE under the Analysis of Variance column headed SS and in the row identified as Error. The value of s^2 is shown in the same row under the column headed MS. The degrees of freedom DF appear in the same row. The corresponding values are shown at the top of the SAS printout in the row labeled ERROR and in the columns designated as SUM OF SQUARES,

*The SAS program is the only one of the four packages that can be automatically instructed to include these terms, and they would appear in the printout with a star (*) that indicates multiplication. In the SAS printout, x_1x_2 is printed as X1*X2 when this option is selected.

†See Chapter 5, page 234 for a footnote on standardized β estimates.

MEAN SQUARE, and DF, respectively. These quantities appear with the same headings in the SPSS^x and BMDP printouts in the row labeled RESIDUAL.

R^2

The value of R^2, as defined in Section 4.6, is given in the Minitab printout below the parameter estimates as 95.4% (we defined this quantity as a ratio where $0 \leq R^2 \leq 1$). It is given in the middle of the SAS printout as .9544, in the left column of the SPSS^x printout as .95436, and in the left column of the BMDP printout as .9544. (The quantities shown in the SAS printout as ADJ R-SQ, in the Minitab as R-sq(adj), and in the SPSS^x printout as ADJUSTED R SQUARE are adjusted for the degrees of freedom associated with the total SS and SSE. The adjusted R^2 will be discussed in Section 4.13.)

F-TEST FOR OVERALL MODEL UTILITY

The F-statistic for testing the utility of the model (Section 4.6), i.e., testing the null hypothesis that all model parameters (except β_0) equal 0, is shown under the title F VALUE in the top center of the SAS printout. In addition, the SAS printout gives the observed significance level of this F-test under PROB > F. This F-value is also printed at the right side of the SPSS^x printout, followed by the observed significance level given as SIGNIF F. BMDP gives the F-value and corresponding observed significance level on the right side of the printout as F RATIO and P(TAIL), respectively. The F-statistic for testing the utility of the model is *not* given in the Minitab printout. If you are using Minitab and want to obtain the value of this statistic you must compute it using one of the formulas (given in the box in Section 4.6):

$$F = \frac{R^2/k}{(1 - R^2)/[n - (k + 1)]}$$

or

$$F = \frac{\text{Mean square for model (or regression)}}{\text{Mean square for error (or residuals)}}$$

$$= \frac{\text{Mean square for regression}}{s^2}$$

These quantities are given in the Minitab printout under the column marked MS = SS/DF. Thus,

$$F = \frac{1,524,174}{7,809} = 195.18$$

a value that agrees with the values given in the SAS, SPSS^x, and BMDP printouts.

We will not comment on the advantages or disadvantages of the various packages because you will have to use and become familiar with the output of the package(s) available at your computer center. (All four packages are also available on floppy diskettes for use on a PC.) Most of the computer printouts are similar and it is relatively easy to learn how to read one output after you have become familiar with another. We have used different packages in the solutions of the examples to help you with this problem. Instructions on how to run multiple regression models with all four of these packages are provided in Appendix C.

In the preceding sections, we have demonstrated the methods of multiple regression analysis by fitting several different models, including a quadratic model and an interaction model. In this section we formally introduce other types of general linear models that are useful for relating a response variable y to a set of data.* We begin with a discussion of models using **quantitative** (numerical) independent variables.

Suppose that the mean value $E(y)$ of a response y is related to two quantitative variables, x_1 and x_2, by the model

$$E(y) = 1 + 2x_1 - x_2$$

Note that when $x_2 = 0$, the relationship between $E(y)$ and x_1 is given by

$$E(y) = 1 + 2x_1 - (0) = 1 + 2x_1$$

A graph of this relationship (a straight line) is shown in Figure 4.21. Similar graphs of the relationship between $E(y)$ and x_1 for $x_2 = 1$,

$$E(y) = 1 + 2x_1 - (1) = 2x_1$$

and for $x_2 = 2$,

$$E(y) = 1 + 2x_1 - (2) = -1 + 2x_1$$

are also shown in Figure 4.21.

FIGURE 4.21
Graphs of
$E(y) = 1 + 2x_1 - x_2$
for $x_2 = 0, 1, 2$

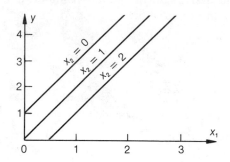

The model $E(y) = 1 + 2x_1 - x_2$ is an example of a **first-order linear model** in two quantitative independent variables, x_1 and x_2. A first-order linear model in five quantitative independent variables is shown in the box.

A FIRST-ORDER LINEAR MODEL RELATING $E(y)$ to $x_1, x_2, \ldots, x_5$

$$E(y) = \beta_0 + \beta_1 x_1 + \beta_2 x_2 + \cdots + \beta_5 x_5$$

Figure 4.21 exhibits a characteristic of all first-order models: If you graph $E(y)$ versus any one variable—say, x_1—for fixed values of the other variables, the

*A more complete discussion of general linear models and their role in model-building is provided in optional Chapter 7.

response curve will always be a *straight line*. If you repeat the process for other values of the fixed independent variables, you will obtain a set of *parallel* straight lines. This indicates that the effect on $E(y)$ of a change in x_1 is independent of the other variables in the model. When this situation occurs (as it always does for a first-order model), we say that the independent variables in the model **do not interact**.

Now, suppose that the mean value $E(y)$ of a response y is related to two quantitative variables, x_1 and x_2, by the model

$$E(y) = 1 + 2x_1 - x_2 + x_1x_2$$

This model contains the second-order cross product term, x_1x_2, in addition to all the terms of the first-order model. Figure 4.22 shows the graphs of the relationship between $E(y)$ and x_1 for $x_2 = 0, 1,$ and 2. The straight-line equations relating $E(y)$ to x_1 are:

For $x_2 = 0$: $E(y) = 1 + 2x_1 - (0) + x_1(0) = 1 + 2x_1$

For $x_2 = 1$: $E(y) = 1 + 2x_1 - (1) + x_1(1) = 3x_1$

For $x_2 = 2$: $E(y) = 1 + 2x_1 - (2) + x_1(2) = -1 + 4x_1$

FIGURE 4.22

Graphs of $E(y) = 1 + 2x_1 - x_2 + x_1x_2$ for $x_2 = 0, 1, 2$

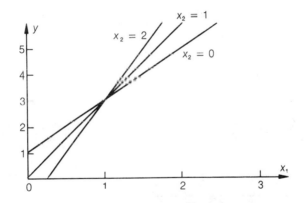

The effect of adding a term involving the cross product x_1x_2 can be seen in Figure 4.22. In·contrast to Figure 4.21, the lines relating $E(y)$ to x_1 are no longer parallel. The effect on $E(y)$ of a change in x_1 is now dependent on the value of x_2. When this situation occurs, we say that x_1 and x_2 **interact**. An **interaction model** is a model that contains first-order terms in the independent variables as well as two-way cross product terms such as x_1x_2, x_1x_3, ... An interaction model with three quantitative independent variables is shown in the box.

AN INTERACTION MODEL RELATING $E(y)$ TO THREE QUANTITATIVE INDEPENDENT VARIABLES

$$E(y) = \beta_0 + \beta_1x_1 + \beta_2x_2 + \beta_3x_3 + \beta_4x_1x_2 + \beta_5x_1x_3 + \beta_6x_2x_3$$

Finally, consider relating the mean value $E(y)$ of a response y to two quantitative independent variables, x_1 and x_2, by the model

$$E(y) = 1 + 2x_1 - x_2 + x_1x_2 + x_1^2 + 3x_2^2$$

This model contains all of the terms contained in the interaction model plus the second-order terms x_1^2 and x_2^2. Figure 4.23 shows a computer-generated graph of the relationship between $E(y)$ and x_1 for $x_2 = 0, 1,$ and 2.

FIGURE 4.23 Computer Graph of $E(y) = 1 + 2x_1 - x_2 + x_1x_2 + x_1^2 + 3x_2^2$ for $x_2 = 0, 1, 2$

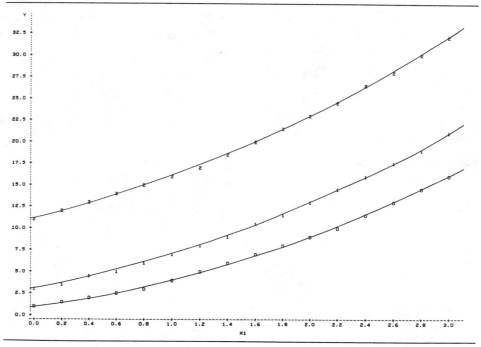

The response curves in Figure 4.23 rise (or fall) in the same manner as the lines shown in Figure 4.22. However, the graphs are curvilinear and the spacing between the curves has changed. These changes were produced by adding the second-order terms (those involving x_1^2 and x_2^2) to the model.

The model $E(y) = 1 + 2x_1 - x_2 - x_1x_2 + x_1^2 + 3x_2^2$ is an example of a **second-order model** in two quantitative independent variables. A second-order model contains all of the terms in a first-order model and, in addition, the second-order terms involving cross products (interaction terms) and squares of the independent variables. A second-order model with three quantitative independent variables is shown in the box. (Note that an interaction model is a special case of a second-order model, where the β coefficients of $x_1^2, x_2^2, \ldots,$ are all equal to 0.)

A SECOND-ORDER MODEL WITH THREE QUANTITATIVE INDEPENDENT VARIABLES

$$E(y) = \beta_0 + \beta_1 x_1 + \beta_2 x_2 + \beta_3 x_3 + \beta_4 x_1 x_2 + \beta_5 x_1 x_3$$
$$+ \beta_6 x_2 x_3 + \beta_7 x_1^2 + \beta_8 x_2^2 + \beta_9 x_3^2$$

How can you choose an appropriate linear model to fit a set of data? Since most relationships in the real world are curvilinear (at least to some extent), a good first choice would be a second-order linear model. If you are fairly certain that the relationships between $E(y)$ and the individual independent variables are approximately first-order and that the independent variables do not interact, you could select a first-order model for the data. If you have prior information that suggests there is moderate or very little curvature over the region in which the independent variables are measured, you could use the interaction model described earlier. However, keep in mind that for all multiple regression models, the number of data points must exceed the number of parameters in the model. Thus, you may be forced to use a first-order model rather than a second-order model simply because you do not have sufficient data to estimate all of the parameters in the second-order model.

A practical example of choosing and fitting a linear model with two quantitative independent variables follows.

EXAMPLE 4.4

Although a regional express delivery service bases the charge for shipping a package on the package weight and distance shipped, its profit per package depends on the package size (volume of space that it occupies) and the size and nature of the load on the delivery truck. The company recently conducted a study to investigate the relationship between the cost, y, of shipment (in dollars) and the variables that control the shipping charge—package weight, x_1 (in pounds), and distance shipped, x_2 (in miles). Twenty packages were randomly selected from among the large number received for shipment and a detailed analysis of the cost of shipment was made for each package, with the results shown in Table 4.5.

TABLE 4.5
Cost of Shipment Data for Example 4.4

PACKAGE	x_1	x_2	y	PACKAGE	x_1	x_2	y
1	5.9	47	2.60	11	5.1	240	11.00
2	3.2	145	3.90	12	2.4	209	5.00
3	4.4	202	8.00	13	.3	160	2.00
4	6.6	160	9.20	14	6.2	115	6.00
5	.75	280	4.40	15	2.7	45	1.10
6	.7	80	1.50	16	3.5	250	8.00
7	6.5	240	14.50	17	4.1	95	3.30
8	4.5	53	1.90	18	8.1	160	12.10
9	.60	100	1.00	19	7.0	260	15.50
10	7.5	190	14.00	20	1.1	90	1.70

a. Give an appropriate linear model for the data.
b. Fit the model to the data and give the prediction equation.
c. Find the value of SSE and specify its number of degrees of freedom.
d. Find the value of R^2 and interpret it.
e. Is the model useful for the prediction of shipping cost y? Find the value of the F-statistic on the printout and give the observed significance level (p-value) for the test.
f. Find a 95% prediction interval for the cost of shipping a 5-pound package a distance of 100 miles.

SOLUTION

a. Since we have no reason to expect that the relationship between y and x_1 and x_2 would be first-order, we will allow for curvature in the response surface and fit the second-order model

$$y = \beta_0 + \beta_1 x_1 + \beta_2 x_2 + \beta_3 x_1 x_2 + \beta_4 x_1^2 + \beta_5 x_2^2 + \varepsilon$$

The mean value of the random error term ε is assumed to equal 0. Therefore, the mean value of y is

$$E(y) = \beta_0 + \beta_1 x_1 + \beta_2 x_2 + \beta_3 x_1 x_2 + \beta_4 x_1^2 + \beta_5 x_2^2$$

b. The SAS computer printout for fitting the model to the $n = 20$ data points is shown in Figure 4.24. You can see from the printout that the parameter estimates (shaded in Figure 4.24) are:

$$\hat{\beta}_0 = .82701584 \qquad \hat{\beta}_1 = -.60913729 \qquad \hat{\beta}_2 = .00402071$$
$$\hat{\beta}_3 = .00732710 \qquad \hat{\beta}_4 = .08975085 \qquad \hat{\beta}_5 = .00001507$$

Therefore, the prediction equation that relates the predicted shipping cost $\hat{y}$ to weight of a package, x_1, and distance shipped, x_2, is

$$\hat{y} = .82701584 - .60913729x_1 + .00402071x_2 + .00732710x_1 x_2 + .08975085x_1^2 + .00001507x_2^2$$

c. The SUM OF SQUARES for ERROR, shaded on the printout, is

SSE $= 2.74473841$

based on 14 degrees of freedom (DF for ERROR in the printout). Recall that SSE has degrees of freedom equal to $n -$ (Number of estimated β parameters).

d. The value of R^2 shown in Figure 4.24 is R-SQUARE $= 0.9939$. This means that 99.4% of the total variation, SS(Total) $= SS_{yy} = \Sigma(y - \bar{y})^2$, is explained by the model; the remainder is explained by random error.

e. The test statistic for testing whether the model is useful for predicting shipping cost is

$$F = \frac{\text{Mean square for model}}{\text{Mean square for error}} = \frac{\text{SS(Model)}/k}{\text{SSE}/[n - (k + 1)]}$$

where $n = 20$ is the number of data points and $k = 5$ is the number of parameters (excluding β_0) contained in the model.

FIGURE 4.24 SAS Printout for the Multiple Regression Analysis of Example 4.4

ANALYSIS OF VARIANCE

SOURCE	DF	SUM OF SQUARES	MEAN SQUARE	F VALUE	PROB>F
MODEL	5	449.34076	89.86815232	458.388	0.0001
ERROR	14	2.74473841	0.19605274		
C TOTAL	19	452.08550			

ROOT MSE	0.4427784	R-SQUARE	0.9939	
DEP MEAN	6.335	ADJ R-SQ	0.9918	
C.V.	6.989399			

PARAMETER ESTIMATES

VARIABLE	DF	PARAMETER ESTIMATE	STANDARD ERROR	T FOR H0: PARAMETER=0	PROB > |T|
INTERCEP	1	0.82701584	0.70228935	1.178	0.2586
X1	1	-0.60913729	0.17990408	-3.386	0.0044
X2	1	0.40207076	0.79984198	0.503	0.6230
X1X2	1	0.73270972	0.06374329	11.495	0.0001
X1SQ	1	0.08975085	0.02020542	4.442	0.0006
X2SQ	1	0.15069916	0.22432800	0.672	0.5127

OBS	ID	ACTUAL	PREDICT VALUE	STD ERR PREDICT	LOWER95% PREDICT	UPPER95% PREDICT	RESIDUAL
1	5.9	2.6000	2.6114	0.2990	1.4655	3.7573	-0.0114
2	3.2	3.9000	4.0964	0.2111	3.0443	5.1486	-0.1964
3	4.4	8.0000	7.8238	0.1994	6.7823	8.8654	0.1762
4	6.6	9.2000	9.4828	0.1818	8.4562	10.5093	-0.2828
5	0.75	4.4000	4.2666	0.3992	2.9880	5.5452	0.1334
6	0.7	1.5000	1.2730	0.2453	0.1874	2.3587	0.2270
7	6.5	14.5000	13.9229	0.2127	12.8693	14.9764	0.5771
8	4.5	1.9000	1.9063	0.2284	0.8378	2.9748	-.006294
9	0.6	1.0000	1.4862	0.2259	0.4202	2.5523	-0.4862
10	7.5	14.0000	13.0560	0.2155	11.9998	14.1122	0.9440
11	5.1	11.0000	10.8562	0.1962	9.8175	11.8949	0.1438
12	2.4	5.0000	5.0559	0.1999	4.0140	6.0979	-0.0559
13	0.3	2.0000	2.0332	0.2602	0.9316	3.1347	-0.0332
14	6.2	6.0000	6.3863	0.1954	5.3482	7.4243	-0.3863
15	2.7	1.1000	0.9383	0.2572	-0.1600	2.0366	0.1617
16	3.5	8.0000	8.1527	0.2117	7.1001	9.2054	-0.1527
17	4.1	3.3000	3.2101	0.1803	2.1847	4.2356	0.0899
18	8.1	12.1000	12.3066	0.3015	11.1577	13.4555	-0.2066
19	7	15.5000	16.3603	0.3109	15.1999	17.5207	-0.8603
20	1.1	1.7000	1.4749	0.1993	0.4335	2.5163	0.2251
21	5	.	4.2414	0.1837	3.2133	5.2695	.

SUM OF RESIDUALS	-3.92325E-14	
SUM OF SQUARED RESIDUALS	2.744738	
PREDICTED RESID SS (PRESS)	6.856756	

This value of F has been computed for us as $F = 458.388$ and is shaded on the printout (Figure 4.24) in the row corresponding to MODEL. The observed significance level (p-value) for the test is shown to the right of the F-value on the printout as PROB > F = .0001. This means that if the model contributed no information for the prediction of y, the probability of observing a value of the F-statistic as large as 458.39 would be only .0001. Thus, we would reject the null hypothesis that the model contributes no information for the prediction of y for all values of α larger than .0001.

f. The predicted value of y for $x_1 = 5.0$ pounds (ID = 5) and $x_2 = 100$ miles is

$$\hat{y} = .82701584 - .60913729(5.0) + .00402071(100) +$$
$$.00732710(5.0)(100) + .08975085(5.0)^2 + .00001507(100)^2$$

This quantity has been computed and is shown (shaded) on the printout as $\hat{y} = 4.2414$. The corresponding 95% prediction interval (shaded) is given as 3.2133 to 5.2695. Therefore, if we were to select a 5-pound package and ship it 100 miles, we would expect the actual cost to fall between \$3.21 and \$5.27.

∎

Linear models can also be written to include **qualitative** (or **categorical**) independent variables. Qualitative variables, unlike quantitative variables, cannot be measured on a numerical scale. Therefore, we need to code the values of the qualitative variable (called **levels**) as numbers before we can fit the model. These coded qualitative variables are called **dummy variables** since the numbers assigned to the various levels are arbitrarily selected.

We gave an example of a dummy variable for a qualitative variable in Section 4.8. Recall that one of the independent variables used in the multiple regression model to project executive salaries was the sex of an executive. The dummy variable used to describe sex was coded as

$$x_3 = \begin{cases} 1 & \text{if male} \\ 0 & \text{if female} \end{cases}$$

The advantage of using a 0–1 coding scheme is that the β coefficients associated with the dummy variables are easily interpreted. For example, consider the following model for executive salary y:

$$E(y) = \beta_0 + \beta_1 x$$

where

$$x = \begin{cases} 1 & \text{if male} \\ 0 & \text{if female} \end{cases}$$

This model allows us to compare the mean executive salary $E(y)$ for males with the corresponding mean for females:

Males $(x = 1)$: $E(y) = \beta_0 + \beta_1(1) = \beta_0 + \beta_1$
Females $(x = 0)$: $E(y) = \beta_0 + \beta_1(0) = \beta_0$

First note that β_0 represents the mean salary for females (say, μ_F). When using a 0–1 coding convention, β_0 will always represent the mean response associated with the level of the qualitative variable assigned the value 0 (called the **base level**). The difference between the mean salary for males and the mean salary for females, $\mu_M - \mu_F$, is represented by β_1—that is,

$$\mu_M - \mu_F = (\beta_0 + \beta_1) - (\beta_0) = \beta_1$$

Therefore, with the 0–1 coding convention, β_1 will always represent the difference between the mean response for the level assigned the value 1 and the mean for the base level. Thus, for the executive salary model we have

$$\beta_0 = \mu_F$$
$$\beta_1 = \mu_M - \mu_F$$

The model relating a mean response $E(y)$ to a qualitative independent variable at two levels is shown in the box.

A MODEL RELATING $E(y)$ TO A QUALITATIVE INDEPENDENT VARIABLE WITH TWO LEVELS

$$E(y) = \beta_0 + \beta_1 x$$

where

$$x = \begin{cases} 1 & \text{if level A} \\ 0 & \text{if level B} \end{cases}$$

Interpretation of β's:

$\beta_0 = \mu_B$ (Mean for base level)

$\beta_1 = \mu_A - \mu_B$

For models that involve qualitative independent variables at more than two levels, additional dummy variables must be created. In general, the number of dummy variables used to describe a qualitative variable will be one less than the number of levels of the qualitative variable. The accompanying box presents a model that includes a qualitative independent variable at three levels.

A MODEL RELATING $E(y)$ TO A QUALITATIVE INDEPENDENT VARIABLE WITH THREE LEVELS

$$E(y) = \beta_0 + \beta_1 x_1 + \beta_2 x_2$$

where

$$x_1 = \begin{cases} 1 & \text{if level A} \\ 0 & \text{if not} \end{cases} \qquad x_2 = \begin{cases} 1 & \text{if level B} \\ 0 & \text{if not} \end{cases} \qquad \text{Base level = Level C}$$

Interpretation of β's:

$\beta_0 = \mu_C$ (Mean for base level)

$\beta_1 = \mu_A - \mu_C$

$\beta_2 = \mu_B - \mu_C$

EXAMPLE 4.5

Refer to the problem of modeling the shipment cost, y, of a regional express delivery service, described in Example 4.4. Suppose we want to model $E(y)$ as a function of cargo type, where cargo type has three levels—fragile, semifragile, and durable.

a. Write a linear model relating $E(y)$ to cargo type.
b. Interpret the β coefficients in the model.
c. Explain the practical significance of the F-test for overall model utility.

SOLUTION

a. Since the qualitative variable of interest, cargo type, has three levels, we need to create $(3 - 1) = 2$ dummy variables. First, select (arbitrarily) one of the levels to be the base level—say, durable cargo. Then each of the remaining levels is assigned the value 1 in one of the two dummy variables as follows:

$$x_1 = \begin{cases} 1 & \text{if fragile} \\ 0 & \text{if not} \end{cases} \qquad x_2 = \begin{cases} 1 & \text{if semifragile} \\ 0 & \text{if not} \end{cases}$$

(Note that for the base level, durable cargo, $x_1 = x_2 = 0$.) Then, the appropriate model is

$$E(y) = \beta_0 + \beta_1 x_1 + \beta_2 x_2$$

b. To interpret the β's, first write the mean shipment cost $E(y)$ for each of the three cargo types as a function of the β's:

Fragile $(x_1 = 1, x_2 = 0)$:

$$E(y) = \beta_0 + \beta_1(1) + \beta_2(0) = \beta_0 + \beta_1 = \mu_F$$

Semifragile $(x_1 = 0, x_2 = 1)$:

$$E(y) = \beta_0 + \beta_1(0) + \beta_2(1) = \beta_0 + \beta_2 = \mu_S$$

Durable $(x_1 = 0, x_2 = 0)$:

$$E(y) = \beta_0 + \beta_1(0) + \beta_2(0) = \beta_0 = \mu_D$$

Then we have

$$\beta_0 = \mu_D \quad \text{(Mean of the base level)}$$
$$\beta_1 = \mu_F - \mu_D$$
$$\beta_2 = \mu_S - \mu_D$$

Note that the β's associated with the non-base levels of cargo type (fragile and semifragile) represent differences between a pair of means. As always, β_0 represents a single mean—the mean response for the base level (durable).

c. The F-test for overall model utility tests the null hypothesis

$$H_0 = \beta_1 = \beta_2 = 0$$

Note that $\beta_1 = 0$ implies that $\mu_F = \mu_D$ and $\beta_2 = 0$ implies that $\mu_S = \mu_D$. Therefore, $\beta_1 = \beta_2 = 0$ implies that $\mu_F = \mu_S = \mu_D$. Thus, a test for model utility is equivalent to a test for equality of means, i.e.,

$$H_0: \quad \mu_F = \mu_S = \mu_D$$

If there is evidence of a difference between any two of the three mean shipment costs, then cargo type is a useful predictor of shipment cost y. ■

The linear models described in this section form the basis for building models with quantitative independent variables and models with qualitative independent variables. More complex models, such as those with interactions between qualitative variables and those with both quantitative and qualitative variables (including interactions) may be required in practice, however. Optional Chapter 7 presents a detailed discussion of model-building with both quantitative and qualitative variables.

$|\ |\ |\ |\ |\ |\ |\ |\ |\ |\ |\ |$

EXERCISES 4.18–4.31

4.18 Write a first-order linear model relating the mean value of y, $E(y)$, to two quantitative independent variables.

4.19 Write a first-order linear model relating the mean value of y, $E(y)$, to four quantitative independent variables.

4.20 Write a second-order linear model relating the mean value of y, $E(y)$, to two quantitative independent variables.

4.21 Write a second-order linear model relating the mean value of y, $E(y)$, to three quantitative independent variables.

4.22 Write a model relating $E(y)$ to a qualitative independent variable with two levels, A and B. Interpret the β parameters.

4.23 Write a model relating $E(y)$ to a qualitative independent variable with four levels, A, B, C, and D. Interpret the β parameters.

4.24 Consider the first-order equation

$$y = 1 + 2x_1 + x_2$$

a. Graph the relationship between y and x_1 for $x_2 = 0$, 1, and 2.
b. Are the graphed curves in part **a** first-order or second-order?
c. How do the graphed curves in part **a** relate to each other?
d. If a linear model is first-order in two independent variables, what type of geometric relationship will you obtain when $E(y)$ is graphed as a function of one of the independent variables for various values of the other independent variable?

4.25 Consider the first-order equation

$$y = 1 + 2x_1 + x_2 - 3x_3$$

a. Graph the relationship between y and x_1 for $x_2 = 1$ and $x_3 = 3$.
b. Repeat part **a** for $x_2 = -1$ and $x_3 = 1$.
c. If a linear model is first-order in three independent variables, what type of geometric relationship will you obtain when $E(y)$ is graphed as a function of one of the independent variables for various values of the other independent variables?

4.26 Consider the second-order model

$$y = 1 + x_1 - x_2 + 2x_1^2 + x_2^2$$

a. Graph the relationship between y and x_1 for $x_2 = 0$, 1, and 2.
b. Are the graphed curves in part **a** first-order or second-order?
c. How do the graphed curves in part **a** relate to each other?
d. Do the independent variables x_1 and x_2 interact? Explain.

4.27 Consider the second-order model

$$y = 1 + x_1 - x_2 + x_1x_2 + 2x_1^2 + x_2^2$$

a. Graph the relationship between y and x_1 for $x_2 = 0$, 1, and 2.
b. Are the graphed curves in part a first-order or second-order?
c. How do the graphed curves in part a relate to each other?
d. Do the independent variables x_1 and x_2 interact? Explain.
e. Note that the model used in this exercise is identical to the noninteraction model of Exercise 4.26, except that it contains the term involving x_1x_2. What does the term x_1x_2 introduce into the model?

4.28 One of the most promising methods for extracting crude oil employs a carbon dioxide (CO_2) flooding technique. CO_2, when flooded into oil pockets, enhances oil recovery by displacing the crude oil. In a microscopic investigation of the CO_2 flooding process, flow tubes were dipped into sample oil pockets containing a known amount of oil. The oil pockets were flooded with CO_2 and the percentage of oil displaced was recorded. The experiment was conducted at three different flow pressures and three different dipping angles. The displacement test data are recorded in the table.

PRESSURE x_1, pounds per square inch	DIPPING ANGLE x_2, degrees	OIL RECOVERY y, percentage
1,000	0	60.58
1,000	15	72.72
1,000	30	79.99
1,500	0	66.83
1,500	15	80.78
1,500	30	89.78
2,000	0	69.18
2,000	15	80.31
2,000	30	91.99

Source: Wang, G. C. "Microscopic investigation of CO_2 flooding process." *Journal of Petroleum Technology*, Vol. 34, No. 8, August 1982, pp. 1789–1797. Copyright © 1982, Society of Petroleum Engineers, American Institute of Mining. First published in the *JPT* August 1982.

a. Write the complete second-order model relating percentage oil recovery y to pressure x_1 and dipping angle x_2.
b. Plot the sample data on a scattergram, with percentage oil recovery y on the vertical axis and pressure x_1 on the horizontal axis. Connect the points corresponding to the same value of dipping angle x_2. Based on the scattergram, do you believe a complete second-order model is appropriate?

SAS Printout for Exercise 4.28

ANALYSIS OF VARIANCE

SOURCE	DF	SUM OF SQUARES	MEAN SQUARE	F VALUE	PROB>F
MODEL	3	843.19083	281.06361	44.670	0.0005
ERROR	5	31.45996667	6.29199333		
C TOTAL	8	874.65080			

			R-SQUARE	0.9640
ROOT MSE	2.508385			
DEP MEAN	76.90667		ADJ R-SQ	0.9425
C.V.	3.261596			

PARAMETER ESTIMATES

VARIABLE	DF	PARAMETER ESTIMATE	STANDARD ERROR	T FOR H0: PARAMETER=0	PROB > |T|
INTERCEP	1	54.50000000	5.03415841	10.826	0.0001
X1	1	0.007696667	0.003238311	2.377	0.0634
X2	1	0.55411111	0.25996282	2.132	0.0862
X1X2	1	0.000113333	0.000167226	0.678	0.5280

c. The SAS printout for the interaction model

$$y = \beta_0 + \beta_1 x_1 + \beta_2 x_2 + \beta_3 x_1 x_2 + \varepsilon$$

is provided here. Give the prediction equation for this model.

d. Construct a plot similar to the scattergram of part b, but use the predicted values from the interaction model on the vertical axis. Compare the two plots. Do you believe the interaction model will provide an adequate fit?

e. Check model adequacy using a statistical test with $\alpha = .05$.

f. Is there evidence of interaction between pressure x_1 and dipping angle x_2? Test using $\alpha = .05$.

4.29 Refer to Exercise 3.15 and the Consumer Attitude (Telephone) Survey conducted by the University of Florida's Bureau of Economic and Business Research (BEBR). Suppose we want to model the refusal rate, y (i.e., the percentage of dialed households in a county that refuse to take part in the survey), as a function of the county's personal income per capita, x_1, and percentage of residents with a college education, x_2.

a. Write a first-order linear model for refusal rate y.

b. Write an interaction model for refusal rate y.

c. Write a second-order model for refusal rate y.

d. The second-order model of part c was fit to data on 12 Florida counties (data supplied by BEBR). The resulting SAS printout is provided on page 194. Find R^2 on the printout and interpret its value.

e. Is there evidence that the model is useful for predicting refusal rate y? Test using $\alpha = .05$.

f. The SAS printout also gives a 95% prediction interval for the refusal rate for each of the 12 counties. County 8 had a per capita income of $x_1 = \$10,000$ (ID) and $x_2 = 29.74\%$ of residents with a college education. Interpret the prediction interval for this county. How do you explain the large width of the interval?

SAS Printout for Exercise 4.29

ANALYSIS OF VARIANCE

SOURCE	DF	SUM OF SQUARES	MEAN SQUARE	F VALUE.	PROB>F
MODEL	5	0.06174787	0.01234957	1.803	0.2465
ERROR	6	0.04109780	0.006849633		
C TOTAL	11	0.10284567			

ROOT MSE	0.08276251	R-SQUARE	0.6004
DEP MEAN	0.3961667	ADJ R-SQ	0.2674
C.V.	20.89083		

PARAMETER ESTIMATES

VARIABLE	DF	PARAMETER ESTIMATE	STANDARD ERROR	T FOR H0: PARAMETER=0	PROB > |T|
INTERCEP	1	1.98524969	0.83441322	2.379	0.0548
X1	1	-0.000431242	0.000255467	-1.688	0.1424
X2	1	0.03724751	0.08995885	0.414	0.6932
X1X2	1	-0.000002983	0.0000069142	-0.431	0.6812
X1SQ	1	2.72363E-08	2.20347E-08	1.236	0.2626
X2SQ	1	-0.000294703	0.000580117	-0.508	0.6296

OBS	ID	ACTUAL	PREDICT VALUE	STD ERR PREDICT	LOWER95% PREDICT	UPPER95% PREDICT	RESIDUAL
1	7737	0.2960	0.2630	0.0814	-0.0211	0.5470	0.0330
2	12330	0.4980	0.5042	0.0748	0.2313	0.7771	-.006186
3	12058	0.3860	0.4166	0.0712	0.1495	0.6838	-0.0306
4	9927	0.3270	0.3410	0.0400	0.1161	0.5659	-0.0140
5	6904	0.5000	0.5234	0.0690	0.2598	0.7871	-0.0234
6	9463	0.3330	0.4112	0.0757	0.1366	0.6857	-0.0782
7	11466	0.4290	0.4034	0.0368	0.1818	0.6249	0.0256
8	10000	0.4220	0.3564	0.0361	0.1354	0.5774	0.0656
9	10052	0.4410	0.3700	0.0378	0.1474	0.5927	0.0710
10	8636	0.1910	0.3234	0.0498	0.0871	0.5597	-0.1324
11	7445	0.5260	0.4634	0.0598	0.2135	0.7133	0.0626
12	9059	0.4050	0.3780	0.0397	0.1534	0.6026	0.0270

SUM OF RESIDUALS	-1.94289E-16
SUM OF SQUARED RESIDUALS	0.0410978
PREDICTED RESID SS (PRESS)	1.363713

4.30 In recent years, many companies have converted to the metric system of measurement. In an attempt to quantify some of the characteristics of companies that have converted to metric production, B. D. Phillips, H. A. G. Lakhani, and S. L. George analyzed data on 350 small manufacturers collected for a U.S. Metric Board study (*Technological Forecasting and Social Change*, April 1984). One of the research objectives was to investigate the relationship between the percentage y of metric work performed by a company, age x_1 of the company (in years), and cost x_2 of metric conversion, where

$$x_2 = \begin{cases} 1 & \text{if cost over \$10,000} \\ 0 & \text{if not} \end{cases}$$

A first-order linear model was fit to the $n = 350$ data points with the following results:

$$\hat{y} = 70.9770 - .2167x_1 - 13.2768x_2$$

$$R^2 = .0576 \qquad t \text{ (for } H_0: \beta_1 = 0) = -1.56 \qquad t \text{ (for } H_0: \beta_2 = 0) = -2.71$$

a. Sketch the least squares relationship between $\hat{y}$ and x_1 for the two levels of metric conversion cost.

b. Is there evidence that the model is useful for predicting percentage y of metric work performed? Test using $\alpha = .01$.

c. Is there evidence that percentage y of metric work performed decreases as age x_1 of the company increases, for companies with the same cost x_2 of conversion? Test using $\alpha = .05$.

d. Write an interaction model for percentage y of metric work performed.

e. How will interaction between age x_1 and cost x_2, if determined to be significant, affect the graphs constructed in part **a**?

4.31 Researchers recently conducted an analysis of bus travel demand in Albuquerque, New Mexico, a city selected because of its unique multicentered "Sun Tran" public transit system. One aspect of the study involved the development of a multiple regression model for predicting y, the home-origin trip rate (that is, the number of home-origin trips per 1,000 residents) of a Sun Tran subzone urban area. The following five independent variables, all designed to measure transit level of service (travel time), were entered into the model:

x_1 = Composite in-vehicle travel time to reach major destination (minutes)

x_2 = Composite transit wait time (minutes)

x_3 = Composite number of transfers required to reach major destination

x_4 = Number of transit routes serving the Sun Tran zone

$x_5 = \begin{cases} 1 & \text{if Sun Tran zone at end of major regional transportation corridor} \\ 0 & \text{if not} \end{cases}$

Data collected from a survey of the city's bus passengers in each of 298 Sun Tran planning analysis zones were used to fit the first-order model

$$E(y) = \beta_0 + \beta_1 x_1 + \beta_2 x_2 + \beta_3 x_3 + \beta_4 x_4 + \beta_5 x_5$$

with the results shown in the accompanying printout.

Computer Printout for Exercise 4.31

SOURCE	DF	SS	MS	F
MODEL	5	32774	6554.8	87.45
ERROR	292	21886	74.95	
TOTAL	297	54660		

VARIABLE	PARAMETER ESTIMATE	STANDARD ERROR
INTERCEPT	22.0189	-
X1	-0.1807	0.0389
X2	-0.2498	0.1207
X3	-4.6910	1.7020
X4	3.6745	0.4027
X5	22.5201	3.5959

Source: Adapted from Nelson, D. and O'Neil, K. "Analyzing demand for grid system transit," *Transportation Quarterly*, Vol. 37, No. 1, January 1983, pp. 41–56.

a. Write the least squares prediction equation.

b. Compute and interpret the value of R^2.

c. Compute and interpret the value of s.

d. Is the model useful for predicting home-origin trip rate y? Test using $\alpha = .05$.

e. Is there evidence that home-origin trip rate y decreases as in-vehicle travel time x_1 increases and the remaining independent variables are held constant? Test using $\alpha = .05$.

f. Construct a 95% confidence interval for β_4. Interpret the interval.

g. Construct a 95% confidence interval for β_5. Interpret the inverval.

SECTION 4.11

TESTING PORTIONS OF A MODEL

In Example 4.4, we fit a second-order model for the mean shipment cost $E(y)$ of an express delivery package as a function of two quantitative variables— package weight, x_1, and distance shipped, x_2. The second-order model was

$$E(y) = \beta_0 + \beta_1 x_1 + \beta_2 x_2 + \beta_3 x_1 x_2 + \beta_4 x_1^2 + \beta_5 x_2^2$$

If we assume that the relationship between shipment cost y, package weight x_1, and distance shipped x_2 is first-order, then the interaction model is appropriate:

$$E(y) = \beta_0 + \beta_1 x_1 + \beta_2 x_2 + \beta_3 x_1 x_2$$

We will call the second-order model the **complete model**, and the straight-line interaction model the **reduced model**. Now, suppose we wanted to test whether the second-order terms contribute information for the prediction of y. This can be done by testing the hypothesis that the parameters for the curvature terms, x_1^2 and x_2^2, equal 0:

H_0: $\beta_4 = \beta_5 = 0$

H_a: At least one of the two parameters, β_4 and β_5, is nonzero

In Section 3.6 we presented the t-test for a single coefficient, and in Section 4.6 we gave the F-test for *all* the β parameters (except β_0) in the model. We now need a test for *some* of the β parameters in the model. The test procedure is intuitive. First, we use the method of least squares to fit the interaction model, and calculate the corresponding sum of squares for error, SSE_1 (the sum of squares of the deviations between observed and predicted y-values). Next, we fit the second-order model, and calculate its sum of squares for error, SSE_2. Then, we compare SSE_1 to SSE_2 by calculating the difference $\text{SSE}_1 - \text{SSE}_2$. If the curvature terms contribute to the model, then SSE_2 should be much smaller than SSE_1, and the difference $\text{SSE}_1 - \text{SSE}_2$ will be large. The larger the difference, the greater the weight of evidence that the variables package weight and distance shipped affect the mean shipment cost in a curvilinear manner.

The sum of squares for error will always decrease when new terms are added to the model since the total sum of squares, $\text{SS}_{yy} = \Sigma(y - \bar{y})^2$, remains the same.

The question is whether this decrease is large enough to conclude that it is due to more than just an increase in the number of model terms and to chance. To test the null hypothesis that the curvature coefficients β_4 and β_5 simultaneously equal 0, we use an F-statistic calculated as follows:

$$F = \frac{\text{Drop in SSE/Number of } \beta \text{ parameters being tested}}{s^2 \text{ for complete model}}$$

$$= \frac{(\text{SSE}_1 - \text{SSE}_2)/2}{\text{SSE}_2/[n - (5 + 1)]}$$

When the assumptions listed in Sections 3.4 and 4.2 about the error term ε are satisfied and the β parameters for curvature are all 0 (H_0 is true), this F-statistic has an F-distribution with $\nu_1 = 2$ and $\nu_2 = n - 6$ df. Note that ν_1 is the number of β parameters being tested and ν_2 is the number of degrees of freedom associated with s^2 in the complete model.

If the quadratic terms *do* contribute to the model (H_a is true), we expect the F-statistic to be large. Thus, we use a one-tailed test and reject H_0 when F exceeds some critical value, F_α, as shown in Figure 4.25.

FIGURE 4.25
Rejection Region for the
F-Test H_0: $\beta_4 = \beta_5 = 0$

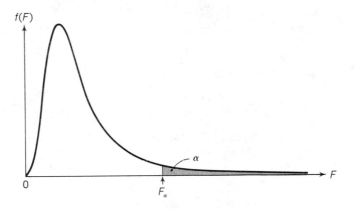

EXAMPLE 4.6

In Example 4.4, we fit the complete second-order model for a set of $n = 20$ data points relating shipment cost to package weight and distance shipped. The SAS printout for this model is shown in Figure 4.24 (page 187). Figure 4.26 (page 198) shows the SAS printout for the straight-line interaction model fit to the same $n = 20$ data points. Referring to the printouts, we find the following:

Straight-line interaction model:
    $\text{SSE}_1 = 6.63330508$ (see Figure 4.26)
Second-order model:
    $\text{SSE}_2 = 2.74473841$ (see Figure 4.24)

Test the hypothesis that the quadratic terms do not contribute information for the prediction of y.

FIGURE 4.26   SAS Printout for Straight-Line Interaction Model of Example 4.6

ANALYSIS OF VARIANCE

SOURCE	DF	SUM OF SQUARES	MEAN SQUARE	F VALUE	PROB>F
MODEL	3	445.45219	148.48406	358.154	0.0001
ERROR	16	6.63330508	0.41458157		
C TOTAL	19	452.08550			

ROOT MSE		0.6438801	R-SQUARE	0.9853	
DEP MEAN		6.335	ADJ R-SQ	0.9826	
C.V.		10.16385			

PARAMETER ESTIMATES

VARIABLE	DF	PARAMETER ESTIMATE	STANDARD ERROR	T FOR H0: PARAMETER=0	PROB > |T|
INTERCEP	1	-0.14050074	0.64810001	-0.217	0.8311
X1	1	0.01908803	0.15821160	0.121	0.9055
X2	1	0.77208456	0.39056785	1.977	0.0656
X1X2	1	0.77957444	0.08976644	8.684	0.0001

SOLUTION

The test statistic is

$$F = \frac{(SSE_1 - SSE_2)/2}{SSE_2/(20 - 6)} = \frac{(6.63330508 - 2.74473841)/2}{2.74473841/14} = \frac{1.944283}{.196053} = 9.92$$

The critical value of F for $\alpha = .05$, $\nu_1 = 2$, and $\nu_2 = 14$ is found in Table 4 (Appendix D) to be

$$F_{.05} = 3.74$$

Since the calculated $F = 9.92$ exceeds 3.74, we are confident in concluding that the quadratic terms contribute to the prediction of y, shipment cost per package. The curvature terms should be retained in the model.   ∎

The F-test can be used to determine whether *any* set of terms should be included in a model by testing the null hypothesis that a particular set of β parameters simultaneously equal 0. For example, we may want to test to determine whether a set of interaction terms for quantitative variables or a set of main effect terms for a qualitative variable should be included in a model. The F-test appropriate for testing the null hypothesis that all of a set of β parameters are equal to 0 is summarized in the box.

F-TEST FOR TESTING THE NULL HYPOTHESIS: SET OF β PARAMETERS EQUAL ZERO

Reduced model: $E(y) = \beta_0 + \beta_1 x_1 + \cdots + \beta_g x_g$

Complete model: $E(y) = \beta_0 + \beta_1 x_1 + \cdots + \beta_g x_g$
$$+ \beta_{g+1} x_{g+1} + \cdots + \beta_k x_k$$

H_0: $\beta_{g+1} = \beta_{g+2} = \cdots = \beta_k = 0$

H_a: At least one of the β parameters under test is nonzero

Test statistic: $F = \dfrac{(SSE_1 - SSE_2)/(k - g)}{SSE_2/[n - (k + 1)]}$

where

SSE_1 = Sum of squared errors for the reduced model

SSE_2 = Sum of squared errors for the complete model

$k - g$ = Number of β parameters specified in H_0

$k + 1$ = Number of β parameters in the complete model (including β_0)

n = Total sample size

Rejection region: $F > F_\alpha$

and

$\nu_1 = k - g$ = Degrees of freedom for the numerator

$\nu_2 = n - (k + 1)$ = Degrees of freedom for the denominator

---

EXERCISES 4.32–4.37

4.32 Consider the second-order model relating $E(y)$ to three quantitative independent variables, x_1, x_2, and x_3.

$$E(y) = \beta_0 + \beta_1 x_1 + \beta_2 x_2 + \beta_3 x_3 + \beta_4 x_1 x_2 + \beta_5 x_1 x_3 + \beta_6 x_2 x_3 + \beta_7 x_1^2 + \beta_8 x_2^2 + \beta_9 x_3^2$$

a. Specify the parameters involved in a test of the hypothesis that no curvature exists in the response surface.

b. State the hypothesis of part **a** in terms of the model parameters.

c. What hypothesis would you test to determine whether x_3 is useful for the prediction of $E(y)$?

4.33 Research was conducted at Temple University to examine the effect of several factors on managerial performance (*Journal of Vocational Behavior*, October 1986). A sample of 100 management personnel from several divisions within a government agency took part in the study. Each manager completed a questionnaire designed to measure the following variables:

y = Performance rating (1 = unacceptable, 5 = outstanding)

$x_1 = \begin{cases} 1 & \text{if male} \\ 0 & \text{if female} \end{cases}$

$x_2 = $ Job tenure (years)

$x_3 = $ Manager–subordinate work relationship rating
(1 = unsatisfactory, 5 = excellent)

$x_4 = $ Effort level (average number of hours per week invested in job)

$x_5 = \begin{cases} 1 & \text{if middle/upper-level manager} \\ 0 & \text{if lower-level manager} \end{cases}$

$x_6 = $ Subordinate-related managerial behavior score (low scores indicate little or no effort spent on counseling, evaluating, and training subordinates)

The data collected on the 100 managers were then used to fit several regression models of managerial performance.

a. Initially, the model

$$E(y) = \beta_0 + \beta_1 x_1 + \beta_2 x_2 + \beta_3 x_3 + \beta_4 x_4$$

was considered to account for the influence of sex, job tenure, manager–subordinate work relationship and effort level on performance rating. For this model, SSE = 352 and R^2 = .11. Calculate the F-statistic for testing model adequacy. Is the model useful for predicting performance rating y? Use $\alpha = .05$.

b. Terms for managerial level and subordinate-related behavior (i.e., $\beta_5 x_5 + \beta_6 x_6$) were added to the model in part **a**, resulting in SSE = 341 and R^2 = .14. Do these terms contribute additional information for the prediction of performance rating y? Test using $\alpha = .05$.

c. A third model was also considered:

$$E(y) = \beta_0 + \beta_1 x_1 + \beta_2 x_2 + \beta_3 x_3 + \beta_4 x_4 + \beta_5 x_5 + \beta_6 x_6 + \beta_7 x_5 x_6$$

This model resulted in SSE = 321 and R^2 = .19. Test the hypothesis that the interaction between managerial level x_5 and subordinate-related behavior x_6 is not important, i.e., test $H_0: \beta_7 = 0$. Use $\alpha = .05$.

d. Interpret the result of part **c** in terms of the problem.

4.34 An insurance company is experimenting with three different training programs, A, B, and C, for their salespeople. The following first-order model is proposed:

$$E(y) = \beta_0 + \beta_1 x_1 + \beta_2 x_2 + \beta_3 x_3$$

where

$y = $ Monthly sales (in thousands of dollars)

$x_1 = $ Number of months experience

$x_2 = \begin{cases} 1 & \text{if training program B was used} \\ 0 & \text{otherwise} \end{cases}$

$x_3 = \begin{cases} 1 & \text{if training program C was used} \\ 0 & \text{otherwise} \end{cases}$

Training program A is the base level.

a. What hypothesis would you test to determine whether the mean monthly sales differ for salespeople trained by the three programs?

b. After experimenting with 50 salespeople over a 5-year period, the complete first-order model is fit, with the following results:

$$\hat{y} = 10 + .5x_1 + 1.2x_2 - .4x_3 \qquad SSE = 140.5$$

Then the reduced model $E(y) = \beta_0 + \beta_1 x_1$ is fit to the same data, with the following results:

$$\hat{y} = 11.4 + .4x_1 \qquad SSE = 183.2$$

Test the hypothesis formulated in part a. Use $\alpha = .05$.

4.35 The following model was proposed for testing salary discrimination against women in a state university system:

$$E(y) = \beta_0 + \beta_1 x_1 + \beta_2 x_2 + \beta_3 x_1 x_2 + \beta_4 x_2^2$$

where

y = Annual salary (in thousands of dollars)

$x_1 = \begin{cases} 1 & \text{if female} \\ 0 & \text{if male} \end{cases}$

x_2 = Experience (years)

A portion of the computer printout that results from fitting this model to a sample of 200 faculty members in the university system is given here.

Printout for Exercise 4.35: Complete Model

SOURCE	DF	SUM OF SQUARES	MEAN SQUARE
MODEL	4	2351.70	587.92
ERROR	195	783.90	4.02
TOTAL	199	3135.60	R-SQUARE
			0.7500

The reduced model $E(y) = \beta_0 + \beta_2 x_2 + \beta_4 x_2^2$ is fit to the same data, and the resulting computer printout is partially reproduced here.

Printout for Exercise 4.35: Complete Model

SOURCE	DF	SUM OF SQUARES	MEAN SQUARE
MODEL	2	2340.37	1170.185
ERROR	197	795.23	4.04
TOTAL	199	3135.60	R-SQUARE
			0.7464

Do the data provide sufficient evidence to support the claim that the mean salary of faculty members is dependent on sex? Use $\alpha = .05$.

4.36 In an attempt to reduce lost work-hours due to accidents, a company tested three safety programs, A, B, and C, each at three of the company's nine factories. The proposed complete first-order model is

$$E(y) = \beta_0 + \beta_1 x_1 + \beta_2 x_2 + \beta_3 x_3$$

where

y = Total work-hours lost due to accidents for a 1-year period beginning 6 months after the plan is instituted

x_1 = Total work-hours lost due to accidents during the year before the plan was instituted

$$x_2 = \begin{cases} 1 & \text{if program B is in effect} \\ 0 & \text{otherwise} \end{cases}$$

$$x_3 = \begin{cases} 1 & \text{if program C is in effect} \\ 0 & \text{otherwise} \end{cases}$$

After the programs have been in effect for 18 months, the complete model is fit to the $n = 9$ data points, with the result

$$\hat{y} = -2.1 + .88x_1 - 150x_2 + 35x_3 \qquad \text{SSE} = 1{,}527.27$$

Then the reduced model $E(y) = \beta_0 + \beta_1 x_1$ is fit, with the result

$$\hat{y} = 15.3 + .84x_1 \qquad \text{SSE} = 3{,}113.14$$

Test to determine whether the mean work-hours lost differ for the three programs. Use $\alpha = .05$.

4.37 A chain of drug stores wants to model mean profit per week, $E(y)$, as a function of three advertising factors: type of design, choice of newspaper, and percent discount offered on sale items. A first proposal is the model

$$E(y) = \beta_0 + \beta_1 x_1 + \beta_2 x_1^2 + \beta_3 x_2 + \beta_4 x_3 + \beta_5 x_2 x_3 + \beta_6 x_1 x_2 + \beta_7 x_1 x_3$$
$$+ \beta_8 x_1 x_2 x_3 + \beta_9 x_1^2 x_2 + \beta_{10} x_1^2 x_3 + \beta_{11} x_1^2 x_2 x_3$$

where

x_1 = Percent discount

$$x_2 = \begin{cases} 1 & \text{if design } D_1 \\ 0 & \text{if design } D_2 \end{cases}$$

$$x_3 = \begin{cases} 1 & \text{if newspaper } N_1 \\ 0 & \text{if newspaper } N_2 \end{cases}$$

a. Specify the parameters that would be involved in a test of the hypothesis, "The design of the advertising and the choice of newspaper have no effect on the mean value of the weekly profits."

b. Refer to part **a**. State the hypothesis that you would make regarding the parameter values.

c. Give the parameters that would be involved in a test of the hypothesis, "The relationship between mean weekly profit and percent discount for each of the combinations of newspaper and design is first-order (i.e., a straight line)."

SECTION 4.12

STEPWISE REGRESSION

The problem of predicting executive salaries was discussed in Section 4.8. Perhaps the biggest problem in building a model to describe executive salaries is choosing the important independent variables to be included in the model. The list of potentially important independent variables is extremely long, and we need some objective method of screening out those that are not important.

The problem of deciding which of a large set of independent variables to include in a model is common. Trying to determine which variables influence the profit of a firm, affect product quality, or are related to the state of the economy are only a few examples.

A systematic approach to building a model with a large number of independent variables is difficult because the interpretation of multivariable interactions and higher-order polynomials (squared terms, cubic terms, and so forth) is tedious. We therefore turn to a screening procedure known as **stepwise regression**.

The most commonly used stepwise regression procedure, available in most popular computer packages, works as follows. The user first identifies the response, y, and the set of potentially important independent variables, x_1, $x_2, \ldots, x_k$, where k will generally be large. (Note that this set of variables could represent both first- and higher-order terms, as well as any interaction terms that might be important information contributors.) The response and independent variables are then entered into the computer, and the stepwise procedure begins.

STEP 1 The computer fits all possible one-variable models of the form

$$E(y) = \beta_0 + \beta_1 x_1$$

to the data. For each model, the test of the null hypothesis

$$H_0: \quad \beta_1 = 0$$

against the alternative hypothesis

$$H_a: \quad \beta_1 \neq 0$$

is conducted using the t- (or the equivalent F-) test for a single β parameter. The independent variable that produces the largest (absolute) t-value is declared the best one variable predictor of y.* Call this independent variable x_1.

STEP 2 The stepwise program now begins to search through the remaining $(k - 1)$ independent variables for the best two-variable model of the form

$$E(y) = \beta_0 + \beta_1 x_1 + \beta_2 x_i$$

This is done by fitting all two-variable models containing x_1 and each of the other $(k - 1)$ options for the second variable x_i. The t-values for the test H_0: $\beta_2 = 0$ are computed for each of the $(k - 1)$ models (corresponding to the remaining independent variables x_i, $i = 2, 3, \ldots, k$), and the variable having the largest t is retained. Call this variable x_2.

At this point, some computer packages diverge in methodology. The better packages now go back and check the t-value of $\hat{\beta}_1$ after $\hat{\beta}_2 x_2$ has been added to the model. If the t-value has become nonsignificant at some specified α level (say $\alpha = .10$), the variable x_1 is removed and a search is made for the independent variable with a β parameter that will yield the most significant t-value in the presence of $\hat{\beta}_2 x_2$. Other packages do not recheck $\hat{\beta}_1$, but proceed directly to step 3.

The best-fitting model may yield a different value for $\hat{\beta}_1$ than that obtained in step 1, because $\hat{\beta}_1$ and $\hat{\beta}_2$ may be correlated. Thus, both the value of $\hat{\beta}_1$

*Note that the variable with the largest t-value will also be the one with the largest Pearson product moment correlation, r (Section 3.7), with y.

and, therefore, its significance will usually change from step 1 to step 2. For this reason, the computer packages that recheck the t-values at each step are preferred.

STEP 3 The stepwise procedure now checks for a third independent variable to include in the model with x_1 and x_2. That is, we seek the best model of the form

$$E(y) = \beta_0 + \beta_1 x_1 + \beta_2 x_2 + \beta_3 x_i$$

To do this, we fit all the $(k - 2)$ models using x_1, x_2, and each of the $(k - 2)$ remaining variables, x_i, as a possible x_3. The criterion is again to include the independent variable with the largest t-value. Call this best third variable x_3.

The better programs now recheck the t-values corresponding to the x_1 and x_2 coefficients, replacing the variables that have t-values that have become nonsignificant. This procedure is continued until no further independent variables can be found that yield significant t-values (at the specified α level) in the presence of the variables already in the model.

The result of the stepwise procedure is a model containing only the main effects with t-values that are significant at the specified α level. Thus, in most practical situations, only several of the large number of independent variables will remain. However, it is very important *not* to jump to the conclusion that all the independent variables important for predicting y have been identified or that the unimportant independent variables have been eliminated. Remember, the stepwise procedure is using only *sample estimates* of the true model coefficients (β's) to select the important variables. An extremely large number of single β parameter t-tests have been conducted, and the probability is very high that one or more errors have been made in including or excluding variables. That is, we have very probably included some unimportant independent variables in the model (Type I errors) and eliminated some important ones (Type II errors).

There is a second reason why we might not have arrived at a good model. When we choose the variables to be included in the stepwise regression, we may often omit high-order terms (to keep the number of variables manageable). Consequently, we may have initially omitted several important terms from the model. Thus, we should recognize stepwise regression for what it is—an objective screening procedure.

Now, we will consider interactions and quadratic terms (for quantitative variables) among variables screened by the stepwise procedure. It would be best to develop this response surface model with a second set of data independent of that used for the screening, so the results of the stepwise procedure can be partially verified with new data. However, this is not always possible, because in many business modeling situations only a small amount of data is available.

Remember, do not be deceived by the impressive-looking t-values that result from the stepwise procedure—it has retained only the independent variables with the largest t-values. Also, be certain to consider second-order terms in systematically developing the prediction model. The first-order model given by the stepwise procedure may be greatly improved by the addition of interaction and quadratic terms.

EXAMPLE 4.7

In Section 4.8 we fit a multiple regression model for executive salaries as a function of experience, education, sex, etc. A preliminary step in the construction of this model was the determination of the most important independent variables. Ten independent variables (seven quantitative and three qualitative) were considered, as shown in Table 4.6. It would be very difficult to construct a second-order model with ten independent variables. Therefore, use the sample of 100 executives from Section 4.8 to decide which of the ten variables should be included in the construction of the final model for executive salaries.

TABLE 4.6

Independent Variables in the Executive Salary Example

INDEPENDENT VARIABLE	DESCRIPTION
x_1	Experience (years)—quantitative
x_2	Education (years)—quantitative
x_3	Sex (1 if male, 0 if female)—qualitative
x_4	Number of employees supervised—quantitative
x_5	Corporate assets (millions of dollars)—quantitative
x_6	Board member (1 if yes, 0 if no)—qualitative
x_7	Age (years)—quantitative
x_8	Company profits (past 12 months, millions of dollars)—quantitative
x_9	Has international responsibility (1 if yes, 0 if no)—qualitative
x_{10}	Company's total sales (past 12 months, millions of dollars)—quantitative

SOLUTION

We will use stepwise regression with the first-order terms of the seven quantitative independent variables and the main effects of the three qualitative independent variables to identify the most important variables. The dependent variable y is the natural logarithm of the executive salaries. The SAS stepwise regression printout is shown in Figure 4.27 (pages 206–208). Note that the first variable included in the model is x_4, number of employees supervised by the executive. At the second step, x_5, corporate assets, enters the model. At the sixth step, x_6, a dummy variable for the qualitative variable board member or not, is brought into the model. However, because the significance (.2295) of the F-statistic (SAS uses the $F = t^2$-statistic in the stepwise procedure rather than the t-statistic) for x_6 is above the preassigned $\alpha = .10$, x_6 is then removed from the model. Thus, at step 7 the procedure indicates that the five-variable model including x_1, x_2, x_3, x_4, and x_5 is best. That is, none of the other independent variables can meet the $\alpha = .10$ criterion for admission to the model.

Thus, in our final modeling effort (Section 4.8) we concentrated on these five independent variables, and determined that several second-order terms were important in the prediction of executive salaries. ■

FIGURE 4.27 Stepwise Regression Computer Printout for Example 4.7

STEP 1
VARIABLE X4 ENTERED   R-SQUARE = 0.42071677   C(P) = 1274.7576

	DF	SUM OF SQUARES	MEAN SQUARE	F	PROB > F
REGRESSION	1	11.46854285	11.46854285	71.17	0.0001
ERROR	98	15.79112802	0.16113396		
TOTAL	99	27.25977087			

	B VALUE	STD ERROR	F	PROB > F
INTERCEPT	10.20077500			
X4 (EMPLOYEES SUPERVISED)	0.00057284	0.00006790	71.17	0.0001

STEP 2
VARIABLE X5 ENTERED   R-SQUARE = 0.78299675   C(P) = 419.4947

	DF	SUM OF SQUARES	MEAN SQUARE	F	PROB > F
REGRESSION	2	21.34431198	10.67215599	175.00	0.0001
ERROR	97	5.91545889	0.06098411		
TOTAL	99	27.25977087			

	B VALUE	STD ERROR	F	PROB > F
INTERCEPT	9.87702903			
X4 (EMPLOYEES SUPERVISED)	0.00058353	0.00004178	195.06	0.0001
X5 (ASSETS)	0.00183730	0.00014438	161.94	0.0001

STEP 3
VARIABLE X1 ENTERED   R-SQUARE = 0.89667614   C(P) = 152.4952

	DF	SUM OF SQUARES	MEAN SQUARE	F	PROB > F
REGRESSION	3	24.44318616	8.14772872	277.71	0.0001
ERROR	96	2.81658471	0.02933942		
TOTAL	99	27.25977087			

	B VALUE	STD ERROR	F	PROB > F
INTERCEPT	9.66449288			
X1 (EXPERIENCE)	0.01870784	0.00182032	105.62	0.0001
X4 (EMPLOYEES SUPERVISED)	0.00055251	0.00002914	359.59	0.0001
X5 (ASSETS)	0.00191195	0.00010041	362.60	0.0001

FIGURE 4.27 (Continued)

STEP 4
VARIABLE X3 ENTERED   R-SQUARE = 0.94815717   C(P) = 32.6757

	DF	SUM OF SQUARES	MEAN SQUARE	F	PROB > F
REGRESSION	4	25.84654710	6.46163678	434.37	0.0001
ERROR	95	1.41322377	0.01487604		
TOTAL	99	27.25977087			

	B VALUE	STD ERROR	F	PROB > F
INTERCEPT	3.40077349			
X1 (EXPERIENCE)	0.02074868	0.00131310	249.68	0.0001
X3 (SEX)	0.30011726	0.03089939	94.34	0.0001
X4 (EMPLOYEES SUPERVISED)	0.00055288	0.00002075	710.15	0.0001
X5 (ASSETS)	0.00190876	0.00007150	712.74	0.0001

STEP 5
VARIABLE X2 ENTERED   R-SQUARE = 0.96039323   C(P) = 5.7215

	DF	SUM OF SQUARES	MEAN SQUARE	F	PROB > F
REGRESSION	5	26.18009940	5.23601988	455.87	0.0001
ERROR	94	1.07967147	0.01148587		
TOTAL	99	27.25977087			

	B VALUE	STD ERROR	F	PROB > F
INTERCEPT	8.85387930			
X1 (EXPERIENCE)	0.02141724	0.00116047	340.61	0.0001
X2 (EDUCATION)	0.03315807	0.00615303	29.04	0.0001
X3 (SEX)	0.31927842	0.02738298	135.95	0.0001
X4 (EMPLOYEES SUPERVISED)	0.00056061	0.00001829	939.84	0.0001
X5 (ASSETS)	0.00193584	0.00006504	943.98	0.0001

(continued)

FIGURE 4.27 (Continued)

STEP 6   VARIABLE X6 ENTERED   R-SQUARE = 0.96100666   C(P) = 6.2699

	DF	SUM OF SQUARES	MEAN SQUARE	F	PROB > F
REGRESSION	6	26.19682148	4.36613691	382.00	0.0001
ERROR	93	1.06294939	0.01142956		
TOTAL	99	27.25977087			

	B VALUE	STD ERROR	F	PROB > F
INTERCEPT	8.87509152			
X1 (EXPERIENCE)	0.02133460	0.00115963	338.48	0.0001
X2 (EDUCATION)	0.03272195	0.00614851	28.32	0.0001
X3 (SEX)	0.31093801	0.02817264	121.81	0.0001
X4 (EMPLOYEES SUPERVISED)	0.00055820	0.00001835	925.32	0.0001
X5 (ASSETS)	0.00193764	0.00006289	949.31	0.0001
X6 (BOARD)	0.03866226	0.03196369	1.46	0.2295

STEP 7   VARIABLE X6 REMOVED   R-SQUARE = 0.96039323   C(P) = 5.7215

	DF	SUM OF SQUARES	MEAN SQUARE	F	PROB > F
REGRESSION	5	26.18009940	5.23601988	455.87	0.0001
ERROR	94	1.07967147	0.01148587		
TOTAL	99	27.25977087			

	B VALUE	STD ERROR	F	PROB > F
INTERCEPT	8.85387930			
X1 (EXPERIENCE)	0.02141724	0.00116047	340.61	0.0001
X2 (EDUCATION)	0.03315807	0.00615303	29.04	0.0001
X3 (SEX)	0.31927842	0.02738298	135.95	0.0001
X4 (EMPLOYEES SUPERVISED)	0.00056061	0.00001829	939.84	0.0001
X5 (ASSETS)	0.00193684	0.00006304	943.98	0.0001

CASE STUDY 4.2

New factors and a lack of knowledge about the importance of factors that affect value continue to complicate the job of the rural appraiser. In order to provide knowledge on the subject, this article reports on and evaluates a study in which multiple linear regression equations were used to evaluate and quantify factors affecting value . . . . It is believed that the findings obtained with these equations, and the relationships they indicate, will be of value to the appraiser.

The authors of this statement, James O. Wise and H. Jackson Dover (1974), use stepwise regression to identify a number of important factors (variables) that can be used to predict rural property values. They obtained their results by analyzing a sample of 105 cases from seven counties in the state of Georgia. A part of their findings are duplicated in Table 4.7. The variable names are listed in the order in which the stepwise regression procedure identified their importance, and the t-values found at each step are given for each variable. Note that both qualitative and quantitative variables have been included. Since each qualitative variable is at two levels, only one main effect term (i.e., dummy variable) could be included in the model for each factor.

TABLE 4.7

Stepwise Regression Analysis of Price per Acre

VARIABLE NAME	t-VALUES	VARIABLE NAME	t-VALUES
Residential land (yes–no)	10.466	Branches or springs (yes–no)	3.855
Seedlings and saplings (number)	6.692	Site index (ratio)	3.160
Percent ponds (percent)	4.141	Size (acres)	1.142
Distance to state park (miles)	3.985	Farm land (yes–no)	2.288

Since there were 105 cases used in the study, a large number of degrees of freedom are associated with each t-statistic (first 103, then 102, etc.). Thus, we can compare the value of the test statistic to a corresponding z-value (1.645 for $\alpha = .10$ and the two-sided alternative $H_a: \beta_i \neq 0$) when we judge the importance of each variable. Although Wise and Dover imply that the size variable is important, we might not include it, since the t-value is only 1.142.

Finally, we have now only isolated a set of important variables. We still must decide exactly how each should be entered into our prediction equation, remembering that interactions and quadratic terms often improve the model.

There are several other stepwise regression techniques designed to select the "most important" independent variables. One of these, called **forward selection**, is nearly identical to the stepwise procedure outlined above. The only difference is that the forward selection technique provides no option for rechecking the t-values corresponding to the x's that have entered the model in an earlier step. Thus, stepwise regression is preferred to forward selection in practice.

Another technique, called **backward elimination**, initially fits a model containing terms for all potential independent variables. That is, for k independent variables, the model $E(y) = \beta_0 + \beta_1 x_1 + \beta_2 x_2 + \cdots + \beta_k x_k$ is fit in step 1. The variable with the smallest t (or F-) statistic for testing $H_0: \beta_i = 0$ is identified and dropped from the model if the t-value is less than some specified critical value. The model with the remaining $(k-1)$ independent variables is fit in step

2, and again, the variable associated with the smallest nonsignificant t-value is dropped. This process is repeated until no further nonsignificant independent variables can be found. The real disadvantage of using the backward elimination technique is that you need a sufficiently large number of data points to fit the initial model in step 1.

| | | | | | | | | | | |

EXERCISE 4.38

4.38 In any production process in which one or more workers are engaged in a variety of tasks, the total time spent in production varies as a function of the size of the workpool and the level of output of the various activities. For example, in a large metropolitan department store, the number of hours worked (y) per day by the clerical staff may depend on the following variables:

x_1 = Number of pieces of mail processed (open, sort, etc.)

x_2 = Number of money orders and gift certificates sold

x_3 = Number of window payments (customer charge accounts) transacted

x_4 = Number of change order transactions processed

x_5 = Number of checks cashed

x_6 = Number of pieces of miscellaneous mail processed on an "as available" basis

x_7 = Number of bus tickets sold

The accompanying table of observations gives the output counts for these activities on each of 52 working days.

OBS.	DAY OF WEEK	y	x_1	x_2	x_3	x_4	x_5	x_6	x_7
1	M	128.5	7781	100	886	235	644	56	737
2	T	113.6	7004	110	962	388	589	57	1029
3	W	146.6	7267	61	1342	398	1081	59	830
4	Th	124.3	2129	102	1153	457	891	57	1468
5	F	100.4	4878	45	803	577	537	49	335
6	S	119.2	3999	144	1127	345	563	64	918
7	M	109.5	11777	123	627	326	402	60	335
8	T	128.5	5764	78	748	161	495	57	962
9	W	131.2	7392	172	876	219	823	62	665
10	Th	112.2	8100	126	685	287	555	86	577
11	F	95.4	4736	115	436	235	456	38	214
12	S	124.6	4337	110	899	127	573	73	484
13	M	103.7	3079	96	570	180	428	59	456
14	T	103.6	7273	51	826	118	463	53	907
15	W	133.2	4091	116	1060	206	961	67	951
16	Th	111.4	3390	70	957	284	745	77	1446
17	F	97.7	6319	58	559	220	539	41	440
18	S	132.1	7447	83	1050	174	553	63	1133
19	M	135.9	7100	80	568	124	428	55	456
20	T	131.3	8035	115	709	174	498	78	968
21	W	150.4	5579	83	568	223	683	79	660
22	Th	124.9	4338	78	900	115	556	84	555
23	F	97.0	6895	18	442	118	479	41	203

OBS.	DAY OF WEEK	y	x_1	x_2	x_3	x_4	x_5	x_6	x_7
24	S	114.1	3629	133	644	155	505	57	781
25	M	88.3	5149	92	389	124	405	59	236
26	T	117.6	5241	110	612	222	477	55	616
27	W	128.2	2917	69	1057	378	970	80	1210
28	Th	138.8	4390	70	974	195	1027	81	1452
29	F	109.5	4957	24	783	358	893	51	616
30	S	118.9	7099	130	1419	374	609	62	957
31	M	122.2	7337	128	1137	238	461	51	968
32	T	142.8	8301	115	946	191	771	74	719
33	W	133.9	4889	86	750	214	513	69	489
34	Th	100.2	6308	81	461	132	430	49	341
35	F	116.8	6908	145	864	164	549	57	902
36	S	97.3	5345	116	604	127	360	48	126
37	M	98.0	6994	59	714	107	473	53	726
38	T	136.5	6781	78	917	171	805	74	1100
39	W	111.7	3142	106	809	335	702	70	1721
40	Th	98.6	5738	27	546	126	455	52	502
41	F	116.2	4931	174	891	129	481	71	737
42	S	108.9	6501	69	643	129	334	47	473
43	M	120.6	5678	94	828	107	384	52	1083
44	T	131.8	4619	100	777	164	834	67	841
45	W	112.4	1832	124	626	158	571	71	627
46	Th	92.5	5445	52	432	121	458	42	313
47	F	120.0	4123	84	432	153	544	42	654
48	S	112.2	5884	89	1061	100	391	31	280
49	M	113.0	5505	45	562	84	444	36	814
50	T	138.7	2882	94	601	139	799	44	907
51	W	122.1	2395	89	637	201	747	30	1666
52	Th	86.6	6847	14	810	230	547	40	614

Source: Adapted from *Work Measurement*, by G. L. Smith, Grid Publishing Co., Columbus, Ohio, 1978 (Table 3-1).

a. Conduct a stepwise regression analysis of the data using an available statistical computer program package.

b. Interpret the β estimates in the resulting stepwise model.

c. What are the dangers associated with drawing inferences from the stepwise model?

S E C T I O N 4.13

**OTHER VARIABLE
SELECTION
TECHNIQUES
(OPTIONAL)**

In Section 4.12, we presented stepwise regression as an objective screening procedure for selecting the most important predictors of y. Other, more subjective, variable selection techniques have been developed in the literature for the purpose of identifying important independent variables. The most popular of these procedures are those that consider all possible regression models given the set of potentially important predictors. (Such a procedure is commonly known as an **all-possible-regressions selection procedure**.) The techniques differ with respect to the criteria for selecting the "best" subset of variables. In this section we describe three criteria widely used in practice, then give an example illustrating the three techniques.

R^2 CRITERION

Consider the set of potentially important variables, $x_1, x_2, x_3, \ldots, x_k$. We learned in Section 4.6 that the multiple coefficient of determination,

$$R^2 = 1 - \frac{\text{SSE}}{\text{SS(Total)}}$$

will increase when independent variables are added to the model. Therefore, the model that includes all k independent variables

$$E(y) = \beta_0 + \beta_1 x_1 + \beta_2 x_2 + \cdots + \beta_k x_k$$

will yield the largest R^2. Yet, we have seen examples (in Section 4.11) where adding terms to the model does not yield a significantly better prediction equation. The objective of the R^2 criterion is to find a subset model (i.e., a model containing a subset of the k independent variables) so that adding more variables to the model will yield only small increases in R^2. In practice, the best model found by the R^2 criterion will rarely be the model with the largest R^2. Generally, you are looking for a simple model that is as good as, or nearly as good as, the model with all k independent variables. Unlike stepwise regression, however, the decision about when to stop adding variables to the model is a subjective one.

ADJUSTED R^2 OR MSE CRITERION

One drawback to using the R^2 criterion is that the value of R^2 does not account for the number of β parameters in the model. If enough variables are added to the model so that the sample size n equals the total number of β's in the model, you will force R^2 to equal 1. As an alternative to the R^2 criterion, the **adjusted multiple coefficient of determination**, R_a^2, is often used, where

$$R_a^2 = 1 - \left(\frac{\text{df(Total)}}{\text{df(Error)}}\right)\frac{\text{SSE}}{\text{SS(Total)}} = 1 - (n-1)\frac{\text{MSE}}{\text{SS(Total)}}$$

Note that R_a^2 increases only if MSE decreases [since SS(Total) remains constant for all models]. Thus, an equivalent procedure is to search for the model with the minimum, or near minimum, MSE.

C_p CRITERION

A recently developed procedure, which is gaining increasing acceptance among regression analysts, is based on a quantity called the **total mean square error (TMSE)** for the fitted regression model, where

$$\text{TMSE} = E\left\{\sum_{i=1}^{n} [y_i - E(y_i)]^2\right\}$$

The objective is to compare the TMSE for a regression model with σ^2, the variance of the random error for the true model, using the ratio

$$\Gamma = \frac{\text{TMSE}}{\sigma^2}$$

Small values of Γ imply that the subset regression model has a small total mean square error relative to σ^2. Unfortunately, both TMSE and σ^2 are unknown and we must rely on sample estimates of these quantities. It can be shown (proof omitted) that an estimator of the ratio Γ is given by

$$C_p = \frac{SSE_p}{MSE_k} + 2(p+1) - n$$

where n is the sample size, p is the number of independent variables in the subset model, k is the total number of potential independent variables, SSE_p is the SSE for the subset model, and MSE_k is the MSE for the model containing all k independent variables. In their latest releases, the statistical computer packages discussed in this text have routines that calculate the C_p statistic. In fact, the C_p value is now automatically printed at each step in the SAS stepwise regression printout (see Figure 4.27 on pages 206–208).

The C_p criterion selects as the best model the subset model with (1) a small value of C_p (i.e., a small total mean square error) and (2) a value of C_p near $p + 1$, a property that indicates that slight or no bias exists in the subset regression model.*

Thus, the C_p criterion focuses on minimizing total mean square error and the regression bias. If you are mainly concerned with minimizing total mean square error, you will want to choose the model with the smallest C_p-value, as long as the bias is not large. On the other hand, you may prefer a model that yields a C_p-value slightly larger than the minimum but which has slight (or no) bias.

Plots aid in the selection of the "best" subset regression model using the all-possible-regressions procedure. The criterion measure, either R^2, MSE, or C_p, is plotted on the vertical axis against p, the number of independent variables in the subset model, on the horizontal axis. We illustrate all three variable selection techniques in Example 4.8.

EXAMPLE 4.8

In Example 4.7 we applied stepwise regression to identify the most important variables for predicting executive salary from the list of ten variables given in Table 4.6. Since it is impractical to fit all possible regression models involving subsets of these independent variables (for $k = 10$, there exist 1,023 possible subset models), for this example we fit only the first-order models shown in Table 4.8 (page 214). These models were selected based, in part, on the results of the stepwise regression. The values of R^2, MSE, and C_p are calculated for each model and appear in the appropriate column in Table 4.8. Plot each of these three quantities against p, the number of predictors in the subset model. Interpret the plots.

---

*A model is said to be *unbiased* if $E(\hat{y}) = E(y)$. We state (without proof) that for an unbiased regression model, $E(C_p) \approx p + 1$. In general, subset models will be biased since $k - p$ independent variables are omitted from the fitted model. However, when C_p is near $p + 1$, the bias is small and can essentially be ignored.

TABLE 4.8 Subset Models for the Executive Salary Example

NUMBER OF PREDICTORS, p	VARIABLES IN THE MODEL	R^2	MSE	C_p
1	x_4	.421	.1611	1,339,6
2	x_4, x_5	.783	.0610	443.8
3	x_1, x_4, x_5	.897	.0293	164.1
4	x_1, x_3, x_4, x_5	.948	.0149	38.5
5	x_1, x_2, x_3, x_4, x_5	.960	.0115	10.2
6	$x_1, x_2, x_3, x_4, x_5, x_8$	.962	.0111	8.2
7	$x_1, x_2, x_4, x_5, x_8, x_9, x_{10}$	.963	.0110	7.7
8	$x_1, x_2, x_3, x_4, x_5, x_8, x_9, x_{10}$	.963	.0111	9.7
9	$x_1, x_2, x_3, x_4, x_5, x_6, x_8, x_9, x_{10}$	.964	.0109	9.2
10	$x_1, x_2, x_3, x_4, x_5, x_6, x_7, x_8, x_9, x_{10}$	.964	.0110	11.2

SOLUTION

Plots of R^2, MSE, and C_p against p are shown in Figures 4.28(a)–(c), respectively. Visually examining Figure 4.28(a), we see that the R^2-values tend to increase in very small amounts for models with more than $p = 4$ predictors. Similarly, Figure 4.28(b) shows that the MSE decreases by negligible amounts for models with more than $p = 4$ predictors. Thus, both the R^2 and MSE criteria suggest that the model containing the four predictors x_1, x_3, x_4, and x_5 is a good candidate for the "best" subset regression model.

Figure 4.28(c) shows the plotted C_p-values and the line $C_p = p + 1$. Notice that the subset models with $p \geq 5$ independent variables all have relatively small C_p-values and vary tightly about the line $C_p = p + 1$. This implies that these models have a small total mean square error and a negligible bias. The model corresponding to $p = 4$, while certainly outperforming the models with $p \leq 3$,

FIGURE 4.28 Plots of R^2, MSE, and C_p for Subset Regression Models of Example 4.8

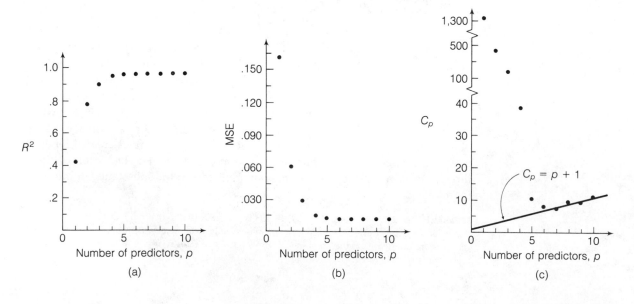

appears to fall short of the larger models according to the C_p criterion. The C_p-value for this model is $C_p = 38.5$ (compared to C_p-values ranging from 7.7 to 11.2 for models with $p \geq 5$) and the model has a slight bias.

According to all three criteria, the variables $x_1, x_3, x_4,$ and x_5 should be included in the group of the most important predictors. The decision of whether to include another variable (e.g., x_2) in this group is left to the data analyst.   ■

In summary, variable selection procedures based on the R^2, MSE, or C_p criterion will assist you in identifying the most important independent variables for predicting y. Keep in mind, however, that these techniques lack the objectivity of a stepwise regression procedure. Furthermore, you should be wary of concluding that the best model for predicting y has been found, since, in practice, interactions and higher-order terms are typically omitted from the list of potential important predictors.

| | | | | | | | | | | | | |

EXERCISE 4.39

4.39 Refer to the data on units of production and time worked for a department store clerical staff in Exercise 4.38. For this exercise, consider only the independent variables $x_1, x_2, x_3,$ and x_4 in an all-possible-regressions selection procedure.
   a. How many models for $E(y)$ are possible, if the model includes (i) 1 variable (ii) 2 variables (iii) 3 variables (iv) 4 variables?
   b. For each case in part **a**, use a statistical computer program package to find the maximum R^2, minimum MSE, and minimum C_p.
   c. Plot each of the quantities R^2, MSE, and C_p in part **b** against p, the number of predictors in the subset model
   d. Based on the plots in part **c**, which variables would you select for predicting total hours worked, y?

| | | | | | | | | | | | | |

S E C T I O N   4.14

A SUMMARY OF THE STEPS TO FOLLOW IN A MULTIPLE REGRESSION ANALYSIS

We have discussed some of the methodology of **multiple regression analysis**, a technique for modeling a dependent variable y as a function of several independent variables $x_1, x_2, \ldots, x_k$. The steps we follow in constructing and using multiple regression models, which are much the same as those for the simple straight-line models, are listed here:

1. **The set of independent variables to be included in the model is identified.** If the number of independent variables is large, you may want to use a variable selection technique such as **stepwise regression** to screen out those that do not seem important for the prediction of y.
2. **The form of the probabilistic model is hypothesized.** The model may include **second-order** terms for quantitative variables, **interaction** terms, and **dummy variables** for qualitative variables. Remember that a model with no interaction terms implies that each of the independent variables affects the response y independently of the other independent variables. **Quadratic** (or **second-order**) terms add curvature to the response curve when $E(y)$ is plotted as a function of the quantitative independent variable. Dummy variables allow the mean response to differ for the different levels of the qualitative variable.

3. The model coefficients are estimated using the method of least squares.

4. The probability distribution of ε is specified and σ^2 is estimated.

5. The utility of the model is checked using the analysis of variance F-test and the multiple coefficient of determination R^2. The F-test for testing a partial set of β parameters and t-tests on individual β parameters aid in deciding the final form of the model.

6. If the model is deemed useful, it may be used to make estimates of $E(y)$ and to predict future values of y.

Subsequent chapters extend the methods of this chapter to special applications and problems encountered during a regression analysis. One of the most common problems faced by regression analysts is the problem of multicollinearity, i.e., intercorrelations among the independent variables. Methods for detecting and overcoming multicollinearity, as well as other problems, are discussed in Chapter 5. Another aspect of regression analysis is the analysis of the residuals, i.e., the deviations between the observed and the predicted values of y. An analysis of residuals (Chapter 6) may indicate that the data do not comply with the assumptions of Section 4.2 and may suggest appropriate procedures for modifying the data analysis.

SUPPLEMENTARY EXERCISES 4.40–4.55

4.40 After a regression model is fit to a set of data, a confidence interval for the mean value of y at a given setting of the independent variables will *always* be narrower than the corresponding prediction interval for a particular value of y at the same setting of the independent variables. Why?

4.41 Most homes in the United States are sold through a real estate broker, who receives a commission on each sale. As a future homebuyer, you may want to know the extent to which brokerage commissions are incorporated into the selling price of the home. Stated more simply, is the price paid by a homebuyer higher when property is sold through a real estate broker than when sold directly by an owner? Evidence collected by researchers at Louisiana State University suggests that homebuyers may, in fact, save in costs by purchasing owner-offered properties (*The Real Estate Appraiser and Analyst*, Winter 1985). Data on the sample of 111 transactions were used to fit the model

$$E(y) = \beta_0 + \beta_1 x_1 + \beta_2 x_2 + \beta_3 x_3 + \beta_4 x_4 + \beta_5 x_5 + \beta_6 x_6$$

where

y = Sale price (in dollars)

x_1 = Age of home (years)

x_2 = Living area (square feet)

x_3 = Area (square feet) of other improvements (e.g., garage, porch, etc.)

x_4 = Selling date

$x_5 = \begin{cases} 1 & \text{if home was conventionally financed} \\ 0 & \text{if not} \end{cases}$

$$x_6 = \begin{cases} \text{Brokerage commission (in dollars)} & \text{if home listed with real estate broker} \\ 0 & \text{if home not listed with real estate broker} \end{cases}$$

The regression results are summarized here. (The standard errors of the β estimates are given in parentheses.)

$$\hat{y} = 10.67 - 725x_1 + 27.50x_2 + 10.15x_3 + 470.12x_4 - 1{,}450x_5 + .43x_6 \qquad R^2 = .86$$
$$\phantom{\hat{y} = 10.67} (120.8) \quad (1.73) \qquad (2.67) \qquad (940.24) \quad (1{,}208.3) \quad (.25)$$

a. Is the model adequate for predicting the sale price of a home? Test using $\alpha = .10$.

b. Calculate a 90% confidence interval for β_1. Interpret your result.

c. Interpret the estimate of β_5.

d. Test the hypothesis that a brokerage commission increases the mean selling price of a home. Test using $\alpha = .10$.

4.42 Most companies institute rigorous safety programs in order to assure employee safety. Suppose accident reports over the last year at a company are sampled, and the number of hours the employee had worked before the accident occurred, x, and the amount of time the employee lost from work, y, are recorded. A second-order model is proposed to investigate a fatigue hypothesis that more serious accidents occur near the end of workdays than near the beginning. Thus, the proposed model is

$$E(y) = \beta_0 + \beta_1 x + \beta_2 x^2$$

A total of 60 accident reports are examined and part of the computer printout appears as shown here.

Printout for Exercise 4.42

SOURCE	DF	SUM OF SQUARES	MEAN SQUARE	F VALUE
MODEL	2	112.110	56.055	1.28
ERROR	57	2496.201	43.793	R-SQUARE
TOTAL	59	2608.311		.0430

a. Do the data support the fatigue hypothesis? Use $\alpha = .05$ to test this hypothesis by testing whether the proposed model is useful in predicting the lost work time, y.

b. Does the result of the test in part **a** necessarily mean that no fatigue factor exists? Explain.

4.43 Refer to Exercise 4.42. Suppose the company persists in using the second-order model despite its apparent lack of utility. The fitted model is

$$\hat{y} = 12.3 + .25x - .0033x^2$$

where $\hat{y}$ is the predicted time lost (days) and x is the number of hours worked prior to an accident.

a. Use the model to predict the number of days missed by an employee who has an accident after 6 hours of work.

b. Suppose the 95% confidence interval for the estimated mean in part **a** is calculated to be (1.35, 26.01). What is the interpretation of this interval? Does this interval reconfirm your conclusion about this model in Exercise 4.42?

4.44 A company that relies on door-to-door sales wants to determine the relationship, if any, between the proportion of customers who buy its product, y, and two independent variables: price, x_1, and years of experience of the salesperson, x_2. Twenty salespeople employed by the company are randomly assigned to sell the products, five to each of four prices, ranging from $1.98 to $5.98. Each salesperson makes a sales presentation to 30 prospects, and the percentage of sales is recorded. The 20 observations (five salespeople for each of four prices) are used to fit the model

$$y = \beta_0 + \beta_1 x_1 + \beta_2 x_2 + \varepsilon$$

The least squares prediction equation is

$$\hat{y} = -.30 - .010x_1 + .10x_2$$

with $s_{\hat{\beta}_1} = .0030$, $s_{\hat{\beta}_2} = .025$, and $R^2 = .86$.

a. Interpret the values of $\hat{\beta}_1$ and $\hat{\beta}_2$.

b. Test the null hypothesis H_0: $\beta_1 = \beta_2 = 0$ that the model is not useful for predicting y. Use $\alpha = .05$.

c. Do the data support the research hypothesis that as the price of the product is increased the mean proportion of buyers will decrease, when x_2 is held fixed?

d. Is there evidence that as the experience of the salesperson increases the mean proportion of buyers increases, when x_1 is fixed?

e. Suppose it is claimed that the least squares model cannot be correct, since the value of $\hat{\beta}_0 = -.30$, and a negative proportion of buyers is clearly impossible. How do you refute this argument?

4.45 To increase the motivation and productivity of workers, an electronics manufacturer decides to experiment with a new pay incentive structure at one of two plants. The experimental plan will be tried at plant A for 6 months, while workers at plant B will remain on the original pay plan. To evaluate the effectiveness of the new plan, the average assembly time for part of an electronic system was measured for employees at both plants at the beginning and end of the 6-month period. Suppose the following model was proposed:

$$y = \beta_0 + \beta_1 x_1 + \beta_2 x_2 + \varepsilon$$

where

y = Assembly time (hours) at end of 6-month period

x_1 = Assembly time (hours) at beginning of 6-month period

$x_2 = \begin{cases} 1 & \text{if plant A} \\ 0 & \text{if plant B} \end{cases}$ (dummy variable)

A sample of $n = 42$ observations yielded

$$\hat{y} = .11 + .98x_1 - .53x_2$$

where

$s_{\hat{\beta}_1} = .231$

$s_{\hat{\beta}_2} = .48$

Test to see whether, after allowing for the effect of initial assembly time, plant A had a lower mean assembly time than plant B. Use $\alpha = .01$.

4.46 The Environmental Protection Agency wants to model the gas mileage ratings, y, of automobiles as a function of their engine size, x. A second-order model,

$$E(y) = \beta_0 + \beta_1 x + \beta_2 x^2$$

is proposed. A sample of 50 engines of varying size is selected and the miles per gallon rating of each is determined. The least squares prediction equation is

$$\hat{y} = 51.3 - 10.1x + .15x^2$$

The size, x, of the engine is measured in hundreds of cubic inches. Also, $s_{\hat{\beta}_2} = .0037$ and $R^2 = .93$.
a. Sketch this model over the interval $x = 1$ to $x = 4$.
b. Is there evidence that the second-order term in the model is contributing to the prediction of the miles per gallon rating, y? Use $\alpha = .05$.
c. Use the model to estimate the mean miles per gallon rating for all cars with 350-cubic-inch engines ($x = 3.5$).
d. Suppose a 95% confidence interval for the quantity estimated in part c is calculated to be (17.2, 18.4). Interpret this interval.
e. Suppose you purchase an automobile with a 350-cubic-inch engine and determine that the miles per gallon rating is 14.7. Is the fact that this value lies outside the confidence interval given in part d surprising? Explain.

4.47 How satisfied are you with your auto repair service? G. J. Biehal (University of Houston) examined the factors that influence the satisfaction level for auto repair services in California. One objective of the study was to show that the extent of the information search conducted by the customer prior to selecting an auto repair service affects the customer's satisfaction with the service. A sample of 208 households in Palo Alto and Menlo Park, California, were questioned about their most recent auto repair experiences. All respondents reported at least one auto repair expense that exceeded $100 during the previous year. The dependent variable, level of dissatis-faction y, was measured on a 7-point Likert scale (0 = very satisfied to 7 = very dissatisfied). The following linear model was proposed:

$$E(y) = \beta_0 + \beta_1 x_1 + \beta_2 x_2 + \beta_3 x_3 + \beta_4 x_4 + \beta_5 x_5 + \beta_6 x_6$$

where

Level of external information search: x_1 (0 = none to 7 = very high)

Repair cost: x_2 (dollars)

Problem corrected: $x_3 = \begin{cases} 1 & \text{if more than one visit required to correct the problem} \\ 0 & \text{if the problem was corrected on the first visit} \end{cases}$

Prior experience with service:

$x_4 = \begin{cases} 1 & \text{if 1--2 times} \\ 0 & \text{if not} \end{cases}$ $\quad x_5 = \begin{cases} 1 & \text{if 3--5 times} \\ 0 & \text{if not} \end{cases}$ $\quad x_6 = \begin{cases} 1 & \text{if more than 5 times} \\ 0 & \text{if not} \end{cases}$

The results of the multiple regression analysis are summarized in the table.

PARAMETER	ESTIMATE	p-VALUE
β_0	2.14	—
β_1	−.15	$p < .05$
β_2	.03	$p > .05$
β_3	2.54	$p < .01$
β_4	−.34	$p > .05$
β_5	−.26	$p > .05$
β_6	−.72	$p < .05$
$R^2 = .43$		

Source: Biehal, G. J. "Consumers' prior experiences and perceptions in auto repair choice," *Journal of Marketing,* Summer 1983, Vol. 47, No. 3, pp. 82–91. Reprinted by permission of the American Marketing Association.

a. Write the least squares prediction equation.
b. Is the model useful for predicting the dissatisfaction level for auto repair services? Test using $\alpha = .05$.
c. Is there evidence to indicate that the dissatisfaction level for auto repair service declines as the amount of external information search increases? Test using $\alpha = .05$.
d. Is there evidence to indicate that repair cost is a useful predictor of the dissatisfaction level for auto repair service? Test using $\alpha = .05$.
e. What hypothesis would you test to determine whether prior experience with service has no effect on the level of dissatisfaction?

4.48 To determine whether extra personnel are needed for the day, the owners of a water adventure park would like to find a model that would allow them to predict the day's attendance each morning before opening based on the day of the week and weather conditions. The model is of the form

$$E(y) = \beta_0 + \beta_1 x_1 + \beta_2 x_2 + \beta_3 x_3$$

where

y = Daily admissions

$x_1 = \begin{cases} 1 & \text{if weekend} \\ 0 & \text{otherwise} \end{cases}$ (dummy variable)

$x_3 = \begin{cases} 1 & \text{if sunny} \\ 0 & \text{if overcast} \end{cases}$ (dummy variable)

x_3 = Predicted daily high temperature (°F)

After collecting 30 days of data, the following least squares model is obtained:

$$\hat{y} = -105 + 25x_1 + 100x_2 + 10x_3$$

with $s_{\hat{\beta}_1} = 10$, $s_{\hat{\beta}_2} = 30$, and $s_{\hat{\beta}_3} = 4$. Also, $R^2 = .65$.

a. Interpret the model coefficients.

b. Is there sufficient evidence to conclude that this model is useful in the prediction of daily attendance? Use $\alpha = .05$.

c. Is there sufficient evidence to conclude that mean attendance increases on weekends? Use $\alpha = .10$.

d. Use the model to predict the attendance on a sunny weekend with a predicted high temperature of 95°F.

e. Suppose the 90% prediction interval for part d is (645, 1,245). Interpret this interval.

4.49 Refer to Exercise 4.48. The owners of the water adventure park are advised that the prediction model could probably be improved if interaction terms were added. In particular, it is thought that the *rate* of increase in mean attendance with increases in predicted high temperature will be greater on weekends than on weekdays. The following model is therefore proposed:

$$E(y) = \beta_0 + \beta_1 x_1 + \beta_2 x_2 + \beta_3 x_3 + \beta_4 x_1 x_3$$

The same 30 days of data as were used in Exercise 4.48 are again used to obtain the least squares model

$$\hat{y} = 250 - 700 x_1 + 100 x_2 + 5 x_3 + 15 x_1 x_3$$

with $s_{\hat{\beta}_4} = 3.0$ and $R^2 = .96$.

a. Graph the predicted day's attendance, $\hat{y}$, against the day's predicted high temperature, x_3, for a sunny weekday and for a sunny weekend day. Graph both on the same paper for x_3 between 70°F and 100°F. Note the increase in slope for the weekend day.

b. Do the data indicate that the interaction term is a useful addition to the model? Use $\alpha = .05$.

c. Use this model to predict the attendance on a sunny weekday with a predicted high temperature of 95°F.

d. Suppose the 90% prediction interval for part c is (800, 850). Compare this with the prediction interval for the model without interaction in Exercise 4.48, part e. Do the relative widths of the confidence intervals support or refute your conclusion about the utility of the interaction term (part b)?

e. The owners, noting that the coefficient $\hat{\beta}_1 = -700$, conclude the model is ridiculous, because it seems to imply that the mean attendance will be 700 less on weekends than on weekdays. Refute their argument.

4.50 Refer to Exercise 4.49. Suppose the second-order model

$$E(y) = \beta_0 + \beta_1 x_1 + \beta_2 x_2 + \beta_3 x_3 + \beta_4 x_1 x_3 + \beta_5 x_3^2 + \beta_6 x_1 x_3^2$$

is fit to the $n = 30$ observations on daily admissions.

a. What hypothesis would you test to determine whether the quadratic terms for predicted daily high temperature x_3 are important?

b. Use the SSE's for the interaction model (Exercise 4.49) and the second-order model given below to test the hypothesis of part a. Use $\alpha = .05$.

Interaction model: $\text{SSE}_1 = 585,000$
Second-order model: $\text{SSE}_2 = 530,000$

4.51 R. N. Horn (James Madison University) and W. J. McGuire (Eastern Kentucky University) used multiple regression analysis to model academic year salaries for secondary-school teachers in Philadelphia during 1981–1982. Information on a sample of 4,316 secondary-school teachers was obtained from records compiled by the Pennsylvania Department of Education. The total sample was divided by race and sex to form four subsamples: black females, black males, white females, and white males. The data for each subsample were used to fit a multiple regression model of the form

$$E(y) = \beta_0 + \beta_1 x_1 + \beta_2 x_2 + \beta_3 x_3 + \cdots + \beta_{30} x_{30}$$

Years of experience, educational level, age, marital status, and major teaching field were among the 30 independent variables in the model. The results for each subsample are summarized in the table.

	BLACK MALES	BLACK FEMALES	WHITE MALES	WHITE FEMALES
n	526	862	1,848	1,080
R^2	.8074	.7674	.7724	.7879

Source: Horn, R. N. and McGuire, W. J. "Determinants of secondary school teacher salaries in a large urban school district," *Southern Economic Journal*, October 1984, pp. 481–493.

a. Is there sufficient evidence to indicate that the model is useful for predicting academic year salaries for black male secondary-school teachers in Philadelphia? Test using $\alpha = .05$.

b. Is there sufficient evidence to indicate that the model is useful for predicting academic year salaries for black female secondary-school teachers in Philadelphia? Test using $\alpha = .05$.

c. Is there sufficient evidence to indicate that the model is useful for predicting academic year salaries for white male secondary-school teachers in Philadelphia? Test using $\alpha = .05$.

d. Is there sufficient evidence to indicate that the model is useful for predicting academic year salaries for white female secondary-school teachers in Philadelphia? Test using $\alpha = .05$.

4.52 Many students must work part-time to help finance their college education. A survey of 100 students was completed at a university to see if the number of hours worked per week, x, was affecting their grade-point averages, y. A second-order model was proposed:

$$y = \beta_0 + \beta_1 x + \beta_2 x^2 + \varepsilon$$

The 100 observations yielded the least squares model

$$\hat{y} = 2.8 - .005x - .0002x^2$$

with $R^2 = .12$.

a. Do these statistics indicate that the model is useful in explaining grade-point averages? Use $\alpha = .05$.

b. Interpret the value $R^2 = .12$. Do you think the relationship between x and y is strong? Would you expect to be able to predict grade-point averages precisely (narrow prediction interval) if you knew how many hours students work per week?

4.53 Recent increases in gasoline prices have increased interest in modes of transportation other than the automobile. A metropolitan bus company wants to know if changes in numbers of bus riders are related to changes in gasoline prices. By using information contained in the company files and gasoline price information obtained from fuel distributors, the company planned to fit the following model:

$$y = \beta_0 + \beta_1 x_1 + \beta_2 x_2 + \beta_3 x_1 x_2 + \varepsilon$$

where

x_1 = Average wholesale price for regular gas in a given month

$x_2 = \begin{cases} 1 & \text{if the bus travels a city route only} \\ 0 & \text{if the bus travels a suburb–city route} \end{cases}$

y = Total number of riders in a bus over the month

a. For the above model, how would you test to determine whether the relationship between gasoline price and the mean number of riders is different for the two different types of bus routes?

b. Suppose 12 months of data are kept, and the least squares model is

$$\hat{y} = 500 + 50x_1 + 5x_2 - 10x_1 x_2$$

Graph the predicted relationship between number of riders and gas price for city buses and for suburb–city buses. Compare the slopes.

c. If $s_{\hat{\beta}_3} = 3.0$, do the data indicate that gas price affects the number of riders differently for city and suburb–city buses? Use $\alpha = .05$.

4.54 A naval base is considering modifying or adding to its fleet of 48 standard aircraft. The final decision regarding the type and number of aircraft to be added depends on a comparison of cost versus effectiveness of the modified fleet. Consequently, the naval base would like to model the projected percentage increase y in fleet effectiveness by the end of the decade as a function of the cost x of modifying the fleet. A first proposal is the quadratic model

$$E(y) = \beta_0 + \beta_1 x + \beta_2 x^2$$

The data provided in the table were collected on ten naval bases of a similar size that recently expanded their fleets. The data were used to fit the model, and the SAS printout of the multiple regression analysis is also reproduced on page 224.

PERCENTAGE IMPROVEMENT AT END OF DECADE y	COST OF MODIFYING FLEET x, millions of dollars
18	125
32	160
9	80
37	162
6	110
3	90
30	140
10	85
25	150
2	50

SAS Printout for Exercise 4.54

ANALYSIS OF VARIANCE

SOURCE	DF	SUM OF SQUARES	MEAN SQUARE	F VALUE	PROB>F
MODEL	2	1368.77501	684.38750	33.079	0.0003
ERROR	7	144.82499	20.68928481		
C TOTAL	9	1513.60000			

ROOT MSE	4.548548	R-SQUARE	0.9043
DEP MEAN	17.2	ADJ R-SQ	0.8770
C.V.	26.44504		

PARAMETER ESTIMATES

VARIABLE	DF	PARAMETER ESTIMATE	STANDARD ERROR	T FOR H0: PARAMETER=0	PROB > ∣T∣
INTERCEP	1	10.65903604	14.55009061	0.733	0.4876
X	1	-0.28160568	0.28087588	-1.003	0.3494
XX	1	0.002671936	0.001253832	2.131	0.0706

a. Interpret the value of R^2 on the printout.

b. Find the value of s and interpret it.

c. Perform a test of overall model adequacy. Use $\alpha = .05$.

d. Is there sufficient evidence to conclude that the percentage improvement y increases more quickly for more costly fleet modifications than for less costly fleet modifications? Test with $\alpha = .05$.

e. Now consider the model

$$E(y) = \beta_0 + \beta_1 x_1 + \beta_2 x_1^2 + \beta_3 x_2 + \beta_4 x_1 x_2$$

where

x_1 = Cost of modifying the fleet

$x_2 = \begin{cases} 1 & \text{if American base} \\ 0 & \text{if foreign base} \end{cases}$

The model is fit to the $n = 10$ data points and results in SSE = 97.645. Is there sufficient evidence to indicate that type of base (American or foreign) is a useful predictor of percentage improvement y? Test using $\alpha = .05$.

4.55 [*Note:* You will need a computer for this exercise.] Refer to the data given in Appendix E. Let

y = Price of residential property

x_1 = Appraised land value

x_2 = Appraised improvement value

a. Fit the model

$$E(y) = \beta_0 + \beta_1 x_1 + \beta_2 x_2 + \beta_3 x_1 x_2 + \beta_4 x_1^2 + \beta_5 x_2^2$$

to the data for neighborhoods A and B.

b. Does the model provide sufficient information for the prediction of sale price?

c. Find R^2 and interpret its value.

d. Now fit the model

$$E(y) = \beta_0 + \beta_1 x_1 + \beta_2 x_2 + \beta_3 x_1 x_2 + \beta_4 x_1^2 + \beta_5 x_2^2 + \beta_6 x_3 + \beta_7 x_1 x_3 + \beta_8 x_2 x_3$$
$$+ \beta_9 x_1 x_2 x_3 + \beta_{10} x_1^2 x_3 + \beta_{11} x_2^2 x_3$$

where

$$x_3 = \begin{cases} 1 & \text{if neighborhood A} \\ 0 & \text{if neighborhood B} \end{cases}$$

e. Is there evidence that the relationship between sale price and appraised value differs for the two neighborhoods? Test using $\alpha = .05$.

ON YOUR OWN . . .

[*Note:* The use of a computer is required for this study.]

This is a continuation of the "On Your Own" presented in Chapter 3, in which you selected three independent variables as predictors of the Gross National Product and obtained 10 years of data for each. Now fit the multiple regression model (use an available computer package, if possible)

$$y = \beta_0 + \beta_1 x_1 + \beta_2 x_2 + \beta_3 x_3 + \varepsilon$$

where

y = Gross National Product

x_1 = First variable you chose

x_2 = Second variable you chose

x_3 = Third variable you chose

a. Compare the coefficients $\hat{\beta}_1$, $\hat{\beta}_2$, and $\hat{\beta}_3$ to their corresponding slope coefficients in the Chapter 3 "On Your Own," where you fit three separate straight-line models. How do you account for the differences?

b. Calculate the coefficient of determination R^2, and conduct the F-test of the null hypothesis $H_0: \beta_1 = \beta_2 = \beta_3 = 0$. What is your conclusion?

c. Now, increase your list of three variables to include approximately ten that you feel would be useful in predicting the GNP. Obtain data for as many years as possible for the new list of variables and the GNP. With the aid of a computer analysis package, use a stepwise regression program to choose the important variables among those you have listed. To test your intuition, list the variables in the order you think they will be selected before you conduct the analysis. How does your list compare with the stepwise regression results?

d. After the group of ten variables has been narrowed to a smaller group of variables by the stepwise analysis, try to improve the model by including interactions and quadratic terms. Be sure to consider the meaning of each interaction or quadratic term before adding it to the model—a quick sketch can be very helpful. See if you can systematically construct a useful model for predicting the GNP. You might want to hold out the last several years of data to test the predictive ability of your model after it is constructed. (As noted in Section 4.11, using the same data to construct *and* to evaluate predictive ability can lead to invalid statistical tests and a false sense of security.) If the independent variables you chose are themselves highly correlated, you may encounter some results that are difficult to explain. For example, the coefficients $\hat{\beta}_1$, $\hat{\beta}_2$, and $\hat{\beta}_3$ in the first-order model may assume signs that run counter to what you expected. Or you may get a highly significant F-value in part **b**, but the individual t-statistics for x_1, x_2, and x_3 may all be nonsignificant. This phenomenon—a high correlation between the independent variables in a regression model—is known as **multicollinearity**. We discuss remedial measures for multicollinearity in Chapter 5.

REFERENCES

Baltagi, B. H. and Levin, D. "Estimating dynamic demand for cigarettes using panel data: The effects of bootlegging, taxation, and advertising reconsidered," *The Review of Economics and Statistics*, Vol. 68, No. 1, February 1986, pp. 148–155.

Blau, G. J. "The relationship of management level to effort level, direction of effort, and managerial performance," *Journal of Vocational Behavior*, Vol. 29, October 1986, pp. 226–239.

Chou, Ya-lun. *Statistical Analysis with Business and Economic Applications*, 2nd ed. New York: Holt, Rinehart, and Winston, 1975. Chapter 18.

Draper, N. and Smith, H. *Applied Regression Analysis*, 2nd ed. New York: Wiley, 1981.

Hamburg, M. *Statistical Analysis for Decision Making*, 2nd ed. New York: Harcourt Brace Jovanovich, 1977. Chapter 9.

Hergert, M. "Market share and profitability: Is bigger really better?" *Business Economics*, October 1984, pp. 45–48.

Kearl, J. R. and Pope, C. L. "Mobility and distribution." *The Review of Economics and Statistics*, May 1984, Vol. 66, No. 2, pp. 192–199.

Miller, R. B. and Wichern, D. W. *Intermediate Business Statistics: Analysis of Variance, Regression, and Time Series*. New York: Holt, Rinehart, and Winston, 1977. Chapters 6–8.

Neter, J., Wasserman, W., and Kutner, M. H. *Applied Linear Statistical Models*. Homewood, Ill.: Richard D. Irwin, 1985. Chapters 7–10.

Phillips, B. D., Lakhani, H. A. G., and George, S. L. "The economics of metric conversion for small manufacturing firms in the U.S.," *Technological Forecasting and Social Change*, April 1984, Vol. 25, No. 2, pp. 109–121.

Weisberg, S. *Applied Linear Regression*. New York: Wiley, 1980.

Winkler, R. L. and Hays, W. L. *Statistics: Probability, Inference, and Decision*, 2nd ed. New York: Holt, Rinehart, and Winston, 1975. Chapter 10.

Wise, J. O. and Dover, H. J. "An evaluation of a statistical method of appraising rural property," *Appraisal Journal*, January 1974, Vol. 42, pp. 103–113.

Younger, M. S. *A First Course in Linear Regression*, 2nd ed. Boston: Duxbury, 1985.

C H A P T E R 5

OBJECTIVE

To identify several potential problems that may be encountered when constructing a model for a response y; to help you recognize when these problems exist so that you can avoid some of the pitfalls of multiple regression analysis

CONTENTS

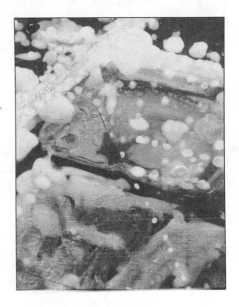

SOME PROBLEMS ENCOUNTERED IN PERFORMING A MULTIPLE REGRESSION ANALYSIS

SECTION 5.1

INTRODUCTION

Multiple regression analysis is recognized by practitioners as a powerful tool for modeling a response y. Because it is so widely used in practice, it is also one of the most abused statistical techniques. The ease with which a multiple regression analysis can be run on the computer has opened the doors to many data analysts who have only a limited knowledge of multiple regression and statistics. In practice, building a model for some response y is rarely a simple, straightforward process. There are a number of pitfalls which trap the unwary analyst. In this chapter we discuss several problems that you should be aware of when constructing a multiple regression model.

SECTION 5.2

OBSERVATIONAL DATA VERSUS DESIGNED EXPERIMENTS

One problem encountered in the use of a regression analysis for business or economic data is caused by the type of data that the analyst is often forced to collect. The sampling situation appropriate for a regression analysis is one in which you have a set of n experimental units and, for each unit, you record a value of y and the corresponding values of the independent (predictor) variables $x_1, x_2, \ldots, x_k$. If the values of $x_1, x_2, \ldots, x_k$ are uncontrolled but are measured without error, the data are said to be **observational data**. For example, suppose you want to relate the selling price y of real estate to a set of predictors and that you randomly select $n = 100$ sales and record the value of y and the observed value for each of the predictor variables. Note that in this experiment you did not specify the x-values, such as location, square footage, etc.; that is, they were uncontrolled. Therefore, the real estate data are observational.

DEFINITION 5.1

If the values of the independent variables (x's) in regression are uncontrolled (i.e., not set in advance before the value of y is observed) but are measured without error, the data are **observational**.

Data that are not observational are generated by **designed experiments** where the values of the independent variables are set in advance before the value of y is observed. For example, if a production supervisor wants to investigate the effect of two quantitative independent variables, say temperature x_1 and pressure x_2, on the purity of batches of a chemical, the supervisor might decide to employ three values of temperature ($100°C$, $125°C$, and $150°C$) and three values of pressure (50, 60, and 70 pounds per square inch) and to produce and measure the impurity y in one batch of chemical for each of the $3 \times 3 = 9$ temperature–pressure combinations. For this experiment, the settings of the independent variables are controlled, in contrast to the uncontrolled nature of observational data in the real estate sales example.

Most data collected in business studies are observational. Whether data are observational is important for the following reasons. First, as you will subsequently learn in Chapter 8, the quantity of information in an experiment is controlled not only by the *amount of data* but also by the *values of the predictor variables* $x_1, x_2, \ldots, x_k$. Consequently, if you can design the experiment (sometimes this is physically impossible), you may be able to increase greatly the amount of information in the data at no additional cost.

Second, the use of observational data creates a problem involving randomization. When an experiment has been designed and we have decided on the various settings of the independent variables to be used, the experimental units are then randomly assigned in such a way that each combination of the independent variables has an equal chance of receiving experimental units with unusually high (or low) readings. (We will illustrate this method of randomization in Chapter 8.) This procedure tends to average out any variation within the experimental units. The result is that if the difference between two sample means is statistically significant, then you can infer (with probability of Type I error equal to α) that the population means differ. But more important, you can infer that this difference was due to the settings of the predictor variables, which is what you did to make the two populations different. Thus, you can infer a cause-and-effect relationship.

If the data are observational, a statistically significant relationship between a response y and a predictor variable x does not imply a cause-and-effect relationship. It simply means that x contributes information for the prediction of y, and nothing more. This point is aptly illustrated by a socioeconomic research project reported in the *Orlando Sentinel Star* (February 28, 1979), in an article entitled "Couples who marry with child on way end up far poorer." The article states that "White suburban couples beginning marriage with the bride already pregnant face lower income and living standards and 22 percent fewer assets by age 40 than couples with no premarital pregnancy." This conclusion was based on interviews with approximately 1,000 white suburban married women over the period 1962–1977. Note that although the quotation does not use the word *cause*, it certainly implies that the researchers have concluded that lower income can be expected to follow premarital pregnancy.

The pitfalls provided by the researcher's observational data are apparent. The response y (financial reward at age 40) is related to a single variable, the presence or absence of premarital pregnancy. The women questioned in the experiment were *not* randomly assigned to the two groups—an obviously impossible task—so a real possibility exists that lower prospective economic achievers tended to fall in the premarital pregnancy group and higher prospective economic achievers tended to fall in the nonpremarital pregnancy group. In other words, perhaps the survey is showing that women from lower socioeconomic groups are more likely to be subject to both premarital pregnancy *and* lower long-term economic rewards. Whether this explanation or the researcher's explanation of the relationship between long-term economic gain and premarital pregnancy is valid is impossible to decide based on observational data, and this demonstrates the primary weakness of observational experiments.

The point of the previous discussion is twofold. If you can control the values of the independent variables in an experiment, it pays to do so. If you cannot control them, you can still learn much from a regression analysis about the relationship between a response y and a set of predictors. In particular, a prediction equation that provides a good fit to your data will almost always be useful. But, you must be careful about deducing cause-and-effect relationships between the response and the predictors in an observational experiment.

WARNING

With observational data, a statistically significant relationship between a response y and a predictor variable x *does not* imply a cause-and-effect relationship.

Learning about the design of experiments is useful even if most of your applications of regression analysis involve observational data. Learning how to design an experiment and control the information in the data will improve your ability to assess the quality of observational data. We introduce experimental design and methods for analyzing the data in Chapter 8.

S E C T I O N 5.3

DEVIATING FROM THE ASSUMPTIONS

When we apply a regression analysis to a set of data, we never know for certain that the assumptions about the random error term ε are satisfied. How far can we deviate from the assumptions and still expect a multiple regression analysis to yield results that will have the reliability stated in Chapter 4? How can we detect departures (if they exist) from the assumptions and what can we do about them? We will provide some partial answers to these questions in this section and direct you to further discussion in succeeding chapters.

Remember (from Section 4.2) that for a given set of values of $x_1, x_2, \ldots, x_k$,

$$y = \beta_0 + \beta_1 x_1 + \beta_2 x_2 + \cdots + \beta_k x_k + \varepsilon$$

where ε is a random error. The first assumption that we made was that the mean value of the random error for *any* given set of values of $x_1, x_2, \ldots, x_k$ is $E(\varepsilon) = 0$.

One consequence of this assumption is that the mean $E(y)$ for a specific set of values of $x_1, x_2, \ldots, x_k$ is

$$E(y) = \beta_0 + \beta_1 x_1 + \beta_2 x_2 + \cdots + \beta_k x_k$$

That is,

$$
\underbrace{
\begin{array}{c}
\text{Mean value of } y \\
\text{for specific values} \\
\text{of } x_1, x_2, \ldots, x_k
\end{array}
}_{y = E(y)}
\quad + \quad
\underbrace{
\begin{array}{c}
\text{Random} \\
\text{error}
\end{array}
}_{\varepsilon}
$$

The second consequence of this assumption (which we will state without proof) is that the least squares estimators of the model parameters, $\beta_0, \beta_1, \beta_2, \ldots, \beta_k$, will be unbiased regardless of the remaining assumptions that we attribute to the random errors and their probability distributions.

The properties of the sampling distributions of the parameter estimators $\hat{\beta}_0$, $\hat{\beta}_1, \ldots, \hat{\beta}_k$ will depend on the remaining assumptions that we specify concerning the probability distributions of the random errors. You will recall that we assumed that for any given set of values of $x_1, x_2, \ldots, x_k$, ε has a normal probability distribution with mean equal to 0 and variance equal to σ^2. Also, we assumed that the random errors are independent (in a probabilistic sense).

It is unlikely that the assumptions stated above are satisfied exactly for many practical situations. If departures from the assumptions are not too great, experience has shown that a least squares regression analysis produces estimates—predictions and statistical test results—that have, for all practical purposes, the properties specified in Chapter 4. On the other hand, if the assumptions are flagrantly violated, any inferences derived from the regression analysis are suspect.

If the observations are likely to be correlated (as in the case of **time series data**—that is, data collected over time), we must check for correlation between the random errors (a topic to be discussed in Chapter 6) and may have to modify our methodology if correlation exists. The solution to this problem is to construct a time series model; this will be the subject of Chapter 9. If the variance of the random error ε changes from one setting of the independent x variables to another, we can sometimes transform the data so that the standard least squares methodology will be appropriate. Some techniques for detecting nonhomogeneous variances of the random errors (a condition called **heteroscedasticity**) and some methods for treating this type of data are discussed in Chapter 6. The normality assumption is the least restrictive of the assumptions when regression analysis is applied in practice. However, nonnormality can result in predicted values that deviate greatly from the observed values. A careful analysis of these extreme values, or **outliers**, is an important component of regression analysis. In Chapter 6 we give some methods for detecting outliers and determining their influence on the prediction equation.

Frequently, the data $(y, x_1, x_2, \ldots, x_k)$ are observational, i.e., we just observe an experimental unit and record values for $y, x_1, x_2, \ldots, x_k$. Do these data violate the assumption that $x_1, x_2, \ldots, x_k$ are fixed? For this particular case, if we can assume that $x_1, x_2, \ldots, x_k$ are *measured without error*, the mean value $E(y)$ can be viewed as a conditional mean. That is, it gives the mean value of y, *given* that the x variables assume a specific set of values. With this modification in our thinking, the least squares regression analysis is applicable to observational data. Keep in mind, however, that inferences about $E(y)$ have the reliability stated in Chapter 4 only for the given set of x-values.

To conclude, remember that when you perform a regression analysis the reliability you can place in your inferences depends on having satisfied the assumptions prescribed in Section 4.2. We will never know for certain that the random errors satisfy these assumptions, but we will examine the residuals [the deviations $(y_i - \hat{y}_i)$ between the observed and the corresponding predicted values of y]

in Chapter 6 to see if we can discover patterns that suggest correlation, heteroscedasticity, nonnormality, or an improper choice for the deterministic portion of the model. An examination of the residuals will also have another beneficial effect: The magnitudes of the residuals will give you an idea of how well the model is predicting. This should convince you that although the assumptions may not always be satisfied exactly, a multiple regression analysis is a powerful statistical tool.

| | | | | | | | | | | | | | |

S E C T I O N 5.4

PARAMETER ESTIMABILITY AND INTERPRETATION

Suppose we want to fit the first-order model

$$E(y) = \beta_0 + \beta_1 x$$

to relate a firm's monthly profit y to advertising expenditure x. Now, suppose we have 3 months of data, and the firm spent \$1,000 on advertising during each month. The data are shown in Figure 5.1. You can see the problem: The parameters of the straight-line model cannot be estimated when all the data are concentrated at a single x-value. Recall that it takes two points (x-values) to fit a straight line. Thus, the parameters are not estimable when only one x-value is observed.

FIGURE 5.1
Profit and Advertising
Expenditure Data:
3 Months

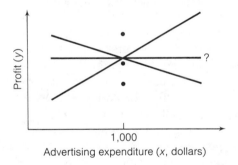

A similar problem would occur if we attempted to fit the second-order model

$$E(y) = \beta_0 + \beta_1 x + \beta_2 x^2$$

to a set of data for which only one *or two* different x-values were observed (see Figure 5.2). At least three different x-values must be observed before a second-order model can be fit to a set of data (that is, before all three parameters are estimable). In general, the number of levels of a quantitative independent variable x must be at least one more than the order of the polynomial in x that you want to fit. If two values of x are too close together, you may not be able to estimate a parameter because of rounding error encountered in fitting the model. Remember, also, that the sample size n must be sufficiently large to allow degrees of freedom for estimating σ^2.

FIGURE 5.2

Only Two Different
x-Values Observed—
the Second-Order Model
Is Not Estimable

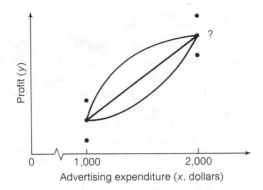

Since most business variables are not controlled by the researcher, the independent variables will almost always be observed at a sufficient number of levels to permit estimation of the model parameters. However, when the computer program you use suddenly refuses to fit a model, the problem is probably inestimable parameters.

REQUIREMENTS FOR FITTING A pTH-ORDER POLYNOMIAL REGRESSION MODEL

$$E(y) = \beta_0 + \beta_1 x + \beta_2 x^2 + \cdots + \beta_p x^p$$

1. The number of levels of x must be greater than or equal to $(p + 1)$.
2. The sample size n must be greater than $(p + 1)$ in order to allow sufficient degrees of freedom for estimating σ^2.

Given that the parameters of the model are estimable, it is important to interpret the parameter estimates correctly. A typical misconception is that $\hat{\beta}_i$ always measures the effect of x_i on $E(y)$, *independent* of the other x variables in the model. This may be true for some models, but it is not true in general. We will see in Section 5.5 that when the independent variables are correlated, the values of the estimated β coefficients are often misleading. Even if the independent variables are uncorrelated, the presence of interaction changes the meaning of the parameters. For example, the underlying assumption of the first-order model

$$E(y) = \beta_0 + \beta_1 x_1 + \beta_2 x_2$$

is, in fact, that x_1 and x_2 affect the mean response $E(y)$ independently. Recall from Section 4.5 that the slope parameter β_1 measures the rate of change of y for a 1-unit increase in x_1, for any given value of x_2. However, if the relationship between $E(y)$ and x_1 depends on x_2 (i.e., if x_1 and x_2 interact), then the interaction model

$$E(y) = \beta_0 + \beta_1 x_1 + \beta_2 x_2 + \beta_3 x_1 x_2$$

is more appropriate. For the interaction model, we showed that the effect of x_1 on $E(y)$, i.e., the slope, is not measured by a single β parameter, but by $\beta_1 + \beta_3 x_2$.

Generally, the interpretation of an individual β parameter becomes increasingly more difficult as the model becomes more complex. In Chapter 7, we give the interpretations for a number of different multiple regression models. For higher-order models, the individual β's usually have no practical interpretation.

Another misconception about the parameter estimates is that the magnitude of $\hat{\beta}_i$ determines the importance of x_i, that is, the larger (in absolute value) the $\hat{\beta}_i$, the more important the independent variable x_i is as a predictor of y. We learned in Chapter 4, however, that the standard error of the estimate $s_{\hat{\beta}_i}$ is critical in making inferences about the true parameter value. To reliably assess the importance of an individual term in the model we conduct a test of H_0: $\beta_i = 0$ or construct a confidence interval for β_i using formulas that reflect the magnitude of $s_{\hat{\beta}_i}$.*

SECTION 5.5

MULTICOLLINEARITY

Often, two or more of the independent variables used in the model for $E(y)$ will contribute redundant information. That is, the independent variables will be correlated with each other. For example, suppose we want to construct a model to predict the gasoline mileage rating, y, of a truck as a function of its load, x_1, and the horsepower, x_2, of its engine. In general, you would expect heavier loads to require greater horsepower and to result in lower mileage ratings. Thus, although both x_1 and x_2 contribute information for the prediction of mileage rating, some of the information is overlapping, because x_1 and x_2 are correlated. When the independent variables are correlated, we say that **multicollinearity** exists. In practice, it is not uncommon to observe correlations among the independent variables. However, a few problems arise when serious multicollinearity is present in the regression analysis.

First, high correlations among the independent variables increase the likelihood of rounding errors in the calculations of the β estimates, standard errors, and so forth.[†] Second, the regression results may be confusing and misleading.

---

*In addition to the parameter estimates, $\hat{\beta}_i$, some multiple regression computer packages report the **standardized regression coefficients**, $\hat{\beta}_i^* = \hat{\beta}_i(s_{x_i}/s_y)$, where s_{x_i} and s_y are the standard deviations of the x_i- and y-values, respectively, in the sample. Unlike $\hat{\beta}_i$, $\hat{\beta}_i^*$ is scaleless. These standardized regression coefficients make it more feasible to compare parameter estimates since the units are the same. However, the problems with interpreting standardized regression coefficients are much the same as those mentioned above. Therefore, you should be wary of using a standardized regression coefficient as the sole determinant of an x variable's importance.

[†]The result is due to the fact that, in the presence of severe multicollinearity, the computer has difficulty inverting the information matrix $(X'X)$. See Appendix A for a discussion of the $(X'X)$ matrix and the mechanics of a regression analysis.

To illustrate, if the gasoline mileage rating model

$$E(y) = \beta_0 + \beta_1 x_1 + \beta_2 x_2$$

were fit to a set of data, we might find that the t-values for both $\hat{\beta}_1$ and $\hat{\beta}_2$ (the least squares estimates) are nonsignificant. However, the F-test for H_0: $\beta_1 = \beta_2 = 0$ would probably be highly significant. The tests may seem to be contradictory, but really they are not. The t-tests indicate that the contribution of one variable, say x_1 = load, is not significant after the effect of x_2 = horsepower has been discounted (because x_2 is also in the model). The significant F-test, on the other hand, tells us that at least one of the two variables is making a contribution to the prediction of y (i.e., either β_1, β_2, or both differ from 0). In fact, both are probably contributing, but the contribution of one overlaps with that of the other.

Multicollinearity can also have an effect on the signs of the parameter estimates. More specifically, a value of $\hat{\beta}_i$ may have the opposite sign from what is expected. For example, we expect the signs of both of the parameter estimates for the gasoline mileage rating model to be negative, yet the regression analysis for the model might yield the estimates $\hat{\beta}_1 = .2$ and $\hat{\beta}_2 = -.7$. The positive value of $\hat{\beta}_1$ seems to contradict our expectation that heavy loads will result in lower mileage ratings. We mentioned in the previous section, however, that it is dangerous to interpret a β coefficient when the independent variables are correlated. Because the variables contribute redundant information, the effect of load x_1 on mileage rating is measured only partially by $\hat{\beta}_1$. Also, we warned in Section 5.2 that we cannot establish a cause and-effect relationship between y and the predictor variables based on observational data. By attempting to interpret the value $\hat{\beta}_1$, we are really trying to establish a cause-and-effect relationship between y and x_1 (by suggesting that a heavy load x_1 will *cause* a lower mileage rating y).

How can you avoid the problems of multicollinearity in regression analysis? One way is to conduct a designed experiment so that the levels of the x variables are uncorrelated (see Section 5.2). Unfortunately, time and cost constraints may prevent you from collecting data in this manner. For these and other reasons, most data collected in business studies are observational. Since observational data frequently consist of correlated independent variables, you will need to recognize when multicollinearity is present and, if necessary, make modifications in the analysis.

Several methods are available for detecting multicollinearity in regression. A simple technique is to calculate the coefficient of correlation r between each pair of independent variables in the model and use the procedure outlined in Section 3.7 to test for evidence of positive or negative correlation. If one or more of the r-values is statistically different from 0, the variables in question are correlated and a severe multicollinearity problem may exist.* Other indications of the presence of multicollinearity include those mentioned in the beginning of this sec-

---

*Remember that r measures only the pairwise correlation between x-values. Three variables, x_1, x_2, and x_3, may be highly correlated as a group, but may not exhibit large pairwise correlations. Thus, multicollinearity may be present even when all pairwise correlations are not significantly different from 0.

tion—namely, nonsignificant t-tests for the individual β parameters when the F-test for overall model adequacy is significant, and estimates with opposite signs from what is expected.

A more formal method for detecting multicollinearity involves the calculation of **variance inflation factors** for the individual β parameters. One reason why the t-tests on the individual β parameters are nonsignificant is because the standard errors of the estimates, $s_{\hat{\beta}_i}$, are inflated in the presence of multicollinearity. When the dependent and independent variables are appropriately transformed,* it can be shown that

$$s_{\hat{\beta}_i}^2 = s^2 \left(\frac{1}{1 - R_i^2} \right)$$

where s^2 is the estimate of σ^2, the variance of ε, and R_i^2 is the multiple coefficient of determination for the model which regresses the independent variable x_i on the remaining independent variables $x_1, x_2, \ldots, x_{i-1}, x_{i+1}, \ldots, x_k$. The quantity $1/(1 - R_i^2)$ is called the *variance inflation factor* for the parameter β_i, and is denoted $(VIF)_i$. Note that $(VIF)_i$ will be large when R_i^2 is large—that is, when the independent variable x_i is strongly related to the other independent variables.

Various authors maintain that, in practice, a severe multicollinearity problem exists if the largest of the variance inflation factors for the β's is greater than 10 or, equivalently, if the largest multiple coefficient of determination R_i^2, is greater than .90.[†] Several of the multiple regression computer packages discussed in Chapter 4 have options for calculating variance inflation factors.[‡]

The methods for detecting multicollinearity are summarized in the accompanying box. We illustrate the use of these statistics in Example 5.1.

EXAMPLE 5.1

The Federal Trade Commission annually ranks varieties of domestic cigarettes according to their tar, nicotine, and carbon monoxide contents. The U.S. surgeon general considers each of these three substances hazardous to a smoker's health. Past studies have shown that increases in the tar and nicotine contents of a cigarette are accompanied by an increase in the carbon monoxide emitted from the cigarette smoke. Table 5.1 contains tar, nicotine, and carbon monoxide con-

*The transformed variables are obtained as

$$y_i^* = (y_i - \bar{y})/s_y \qquad x_{1i}^* = (x_{1i} - \bar{x}_1)/s_1 \qquad x_{2i}^* = (x_{2i} - \bar{x}_2)/s_2$$

and so on, where $\bar{y}, \bar{x}_1, \bar{x}_2, \ldots$, and $s_y, s_1, s_2, \ldots$ are the sample means and standard deviations, respectively, of the original variables.

[†]See, for example, Montgomery and Peck (1982) or Neter, Wasserman, and Kutner (1985).

[‡]Some computer packages calculate an equivalent statistic, called the **tolerance**. The tolerance for a β coefficient is the reciprocal of the variance inflation factor, i.e.,

$$(TOL)_i = \frac{1}{(VIF)_i} = 1 - R_i^2$$

For $R_i^2 > .90$ (the extreme multicollinearity case), $(TOL)_i < .10$. These computer packages allow the user to set tolerance limits, so that any independent variable with a value of $(TOL)_i$ below the tolerance limit will not be allowed to enter into the model.

DETECTING MULTICOLLINEARITY IN THE REGRESSION MODEL

$$E(y) = \beta_0 + \beta_1 x_1 + \beta_2 x_2 + \cdots + \beta_k x_k$$

The following are indicators of multicollinearity:

1. Significant correlations between pairs of independent variables in the model
2. Nonsignificant *t*-tests for the individual β parameters when the *F*-test for overall model adequacy H_0: $\beta_1 = \beta_2 = \cdots = \beta_k = 0$ is significant
3. Opposite signs (from what is expected) in the estimated parameters
4. A variance inflation factor (VIF) for a β parameter greater than 10, where

$$(\text{VIF})_i = \frac{1}{1 - R_i^2}, \quad i = 1, 2, \ldots, k$$

and R_i^2 is the multiple coefficient of determination for the model

$$E(x_i) = \alpha_0 + \alpha_1 x_1 + \alpha_2 x_2 + \cdots + \alpha_{i-1} x_{i-1} + \alpha_{i+1} x_{i+1} + \cdots + \alpha_k x_k$$

TABLE 5.1

FTC Cigarette Data for Example 5.1

BRAND	TAR x_1, milligrams	NICOTINE x_2, milligrams	WEIGHT x_3, grams	CARBON MONOXIDE y, milligrams
Alpine	14.1	.86	.9853	13.6
Benson & Hedges	16.0	1.06	1.0938	16.6
Bull Durham	29.8	2.03	1.1650	23.5
Camel Lights	8.0	.67	.9280	10.2
Carlton	4.1	.40	.9462	5.4
Chesterfield	15.0	1.04	.8885	15.0
Golden Lights	8.8	.76	1.0267	9.0
Kent	12.4	.95	.9225	12.3
Kool	16.6	1.12	.9372	16.3
L&M	14.9	1.02	.8858	15.4
Lark Lights	13.7	1.01	.9643	13.0
Marlboro	15.1	.90	.9316	14.4
Merit	7.8	.57	.9705	10.0
Multifilter	11.4	.78	1.1240	10.2
Newport Lights	9.0	.74	.8517	9.5
Now	1.0	.13	.7851	1.5
Old Gold	17.0	1.26	.9186	18.5
Pall Mall Light	12.8	1.08	1.0395	12.6
Raleigh	15.8	.96	.9573	17.5
Salem Ultra	4.5	.42	.9106	4.9
Tareyton	14.5	1.01	1.0070	15.9
True	7.3	.61	.9806	8.5
Viceroy Rich Lights	8.6	.69	.9693	10.6
Virginia Slims	15.2	1.02	.9496	13.9
Winston Lights	12.0	.82	1.1184	14.9

Source: Federal Trade Commission.

tents (in milligrams) and weight (in grams) for a sample of 25 (filter) brands tested in a recent year. Suppose we want to model carbon monoxide content, y, as a function of tar content, x_1, nicotine content, x_2, and weight, x_3, using the model

$$E(y) = \beta_0 + \beta_1 x_1 + \beta_2 x_2 + \beta_3 x_3$$

The model is fit to the 25 data points in Table 5.1 and a portion of the resulting SAS printout is shown in Figure 5.3. Examine the printout. Do you detect any signs of multicollinearity?

SOLUTION

First, notice that a test of

$$H_0: \quad \beta_1 = \beta_2 = \beta_3 = 0$$

is highly significant. The F-value (shaded on the printout) is very large ($F = 78.984$) and the observed significance level of the test (also shaded) is small ($p = .0001$). Therefore, we can reject H_0 for any α greater than .0001 and conclude that at least one of the parameters, β_1, β_2, and β_3, is nonzero. The t-tests for two of the three individual β's, however, are nonsignificant. (The p-values for these tests are shaded on the printout.) Unless tar is the only one of the three variables useful for predicting carbon monoxide content, these results are the first indication of a potential multicollinearity problem.

A second clue to the presence of multicollinearity is the negative value for $\hat{\beta}_2$ (shaded on the printout),

$$\hat{\beta}_2 = -2.63$$

From past studies, we expect carbon monoxide content y to increase when nicotine content x_2 increases—that is, we expect a *positive* relationship between y and x_2, not a negative one.

A more formal procedure for detecting multicollinearity is to examine the variance inflation factors. Figure 5.3 shows the variance inflation factors (shaded) for each of the three parameters under the column labeled VARIANCE INFLA-TION. Note that the variance inflation factors for both the tar and nicotine parameters are greater than 10. The variance inflation factor for the tar parameter, $(VIF)_1 = 21.63$, implies that a model relating tar content x_1 to the remaining two independent variables, nicotine content x_2 and weight x_3, resulted in a coefficient of determination

$$R_1^2 = 1 - \frac{1}{(VIF)_1}$$

$$= 1 - \frac{1}{21.63} = .954$$

FIGURE 5.3 Portion of the SAS Printout for Example 5.1

```
DEP VARIABLE: CO                        ANALYSIS OF VARIANCE

                              SUM OF        MEAN
             SOURCE    DF     SQUARES      SQUARE     F VALUE    PROB>F

             MODEL      3    495.25781    165.08594    78.984    0.0001
             ERROR     21     43.89258562  2.09012312
             C TOTAL   24    539.15040

             ROOT MSE        1.445726    R-SQUARE     0.9186
             DEP MEAN          12.528    ADJ R-SQ     0.9070
             C.V.           11.53996

                              PARAMETER ESTIMATES

                     PARAMETER      STANDARD    T FOR HO:                                VARIANCE
   VARIABLE    DF    ESTIMATE       ERROR       PARAMETER=0   PROB > :T:    TOLERANCE    INFLATION

   INTERCEP    1    3.20219002     3.46175473     0.925        0.3655                           0
   TAR         1    0.96257386     0.24224436     3.974        0.0007      0.04623058   21.63070592
   NICOTINE    1   -2.63166111     3.90055745    -0.675        0.5072      0.04566227   21.89991722
   WEIGHT      1   -0.13048185     3.88534182    -0.034        0.9735      0.74970451    1.33385886
```

All signs indicate that a serious multicollinearity problem exists. To confirm our suspicions, we calculated the coefficient of correlation r for each of the three pairs of independent variables in the model. These values are given in Table 5.2. You can see that tar content x_1 and nicotine content x_2 appear to be highly correlated ($r = .977$), while weight x_3 appears to be moderately correlated with both tar content ($r = .491$) and nicotine content ($r = .500$). In fact, all three sample correlations exceed the critical value, $r_{.025} = .423$, for a two-tailed test of $H_0: \rho = 0$ conducted at $\alpha = .05$ with $n - 2 = 23$ df.

TABLE 5.2

Correlation Coefficients for the Three Pairs of Independent Variables in Example 5.1

PAIR	r
x_1, x_2	.977
x_1, x_3	.491
x_2, x_3	.500

■

Once you have detected that a multicollinearity problem exists, there are several alternative measures available for solving the problem. The appropriate measure to take depends on the severity of the multicollinearity and the ultimate goal of the regression analysis.

Some researchers, when confronted with highly correlated independent variables, choose to include only one of the correlated variables in the final model. One way of deciding which variable to include is by using **stepwise regression**, a topic discussed in Chapter 4. Generally, only one (or a small number) of a set of multicollinear independent variables will be included in the regression model by the stepwise regression procedure since this procedure tests the parameter associated with each variable in the presence of all the variables already in the model. For example, in fitting the gasoline mileage rating model introduced earlier, if at one step the variable representing truck load is included as a significant variable in the prediction of the mileage rating, the variable representing horsepower will probably never be added in a future step. Thus, if a set of independent variables is thought to be multicollinear, some screening by stepwise regression may be helpful.

If you are interested only in using the model for estimation and prediction, you may decide not to drop any of the independent variables from the model. In the presence of multicollinearity, we have seen that it is dangerous to interpret the individual β's for the purpose of establishing cause and effect. However, confidence intervals for $E(y)$ and prediction intervals for y generally remain unaffected *as long as the values of the independent variables used to predict y follow the same pattern of multicollinearity exhibited in the sample data.* That is, you must take strict care to ensure that the values of the x variables fall within the experimental region. (We will discuss this problem in further detail in Section 5.6.) Alternatively, if your goal is to establish a cause-and-effect relationship between y and the independent variables, you will need to conduct a designed experiment to break up the pattern of multicollinearity.

SOLUTIONS TO SOME PROBLEMS CREATED BY MULTICOLLINEARITY

1. Drop one or more of the correlated independent variables from the final model. A screening procedure such as stepwise regression is helpful in determining which variables to drop.
2. If you decide to keep all the independent variables in the model:
   a. Avoid making inferences about the individual β parameters (such as establishing a cause-and-effect relationship between y and the predictor variables).
   b. Restrict inferences about $E(y)$ and future y-values to values of the independent variables that fall within the experimental region (see Section 5.6).
3. If your ultimate objective is to establish a cause-and-effect relationship between y and the predictor variables, use a designed experiment (see Chapter 8).
4. To reduce rounding errors in polynomial regression models, code the independent variables so that first-, second-, and higher-order terms for a particular x variable are not highly correlated (see Section 7.6).
5. To reduce rounding errors and stabilize the regression coefficients, use ridge regression to estimate the β parameters (see Section 10.7).

When fitting a polynomial regression model [for example, the second-order model $E(y) = \beta_0 + \beta_1 x + \beta_2 x^2$], the independent variables $x_1 = x$ and $x_2 = x^2$ will often be correlated. If the correlation is high, the computer solution may result in extreme rounding errors. For this model, the solution is not to drop one of the independent variables but to transform the x variable in such a way that the correlation between the coded x- and x^2-values is substantially reduced. Coding the independent quantitative variables in polynomial regression models is a topic discussed in optional Chapter 7.

Another, more complex, procedure for reducing the rounding errors caused by multicollinearity involves a modification of the least squares method, called **ridge regression**. In ridge regression, the estimates of the β coefficients are biased [that is, $E(\hat{\beta}_i) \neq \beta_i$] but have significantly smaller standard errors than the unbiased β estimates yielded by the least squares method. Thus, the β estimates for the ridge regression are more stable than the corresponding least squares estimates. Ridge regression is a topic discussed in optional Chapter 10.

| | | | | | | | | | | | |

SECTION 5.6

EXTRAPOLATION: PREDICTING OUTSIDE THE EXPERIMENTAL REGION

By the late 1960's many research economists had developed highly technical models to relate the state of the economy to various economic indices and other independent variables. Many of these models were multiple regression models, where, for example, the dependent variable y might be next year's growth in GNP and the independent variables might include this year's rate of inflation, this year's Consumer Price Index, and so forth. In other words, the model might be constructed to predict next year's economy using this year's knowledge.

Unfortunately, these models were almost unanimously unsuccessful in predicting the recession in the early 1970's. What went wrong? Well, one of the problems was that regression models were used to predict y for values of the independent variables that were outside the region in which the model was developed. For example, the inflation rate in the late 1960's, when the models were developed, ranged from 6 to 8%. When the double-digit inflation of the early 1970's became a reality, some researchers attempted to use the same models to predict the growth in GNP 1 year hence. As you can see in Figure 5.4, the model may be very accurate for predicting y when x is in the range of experimentation, but the use of the model outside that range is a dangerous practice. A $100(1 - \alpha)\%$ prediction interval for GNP when the inflation rate is, say 10%, will be less reliable than the stated confidence coefficient $(1 - \alpha)$. How much less is unknown.

FIGURE 5.4

Using a Regression Model Outside the Experimental Region

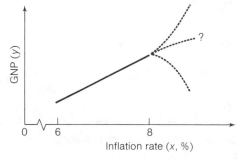

For a single independent variable x, the experimental region is simply the range of the values of x in the sample. Establishing the experimental region for a multiple regression model that includes a number of independent variables may be more difficult. For example, consider a model for GNP (y) using inflation rate (x_1) and prime interest rate (x_2) as predictor variables. Suppose a sample of size $n = 5$ was observed, and the values of x_1 and x_2 corresponding to the five values for GNP were (6, 10), (6.25, 12), (7.25, 10.25), (7.5, 13), and (8, 11.5). Notice that x_1 ranges from 6% to 8% and x_2 ranges from 10% to 13% in the sample data. You may think that the experimental region is defined by the ranges of the individual variables, i.e., $6 \le x_1 \le 8$ and $10 \le x_2 \le 13$. However, the levels of x_1 and x_2 *jointly* define the region. Figure 5.5 shows the experimental region for our hypothetical data. You can see that an observation with levels $x_1 = 8$ and $x_2 = 10$ clearly falls outside the experimental region, yet is within the ranges of the individual x-values. Using the model to predict GNP for this observation may lead to unreliable results.

FIGURE 5.5

Experimental Region for Modeling GNP (y) as a Function of Inflation Rate (x_1) and Prime Interest Rate (x_2)

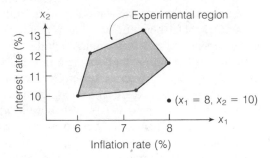

The word *transform* means to change the form of some object or thing. Consequently, the phrase *data transformation* means that we have done, or plan to do, something to change the form of the data. For example, if one of the independent variables in a model is the price p of a commodity, we might choose to introduce this variable into the model as $x = 1/p$, $x = \sqrt{p}$, or $x = e^{-p}$. Thus, if we were to let $x = \sqrt{p}$, we would compute the square root of each price value, and these square roots would be the values of x that would be used in the regression analysis.

Data transformations are performed on the y-values to make them more nearly satisfy the assumptions of Section 4.2 and, sometimes, to make the deterministic portion of the model a better approximation to the mean value of the transformed response. Transformations of the values of the independent variables are performed solely for the latter reason—that is, to achieve a model that provides a better approximation to $E(y)$. The purpose of this section is to discuss transformations on the independent variables. In particular, we want you to see why these transformations are sometimes useful and to see how you might use them. (Transformations on the y-values for the purpose of satisfying the assumptions will be discussed in Chapter 6.)

Suppose you want to fit a model relating the demand y for a product to its price p. Also, suppose the product is a nonessential item, and you expect the

SECTION 5.7

DATA TRANSFORMATIONS

mean demand to decrease as price p increases and then to decrease more slowly as p gets larger (see Figure 5.6). What function of p will provide a good approximation to $E(y)$?

FIGURE 5.6

Hypothetical Relation Between Demand y and Price p

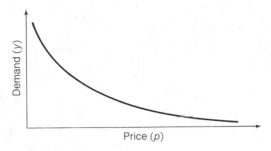

To answer this question, you need to know the graphs of some elementary mathematical functions—there is a one-to-one relationship between mathematical functions and graphs. If we want to model a relationship similar to the one indicated in Figure 5.6, we need to be able to select a mathematical function that will possess a graph similar to the curve shown.

Portions of some curves corresponding to mathematical functions that decrease as p increases are shown in Figure 5.7. Of the four functions shown, the curves in Figures 5.7(c) and (d) will probably provide the best approximations to $E(y)$. This is because both provide graphs that show $E(y)$ decreasing and approaching

FIGURE 5.7

Graphs of Some Mathematical Functions Relating $E(y)$ to p

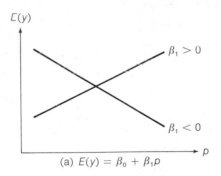

(a) $E(y) = \beta_0 + \beta_1 p$

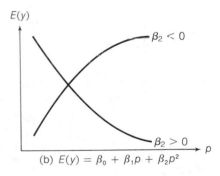

(b) $E(y) = \beta_0 + \beta_1 p + \beta_2 p^2$

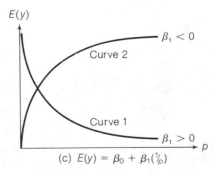

(c) $E(y) = \beta_0 + \beta_1 (1/p)$

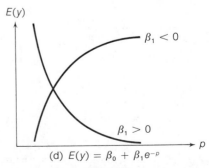

(d) $E(y) = \beta_0 + \beta_1 e^{-p}$

(but never reaching) 0 as p increases. This suggests that the independent variable, price, should be transformed using either $x = 1/p$ or $x = e^{-p}$. Then you might try fitting the model

$$E(y) = \beta_0 + \beta_1 x$$

using the transformed data.

The functions shown in Figure 5.7 produce curves that either rise or fall depending on the sign of the parameter β_1 in parts (a), (c), and (d), and on β_2 and the portion of the curve used in part (b). When you choose a model for a regression analysis, you do not have to specify the sign of the parameter(s). The least squares procedure will choose as estimates of the parameters those that minimize the sum of squares of the residuals. Consequently, if you were to fit the model shown in Figure 5.7(c) to set of y-values that increase in value as p increases, your least squares estimate of β_1 would be negative, and a graph of y would produce a curve similar to curve 2 in Figure 5.7(c). If the y-values decrease as p increases, your estimate of β_1 will be positive and the curve will be similar to curve 1 in Figure 5.7(c). All the curves in Figure 5.7 shift upward or downward depending on the value of β_0.

EXAMPLE 5.2

A supermarket chain conducted an experiment to investigate the effect of price p on the weekly demand (in pounds) for a house brand of coffee. Eight supermarket stores that had nearly equal past records of demand for the product were used in the experiment. Eight prices were randomly assigned to the stores and were advertised using the same procedures. The number of pounds of coffee sold during the following week was recorded for each of the stores and is shown in the table.

DEMAND y, pounds	PRICE p, dollars
1,120	3.00
999	3.10
932	3.20
884	3.30
807	3.40
760	3.50
701	3.60
688	3.70

a. Fit the model

$$E(y) = \beta_0 + \beta_1 x$$

to the data, letting $x = 1/p$.
b. Do the data provide sufficient evidence to indicate that the model contributes information for the prediction of demand?
c. Find a 90% confidence interval for the mean demand when the price is set at $3.20 per pound. Interpret this interval.

SOLUTION

y	$x = 1/p$
1,120	.333
999	.323
932	.313
884	.303
807	.294
760	.286
701	.278
688	.270

a. The first step is to calculate $x = 1/p$ for each data point. These values are given in the table. The Minitab computer printout* (Figure 5.8) gives

$$\hat{\beta}_0 = -1,180 \qquad \hat{\beta}_1 = 6,808$$

and

$$\hat{y} = -1,180 + 6,808x$$
$$= -1,180 + 6,808\left(\frac{1}{p}\right)$$

*The Minitab program uses full decimal accuracy for $x = 1/p$. Hence, the results shown in Figure 5.8 differ from results that would be calculated using the three-decimal values for $x = 1/p$ shown in the table.

(You can verify that the formulas of Section 3.3 give the same answers.) A graph of this prediction equation is shown in Figure 5.9.

FIGURE 5.8

Minitab Printout for Example 5.2

```
The regression equation is
Y = - 1180 + 6808 X

Predictor        Coef        Stdev      t-ratio
Constant       -1180.5       107.7      -10.96
X               6808.1       358.4       19.00

s = 20.90       R-sq = 98.4%      R-sq(adj) = 98.1%

Analysis of Variance

SOURCE          DF          SS            MS
Regression       1        157718        157718
Error            6          2622           437
Total            7        160340
```

FIGURE 5.9

Graph of the Demand–Price Curve for Example 5.2

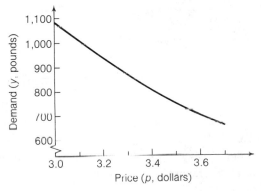

b. To determine whether x contributes information for the prediction of y, we test $H_0: \beta_1 = 0$ against the alternative hypothesis $H_a: \beta_1 \neq 0$. The test statistic is

$$t = \frac{\hat{\beta}_1 - 0}{s_{\hat{\beta}_1}} = \frac{\hat{\beta}_1}{s/\sqrt{SS_{xx}}}$$

The computed value of this t-statistic, given in Figure 5.8 (you can verify the computations), is $t = 19.00$. Since we wish to detect either $\beta_1 > 0$ or $\beta_1 < 0$, we will use a two-tailed test and will reject H_0 when t exceeds $t_{\alpha/2}$ based on $(n - 2) = 6$ df. Using $\alpha = .05$, $t_{.025} = 2.447$. Since the computed value of t exceeds $t_{.025}$, we reject $H_0: \beta_1 = 0$ and conclude that $x = 1/p$ contributes information for the prediction of demand y.

c. Note on the printout that $s = \sqrt{MSE} = \sqrt{437} = 20.90$, and from Table 2 of Appendix D, we have $t_{\alpha/2} = t_{.05} = 1.943$. Then the 90% confidence interval for mean demand $E(y)$ when the price per pound is $p = \$3.20$ is

$$\hat{y} \pm t_{\alpha/2}\, s\sqrt{\frac{1}{n} + \frac{(x_p - \bar{x})^2}{\sum (x_i - \bar{x})^2}}$$

where

$$\hat{y} = \hat{\beta}_0 + \hat{\beta}_1 x = -1{,}180 + 6{,}808\left(\frac{1}{p}\right)$$

$$= -1{,}180 + 6{,}808\left(\frac{1}{3.20}\right)$$

$$= 947$$

$$\bar{x} = \frac{\sum x_i}{n} = \frac{2.400}{8} = .300$$

$$\sum (x_i - \bar{x})^2 = \sum x_i^2 - \frac{\left(\sum x_i\right)^2}{n} = .723412 - \frac{(2.4)^2}{8}$$

$$= .003412$$

Substituting all these values into the formula for the confidence interval, we estimate the mean demand to be enclosed by the interval:

$$\hat{y} \pm t_{\alpha/2} s \sqrt{\frac{1}{n} + \frac{(x_p - \bar{x})^2}{\sum (x_i - \bar{x})^2}}$$

$$947 \pm (1.943)(20.90)\sqrt{\frac{1}{8} + \frac{(.313 - .300)^2}{.003412}}$$

or

930 to 964

with confidence coefficient .90. Thus, we are reasonably certain that this confidence interval encloses the mean demand $E(y)$. ∎

This discussion has been intended to emphasize the importance of data transformation and to explain its role in model building, which is presented later in optional Chapter 7. As you proceed through that chapter, remember that the symbols, $x_1, x_2, \ldots, x_k$ that appear in the linear models can be transformations on the independent variables you have observed. These transformations, coupled with the methods of optional Chapter 7, allow you to use a great variety of mathematical functions to model the mean $E(y)$ for data.

SECTION 5.8

SUMMARY

There are several problems that you should be aware of when constructing a model for a response y. In this chapter we have identified a few of the most important of these problem areas:

1. *Establishing cause and effect* When the data used in the regression analysis are observational (i.e., uncontrolled), it is dangerous to deduce a cause-and-effect relationship between y and the independent (predictor) variables. Only

when the experiment has been designed properly can you be certain that any changes in y are due solely to the different settings of the predictor variables.

2. *Departures from the assumptions* In a practical setting, it is unlikely that the standard least squares assumptions about the error term are satisfied exactly. When departures from the assumptions are slight, the model remains a powerful predictor of the response y. On the other hand, the model performs poorly when the assumptions of equal variances and uncorrelated errors are violated.

3. *Parameter estimability and interpretation* In order to estimate all the parameters in a pth-order model, the number of levels of x must be greater than or equal to $(p + 1)$. For any model, the sample size n must be sufficiently large to allow degrees of freedom for estimating σ^2 so that a test of model adequacy can be performed. Be wary of interpreting the individual β parameters in the presence of interaction or highly correlated independent variables.

4. *Multicollinearity* When highly correlated independent variables are present in a regression model, the results may be confusing; the t-tests on the individual β's may be nonsignificant even though the F-test for overall model adequacy is significant, and the β's may have signs opposite from what is expected. Also, there may be extreme rounding errors in the computation of the β estimates. Variance inflation factors aid in determining whether a serious multicollinearity problem exists. The solution to the problem depends on the severity of the multicollinearity and the ultimate goal of the regression analyst.

5. *Extrapolation* Predicting y when the x-values are outside the range of experimentation is a dangerous practice. The level of reliability associated with any inference derived from the model will be less than the stated level of confidence $(1 - \alpha)$ since the adequacy of the model outside the experimental region is unknown.

6. *Data transformations* In order to achieve a model that provides a better approximation to $E(y)$, you may need to transform the values of the independent variables or the value of y. The type of transformation you should make depends on the theoretical relationships between $E(y)$ and the independent variables.

EXERCISES 5.1–5.19

5.1 Discuss the consequences of fitting multiple regression models when the assumptions of Section 4.2 are violated.

5.2 Why is it dangerous to predict y for values of independent variables that fall outside the experimental region?

5.3 Discuss the problems that result when multicollinearity is present in a regression analysis.

5.4 How can you detect multicollinearity?

5.5 What remedial measures are available when multicollinearity is detected?

5.6 Refer to Example 5.2. Can you think of any other transformations on price that might provide a good fit to the data? Try them and answer the questions of Example 5.2 again.

5.7 The management of a manufacturing firm is considering the possibility of setting up its own market research department rather .than continuing to use the services of a market research firm. Management wants to know what salary should be paid to a market researcher, based on years of experience. An independent consultant has proposed the quadratic model

$$E(y) = \beta_0 + \beta_1 x + \beta_2 x^2$$

where

y = Annual salary (thousands of dollars)

x = Years of experience

In order to fit the model, the consultant randomly sampled three market researchers at other firms and recorded the information given in the accompanying table. Give your opinion regarding the adequacy of the proposed model.

	y	x
Researcher 1	40	2
Researcher 2	25	1
Researcher 3	42	3

5.8 A firm that sells a special skin cream exclusively through drug stores currently operates in 15 marketing districts. As part of an expansion feasibility study, the company wants to model district sales (y) as a function of target population (x_1), per capita income (x_2), and number of drug stores (x_3) in the district. Data collected for each of the 15 districts were used to fit the first-order model

$$E(y) = \beta_0 + \beta_1 x_1 + \beta_2 x_2 + \beta_3 x_3$$

A summary of the regression results follows:

$$\hat{y} = -3{,}000 + 3.2x_1 - .4x_2 - 1.1x_3 \qquad R^2 = .93$$

$$s_{\hat{\beta}_1} = 2.4 \qquad s_{\hat{\beta}_2} = .6 \qquad s_{\hat{\beta}_3} = .8$$

$$r_{12} = .92 \qquad r_{13} = .87 \qquad r_{23} = .81$$

Based on these results, the company concludes that none of the three independent variables x_1, x_2, and x_3, is a useful predictor of district sales, y. Do you agree with this statement? Explain.

5.9 A particular meat-processing plant slaughters steers and cuts and wraps the beef for its customers. Suppose a complaint has been filed with the Food and Drug Administration (FDA) against the processing plant. The complaint alleges that the consumer does not get all the beef from the steer he purchases. In particular, one consumer purchased a 300-pound steer but received only 150 pounds of cut and wrapped beef. In order to settle the complaint, the FDA collected data on the live weights and dressed weights of nine steers processed by a reputable meat-processing plant (not the firm in question). The results are listed in the table.

LIVE WEIGHT x, pounds	DRESSED WEIGHT y, pounds
420	280
380	250
480	310
340	210
450	290
460	280
430	270
370	240
390	250

a. Fit the model $E(y) = \beta_0 + \beta_1 x$ to the data.
b. Construct a 95% prediction interval for the dressed weight y of a 300-pound steer.
c. Would you recommend that the FDA use the interval obtained in part **b** to determine whether the dressed weight of 150 pounds is a reasonable amount to receive from a 300-pound steer? Explain.

5.10 Refer to the FTC cigarette data of Example 5.1. The data of Table 5.1 are reproduced here for convenience.

BRAND	TAR x_1, milligrams	NICOTINE x_2, milligrams	WEIGHT x_3, grams	CARBON MONOXIDE y, milligrams
Alpine	14.1	.86	.9853	13.6
Benson & Hedges	16.0	1.06	1.0938	16.6
Bull Durham	29.8	2.03	1.1650	23.5
Camel Lights	8.0	.67	.9280	10.2
Carlton	4.1	.40	.9462	5.4
Chesterfield	15.0	1.04	.8885	15.0
Golden Lights	8.8	.76	1.0267	9.0
Kent	12.4	.95	.9225	12.3
Kool	16.6	1.12	.9372	16.3
L&M	14.9	1.02	.8858	15.4
Lark Lights	13.7	1.01	.9643	13.0
Marlboro	15.1	.90	.9316	14.4
Merit	7.8	.57	.9705	10.0
Multifilter	11.4	.78	1.1240	10.2
Newport Lights	9.0	.74	.8517	9.5
Now	1.0	.13	.7851	1.5
Old Gold	17.0	1.26	.9186	18.5
Pall Mall Light	12.8	1.08	1.0395	12.6
Raleigh	15.8	.96	.9573	17.5
Salem Ultra	4.5	.42	.9106	4.9
Tareyton	14.5	1.01	1.0070	15.9
True	7.3	.61	.9806	8.5
Viceroy Rich Lights	8.6	.69	.9693	10.6
Virginia Slims	15.2	1.02	.9496	13.9
Winston Lights	12.0	.82	1.1184	14.9

Source: Federal Trade Commission.

a. Fit the model $E(y) = \beta_0 + \beta_1 x_1$ to the data. Is there evidence that tar content x_1 is useful for predicting carbon monoxide content y?

b. Fit the model $E(y) = \beta_0 + \beta_2 x_2$ to the data. Is there evidence that nicotine content x_2 is useful for predicting carbon monoxide content y?

c. Fit the model $E(y) = \beta_0 + \beta_3 x_3$ to the data. Is there evidence that weight x_3 is useful for predicting carbon monoxide content y?

d. Compare the signs of $\hat{\beta}_1$, $\hat{\beta}_2$, and $\hat{\beta}_3$ in the models of parts **a**, **b**, and **c**, respectively, to the signs of the $\hat{\beta}$'s in the multiple regression model fit in Example 5.1. The fact that the $\hat{\beta}$'s change dramatically when the independent variables are removed from the model is another indication of a serious multicollinearity problem.

5.11 An economist wants to model annual per capita demand, y, for passenger car motor fuel in the United States as a function of the two quantitative independent variables, average real weekly earnings (x_1) and average price of regular gasoline (x_2). Data on these three variables for the years 1965–1980 are shown in the table. Suppose the economist fits the model $E(y) = \beta_0 + \beta_1 x_1 + \beta_2 x_2$ to the data for the years 1965–1979. Would you recommend that the economist use the least squares prediction equation to predict per capita consumption of motor fuel in 1980? Explain.

YEAR	PER CAPITA CONSUMPTION OF MOTOR FUEL y, gallons	AVERAGE REAL WEEKLY EARNINGS x_1, 1967 dollars	AVERAGE PRICE OF GASOLINE x_2, dollars
1965	258.88	101.01	.32
1966	271.16	101.67	.33
1967	277.35	101.84	.34
1968	291.58	103.39	.34
1969	308.09	104.38	.36
1970	320.82	103.04	.36
1971	335.63	104.96	.36
1972	350.17	109.26	.37
1973	368.10	109.23	.40
1974	346.89	104.78	.53
1975	354.17	101.45	.57
1976	361.47	102.90	.59
1977	366.49	104.13	.62
1978	376.46	104.30	.63
1979	356.29	101.02	.86
1980	323.67	95.18	1.19

Source: Per capita consumption and average price of gasoline from *Statistical Abstract of the United States.* Average real weekly earnings from *Employment and Earnings,* US. Department of Labor, Bureau of Labor Statistics, October 1983, p. 109.

5.12 A firm wants to use multiple regression in a cost analysis of its shipping department. Since most of the costs incurred by shipping result from direct labor, the firm will model weekly hours of labor (y) as a function of total weight shipped (x_1), percentage of units shipped by truck (x_2), and average weight per shipment (x_3). Data collected from the firm's accounting and production records for a 20-week period are shown in the table.

WEEK	HOURS OF LABOR y	THOUSANDS OF POUNDS SHIPPED x_1	PERCENTAGE OF UNITS SHIPPED BY TRUCK x_2	AVERAGE NUMBER OF POUNDS PER SHIPMENT x_3
1	100	5.1	90	20
2	85	3.8	99	22
3	108	5.3	58	19
4	116	7.5	16	15
5	92	4.5	54	20
6	63	3.3	42	26
7	79	5.3	12	25
8	101	5.9	32	21
9	88	4.0	56	24
10	71	4.2	64	29
11	122	6.8	78	10
12	85	3.9	90	30
13	50	3.8	74	28
14	114	7.5	89	14
15	104	4.5	90	21
16	111	6.0	40	20
17	110	8.1	55	16
18	100	2.9	64	19
19	82	4.0	35	23
20	85	4.8	58	25

The SAS computer printout for the model $E(y) = \beta_0 + \beta_1 x_1 + \beta_2 x_2 + \beta_3 x_3$ is shown here. The firm is concerned about the problems that occur in regression analysis when multicollinearity is present. Examine the SAS printout. Do you detect any signs of multicollinearity?

SAS Printout for Exercise 5.12

DEP VARIABLE: LABOR

ANALYSIS OF VARIANCE

SOURCE	DF	SUM OF SQUARES	MEAN SQUARE	F VALUE	PROB>F
MODEL	3	5158.31383	1719.43794	17.866	0.0001
ERROR	16	1539.88617	96.24288576		
C TOTAL	19	6698.20000			

ROOT MSE	9.810346	R-SQUARE 0.7701
DEP MEAN	93.3	ADJ R-SQ 0.7270
C.V.	10.51484	

PARAMETER ESTIMATES

VARIABLE	DF	PARAMETER ESTIMATE	STANDARD ERROR	T FOR H0: PARAMETER=0	PROB > :T:	TOLERANCE	VARIANCE INFLATION
INTERCEP	1	131.92425	25.69321439	5.135	0.0001	.	0
WEIGHT	1	2.72608977	2.27500488	1.198	0.2483	0.44435417	2.25045709
TRUCK	1	0.04721841	0.09334856	0.506	0.6199	0.91496179	1.09294182
AVGSHIP	1	-2.58744391	0.64281819	-4.025	0.0010	0.46162372	2.16626650

5.13 How many levels of x are required to fit the model $E(y) = \beta_0 + \beta_1 x + \beta_2 x^2$? How large a sample size is required to have sufficient degrees of freedom for estimating σ^2?

5.14 How many levels of x_1 and x_2 are required to fit the model $E(y) = \beta_0 + \beta_1 x_1 + \beta_2 x_2 + \beta_3 x_1 x_2$? How large a sample size is required to have sufficient degrees of freedom for estimating σ^2?

5.15 How many levels of x_1 and x_2 are required to fit the model $E(y) = \beta_0 + \beta_1 x_1 + \beta_2 x_2 + \beta_3 x_1 x_2 + \beta_4 x_1^2 + \beta_5 x_2^2$? How large a sample is required to have sufficient degrees of freedom for estimating σ^2?

5.16 A physiologist wanted to investigate the relationship between the physical characteristics of preadolescent boys and their maximal oxygen uptake (measured in milliliters of oxygen per kilogram of body weight). The data shown in the table were collected on a random sample of ten preadolescent boys. As a first step in the data analysis, the researcher fit the regression model

$$y = \beta_0 + \beta_1 x_1 + \beta_2 x_2 + \beta_3 x_3 + \beta_4 x_4 + \varepsilon$$

to the data. The output for a SAS regression analysis is given.

MAXIMAL OXYGEN UPTAKE y	AGE x_1, years	HEIGHT x_2, centimeters	WEIGHT x_3, kilograms	CHEST DEPTH x_4, centimeters
1.54	8.4	132.0	29.1	14.4
1.74	8.7	135.5	29.7	14.5
1.32	8.9	127.7	28.4	14.0
1.50	9.9	131.1	28.8	14.2
1.46	9.0	130.0	25.9	13.6
1.35	7.7	127.6	27.6	13.9
1.53	7.3	129.9	29.0	14.0
1.71	9.9	138.1	33.6	14.6
1.27	9.3	126.6	27.7	13.9
1.50	8.1	131.8	30.8	14.5

SAS Printout for Exercise 5.16

SOURCE	DF	SUM OF SQUARES	MEAN SQUARE	F VALUE	PROB > F
MODEL	4	0.20206274	0.05051568	23.18	0.0020
ERROR	5	0.01089726	0.00217945		
TOTAL	9	0.21296000		R-SQUARE	ROOT MSE
				0.948830	0.04668461

VARIABLE	PARAMETER ESTIMATE	STANDARD ERROR	T FOR H0: PARAMETER = 0	PROB > ¦T¦
INTERCEPT	-3.42634621	0.84078488	-4.08	0.0096
AGE	-0.03999447	0.02065180	-1.94	0.1105
HEIGHT	0.04814593	0.00759695	6.34	0.0014
WEIGHT	0.00179116	0.00544641	0.33	0.7556
CHEST	-0.07709906	0.08343441	-0.92	0.3979

a. Is the model adequate for predicting maximal oxygen uptake?

b. It seems reasonable to assume that the greater a child's chest depth, the greater should be the maximal oxygen uptake. But note that $\hat{\beta}_4$, the estimated coefficient of chest depth, x_4, is negative. Give an explanation for this result.

c. It would seem that the weight of a child should be positively correlated to lung volume and hence to maximal oxygen uptake. Can you explain the small t-value associated with $\hat{\beta}_3$?

d. Calculate the coefficient of correlation r for each pair of independent variables. Does this information confirm your suspicions in parts b and c?

5.17 Consider the data shown in the table.

x	54	42	28	38	25	70	48	41	20	52	65
y	6	16	33	18	41	3	10	14	45	9	5

a. Plot the points on a scattergram. What type of relationship appears to exist between x and y?

b. For each observation calculate $\log x$ and $\log y$. Plot the log-transformed data points on a scattergram. What type of relationship appears to exist between $\log x$ and $\log y$?

c. The scattergram from part b suggests that the transformed model

$$\log y = \beta_0 + \beta_1 \log x + \varepsilon$$

may be appropriate. Fit the transformed model to the data. Is the model adequate? Test using $\alpha = .05$.

d. Use the transformed model to predict the value of y when $x = 30$. [*Hint:* Use the inverse transformation $y = e^{\log y}$.]

5.18 Hamilton (1987) illustrated the multicollinearity problem with an example using the data shown in the accompanying table. The values of x_1, x_2, and y in the table represent appraised land value, appraised improvements value, and sale price, respectively, of a randomly selected residential property. (All measurements are in thousands of dollars.)

x_1	x_2	y
22.3	96.6	123.7
25.7	89.4	126.6
38.7	44.0	120.0
31.0	66.4	119.3
33.9	49.1	110.6
28.3	85.2	130.3
30.2	80.4	131.3
21.4	90.5	114.4
30.4	77.1	128.6
32.6	51.1	108.4
33.9	50.5	112.0
23.5	85.1	115.6
27.6	65.9	108.3
39.0	49.0	126.3
31.6	69.6	124.6

Source: Hamilton, D. "Sometimes $R^2 > r_{yx_1}^2 + r_{yx_2}^2$: Correlated variables are not always redundant," *The American Statistician*, Vol. 41, No. 2, May 1987, pp. 129–132.

a. Calculate the coefficient of correlation between y and x_1. Is there evidence of a linear relationship between sale price and appraised land value?

b. Calculate the coefficient of correlation between y and x_2. Is there evidence of a linear relationship between sale price and appraised improvements?

c. Based on the results in parts a and b, do you think the model $E(y) = \beta_0 + \beta_1 x_1 + \beta_2 x_2$ will be useful for predicting sale price?

d. Use a statistical computer program package to fit the model in part c, and conduct a test of model adequacy. In particular, note the value of R^2. Does the result agree with your answer to part c?

e. Calculate the coefficient of correlation between x_1 and x_2. What does the result imply?

f. Many researchers avoid the problems of multicollinearity by always omitting all but one of the "redundant" variables from the model. Would you recommend this strategy for this example? Explain. (Hamilton notes that in this case, such a strategy "can amount to throwing out the baby with the bathwater.")

5.19 Neil A. Palomba used multiple regression to relate the strike activity in a state (the percentage y of total working hours lost due to strikes) to three independent variables:

x_1 = Percentage of union members in nonagricultural establishments

x_2 = Percentage of all nonagricultural employment that is manufacturing

x_3 = Hourly earnings of workers on manufacturing payrolls

The data for the analysis are reproduced in the table.

STATE	y	x_1	x_2	x_3	STATE	y	x_1	x_2	x_3
Alabama	.14	18.0	30.5	2.17	Nebraska	.05	19.3	16.6	2.36
Alaska	.11	32.2	8.9	3.54	Nevada	.36	32.8	4.6	3.16
Arizona	.09	20.08	15.3	2.72	New Hampshire	.03	20.9	40.9	2.00
Arkansas	.10	26.2	29.2	1.78	New Jersey	.27	37.7	37.1	2.67
California	.16	33.8	24.9	2.96	New Mexico	.09	13.4	6.8	2.29
Colorado	.04	21.6	15.8	2.74	New York	.11	39.4	28.2	2.60
Connecticut	.08	24.6	42.5	2.62	North Carolina	.01	6.7	41.6	1.75
Delaware	.41	21.5	36.1	2.65	North Dakota	.03	14.8	5.8	2.28
Florida	.20	13.1	15.5	2.11	Ohio	.38	35.7	39.0	2.91
Georgia	.13	12.7	31.8	1.92	Oklahoma	.01	13.7	15.5	2.35
Hawaii	.02	24.2	12.1	2.14	Oregon	.12	34.8	26.5	2.85
Idaho	.11	19.2	18.8	2.50	Pennsylvania	.14	38.4	37.8	2.55
Illinois	.18	37.9	33.5	2.76	Rhode Island	.09	29.6	38.2	2.11
Indiana	.16	34.1	40.8	2.81	South Carolina	.01	7.9	42.7	1.80
Iowa	.16	20.8	25.4	2.71	South Dakota	.16	9.5	8.8	2.34
Kansas	.11	18.8	20.6	2.65	Tennessee	.23	17.6	34.6	2.03
Kentucky	.17	25.7	26.6	2.43	Texas	.06	13.3	19.4	2.42
Louisiana	.10	17.1	17.8	2.49	Utah	.66	19.7	17.6	2.77
Maine	.15	20.3	36.6	2.00	Vermont	.26	19.3	30.9	2.08
Massachusetts	.07	29.1	33.0	2.37	Virginia	.04	15.5	26.5	2.04
Michigan	.83	38.9	40.7	3.11	Washington	.16	43.1	25.6	2.98
Minnesota	.02	33.0	24.0	2.64	West Virginia	.45	42.0	27.4	2.67
Mississippi	.14	11.6	30.5	1.76	Wisconsin	.21	31.5	37.0	2.66
Missouri	.14	39.8	28.5	2.53	Wyoming	.01	19.2	7.7	2.82
Montana	.28	36.2	12.2	2.71	Maryland	—	—	—	—

Source: Palomba, N. A. "Strike activity and union membership: An empirical approach," *University of Washington Business Review*, Winter 1969.

a. Palomba initially considered the straight-line model

$$y = \beta_0 + \beta_1 x_1 + \varepsilon$$

Fit the model to the data and give your opinion on the adequacy of the model. In particular, interpret the value of $\hat{\beta}_1$.

b. Now consider the first-order model

$$y = \beta_0 + \beta_1 x_1 + \beta_2 x_2 + \beta_3 x_3 + \varepsilon$$

Fit the model to the data using an available multiple regression package and give your opinion on the adequacy of the model.

    c. Refer to part **b**. Conduct a test to determine whether the percentage of union members x_1 has a positive effect on the level of strike activity (when the other independent variables, x_2 and x_3 are held constant). Does this result contradict your result in part **a**? Explain.

    d. Calculate the variance inflation factor (VIF) for β_1. [*Hint:* You will need to fit the model $x_1 = \alpha_0 + \alpha_1 x_2 + \alpha_2 x_3 + \varepsilon$.] Interpret your result.

REFERENCES

Draper, N. and Smith, H. *Applied Regression Analysis*, 2nd ed. New York: Wiley, 1981.

Montgomery, D. C., and Peck, E. A. *Introduction to Linear Regression Analysis*. New York: Wiley, 1982. Chapters 3, 4, 8, and 9.

Mosteller, F. and Tukey, J. W. *Data Analysis and Regression: A Second Course in Statistics*. Reading, Mass.: Addison-Wesley, 1977. Chapters 4–6, 12–13.

Neter, J., Wasserman, W., and Kutner, M. H. *Applied Linear Statistical Models*. 2nd ed. Homewood, Ill.: Richard D. Irwin, 1985. Chapters 4, 8, and 11.

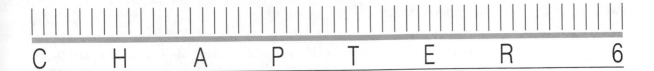

C H A P T E R 6

OBJECTIVE

To show how residuals can be used to detect departures from the model assumptions and to suggest some procedures for coping with these problems

CONTENTS

RESIDUAL ANALYSIS

SECTION 6.1

INTRODUCTION

We have repeatedly stated that the validity of many of the inferences associated with a regression analysis depends on the error term, ε, satisfying certain assumptions. Thus, when we test a hypothesis about a regression coefficient or a set of regression coefficients, or when we form a prediction interval for a future value of y, we must assume (1) that ε is normally distributed with a mean of 0, (2) constant variance σ^2, and (3) that all pairs of error terms are uncorrelated.* The objective of this chapter is to provide you with both graphical tools and statistical tests that will aid in checking the validity of these assumptions. In addition, these tools will help you evaluate the utility of the model and, in some cases, may suggest modifications to the model that will allow us to better describe the mean response.

In Section 6.2 we will show how to plot the residuals to reveal model inadequacies. The use of these plots and a simple test to detect unequal variances at different levels of the independent variable(s) is presented in Section 6.3. In Section 6.5 residual plots are used to detect outliers, i.e., observations that are unusually large or small relative to the others; procedures for measuring the influence these outliers may have on the fitted regression model are also presented. Finally, we discuss the use of residuals to test for time series correlation of the error term in Section 6.6.

SECTION 6.2

PLOTTING RESIDUALS AND DETECTING LACK OF FIT

The error term in a multiple regression model is, in general, not observable. To see this, consider the model

$$y = \beta_0 + \beta_1 x_1 + \cdots + \beta_k x_k + \varepsilon$$

and solve for the error term:

$$\varepsilon = y - (\beta_0 + \beta_1 x_1 + \cdots + \beta_k x_k)$$

Although you will observe values of the dependent variable and the independent variables $x_1, x_2, \ldots, x_k$, you will not know the true values of the regression coefficients $\beta_0, \beta_1, \ldots, \beta_k$. Therefore, the exact value of ε cannot be calculated.

After the data have been used to obtain least squares estimates $\hat{\beta}_0, \hat{\beta}_1, \ldots, \hat{\beta}_k$ of the regression coefficients, we can estimate the value of ε associated with each y-value using the corresponding **regression residual**, i.e., the deviation between the observed and the predicted value of y:

$$\hat{\varepsilon}_i = y_i - \hat{y}_i$$

To accomplish this, we must substitute the values of $x_1, x_2, \ldots, x_k$ into the prediction equation for each data point to obtain $\hat{y}$, and then this value must be subtracted from the observed value of y. Remember that you encountered the

---

*We assumed (Section 4.2) that the random errors associated with the linear model were independent. If two random variables are independent, it follows (proof omitted) that they will be uncorrelated. The reverse is generally untrue, except for normally distributed random variables. If two normally distributed random variables are uncorrelated, it can be shown that they are also independent.

regression residual in Chapters 3 and 4. In particular, the least squares estimates of $\beta_0, \beta_1, \beta_2, \ldots, \beta_k$ are those that minimize the sum of squares of the residuals,

$$\sum_{i=1}^{n} \hat{\varepsilon}_i^2 = \sum_{i=1}^{n} (y_i - \hat{y}_i)^2$$

DEFINITION 6.1

The **regression residual** is the observed value of the dependent variable minus the predicted value, or

$$\hat{\varepsilon} = y - \hat{y} = y - (\hat{\beta}_0 + \hat{\beta}_1 x_1 + \cdots + \hat{\beta}_k x_k)$$

EXAMPLE 6.1

The data in Table 6.1 show the monthly usage of electricity and the size of the home for a sample of ten homes (the same data were presented earlier in Table 4.1). Calculate the regression residuals for both the straight-line (first-order) model and the quadratic (second-order) model.

TABLE 6.1
Data for Monthly Electrical Usage Example

SIZE OF HOME x, square feet	MONTHLY USAGE y, kilowatt-hours
1,290	1,182
1,350	1,172
1,470	1,264
1,600	1,493
1,710	1,571
1,840	1,711
1,980	1,804
2,230	1,840
2,400	1,956
2,930	1,954

SOLUTION

The computer printout for the regression analysis of the first-order model,

$$y = \beta_0 + \beta_1 x + \varepsilon$$

is shown in Figure 6.1 (page 260). The least squares model is

$$\hat{y} = 578.928 + .540304x$$

Thus, the residual for the first observation, $x = 1,290$ and $y = 1,182$, is obtained by first calculating the predicted value

$$\hat{y} = 578.928 + .540304(1,290) = 1,275.92$$

and then subtracting from the observed value:

$$\hat{\varepsilon} = y - \hat{y} = 1,182 - 1,275.92 = -93.92$$

FIGURE 6.1   SAS Computer Printout for First-Order Model: Electrical Usage Example

ANALYSIS OF VARIANCE

SOURCE	DF	SUM OF SQUARES	MEAN SQUARE	F VALUE	PROB>F
MODEL	1	703957.18	703957.18	39.536	0.0002
ERROR	8	142444.92	17805.61457		
C TOTAL	9	846402.10			

ROOT MSE	133.4377	R-SQUARE	0.8317
DEP MEAN	1594.7	ADJ R-SQ	0.8107
C.V.	8.367573		

PARAMETER ESTIMATES

VARIABLE	DF	PARAMETER ESTIMATE	STANDARD ERROR	T FOR H0: PARAMETER=0	PROB > |T|
INTERCEP	1	578.92775	166.96806	3.467	0.0085
X	1	0.54030439	0.08592981	6.288	0.0002

Similar calculations for the other nine observations produce the residuals shown in Table 6.2.

TABLE 6.2
Regression Residuals for First-Order Model: Example 6.1

x	y	$\hat{y}$	$\hat{\varepsilon} = y - \hat{y}$
1,290	1,182	1,275.92	−93.92
1,350	1,172	1,308.34	−136.34
1,470	1,264	1,373.18	−109.18
1,600	1,493	1,443.41	49.59
1,710	1,571	1,502.85	68.15
1,840	1,711	1,573.09	137.91
1,980	1,804	1,648.73	155.27
2,230	1,840	1,783.81	56.19
2,400	1,956	1,875.66	80.34
2,930	1,954	2,162.02	−208.02

The printout for the second-order model

$$y = \beta_0 + \beta_1 x + \beta_2 x^2 + \varepsilon$$

is shown in Figure 6.2. The least squares model is

$$\hat{y} = -1,216.14 + 2.39893x - .00045004x^2$$

For the first observation, $x = 1,290$ and $y = 1,182$, the predicted electrical usage is

$$\hat{y} = -1,216.14 + 2.39893(1,290) - .00045004(1,290)^2 = 1,129.56*$$

*The residuals in Tables 6.2 and 6.3 have been generated using a computer program. Therefore, the results reported here will differ slightly from hand-calculated residuals due to rounding error.

and the regression residual is

$$\hat{\varepsilon} = y - \hat{y} = 1{,}182 - 1{,}129.56 = 52.44$$

All the regression residuals for the second-order model are given in Table 6.3.

FIGURE 6.2   SAS Regression Printout for Second-Order Model: Example 6.1

ANALYSIS OF VARIANCE

SOURCE	DF	SUM OF SQUARES	MEAN SQUARE	F VALUE	PROB>F
MODEL	2	831069.55	415534.77	189.710	0.0001
ERROR	7	15332.55363	2190.36480		
C TOTAL	9	846402.10			

ROOT MSE	46.80133	R-SQUARE	0.9819	
DEP MEAN	1594.7	ADJ R-SQ	0.9767	
C.V.	2.934805			

PARAMETER ESTIMATES

VARIABLE	DF	PARAMETER ESTIMATE	STANDARD ERROR	T FOR H0: PARAMETER=0	PROB > |T|
INTERCEP	1	-1216.14389	242.80637	-5.009	0.0016
X	1	2.39893018	0.24583560	9.758	0.0001
XX	1	-0.000450040	0.000059077	-7.618	0.0001

TABLE 6.3

Regression Residuals for Second-Order Model: Example 6.1

x	y	$\hat{y}$	$\hat{\varepsilon} = y - \hat{y}$
1,290	1,182	1,129.56	52.44
1,350	1,172	1,202.21	−30.21
1,470	1,264	1,337.79	−73.79
1,600	1,493	1,470.04	22.96
1,710	1,571	1,570.06	.94
1,840	1,711	1,674.23	36.77
1,980	1,804	1,769.40	34.60
2,230	1,840	1,895.47	−55.47
2,400	1,956	1,949.06	6.94
2,930	1,954	1,949.17	4.83

Graphical displays of regression residuals are useful aids to their interpretation. For example, the regression residual can be plotted on the vertical axis against one of the independent variables on the horizontal axis, or against the predicted value $\hat{y}$ (which is a linear function of the independent variables). If the assumptions concerning the error term ε are satisfied, we would expect to see residual plots that have no trends, no dramatic increases or decreases in variability, and only a few residuals (about 5%) more than two estimated standard deviations $(2s)$ of ε above or below 0. It is a property of the least squares prediction equation that the mean of the regression residuals will always be 0. That is, the least squares prediction equation not only minimizes the SSE, the sum of squared errors (residuals), but also produces residuals that have mean 0.

To illustrate, the residuals $\hat{\varepsilon}$ obtained by fitting the second-order model to the electrical usage data (Table 6.3) are plotted against the size of the home, x, in Figure 6.3. The middle, upper, and lower horizontal lines in Figure 6.3 locate the mean (0), $+2s$, and $-2s$, respectively, for the residuals. We detect no distinctive patterns or trends in this plot. All the residuals lie within $2s$ of the mean (0), and the variability around the mean is consistent for small and large homes.

FIGURE 6.3

SAS Plot of Regression Residuals for the Second-Order Model: Electrical Usage Example

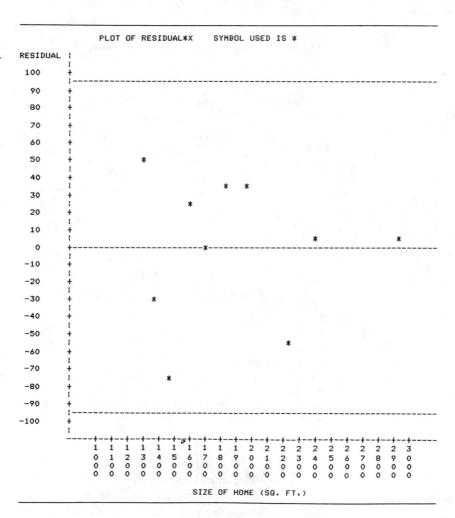

EXAMPLE 6.2

Example 6.1 gives the residuals obtained from fitting a first-order model to the electrical usage data. Plot the residuals obtained in this regression analysis against the home size, x (x is recorded on the horizontal axis). Does the plot suggest model inadequacy, or departure from the usual assumptions made about the error term ε?

SOLUTION

The residuals for the first-order model were given in Table 6.2. The plot of these residuals versus home size is shown in Figure 6.4. The distinctive aspect of this plot is the parabolic distribution of the residuals about their mean, i.e., all residuals tend to be positive for the homes of intermediate size and negative for the homes of either small or large size. This parabolic appearance of the trend in the residuals suggests that a second-order term may improve the model. As we know, the addition of a second-order term does improve the model, and the residual plot for the second-order model no longer shows an observable pattern (see Figure 6.3).

FIGURE 6.4

SAS Plot of Regression Residuals for the First-Order Model: Electrical Usage Example

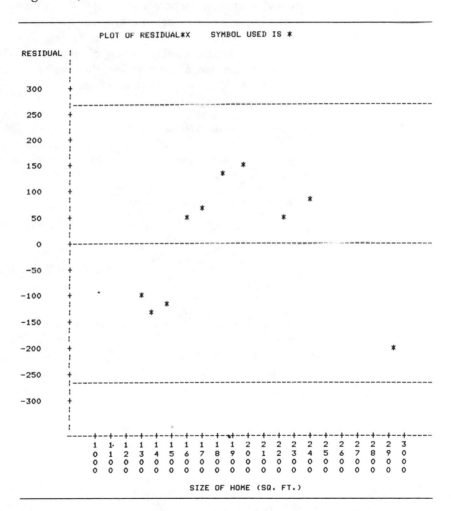

An alternative method of detecting lack of fit in models with more than one independent variable is to construct a partial residual plot. The **partial residuals** for the jth independent variable, x_j, in the model are calculated as follows:

$$\hat{\varepsilon}^* = y - (\hat{\beta}_0 + \hat{\beta}_1 x_1 + \hat{\beta}_2 x_2 + \cdots + \hat{\beta}_{j-1} x_{j-1} + \hat{\beta}_{j+1} x_{j+1} + \cdots + \hat{\beta}_k x_k)$$
$$= \hat{\varepsilon} + \hat{\beta}_j x_j$$

where $\hat{\varepsilon}$ is the usual regression residual.

Partial residuals measure the influence of x_j on the independent variable y *after the effects of the other independent variables* $(x_1, x_2, \ldots, x_{j-1}, x_{j+1}, \ldots, x_k)$ *have been removed, or accounted for*. If the partial residuals $\hat{\varepsilon}^*$ are regressed against x_j in a straight-line model, the resulting least squares slope is equal to $\hat{\beta}_j$—the β estimate obtained from the full model. Therefore, when the partial residuals are plotted against x_j, the points are scattered around a line with slope equal to $\hat{\beta}_j$. Unusual deviations or patterns around this line indicate lack of fit for the variable x_j.

A plot of the partial residuals versus x_j often reveals more information about the relationship between y and x_j than the usual residual plot. In particular, a partial residual plot usually indicates more precisely how to modify the model,[†] as the next example illustrates.

DEFINITION 6.2

The set of **partial regression residuals** for the jth independent variable x_j are calculated as follows:

$$\hat{\varepsilon}^* = y - (\hat{\beta}_0 + \hat{\beta}_1 x_1 + \hat{\beta}_2 x_2 + \cdots$$
$$+ \hat{\beta}_{j-1} x_{j-1} + \hat{\beta}_{j+1} x_{j+1} + \cdots + \hat{\beta}_k x_k)$$
$$= \hat{\varepsilon} + \hat{\beta}_j x_j$$

where $\hat{\varepsilon} = y - \hat{y}$ is the usual regression residual (see Definition 6.1).

EXAMPLE 6.3

Refer to the supermarket chain experiment in Example 5.2. Recall that the chain wanted to investigate the effect of price x_1 on the weekly demand y for a house brand of coffee at eight of its stores. Eight prices were randomly assigned to the stores and were advertised using the same procedures. A few weeks later, the chain conducted the same experiment using no advertisements. The data for the entire study are shown in Table 6.4.

Consider the model

$$E(y) = \beta_0 + \beta_1 p + \beta_2 x_2$$

where

$$x_2 = \begin{cases} 1 & \text{if advertisement used} \\ 0 & \text{if not} \end{cases}$$

[†]Partial residual plots display the correct functional form of the predictor variables across the relevant range of interest, except in cases where severe multicollinearity exists. See Mansfield and Conerly (1987) for an excellent discussion of the use of residual and partial residual plots.

	WEEKLY DEMAND y, pounds	PRICE p, dollars per pound	ADVERTISEMENT x_2
TABLE 6.4 Data for Example 6.3	1,120	3.00	1
	999	3.10	1
	932	3.20	1
	884	3.30	1
	807	3.40	1
	760	3.50	1
	701	3.60	1
	688	3.70	1
	1,037	3.00	0
	962	3.10	0
	904	3.20	0
	827	3.30	0
	775	3.40	0
	715	3.50	0
	666	3.60	0
	607	3.70	0

a. Fit the model to the data. Is the model adequate for predicting weekly demand y?

b. Calculate and plot the residuals versus p. Do you detect any trends?

c. Calculate the partial residuals for the independent variable p, and construct the corresponding partial residual plot. What does the plot reveal?

d. Fit the model $E(y) = \beta_0 + \beta_1 x_1 + \beta_2 x_2$, where $x_1 = 1/p$. Has the predictive ability of the model improved?

SOLUTION

a. The SAS printout for the regression analysis is shown in Figure 6.5 (page 266). The F-value for testing model adequacy, i.e., H_0: $\beta_1 = \beta_2 = 0$, is given on the printout (shaded) as $F = 368.298$ with a corresponding p-value (also shaded) of $p = .0001$. Thus, there is sufficient evidence (at any α of .0001 or greater) that the model contributes information for the prediction of weekly demand, y. Also, the coefficient of determination is $R^2 = .9827$, meaning that the model explains approximately 98% of the sample variation in weekly demand.

Recall from Example 5.2, however, that we fit a model with the transformed independent variable $x_1 = 1/p$. That is, we expect the relationship between weekly demand y and price p to be decreasing in a curvilinear fashion and approaching (but never reaching) 0 as p increases. (See curve 1 in Figure 5.7.) If such a relationship exists, the model (with untransformed price), although statistically useful for predicting demand y, will be inadequate in a practical setting.

b. The regression residuals for the model in part **a** are shown in the bottom portion of Figure 6.5. A computer-generated (SAS) plot of these residuals against price p is portrayed in Figure 6.6 (page 267). Notice that the plot

FIGURE 6.5   SAS Printout for Example 6.3

ANALYSIS OF VARIANCE

SOURCE	DF	SUM OF SQUARES	MEAN SQUARE	F VALUE	PROB>F
MODEL	2	320515.30	160257.65	368.298	0.0001
ERROR	13	5656.70238	435.13095		
C TOTAL	15	326172.00			

ROOT MSE	20.85979	R-SQUARE	0.9827	
DEP MEAN	836.5	ADJ R-SQ	0.9800	
C.V.	2.493699			

PARAMETER ESTIMATES

VARIABLE	DF	PARAMETER ESTIMATE	STANDARD ERROR	T FOR H0: PARAMETER=0	PROB > |T|
INTERCEP	1	2848.74405	76.60151888	37.189	0.0001
P	1	-608.09524	22.75989979	-26.718	0.0001
X2	1	49.75000000	10.42989636	4.770	0.0004

OBS	ID	ACTUAL	PREDICT VALUE	RESIDUAL
1	3	1120.0	1074.2	45.7917
2	3.1	999.0	1013.4	-14.3988
3	3.2	932.0	952.6	-20.5893
4	3.3	884.0	891.8	-7.7798
5	3.4	807.0	831.0	-23.9702
6	3.5	760.0	770.2	-10.1607
7	3.6	701.0	709.4	-8.3512
8	3.7	688.0	648.5	39.4583
9	3	1037.0	1024.5	12.5417
10	3.1	962.0	963.6	-1.6488
11	3.2	904.0	902.8	1.1607
12	3.3	827.0	842.0	-15.0298
13	3.4	775.0	781.2	-6.2202
14	3.5	715.0	720.4	-5.4107
15	3.6	666.0	659.6	6.3988
16	3.7	607.0	598.8	8.2083

SUM OF RESIDUALS   8.41283E-12
SUM OF SQUARED RESIDUALS   5656.702

reveals a parabolic trend, implying a lack of fit. Thus, the residual plot supports our hypothesis that the weekly demand–price relationship is curvilinear, not linear. However, the appropriate transformation on price (i.e., $1/p$) is not evident from the plot. In fact, the nature of the curvature in Figure 6.6 may lead you to conclude that the addition of the quadratic term, $\beta_3 p^2$, to the model will solve the problem. In general, a residual plot will detect curvature if it exists, but may not reveal the appropriate transformation.

c. The partial residuals (denoted $\hat{\varepsilon}^*$) for the independent variable p are calculated using the formula given in Definition 6.2:

$$\hat{\varepsilon}^* = \hat{\varepsilon} + \hat{\beta}_1 p$$

where $\hat{\beta}_1 = -608.1$ (from Figure 6.5). For example, the partial residual for the first observation is calculated by substituting $\hat{\varepsilon} = 45.79$ and $p = 3.00$ into the equation:

$$\hat{\varepsilon}^* = \hat{\varepsilon} - 608.1p$$
$$= 45.79 - 608.1(3) = -1,778.5$$

FIGURE 6.6 SAS Printout of Residuals Against Price for Example 6.3

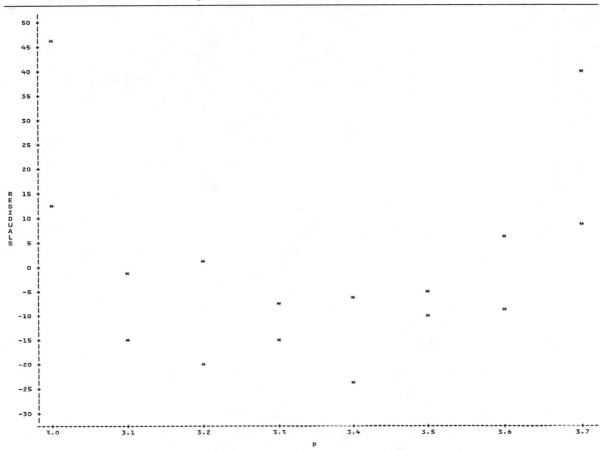

The complete set of partial residuals for p is shown in Table 6.5, accompanied by a computer-generated (SAS) partial residual plot in Figure 6.7 (page 268). You can see that the partial residual plot also reveals a curvilinear trend but, in addition, displays the correct functional form of weekly demand–price

TABLE 6.5
Partial Residuals for Price p in First-Order Model for Example 6.3

PRICE p	RESIDUAL $\hat{\varepsilon}$	PARTIAL RESIDUAL $\hat{\varepsilon}^*$	PRICE p	RESIDUAL $\hat{\varepsilon}$	PARTIAL RESIDUAL $\hat{\varepsilon}^*$
3.0	45.792	−1,778.5	3.0	12.542	−1,811.7
3.1	−14.399	−1,899.5	3.1	−1.649	−1,886.7
3.2	−20.589	−1,966.5	3.2	1.161	−1,944.7
3.3	−7.780	−2,014.5	3.3	−15.030	−2,021.7
3.4	−23.970	−2,091.5	3.4	−6.220	−2,073.7
3.5	−10.161	−2,138.5	3.5	−5.411	−2,133.7
3.6	−8.351	−2,197.5	3.6	6.399	−2,182.7
3.7	39.458	−2,210.5	3.7	8.208	−2,241.7

FIGURE 6.7   Printout of Partial Residuals Against Price for Example 6.3

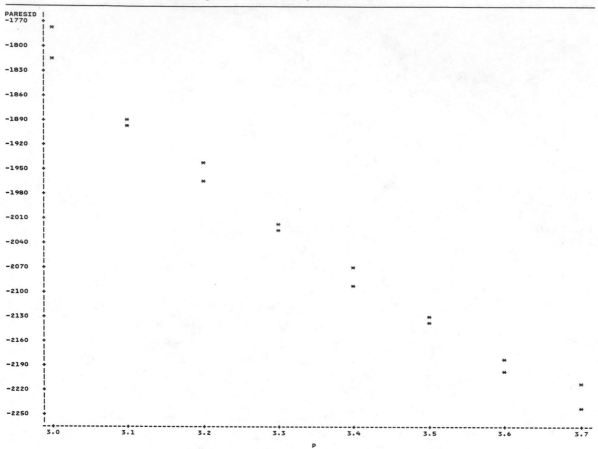

relationship. Notice that the curve is decreasing and approaching (but never reaching) 0 as p increases. This suggests that the appropriate transformation on price is either $1/p$ or e^{-p} (see Figure 5.7).

d. Using the transformation $x_1 = 1/p$, we refit the model to the data, and the resulting SAS printout is shown in Figure 6.8. The small p-value ($p = .0001$) for testing H_0: $\beta_1 = \beta_2 = 0$ indicates that the model is adequate for predicting y. Although the coefficient of determination increased only slightly (from $R^2 = .98$ to $R^2 = .99$), the model standard deviation (ROOT MSE) decreased significantly (from $s = 20.9$ to $s = 14.6$). Thus, whereas the model with untransformed price can predict weekly demand for coffee to within $2s = 2(20.9) = 41.8$ pounds, the transformed model can predict demand to within $2(14.6) = 29.2$ pounds.

FIGURE 6.8 SAS Printout for Example 6.3

```
                         ANALYSIS OF VARIANCE

                        SUM OF          MEAN
       SOURCE    DF     SQUARES        SQUARE      F VALUE     PROB>F

       MODEL      2    323412.38     161706.19     761.765     0.0001
       ERROR     13   2759.61908     212.27839
       C TOTAL   15    326172.00

            ROOT MSE      14.56978      R-SQUARE       0.9915
            DEP MEAN         836.5      ADJ R-SQ       0.9902
            C.V.         1.741755

                        PARAMETER ESTIMATES

                     PARAMETER      STANDARD      T FOR H0:
    VARIABLE    DF    ESTIMATE       ERROR      PARAMETER=0    PROB > |T|

    INTERCEP     1   -1223.99520   53.21896999    -22.999       0.0001
    X1           1    6787.31169   176.61334       38.430       0.0001
    X2           1    49.75000000    7.28488832      6.829       0.0001
```

■

Residual (or partial residual) plots are useful for indicating potential model improvements but they are no substitute for formal statistical tests of model terms to determine their importance. Thus, a true test of whether the second-order term contributes to the electrical usage model (Example 6.1) is the t-test of the null hypothesis H_0: $\beta_2 = 0$. The appropriate test statistic, shown in the printout of Figure 6.2, indicates that the second-order term does contribute information for the prediction of electrical usage y. We have confidence in this statistical inference because we know the probability α of committing a Type I error (concluding a term is important when, in fact, it is not). In contrast, decisions based on residual plots are subjective, and their reliability cannot be measured. Therefore, we suggest that such plots only be used as indicators of *potential* problems. The final judgment on model adequacy should be based on appropriate statistical tests.*

EXERCISES 6.1–6.6

6.1 A first-order model is fit to the data shown in the table with the following result:

$$\hat{y} = 2.588 + .541x$$

x	−2	−2	−1	−1	0	0	1	1	2	2	3	3
y	1.1	1.3	2.0	2.1	2.7	2.8	3.4	3.6	4.0	3.9	3.8	3.6

*A more general procedure for determining whether the straight-line model adequately fits the data tests the null hypothesis H_0: $E(y) = \beta_0 + \beta_1 x$ against the alternative H_a: $E(y) \neq \beta_0 + \beta_1 x$. You can see that this test, called a test for *lack of fit*, does not restrict the alternative hypothesis to second-order models. Lack-of-fit tests are appropriate when the x-values are replicated, i.e., when the sample data include two or more observations for several different levels of x. When the data are observational, however, replication rarely occurs. (Note that none of the values of x are repeated in Table 6.1.) For details on how to conduct tests for lack of fit, consult the references given at the end of this chapter.

a. Calculate the residuals for the model.

b. Plot the residuals versus x. Do you detect any trends? If so, what does the pattern suggest about the model?

6.2 A first-order model is fit to the data shown in the table with the following result:

$$\hat{y} = -3.179 + 2.491x$$

x	2	4	7	10	12	15	18	20	21	25
y	5	10	12	22	25	27	39	50	47	65

a. Calculate the residuals for the model.

b. Plot the residuals versus x. Do you detect any trends? If so, what does the pattern suggest about the model?

6.3 Underinflated or overinflated tires can increase tire wear and decrease gas mileage. A new tire was tested for wear at different pressures with the results shown in the accompanying table. Suppose you are interested in modeling the relationship between y and x.

PRESSURE x, pounds per square inch	MILEAGE y, thousands
30	29
31	32
32	36
33	38
34	37
35	33
36	26

a. Fit the straight-line model $y = \beta_0 + \beta_1 x + \varepsilon$ to the data.

b. Calculate the residuals for the model.

c. Plot the residuals versus x. Do you detect any trends? If so, what does the pattern suggest about the model?

d. Fit the quadratic model $y = \beta_0 + \beta_1 x + \beta_2 x^2 + \varepsilon$ to the data using an available multiple regression computer program package. Has the addition of the quadratic term improved model adequacy?

6.4 The 1986 EPA gas mileage guide gives the engine size and estimated city miles per gallon ratings for the 11 gasoline-fueled subcompact and compact cars shown in the table. (The engine sizes are in total cubic inches of cylinder volume.) In an attempt to predict gas mileage from the engine size of subcompact and compact cars, the first-order model

$$y = \beta_0 + \beta_1 x + \varepsilon$$

is fit to the data. The resulting least squares model is $\hat{y} = 37.677 - .0724x$.

CAR	CYLINDER VOLUME x	MILES PER GALLON y
VW Golf	97	37
Chevy Cavalier	173	19
Plymouth Horizon	97	31
Pontiac Firebird	151	23
Corvette	350	17
Honda Accord	119	27
Dodge Omni	97	31
Renault Alliance	85	35
Olds Firenza	173	19
Nissan Sentra	97	31
Ford Escort	114	32

Sources: 1986 Gas Mileage Guide, EPA Fuel Economy Estimates.
U.S. Department of Energy. *Wards Automotive Yearbook,* 1986.

a. Calculate the regression residuals for this model.
b. Verify that the sum of the residuals is 0.
c. Plot these residuals against cylinder volume, x.
d. Do you detect any distinctive patterns or trends in this plot?
e. What does your answer to part c suggest about model adequacy or the usual assumptions made about the error term ε?

6.5 The real estate data for Exercise 4.3 are reproduced below. Recall that the property appraisers fit the model

$$E(y) = \beta_0 + \beta_1 x_1 + \beta_2 x_2$$

SALE PRICE y, \$ thousands	HOME SIZE x_1, hundreds of sq. ft.	CONDITION RATING x_2, 1 to 10
60.0	23	5
32.7	11	2
57.7	20	9
45.5	17	3
47.0	15	8
55.3	21	4
64.5	24	7
42.6	13	6
54.5	19	7
57.5	25	2

Source: Andrews, R. L. and Ferguson, J. T. "Integrating judgment with a regression appraisal," *The Real Estate Appraiser and Analyst,* Vol. 52, No. 2, Spring 1986 (Table I).

where

y = Sale price (in $ thousands)

x_1 = Home size (in square feet)

x_2 = Condition rating (1 to 10)

The resulting least squares prediction equation is

$$\hat{y} = 9.782 + 1.871x_1 + 1.278x_2$$

a. Calculate the residuals for the model.

b. Plot the residuals versus x_1. Do you detect any trends? If so, what does the pattern suggest about the model?

c. Plot the residuals versus x_2. Do you detect any trends? If so, what does the pattern suggest about the model?

d. Calculate and plot the partial residuals for x_1. Interpret the result.

e. Calculate and plot the partial residuals for x_2. Interpret the result.

6.6 Taxes, a major source of income for the state of Florida, have grown steadily since 1970. The data in the table give the total state tax collections for the years 1971–1985. A tax economist fit the model

$$y = \beta_0 + \beta_1 x + \beta_2 x^2 + \varepsilon$$

to the data. Results from the SAS printout gave the least squares model

$$\hat{y} = 1{,}772.39 + 88.28x + 20.85x^2$$

YEAR	YEAR – 1970 x	TOTAL TAX COLLECTIONS y, million dollars
1971	1	1,587
1972	2	1,996
1973	3	2,488
1974	4	2,794
1975	5	2,791
1976	6	2,936
1977	7	3,275
1978	8	3,764
1979	9	4,291
1980	10	4,804
1981	11	5,314
1982	12	5,556
1983	13	6,225
1984	14	7,329
1985	15	7,883

Source: United States Bureau of the Census, *Statistical Abstracts of the United States 1972–1987.*

a. Calculate the regression residuals for this second-order (quadratic) model.

b. As a check on your calculations, compute the sum of the residuals. (The sum should be very close to 0.)

c. Graph the residuals against the independent variable x. What does the residual plot suggest about the validity of the usual assumptions about the error term ε?

SECTION 6.3

**DETECTING UNEQUAL
VARIANCES**

Recall that one of the assumptions necessary for the validity of regression inferences is that the error term ε have constant variance σ^2 for all levels of the independent variable(s). Variances that satisfy this property are called **homoscedastic**. Unequal variances for different settings of the independent variable(s) are said to be **heteroscedastic**. Various statistical tests for heteroscedasticity have been developed. However, plots of the residuals will frequently reveal the presence of heteroscedasticity. In this section we will show how residual plots can be used to detect departures from the assumption of equal variances, and then give a simple test for heteroscedasticity. In addition, we will suggest some modifications to the model that may remedy the situation.

When business and economic data fail to be homoscedastic, the reason is often that the variance of the response y is a function of its mean $E(y)$. Some examples follow:

1. If the response y is a count that has a Poisson distribution, the variance will be equal to the mean $E(y)$. Poisson data are usually counts per unit volume, area, time, etc. For example, the number of sick days per month for an employee would very likely be a Poisson random variable. If the variance of a response is proportional to $E(y)$, the regression residuals produce a pattern about $\hat{y}$, the least squares estimate of $E(y)$, like that shown in Figure 6.9.

FIGURE 6.9

A Plot of Residuals
for Poisson Data

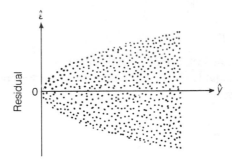

2. Many responses are proportions (or percentages) generated by **binomial experiments**. For example, the proportion of a random sample of 100 economists who forecast a recession within the next year is an example of a binomial response. Binomial proportions have variances that are functions of both the true proportion (the mean) and the sample size. In fact, if the observed proportion $y_i = \hat{p}_i$ is generated by a binomial distribution with sample size n_i and true probability p_i, the variance of y_i is

$$\text{Var}(y_i) = \frac{p_i(1 - p_i)}{n_i} = \frac{E(y_i)[1 - E(y_i)]}{n_i}$$

Residuals for binomial data produce a pattern about $\hat{y}$ like that shown in Figure 6.10 (page 274).

FIGURE 6.10
A Plot of Residuals
for Binomial Data
(Proportions or
Percentages)

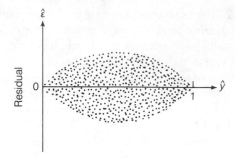

3. The random error component has been assumed to be **additive** in all the models we have constructed. An additive error is one for which the response is equal to the mean $E(y)$ *plus* random error,

$$y = E(y) + \varepsilon$$

Another useful type of model for business and economic data is the **multiplicative** model. In this model, the response is written as the *product* of its mean and the random error component, i.e.,

$$y = [E(y)]\varepsilon$$

The variance of this response will grow proportionally to the *square* of the mean, i.e.,

$$\text{Var}(y) = [E(y)]^2 \sigma^2$$

where σ^2 is the variance of ε. Data subject to multiplicative errors produce a pattern of residuals about $\hat{y}$ like that shown in Figure 6.11.

FIGURE 6.11
A Plot of Residuals
for Data Subject to
Multiplicative Errors

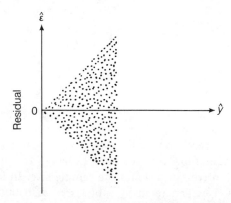

When the variance of y is a function of its mean, we can often satisfy the least squares assumption of homoscedasticity by transforming the response to some new response that has a constant variance. These are called **variance-stabilizing transformations**. For example, if the response y is a count that follows a Poisson distribution, the square root transform $\sqrt{y}$ can be shown to have approximately constant variance. Consequently, if the response is a Poisson random variable, we would let

$$y^* = \sqrt{y}$$

and fit the model

$$y^* = \beta_0 + \beta_1 x_1 + \cdots + \beta_k x_k + \varepsilon$$

This model will satisfy approximately the least squares assumption of homoscedasticity.

Similar transformations that are appropriate for percentages and proportions (binomial data) or for data subject to multiplicative errors are shown in Table 6.6. The transformed responses will satisfy (at least approximately) the assumption of homoscedasticity.

TABLE 6.6
Stabilizing Transformations for Heteroscedastic Responses

TYPE OF RESPONSE	VARIANCE	STABILIZING TRANSFORMATION
Poisson	$E(y)$	$\sqrt{y}$
Binomial proportion	$\dfrac{E(y)[1 - E(y)]}{n}$	$\sin^{-1}\sqrt{y}$
Multiplicative	$[E(y)]^2\sigma^2$	$\log y^a$

[a]Unless otherwise noted, natural logarithms (to the base e) will be used.

The data in Table 6.7 (page 276) are the executive salaries, y, and years of experience, x, for a sample of 50 executives from a major industry. If we fit the second-order model $E(y) = \beta_0 + \beta_1 x + \beta_2 x^2$ to the data, we obtain the Minitab computer printout for the regression analysis shown in Figure 6.12 and the prediction equation

$$\hat{y} = 20{,}242 + 522x + 53x^2$$

FIGURE 6.12
Minitab Regression Analysis Printout for Second-Order Model: Executive Salaries

```
The regression equation is
Y = 20242 + 522 X + 53.0 XX

Predictor        Coef        Stdev      t-ratio
Constant        20242        4423         4.58
X               522.4        616.7        0.85
XX              53.00        19.57        2.71

s = 8123        R-sq = 81.6%      R-sq(adj) = 80.8%

Analysis of Variance

SOURCE        DF          SS              MS
Regression     2     13722605568      6861302784
Error         47      3101243136        65983896
Total         49     16823848960
```

TABLE 6.7

Salary and Experience Data for 50 Executives from a Major Industry

YEARS OF EXPERIENCE x	SALARY y	YEARS OF EXPERIENCE x	SALARY y	YEARS OF EXPERIENCE x	SALARY y
7	$26,075	21	$43,628	28	$99,139
28	79,370	4	16,105	23	52,624
23	65,726	24	65,644	17	50,594
18	41,983	20	63,022	25	53,272
19	62,309	20	47,780	26	65,343
15	41,154	15	38,853	19	46,216
24	53,610	25	66,537	16	54,288
13	33,697	25	67,447	3	20,844
2	22,444	28	64,785	12	32,586
8	32,562	26	61,581	23	71,235
20	43,076	27	70,678	20	36,530
21	56,000	20	51,301	19	52,745
18	58,667	18	39,346	27	67,282
7	22,210	1	24,833	25	80,931
2	20,521	26	65,929	12	32,303
18	49,727	20	41,721	11	38,371
11	33,233	26	82,641		

The computer printout suggests that the second-order model provides an adequate fit to the data. The R^2-value, .816, indicates that the model explains 81.6% of the total variation of the y-values about $\bar{y}$. The F-value, the ratio of mean square regression (6,861,302,784) to mean square error (65,983,896), $F = 103.98$, is highly significant and indicates that the model contributes information for the prediction of y. However, an examination of the salary residuals plotted against the estimated mean salary, $\hat{y}$, as shown in Figure 6.13, reveals a potential problem. Note the "cone" shape of the residual variability; the size of the residuals* increases as the estimated mean salary increases. This residual plot indicates the possibility of a multiplicative model. We will explore this possibility further in Example 6.4.

EXAMPLE 6.4

Consider the salary and experience data in Table 6.7. Use the logarithmic transformation on the dependent variable, and relate log y to years of experience, x, using the second-order model

$$\log y = \beta_0 + \beta_1 x + \beta_2 x^2 + \varepsilon$$

Evaluate the adequacy of the model.

*The vertical axis of the Minitab residual plot in Figure 6.13 gives the residuals in terms of number of standard deviations (positive or negative) from the mean 0. For example, the residual value $\hat{\varepsilon} = -10,161$ is reported on the plot as $-10,161/8,123 = -1.25$, where $s = 8,123$ is the estimated standard deviation given on the printout.

FIGURE 6.13

Executive Salary Versus Experience Example: Minitab Residual Plot for Second-Order Model with Dependent Variable, Salary

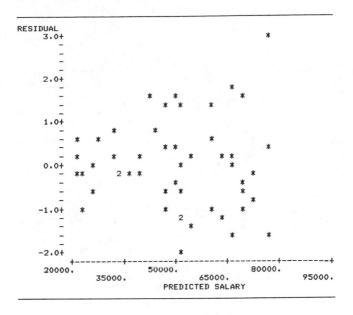

SOLUTION

The Minitab computer printout in Figure 6.14 gives the regression analysis for the $n = 50$ measurements. The prediction equation used in computing the residuals is

$$\widehat{\log y} = 9.8429 + .0497x + .000009x^2$$

The residual plot in Figure 6.15 (page 278) indicates that the logarithmic transformation has significantly reduced the heteroscedasticity.* Note that the cone shape is gone; there is no apparent tendency of the residual variance to increase as mean salary increases. We therefore are confident that inferences using the logarithmic model are more reliable than those using the untransformed model.

FIGURE 6.14

Minitab Regression Analysis for Logarithmic Transform of Salary Data: Second-Order Model

```
The regression equation is
LOGY = 9.84 + 0.0497 X +0.000009 XX

Predictor       Coef        Stdev      t-ratio
Constant     9.84289      0.08479      116.08
X            0.04969      0.01182        4.20
XX         0.0000093      0.0003753       0.02

s = 0.1557       R-sq = 86.4%      R-sq(adj) = 85.8%

Analysis of Variance

SOURCE       DF          SS           MS
Regression    2       7.2119       3.6059
Error        47       1.1400       0.0243
Total        49       8.3519
```

*A printout of the residuals is omitted.

FIGURE 6.15
Executive Salary Versus
Experience Example:
Minitab Residual Plot for
Second-Order Model with
Dependent Variable,
log(Salary)

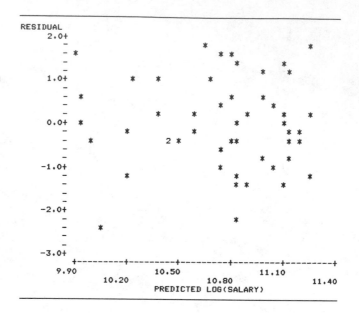

To evaluate model adequacy, we first note that $R^2 = .864$ and that 86.4% of the variation in log(salary) is accounted for by the model (which includes first- and second-order terms for years of experience). The computed F-value, the ratio of mean square regression (3.6059) to mean square error (.0243), $F = 148.39$, indicates that the model significantly improves upon the sample mean as a predictor of log(salary).

Although the estimate $\hat{\beta}_2$ of β_2 is very small, we should check to determine whether the data provide sufficient evidence to indicate that the second-order term contributes information for the prediction of log(salary). The test of

$$H_0: \quad \beta_2 = 0 \qquad H_a: \quad \beta_2 \neq 0$$

is conducted using the t-statistic shown in Figure 6.14, $t = .02$. The absolute value of this computed t-value is extremely small and, clearly, does not exceed the tabulated t-value for any reasonable value of α ($t_{.05} \approx 1.645$). Consequently, there is insufficient evidence to indicate that the second-order term contributes to the prediction of log(salary). There is no indication that the second-order model is an improvement over the straight-line model,

$$\log y = \beta_0 + \beta_1 x + \varepsilon$$

for predicting log(salary).

The Minitab computer printout for the first-order model (Figure 6.16) shows that the prediction equation for the first-order model is

$$\widehat{\log y} = 9.8413 + .04998x$$

The value of R^2, .863, is approximately the same as the value of R^2 obtained for the second-order model. The F-statistic, computed from the mean squares in

Figure 6.16, $F = 303.0$, indicates that the model contributes significantly to the prediction of log y.

FIGURE 6.16

Minitab Regression
Analysis for Logarithmic
Transform of Salary Data:
First-Order Model

```
The regression equation is
LOGY = 9.84 + 0.0500 X

Predictor          Coef        Stdev      t-ratio
Constant        9.84133      0.05636       174.63
X               0.049978     0.002868       17.43

s = 0.1541      R-sq = 86.3%      R-sq(adj) = 86.1%

Analysis of Variance

SOURCE          DF          SS           MS
Regression       1       7.2118       7.2118
Error           48       1.1400       0.0238
Total           49       8.3519
```

When the transformed model of Example 6.4 is used to predict the value of log y, the predicted value of y is the antilog, $\hat{y} = e^{\widehat{\log y}}$. The endpoints of the prediction interval are similarly transformed back to the original scale, and the interval will retain its meaning. In repeated use, the intervals will contain the observed y-value $100(1 - \alpha)\%$ of the time.

Unfortunately, you cannot take antilogs to find the confidence interval for the mean value $E(y)$. The reason for this is that the mean value of log y is not equal to the logarithm of the mean of y. In fact, the antilog of the logarithmic mean of a random variable y is called its **geometric mean**. Thus, the antilogs of the endpoints of the confidence interval for the mean of the transformed response will give a confidence interval for the geometric mean. Similar care must be exercised with other types of transformations. In general, prediction intervals can be transformed back to the original scale without losing their meaning, but confidence intervals for the mean of a transformed response cannot.

The preceding examples illustrate that, in practice, residual plots can be a powerful technique for detecting heteroscedasticity. Furthermore, the pattern of the residuals often suggests the appropriate variance-stabilizing transformation to use. Keep in mind, however, that no measure of reliability can be attached to inferences derived from a graphical technique. For this reason, you may want to rely on a statistical test.

Various tests for heteroscedasticity in regression have been developed. One of the simpler techniques utilizes the F-test (discussed in Chapter 2) for comparing population variances. The procedure requires that you divide the sample data in half and fit the regression model to each half. If the regression model fit to one-half the observations yields a significantly smaller or larger MSE than the model fit to the other half, there is evidence to indicate that the assumption of equal variances for all levels of the x variables in the model is being violated. (Recall that MSE, or mean square for error, estimates σ^2, the variance of the random error term.) Where you divide the data depends on where you suspect the differences in variances to be. We illustrate this procedure with an example.

EXAMPLE 6.5

Refer to the executive salary (y) versus years of experience (x) example. The residual plot for the quadratic model

$$E(y) = \beta_0 + \beta_1 x + \beta_2 x^2$$

indicates that the assumption of equal variances may be violated (see Figure 6.13). Conduct a statistical test of hypothesis to determine whether heteroscedasticity exists. Use $\alpha = .05$.

SOLUTION

The residual plot shown in Figure 6.13 reveals that the residuals associated with larger values of predicted salary tend to be more variable than the residuals associated with smaller values of predicted salary. Therefore, we will divide the sample observations based on the values of $\hat{y}$, or, equivalently, the value of x (since, for the fitted model, $\hat{y}$ increases as x increases). An examination of the data in Table 6.7 reveals that approximately one-half of the 50 observed values of years of experience, x, fall below $x = 20$. Thus, we will divide the data into two subsamples as follows:

SUBSAMPLE 1	SUBSAMPLE 2
$x < 20$	$x \geq 20$
$n_1 = 24$	$n_2 = 26$

Figures 6.17(a) and (b) give the SAS printouts for the quadratic model fit to subsample 1 and subsample 2, respectively. The value of MSE is shaded in each printout.

The null and alternative hypotheses to be tested are:

H_0: $\dfrac{\sigma_1^2}{\sigma_2^2} = 1$ (Assumption of equal variances satisfied)

H_a: $\dfrac{\sigma_1^2}{\sigma_2^2} \neq 1$ (Assumption of equal variances violated)

where

σ_1^2 = Variance of the random error term, ε, for subpopulation 1 (i.e., $x < 20$)
σ_2^2 = Variance of the random error term, ε, for subpopulation 2 (i.e., $x \geq 20$)

The test statistic for a two-tailed test is given by:

$$F = \frac{\text{Larger } s^2}{\text{Smaller } s^2} = \frac{\text{Larger MSE}}{\text{Smaller MSE}} \quad \text{(see Section 2.9)}$$

where the distribution of F is based on ν_1 = df(error) associated with the larger MSE and ν_2 = df(error) associated with the smaller MSE. Recall that for a quadratic model, df(error) = $n - 3$.

From the printouts shown in Figures 6.17(a) and (b), we have

$$MSE_1 = 31{,}577{,}876 \quad \text{and} \quad MSE_2 = 94{,}708{,}058$$

FIGURE 6.17(a) SAS Regression Analysis for Second-Order Model: Subsample 1
(Years of Experience < 20)

ANALYSIS OF VARIANCE

SOURCE	DF	SUM OF SQUARES	MEAN SQUARE	F VALUE	PROB>F
MODEL	2	3231228653	1615614327	51.163	0.0001
ERROR	21	663135395	31577875.97		
C TOTAL	23	3894364049			

ROOT MSE	5619.42	R-SQUARE	0.8297	
DEP MEAN	37152.75	ADJ R-SQ	0.8135	
C.V.	15.12518			

PARAMETER ESTIMATES

VARIABLE	DF	PARAMETER ESTIMATE	STANDARD ERROR	T FOR H0: PARAMETER=0	PROB > |T|
INTERCEP	1	20372.31906	3817.93090	5.336	0.0001
X	1	263.14290	861.88016	0.305	0.7631
XX	1	76.77081657	40.01457868	1.919	0.0687

(b) SAS Regression Analysis for Second-Order Model: Subsample 2 (Years of Experience ≥ 20)

ANALYSIS OF VARIANCE

SOURCE	DF	SUM OF SQUARES	MEAN SQUARE	F VALUE	PROB>F
MODEL	2	2930533276	1465266638	15.471	0.0001
ERROR	23	2178285323	94708057.53		
C TOTAL	25	5108818599			

ROOT MSE	9731.806	R-SQUARE	0.5736	
DEP MEAN	62185.85	ADJ R-SQ	0.5365	
C.V.	15.64955			

PARAMETER ESTIMATES

VARIABLE	DF	PARAMETER ESTIMATE	STANDARD ERROR	T FOR H0: PARAMETER=0	PROB > |T|
INTERCEP	1	-19228.62348	168795.40	-0.114	0.9103
X	1	2996.18859	14415.13670	0.208	0.8372
XX	1	17.03645465	304.21753	0.056	0.9558

Therefore, the test statistic is

$$F = \frac{MSE_2}{MSE_1} = \frac{94,708,058}{31,577,876} = 3.00$$

Since the MSE for subsample 2 is placed in the numerator of the test statistic, this F-value is based on $n_2 - 3 = 26 - 3 = 23$ numerator df and $n_1 - 3 = 24 - 3 = 21$ denominator df. For a two-tailed test at $\alpha = .05$, the critical value for $\nu_1 = 23$ and $\nu_2 = 21$ (found in Table 5 of Appendix D) is approximately $F_{.025} = 2.37$.

Since the observed value, $F = 3.00$, exceeds the critical value, there is sufficient evidence (at $\alpha = .05$) to indicate that the error variances differ.* Thus, this test

*Most statistical tests require that the observations in the sample be independent. For this F-test, the observations are the residuals. Even if the standard least squares assumption of independent errors is satisfied, the regression residuals will be correlated. Fortunately, when n is large compared to the number of β parameters in the model, the correlation among the residuals is reduced and, in most cases, can be ignored.

supports the conclusions reached by using the residual plots in the preceding examples.   ∎

The test for heteroscedasticity outlined in Example 6.5 is easy to apply when only a single independent variable appears in the model. For a multiple regression model which contains several different independent variables, the choice of the levels of the x variables for dividing the data is more difficult, if not impossible. If you require a statistical test for heteroscedasticity in a multiple regression model, you may need to resort to other, more complex, tests. Consult the references at the end of this chapter for details on how to conduct these tests.

EXERCISES 6.7–6.12

6.7 Refer to Exercise 6.1. Plot the residuals for the first-order model versus $\hat{y}$. Do you detect any trends? If so, what does the pattern suggest about the model?

6.8 Refer to Exercise 6.1. Plot the residuals for the first-order model versus $\hat{y}$. Do you detect any trends? If so, what does the pattern suggest about the model?

6.9 Breakdowns of machines that produce steel cans are very costly. The more breakdowns, the fewer cans produced, and the smaller the company's profits. To help anticipate profit loss, the owners of a can company would like to find a model that will predict the number of breakdowns on the assembly line. The model proposed by the company's statisticians is the following:

$$y = \beta_0 + \beta_1 x_1 + \beta_2 x_2 + \beta_3 x_3 + \beta_4 x_4 + \varepsilon$$

where y is the number of breakdowns per 8-hour shift,

$$x_1 = \begin{cases} 1 & \text{if afternoon shift} \\ 0 & \text{otherwise} \end{cases} \qquad x_2 = \begin{cases} 1 & \text{if midnight shift} \\ 0 & \text{otherwise} \end{cases}$$

x_3 is the temperature of the plant (°F), and x_4 is the number of inexperienced personnel working on the assembly line. After the model is fit using the least squares procedure, the residuals are plotted. The trend in the plot reveals that the response variable y may be of the Poisson type, and hence that a square root transformation is necessary in order to achieve a variance that is approximately the same for all settings of the independent variables. The regression analysis for the transformed model

$$y^* = \sqrt{y} = \beta_0 + \beta_1 x_1 + \beta_2 x_2 + \beta_3 x_3 + \beta_4 x_4 + \varepsilon$$

produces the prediction equation

$$\hat{y}^* = 1.3 + .008 x_1 - .13 x_2 + .0025 x_3 + .26 x_4$$

a. Use the equation to predict the number of breakdowns during the midnight shift if the temperature of the plant at that time is 87°F and if there is only one inexperienced worker on the assembly line.

b. A 95% prediction interval for y^* when $x_1 = 0$, $x_2 = 0$, $x_3 = 90$°F, and $x_4 = 2$ is (1.965, 2.125). For those same values of the independent variables, find a 95% prediction interval for y, the number of breakdowns per 8-hour shift.

c. A 95% confidence interval for $E(y^*)$ when $x_1 = 0$, $x_2 = 0$, $x_3 = 90$°F, and $x_4 = 2$ is (1.987, 2.107). Using only the information given in this problem, is it possible to find a 95% confidence interval for $E(y)$? Explain.

6.10 The data in the table are the monthly market shares for a product over most of the past year. The least squares line relating market share to television advertising expenditure is found to be

$$\hat{y} = -1.56 + .687x$$

MONTH	MARKET SHARE y, %	TELEVISION ADVERTISING EXPENDITURE x, thousands of dollars
January	15	23
February	17	27
March	17	25
May	13	21
June	12	20
July	14	24
September	16	26
October	14	23
December	15	25

a. Calculate and plot the regression residuals in the manner outlined in this section.
b. The response variable y, market share, is recorded as a percentage. What does this lead you to believe about the least squares assumption of homoscedasticity? Does the residual plot substantiate this belief?
c. What variance-stabilizing transformation is suggested by the trend in the residual plot? If you have access to a computer package, refit the first-order model using the transformed responses. Calculate and plot these new regression residuals. Is there evidence that the transformation has been successful in stabilizing the variance of the error term, ε?

6.11 The manager of a retail appliance store wants to model the proportion of appliance owners who decide to purchase a service contract for a specific major appliance. Since the manager believes that the proportion y decreases with age x of the appliance (in years), he will fit the first-order model

$$E(y) = \beta_0 + \beta_1 x$$

A sample of 50 purchasers of new appliances are contacted about the possibility of purchasing a service contract. Fifty owners of 1-year-old machines, and 50 owners each of 2-, 3-, and 4-year-old machines are also contacted. One year later, another survey is conducted in a similar manner. The proportion y of owners deciding to purchase the service policy is shown in the table.

AGE OF APPLIANCE x, years	0	0	1	1	2	2	3	3	4	4
PROPORTION BUYING SERVICE CONTRACT, y	.94	.96	.7	.76	.6	.4	.24	.3	.12	.1

a. Fit the first-order model to the data.
b. Calculate the residuals and construct a residual plot versus $\hat{y}$.

c. What does the plot from part **b** suggest about the variance of *y*?

d. Explain how you could stabilize the variances.

e. Refit the model using the appropriate variance-stabilizing transformation. Plot the residuals for the transformed model and compare to the plot obtained in part **b**. Does the assumption of homoscedasticity appear to be satisfied?

6.12 In Hawaii, condemnation proceedings are underway to enable private citizens to own the property that their homes are built on. Prior to 1980, only estates were permitted to own land and homeowners leased the land from the estate (a law that dates back to the feudal period in Hawaii). In order to comply with the new law, a large Hawaiian estate wants to use regression analysis to estimate the fair market value of its land. A first proposal is the quadratic model

$$E(y) = \beta_0 + \beta_1 x + \beta_2 x^2$$

where

　　y = Leased fee value (i.e., sale price of property)

　　x = Size of property in square feet

Data collected for 20 property sales in a particular neighborhood, given in the accompanying table, were used to fit the model. The least squares prediction equation is

$$\hat{y} = -44.0947 + 11.5339x - .06378x^2$$

PROPERTY	LEASED FEE VALUE y, thousands of dollars	SIZE x, thousands	PROPERTY	LEASED FEE VALUE y, thousands of dollars	SIZE x, thousands
1	70.7	13.5	11	148.0	14.5
2	52.7	9.6	12	85.0	10.2
3	87.6	17.6	13	171.2	18.7
4	43.2	7.9	14	97.5	13.2
5	103.8	11.5	15	158.1	16.3
6	45.1	8.2	16	74.2	12.3
7	86.8	15.2	17	47.0	7.7
8	73.3	12.0	18	54.7	9.9
9	144.3	13.8	19	68.0	11.2
10	61.3	10.0	20	75.2	12.4

a. Calculate the predicted values and corresponding residuals for the model.

b. Plot the residuals versus $\hat{y}$. Do you detect any trends? If so, what does the pattern suggest about the model?

c. Conduct a test for heteroscedasticity. [*Hint:*　Divide the data into two subsamples, $x \leq 12$ and $x > 12$, and fit the model to both subsamples.]

d. Based on your results from parts **b** and **c**, how should the estate proceed?

|||||||||||||

S E C T I O N　6.4

CHECKING THE NORMALITY ASSUMPTION

Recall from Section 4.2 that all the inferential procedures associated with a regression analysis are based on the assumptions that, for any setting of the independent variables, the random error ε is normally distributed with mean 0 and variance σ^2, and all pairs of errors are independent. Of these assumptions, the normality assumption is the least restrictive when we apply regression analysis in practice. That is, moderate departures from the assumption of normality have very little effect on error rates associated with the statistical tests and on the confidence coefficients associated with the confidence intervals.

Although tests are available to check the normality assumption (see, for example, Stephens, 1974), we discuss only graphical techniques in this section. The simplest way to determine whether the data violate the assumption of normality is to construct a frequency or relative frequency distribution for the residuals using the computer. If this distribution is not badly skewed, you can feel reasonably confident that the measures of reliability associated with your inferences are as stated in Chapter 4. This visual check is not foolproof because we are lumping the residuals together for all settings of the independent variables. It is conceivable (but not likely) that the distribution of residuals might be skewed to the left for some values of the independent variables and skewed to the right for others. Combining these residuals into a single relative frequency distribution could produce a distribution that is relatively symmetric. But, as noted above, we think that this situation is unlikely and that this graphical check is very useful.

To illustrate, a computer-generated frequency distribution for the $n = 50$ residuals of Example 6.4* is shown in Figure 6.18. You can see that this distribution is mound-shaped and reasonably symmetric. Consequently, it is unlikely that the normality assumption would be violated using these data.

FIGURE 6.18   A Relative Frequency Distribution for the $n = 50$ Residuals of Example 6.4

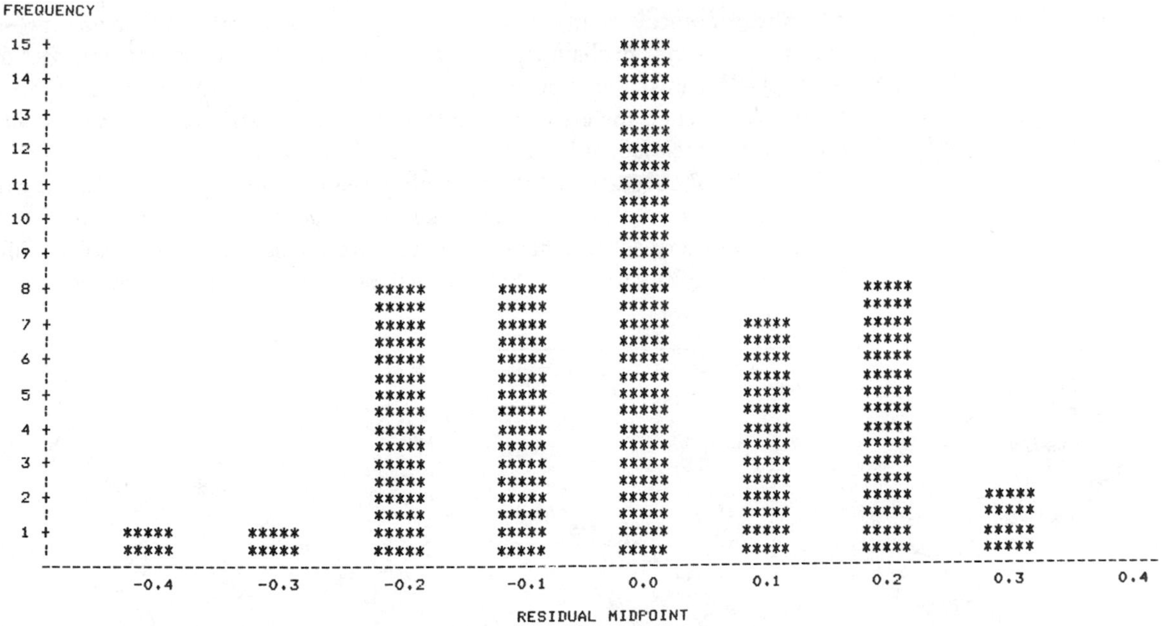

*A printout of the residuals is omitted.

If the number of observations is small or if you do not have access to a computer, you may want to construct a **stem-and-leaf display** for the residuals. A stem-and-leaf display organizes the residuals in much the same way as a relative frequency distribution. Figure 6.19 shows a stem-and-leaf display for the 50 residuals of Example 6.4. To construct this display, each residual value is partitioned into a **stem** and a **leaf**. For our display, we chose the *stem* portion of a residual as the first digit to the right of the decimal point. The remaining portion of the residual, to the right of the stem, is called the *leaf*. For example, a residual value of .13 has a stem of .1 and a leaf of 3. The list of possible stems for the set of residuals is listed in a column from the smallest (−.3) to the largest (.2), then the leaf for each residual is recorded in the row corresponding to the residual's stem. If you turn the stem-and-leaf display (Figure 6.19) on its side, it will look very much like the relative frequency distribution (Figure 6.18), mound-shaped and reasonably symmetric.

FIGURE 6.19

Stem-and-Leaf Display for the *n* = 50 Residuals of Example 6.4

STEM	LEAF
−.3	3, 5
−.2	0, 1, 1
−.1	0, 1, 2, 5, 6, 6, 7, 8
−.0	0, 1, 2, 2, 3, 5, 5, 5, 5, 5, 6, 7, 7, 8
.0	1, 2, 3, 3, 4, 4, 5, 7, 8, 8
.1	0, 4, 5, 6, 8, 8
.2	1, 1, 3, 4, 5, 6, 6

A third graphical technique for checking the assumption of normality is to construct a **normal probability plot**. In a normal probability plot, the residuals are graphed against the expected values of the residuals under the assumption of normality. When the errors are, in fact, normally distributed, a residual value will approximately equal its expected value. Thus, a linear trend on the normal probability plot suggests that the normality assumption is nearly satisfied, while a nonlinear trend indicates that the assumption is most likely violated.

Most computer packages have procedures for constructing normal probability plots. Figure 6.20 shows the SAS normal probability plot for the residuals of

FIGURE 6.20

Normal Probability Plot for *n* = 50 Residuals of Example 6.4

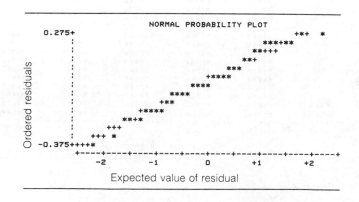

Example 6.4.* Notice that the points (represented by the plotting symbol "*") fall reasonably close to a straight line (plotting symbol "+"), indicating that the normality assumption is most likely satisfied. If you do not have access to a computer package that contains a normal probability plot option, you can calculate the expected values of the residuals (under normality) using the procedure outlined in the next box.

CONSTRUCTING A NORMAL PROBABILITY PLOT FOR REGRESSION RESIDUALS

1. List the residuals in ascending order, where $\hat{\varepsilon}_i$ represents the ith ordered residual.
2. For each residual, calculate the corresponding tail area (of the standard normal distribution),

$$A = \frac{i - .375}{n + 25}$$

where n is the sample size.
3. Calculate the estimated value of $\hat{\varepsilon}_i$ under normality using the following formula:

$$E(\hat{\varepsilon}_i) \approx \sqrt{\text{MSE}}[Z(A)]$$

where

MSE = Mean square error for the fitted model
$Z(A)$ = Value of the standard normal distribution (z-value) that cuts off an area of A in the lower tail of the distribution.

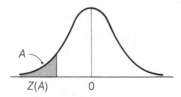

4. Plot the residuals $\hat{\varepsilon}_i$ against the estimated expected residuals, $i = 1, 2, \ldots, n$.

Nonnormality of the distribution of the random error ε is often accompanied by heteroscedasticity. Both these situations can frequently be rectified by applying the variance-stabilizing transformations of Section 6.3. For example, if the relative frequency distribution (or stem-and-leaf display) of the residuals is highly skewed to the right (as it would be for Poisson data), the square-root transformation on y will stabilize (approximately) the variance and, at the same time, will reduce

---

*The horizontal axis of the SAS normal probability plot (Figure 6.20) gives the expected values of the residuals in terms of number of standard deviations (positive or negative) from the mean 0.

skewness in the distribution of the residuals. Thus, for any given setting of the independent variables, the square-root transformation will reduce the larger values of y to a greater extent than the smaller ones. This has the effect of reducing or eliminating the positive skewness.

| | | | | | | | | | | | | | |

EXERCISES 6.13–6.16

6.13 Refer to Exercise 3.11. Use one of the graphical techniques described in this section to check the normality assumption.

6.14 Refer to Exercise 4.28. Use one of the graphical techniques described in this section to check the normality assumption.

6.15 Refer to Exercise 6.12. Use one of the graphical techniques described in this section to check the normality assumption.

6.16 L. De Cola conducted an extensive investigation of the geopolitical and socioeconomic processes that shape the urban size distributions of the world's nations. One of the goals of the study was to determine the factors that influence population size in each nation's largest city. Using data collected for a sample of 126 countries, De Cola fit the following log model:

$$E(y) = \beta_0 + \beta_1 x_1 + \beta_2 x_2 + \beta_3 x_3 + \beta_4 x_4 + \beta_5 x_5 + \beta_6 x_6 + \beta_7 x_7 + \beta_8 x_8 + \beta_9 x_9 + \beta_{10} x_{10}$$

where

y = Log of population (in thousands) of largest city in country

x_1 = Log of area (in thousands of square kilometers) of country

x_2 = Log of radius (in hundred kilometers) of country

x_3 = Log of national population (in thousands)

x_4 = Percentage annual change in national population (1960–1970)

x_5 = Log of energy consumption per capita (in kilograms of coal equivalent)

x_6 = Percentage of nation's population in urban areas

x_7 = Log of population (in thousands) of second largest city in country

$x_8 = \begin{cases} 1 & \text{if seaport city} \\ 0 & \text{if not} \end{cases}$

$x_9 = \begin{cases} 1 & \text{if capital city} \\ 0 & \text{if not} \end{cases}$

$x_{10} = \begin{cases} 1 & \text{if city data are for metropolitan area} \\ 0 & \text{if not} \end{cases}$

[*Note:* All logarithms are to the base 10.]

The regression resulted in

$$R^2 = .879 \quad \text{and} \quad MSE = .036$$

a. Conduct a test for model adequacy. Use $\alpha = .05$.

b. The residuals for five cities selected from the total sample are given in the table. For each of these cities, calculate the estimated expected residuals under the assumption of normality.

CITY	RESIDUAL	RANK
Bangkok	.510	126
Paris	.228	110
London	.033	78
Warsaw	−.132	32
Lagos	−.392	2

Source: De Cola, L. "Statistical determinants of the population of a nation's largest city," *Economic Development and Cultural Change*, Vol. 3, No. 1, October 1984, pp. 71–98.

c. A computer-generated (SAS) stem-and-leaf plot of all the residuals is shown here. Does it appear that the assumption of normal errors is satisfied?

SAS Stem-and-Leaf Plot of Residuals for Exercise 6.16

```
STEM LEAF                            #
   5 1                               1
   4 5                               1
   4
   3 79                              2
   3 1224                            4
   2 5899                            4
   2 000012233344                   12
   1 5789                            4
   1 00001113444                    11
   0 5566888                         7
   0 11111122223334                 14
  -0 4433333322111000               16
  -0 9999988665                     10
  -1 44433332221110                 14
  -1 88865                           5
  -2 444443100                       9
  -2 77665                           5
  -3 3000                            4
  -3 97                             2
  -4
  -4 6                              1
     ----+----+----+----+
MULTIPLY STEM.LEAF BY 10**-01
```

| | | | | | | | | | | | | |

S E C T I O N 6.5

DETECTING OUTLIERS AND IDENTIFYING INFLUENTIAL OBSERVATIONS

Although we expect almost all the regression residuals to fall within 3 standard deviations of their mean (0), sometimes one or several residuals fall outside this interval. Observations with residuals that are extremely large or small (say, more than 3 standard deviations from 0) are called **outliers**.

Outliers are usually attributable to one of several causes. The measurement associated with the outlier may be invalid. For example, the experimental procedure used to generate the measurement may have malfunctioned, the experimenter may have misrecorded the measurement, or the data might have been coded incorrectly for entry into the computer. Careful checks of the experimental and coding procedures should reveal this type of problem if it exists, so that we can eliminate erroneous observations from a data set.

TABLE 6.8   Data for Fast-Food Sales

CITY	TRAFFIC FLOW thousands of cars	WEEKLY SALES y, thousands of dollars	CITY	TRAFFIC FLOW thousands of cars	WEEKLY SALES y, thousands of dollars
1	59.3	6.3	3	75.8	8.2
1	60.3	6.6	3	48.3	5.0
1	82.1	7.6	3	41.4	3.9
1	32.3	3.0	3	52.5	5.4
1	98.0	9.5	3	41.0	4.1
1	54.1	5.9	3	29.6	3.1
1	54.4	6.1	3	49.5	5.4
1	51.3	5.0	4	73.1	8.4
1	36.7	3.6	4	81.3	9.5
2	23.6	2.8	4	72.4	8.7
2	57.6	6.7	4	88.4	10.6
2	44.6	5.2	4	23.2	3.3

For example, Table 6.8 presents the sales, y, in thousands of dollars per week, for fast-food outlets in each of four cities. The objective is to model sales, y, as a function of traffic flow, adjusting for city-to-city variations which might be due to size or other market conditions. We expect a first-order (linear) relationship to exist between mean sales, $E(y)$, and traffic flow. Further, we believe that the level of mean sales will differ from city to city, but that the change in mean sales per unit increase in traffic flow will remain the same for all cities, i.e., that the factors traffic flow and cities do not interact. The model is therefore

$$E(y) = \beta_0 + \beta_1 x_1 + \beta_2 x_2 + \beta_3 x_3 + \beta_4 x_4$$

where

$$x_1 = \begin{cases} 1 & \text{if city 1} \\ 0 & \text{other} \end{cases} \qquad x_2 = \begin{cases} 1 & \text{if city 2} \\ 0 & \text{other} \end{cases}$$

$$x_3 = \begin{cases} 1 & \text{if city 3} \\ 0 & \text{other} \end{cases} \qquad x_4 = \text{Traffic flow}$$

The computer printout for the regression analysis is shown in Figure 6.21. The regression analysis indicates that the first-order model in traffic flow is inadequate for explaining mean sales. The coefficient of determination, R^2, is .259, indicating that only 25.9% of the total sum of squares of deviations of the sales y about their mean $\bar{y}$ is accounted for by the model. The F-value, 1.67, which tests the adequacy of the model, does not indicate that the model is useful for predicting sales. The observed significance level is only .1996.

Plots of the residuals against traffic flow and city are shown in Figures 6.22 and 6.23 (pages 292–293), respectively. The dashed horizontal lines locate the mean (0), $+2s$, and $-2s$ for the residuals. As you can see, the plots of the residuals

FIGURE 6.21 Regression Analysis: SAS Computer Printout for Fast-Food Sales (First-Order Model)

ANALYSIS OF VARIANCE

SOURCE	DF	SUM OF SQUARES	MEAN SQUARE	F VALUE	PROB>F
MODEL	4	1469.76287	367.44072	1.665	0.1996
ERROR	19	4194.22671	220.74877		
C TOTAL	23	5663.98958			

ROOT MSE	14.85762	R-SQUARE	0.2595	
DEP MEAN	9.070833	ADJ R-SQ	0.1036	
C.V.	163.7955			

PARAMETER ESTIMATES

VARIABLE	DF	PARAMETER ESTIMATE	STANDARD ERROR	T FOR H0: PARAMETER=0	PROB > |T|
INTERCEP	1	-16.45924803	13.16399794	-1.250	0.2264
X1	1	1.10609170	8.42256884	0.131	0.8969
X2	1	6.14277146	11.67996860	0.526	0.6050
X3	1	14.48962257	9.28839086	1.560	0.1353
X4	1	0.36287305	0.16790819	2.161	0.0437

OBS	ACTUAL	PREDICT VALUE	RESIDUAL
1	6.3000	6.1652	0.1348
2	6.6000	6.5281	0.0719
3	7.6000	14.4387	-6.8387
4	3.0000	-3.6324	6.6324
5	9.5000	20.2084	-10.7084
6	5.9000	4.2783	1.6217
7	6.1000	4.3871	1.7129
8	5.0000	3.2622	1.7378
9	3.6000	-2.0357	5.6357
10	2.8000	-1.7527	4.5527
11	6.7000	10.5850	-3.8850
12	5.2000	5.8677	-0.6677
13	82.0000	25.5362	56.4638
14	5.0000	15.5571	-10.5571
15	3.9000	13.0533	-9.1533
16	5.4000	17.0812	-11.6812
17	4.1000	12.9082	-8.8082
18	3.1000	8.7714	-5.6714
19	5.4000	15.9926	-10.5926
20	8.4000	10.0668	-1.6668
21	9.5000	13.0423	-3.5423
22	8.7000	9.8128	-1.1128
23	10.6000	15.6187	-5.0187
24	3.3000	-8.0406	11.3406

SUM OF RESIDUALS   -2.87992E-13
SUM OF SQUARED RESIDUALS   4194.227

are very revealing. Both the plot of residuals against traffic flow in Figure 6.22 and the plot of residuals against city in Figure 6.23 indicate the presence of an outlier. One observation in city 3, with traffic flow of 75.8, is approximately 4 standard deviations from 0. (Note in Figure 6.21 that the standard deviation is 14.86, and the outlier residual is 56.46.) A further check of the observation associated with this residual reveals that the sales value entered into the computer, 82.0, does not agree with the corresponding value of sales, 8.2, that appears in Table 6.8. The decimal point was evidently dropped when the data were entered into the computer.

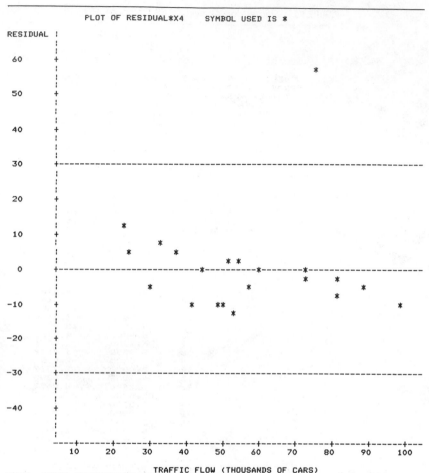

NOTE: THREE OBSERVATIONS ARE NOT SHOWN BECAUSE THEY COINCIDED WITH OTHERS
THAT APPEAR ON THE PLOT. THEIR OMISSION DOES NOT DETRACT FROM THE
INFORMATION CONTRIBUTED BY THE PLOT.

If the correct y-value, 8.2, is substituted for the 82.0, we obtain the regression analysis shown in Figure 6.24 (page 294). Plots of the residuals against traffic flow and city are shown in Figures 6.25 and 6.26 (pages 295–296), respectively. The corrected computer printout indicates the dramatic effect that a single outlier can have on a regression analysis. The value of R^2 is now .979, and the F-value that tests the adequacy of the model, 222.17, verifies the strong predictive capability of the model. Further analysis reveals that significant differences exist in the mean sales among cities, and that the estimated mean weekly sales increase by \$104 for every 1,000-car increase in traffic flow ($\hat{\beta}_4 = .104$). The 95% confidence interval for β_4 is

$$\hat{\beta}_4 \pm t_{.025}s_{\hat{\beta}_4} = .104 \pm (2.093)(.004094) = .104 \pm .009$$

Thus, a 95% confidence interval for the mean increase in sales per 1,000-car increase in traffic flow is \$95 to \$113.

FIGURE 6.23

SAS Plot of Residuals
Versus City

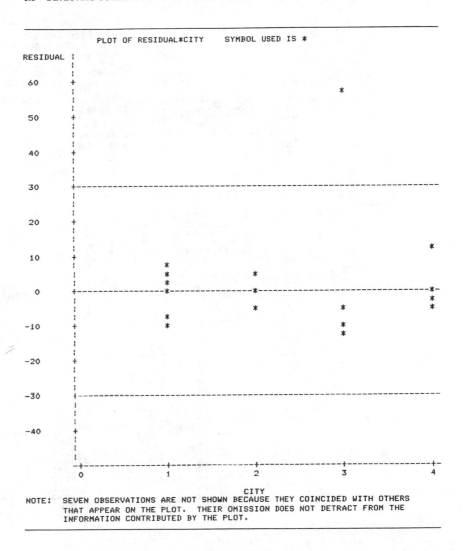

NOTE: SEVEN OBSERVATIONS ARE NOT SHOWN BECAUSE THEY COINCIDED WITH OTHERS
        THAT APPEAR ON THE PLOT.   THEIR OMISSION DOES NOT DETRACT FROM THE
        INFORMATION CONTRIBUTED BY THE PLOT.

Outliers cannot always be explained by computer or recording errors. Extremely large or small residuals may be attributable to skewness (nonnormality) of the probability distribution of the random error, chance, or unassignable causes. Although some analysts advocate elimination of outliers, regardless of whether cause can be assigned, others encourage the correction of only those outliers which can be traced to specific causes. The best philosophy is probably a compromise between these extremes. For example, before deciding the fate of an outlier you may want to determine how much influence it has on the regression analysis. When an accurate outlier (i.e., an outlier that is not due to recording or measurement error) is found to have a dramatic effect on the regression analysis, it may be the model and not the outlier that is suspect. Omission of important independent variables or higher-order terms could be the reason why the model is not predicting well for the outlying observation. Several sophisticated

FIGURE 6.24   Corrected SAS Computer Printout for Fast-Food Example

ANALYSIS OF VARIANCE

SOURCE	DF	SUM OF SQUARES	MEAN SQUARE	F VALUE	PROB>F
MODEL	4	116.65552	29.16387942	222.173	0.0001
ERROR	19	2.49406564	0.13126661		
C TOTAL	23	119.14958			

ROOT MSE	0.3623073	R-SQUARE	0.9791	
DEP MEAN	5.995833	ADJ R-SQ	0.9747	
C.V.	6.042652			

PARAMETER ESTIMATES

VARIABLE	DF	PARAMETER ESTIMATE	STANDARD ERROR	T FOR H0: PARAMETER=0	PROB > |T|
INTERCEP	1	1.08338756	0.32100795	3.375	0.0032
X1	1	-1.21576160	0.20538681	-5.919	0.0001
X2	1	-0.53075677	0.28481946	-1.863	0.0779
X3	1	-1.07652473	0.22650014	-4.753	0.0001
X4	1	0.10367335	0.004094490	25.320	0.0001

OBS	ACTUAL	PREDICT VALUE	RESIDUAL
1	6.3000	6.0155	0.2845
2	6.6000	6.1191	0.4809
3	7.6000	8.3792	-0.7792
4	3.0000	3.2163	-0.2163
5	9.5000	10.0276	-0.5276
6	5.9000	5.4764	0.4236
7	6.1000	5.5075	0.5925
8	5.0000	5.1861	-0.1861
9	3.6000	3.6724	-0.0724
10	2.8000	2.9993	-0.1993
11	6.7000	6.5242	0.1758
12	5.2000	5.1765	0.0235
13	8.2000	7.8653	0.3347
14	5.0000	5.0143	-0.0143
15	3.9000	4.2989	-0.3989
16	5.4000	5.4497	-0.0497
17	4.1000	4.2575	-0.1575
18	3.1000	3.0756	0.0244
19	5.4000	5.1387	0.2613
20	8.4000	8.6619	-0.2619
21	9.5000	9.5120	-0.0120
22	8.7000	8.5893	0.1107
23	10.6000	10.2481	0.3519
24	3.3000	3.4886	-0.1886

SUM OF RESIDUALS	-2.42029E-14
SUM OF SQUARED RESIDUALS	2.494066

numerical techniques are available for identifying outlying influential observations. We conclude this section with a brief discussion of some of these methods and an example.

LEVERAGE

This procedure is based on a result (proof omitted) in regression analysis which states that the predicted value for the ith observation, $\hat{y}_i$, can be written as a linear combination of the n observed values $y_1, y_2, \ldots, y_n$:

$$\hat{y}_i = h_1 y_1 + h_2 y_2 + \cdots + h_i y_i + \cdots + h_n y_n, \quad i = 1, 2, \ldots, n$$

where the weights $h_1, h_2, \ldots, h_n$ of the observed values are functions of the independent variables. In particular, the coefficient h_i measures the influence of

FIGURE 6.25

Corrected SAS Plot of
Residuals Versus Traffic
Flow

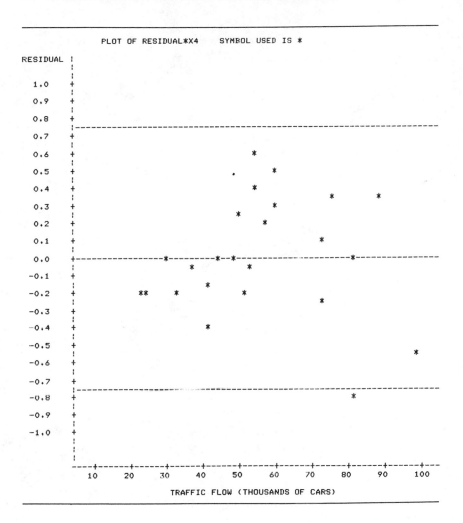

the observed value y_i on its own predicted value $\hat{y}_i$. This value, h_i, is called the
leverage of the ith observation (with respect to the values of the independent
variables). Thus, leverage values can be used to identify influential observations—
the larger the leverage value, the more influence the observed y-value has on its
predicted value.

Leverage values are extremely difficult to calculate without the aid of a computer.* Fortunately, most of the statistical computer program packages discussed
in this text have options that give the leverage associated with each observation.
The leverage value for an observation is usually compared with the average
leverage value of all n observations, $\bar{h}$, where

*In matrix notation, the leverage values are the diagonals of the **H** matrix (called the "hat" matrix),
where $\mathbf{H} = \mathbf{X}(\mathbf{X}'\mathbf{X})^{-1}\mathbf{X}'$. See Appendix A for details on matrix multiplication and definition of the
X matrix in regression.

FIGURE 6.26
Corrected SAS Plot of
Residuals Versus City

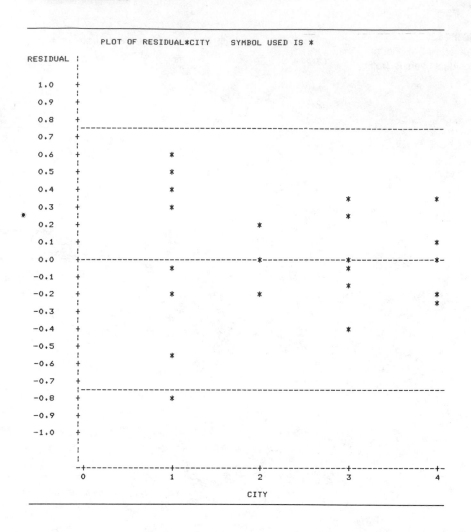

$$\bar{h} = \frac{k + 1}{n} = \frac{\text{Number of } \beta \text{ parameters in the model, including } \beta_0{}^*}{n}$$

A good rule of thumb identifies an observation y_i as influential if its leverage value h_i is more than twice as large as $\bar{h}$, that is, if

$$h_i > \frac{2(k + 1)}{n}$$

THE JACKKNIFE

Another technique for identifying influential observations requires that you delete the observations one at a time, each time refitting the regression model based on only the remaining $n - 1$ observations. This method is based on a statistical

---

*The proof of this result is beyond the scope of this text. Consult the references given at the end of this chapter. [See Neter, Wasserman, and Kutner (1985).]

procedure that is gaining increasing acceptance among practitioners, called the **jackknife.*** The basic principle of the jackknife when applied to regression is to compare the regression results using all n observations to the results with the ith observation deleted, in order to ascertain how much influence a particular observation has on the analysis. Using the jackknife, several alternative influence measures can be calculated.

The **deleted residual**, $d_i = y_i - \hat{y}_{(i)}$, measures the difference between the observed value y_i and the predicted value $\hat{y}_{(i)}$ based on the model with the ith observation deleted. [The notation (i) is generally used to indicate that the observed value y_i was deleted from the regression analysis.] An observation with an unusually large (in absolute value) deleted residual is considered to have large influence on the fitted model.

A measure closely related to the deleted residual is the difference between the predicted value based on the model fit to all n observations and the predicted value obtained when y_i is deleted, i.e., $\hat{y}_i - \hat{y}_{(i)}$. When the difference $\hat{y}_i - \hat{y}_{(i)}$ is large relative to the predicted value $\hat{y}_i$, the observation y_i is said to influence the regression fit.

A third way to identify an influential observation using the jackknife is to calculate, for each β parameter in the model, the difference between the parameter estimate based on all n observations and the estimate based on only $n - 1$ observations (with the observation in question deleted). Consider, for example, the straight-line model $E(y) = \beta_0 + \beta_1 x$. The differences $\hat{\beta}_0 - \hat{\beta}_0^{(i)}$ and $\hat{\beta}_1 - \hat{\beta}_1^{(i)}$ measure how influential the ith observation y_i is on the parameter estimates. [Using the (i) notation defined earlier, $\hat{\beta}^{(i)}$ represents the estimate of the β coefficient when the ith observation is omitted from the analysis.] If the parameter estimates change drastically, i.e., if the absolute differences are large, y_i is deemed an influential observation.

Each of the four computer packages discussed in this text has a jackknife routine that produces one or more of the measures described above.

COOK'S DISTANCE

A measure of the overall influence an outlying observation has on the estimated β coefficients was proposed by Cook (1979). Cook's distance, D_i, is calculated for the ith observation as follows:

$$D_i = \frac{y_i - \hat{y}_i}{(k + 1)\text{MSE}} \left[\frac{h_i}{(1 - h_i)^2} \right]$$

Note that D_i depends on both the residual ($y_i - \hat{y}_i$) and the leverage h_i for the ith observation. A large value of D_i indicates that the observed y_i-value has strong influence on the estimated β coefficients (since either the residual or the leverage or both will be large). Values of D_i can be compared to the values of the F-distribution with $\nu_1 = k + 1$ and $\nu_2 = n - (k + 1)$ degrees of freedom. Usually,

---

*The procedure derives its name from the Boy Scout jackknife, which serves as a handy tool in a variety of situations when specialized techniques may not be applicable. [See Belsley, Kuh, and Welsch (1980).]

an observation with a value of D_i that falls at or above the 50th percentile of the F-distribution is considered to be an influential observation. Like the other numerical measures of influence, options for calculating Cook's distance are available in most statistical computer program packages.

EXAMPLE 6.6

We now return to the fast-food sales example in which we detected an outlier using residual plots. Recall that the outlier was due to an error in coding the weekly sales value for observation 13 (denoted y_{13}). The SAS regression analysis is rerun with options for producing influence diagnostics. (An **influence diagnostic** is a number that measures how much influence an observation has on the regression analysis.) The resulting SAS printout is shown in Figure 6.27. Locate and interpret the measures of influence for y_{13} on the printout.

SOLUTION

The influence diagnostics are shown in the last portion of the SAS printout in Figure 6.27. Leverage values for each observation are given under the column heading HAT DIAG H. The leverage value for y_{13} (shaded on the printout) is h_{13} = .2431 while the average leverage for all n = 24 observations is

$$\bar{h} = \frac{k+1}{n} = \frac{5}{24} = .2083$$

Since the leverage value .2431 does not exceed $2\bar{h}$ = .4166, we would not identify y_{13} as an influential observation. At first, this result may seem confusing since we already know the dramatic effect the incorrectly coded value of y_{13} had on the regression analysis. Remember, however, that the leverage values, h_1, h_2, . . . , h_{24}, are functions of the independent variables only. Since we know the values of x_1, x_2, x_3, and x_4 were coded correctly, the relatively small leverage value of .2431 simply indicates that observation 13 is not an outlier with respect to the values of the independent variables.

A better overall measure of the influence of y_{13} on the fitted regression model is Cook's distance, D_{13}. Recall that Cook's distance is a function of both leverage and the magnitude of the residual. This value, D_{13} = 1.219 (shaded) is given in the column labeled COOK'S D located on the right side of the printout. You can see that the value is extremely large relative to the other values of D_i in the printout. [In fact, D_{13} = 1.219 falls in the 66th percentile of the F-distribution with $\nu_1 = k + 1 = 5$ and $\nu_2 = n - (k + 1) = 24 - 5 = 19$ degrees of freedom.] This implies that the observed value y_{13} has substantial influence on the estimates of the model parameters.

A statistic related to the deleted residual of the jackknife procedure is the **Studentized deleted residual** given under the column heading RSTUDENT. The Studentized deleted residual, denoted d_i^*, is calculated by dividing the deleted residual d_i by its standard error s_{d_i}:

$$d_i^* = \frac{d_i}{s_{d_i}}$$

FIGURE 6.27 Regression Analysis with Influence Diagnostics: SAS Computer Printout for Fast-Food Sales Example

ANALYSIS OF VARIANCE

SOURCE	DF	SUM OF SQUARES	MEAN SQUARE	F VALUE	PROB>F
MODEL	4	1494.97413	373.74353	1.703	0.1908
ERROR	19	4169.01546	219.42187		
C TOTAL	23	5663.98958			

ROOT MSE	14.8129	R-SQUARE	0.2639
DEP MEAN	9.070833	ADJ R-SQ	0.1090
C.V.	163.3025		

PARAMETER ESTIMATES

VARIABLE	DF	PARAMETER ESTIMATE	STANDARD ERROR	T FOR H0: PARAMETER=0	PROB > |T|
INTERCEP	1	-17.22248485	13.30767779	-1.294	0.2111
X1	1	0.87453186	8.37603634	0.104	0.9179
X2	1	6.43312022	11.67487958	0.551	0.5880
X3	1	14.70817354	9.28185800	1.585	0.1296
X4	1	0.37415019	0.17053280	2.194	0.0409

OBS	ACTUAL	PREDICT VALUE	STD ERR PREDICT	RESIDUAL	STD ERR RESIDUAL	STUDENT RESIDUAL	-2-1-0 1 2	COOK'S D
1	6.3000	5.8392	4.9379	0.4608	13.9656	0.0330		0.000
2	6.6000	6.2133	4.9390	0.3667	13.9652	0.0277		0.000
3	7.6000	14.3698	6.2521	-6.7698	13.4288	-0.5041	*|	0.011
4	3.0000	-4.2629	6.7876	7.2629	13.1662	0.5516	|*	0.016
5	9.5000	20.3562	8.2135	-10.8562	12.3272	-0.8807	*|	0.069
6	5.9000	3.8936	5.0263	2.0064	13.9341	0.1440		0.001
7	6.1000	4.0058	5.0170	2.0942	13.9374	0.1503		0.001
8	5.0000	2.8460	5.1370	2.1540	13.8936	0.1550		0.001
9	3.6000	0.3391	5.5618	3.2609	13.7291	0.2375		0.002
10	2.8000	-1.9594	9.1058	4.7594	11.6836	0.4074		0.020
11	6.7000	10.7617	8.9598	-4.0617	11.7959	-0.3443	*|	0.014
12	5.2000	5.8977	8.5643	-0.6977	12.0861	-0.0577		0.000
13	82.0000	25.8463	7.3033	56.1537	12.8873	4.3573	|******	1.219
14	5.0000	15.5571	5.5987	-10.5571	13.7141	-0.7698	*|	0.020
15	3.9000	12.9755	5.7211	-9.0755	13.6635	-0.6642	*|	0.015
16	3.1000	8.5605	6.4432	-5.4605	13.3382	-0.4094	*|	0.008
17	5.4000	16.0061	5.6025	-10.5061	13.7125	-0.7735	*|	0.020
18	5.4000	17.1286	5.6444	-11.7286	13.6954	-0.8564	*|	0.025
19	4.1000	12.8258	5.7355	-8.7258	13.6575	-0.6389	*|	0.020
20	8.4000	10.1279	6.8887	-1.7279	13.2168	-0.1307		0.001
21	9.5000	13.1959	7.0199	-3.6959	13.0439	-0.2833	*|	0.005
22	8.7000	9.8660	6.6732	-1.1660	13.2246	-0.0882		0.000
23	10.6000	15.8524	7.5080	-5.2524	12.7692	-0.4113	*|	0.012
24	3.3000	-8.5422	10.0708	11.8422	10.8628	1.0902	|**	0.204

SUM OF RESIDUALS	-3.09086E-13
SUM OF SQUARED RESIDUALS	4169.015
PREDICTED RESID SS (PRESS)	7356.261

(continued)

FIGURE 6.27 (continued)

OBS	RESIDUAL	RSTUDENT	HAT DIAG H	COV RATIO	DFFITS	INTERCEP DFBETAS	X1 DFBETAS	X2 DFBETAS	X3 DFBETAS	X4 DFBETAS
1	0.4608	0.0321	0.1111	1.4738	0.0114	0.0001	0.0067	-0.0000	-0.0000	-0.0001
2	0.3867	0.0270	0.1112	1.4740	0.0095	-0.0002	0.0057	0.0001	0.0001	0.0002
3	-6.7698	-0.4940	0.1781	1.4906	-0.2300	0.1224	-0.1302	-0.0531	-0.0502	-0.1411
4	7.2629	0.5413	0.2100	1.5300	0.2790	0.1661	0.0882	-0.0720	-0.0682	-0.1915
5	-10.8562	-0.8752	0.3075	1.5362	-0.5832	0.4042	-0.2832	-0.1753	-0.1659	-0.4660
6	2.0064	0.1402	0.1151	1.4728	0.0506	0.0082	0.0277	-0.0036	-0.0034	-0.0095
7	2.0942	0.1463	0.1147	1.4714	0.0527	0.0081	0.0290	-0.0035	-0.0033	-0.0093
8	2.1540	0.1510	0.1203	1.4801	0.0558	0.0134	0.0291	-0.0058	-0.0055	-0.0154
9	3.2609	0.2315	0.1410	1.5029	0.0938	0.0374	0.0420	-0.0162	-0.0154	-0.0432
10	4.7594	0.3982	0.3779	2.0160	0.3104	0.0924	-0.0175	0.1735	-0.0379	-0.1066
11	-4.0617	-0.3362	0.3659	2.0028	-0.2554	0.0660	-0.0125	-0.2072	-0.0271	-0.0761
12	-0.6977	-0.0562	0.3343	1.9667	-0.0398	0.0018	-0.0003	-0.0299	-0.0008	-0.0021
13	56.1537	155.8461	0.2431	0.0000	88.3192	-49.1858	9.3166	21.3280	61.0325	56.7119
14	-10.5371	-0.7612	0.1429	1.3048	-0.3108	0.0000	-0.0000	0.0000	-0.1875	-0.0000
15	-9.0755	-0.6541	0.1492	1.3694	-0.2739	-0.0489	0.0093	0.0212	-0.1416	0.0563
16	-5.4605	-0.4002	0.1892	1.5462	-0.1933	-0.0830	0.0157	0.0360	-0.0673	0.0957
17	-10.6061	-0.7650	0.1430	1.3031	-0.3125	0.0099	-0.0019	-0.0043	-0.1925	-0.0114
18	-11.7286	-0.8501	0.1452	1.2591	-0.3504	0.0386	-0.0073	-0.0167	-0.2255	-0.0445
19	-8.7258	-0.6287	0.1499	1.3829	-0.2640	-0.0497	0.0094	0.0216	-0.1350	0.0573
20	-1.7279	-0.1273	0.2039	1.6386	-0.0644	-0.0240	0.0490	0.0329	0.0424	-0.0089
21	-3.6959	-0.2764	0.2246	1.6545	-0.1487	-0.0272	0.1029	0.0611	0.0827	-0.0492
22	-1.1660	-0.0858	0.2030	1.6407	-0.0433	-0.0169	0.0331	0.0224	0.0288	-0.0052
23	-5.2524	-0.4022	0.2569	1.6863	-0.2365	-0.0073	0.1467	0.0765	0.1093	-0.1113
24	11.8422	1.0959	0.4622	1.7642	1.0160	0.9964	-0.6543	-0.6670	-0.7495	-0.7653

The Studentized deleted residual for y_{13} (shaded on the printout) is $d_{13}^* = 155.8461$. This extremely large value* is another indication that y_{13} is an influential observation.

The DFFITS column gives the difference between the predicted value when all 24 observations are used and when the ith observation is deleted. The difference, $\hat{y}_i - \hat{y}_{(i)}$, is divided by its standard error so that the differences can be compared more easily. For observation 13, this scaled difference (shaded on the printout) is $\hat{y}_{13} - \hat{y}_{(13)} = 88.3192$, an extremely large value relative to the other differences in predicted values. Similarly, the changes in the parameter estimates when observation 13 is deleted are given in the DFBETAS columns (shaded) immediately to the right of DFFITS on the printout. (Each difference is also divided by the appropriate standard error.) The large magnitude of these differences provides further evidence that y_{13} is very influential on the regression analysis. ∎

Several techniques designed to dampen the influence an outlying observation has on the regression analysis are available. One method produces estimates of the β's which minimize the sum of the absolute deviations, $\sum_{i=1}^{n} |y_i - \hat{y}_i|$.† Because the deviations $(y_i - \hat{y}_i)$ are not squared, this method places less emphasis on outliers than the method of least squares. Regardless of whether you choose to eliminate an outlier or dampen its influence, careful study of residual plots and influence diagnostics are essential for outlier detection.

EXERCISES 6.17–6.22

6.17 Refer to the data and model of Exercise 6.1. The MSE for the model is .1267. Plot the residuals versus $\hat{y}$. Identify any outliers on the plot.

6.18 Refer to the data and model of Exercise 6.2. The MSE for the model is 17.2557. Plot the residuals versus $\hat{y}$. Identify any outliers on the plot.

6.19 Refer to the data and model of Exercise 6.4. The MSE for the model is 19.45. Plot the residuals versus $\hat{y}$. Identify any outliers on the plot.

6.20 Refer to the grandfather clock example (Example 4.1) in Chapter 4. The least squares model used to predict auction price, y, from age of the clock, x_1, and number of bidders, x_2, was determined to be

$$\hat{y} = -1,336.722 + 12.7362x_1 + 85.8151x_2$$

a. Use this equation to calculate the residuals of each of the prices given in Table 4.2.

*Under the assumptions of Section 4.2, the Studentized deleted residual d_i^* has a sampling distribution that is approximated by a Student's t-distribution with $(n - 1) - (k + 1)$ df.

†The method of absolute deviations requires linear programming techniques which are beyond the scope of this text. Consult the references given at the end of the chapter for details on how to apply this method.

b. Calculate the mean and the variance of the residuals. The mean should equal 0 and the variance should be close to the value of MSE given in the portion of the SAS printout shown in Figure 4.7.

c. Determine the proportion of the residuals that fall outside two estimated standard deviations ($2s$) of 0.

d. If you have access to a multiple regression computer package, rerun the analysis and request influence diagnostics. Interpret the measures of influence given on the printout.

6.21 Refer to the study of the population of the world's largest cities in Exercise 6.16. A multiple regression model for log of the population (in thousands) of each country's largest city was fit to data collected on 126 nations and resulted in MSE = .036. A computer-generated (SAS) plot of the regression residuals versus $\hat{y}$ is shown here. Identify any outliers on the plot.

SAS Residual Plot for Exercise 6.21

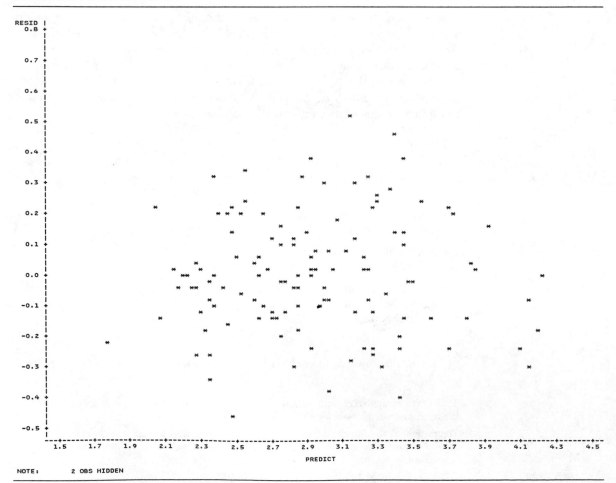

NOTE: 2 OBS HIDDEN

6.22 A large manufacturing firm wants to determine whether a relationship exists between the number of work-hours an employee misses per year and the employee's annual wages. A sample of 15 employees produced the data in the accompanying table. A first-order model was fit to the data with the following results:

$$\hat{y} = 222.64 - 9.60x \qquad r^2 = .073$$

EMPLOYEE	WORK-HOURS MISSED y	ANNUAL WAGES x, thousands of dollars
1	49	12.8
2	36	14.5
3	127	8.3
4	91	10.2
5	72	10.0
6	34	11.5
7	155	8.8
8	11	17.2
9	191	7.8
10	6	15.8
11	63	10.8
12	79	9.7
13	543	12.1
14	57	21.2
15	82	10.9

a. Interpret the value of r^2
b. Calculate and plot the regression residuals. What do you notice?
c. After searching through its employees' files, the firm has found that employee #13 had been fired but that his name had not been removed from the active employee payroll. This explains the large accumulation of work-hours missed (543) by that employee. In view of this fact, what is your recommendation concerning this outlier?
d. Use an available multiple regression program to measure how influential the observation for employee #13 is on the regression analysis.
e. Refit the model to the data, excluding the outlier, and find the least squares line. Calculate r^2 and comment on model adequacy.

SECTION 6.6

DETECTION OF RESIDUAL CORRELATION: THE DURBIN–WATSON TEST

Many types of business data are observed at regular time intervals. The Consumer Price Index (CPI) is computed and published monthly, the profits of most major corporations are published quarterly, and the *Fortune* 500 list of largest corporations is published annually. Data like these, which are observed over time, are called **time series.** We will often want to construct regression models where the data for the dependent and independent variables are time series.

Regression models of time series may pose a special problem. Because business time series tend to follow economic trends and seasonal cycles, the value of a time series at time t is often indicative of its value at time $(t + 1)$. That is, the value of a time series at time t is **correlated** with its value at time $(t + 1)$. If such a series is used as the dependent variable in a regression analysis, the result is that the random errors are correlated, and this violates one of the assumptions basic to the least squares inferential procedures. Consequently, we cannot apply the standard least squares inference-making tools and have confidence in their validity. Modifications of the methods, which allow for correlated residuals in time series regression models, will be presented in Chapter 9. In this section we present a method of testing for the presence of residual correlation.

Consider the time series data in Table 6.9 which gives sales data for the 35-year history of a company. The computer printout shown in Figure 6.28 gives the regression analysis for the first-order linear model

$$y = \beta_0 + \beta_1 t + \varepsilon$$

You will note that this model seems to fit the data very well, since $R^2 = .98$ and the F-value (1,615.72) that tests the adequacy of the model is significant. The hypothesis that the coefficient β_1 is positive is accepted at almost any α level ($t = 40.2$ with 33 df).

TABLE 6.9

A Firm's Annual Sales Revenue (thousands of dollars)

YEAR t	SALES y	YEAR t	SALES y	YEAR t	SALES y
1	4.8	13	48.4	25	100.3
2	4.0	14	61.6	26	111.7
3	5.5	15	65.6	27	108.2
4	15.6	16	71.4	28	115.5
5	23.1	17	83.4	29	119.2
6	23.3	18	93.6	30	125.2
7	31.4	19	94.2	31	136.3
8	46.0	20	85.4	32	146.8
9	46.1	21	86.2	33	146.1
10	41.9	22	89.9	34	151.4
11	45.5	23	89.2	35	150.9
12	53.5	24	99.1		

The residuals $\hat{\varepsilon} = y - (\hat{\beta}_0 + \hat{\beta}_1 t)$ are plotted in Figure 6.29 (page 306). Note that there is a distinct tendency for the residuals to have long positive and negative runs. That is, if the residual for year t is positive, there is a tendency for the residual for year $(t + 1)$ to be positive. These cycles are indicative of possible positive correlation between residuals. For most economic time series models, we want to test the null hypothesis

H_0:   No residual correlation

against the alternative

H_a: Positive residual correlation

since the hypothesis of positive residual correlation is consistent with economic trends and seasonal cycles.

FIGURE 6.28 SAS Printout for the Regression Analysis: Annual Sales Data (First-Order Model)

ANALYSIS OF VARIANCE

SOURCE	DF	SUM OF SQUARES	MEAN SQUARE	F VALUE	PROB>F
MODEL	1	65875.20817	65875.20817	1615.724	0.0001
ERROR	33	1345.45355	40.77131958		
C TOTAL	34	67220.66171			

ROOT MSE	6.385242	R-SQUARE	0.9800	
DEP MEAN	77.72286	ADJ R-SQ	0.9794	
C.V.	8.215398			

PARAMETER ESTIMATES

VARIABLE	DF	PARAMETER ESTIMATE	STANDARD ERROR	T FOR H0: PARAMETER=0	PROB > |T|
INTERCEP	1	0.40151261	2.20570829	0.182	0.8567
T	1	4.29563025	0.10686692	40.196	0.0001

OBS	ACTUAL	PREDICT VALUE	RESIDUAL
1	4.8000	4.6971	0.1029
2	4.0000	8.9928	-4.9928
3	5.5000	13.2884	-7.7884
4	15.6000	17.5840	-1.9840
5	23.1000	21.8797	1.2203
6	23.3000	26.1753	-2.8753
7	31.4000	30.4709	0.9291
8	46.0000	34.7666	11.2334
9	46.1000	39.0622	7.0378
10	41.9000	43.3578	-1.4578
11	45.5000	47.6534	-2.1554
12	53.5000	51.9491	1.5509
13	48.4000	56.2447	-7.8447
14	61.6000	60.5403	1.0597
15	65.6000	64.8360	0.7640
16	71.4000	69.1316	2.2684
17	83.4000	73.4272	9.9728
18	93.6000	77.7229	15.8771
19	94.2000	82.0185	12.1815
20	85.4000	86.3141	-0.9141
21	86.2000	90.6097	-4.4097
22	89.9000	94.9054	-5.0054
23	89.2000	99.2010	-10.0010
24	99.1000	103.5	-4.3966
25	100.3	107.8	-7.4923
26	111.7	112.1	-0.3879
27	108.2	116.4	-8.1835
28	115.5	120.7	-5.1792
29	119.2	125.0	-5.7748
30	125.2	129.3	-4.0704
31	136.3	133.6	2.7339
32	146.8	137.9	8.9383
33	146.1	142.2	3.9427
34	151.4	146.5	4.9471
35	150.9	150.7	0.1514

SUM OF RESIDUALS	-7.10543E-14
SUM OF SQUARED RESIDUALS	1345.454

DURBIN–WATSON D	0.821
(FOR NUMBER OF OBS.)	35
1ST ORDER AUTOCORRELATION	0.590

FIGURE 6.29
SAS Plot of Residuals for
the Sales Data: Least
Squares Model

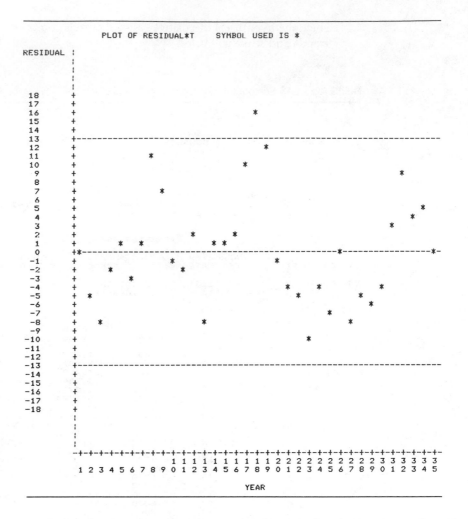

The **Durbin–Watson d-statistic** is used to test for the presence of residual correlation. This statistic is given by the formula

$$d = \frac{\sum_{t=2}^{n} (\hat{\varepsilon}_t - \hat{\varepsilon}_{t-1})^2}{\sum_{t=1}^{n} \hat{\varepsilon}_t^2}$$

where n is the number of observations and $(\hat{\varepsilon}_t - \hat{\varepsilon}_{t-1})$ represents the difference between a pair of successive residuals. By expanding the numerator of d, we can also write

$$d = \frac{\sum_{t=2}^{n} \hat{\varepsilon}_t^2}{\sum_{t=1}^{n} \hat{\varepsilon}_t^2} + \frac{\sum_{t=2}^{n} \hat{\varepsilon}_{t-1}^2}{\sum_{t=1}^{n} \hat{\varepsilon}_t^2} - \frac{2\sum_{t=2}^{n} \hat{\varepsilon}_t \hat{\varepsilon}_{t-1}}{\sum_{t=1}^{n} \hat{\varepsilon}_t^2} \approx 2 - \frac{2\sum_{t=2}^{n} \hat{\varepsilon}_t \hat{\varepsilon}_{t-1}}{\sum_{t=1}^{n} \hat{\varepsilon}_t^2}$$

If the residuals are uncorrelated,

$$\sum_{t=2}^{n} \hat{\varepsilon}_t \hat{\varepsilon}_{t-1} \approx 0$$

indicating no relationship between $\hat{\varepsilon}_t$ and $\hat{\varepsilon}_{t-1}$, the value of d will be close to 2. If the residuals are highly positively correlated,

$$\sum_{t=2}^{n} \hat{\varepsilon}_t \hat{\varepsilon}_{t-1} \approx \sum_{t=2}^{n} \hat{\varepsilon}_t^2$$

(since $\hat{\varepsilon}_t \approx \hat{\varepsilon}_{t-1}$), and the value of d will be near 0:

$$d \approx 2 - \frac{2 \sum_{t=2}^{n} \hat{\varepsilon}_t \hat{\varepsilon}_{t-1}}{\sum_{t=1}^{n} \hat{\varepsilon}_t^2} \approx 2 - \frac{2 \sum_{t=2}^{n} \hat{\varepsilon}_t^2}{\sum_{t=1}^{n} \hat{\varepsilon}_t^2} \approx 2 - 2 = 0$$

If the residuals are very negatively correlated, then $\hat{\varepsilon}_t \approx -\hat{\varepsilon}_{t-1}$, so that

$$\sum_{t=2}^{n} \hat{\varepsilon}_t \hat{\varepsilon}_{t-1} \approx -\sum_{t=2}^{n} \hat{\varepsilon}_t^2$$

and d will be approximately equal to 4. Thus, d ranges from 0 to 4, with interpretations as summarized in the box.

INTERPRETATION OF DURBIN–WATSON d-STATISTIC

DEFINITION

$$d = \frac{\sum_{t=2}^{n} (\hat{\varepsilon}_t - \hat{\varepsilon}_{t-1})^2}{\sum_{t=1}^{n} \hat{\varepsilon}_t^2}$$

Range of d: $0 \le d \le 4$

1. If residuals are uncorrelated, $d \approx 2$.
2. If residuals are positively correlated, $d < 2$, and if the correlation is very strong, $d \approx 0$.
3. If residuals are negatively correlated, $d > 2$, and if the correlation is very strong, $d \approx 4$.

Durbin and Watson (1951) have given tables for the lower-tail values of the d-statistic, which we show in Table 9 ($\alpha = .05$) and Table 10 ($\alpha = .01$) of Appendix D. Part of Table 9 is reproduced in Table 6.10 (page 308). For the sales example, we have $k = 1$ independent variable and $n = 35$ observations. Using $\alpha = .05$ for the one-tailed test for positive residual correlation, the table values are $d_L = 1.40$ and $d_U = 1.52$. The meaning of these values is illustrated in Figure 6.30. Because of the complexity of the sampling distribution of d, it is not possible to specify a single point that acts as a boundary between the rejection and nonrejection regions, as we did for the z, t, F, and other test statistics. Instead, an upper (d_U) and lower (d_L) bound are specified so that a d-value less than d_L definitely *does* provide strong evidence of positive residual correlation at $\alpha = .05$ (recall that small d-values indicate positive correlation), a d-value greater than d_U *does not* provide evidence of positive correlation at $\alpha = .05$, but a value of d between d_L and d_U *might* be significant at the $\alpha = .05$ level. If $d_L < d < d_U$, more information is needed before we can reach any conclusion about the presence of residual correlation.

TABLE 6.10 Reproduction of Part of Table 9 of Appendix D ($\alpha = .05$)

n	$k = 1$		$k = 2$		$k = 3$		$k = 4$		$k = 5$	
	d_L	d_U	d_L	d_U	d_L	d_U	d_L	d_U	d_L	d_U
31	1.36	1.50	1.30	1.57	1.23	1.65	1.16	1.74	1.09	1.83
32	1.37	1.50	1.31	1.57	1.24	1.65	1.18	1.73	1.11	1.82
33	1.38	1.51	1.32	1.58	1.26	1.65	1.19	1.73	1.13	1.81
34	1.39	1.51	1.33	1.58	1.27	1.65	1.21	1.73	1.15	1.81
35	1.40	1.52	1.34	1.58	1.28	1.65	1.22	1.73	1.16	1.80
36	1.41	1.52	1.35	1.59	1.29	1.65	1.24	1.73	1.18	1.80
37	1.42	1.53	1.36	1.59	1.31	1.66	1.25	1.72	1.19	1.80
38	1.43	1.54	1.37	1.59	1.32	1.66	1.26	1.72	1.21	1.79
39	1.43	1.54	1.38	1.60	1.33	1.66	1.27	1.72	1.22	1.79
40	1.44	1.54	1.39	1.60	1.34	1.66	1.29	1.72	1.23	1.79

FIGURE 6.30
Rejection Region for the
Durbin–Watson d-Test:
Sales Example

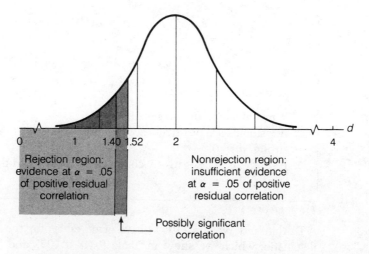

As indicated in the printout for the sales example (Figure 6.28), the computed value of d, .82, is less than the tabulated value of d_L, 1.40. Thus, we conclude that the residuals of the straight-line model for sales are positively correlated.

Tests for negative correlation and two-tailed tests can be conducted by making use of the symmetry of the sampling distribution of the d-statistic about its mean, 2 (see Figure 6.30). That is, we compare $(4 - d)$ to d_L and d_U, and conclude that the residuals are negatively correlated if $(4 - d) < d_L$, that there is insufficient evidence to conclude that the residuals are negatively correlated if $(4 - d) > d_U$, and that the test for negative residual correlation is *possibly* significant if $d_L < (4 - d) < d_U$.

DURBIN–WATSON d-TEST

ONE-TAILED TEST	TWO-TAILED TEST

ONE-TAILED TEST

H_0: No residual correlation

H_a: Positive residual correlation [or H_a: Negative residual correlation]

Test statistic:

$$d = \frac{\sum_{t=2}^{n} (\hat{\varepsilon}_t - \hat{\varepsilon}_{t-1})^2}{\sum_{t=1}^{n} \hat{\varepsilon}_t^2}$$

Rejection region: $d < d_{L,\alpha}$ [or $(4 - d) < d_{L,\alpha}$ if H_a: Negative residual correlation]

Nonrejection region: $d > d_{U,\alpha}$ [or $(4 - d) > d_{U,\alpha}$ if H_a: Negative residual correlation]

Inconclusive ("possibly significant") region:

$d_{L,\alpha} \le d \le d_{U,\alpha}$ (see Figure 6.30) [or $d_{L,\alpha} \le (4 - d) \le d_{U,\alpha}$ if H_a: Negative residual correlation]

where $d_{L,\alpha}$ and $d_{U,\alpha}$ are the lower and upper tabled values, respectively, corresponding to k independent variables and n observations.

TWO-TAILED TEST

H_0: No residual correlation

H_a: Positive or negative residual correlation

Test statistic:

$$d = \frac{\sum_{t=2}^{n} (\hat{\varepsilon}_t - \hat{\varepsilon}_{t-1})^2}{\sum_{t=1}^{n} \hat{\varepsilon}_t^2}$$

Rejection region: $d < d_{L,\alpha/2}$ or $(4 - d) < d_{L,\alpha/2}$

Nonrejection region: $d > d_{U,\alpha/2}$ or $(4 - d) > d_{U,\alpha/2}$

Inconclusive ("possibly significant") region:

Any other result

where $d_{L,\alpha/2}$ and $d_{U,\alpha/2}$ are the lower and upper tabled values, respectively, corresponding to k independent variables and n observations.

Assumption: The residuals are normally distributed.

Once strong evidence of residual correlation has been established, as in the case of the sales example, doubt is cast on the least squares results and any inferences drawn from them. In Chapter 9 we will present a time series model that accounts for the correlation of the random errors. The residual correlation can be taken into account in a time series model and thereby used to improve both the fit of the model and the reliability of model inferences.

EXERCISES 6.23–6.29

6.23 Find the values of d_L and d_U from Tables 9 and 10 of Appendix D for each of the following situations:
  a. $n = 30$, $k = 3$, $\alpha = .05$
  b. $n = 40$, $k = 1$, $\alpha = .01$
  c. $n = 35$, $k = 5$, $\alpha = .05$

6.24 Suppose you fit the time series model

$$E(y_t) = \beta_0 + \beta_1 t + \beta_2 t^2$$

to quarterly time series data collected over a 10-year period ($n = 40$ quarters), where t = quarter ($t = 1, 2, 3, \ldots, 40$).
  a. Set up the test of hypothesis for positively correlated residuals. Specify H_0, H_a, the test statistic, and the rejection region.
  b. Suppose the Durbin–Watson d-statistic is calculated to be 1.14. What is the appropriate conclusion?

6.25 Forecasts of automotive vehicle sales in the United States provide the basis for financial and strategic planning of large automotive corporations. Olson and Janakiraman (1985) developed a forecasting model for y, total monthly passenger car and light truck sales (in thousands):

$$E(y) = \beta_0 + \beta_1 x_1 + \beta_2 x_2 + \beta_3 x_3 + \beta_4 x_4 + \beta_5 x_5$$

where

  x_1 = Average monthly retail price of regular gasoline

  x_2 = Annual percentage change in GNP per quarter

  x_3 = Monthly consumer confidence index

  x_4 = Total number of vehicles scrapped (millions) per month

  x_5 = Vehicle seasonality

The model was fit to monthly data collected over a 12-year period (i.e., $n = 144$ months) with the following results:

$$\hat{y} = -676.42 - 1.93x_1 + 6.54x_2 + 2.02x_3 + .08x_4 + 9.82x_5$$

$$R^2 = .856 \qquad \text{Durbin–Watson } d = 1.01$$

a. Is there sufficient evidence to indicate that the model contributes information for the prediction of y? Test using $\alpha = .05$.

b. Is there sufficient evidence to indicate that the regression errors are positively correlated? Test using $\alpha = .05$.

c. Comment on the validity of the inference concerning model adequacy in the light of the result of part **b**.

6.26 The decrease in the value of the dollar since 1970 is illustrated by the data in the accompanying table. The buying power of the dollar (compared to 1967) is listed for each year. The first-order model $y = \beta_0 + \beta_1 t + \varepsilon$ was fit to the data using the method of least squares. The least squares estimates of β_0 and β_1 were calculated to be

$$\hat{\beta}_0 = 79.834837 \quad \text{and} \quad \hat{\beta}_1 = -.0400956$$

YEAR t	VALUE y	YEAR t	VALUE y
1970	.860	1978	.512
1971	.824	1979	.460
1972	.799	1980	.405
1973	.755	1981	.367
1974	.680	1982	.346
1975	.621	1983	.335
1976	.587	1984	.321
1977	.551	1985	.310

Source: United States Bureau of the Census. *Statistical Abstracts of the United States 1971–1987.*

a. Calculate and plot the regression residuals against t. Is there a tendency for the residuals to have long positive and negative runs? How do you account for this?

b. Calculate the Durbin–Watson d-statistic and test the hypothesis that the random errors, ε, are uncorrelated. Test at $\alpha = .05$.

c. What assumption(s) must be satisfied in order for the test of part **b** to be valid?

6.27 Refer to Exercise 6.6. The tax economist theorizes that the total tax collected in the state of Florida in year t is indicative of the total tax collected in the state in year $(t + 1)$. If this is so, then the least squares assumption that the random errors are uncorrelated is violated. Use the Durbin–Watson procedure to test the null hypothesis

H_0: No residual correlation

against the alternative

H_a: Positive residual correlation

Use $\alpha = .05$.

6.28 The accompanying table gives the retail sales of passenger cars in the United States for the years 1984 and 1985.

MONTH	TIME t	1984 SALES y, thousands	TIME t	1985 SALES y, thousands
January	1	583.4	13	628.0
February	2	655.0	14	645.2
March	3	756.2	15	768.9
April	4	721.1	16	788.2
May	5	803.3	17	807.6
June	6	727.4	18	676.7
July	7	684.1	19	633.5
August	8	603.5	20	744.8
September	9	566.7	21	839.4
October	10	689.6	22	598.4
November	11	600.7	23	515.8
December	12	560.7	24	558.0

Source: Wards Automotive Yearbook, 1986.

a. Fit the first-order time series model $y = \beta_0 + \beta_1 t + \varepsilon$ to the data using the method of least squares.

b. As a first step in a check of the usual least squares assumptions about the random error ε, calculate the regression residuals, $\hat{\varepsilon}$.

c. Calculate the Durbin–Watson d-statistic.

d. Is there evidence at the $\alpha = .05$ level of significance to conclude that the random errors ε are correlated?

e. Do you recommend the use of the least squares model found in part **a** as an inference-making tool? Why or why not?

6.29 B. N. Song compared annual consumption for two lower developed countries (LDCs)—Korea, a poor LDC, and Italy, a rich LDC (*Economic Development and Cultural Change*, April 1981). Using data from the post–Korean War period, Song modeled annual consumption y_t as a function of total labor income x_{1t} and total property income x_{2t}, with the following results (assume data for $n = 40$ years were used in the analysis):

Korea: $\hat{y}_t = 7.81 + .91x_{1t} + .57x_{2t}$

$s = 1.29$

$d = 2.09$

Italy: $\hat{y}_t = 1{,}043.4 + .85x_{1t} + .40x_{2t}$

$s = 290.5$

$d = 1.07$

a. Is there evidence of positively correlated residuals in the consumption model for Korea? Test using $\alpha = .05$.

b. Is there evidence of positively correlated residuals in the consumption model for Italy? Test using $\alpha = .05$.

SECTION 6.7

SUMMARY

An analysis of regression **residuals** can play an important role in the modeling process. Plots of the residuals against the independent variables can suggest modifications that will improve the model. These include the **addition of quadratic terms to allow for curvature** in the response surface, and **transforming the dependent variable to stabilize its variance. Histograms, stem-and-leaf displays,** and **normal probability plots** of residuals give visual clues as to whether the normality assumption is satisfied. **Plots of residuals** are also helpful for identification of **outliers,** which can then be traced to determine the cause of an unusually large or small observation. The influence the outlying observation has on the regression analysis can be measured using **leverage,** Cook's distance D, and **deleted residuals**.

When you use residual plots, you should always be aware that conclusions drawn from them are subjective. Therefore, they cannot substitute for formal tests to detect model inadequacies.

The F-test (Section 2.9) can be used to detect differences in pairs of variances, and F-tests for sets of parameters (Section 4.10) can be used to detect factor interactions and curvature in a response surface. The **Durbin–Watson d-test** can be used to test for the presence of **residual correlation**. We will present methods for constructing time series models that allow for correlation of the random errors in Chapter 9.

SUPPLEMENTARY EXERCISES 6.30–6.39

6.30 A certain type of rare gem serves as a status symbol for many of its owners. As the price of the gem decreases, the demand increases, but then at moderate prices the demand falls off. However, as the price increases, the demand again increases, due to the status the owners believe they gain by obtaining the gem. The model proposed to best explain the demand for the gem by its price is the second-order model

$$y = \beta_0 + \beta_1 x + \beta_2 x^2 + \varepsilon$$

where y = demand (in thousands) and x = retail price per carat (dollars). This model was fit to the data given in the table. The least squares analysis gives the following results: $\hat{\beta}_0 = 200.4$, $\hat{\beta}_1 = -.78$, $\hat{\beta}_2 = .000987$, $F = 171.98$, and $R^2 = .975$, with $s_{\hat{\beta}_1} = .0469$ and $s_{\hat{\beta}_2} = .000055$.

RETAIL PRICE PER CARAT x, dollars	DEMAND y, thousands	RETAIL PRICE PER CARAT x, dollars	DEMAND y, thousands
100	130	70	150
700	150	50	160
450	60	300	50
150	120	350	40
500	50	750	180
800	200	700	130

a. Give the least squares prediction equation. Use the prediction equation to calculate the regression residuals.

b. Plot the residuals against retail price per carat, x. Can you detect any trends in the residual plot?

c. Plot the residuals against $\hat{y}$. Do you detect any trends?

d. Use a graphical technique to check the normality assumption.

e. Use the F- and R^2-values given above, in combination with the results of the residual plot, to make a statement about model adequacy and model assumptions.

6.31 Refer to Exercise 4.11. The first-order least squares model was found to be

$$\hat{y} = 95.75 - .3199x$$

where y is the average utility bill (dollars) for a standard-size home and x is the average monthly temperature (°F).

a. Use the data in the table that appears in Exercise 4.11 to calculate the regression residuals for this first-order model.

b. Display a plot of the residuals versus average monthly temperature, x.

c. Are there any observable trends in this plot? If so, what alternative model is suggested by the trend? Does this agree with your answer to part **c** in Exercise 4.11?

d. Fit the alternative model (part **c**), calculate the new regression residuals, and make a plot similar to part **b** above. Comment on the result of this residual plot.

6.32 A naval base is considering modifying or adding to its fleet of 48 standard aircraft. The final decision regarding the type and number of aircraft to be added depends on a comparison of cost versus effectiveness of the modified fleet. Consequently, the naval base would like to model the projected percentage increase y in fleet effectiveness by the end of the decade as a function of the cost x of modifying the fleet. A first proposal is the quadratic model

$$E(y) = \beta_0 + \beta_1 x + \beta_2 x^2$$

The data provided in the accompanying table were collected on ten naval bases of a similar size that recently expanded their fleets. The data were used to fit the model, and the SAS printout of the multiple regression analysis is reproduced here.

PERCENTAGE IMPROVEMENT AT END OF DECADE y	COST OF MODIFYING FLEET x, millions of dollars
18	125
32	160
9	80
37	162
6	110
3	90
30	140
10	85
25	150
2	50

SAS Printout for Exercise 6.32

DEP VARIABLE: Y

ANALYSIS OF VARIANCE

SOURCE	DF	SUM OF SQUARES	MEAN SQUARE	F VALUE	PROB>F
MODEL	2	1368.77501	684.38750	33.079	0.0003
ERROR	7	144.82499	20.68928481		
C TOTAL	9	1513.60000			

ROOT MSE	4.548548	R-SQUARE	0.9043	
DEP MEAN	17.2	ADJ R-SQ	0.8770	
C.V.	26.44504			

PARAMETER ESTIMATES

VARIABLE	DF	PARAMETER ESTIMATE	STANDARD ERROR	T FOR HO: PARAMETER=0	PROB > :T:
INTERCEP	1	10.65903604	14.55009061	0.733	0.4876
X	1	-0.28160568	0.28087588	-1.003	0.3494
XX	1	0.002671936	0.001253832	2.131	0.0706

OBS	ACTUAL	PREDICT VALUE	STD ERR PREDICT	RESIDUAL	STD ERR RESIDUAL	STUDENT RESIDUAL	-2-1-0 1 2	COOK'S D
1	18.0000	17.2073	2.0627	0.7927	4.0539	0.1955	: : :	0.003
2	32.0000	34.0037	2.6525	-2.0037	3.6951	-0.5423	: *: :	0.051
3	9.0000	5.2310	2.0601	3.7690	4.0553	0.9294	: :* :	0.074
4	37.0000	35.1612	2.8397	1.8388	3.5532	0.5175	: :* :	0.057
5	6.0000	12.0128	2.2177	-6.0128	3.9713	-1.5141	: ***: :	0.238
6	3.0000	6.9572	2.0895	-3.9572	4.0433	-0.9787	: *: :	0.085
7	30.0000	23.6042	1.8468	6.3958	4.1568	1.5387	: :*** :	0.156
8	10.0000	6.0273	2.0492	3.9727	4.0608	0.9783	: :* :	0.081
9	25.0000	28.5367	2.0070	-3.5367	4.0818	-0.8665	: *: :	0.060
10	2.0000	3.2586	4.1920	-1.2586	1.7654	-0.7129	: *: :	0.955

SUM OF RESIDUALS   4.97380E-14
SUM OF SQUARED RESIDUALS   144.825
PREDICTED RESID SS (PRESS) 301.668

OBS	RESIDUAL	RSTUDENT	HAT DIAG H	COV RATIO	DFFITS	INTERCEP DFBETAS	X DFBETAS	XX DFBETAS
1	0.7927	0.1815	0.2057	1.9665	0.0924	-0.0594	0.0657	-0.0639
2	-2.0037	-0.5129	0.3401	2.1156	-0.3682	-0.1192	0.1505	-0.1871
3	3.7690	0.9190	0.2051	1.3457	0.4669	0.0286	0.0632	-0.1088
4	1.8388	0.4886	0.3898	2.3148	0.3904	0.1473	-0.1817	0.2199
5	-6.0128	-1.7093	0.2377	0.6336	-0.9545	0.5920	-0.7014	0.7211
6	-3.9572	-0.9753	0.2098	1.2924	-0.5025	0.1474	-0.2357	0.2724
7	6.3958	1.7511	0.1649	0.5511	0.7780	-0.3140	0.3143	-0.2585
8	3.9727	0.9748	0.2030	1.2818	0.4919	-0.0653	0.1582	-0.2007
9	-3.5367	-0.8490	0.1947	1.4030	-0.4174	0.0129	0.0093	-0.0503
10	-1.2586	-0.6854	0.8494	8.4082	-1.6276	-1.4594	1.2838	-1.1539

a. Calculate the regression residuals and construct a residual plot versus x. Do you detect any trends? Any outliers?

b. Interpret the influence diagnostics shown on the printout. Are there any observations that have large influence on the analysis?

6.33 In 1974, Congress adopted the Federal-Aid Highway Amendments which reduced the highway speed limit to 55 miles per hour (mph). Since that time, controversy over the social efficiency of the decision has grown. University of Colorado Professors T. H. Ferrester, R. F. McNown, and L. D. Singell conducted an analysis to estimate the effect of the 55-mph speed limit on traffic fatailities (*Southern Economic Journal*, January 1984). Time series data for the United States from 1952 to 1979 ($n = 28$ years) were used to fit a regression model relating traffic fatalities y_t at time t to $k = 7$ independent variables:

x_{1t} = Real earned income

x_{2t} = Vehicle miles

x_{3t} = Ratio of number of youths to number of adults

x_{4t} = Percentage of all car purchases that are imported cars

x_{5t} = Average highway speed

x_{6t} = Percentage of cars traveling between 45 and 60 mph

$x_{7t} = \begin{cases} 0 & \text{if 55-mph speed limit imposed} \\ 1 & \text{otherwise} \end{cases}$

The results of the multiple regression are summarized as follows:

$$\hat{y}_t = -20{,}016.4 + 7{,}544.85x_{1t} - .01046x_{2t} - 36{,}758.0x_{3t} - 117.609x_{4t} + 1{,}325.22x_{5t} - 415.742x_{6t} + 9{,}678.08x_{7t}$$

$$R^2 = .987 \qquad F = 217.23 \qquad d = 1.97$$

a. Is there evidence that the model is useful for predicting annual traffic fatalities? Test using $\alpha = .05$.

b. Is there evidence that the regression residuals are correlated? Test using $\alpha = .05$.

6.34 A leading pharmaceutical company that produces a new hypertension pill would like to model annual revenue generated by this product. Company researchers utilized data collected over the past 15 years (1974–1988) to fit the model

$$E(y_t) = \beta_0 + \beta_1 x_t + \beta_2 t$$

where

y_t = Revenue in year t (in millions of dollars)

x_t = Cost per pill in year t

t = Year (1, 2, . . . , 15)

The SAS printout for the regression analysis is reproduced on page 317. A company statistician suspects that the assumption of independent errors may be violated and that, in fact, the regression residuals are positively correlated. Test this claim using $\alpha = .05$.

6.35 A ten-speed bicycle shop is located near a large southern university. The owner of the shop is having difficulty in determining the quantity of bicycles to order each month from the manufacturer. To solve this problem, it is essential that the owner be able to predict the monthly demand for the bikes. The owner proposes the following model:

$$y = \beta_0 + \beta_1 x_1 + \beta_2 x_2 + \beta_3 x_3 + \beta_4 x_4 + \beta_5 x_5 + \varepsilon$$

where y = Monthly demand for ten-speed bicycles, x_1 = Selling price of ten-speed bicycles, x_2 = Average price of lead-free gasoline,

SAS Printout for Exercise 6.34

DEP VARIABLE: Y

ANALYSIS OF VARIANCE

SOURCE	DF	SUM OF SQUARES	MEAN SQUARE	F VALUE	PROB>F
MODEL	2	48.82325339	24.41162670	206.187	0.0001
ERROR	12	1.42074661	0.11839555		
C TOTAL	14	50.24400000			

ROOT MSE	0.3440865	R-SQUARE	0.9717	
DEP MEAN	7.32	ADJ R-SQ	0.9670	
C.V.	4.700636			

PARAMETER ESTIMATES

VARIABLE	DF	PARAMETER ESTIMATE	STANDARD ERROR	T FOR H0: PARAMETER=0	PROB > |T|
INTERCEP	1	3.26119109	1.87880228	1.736	0.1082
T	1	0.39158795	0.07045937	5.558	0.0001
X	1	1.58760907	4.12905034	0.384	0.7073

OBS	ID	ACTUAL	PREDICT VALUE	RESIDUAL
1	0.48	5.0000	4.4148	0.5852
2	0.45	4.9000	4.7588	0.1412
3	0.5	5.3000	5.2298	0.0702
4	0.5	5.7000	5.6213	0.0787
5	0.55	5.9000	6.0923	-0.1923
6	0.57	6.1000	6.5157	-0.4157
7	0.6	6.9000	6.9549	-0.0549
8	0.6	7.0000	7.3465	-0.3465
9	0.62	7.5000	7.7698	-0.2698
10	0.6	7.8000	8.1296	-0.3296
11	0.6	8.3000	8.5212	-0.2212
12	0.62	8.9000	8.9446	-0.0446
13	0.65	9.6000	9.3838	0.2162
14	0.7	10.5000	9.8547	0.6453
15	0.71	10.4000	10.2622	0.1378

SUM OF RESIDUALS	1.82077E-14
SUM OF SQUARED RESIDUALS	1.420747
DURBIN-WATSON D	0.776
(FOR NUMBER OF OBS.)	15
1ST ORDER AUTOCORRELATION	0.485

$$x_3 = \begin{cases} 1 & \text{if fall quarter (September, October, November)} \\ 0 & \text{if not} \end{cases}$$

$$x_4 = \begin{cases} 1 & \text{if winter quarter (December, January, February)} \\ 0 & \text{if not} \end{cases}$$

$$x_5 = \begin{cases} 1 & \text{if spring quarter (March, April, May)} \\ 0 & \text{if not} \end{cases}$$

Data obtained from past records (given in the table on page 318) were used to fit the first-order model, and a portion of the SAS printout is also shown.

YEAR	MONTH	DEMAND y	SELLING PRICE x_1, dollars	AVERAGE PRICE OF LEAD-FREE GASOLINE x_2, dollars
1987	May	50	93	1.22
	June	31	95	1.21
	July	16	85	1.21
	August	22	83	1.19
	September	99	95	1.18
	October	80	103	1.16
	November	37	100	1.17
	December	39	105	1.20
1988	January	51	105	1.20
	February	30	105	1.21
	March	22	115	1.22
	April	37	120	1.24
	May	47	115	1.26
	June	28	110	1.27
	July	20	110	1.28

SAS Printout for Exercise 6.35

DEP VARIABLE: Y   ANALYSIS OF VARIANCE

SOURCE	DF	SUM OF SQUARES	MEAN SQUARE	F VALUE	PROB>F
MODEL	5	4773.85553	954.77111	3.383	0.0538
ERROR	9	2539.74447	282.19383		
C TOTAL	14	7313.60000			

ROOT MSE	16.79863	R-SQUARE	0.6527	
DEP MEAN	40.6	ADJ R-SQ	0.4598	
C.V.	41.37593			

PARAMETER ESTIMATES

VARIABLE	DF	PARAMETER ESTIMATE	STANDARD ERROR	T FOR H0: PARAMETER=0	PROB > |T|
INTERCEP	1	-252.61710	308.22379	-0.820	0.4336
X1	1	-0.78982290	0.76446032	-1.033	0.3285
X2	1	285.96915	292.73704	0.977	0.3542
X3	1	68.48893664	23.28222353	2.942	0.0164
X4	1	31.43229467	18.55433765	1.694	0.1245
X5	1	25.91808652	15.17546604	1.708	0.1218

OBS		ID	ACTUAL	PREDICT VALUE	RESIDUAL
	1	MAY87	50.0000	48.7298	1.2702
	2	JUN87	31.0000	18.3724	12.6276
	3	JUL87	16.0000	26.2706	-10.2706
	4	AUG87	22.0000	22.1309	-0.1309
	5	SEP87	99.0000	78.2823	20.7177
	6	OCT87	80.0000	66.2443	13.7557
	7	NOV87	37.0000	71.4735	-34.4735
	8	DEC87	39.0000	39.0468	-0.0468
	9	JAN88	51.0000	39.0468	11.9532
	10	FEB88	30.0000	41.9065	-11.9065
	11	MAR88	22.0000	31.3537	-9.3537
	12	APR88	37.0000	33.1240	3.8760
	13	MAY88	47.0000	42.7925	4.2075
	14	JUN88	28.0000	23.6832	4.3168
	15	JUL88	20.0000	26.5429	-6.5429

SUM OF RESIDUALS   -1.64491E-12
SUM OF SQUARED RESIDUALS   2539.744

a. Construct and interpret plots of residuals against each of the independent variables, x_1 and x_2.
b. Calculate and plot the partial residuals for x_1. Interpret the plot.
c. Calculate and plot the partial residuals for x_2. Interpret the plot.
d. Check the plots in part **a** for residuals that lie more than 2 estimated standard deviations of ε away from the mean of 0. Can you classify any of the residuals as outliers?
e. Would you recommend that the owner use this model to predict monthly bicycle demand? If not, suggest an alternative model.

6.36 The foreman of a printing shop is scheduling his work load for 1989, and he must estimate the number of employees available for work. He asks the company statistician to forecast the absentee rate for 1989. Since it is known that quarterly fluctuations exist, the following model is proposed:

$$y = \beta_0 + \beta_1 x_1 + \beta_2 x_2 + \beta_3 x_3 + \varepsilon$$

where

$$y = \text{Absentee rate} = \frac{\text{Total employees absent}}{\text{Total employees}}$$

$$x_1 = \begin{cases} 1 & \text{if quarter 1 (January–March)} \\ 0 & \text{if not} \end{cases}$$

$$x_2 = \begin{cases} 1 & \text{if quarter 2 (April–June)} \\ 0 & \text{if not} \end{cases}$$

$$x_3 = \begin{cases} 1 & \text{if quarter 3 (July–September)} \\ 0 & \text{if not} \end{cases}$$

YEAR	QUARTER 1	QUARTER 2	QUARTER 3	QUARTER 4
1984	.06	.13	.28	.07
1985	.12	.09	.19	.09
1986	.08	.18	.41	.07
1987	.05	.13	.23	.08
1988	.06	.07	.30	.05

a. Fit the model to the data given in the table.
b. Consider the nature of the response variable, y. Do you think that there may be possible violations of the usual assumptions about ε? Explain.
c. Suggest an alternative model that will approximately stabilize the variance of the error term ε.
d. Fit the alternative model. Check R^2 to determine whether model adequacy has improved.

6.37 Since the energy shortage, the price of foreign crude oil has skyrocketed. Consequently, crude oil imports into the United States have declined. The data in the table at the top of page 320 are the amounts of crude oil (millions of barrels) imported into the United States from the Organization of Petroleum Exporting Countries (OPEC) for the years 1973–1985.

YEAR	t	IMPORTS y_t	YEAR	t	IMPORTS y_t
1973	1	767	1980	8	1,414
1974	2	926	1981	9	1,067
1975	3	1,171	1982	10	633
1976	4	1,663	1983	11	540
1977	5	2,058	1984	12	553
1978	6	1,892	1985	13	479
1979	7	1,866			

Source: United States Bureau of the Census, *Statistical Abstracts of the United States, 1975–1987.*

a. Fit the model $E(y) = \beta_0 + \beta_1 t$.
b. Calculate the residuals for the model and plot the residuals against t. Do you detect any trends?
c. Test for correlated residuals using $\alpha = .01$.

6.38 The breeding ability of a thoroughbred horse is sometimes a more important consideration to prospective buyers than is racing ability. Usually, the longer a horse lives, the greater its value for breeding purposes. Before marketing a group of horses, a breeder would like to be able to predict their lifelengths. The breeder believes that the gestation period of a thoroughbred horse may be an indicator of its lifelength. The information in the table was supplied to the breeder by various stables in the area. (Note that the horse has the greatest variation of gestation period of any species due to seasonal and feed factors.) The first-order (linear) model

$$y = \beta_0 + \beta_1 x + \varepsilon$$

was fit to the data. A portion of the SAS printout is shown here.

HORSE	GESTATION PERIOD x, days	LIFELENGTH y, years
1	403	30
2	279	22
3	307	7
4	416	31
5	265	21
6	356	27
7	298	25

a. Check model adequacy by interpreting the F- and R^2-statistics.
b. Construct a plot of the residuals versus x, gestation period.
c. Check for residuals that lie outside the interval $0 \pm 2s$ or $0 \pm 3s$.
d. The breeder has been informed that the short life span of horse number 3 (7 years) was due to a very rare disease. Omit the data for horse number 3 and refit the least squares line. Has the omission of this observation improved the model?

SAS Printout for Exercise 6.38

ANALYSIS OF VARIANCE

SOURCE	DF	SUM OF SQUARES	MEAN SQUARE	F VALUE	PROB>F
MODEL	1	146.31556	146.31556	2.960	0.1459
ERROR	5	247.11301	49.42260285		
C TOTAL	6	393.42857			

ROOT MSE	7.030121	R-SQUARE	0.3719	
DEP MEAN	23.28571	ADJ R-SQ	0.2463	
C.V.	30.1907			

PARAMETER ESTIMATES

VARIABLE	DF	PARAMETER ESTIMATE	STANDARD ERROR	T FOR H0: PARAMETER=0	PROB > |T|
INTERCEP	1	-3.94341407	16.04679821	-0.246	0.8156
X	1	0.08201545	0.04766649	1.721	0.1459

OBS	ID	ACTUAL	PREDICT VALUE	RESIDUAL
1	403	30.0000	29.1088	0.8912
2	279	22.0000	18.9389	3.0611
3	307	7.0000	21.2353	-14.2353
4	416	31.0000	30.1750	0.8250
5	265	21.0000	17.7907	3.2093
6	356	27.0000	25.2541	1.7459
7	298	25.0000	20.4972	4.5028

SUM OF RESIDUALS   7.10543E-15
SUM OF SQUARED RESIDUALS   247.113

6.39 Refer to Exercise 3.26. An avid golfer fit the model $y = \beta_0 + \beta_1 x + \varepsilon$ to the data given in the table that appears in Exercise 3.26 (where y = United States divorce rate per 1,000 population and x = number of 18-hole and larger golf courses in the United States). The least squares procedure gave the following results:

$$\hat{\beta}_0 = .108199 \qquad \hat{\beta}_1 = .000708$$

a. Calculate the regression residuals for this first-order model.

b. There is a very good possibility that the United States divorce rate for year t is correlated with the United States divorce rate for year $(t + 1)$. If this is so, then the least squares assumption of uncorrelated random errors would be violated, lending doubt to the inference-making ability of the model. As a check on the assumption in question, calculate the Durbin–Watson d-statistic and test the hypothesis of uncorrelated errors. Use $\alpha = .05$.

REFERENCES

Barnett, V. and Lewis, T. *Outliers in Statistical Data*. New York: Wiley, 1978.

Belsley, D. A., Kuh, E., and Welsch, R. E. *Regression Diagnostics: Identifying Influential Data and Sources of Collinearity*. New York: Wiley, 1980.

Box, G. E. P. and Cox, D. R. "An analysis of transformations," *Journal of the Royal Statistical Society, Series B*, 1964, 26, pp. 211–243.

Cook, R. D. "Influential observations in linear regression," *Journal of the American Statistical Association*, 1979, 74, pp. 169–174.

Draper, N. and Smith, H. *Applied Regression Analysis*, 2nd ed. New York: Wiley, 1981.

Durbin, J. and Watson, G. S. "Testing for serial correlation in least squares regression, I," *Biometrika*, 1950, *37*, pp. 409–428.

Durbin, J. and Watson, G. S. "Testing for serial correlation in least squares regression, II," *Biometrika*, 1951, *38*, pp. 159–178.

Durbin, J. and Watson, G. S. "Testing for serial correlation in least squares regression, III," *Biometrika*, 1971, *58*, pp. 1–19.

Ferrester, T. H., McNown, R. F., and Singell, L. D. "A cost-benefit analysis of the 55-mph speed limit," *Southern Economic Journal*, January 1984, *50*, pp. 631–641.

Granger, C. W. J. and Newbold, P. *Forecasting Economic Time Series*. New York: Academic Press, 1977.

Larsen, W. A. and McCleary, S. J. "The use of partial residual plots in regression analysis," *Technometrics*, Vol. 14, 1972, pp. 781–790.

Mansfield, E. R. and Conerly, M. D. "Diagnostic value of residual and partial residual plots," *The American Statistician*, Vol. 41, No. 2, May 1987, pp. 107–116.

Mendenhall, W. *Introduction to Linear Models and the Design and Analysis of Experiments*. Belmont, Ca.: Wadsworth, 1968.

Montgomery, D. C. and Peck, E. A. *Introduction to Linear Regression Analysis*. New York: Wiley, 1982.

Neter, J., Wasserman, W., and Kutner, M. H. *Applied Linear Statistical Models*, 2nd ed. Homewood, Ill.: Richard D. Irwin, 1985.

Olson, S. J. and Janakiraman, J. "Proposed U.S. passenger car and light truck sales forecast model." Paper presented at SAS Users Group International Conference, Reno, Nevada, 1985.

Song, B. N. "Empirical research on consumption behavior: Evidence from rich and poor LDCs," *Economic Development and Cultural Change*, April 1981, *29*, pp. 597–611.

Stephens, M. A. "EDF statistics for goodness of fit and some comparisons," *Journal of the American Statistical Association*, 1974, *69*, pp. 730–737.

C H A P T E R 7

OBJECTIVE

To show you why the choice of the deterministic portion of a linear model is crucial to the acquisition of a good prediction equation; to present some basic concepts and procedures for constructing good linear models

CONTENTS

MODEL-BUILDING (OPTIONAL)

We have emphasized in both Chapters 3 and 4 that one of the first steps in the construction of a regression model is to hypothesize the form of the deterministic portion of the probabilistic model. This **model-building**, or model-construction, stage is the key to the success (or failure) of the regression analysis. If the hypothesized model does not reflect, at least approximately, the true nature of the relationship between the mean response $E(y)$ and the independent variables $x_1, x_2, \ldots, x_k$, the modeling effort will usually be unrewarded.

By *model-building*, we mean writing a model that will provide a good fit to a set of data and that will give good estimates of the mean value of y and good predictions of future values of y for given values of the independent variables. To illustrate, several years ago, a nationally recognized educational research group issued a report concerning the variables related to academic achievement for a certain type of college student. The researchers selected a random sample of students and recorded a measure of academic achievement, y, at the end of the senior year along with data on an extensive list of independent variables, $x_1, x_2, \ldots, x_k$, that they thought were related to y. Among these independent variables were the student's IQ, scores on mathematics and verbal achievement examinations, rank in class, etc. They fit the model

$$E(y) = \beta_0 + \beta_1 x_1 + \beta_2 x_2 + \cdots + \beta_k x_k$$

to the data, analyzed the results, and reached the conclusion that none of the independent variables are "significantly related" to y. The **goodness of fit** of the model, measured by the coefficient of determination R^2, was not particularly good, and t-tests on individual parameters did not lead to rejection of the null hypotheses that these parameters equaled 0.

How could the researchers have reached the conclusion that there is no significant relationship, when it is evident, just as a matter of experience, that some of the independent variables studied are related to academic achievement? For example, achievement on a college mathematics placement test should be related to achievement in college mathematics. Certainly, many other variables will affect achievement—motivation, environmental conditions, and so forth—but generally speaking, there will be a positive correlation between entrance achievement test scores and college academic achievement. So, what went wrong with the educational researchers' study?

Although you can never discard the possibility of computing error as a reason for erroneous answers, most likely the difficulties in the results of the educational study were caused by the use of an improperly constructed model. For example, the model

$$E(y) = \beta_0 + \beta_1 x_1 + \beta_2 x_2 + \cdots + \beta_k x_k$$

assumes that the independent variables $x_1, x_2, \ldots, x_k$ affect mean achievement $E(y)$ independently of each other.* Thus, if you hold all the other independent variables constant and vary only x_1, $E(y)$ will increase by the amount β_1 for

*Keep in mind that we are discussing the deterministic portion of the model and that the word *independent* is used in a mathematical rather than a probabilistic sense.

every unit increase in x_1. A 1-unit change in any of the other independent variables will increase $E(y)$ by the value of the corresponding β parameter for that variable.

Do the assumptions implied by the model agree with your knowledge about academic achievement? First, is it reasonable to assume that the effect of time spent on study is independent of native intellectual ability? We think not. No matter how much effort some students invest in a particular subject, their rate of achievement is low. For others, it may be high. Therefore, assuming that these two variables—effort and native intellectual ability—affect $E(y)$ independently of each other is likely to be an erroneous assumption. Second, suppose that x_5 is the amount of time a student devotes to study. Is it reasonable to expect that a 1-unit increase in x_5 will always produce the same change β_5 in $E(y)$? The changes in $E(y)$ for a 1-unit increase in x_5 might depend on the value of x_5 (for example, the law of diminishing returns). Consequently, it is quite likely that the assumption of a constant rate of change in $E(y)$ for 1-unit increases in the independent variables will not be satisfied.

Clearly, the model

$$E(y) = \beta_0 + \beta_1 x_1 + \beta_2 x_2 + \cdots + \beta_k x_k$$

was a poor choice in view of the researchers' prior knowledge of some of the variables involved. Terms have to be added to the model to account for interrelationships among the independent variables and for curvature in the response function. Failure to include needed terms causes inflated values of SSE, nonsignificance in statistical tests, and, often, erroneous practical conclusions.

In this chapter we discuss the most difficult part of a multiple regression analysis—the formulation of a good model for $E(y)$. Although many of the models presented in this chapter have already been introduced in Section 4.10, we assume the reader has little or no background in model-building. This chapter serves as a basic reference guide to model-building for teachers, students, and practitioners of multiple regression analysis.

| | | | | | | | | | | | |

S E C T I O N   7.2

THE TWO TYPES OF INDEPENDENT VARIABLES: QUANTITATIVE AND QUALITATIVE

The independent variables that appear in a linear model can be one of two types—either **quantitative** or **qualitative**.

DEFINITION 7.1

A **quantitative** independent variable is one that assumes numerical values corresponding to the points on a line. An independent variable that is not quantitative is called **qualitative**.

The Gross National Product, prime interest rate, number of defects in a product, and kilowatt-hours of electricity used per day are all examples of quantitative independent variables. On the other hand, suppose three different styles of packaging, A, B, and C, are used by a manufacturer. This independent variable, style of packaging, is qualitative, since it is not measured on a numerical scale. Certainly, style of packaging is an independent variable that may affect sales of a product, and we would want to include it in a model describing the product's sales, y.

DEFINITION 7.2

The different intensity settings of an independent variable are called its levels.

For a quantitative variable, the levels correspond to the numerical values it assumes. For example, if the number of defects in a product ranges from 0 to 3, the independent variable assumes four levels: 0, 1, 2, and 3.

The levels of a qualitative variable are not numerical. They can be defined only by describing them. For example, the independent variable style of packaging was observed at three levels: A, B, and C.

EXAMPLE 7.1

In Chapter 4 we considered the problem of predicting executive salary as a function of several independent variables. Consider the following four independent variables that may affect executive salaries:

a. Number of years of experience
b. Sex of the employee
c. Firm's net asset value
d. Rank of the employee

For each of these independent variables, give its type and describe the nature of the levels you would expect to observe.

SOLUTION

a. The independent variable for the number of years of experience is quantitative, since its values are numerical. We would expect to observe levels ranging from 0 to 40 (approximately) years.
b. The independent variable for sex is qualitative, since its levels can only be described by the nonnumerical labels "female" and "male."
c. The independent variable for the firm's net asset value is quantitative, with a very large number of possible levels corresponding to the range of dollar values representing various firms' net asset values.
d. Suppose the independent variable for the rank of the employee is observed at three levels: supervisor, assistant vice president, and vice president. Since we cannot assign a realistic measure of relative importance to each position, rank is a qualitative independent variable.   ■

Quantitative and qualitative independent variables are treated differently in regression modeling. In the next section, we will see how quantitative variables are entered into a regression model.

7.1 Companies keep personnel files that contain important information on each employee's background. The data in these files could be used to predict employee performance ratings. Identify the independent variables listed below as qualitative or quantitative. For qualitative variables, suggest several levels that might be observed. For quantitative variables, give a range of values (levels) for which the variable might be observed.

a. Age
b. Years of experience with the company
c. Highest educational degree
d. Job classification
e. Marital status
f. Religious preference
g. Salary
h. Sex

7.2 Which of the assumptions about ε (Section 4.2) prohibit the use of a qualitative variable as a dependent variable? (We present a technique for modeling a qualitative dependent variable in Chapter 10.)

| | | | | | | | | | | | | |

S E C T I O N   7.3

MODELS WITH A SINGLE QUANTITATIVE INDEPENDENT VARIABLE

To write a prediction equation that provides a good model for a response (one that will eventually yield good predictions), we have to know how the response might vary as the levels of an independent variable change. Then we have to know how to write a mathematical equation to model it. To illustrate (with a simple example), suppose we want to model corporate profit, y, as a function of the single independent variable x, the amount of capital invested. It may be that corporate profit, y, increases in a straight line as the amount of capital invested, x, varies from $1 to 2 million, as shown in Figure 7.1(a). If this were the entire range of x-values for which you wanted to predict y, the model

$$E(y) = \beta_0 + \beta_1 x$$

would be appropriate.

FIGURE 7.1

Modeling Corporate Profit, y, as a Function of Capital Invested, x

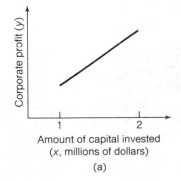

Amount of capital invested
(x, millions of dollars)

(a)

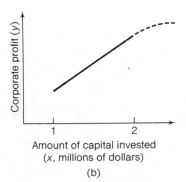

Amount of capital invested
(x, millions of dollars)

(b)

Now, suppose you want to expand the range of values of x to $x = \$3$ or 4 million of capital investment. Will the straight-line model

$$E(y) = \beta_0 + \beta_1 x$$

be satisfactory? Perhaps, but making this assumption could be risky. As the amount of capital invested, x, is increased, sooner or later the point of diminishing returns will be reached. That is, the increase in profit for a unit increase in capital invested will decrease, as shown by the dashed line in Figure 7.1(b). To produce this type of curvature, you need to know the relationship between models and graphs, and how types of terms will change the shape of the curve.

A response that is a function of a single quantitative independent variable can often be modeled by the first few terms of a polynomial algebraic function. The equation relating the mean value of y to a polynomial of order p in one independent variable x is shown in the box. .

FORMULA FOR A pth-ORDER POLYNOMIAL WITH ONE INDEPENDENT VARIABLE

$$E(y) = \beta_0 + \beta_1 x + \beta_2 x^2 + \beta_3 x^3 + \cdots + \beta_p x^p$$

where p is an integer and $\beta_0, \beta_1, \ldots, \beta_p$ are unknown parameters that must be estimated.

As we mentioned in Chapters 3 and 4, a **first-order polynomial** in x (i.e., $p = 1$),

$$E(y) = \beta_0 + \beta_1 x$$

graphs as a straight line. A **second-order polynomial** model ($p = 2$), called a **quadratic**, is given by the equation below:

A SECOND-ORDER (QUADRATIC) MODEL WITH ONE INDEPENDENT VARIABLE

$$E(y) = \beta_0 + \beta_1 x + \beta_2 x^2$$

where β_0, β_1, and β_2 are unknown parameters that must be estimated.

Graphs of two quadratic models are shown in Figure 7.2. The quadratic model is the equation of a **parabola** that opens either upward, as in Figure 7.2(a), or downward, as in Figure 7.2(b). (If the coefficient of x^2 is positive, it opens upward; if it is negative, it opens downward.) The parabola may be shifted upward or downward, left or right. The least squares procedure only uses the portion of the parabola that is needed to model the data. For example, if you fit a parabola to the data points shown in Figure 7.3, the portion shown as a solid curve passes through the data points. The outline of the unused portion of the parabola is indicated by a dashed curve.

FIGURE 7.2

Graphs for Two Second-
Order Polynomial Models

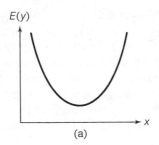
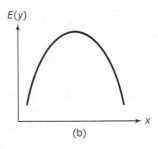

Figure 7.3 illustrates an important limitation on the use of prediction equations—the model is valid only over the range of x-values that were used to fit the model. For example, the response might rise, as shown in the figure, until it reaches a plateau. The second-order model might fit the data very well over the range of x-values shown in Figure 7.3, but would provide a very poor fit if data were collected in the region where the parabola turns downward.

FIGURE 7.3

Example of the Use of
a Quadratic Model

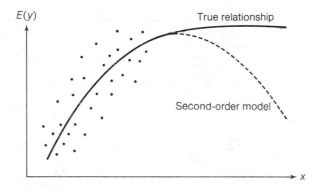

How do you decide the order of the polynomial you should use to model a response if you have no prior information on the relationship between $E(y)$ and x? If you have data, construct a scattergram of the data points and see if you can deduce the nature of a good approximating function. The graphs of most responses as a function of an independent variable x are, in general, curvilinear. Nevertheless, if the rate of curvature of the response curve is very small over the range of x that is of interest to you, a straight line might provide an excellent fit to the response data and function as a very useful prediction equation. If the curvature is not (or may not be) slight, you should try a second-order model. Third- or higher-order models would be used only where you expect more than one reversal in the direction of the curve. These situations are rare, except where the response is a function of time. Models for forecasting over time are presented in Chapter 9.

EXAMPLE 7.2

Power companies have to be able to predict the peak power load at their various stations in order to operate effectively. The peak power load is the maximum amount of power that must be generated each day in order to meet demand.

Suppose a power company located in the southern part of the United States decides to model daily peak power load, y, as a function of the daily high temperature, x, and the model is to be constructed for the summer months when demand is greatest. Although we would expect the peak power load to increase as the high temperature increases, the *rate* of increase in $E(y)$ might also increase as x increases. That is, a 1-unit increase in high temperature from $100°$ to $101°F$ might result in a larger increase in power demand than would a 1-unit increase from $80°$ to $81°F$. Therefore, we postulate the quadratic model

$$E(y) = \beta_0 + \beta_1 x + \beta_2 x^2$$

and we expect β_2 to be positive.

A random sample of 25 summer days is selected, and the data are shown in Table 7.1. Fit a second-order model using the data, and test the hypothesis that the power load increases at an increasing *rate* with temperature, i.e., that $\beta_2 > 0$.

TABLE 7.1   Power Load Data

TEMPERATURE °F	PEAK LOAD megawatts	TEMPERATURE °F	PEAK LOAD megawatts	TEMPERATURE °F	PEAK LOAD megawatts
94	136.0	106	178.2	76	100.9
96	131.7	67	101.6	68	96.3
95	140.7	71	92.5	92	135.1
108	189.3	100	151.9	100	143.6
67	96.5	79	106.2	85	111.4
88	116.4	97	153.2	89	116.5
89	118.5	98	150.1	74	103.9
84	113.4	87	114.7	86	105.1
90	132.0				

SOLUTION

The SAS printout shown in Figure 7.4 gives the least squares fit of the second-order model using the data in Table 7.1. The prediction equation is

$$\hat{y} = 385.048 - 8.293x + .05982x^2$$

A plot of this equation and the observed values is given in Figure 7.5. Note that this curve passes through the data points and seems to produce (by visual examination) a set of deviations that are relatively small.

We now test to determine whether the sample value, $\hat{\beta}_2 = .05982$, is large enough to conclude *in general* that the power load increases at an increasing rate with temperature:

$$H_0: \quad \beta_2 = 0 \qquad H_a: \quad \beta_2 > 0$$

Test statistic: $\quad t = \dfrac{\hat{\beta}_2}{s_{\hat{\beta}_2}}$

FIGURE 7.4 Portion of the SAS Printout for the Second-Order Model of Example 7.2

ANALYSIS OF VARIANCE

SOURCE	DF	SUM OF SQUARES	MEAN SQUARE	F VALUE	PROB>F
MODEL	2	15011.77200	7505.88600	259.687	0.0001
ERROR	22	635.87840	28.90356374		
C TOTAL	24	15647.65040			

ROOT MSE	5.376203	R-SQUARE	0.9594	
DEP MEAN	125.428	ADJ R-SQ	0.9557	
C.V.	4.286287			

PARAMETER ESTIMATES

VARIABLE	DF	PARAMETER ESTIMATE	STANDARD ERROR	T FOR H0: PARAMETER=0	PROB > |T|
INTERCEP	1	385.04809	55.17243578	6.979	0.0001
X	1	-8.29252680	1.29904502	-6.384	0.0001
XX	1	0.05982337	0.007548554	7.925	0.0001

FIGURE 7.5
Plot of the Observations and the Second-Order Least Squares Fit

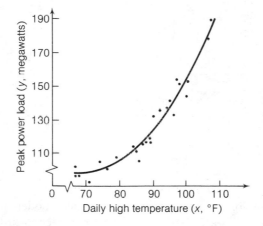

For $\alpha = .05$, $n = 25$, $k = 2$, and df $= n - (k + 1) = 22$, we reject H_0 if

$$t > t_{.05}$$

where $t_{.05} = 1.717$ (from Table 2 of Appendix D). From Figure 7.4, the calculated value of t is 7.925. Since this value exceeds $t_{.05} = 1.717$, we reject H_0 and conclude that the mean power load increases at an increasing rate with temperature. Note that the observed significance level of the test, $p = .0001/2 = .00005$ (shaded on the printout) supports this conclusion. ∎

EXERCISES 7.3–7.4

7.3 A company is considering having the employees on its assembly line work 4 days for 10 hours each instead of 5 days for 8 hours. Management is concerned that the effect of fatigue due to longer afternoons of work might increase assembly times to an unsatisfactory level. An experiment with the 4-day week is planned in which time studies will be conducted on some of the workers during the afternoons. It is believed that an adequate model of the relationship between assembly time, y, and time since lunch, x, should allow for the average assembly time to decrease for a while after lunch (as workers get back in the groove) before it starts to increase as the workers become tired. Write a model to relate $E(y)$ and x that would reflect the management's belief, and sketch the hypothesized shape of the model.

7.4 Underinflated or overinflated tires can increase the tire wear and decrease gas mileage. A new tire was tested for wear at different pressures with the results shown in the table.

PRESSURE x, pounds per square inch	MILEAGE y, thousands
30	29
31	32
32	36
33	38
34	37
35	33
36	26

a. Plot the data on a scattergram.
b. If you were given the information for $x = 30, 31, 32, 33$ only, what kind of model would you suggest? For $x = 33, 34, 35, 36$? For all the data?

SECTION 7.4

FIRST-ORDER MODELS WITH TWO OR MORE QUANTITATIVE INDEPENDENT VARIABLES

Like models for a single independent variable, models with two or more independent variables are classified as first-order, second-order, and so forth, but it is difficult (most often impossible) to graph the response because the plot is in a multi-dimensional space. For example, with one quantitative independent variable, x, the response y traces a curve. But for two quantitative independent variables, x_1 and x_2, the plot of y traces a surface over the x_1, x_2-plane (see Figure 7.6). For three or more quantitative independent variables, the response traces a surface in a four- or higher-dimensional space. For these, we can construct two-dimensional contour curves for one independent variable or three-dimensional plots of response surfaces for two independent variables for fixed levels of the remaining independent variables, but this is the best we can do in providing a graphic description of a response.

A **first-order model** in k quantitative variables is a first-order polynomial in k independent variables. For $k = 1$, the graph is a straight line. For $k = 2$, the response surface is a plane (usually tilted) over the x_1, x_2-plane.

If we use a first-order polynomial to model a response, we are assuming that there is no curvature in the response surface and that the variables affect the

FIGURE 7.6

A Response Surface
for Two Quantitative
Independent Variables

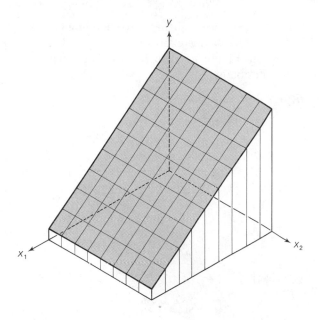

FIRST-ORDER MODEL IN k QUANTITATIVE INDEPENDENT VARIABLES

$$E(y) = \beta_0 + \beta_1 x_1 + \beta_2 x_2 + \cdots + \beta_k x_k$$

where $\beta_0, \beta_1, \ldots, \beta_k$ are unknown parameters that must be estimated.

INTERPRETATION OF MODEL PARAMETERS

β_0: y-intercept of $(k + 1) -$ dimensional surface;
the value of $E(y)$ when $x_1 = x_2 = \cdots = x_k = 0$

β_1: Change in $E(y)$ for a 1-unit increase in x_1, when $x_2, x_3, \ldots, x_k$
are held fixed

β_2: Change in $E(y)$ for a 1-unit increase in x_2, when $x_1, x_3, \ldots, x_k$
are held fixed

$\vdots$

β_k: Change in $E(y)$ for a 1-unit increase in x_k, when $x_1, x_2, \ldots, x_{k-1}$
are held fixed

---

response independently of each other. For example, suppose the true relationship between the mean response and the independent variables x_1 and x_2 is given by the equation

$$E(y) = 1 + 2x_1 + x_2$$

The response surface (a plane) corresponding to this equation is shown in Figure 7.7. The graphs of this expression for $x_2 = 1$, 2, and 3 (called **contour lines**), are shown in Figure 7.8. You can see that when you substitute $x_2 = 1$ into the model, you obtain

$$E(y) = 1 + 2x_1 + x_2$$
$$= 1 + 2x_1 + 1$$
$$= 2 + 2x_1$$

FIGURE 7.7

The Plane $E(y) =$
$1 + 2x_1 + x_2$

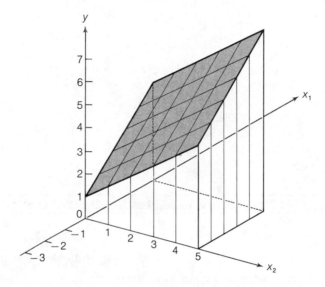

FIGURE 7.8

Contour Lines of $E(y)$
for $x_2 = 1$, 2, 3
(First-Order Model)

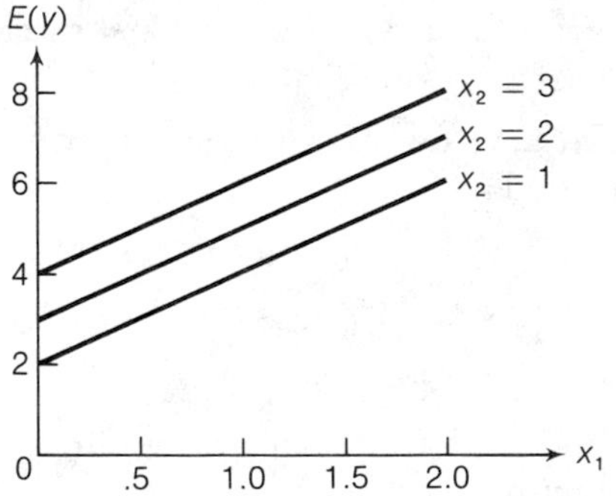

For $x_2 = 2$,

$$E(y) = 1 + 2x_1 + 2$$
$$= 3 + 2x_1$$

And for $x_3 = 3$, $E(y) = 4 + 2x_1$. In other words, regardless of the value of x_2, $E(y)$ graphs as a straight line. Changing x_2 changes only the y-intercept (the constant in the equation). Consequently, assuming that a first-order model will adequately model a response is equivalent to assuming that a 1-unit change in one independent variable will have the same effect on the mean value of y regardless of the levels of the other independent variables. That is, *the contour lines will be parallel.* In Chapter 4, we stated that independent variables that have this property *do not interact.*

Except in cases where the ranges of levels for all independent variables are very small, the implication of no curvature in the response surface and the independence of variable effects on the response restrict the applicability of first-order models.

S E C T I O N 7.5

SECOND-ORDER MODELS WITH TWO OR MORE QUANTITATIVE INDEPENDENT VARIABLES

Second-order models with two or more independent variables permit curvature in the response surface. One important type of second-order term accounts for **interaction** between two variables.* To see the effect of interaction on the model, consider the two-variable model

$$E(y) = \beta_0 + \beta_1 x_1 + \beta_2 x_2 + \beta_3 x_1 x_2$$

This interaction model traces a ruled surface (twisted plane) in a three-dimensional space (see Figure 7.9 on page 336). The second-order term $\beta_3 x_1 x_2$ is called the **interaction term,** and it permits the contour lines to be *nonparallel.* For example, suppose the true equation of the response surface is

$$E(y) = 1 + 2x_1 + x_2 - x_1 x_2$$

We graph the contour lines of this response for $x_2 = 1, 2$, and 3 in Figure 7.10. Note that when we substitute $x_2 = 1$ into the model, we get

$$\begin{aligned} E(y) &= 1 + 2x_1 + x_2 - x_1 x_2 \\ &= 1 + 2x_1 + 1 - x_1(1) \\ &= 2 + x_1 \end{aligned}$$

For $x_2 = 2$,

$$\begin{aligned} E(y) &= 1 + 2x_1 + 2 - x_1(2) \\ &= 3 \end{aligned}$$

Similarly, when $x_2 = 3$, $E(y) = 4 - x_1$. Thus, when interaction is present in the model, both the *y-intercept and the slope* change as x_2 changes. Consequently, *the contour lines are not parallel. The presence of an interaction term implies that the effect of a 1-unit change in one independent variable will depend on the level of the other independent variable.* In our example (Figure 7.10), a 1-unit change in x_1 produces a 1-unit change in $E(y)$ when $x_2 = 1$, but a 1-unit change in x_1 produces *no* change in $E(y)$ when $x_2 = 2$ (i.e., the slope is 0).

*The order of a term involving two or more *quantitative* independent variables is equal to the sum of their exponents. Thus, $\beta_3 x_1 x_2$ is a second-order term, as is $\beta_4 x_1^2$. A term of the form $\beta_i x_1 x_2 x_3$ is a third-order term.

FIGURE 7.9
Computer-Generated
Graph for an Interaction
Model (Second-Order)

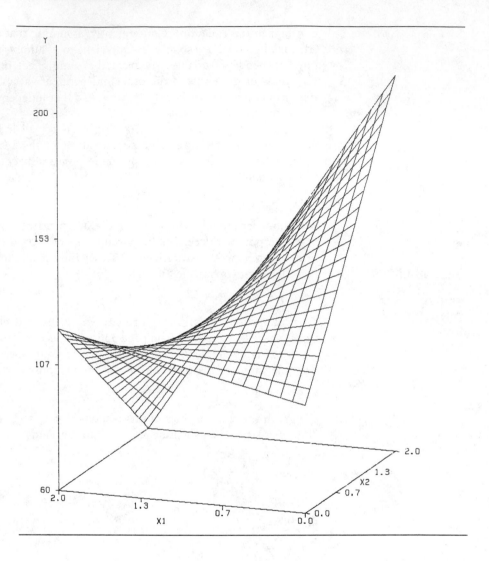

FIGURE 7.10
Contour Lines of $E(y)$ for
$x_2 = 1, 2, 3$ (First-Order
Plus Interaction)

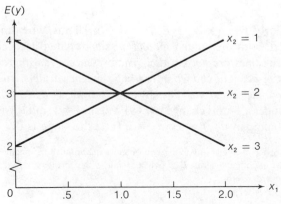

AN INTERACTION MODEL (SECOND-ORDER) WITH TWO QUANTITATIVE INDEPENDENT VARIABLES

$$E(y) = \beta_0 + \beta_1 x_1 + \beta_2 x_2 + \beta_3 x_1 x_2$$

where β_0, β_1, β_2, and β_3 are unknown parameters that must be estimated.

INTERPRETATION OF MODEL PARAMETERS

β_0: y-intercept of three-dimensional surface (see Figure 7.9); the value of $E(y)$ when $x_1 = x_2 = 0$

β_1 and β_2: Changing β_1 and β_2 causes the surface to shift along the x_1- and x_2-axes

β_3: Controls the rate of twist in the ruled surface (see Figure 7.9)

We can introduce even more flexibility into a model by the addition of quadratic terms. *The complete second-order model includes the constant β_0, all linear (first-order) terms, all two-variable interactions, and all quadratic terms.* This complete second-order model for two quantitative independent variables is shown in the box.

COMPLETE SECOND-ORDER MODEL WITH TWO QUANTITATIVE INDEPENDENT VARIABLES

$$E(y) = \beta_0 + \beta_1 x_1 + \beta_2 x_2 + \beta_3 x_1 x_2 + \beta_4 x_1^2 + \beta_5 x_2^2$$

where β_0, β_1, ... , β_5 are unknown parameters that must be estimated.

INTERPRETATION OF MODEL PARAMETERS

β_0: y-intercept of three-dimensional surface (see Figure 7.11 on page 338); the value of $E(y)$ when $x_1 = x_2 = 0$

β_1 and β_2: Changing β_1 and β_2 causes the surface to shift along the x_1- and x_2-axes

β_3: The value of β_3 controls the rotation of the surface

β_4 and β_5: Signs and values of these parameters control the type of surface and the rates of curvature

β_4 and β_5 positive: A paraboloid that opens upward [Figure 7.11(a)]

β_4 and β_5 negative: A paraboloid that opens downward [Figure 7.11(b)]

β_4 and β_5 differ in sign: A saddle-shaped surface [Figure 7.11(c)]

The quadratic terms $\beta_4 x_1^2$ and $\beta_5 x_2^2$ in the second-order model imply that the response surface for $E(y)$ will possess curvature (see Figure 7.11). The interaction term $\beta_3 x_1 x_2$ allows the contours depicting $E(y)$ as a function of x_1 to have different

FIGURE 7.11

Graphs of Three Second-
Order Surfaces

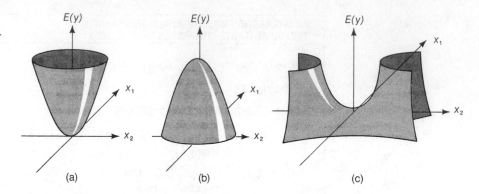

shapes for various values of x_2. For example, suppose the complete second-order model relating $E(y)$ to x_1 and x_2 is

$$E(y) = 1 + 2x_1 + x_2 - 10x_1x_2 + x_1^2 - 2x_2^2$$

Then the contours of $E(y)$ for $x_2 = -1$, 0, and 1 are shown in Figure 7.12. When we substitute $x_2 = -1$ into the model, we get

$$\begin{aligned} E(y) &= 1 + 2x_1 + x_2 - 10x_1x_2 + x_1^2 - 2x_2^2 \\ &= 1 + 2x_1 - 1 - 10x_1(-1) + x_1^2 - 2(-1)^2 \\ &= -2 + 12x_1 + x_1^2 \end{aligned}$$

For $x_2 = 0$

$$\begin{aligned} E(y) &= 1 + 2x_1 + (0) - 10x_1(0) + x_1^2 - 2(0)^2 \\ &= 1 + 2x_1 + x_1^2 \end{aligned}$$

Similarly, for $x_2 = 1$,

$$E(y) = -8x_1 + x_1^2$$

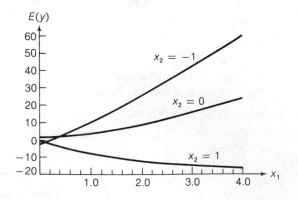

Note how the shapes of the three contour curves in Figure 7.12 differ, thus indicating that the β parameter associated with the x_1x_2 (interaction) term differs from 0.

The complete second-order model for three independent variables is shown in the next box.

COMPLETE SECOND-ORDER MODEL WITH THREE QUANTITATIVE INDEPENDENT VARIABLES

$$E(y) = \beta_0 + \beta_1 x_1 + \beta_2 x_2 + \beta_3 x_3 + \beta_4 x_1 x_2 + \beta_5 x_1 x_3 + \beta_6 x_2 x_3$$
$$+ \beta_7 x_1^2 + \beta_8 x_2^2 + \beta_9 x_3^2$$

where $\beta_0, \beta_1, \beta_2, \ldots, \beta_9$ are unknown parameters that must be estimated.

This second-order model in three independent variables demonstrates how you would write a second-order model for any number of independent variables. Always include the constant β_0 and then all first-order terms corresponding to $x_1, x_2, \ldots$. Then add the interaction terms for all pairs of independent variables $x_1 x_2, x_1 x_3, x_2 x_3, \ldots$. Finally, include the second-order terms $x_1^2, x_2^2, \ldots$.

For any number, say p, of quantitative independent variables, the response traces a surface in a $(p + 1)$-dimensional space, which is impossible to visualize. In spite of this handicap, the prediction equation can still tell us much about the phenomenon being studied.

EXAMPLE 7.3

Many companies manufacture products that are at least partially produced using chemicals (for example, steel, paint, gasoline). In many instances, the quality of the finished product is a function of the temperature and pressure at which the chemical reactions take place.

Suppose you wanted to model the quality, y, of a product as a function of the temperature, x_1, and the pressure, x_2, at which it is produced. Four inspectors independently assign a quality score between 0 and 100 to each product, and then the quality, y, is calculated by averaging the four scores. An experiment is conducted by varying temperature between 80° and 100°F and pressure between 50 and 60 pounds per square inch. The resulting data ($n = 27$) are given in Table 7.2. Fit a complete second-order model to the data and sketch the response surface.

TABLE 7.2
Temperature, Pressure, and Quality of the Finished Product

x_1, °F	x_2, pounds per square inch	y	x_1, °F	x_2, pounds per square inch	y	x_1, °F	x_2, pounds per square inch	y
80	50	50.8	90	50	63.4	100	50	46.6
80	50	50.7	90	50	61.6	100	50	49.1
80	50	49.4	90	50	63.4	100	50	46.4
80	55	93.7	90	55	93.8	100	55	69.8
80	55	90.9	90	55	92.1	100	55	72.5
80	55	90.9	90	55	97.4	100	55	73.2
80	60	74.5	90	60	70.9	100	60	38.7
80	60	73.0	90	60	68.8	100	60	42.5
80	60	71.2	90	60	71.3	100	60	41.4

SOLUTION

The complete second-order model is

$$E(y) = \beta_0 + \beta_1 x_1 + \beta_2 x_2 + \beta_3 x_1 x_2 + \beta_4 x_1^2 + \beta_5 x_2^2$$

The data in Table 7.2 were used to fit this model, and a portion of the SAS output is shown in Figure 7.13.

The least squares model is

$$\hat{y} = -5{,}127.90 + 31.10 x_1 + 139.75 x_2 - .146 x_1 x_2 - .133 x_1^2 - 1.14 x_2^2$$

A three-dimensional graph of this prediction model is shown in Figure 7.14. Note that the mean quality seems to be greatest for temperatures of about 85°–

FIGURE 7.13   SAS Printout for Example 7.3

ANALYSIS OF VARIANCE

SOURCE	DF	SUM OF SQUARES	MEAN SQUARE	F VALUE	PROB>F
MODEL	5	8402.26454	1680.45291	596.324	0.0001
ERROR	21	59.17842620	2.81802030		
C TOTAL	26	8461.44296			

ROOT MSE	1.678696	R-SQUARE	0.9930
DEP MEAN	66.96296	ADJ R-SQ	0.9913
C.V.	2.506902		

PARAMETER ESTIMATES

VARIABLE	DF	PARAMETER ESTIMATE	STANDARD ERROR	T FOR H0: PARAMETER=0	PROB > |T|
INTERCEP	1	-5127.89907	110.29602	-46.492	0.0001
X1	1	31.09638889	1.34441322	23.130	0.0001
X2	1	139.74722	3.14005412	44.505	0.0001
X1X2	1	-0.14550000	0.009691956	-15.012	0.0001
X1X1	1	-0.13338889	0.006853248	-19.464	0.0001
X2X2	1	-1.14422222	0.02741299	-41.740	0.0001

FIGURE 7.14
Plot of Second-Order
Least Squares Model
for Example 7.3

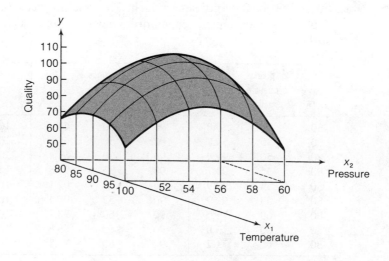

90°F and for pressures of about 55–57 pounds per square inch.* Further experimentation in these ranges might lead to a more precise determination of the optimal temperature–pressure combination.

A look at the coefficient of determination, $R^2 = .993$, the F-value for testing the entire model, $F = 596.32$, and the p-value for the test, $p = .0001$ (in Figure 7.13), leaves little doubt that the complete second-order model is useful for explaining mean quality as a function of temperature and pressure. This, of course, will not always be the case. The additional complexity of second-order models is worthwhile only if a better model results. To determine whether the quadratic terms are important, we would test $H_0: \beta_4 = \beta_5 = 0$ using the partial F-test outlined in Section 4.11. ∎

EXERCISES 7.5–7.9

7.5 Some corporations, instead of owning a fleet of cars, rent cars from a rental agency. A corporation may do this because it is sometimes more economical to rent new cars for a year than to buy new cars each year. A major rental agency wants to develop a model that will allow it to estimate the average annual cost to the prospective customer of renting cars, y, as a function of two independent variables:

x_1 = Number of cars rented

x_2 = Average number of miles driven per car during year (in thousands)

a. Identify the independent variables as quantitative or qualitative.
b. Write the first-order model for $E(y)$.
c. Add an interaction term between x_1 and x_2 to the first-order model. Suppose interaction exists. Graph $E(y)$, the mean cost, versus x_2, the average mileage driven (as you would expect it to appear) for several values of x_1.
d. Write the complete second-order model for $E(y)$.

7.6 Refer to Exercise 7.5. Suppose the model from part c is fit, with the following result:

$$\hat{y} = 1 + .05x_1 + x_2 + .05x_1x_2$$

(The units of $\hat{y}$ are thousands of dollars.) Graph the estimated cost $\hat{y}$ as a function of the average number of miles driven, x_2, over the range $x_2 = 10$ to $x_2 = 50$ (10–50 thousand miles) for $x_1 = 1, 5,$ and 10. Do these functions agree (approximately) with the graphs you drew for Exercise 7.5, part c?

7.7 Refer to Exercise 7.5. Suppose an additional independent variable is considered:

x_3 = Average yearly gas price

a. Write the first-order model plus interaction for $E(y)$ as a function of x_1, x_2, and x_3.
b. Write the complete second-order model for $E(y)$ as a function of x_1, x_2, and x_3.

*Students with knowledge of calculus should note that we can determine the exact temperature and pressure that maximize quality in the least squares model by solving $\partial\hat{y}/\partial x_1 = 0$ and $\partial\hat{y}/\partial x_2 = 0$ for x_1 and x_2. These estimated optimal values are $x_1 = 86.25°F$ and $x_2 = 55.58$ pounds per square inch. Remember, however, that these are only sample estimates of the coordinates for the optimal value.

7.8 *Multinational* is the term given to an industry with foreign investors. A study of 216 manufacturing industries in Mexico found that multinational presence in a firm has a positive influence on market concentration (Blomstrom, 1986). The result was revealed in a multiple regression analysis on the dependent variable y, market concentration index, using the following quantitative independent variables:

x_1 = Market size

x_2 = Market rate of growth

x_3 = Gross production in largest plants (expressed as a percentage of total gross production)

x_4 = Capital intensity (ratio of total assets to total number of employees)

x_5 = Advertising intensity (ratio of advertising to value added)

x_6 = Foreign share (i.e., gross output produced by foreign subsidiaries)

a. Write a first-order model for $E(y)$ as a function of x_1–x_6.
b. Interpret β_6 in the model in part **a**.
c. Based on the results of the study, is β_6 positive or negative?
d. Write a second-order model for $E(y)$ that proposes interaction between the independent variables but with no curvature.
e. Using the model in part **d**, how would you test the hypothesis that effect of a multinational presence on market concentration is independent of the other independent variables in the model?

7.9 Researchers at the Upjohn Company utilized multiple regression analysis in the development of a sustained-release tablet (Klassen, 1986). One of the objectives of the research was to develop a model relating the dissolution y of a tablet (i.e., the percentage of the tablet dissolved over a specified period of time) to the following independent variables:

x_1 = Excipient level (i.e., amount of nondrug ingredient in the tablet)

x_2 = Process variable (e.g., machine setting under which tablet is processed)

a. Write the complete second-order model for $E(y)$.
b. Write a model that hypothesizes straight-line relationships between $E(y)$, x_1, and x_2. Assume that x_1 and x_2 do not interact.
c. Repeat part **b**, but add interaction to the model.
d. For the model in part **c**, what is the slope of the $E(y),x_1$ line for fixed x_2?
e. For the model in part **c**, what is the slope of the $E(y),x_2$ line for fixed x_1?

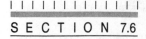

S E C T I O N   7.6

CODING QUANTITATIVE INDEPENDENT VARIABLES

In fitting higher-order polynomial regression models (e.g., second- or third-order models), it is often a good practice to code the quantitative independent variables. For example, suppose one of the independent variables in a regression analysis is advertising expenditure, A, and A is observed at three levels: 500, $1,000$, and $1,500$. We can code (or transform) the advertising expenditure measurements using the formula

$$x = \frac{A - 1,000}{500}$$

Then the coded levels $x = -1$, 0, and 1 correspond to the original levels $500, $1,000, and $1,500.

In a general sense, *coding* means transforming a set of independent variables (qualitative or quantitative) into a new set of independent variables. For example, if we observe two independent variables,

A = Advertising expenditure

I = Prime interest rate

then we can transform the two independent variables, A and I, into two new coded variables, x_1 and x_2, where x_1 and x_2 are related to A and I by two functional equations:

$$x_1 = f_1(A, I) \qquad x_2 = f_2(A, I)$$

The functions f_1 and f_2, which are frequently expressed as equations, establish a one-to-one correspondence between combinations of levels of A and I with combinations of the coded values of x_1 and x_2.

Since qualitative independent variables are not numerical, it is necessary to code their values in order to fit the regression model. However, you might ask why we would bother to code the quantitative independent variables. There are two related reasons for coding quantitative variables. At first glance, it would appear that a computer would be oblivious to the values assumed by the independent variables in a regression analysis, but this is not the case. In order to calculate the estimates of the model parameters using the method of least squares, the computer must invert a matrix of numbers, called the **coefficient (or information) matrix** (see Appendix A). Considerable rounding error may occur during the inversion process if the numbers in the coefficient matrix vary greatly in absolute value. This can produce sizable errors in the computed values of the least squares estimates, $\hat{\beta}_0$, $\hat{\beta}_1$, $\hat{\beta}_2$, ... Coding makes it computationally easier for the computer to invert the matrix, thus leading to more accurate estimates.

A second reason for coding quantitative variables pertains to the problem of multicollinearity discussed in Chapter 5. When polynomial regression models (e.g., second-order models) are fit, the problem of multicollinearity is unavoidable, especially when higher-order terms are fit. For example, consider the quadratic model

$$E(y) = \beta_0 + \beta_1 x + \beta_2 x^2$$

If the range of the values of x is narrow, then the two variables, $x_1 = x$ and $x_2 = x^2$, will generally be highly correlated. As we pointed out in Chapter 5, the likelihood of rounding errors in the regression coefficients is increased in the presence of multicollinearity.

The best way to cope with the rounding error problem is to:

1. Code the quantitative variable so that the new coded origin is in the center of the coded values. For example, by coding advertising expenditure, A, as

$$x = \frac{A - 1,000}{500}$$

we obtain coded values $-1, 0, 1$. This places the coded origin, 0, in the middle of the range of coded values (-1 to 1).

2. Code the quantitative variable so that the range of the coded values is approximately the same for all coded variables. You need not hold exactly to this requirement. The range of values for one independent variable could be double or triple the range of another without causing any difficulty, but it would not be desirable to have a sizable disparity in the ranges, say a ratio of 100 to 1.

When the data are observational (the values assumed by the independent variables are uncontrolled), the coding procedure described in the next box satisfies, reasonably well, these two requirements. The coded variable u is similar to the standardized normal z-statistic of Section 2.4. Thus, the u-value is the deviation (the distance) between an x-value and the mean of the x-values, $\bar{x}$, expressed in units of s_x.* Since we know that most (approximately 95%) measurements in a set will lie within 2 standard deviations of their mean, it follows that most of the coded u-values will lie in the interval -2 to $+2$.

CODING PROCEDURE FOR OBSERVATIONAL DATA

Let

$x = $ Uncoded quantitative independent variable

$u = $ Coded quantitative independent variable

Then if x takes values $x_1, x_2, \ldots, x_n$ for the n data points in the regression analysis, let

$$u_i = \frac{x_i - \bar{x}}{s_x}$$

where s_x is the standard deviation of the x-values, i.e.,

$$s_x = \sqrt{\frac{\sum_{i=1}^{n} (x_i - \bar{x})^2}{n - 1}}$$

If you apply this coding to each quantitative variable, the range of values for each will be approximately -2 to $+2$. The variation in the absolute values of the elements of the coefficient matrix will be moderate, and rounding errors generated in finding the inverse of the matrix will be reduced. Additionally, the correlation between x and x^2 will be reduced.

*The divisor of the deviation, $x - \bar{x}$, need not equal s_x exactly. Any number approximately equal to s_x would suffice.

EXAMPLE 7.4

Table 7.3 gives observational data on the index of building construction costs per month as a function of the index of the cost of construction materials (other components of construction costs would be labor, the cost of money, and so forth). Suppose we are interested in relating monthly construction cost y to monthly index of construction materials x using a quadratic model.

TABLE 7.3
Index of Building
Construction Costs

MONTH	CONSTRUCTION COST[a] y	INDEX OF ALL CONSTRUCTION MATERIALS[b] x
January	193.2	180.0
February	193.1	181.7
March	193.6	184.1
April	195.1	185.3
May	195.6	185.7
June	198.1	185.9
July	200.9	187.7
August	202.7	189.6

[a]*Source:* United States Department of Commerce, Bureau of the Census.
[b]*Source:* United States Department of Labor, Bureau of Labor Statistics.

a. Give the equation relating the coded variable u to the index of construction materials x using the coding system for observational data.
b. Calculate the coded values, u, for the eight x-values.
c. Find the sum of the $n = 8$ values for u.

SOLUTION

a. We first find $\bar{x}$ and s_x:

$$\bar{x} = \frac{\sum_{i=1}^{n} x_i}{n} = \frac{1{,}480.0}{8} = 185.0$$

$$\sum_{i=1}^{n} (x_i - \bar{x})^2 = \sum_{i=1}^{n} x_i^2 - \frac{\left(\sum_{i=1}^{n} x_i\right)^2}{n} = 273{,}866.54 - \frac{(1{,}480.0)^2}{8} = 66.54$$

$$s_x = \sqrt{\frac{\sum_{i=1}^{n} (x_i - \bar{x})^2}{n - 1}} = \sqrt{\frac{66.54}{7}} = 3.08$$

Then the equation relating u and x is

$$u = \frac{x - 185.0}{3.08}$$

b. When $x = 180.0$,

$$u = \frac{x - 185.0}{3.08} = \frac{180.0 - 185.0}{3.08} = -1.62$$

Similarly, when $x = 181.7$,

$$u = \frac{x - 185.0}{3.08} = \frac{181.7 - 185.0}{3.08} = -1.07$$

Table 7.4 gives the coded values for all $n = 8$ observations. [*Note:* You can see that all the $n = 8$ values for u lie in the interval from -2 to $+2$.]

TABLE 7.4

Coded Values of x,
Example 7.4

INDEX	CODED VALUES
x	u
180.0	-1.62
181.7	-1.07
184.1	$-.29$
185.3	.10
185.7	.23
185.9	.29
187.7	.88
189.6	1.49

c. If you ignore rounding error, the sum of the $n = 8$ values for u will equal 0. This is because the sum of the deviations of a set of measurements about their mean is always equal to 0.   ∎

To illustrate the advantage of coding, consider fitting the second-order model

$$E(y) = \beta_0 + \beta_1 x + \beta_2 x^2$$

to the data of Example 7.3. It can be shown that the coefficient of correlation between the two variables, x and x^2, is $r = .999$. However, the coefficient of correlation between the corresponding coded values, u and u^2, is only $r = -.203$. Thus, we can avoid potential rounding error caused by multicollinearity by fitting, instead, the model

$$E(y) = \beta_0^* + \beta_1^* u + \beta_2^* u^2$$

Other methods of coding have been developed to reduce rounding errors and multicollinearity. One of the more complex coding systems involves fitting **orthogonal polynomials**. An orthogonal system of coding guarantees that the coded independent variables will be uncorrelated. For a discussion of orthogonal polynomials, consult the references given at the end of this chapter.

7.10 As part of the first-year evaluation for new salespeople, a large food-processing firm projects the second-year sales for each salesperson based on his or her sales for the first year. Data for $n = 8$ salespeople are shown in the table.

FIRST-YEAR SALES x, thousands of dollars	SECOND-YEAR SALES y, thousands of dollars
75.2	99.3
91.7	125.7
100.3	136.1
64.2	108.6
81.8	102.0
110.2	153.7
77.3	108.8
80.1	105.4

a. Give the equation relating the coded variable u to the first-year sales, x, using the coding system for observational data.

b. Calculate the coded values, u.

c. Calculate the coefficient of correlation r between the variables x and x^2.

d. Calculate the coefficient of correlation r between the variables u and u^2. Compare this value to the value computed in part c.

e. If you have access to a statistical computer package, fit the model

$$E(y) = \beta_0 + \beta_1 u + \beta_2 u^2$$

7.11 Suppose you want to use the coding system for observational data to fit a second-order model to the tire pressure–automobile mileage data of Exercise 7.4, which are repeated in the table.

PRESSURE x, pounds per square inch	MILEAGE y, thousands
30	29
31	32
32	36
33	38
34	37
35	33
36	26

a. Give the equation relating the coded variable u to pressure, x, using the coding system for observational data.

b. Calculate the coded values, u.

c. Calculate the coefficient of correlation r between the variables x and x^2.

d. Calculate the coefficient of correlation r between the variables u and u^2. Compare this value to the value computed in part c.

e. If you have access to a statistical computer package, fit the model

$$E(y) = \beta_0 + \beta_1 u + \beta_2 u^2$$

7.12 Use the coding system for observational data to fit a complete second-order model to the data of Example 7.3, which are repeated in the table.

x_1, °F	x_2, pounds per square inch	y	x_1, °F	x_2, pounds per square inch	y	x_1, °F	x_2, pounds per square inch	y
80	50	50.8	90	50	63.4	100	50	46.6
80	50	50.7	90	50	61.6	100	50	49.1
80	50	49.4	90	50	63.4	100	50	46.4
80	55	93.7	90	55	93.8	100	55	69.8
80	55	90.9	90	55	92.1	100	55	72.5
80	55	90.9	90	55	97.4	100	55	73.2
80	60	74.5	90	60	70.9	100	60	38.7
80	60	73.0	90	60	68.8	100	60	42.5
80	60	71.2	90	60	71.3	100	60	41.4

a. Give the coded values u_1 and u_2 for x_1 and x_2, respectively.
b. Compare the coefficient of correlation between x_1 and x_1^2 with the coefficient of correlation between u_1 and u_1^2.
c. Compare the coefficient of correlation between x_2 and x_2^2 with the coefficient of correlation between u_2 and u_2^2.
d. Give the prediction equation.

SECTION 7.7

MODELS WITH ONE QUALITATIVE INDEPENDENT VARIABLE

Suppose we want to write a model for the mean profit, $E(y)$, per sales dollar of a construction company as a function of the sales engineer who estimates and bids on a job (for purposes of explanation, we will ignore other independent variables that might affect the response). Further, suppose there are three sales engineers, Jones, Smith, and Adams. Then sales engineer is a single qualitative variable with three levels corresponding to Jones, Smith, and Adams. Recall that with a qualitative independent variable, we cannot attach a quantitative meaning to a given level. All we can do is describe it.

To simplify our notation, let μ_A be the mean profit per sales dollar for Jones, and let μ_B and μ_C be the corresponding mean profits for Smith and Adams, respectively. Our objective is to write a single prediction equation that will give the mean value of y for the three sales engineers. This can be done as follows:

$$E(y) = \beta_0 + \beta_1 x_1 + \beta_2 x_2$$

where

$$x_1 = \begin{cases} 1 & \text{if Smith is the sales engineer} \\ 0 & \text{if Smith is not the sales engineer} \end{cases}$$

$$x_2 = \begin{cases} 1 & \text{if Adams is the sales engineer} \\ 0 & \text{if Adams is not the sales engineer} \end{cases}$$

The variables x_1 and x_2 are not meaningful independent variables as in the case of the models containing quantitative independent variables. Instead, they are **dummy** (or **indicator**) **variables** that make the model function. To see how they work, let $x_1 = 0$ and $x_2 = 0$. This condition will apply when we are seeking the mean response for Jones (neither Smith nor Adams will be the sales engineer; hence, it must be Jones). Then the mean value of y when Jones is the sales engineer is

$$\mu_A = E(y) = \beta_0 + \beta_1(0) + \beta_2(0) = \beta_0$$

This tells us that the mean profit per sales dollar for Jones is β_0. Equivalently, it means that $\beta_0 = \mu_A$.

Now suppose we want to represent the mean response, $E(y)$, when Smith is the sales engineer. Checking the dummy variable definitions, we see that we should let $x_1 = 1$ and $x_2 = 0$:

$$\mu_B = E(y) = \beta_0 + \beta_1 x_1 + \beta_2 x_2 = \beta_0 + \beta_1(1) + \beta_2(0) = \beta_0 + \beta_1$$

or, since $\beta_0 = \mu_A$,

$$\mu_B = \mu_A + \beta_1$$

Then it follows that the interpretation of β_1 is

$$\beta_1 = \mu_B - \mu_A$$

which is the difference in the mean profit per sales dollar for Jones and Smith.

Finally, if we want the mean value of y when Adams is the sales engineer, we let $x_1 = 0$ and $x_2 = 1$:

$$\mu_C = E(y) = \beta_0 + \beta_1(0) + \beta_2(1) = \beta_0 + \beta_2$$

or, since $\beta_0 = \mu_A$,

$$\mu_C = \mu_A + \beta_2$$

Then it follows that the interpretation of β_2 is

$$\beta_2 = \mu_C - \mu_A$$

which represents the difference in the mean profit per sales dollar between Jones and Adams.

Note that we were able to describe *three levels* of the qualitative variable with only *two dummy variables*. This is because the mean of the base level (Jones, in this case) is accounted for by the intercept β_0.

Now, carefully examine the model for a single qualitative independent variable three levels, because we will use exactly the same pattern for any number of levels. Also, the interpretation of the parameters will always be the same.

One level is selected as the base level (we used Jones as level A). Then, for the 1–0 system of coding* for the dummy variables,

*We do not have to use a 1–0 system of coding for the dummy variables. Any two-value system will work, but the interpretation given to the model parameters will depend on the code. Using the 1–0 system makes the model parameters easy to interpret.

$$\mu_A = \beta_0$$

The coding for all dummy variables is as follows: To represent the mean value of y for a particular level, let that dummy variable equal 1; otherwise, the dummy variable is set equal to 0. Using this system of coding, we have

$$\mu_B = \beta_0 + \beta_1 \qquad \mu_C = \beta_0 + \beta_2$$

Because $\mu_A = \beta_0$, any other model parameter will represent the difference in means for that level and the base level:

$$\beta_1 = \mu_B - \mu_A \qquad \beta_2 = \mu_C - \mu_A$$

The general procedure is given in the accompanying box.

PROCEDURE FOR WRITING A MODEL WITH ONE QUALITATIVE INDEPENDENT VARIABLE AT k LEVELS

$$E(y) = \beta_0 + \beta_1 x_1 + \beta_2 x_2 + \cdots + \beta_{k-1} x_{k-1}$$

where x_i is the dummy variable for level i and

$$x_i = \begin{cases} 1 & \text{if } E(y) \text{ is the mean for level } i \\ 0 & \text{otherwise} \end{cases}$$

The number of dummy variables for a single qualitative variable is always 1 less than the number of levels for the variable.

Then, for this system of coding,

$$\mu_A = \beta_0$$
$$\mu_B = \beta_0 + \beta_1$$
$$\mu_C = \beta_0 + \beta_2$$
$$\mu_D = \beta_0 + \beta_3$$
$$\vdots$$

Also, note that

$$\beta_1 = \mu_B - \mu_A$$
$$\beta_2 = \mu_C - \mu_A$$
$$\beta_3 = \mu_D - \mu_A$$
$$\vdots$$

Thus, we can use least squares to fit predictive models in which the response y is a function of quantitative or qualitative variables. The interpretation of the parameters will differ for the two types of variables, but the objective is the same no matter what type of variable is used—to obtain a good prediction model for y.

SECTION 7.8

MODELS WITH TWO QUALITATIVE INDEPENDENT VARIABLES

We will demonstrate how to write a model with two qualitative independent variables and then, in Section 7.9, will explain how to use this technique to write models with any number of qualitative independent variables.

Let us return to the example used in Section 7.7, where we wrote a model for the mean profit per sales dollar, $E(y)$, as a function of one qualitative independent variable, sales engineer. Now suppose the mean profit is also a function of the state in which the construction job is located, because of tax differences, different labor conditions, and so forth. Assume that the company operates in two states. Therefore, this second qualitative independent variable, state, will be observed at two levels. To simplify our notation, we will change the symbols for the three levels of sales engineer from A, B, C, to E_1, E_2, E_3, and we will let S_1 and S_2 represent the two states in which the company operates. The six population means of profit per sales dollar measurements (measurements of y) are symbolically represented by the six cells in the two-way table shown in Table 7.5. Each μ subscript corresponds to one sales engineer–state combination.

TABLE 7.5

Table Showing the Six Combinations of Sales Engineer and State

		STATE	
		S_1	S_2
SALES ENGINEER	E_1	μ_{11}	μ_{12}
	E_2	μ_{21}	μ_{22}
	E_3	μ_{31}	μ_{32}

First we will write a model in its simplest form—where the two qualitative variables affect the response independently of each other. To write the model for a mean profit, $E(y)$, we start with a constant β_0 and then add *two* dummy variables for the three levels of sales engineer in the manner explained in Section 7.7. These terms, which are called the **main effect** terms for sales engineer, E, account for the effect of E on $E(y)$ when sales engineer, E, and state, S, affect $E(y)$ independently. Then,

$$E(y) = \beta_0 + \overbrace{\beta_1 x_1 + \beta_2 x_2}^{\substack{\text{Main effect} \\ \text{terms for } E}}$$

where

$$x_1 = \begin{cases} 1 & \text{if } E_2 \text{ was the sales engineer} \\ 0 & \text{if not} \end{cases}$$

$$x_2 = \begin{cases} 1 & \text{if } E_3 \text{ was the sales engineer} \\ 0 & \text{if not} \end{cases}$$

Now let level S_1 be the base level of the state variable. Since there are two levels of this variable, we will need only one dummy variable to include the state in the model:

$$\begin{array}{cc} \overbrace{}^{\substack{\text{Main effect} \\ \text{terms for } E}} & \overbrace{}^{\substack{\text{Main effect} \\ \text{term for } S}} \end{array}$$
$$E(y) = \beta_0 + \beta_1 x_1 + \beta_2 x_2 + \beta_3 x_3$$

where the dummy variables x_1 and x_2 are defined as above, and

$$x_3 = \begin{cases} 1 & \text{if } S_2 \text{ was the state in which the job was located} \\ 0 & \text{if } S_1 \text{ (base level) was the state in which the job was located} \end{cases}$$

If you check the model, you will see that by assigning specific values to x_1, x_2, and x_3, you create a model for the mean value of y corresponding to one of the cells of Table 7.5. We will illustrate with two examples.

EXAMPLE 7.5

Give the values of x_1, x_2, and x_3 and the model for the mean profit per sales dollar, $E(y)$, when E_1 is the sales engineer and S_1 is the state in which the job is located.

SOLUTION

Checking the coding system, you will see that E_1 and S_1 occur when $x_1 = x_2 = x_3 = 0$. Then,

$$E(y) = \beta_0 + \beta_1 x_1 + \beta_2 x_2 + \beta_3 x_3 = \beta_0 + \beta_1(0) + \beta_2(0) + \beta_3(0)$$
$$= \beta_0$$

Therefore, the mean value of y at levels E_1 and S_1, which we represent as μ_{11}, is

$$\mu_{11} = \beta_0 \qquad\qquad\blacksquare$$

EXAMPLE 7.6

Give the values of x_1, x_2, and x_3 and the model for the mean profit per sales dollar, $E(y)$, when E_3 is the sales engineer and S_2 is the state.

SOLUTION

Checking the coding system, you will see that for levels E_3 and S_2,

$$x_1 = 0 \qquad x_2 = 1 \qquad x_3 = 1$$

Then, the mean profit per sales dollar when E_3 is the sales engineer and S_2 is the state, represented by the symbol μ_{32} (see Table 7.5), is

$$\mu_{32} = E(y) = \beta_0 + \beta_1 x_1 + \beta_2 x_2 + \beta_3 x_3 = \beta_0 + \beta_1(0) + \beta_2(1) + \beta_3(1)$$
$$= \beta_0 + \beta_2 + \beta_3 \qquad\qquad\blacksquare$$

Note that in the model described above, we assumed the qualitative independent variables for sales engineer and state affect the mean response, $E(y)$, independently of each other. This type of model is called a **main effects model**. Changing the level of one qualitative variable will have the same effect on $E(y)$ for any level of the second qualitative variable. In other words, the effect of one qualitative variable on $E(y)$ is independent (in a mathematical sense) of the level of the second qualitative variable.

MAIN EFFECTS MODEL WITH TWO QUALITATIVE INDEPENDENT VARIABLES, ONE AT THREE LEVELS (E_1, E_2, E_3) AND THE OTHER AT TWO LEVELS (S_1, S_2)

$$E(y) = \beta_0 + \overbrace{\beta_1 x_1 + \beta_2 x_2}^{\text{Main effect terms for } E} + \overbrace{\beta_3 x_3}^{\text{Main effect term for } S}$$

where

$$x_1 = \begin{cases} 1 & \text{if } E_2 \\ 0 & \text{if not} \end{cases} \qquad x_2 = \begin{cases} 1 & \text{if } E_3 \\ 0 & \text{if not} \end{cases} \qquad (E_1 \text{ is base level})$$

$$x_3 = \begin{cases} 1 & \text{if } S_2 \\ 0 & \text{if } S_1 \end{cases} \quad (\text{base level})$$

INTERPRETATION OF MODEL PARAMETERS

β_0: μ_{11} (Mean of the combination of base levels)

β_1: $\mu_{2j} - \mu_{1j}$, for any level S_j ($j = 1, 2$)

β_2: $\mu_{3j} - \mu_{1j}$, for any level S_j ($j = 1, 2$)

β_3: $\mu_{i2} - \mu_{i1}$, for any level E_i ($i = 1, 2, 3$)

When two independent variables affect the mean response independently of each other, you may obtain the pattern shown in Figure 7.15. Note that the difference in mean profit between any two sales engineers (levels of E) is the same, *regardless* of the state in which the job is located. That is, the main effects model assumes the relative effect of sales engineer on profit is the same in both states.

If E and S do not affect $E(y)$ independently of each other, then the response function might appear as shown in Figure 7.16. Note the difference between the mean response functions for Figures 7.15 and 7.16. When E and S affect the

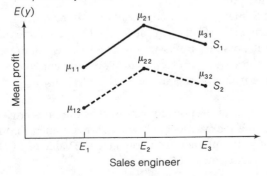

FIGURE 7.15 Main Effects Model: Mean Response as a Function of E and S When E and S Affect $E(y)$ Independently

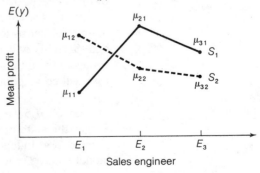

FIGURE 7.16 Interaction Model: Mean Response as a Function of E and S When E and S Interact to Affect $E(y)$

mean response in a dependent manner (Figure 7.16), the response functions differ for each state. This means that you cannot study the effect of one variable on $E(y)$ without considering the level of the other. When this situation occurs, we say that the qualitative independent variables **interact**. In this example, interaction might be expected if one sales engineer tends to develop a knack for bidding on jobs in state S_1, while another becomes adept at bidding in state S_2.

INTERACTION MODEL WITH TWO QUALITATIVE INDEPENDENT VARIABLES, ONE AT THREE LEVELS (E_1, E_2, E_3) AND THE OTHER AT TWO LEVELS (S_1, S_2)

$$E(y) = \beta_0 + \overbrace{\beta_1 x_1 + \beta_2 x_2}^{\substack{\text{Main effect} \\ \text{terms for } E}} + \overbrace{\beta_3 x_3}^{\substack{\text{Main effect} \\ \text{term for } S}} + \overbrace{\beta_4 x_1 x_3 + \beta_5 x_2 x_3}^{\substack{\text{Interaction} \\ \text{terms}}}$$

where the dummy variables x_1, x_2, and x_3 are defined in the same way as for the main effects model.

INTERPRETATION OF MODEL PARAMETERS

β_0: μ_{11} (Mean of the combination of base levels)

β_1: $\mu_{21} - \mu_{11}$ (i.e., for base level S_1 only)

β_2: $\mu_{31} - \mu_{11}$ (i.e., for base level S_1 only)

β_3: $\mu_{12} - \mu_{11}$ (i.e., for base level E_1 only)

β_4: $(\mu_{22} - \mu_{12}) - (\mu_{21} - \mu_{11})$

β_5: $(\mu_{32} - \mu_{12}) - (\mu_{31} - \mu_{11})$

When qualitative independent variables interact, the model for $E(y)$ must be constructed so that it is able (if necessary) to give a different mean value, $E(y)$, for every cell in Table 7.5. We do this by adding **interaction terms** to the main effects model. These terms will involve all possible two-way cross products between each of the two dummy variables for E, x_1, and x_2, and the one dummy variable for S, x_3. The number of interaction terms (for two independent variables) will equal the number of main effect terms for the one variable times the number of main effect terms for the other.

Note that when E and S interact, the model contains six parameters, the two main effect terms for E, one main effect term for S, $(2)(1) = 2$ interaction terms, and β_0. This will make it possible, by assigning the various combinations of values to the dummy variables x_1, x_2, and x_3, to give six different values for $E(y)$ that will correspond to the means of the six cells of Table 7.5.

EXAMPLE 7.7

In Example 7.5 we gave the mean response when E_1 was the sales engineer and S_1 was the state in which the job was located, where we assumed that E and S affected $E(y)$ independently (no interaction). Now give the value of $E(y)$ for the model where E and S interact to affect $E(y)$.

SOLUTION

When E and S interact,

$$E(y) = \beta_0 + \beta_1 x_1 + \beta_2 x_2 + \beta_3 x_3 + \beta_4 x_1 x_3 + \beta_5 x_2 x_3$$

For levels E_1 and S_1, we have agreed (according to our system of coding) to let $x_1 = x_2 = x_3 = 0$. Substituting into the equation for $E(y)$, we have

$$E(y) = \beta_0$$

(the same as for the main effects model). ∎

EXAMPLE 7.8

In Example 7.6 we gave the mean response for sales engineer E_3 and state S_2 when E and S affected $E(y)$ independently. Now assume that E and S interact and write a model for $E(y)$ when E_3 is the sales engineer and S_2 is the state in which the job is located.

SOLUTION

When E and S interact,

$$E(y) = \beta_0 + \beta_1 x_1 + \beta_2 x_2 + \beta_3 x_3 + \beta_4 x_1 x_3 + \beta_5 x_2 x_3$$

To model $E(y)$ for E_3 and S_2, we set $x_1 = 0$, $x_2 = 1$, and $x_3 = 1$:

$$E(y) = \beta_0 + \beta_1(0) + \beta_2(1) + \beta_3(1) + \beta_4(0)(1) + \beta_5(1)(1)$$
$$= \beta_0 + \beta_2 + \beta_3 + \beta_5$$

This is the model for the value of μ_{32} in Table 7.5. Note the difference in $E(y)$ for the model assuming independence between E and S versus this one, which assumes interaction between E and S. The difference is β_5. ∎

EXAMPLE 7.9

The profit per sales dollar, y, for the six combinations of sales engineer and state is shown in Table 7.6. Note that the number of construction jobs per combination varies from one for levels (E_1, S_2) to three for levels (E_1, S_1). A total of twelve jobs are sampled.

TABLE 7.6
Profit Data for
Combinations of Sales
Engineer and State

		STATE	
		S_1	S_2
SALES ENGINEER	E_1	$.065 \atop .073 \atop .068$	\$.036
	E_2	$.078 \atop .082$	$.050 \atop .043$
	E_3	$.048 \atop .046$	$.061 \atop .062$

a. Assume the interaction between E and S is negligible. Fit the model for $E(y)$ with interaction terms omitted.

b. Fit the complete model for $E(y)$ allowing for the fact that interactions might occur.

c. Estimate the mean profit for jobs bid by sales engineer E_3 in state S_2 using the prediction equation $\hat{y}$ for each of the two models in parts **a** and **b**. Then calculate the sample mean for this cell of Table 7.6. Explain the discrepancy between the sample mean for levels (E_3, S_2) and the estimate(s) obtained from one or both of the two prediction equations.

SOLUTION

a. A portion of the SAS printout for the main effects model

$$E(y) = \beta_0 + \overbrace{\beta_1 x_1 + \beta_2 x_2}^{\substack{E \\ \text{main effect}}} + \overbrace{\beta_3 x_3}^{\substack{S \\ \text{main effect}}}$$

is given in Figure 7.17. The least squares prediction equation is

$$\hat{y} = .0645 + .0067x_1 - .00230x_2 - .0158x_3$$

FIGURE 7.17 SAS Printout for Main Effects Model of Example 7.9

```
DEP VARIABLE: Y                         ANALYSIS OF VARIANCE

                        SUM OF          MEAN
            SOURCE   DF  SQUARES        SQUARE       F VALUE   PROB>F

            MODEL     3  0.000858258   0.000286086   1.513     0.2838
            ERROR     8  0.001512409   0.000189051
            C TOTAL  11  0.002370667

            ROOT MSE    0.01374959     R-SQUARE     0.3620
            DEP MEAN    0.05933333     ADJ R-SQ     0.1228
            C.V.          23.17346

                        PARAMETER ESTIMATES

                        PARAMETER       STANDARD     T FOR H0:
   VARIABLE   DF        ESTIMATE        ERROR        PARAMETER=0   PROB > |T|

   INTERCEP    1        0.06445455     0.007180488    8.976        0.0001
   X1          1        0.006704545    0.009940935    0.674        0.5190
   X2          1       -0.002295455    0.009940935   -0.231        0.8232
   X3          1       -0.01581818     0.008291313   -1.908        0.0928


                                 PREDICT    STD ERR    LOWER95%   UPPER95%
   OBS    ID    ACTUAL           VALUE      PREDICT    MEAN       MEAN       RESIDUAL

    1    E1S1   0.0650           0.0645     .0071805   0.0479     0.0810     5.5E-04
    2    E1S1   0.0730           0.0645     .0071805   0.0479     0.0810     .0085455
    3    E1S1   0.0680           0.0645     .0071805   0.0479     0.0810     .0035455
    4    E2S1   0.0780           0.0712     0.008028   0.0526     0.0897     .0068409
    5    E2S1   0.0820           0.0712     0.008028   0.0526     0.0897     0.0108
    6    E3S1   0.0480           0.0622     0.008028   0.0436     0.0807    -0.0142
    7    E3S1   0.0460           0.0622     0.008028   0.0436     0.0807    -0.0162
    8    E1S2   0.0360           0.0486     0.00927    0.0273     0.0700    -0.0126
    9    E2S2   0.0500           0.0553     0.008028   0.0368     0.0739    -.005341
   10    E2S2   0.0430           0.0553     0.008028   0.0368     0.0739    -0.0123
   11    E3S2   0.0610           0.0463     0.008028   0.0278     0.0649     0.0147
   12    E3S2   0.0620           0.0463     0.008028   0.0278     0.0649     0.0157

   SUM OF RESIDUALS                2.29851E-16
   SUM OF SQUARED RESIDUALS        0.001512409
   PREDICTED RESID SS (PRESS)      0.003615375
```

b. The complete model SAS printout is given in Figure 7.18. Recall that the complete model is

$$E(y) = \beta_0 + \beta_1 x_1 + \beta_2 x_2 + \beta_3 x_3 + \beta_4 x_1 x_3 + \beta_5 x_2 x_3$$

The least squares prediction equation is

$$\hat{y} = .0687 + .0113 x_1 - .0217 x_2 - .0327 x_3 - .0008 x_1 x_3 + .0472 x_2 x_3$$

FIGURE 7.18 SAS Printout for Interaction Model of Example 7.9

DEP VARIABLE: Y

ANALYSIS OF VARIANCE

SOURCE	DF	SUM OF SQUARES	MEAN SQUARE	F VALUE	PROB>F
MODEL	5	0.002303000	0.000460600	40.841	0.0001
ERROR	6	0.000067667	0.000011278		
C TOTAL	11	0.002370667			

ROOT MSE	0.00335824	R-SQUARE	0.9715	
DEP MEAN	0.05933333	ADJ R-SQ	0.9477	
C.V.	5.659956			

PARAMETER ESTIMATES

VARIABLE	DF	PARAMETER ESTIMATE	STANDARD ERROR	T FOR H0: PARAMETER=0	PROB > |T|
INTERCEP	1	0.06866667	0.001938881	35.416	0.0001
X1	1	0.01133333	0.003065640	3.697	0.0101
X2	1	-0.02166667	0.003065640	-7.068	0.0004
X3	1	-0.03266667	0.003877762	-8.424	0.0002
X1X3	1	-0.000833333	0.005129797	-0.162	0.8763
X2X3	1	0.04716667	0.005129797	9.195	0.0001

OBS	ID	ACTUAL	PREDICT VALUE	STD ERR PREDICT	LOWER95% MEAN	UPPER95% MEAN	RESIDUAL
1	E1S1	0.0650	0.0687	.0019389	0.0639	0.0734	-.003667
2	E1S1	0.0730	0.0687	.0019389	0.0639	0.0734	.0043333
3	E1S1	0.0680	0.0687	.0019389	0.0639	0.0734	-6.7E-04
4	E2S1	0.0780	0.0800	.0023746	0.0742	0.0858	-0.002
5	E2S1	0.0820	0.0800	.0023746	0.0742	0.0858	0.002
6	E3S1	0.0480	0.0470	.0023746	0.0412	0.0528	0.001
7	E3S1	0.0460	0.0470	.0023746	0.0412	0.0528	-1.0E-03
8	E1S2	0.0360	0.0360	.0033582	0.0278	0.0442	-8.7E-18
9	E2S2	0.0500	0.0465	.0023746	0.0407	0.0523	0.0035
10	E2S2	0.0430	0.0465	.0023746	0.0407	0.0523	-0.0035
11	E3S2	0.0610	0.0615	.0023746	0.0557	0.0673	-5.0E-04
12	E3S2	0.0620	0.0615	.0023746	0.0557	0.0673	5.0E-04

SUM OF RESIDUALS   3.76435E-16
SUM OF SQUARED RESIDUALS   .00006766667
PREDICTED RESID SS (PRESS) 0.0003151194

c. To obtain the estimated mean response for cell (E_3, S_2), we let $x_1 = 0$, $x_2 = 1$, and $x_3 = 1$. Then, for the main effects model, we find

$$\hat{y} = .0645 + .0067(0) - .0023(1) - .0158(1) = .0464$$

The 95% confidence interval for the true mean profit per sales dollar (shown in Figure 7.17) is (.0278, .0649).

For the complete model, we find

$$\hat{y} = .0687 + .0113(0) - .0217(1) - .0327(1) - .0008(0)(1) + .0472(1)(1)$$
$$= .0615$$

The 95% confidence interval for true mean profit (Figure 7.18) is (.0557, .0673).

The mean for the cell (E_3, S_2) in Table 7.6 is

$$\bar{y}_{32} = \frac{.061 + .062}{2} = .0615$$

which is precisely what is estimated by the complete (interaction) model. However, the main effects model yields a different estimate, .0464. The reason for the discrepancy is that the main effects model assumes the two qualitative independent variables affect $E(y)$ independently of each other. That is, the change in $E(y)$ produced by a change in levels of one variable is the same regardless of the level of the other variable. In contrast, the complete model contains six parameters $(\beta_0, \beta_1, \ldots, \beta_5)$ to describe the six cell populations, so that each population cell mean will be estimated by its sample mean. Thus, the complete model estimate for any cell mean is equal to the observed (sample) mean for that cell.   ∎

Example 7.9 demonstrates an important point. If we were to ignore the least squares analysis and calculate the six sample means of Table 7.6 directly, we would obtain exactly the same estimates of $E(y)$ as would be obtained by a least squares analysis for the case where the interaction between E and S is assumed to exist. We would not obtain the same estimates if the model assumes interaction does not exist.

Also, the estimates of means raise important questions. Do the data provide sufficient evidence to indicate that E and S interact? For our example, does the contribution to mean profit per sales dollar for a job in one state depend on which sales engineer estimated and bid the job? The plot of all six sample means, shown in Figure 7.19, seems to indicate interaction, since engineers E_1 and E_2 appear to operate more effectively in state S_1, while the mean profit of E_3 is higher in state S_2. Can these sample facts be reliably generalized to conclusions about the populations?

To answer this question, we will want to perform a test for interaction between the two qualitative independent variables, sales engineer and state. Since allowance for interaction between sales engineer and state in the complete model was provided by the addition of the terms $\beta_4 x_1 x_3$ and $\beta_5 x_2 x_3$, it follows that the null hypothesis that the independent variables sales engineer and state do not interact is equivalent to the hypothesis that the terms $\beta_4 x_1 x_3$ and $\beta_5 x_2 x_3$ are not needed in the model for $E(y)$—or equivalently, that $\beta_4 = \beta_5 = 0$. Conversely, the alternative hypothesis that sales engineer and state do interact is equivalent to stating that at least one of the two parameters, β_4 or β_5, differs from 0.

FIGURE 7.19

Graph of Sample Means for Profit Example

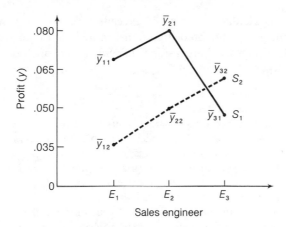

The appropriate procedure for testing a portion of the model parameters, an F-test, was discussed in Section 4.11. The F-test is carried out as follows:

H_0: $\quad \beta_4 = \beta_5 = 0$

H_a: $\quad$ At least one of β_4 and β_5 differs from 0

Test statistic: $\quad F = \dfrac{(SSE_1 - SSE_2)/g}{SSE_2/[n - (k + 1)]}$

where

$$SSE_1 = \text{SSE for reduced model (main effects model)}$$

$$SSE_2 = \text{SSE for complete model (interaction model)}$$

$$g = \text{number of } \beta\text{'s tested}$$

$$= \text{numerator df for the } F\text{-statistic}$$

$$n - (k + 1) = \text{df for error for complete model}$$

$$= \text{denominator df for the } F\text{-statistic}$$

For this example we have

$SSE_1 = .00151241 \quad$ (see Figure 7.17)

$SSE_2 = .00006767 \quad$ (see Figure 7.18)

$g = 2 \quad$ and $\quad n - (k + 1) = 6$

Then

$$F = \frac{(.00151241 - .00006767)/2}{.00006767/6}$$

$$= \frac{.00072237}{.00001128} = 64.04$$

The critical value of F for $\alpha = .05$, $\nu_1 = 2$, and $\nu_2 = 6$ is (from Table 4 of Appendix D) $F_{.05} = 5.14$. Therefore, the rejection region is

Rejection region: $\quad F > 5.14$

Since the calculated $F = 64.04$ exceeds 5.14, we are confident (at $\alpha = .05$) in concluding that the interaction terms contribute to the prediction of y, profit per sales. Equivalently, there is sufficient evidence to conclude that E and S do interact.

EXERCISES 7.13–7.17

7.13 The manager of a supermarket wants to model the total weekly sales of beer, y, as a function of brand. (This model will enable the manager to plan the store's inventory.) The market carries three brands, B_1, B_2, and B_3.
  a. What type of independent variable is brand of beer?
  b. Write the model relating mean weekly beer sales, $E(y)$, to brand of beer. Be sure to explain any dummy variables you use.
  c. Interpret the parameters (β's) of your model in part **b**.
  d. In terms of the model parameters, what is the mean weekly sales for brand B_3?

7.14 Refer to Exercise 7.13. Suppose the manager uses brand B_1 as the base level and obtains the model

$$\hat{y} = 450 + 60x_1 - 30x_2$$

where

$$x_1 = \begin{cases} 1 & \text{if brand } B_2 \\ 0 & \text{otherwise} \end{cases} \qquad x_2 = \begin{cases} 1 & \text{if brand } B_3 \\ 0 & \text{otherwise} \end{cases}$$

  a. What is the difference between the estimated mean* weekly sales for brands B_2 and B_1?
  b. What is the estimated mean weekly sales for B_2?

7.15 The performance of an industry is often measured by the level of excess (or unutilized) capacity within the firm. Esposito and Esposito (1986) examined the relationship between excess capacity y and several market variables in 273 U.S. manufacturing industries. Two qualitative independent variables considered in the study were

  Market concentration (Low, Moderate, and High)
  Industry type (Producer or Consumer)

  a. Write the main effects model for $E(y)$ as a function of the two qualitative variables.
  b. Interpret the β coefficients in the main effects model.
  c. Write the model for $E(y)$ that includes interaction between market concentration and industry type.
  d. Interpret the β coefficients in the interaction model.
  e. How would you test the hypothesis that the difference between the mean excess capacity levels of producer and consumer industry types is the same across all three market concentrations?

7.16 Due to the increase in gasoline prices, many service stations are offering self-service gasoline at reduced prices. Suppose an oil company wants to model the mean monthly gasoline sales, $E(y)$, of its affiliated stations as a function of type of gasoline pur-

---

*We would generally form confidence intervals for the true means in order to assess the reliability of these estimates. Our objective in these exercises is to develop the ability to use the models to obtain the estimates. The corresponding confidence intervals can be obtained using the methods of Chapter 4.

chased—regular, premium, or lead-free—and of type of service—self-service or full-service.

a. How many dummy variables will be needed to describe each of the qualitative variables, type of gasoline and type of service?

b. Write the main effects model relating $E(y)$ to type of gasoline and type of service. Be sure to code the dummy variables.

c. Write a model for $E(y)$ that includes interaction between type of gasoline and type of service.

d. Do you think interaction would be important in this model? A plot of your intuitive estimates of the mean will help you decide.

7.17 Suppose the interaction model of Exercise 7.16 part c, is used, and the following least squares model is obtained (units of sales, y, are millions of dollars):

$$\hat{y} = 4 - 2x_1 - x_2 - x_3 + 2x_1x_3 + 3x_2x_3$$

where

$$x_1 = \begin{cases} 1 & \text{if premium} \\ 0 & \text{otherwise} \end{cases} \qquad x_2 = \begin{cases} 1 & \text{if lead-free} \\ 0 & \text{otherwise} \end{cases} \qquad x_3 = \begin{cases} 1 & \text{if full-service} \\ 0 & \text{if self-service} \end{cases}$$

a. What is the estimate of mean sales for lead-free gasoline at the full-service pumps?

b. What is the estimated difference between mean sales of regular gasoline at full-service and self-service pumps?

c. To see the effects of interaction, compare the difference between the regular self-service and full-service mean sales to the difference between the premium self-service and full-service mean sales.

SECTION 7.9

MODELS WITH THREE OR MORE QUALITATIVE INDEPENDENT VARIABLES

Models with three or more qualitative independent variables are constructed in the same manner as for two qualitative independent variables, except that you must add three-way interaction terms if you have three qualitative independent variables, three-way and four-way interaction terms for four independent variables, and so on. In this section we will explain what we mean by three-way, four-way, etc., interactions and will demonstrate the procedure for writing the model for any number, say k, of qualitative independent variables. The pattern employed in writing the model is shown in the box.

PATTERN OF THE MODEL RELATING $E(y)$ TO k QUALITATIVE INDEPENDENT VARIABLES

$E(y) = \beta_0 +$ Main effect terms for all independent variables

+ All two-way interaction terms between pairs of independent variables

+ All three-way interaction terms between different groups of three independent variables

+

⋮

+ All k-way interaction terms for the k independent variables

---

Recall that a two-way interaction term was formed by multiplying the dummy variable associated with one of the main effect terms of one (call it the first) independent variable by the dummy variable from a main effect term of another (the second) independent variable. Three-way interaction terms are formed in a similar way, by forming the product of three dummy variables, one from a main effect term from each of the three independent variables. Similarly, four-way interaction terms are formed by taking the product of four dummy variables, one from a main effect term from each of four independent variables. We will illustrate with three examples.

EXAMPLE 7.10

Suppose you have three qualitative independent variables, the first at three levels, A_1, A_2, and A_3, the second at three levels, B_1, B_2, and B_3, and the third at two levels, C_1 and C_2. Write a model for $E(y)$ that includes all main effect and interaction terms for the independent variables.

SOLUTION

First write a model containing the main effect terms for the three variables:

$$E(y) = \beta_0 + \overbrace{\beta_1 x_1 + \beta_2 x_2}^{\substack{\text{Main effect} \\ \text{terms for } A}} + \overbrace{\beta_3 x_3 + \beta_4 x_4}^{\substack{\text{Main effect} \\ \text{terms for } B}} + \overbrace{\beta_5 x_5}^{\substack{\text{Main effect} \\ \text{term for } C}}$$

where

$$x_1 = \begin{cases} 1 & \text{if level } A_2 \\ 0 & \text{if not} \end{cases} \qquad x_3 = \begin{cases} 1 & \text{if level } B_2 \\ 0 & \text{if not} \end{cases} \qquad x_5 = \begin{cases} 1 & \text{if level } C_2 \\ 0 & \text{if not} \end{cases}$$

$$x_2 = \begin{cases} 1 & \text{if level } A_3 \\ 0 & \text{if not} \end{cases} \qquad x_4 = \begin{cases} 1 & \text{if level } B_3 \\ 0 & \text{if not} \end{cases}$$

The next step is to add two-way interaction terms. These will be of three types—those for the interaction between A and B, between A and C, and between B and C. Thus,

$$E(y) = \beta_0 + \overbrace{\beta_1 x_1 + \beta_2 x_2}^{\substack{\text{Main effect} \\ A}} + \overbrace{\beta_3 x_3 + \beta_4 x_4}^{\substack{\text{Main effect} \\ B}} + \overbrace{\beta_5 x_5}^{\substack{\text{Main effect} \\ C}}$$

$$+ \overbrace{\beta_6 x_1 x_3 + \beta_7 x_1 x_4 + \beta_8 x_2 x_3 + \beta_9 x_2 x_4}^{AB \text{ interaction terms}}$$

$$+ \overbrace{\beta_{10} x_1 x_5 + \beta_{11} x_2 x_5}^{AC \text{ interaction terms}}$$

$$+ \overbrace{\beta_{12} x_3 x_5 + \beta_{13} x_4 x_5}^{BC \text{ interaction terms}}$$

Finally, since there are three independent variables, we must include terms for the interaction of A, B, and C. These terms are formed as the products of dummy variables, one from each of the A, B, and C main effect terms. The complete model for $E(y)$ is

$$
E(y) = \beta_0 + \overbrace{\beta_1 x_1 + \beta_2 x_2}^{\substack{\text{Main effect}\\ A}} + \overbrace{\beta_3 x_3 + \beta_4 x_4}^{\substack{\text{Main effect}\\ B}} + \overbrace{\beta_5 x_5}^{\substack{\text{Main effect}\\ C}}
$$

$$
+ \overbrace{\beta_6 x_1 x_3 + \beta_7 x_1 x_4 + \beta_8 x_2 x_3 + \beta_9 x_2 x_4}^{AB \text{ interaction terms}}
$$

$$
+ \overbrace{\beta_{10} x_1 x_5 + \beta_{11} x_2 x_5}^{AC \text{ interaction terms}}
$$

$$
+ \overbrace{\beta_{12} x_3 x_5 + \beta_{13} x_4 x_5}^{BC \text{ interaction terms}}
$$

$$
+ \overbrace{\beta_{14} x_1 x_3 x_5 + \beta_{15} x_1 x_4 x_5 + \beta_{16} x_2 x_3 x_5 + \beta_{17} x_2 x_4 x_5}^{\text{Three-way } ABC \text{ interaction terms}} \qquad ■
$$

Note that the complete model in Example 7.10 contains 18 parameters, one for each of the $3 \times 3 \times 2$ combinations of levels for A, B, and C. There are 18 linearly independent linear combinations of these parameters, *one corresponding to each of the means of the $3 \times 3 \times 2$ combinations of levels of A, B, and C.* We will illustrate with an example.

EXAMPLE 7.11

Refer to Example 7.10 and give the expression for the mean value of y for observations taken at the second level of A, the first level of B, and the second level of C, i.e., at (A_2, B_1, C_2).

SOLUTION

Check the coding for the dummy variables (given in Example 7.10) and you will see that they assume the following values:

> *For level A_2:* $x_1 = 1$, $x_2 = 0$
> *For level B_1:* $x_3 = 0$, $x_4 = 0$
> *For level C_2:* $x_5 = 1$

Substituting these values into the expression for $E(y)$, we obtain

$$
\begin{aligned}
E(y) &= \beta_0 + \beta_1(1) + \beta_2(0) + \beta_3(0) + \beta_4(0) + \beta_5(1) \\
&\quad + \beta_6(1)(0) + \beta_7(1)(0) + \beta_8(0)(0) + \beta_9(0)(0) \\
&\quad + \beta_{10}(1)(1) + \beta_{11}(0)(1) + \beta_{12}(0)(1) + \beta_{13}(0)(1) \\
&\quad + \beta_{14}(1)(0)(1) + \beta_{15}(1)(0)(1) + \beta_{16}(0)(0)(1) + \beta_{17}(0)(0)(1) \\
&= \beta_0 + \beta_1 + \beta_5 + \beta_{10}
\end{aligned}
$$

Thus, the mean value of y observed at levels A_2, B_1, and C_2 is $\beta_0 + \beta_1 + \beta_5 + \beta_{10}$. You could find the mean values of y for the other 17 combinations of levels of A, B, and C by substituting the appropriate values of the dummy variables into the expression for $E(y)$ in the same manner. Each of the 18 means is a unique linear combination of the 18 β parameters in the model. ∎

EXAMPLE 7.12

Suppose you want to test the hypothesis that the three qualitative independent variables discussed in Example 7.10 do not interact, i.e., the hypothesis that the effect of any one of the variables on $E(y)$ is independent of the level settings of the other two variables. Formulate the appropriate test of hypothesis about the model parameters.

SOLUTION

No interaction among the three qualitative independent variables implies that the main effects model,

$$E(y) = \beta_0 + \overbrace{\beta_1 x_1 + \beta_2 x_2}^{\substack{\text{Main effect}\\A}} + \overbrace{\beta_3 x_3 + \beta_4 x_4}^{\substack{\text{Main effect}\\B}} + \overbrace{\beta_5 x_5}^{\substack{\text{Main effect}\\C}}$$

is appropriate for modeling $E(y)$ or, equivalently, that all interaction terms should be excluded from the model. This situation will occur if

$$\beta_6 = \beta_7 = \cdots = \beta_{17} = 0$$

Consequently, we will test the null hypothesis

$$H_0: \quad \beta_6 = \beta_7 = \cdots = \beta_{17} = 0$$

against the alternative hypothesis that at least one of these β parameters differs from 0, or equivalently, that some interaction among the independent variables exists. This statistical test was described in Section 4.11. ∎

Of what value is this section? If you are modeling a response, say profit of a corporation, and you believe that several qualitative independent variables affect the response, then you must know how to enter these variables into your model. You must understand the implication of the interaction (or lack of it) among a subset of independent variables and how to write the appropriate terms in the model to account for it. Failure to write a good model for your response will usually lead to inflated values of the SSE and s^2 (with a consequent loss of information), and it also can lead to biased estimates of $E(y)$ and biased predictions of y.

S E C T I O N 7.10

MODELS WITH BOTH QUANTITATIVE AND QUALITATIVE INDEPENDENT VARIABLES

Perhaps the most interesting data analysis problems are those that involve both quantitative and qualitative independent variables. For example, suppose mean profit per sales dollar of the construction company is a function of one qualitative independent variable, sales engineer, at levels E_1, E_2, and E_3, and one quantitative independent variable, job size in millions of dollars. The company might be expected to make more profit per sales dollar on small jobs than on very large jobs due to increased competition for the larger jobs. We will proceed to build a model in stages, showing graphically the interpretation that we would give to the model at each stage. This will help you see the contribution of various terms in the model.

At first we assume that the qualitative independent variable has no effect on the response (i.e., the mean contribution to the response is the same for all three sales engineers), but the mean profit per sales dollar, $E(y)$, is related to job size. In this case, one response curve, which might appear as shown in Figure 7.20, would be sufficient to characterize $E(y)$ for all three sales engineers. The following second-order model would likely provide a good approximation to $E(y)$:

$$E(y) = \beta_0 + \beta_1 x_1 + \beta_2 x_1^2$$

where x_1 is job size in millions of dollars. This model has some distinct disadvantages. If differences in mean profit exist for the three sales engineers, they cannot be detected (because the model does not contain any parameters representing differences among sales engineers). Also, the differences would inflate the SSE associated with the fitted model and consequently would increase errors of estimation and prediction.

FIGURE 7.20

Model for $E(y)$ as a Function of Job Size

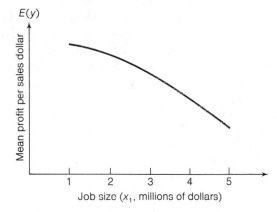

The next stage in developing a model for $E(y)$ is to assume the qualitative independent variable, sales engineer, does affect mean profit, but the effect on $E(y)$ is independent of job size. In other words, the assumption is that the two independent variables do not interact. This model is obtained by adding main effect terms for sales engineer to the second-order model we used in the first stage. Therefore, using the methods of Sections 7.7 and 7.8, we choose E_1 as the base level and add two terms to the model corresponding to levels E_2 and E_3:

$$E(y) = \beta_0 + \beta_1 x_1 + \beta_2 x_1^2 + \beta_3 x_2 + \beta_4 x_3$$

where

$$x_1 = \text{Job size} \qquad x_2 = \begin{cases} 1 & \text{if } E_2 \\ 0 & \text{if not} \end{cases} \qquad x_3 = \begin{cases} 1 & \text{if } E_3 \\ 0 & \text{if not} \end{cases}$$

What effect do these terms have on the graph for the response curve(s)? Suppose we want to model $E(y)$ for level E_1. Then we let $x_2 = 0$ and $x_3 = 0$.

Substituting into the model equation, we have

$$E(y) = \beta_0 + \beta_1 x_1 + \beta_2 x_1^2 + \beta_3(0) + \beta_4(0)$$
$$= \beta_0 + \beta_1 x_1 + \beta_2 x_1^2$$

which would graph as a second-order curve similar to the one shown in Figure 7.20.

Now suppose that one of the other two sales engineers bids a job, for example, E_2. Then $x_2 = 1$, $x_3 = 0$, and

$$E(y) = \beta_0 + \beta_1 x_1 + \beta_2 x_1^2 + \beta_3(1) + \beta_4(0)$$
$$= (\beta_0 + \beta_3) + \beta_1 x_1 + \beta_2 x_1^2$$

This is the equation of exactly the same parabola that we obtained for sales engineer E_1 except that the y-intercept has changed from β_0 to $(\beta_0 + \beta_3)$. Similarly, the response curve for E_3 is

$$E(y) = (\beta_0 + \beta_4) + \beta_1 x_1 + \beta_2 x_1^2$$

Therefore, the three response curves for levels E_1, E_2, and E_3 (shown in Figure 7.21) are identical except that they are shifted vertically upward or downward in relation to each other. The curves depict the situation when the two independent variables do not interact; that is, the effect of job size on mean profit is the same regardless of the sales engineer, and the effect of sales engineer on mean profit is the same for all job sizes (the relative distances between the curves is constant).

FIGURE 7.21

Model for $E(y)$ as a Function of Sales Engineer and Job Size (No Interaction)

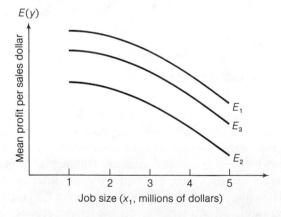

This noninteractive second-stage model has drawbacks similar to those of the simple first-stage model. It is highly unlikely that the response curves for the three sales engineers would be identical except for differing y-intercepts. Because the model does not contain parameters that measure interaction between job size and sales engineer, we cannot test to see if a relationship exists. Also, if interaction does exist, it will cause the SSE for the fitted model to be inflated and will consequently increase the errors of estimating model parameters $E(y)$.

This leads us to the final stage of the model-building process—adding interaction terms to allow the three response curves to differ in shape:

$$E(y) = \beta_0 + \overbrace{\beta_1 x_1 + \beta_2 x_1^2}^{\substack{\text{Main effect} \\ \text{terms for} \\ \text{job size}}} + \overbrace{\beta_3 x_2 + \beta_4 x_3}^{\substack{\text{Main effect} \\ \text{terms for} \\ \text{sales engineer}}}$$

$$+ \overbrace{\beta_5 x_1 x_2 + \beta_6 x_1 x_3 + \beta_7 x_1^2 x_2 + \beta_8 x_1^2 x_3}^{\text{Interaction terms}}$$

where

$$x_1 = \text{Job size} \qquad x_2 = \begin{cases} 1 & \text{if } E_2 \\ 0 & \text{if not} \end{cases} \qquad x_3 = \begin{cases} 1 & \text{if } E_3 \\ 0 & \text{if not} \end{cases}$$

Notice that this model graphs as three different second-order curves.* If E_1 is the sales engineer, we substitute $x_2 = x_3 = 0$ into the formula for $E(y)$, and all but the first three terms equal 0. The result is

$$E(y) = \beta_0 + \beta_1 x_1 + \beta_2 x_1^2$$

If E_2 is the sales engineer, $x_2 = 1$, $x_3 = 0$, and

$$E(y) = \beta_0 + \beta_1 x_1 + \beta_2 x_1^2 + \beta_3(1) + \beta_4(0) + \beta_5 x_1(1) + \beta_6 x_1(0) + \beta_7 x_1^2(1) + \beta_8 x_1^2(0)$$
$$= (\beta_0 + \beta_3) + (\beta_1 + \beta_5)x_1 + (\beta_2 + \beta_7)x_1^2$$

The y-intercept, the coefficient of x_1, and the coefficient of x_1^2 differ from the corresponding coefficients in $E(y)$ at level E_1. Finally, when E_3 is the sales engineer, $x_2 = 0$, $x_3 = 1$, and the result is

$$E(y) = (\beta_0 + \beta_4) + (\beta_1 + \beta_6)x_1 + (\beta_2 + \beta_8)x_1^2$$

A graph of the model for $E(y)$ might appear as shown in Figure 7.22 (page 368). Compare this figure with Figure 7.20, where we assumed the response curves were identical for all three sales engineers, and with Figure 7.21, where we assumed no interaction between the independent variables. Note in Figure 7.22 that the second-order curves may be completely different.

Now that you know how to write a model for two independent variables—one qualitative and one quantitative—we ask a question. Why do it? Why not write a separate second-order model for each level of sales engineer where $E(y)$ is a function of only job size? *One reason we wrote the single model representing all three response curves is so that we can test to determine whether the curves are different.* For example, we might want to know whether the effect of sales engineer depends on job size. Thus, one sales engineer might be especially good on small jobs, but poor on large jobs. The reverse might be true for one of the other two

*Note that the model remains a second-order model for the quantitative independent variable x_1. The terms involving $x_1^2 x_2$ and $x_1^2 x_3$ appear to be third-order terms, but they are not because x_2 and x_3 are dummy variables.

FIGURE 7.22

Graph of $E(y)$ as a
Function of Sales
Engineer and Job Size
(Interaction)

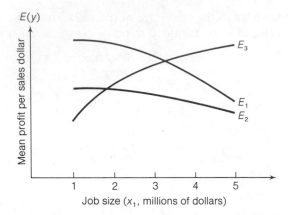

sales engineers. The hypothesis that the independent variables, sales engineer and job size, affect the response independently of one another (a case of no interaction) is equivalent to testing the hypothesis that $\beta_5 = \beta_6 = \beta_7 = \beta_8 = 0$ [i.e., that the model in Figure 7.21 adequately characterizes $E(y)$] using the F-test discussed in Section 4.11. *A second reason for writing a single model is that we obtain a pooled estimate of σ^2, the variance of the random error component ε.* If the variance of ε is truly the same for each sales engineer, the pooled estimate is superior to calculating three estimates by fitting a separate model for each sales engineer.

In conclusion, suppose you want to write a model relating $E(y)$ to several quantitative and qualitative independent variables. Proceed in exactly the same manner as for two independent variables, one qualitative and one quantitative. First, write the model (using the methods of Sections 7.4 and 7.5) that you want to use to describe the quantitative independent variables. Then introduce the main effect and interaction terms for the qualitative independent variables. This gives a model that represents a set of identically shaped response surfaces, one corresponding to each combination of levels of the qualitative independent variables. If you could imagine surfaces in a multidimensional space, their appearance would be analogous to the response curves of Figure 7.21. To complete the model, add terms for the interaction between the quantitative and qualitative variables. This is done by interacting *each* qualitative variable term with every quantitative variable term. We will demonstrate with an example.

EXAMPLE 7.13

A chain of drug stores wished to investigate the effects of three factors on its weekly profit. The factors were:

1. Newspaper selected for the chain's weekly advertisement (two levels)
2. Design of the advertisement (two levels)
3. Percent discount of special sale items (five levels)

The different combinations of newspaper, design, and percent discount were employed in a random sequence, one combination per week, during a period of normal sales activity (holiday periods were excluded). Write a model for the profit per week, y.

SOLUTION

The response y is affected by two qualitative factors (newspaper and design), each at two levels, and one quantitative factor (percent discount), with five levels. Each of the two advertising designs, D_1 and D_2, could be used with each of the two newspapers, N_1 and N_2, giving $2 \times 2 = 4$ possible advertising media—call them (D_1, N_1), (D_1, N_2), (D_2, N_1), (D_2, N_2). For each of these combinations you would obtain a curve that graphs profit as a function of the quantitative factor x_1, percent discount (see Figure 7.23). The stages in writing the model for the response y shown in Figure 7.23 are given below.

FIGURE 7.23

A Graphical Portrayal of Three Factors—Two Qualitative and One Quantitative—on Profit y

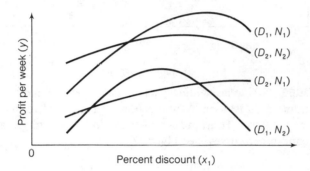

STAGE 1 *Write a model relating y to the quantitative factor(s)*   It is likely that increasing the value of the single quantitative factor x_1, percent discount, will increase sales. This should initially increase profit, but, eventually, when the discount is sufficiently large, the profit should decrease, thus producing curvature in the profit curves of Figure 7.23. Consequently, we will model the mean profit per week, $E(y)$, with the second-order model

$$E(y) = \beta_0 + \beta_1 x_1 + \beta_2 x_1^2$$

This is the model we would use if we were certain that the profit curves were identical for all design–newspaper combinations (D_i, N_j). The model would appear as shown in Figure 7.24(a) (page 370).

STAGE 2 *Add the terms, both main effect and interaction, for the qualitative factors* Thus,

$$E(y) = \beta_0 + \overbrace{\beta_1 x_1 + \beta_2 x_1^2}^{\text{Terms for quantitative factor}}$$

$$+ \underbrace{\beta_3 x_2}_{\substack{\text{Main effect} \\ D}} + \underbrace{\beta_4 x_3}_{\substack{\text{Main effect} \\ N}} + \underbrace{\beta_5 x_2 x_3}_{DN \text{ interaction}}$$

$$x_2 = \begin{cases} 1 & \text{if } D_2 \text{ is employed} \\ 0 & \text{if not} \end{cases} \qquad x_3 = \begin{cases} 1 & \text{if } N_2 \text{ is employed} \\ 0 & \text{if not} \end{cases}$$

This model implies that the profit curves are identically shaped for each of the (D_i, N_j) combinations but that they possess different y-intercepts, as shown in Figure 7.24(b).

FIGURE 7.24
Profit Curves for
Stages 1 and 2

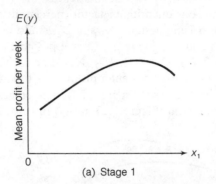

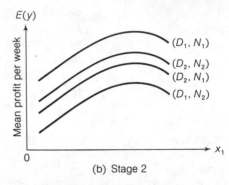

STAGE 3 *Add terms to allow for interaction between the quantitative and qualitative factors* This is done by interacting every pair of terms—one quantitative and one qualitative. Thus, the complete model, which graphs as four different-shaped second-order curves (see Figure 7.23), is

$$E(y) = \overbrace{\beta_0 + \beta_1 x_1 + \beta_2 x_1^2}^{\text{First-stage terms}}$$

$$+ \overbrace{\beta_3 x_2 + \beta_4 x_3 + \beta_5 x_2 x_3}^{\text{Second-stage terms}}$$

$$+ \overbrace{\beta_6 x_1 x_2 + \beta_7 x_1 x_3 + \beta_8 x_1 x_2 x_3 + \beta_9 x_1^2 x_2 + \beta_{10} x_1^2 x_3 + \beta_{11} x_1^2 x_2 x_3}^{\text{Third-stage terms}} \; \blacksquare$$

EXAMPLE 7.14

Use the model of Example 7.13 to find the equation relating $E(y)$ to x_1 for design D_1 and newspaper N_2.

SOLUTION

Checking the coding for the model, we see (noted at the second stage) that when a profit observation y is made during a week when design D_1 and newspaper N_2 were used, we set $x_2 = 0$ and $x_3 = 1$. Substituting these values into the complete model, we obtain

$$\begin{aligned} E(y) &= \beta_0 + \beta_1 x_1 + \beta_2 x_1^2 \\ &\quad + \beta_3 x_2 + \beta_4 x_3 + \beta_5 x_2 x_3 \\ &\quad + \beta_6 x_1 x_2 + \beta_7 x_1 x_3 + \beta_8 x_1 x_2 x_3 + \beta_9 x_1^2 x_2 + \beta_{10} x_1^2 x_3 + \beta_{11} x_1^2 x_2 x_3 \\ &= \beta_0 + \beta_1 x_1 + \beta_2 x_1^2 \\ &\quad + \beta_3(0) + \beta_4(1) + \beta_5(0)(1) \end{aligned}$$

$$+ \beta_6 x_1(0) + \beta_7 x_1(1) + \beta_8 x_1(0)(1)$$
$$+ \beta_9 x_1^2(0) + \beta_{10} x_1^2(1) + \beta_{11} x_1^2(0)(1)$$
$$= (\beta_0 + \beta_4) + (\beta_1 + \beta_7)x_1 + (\beta_2 + \beta_{10})x_1^2$$

Note that this equation graphs as a portion of a parabola with y-intercept equal to $(\beta_0 + \beta_4)$. The coefficient of x_1 is $(\beta_1 + \beta_7)$, and the curvature coefficient (the coefficient of x_1^2) is $(\beta_2 + \beta_{10})$.   ∎

EXAMPLE 7.15

Suppose you have two qualitative independent variables, A and B, and A is at two levels and B is at three levels. You also have two quantitative independent variables, C and D, each at three levels. Further, suppose you plan to fit a second-order response surface as a function of the quantitative independent variables C and D, and that you want your model for $E(y)$ to allow for different shapes of the second-order surfaces for the six (2×3) combinations of levels corresponding to the qualitative independent variables A and B. Write a model for $E(y)$.

SOLUTION

STAGE 1   *Write the second-order model corresponding to the two quantitative independent variables*   If we let

x_1 = Level for independent variable C

x_2 = Level for independent variable D

then

$$E(y) = \beta_0 + \beta_1 x_1 + \beta_2 x_2 + \beta_3 x_1 x_2 + \beta_4 x_1^2 + \beta_5 x_2^2$$

This is the model you would use if you believed that the six response surfaces, corresponding to the six combinations of levels of A and B, were identical.

STAGE 2   *Add the main effect and interaction terms for the qualitative independent variables*   These are

Main effect term for A		Main effect terms for B		AB interaction terms

$$\underbrace{\beta_6 x_6}_{\text{Main effect term for } A} + \underbrace{\beta_7 x_7 + \beta_8 x_8}_{\text{Main effect terms for } B} + \underbrace{\beta_9 x_6 x_7 + \beta_{10} x_6 x_8}_{\text{AB interaction terms}}$$

$$x_6 = \begin{cases} 1 & \text{if at level } A_2 \\ 0 & \text{if not} \end{cases} \qquad x_7 = \begin{cases} 1 & \text{if at level } B_2 \\ 0 & \text{if not} \end{cases} \qquad x_8 = \begin{cases} 1 & \text{if at level } B_3 \\ 0 & \text{if not} \end{cases}$$

The addition of these terms to the model produces six identically *shaped* second-order surfaces, one corresponding to each of the six combinations of levels of A and B. They differ only in their y-intercepts.

STAGE 3   *Add terms that allow for interaction between the quantitative and qualitative independent variables*   This is done by interacting each of the five qualitative independent variable terms (both main effect and interaction) with each term (except β_0) of the quantitative first-stage model. Thus,

$$E(y) = \beta_0 + \beta_1 x_1 + \beta_2 x_2 + \beta_3 x_1 x_2 + \beta_4 x_1^2 + \beta_5 x_2^2$$

First-stage model

$$\overbrace{+\ \beta_6 x_6}^{\text{Main effect } A} + \overbrace{\beta_7 x_7 + \beta_8 x_8}^{\text{Main effect } B} + \overbrace{\beta_9 x_6 x_7 + \beta_{10} x_6 x_8}^{AB \text{ interaction}}$$

Portion added to form second-stage model

$$+\ \beta_{11} x_6 x_1 + \beta_{12} x_6 x_2 + \beta_{13} x_6 x_1 x_2 + \beta_{14} x_6 x_1^2 + \beta_{15} x_6 x_2^2$$

Interacting x_6 with the quantitative terms

$$+\ \beta_{16} x_7 x_1 + \beta_{17} x_7 x_2 + \beta_{18} x_7 x_1 x_2 + \beta_{19} x_7 x_1^2 + \beta_{20} x_7 x_2^2$$

Interacting x_7 with the quantitative terms

$$+\ \cdots$$

$\cdots$

$$+\ \beta_{31} x_6 x_8 x_1 + \beta_{32} x_6 x_8 x_2 + \beta_{33} x_6 x_8 x_1 x_2 + \beta_{34} x_6 x_8 x_1^2 + \beta_{35} x_6 x_8 x_2^2$$

Interacting $x_6 x_8$ with the quantitative terms

Note that the complete model contains 36 terms, one for β_0, five needed to complete the second-order model in the two quantitative variables, five for the two qualitative variables, and $5 \times 5 = 25$ terms for the interactions between the quantitative and qualitative variables. ∎

To see how the model gives different second-order surfaces—one for each combination of the levels of variables A and B—consider the next example.

EXAMPLE 7.16

Refer to Example 7.15. Find the response surface that portrays $E(y)$ as a function of the two quantitative independent variables C and D for the (A_1, B_2) combination of levels of the qualitative independent variables.

SOLUTION

Checking the coding, we see that when y is observed at the first level of A (level A_1) and the second level of B (level B_2), the dummy variables take the following values: $x_6 = 0$, $x_7 = 1$, $x_8 = 0$. Substituting these values into the formula for the complete model (and deleting the terms that equal 0), we obtain

$$\begin{aligned} E(y) &= \beta_0 + \beta_1 x_1 + \beta_2 x_2 + \beta_3 x_1 x_2 + \beta_4 x_1^2 + \beta_5 x_2^2 + \beta_7 + \beta_{16} x_1 + \beta_{17} x_2 + \beta_{18} x_1 x_2 + \beta_{19} x_1^2 + \beta_{20} x_2^2 \\ &= (\beta_0 + \beta_7) + (\beta_1 + \beta_{16}) x_1 + (\beta_2 + \beta_{17}) x_2 + (\beta_3 + \beta_{18}) x_1 x_2 + (\beta_4 + \beta_{19}) x_1^2 + (\beta_5 + \beta_{20}) x_2^2 \end{aligned}$$

Note that this is the equation of a second-order model for $E(y)$. It graphs the response surface for $E(y)$ when the qualitative independent variables A and B are at levels A_1 and B_2. ∎

EXERCISES 7.18–7.24

7.18 Suppose the model described in Example 7.13 were fit to a set of weekly profit data and produced the following prediction equation:

$$\hat{y} = 650.0 + 79.0 x_1 - 1.80 x_1^2 + 370.0 x_2 - 170.0 x_3 - 550.0 x_2 x_3$$

$$- 15.6x_1x_2 - 5.6x_1x_3 + 10.3x_1x_2x_3 + .3x_1^2x_2 - .1x_1^2x_3 - .1x_1^2x_2x_3$$

Assume the percent discount levels varied from 5% to 30%. Graph $\hat{y}$ for each of the four combinations of levels of design, D, and newspaper, N.

7.19 Research conducted at Ohio State University focused on the factors that influence the allocation of black and white men in labor market positions (*American Sociological Review*, June 1986). Data collected for each of 837 labor market positions were used to build a regression model for y, defined as the natural logarithm of the ratio of the proportion of blacks employed in a labor market position to the corresponding proportion of whites employed (called the *black–white log odds ratio*). Positive values of y indicate that blacks have a greater likelihood of employment than whites. Several independent variables were considered, including the following:

x_1 = Market power (a quantitative measure of the size and visibility of firms in the labor market)

x_2 = Percentage of workers in the labor market who are union members

$x_3 = \begin{cases} 1 & \text{if labor market position includes craft occupations} \\ 0 & \text{if not} \end{cases}$

a. Write the first-order main effects model for $E(y)$ as a function of x_1, x_2, and x_3.

b. One theory hypothesized by the researchers is that the mean log odds ratio $E(y)$ is smaller for craft occupations than for noncraft occupations. (That is, the likelihood of black employment is less for craft occupations.) Explain how to test this hypothesis using the model in part a.

c. Write the complete second-order model for $E(y)$ as a function of x_1, x_2, and x_3.

d. Using the model in part c, explain how to test the hypothesis that level of market power x_1 has no effect on black–white log odds ratio y.

e. Consider the model

$$E(y) = \beta_0 + \beta_1x_1 + \beta_2x_2 + \beta_3x_3 + \beta_4x_1x_3 + \beta_5x_2x_3$$

Holding the percentage of union members x_2 fixed, sketch the proposed contour lines for the relationship between log odds ratio y and market power x_1.

7.20 Researchers for a dog food company have developed a new puppy food that they hope will compete with the major brands. One premarketing test involved the comparison of the new food with that of two competitors in terms of weight gain. Fifteen 8-week-old German shepherd puppies, each from a different litter, were divided into three groups of five puppies each. Each group was fed one of the three brands of food.

a. Set up a model that assumes the initial weight, x_1, is linearly related to final weight, y, but does not allow for differences among the three brands; i.e., assume the response curve is the same for the three brands of dog food. Sketch the response curve as it might appear.

b. Set up a model that assumes the effect of initial weight is linearly related to final weight, and allows the intercept of the line to differ for the three brands. In other words, assume the initial weight and brand both affect final weight, but in an independent fashion. Sketch typical response curves.

c. Now write the main effects plus interaction model. For this model we assume the initial weight is linearly related to final weight, but both the slope and the intercept of the line depend upon the brand. Sketch typical response curves.

7.21 A company is studying three different safety programs in an attempt to reduce the number of work-hours lost due to accidents. Each program is to be tried at three of the company's nine factories, and the plan is to monitor the lost work-hours, y, for a 1-year period beginning 6 months after the new safety program is instituted.
  a. Write a main effects model relating $E(y)$ to the lost work-hours, x_1, the year before the plan is instituted and to the type of safety program that is instituted.
  b. In terms of the model parameters from part **a**, what hypothesis would you test to determine whether the mean work-hours lost differs for the three safety programs?

7.22 An equal rights group has charged that women are being discriminated against in terms of the salary structure in a state university system. It is thought that a complete second-order model will be adequate to describe the relationship between salary and years of experience for both groups. A sample is to be taken from the records for faculty members (all of equal status) within the system and the following model is to be fit.*

$$E(y) = \beta_0 + \beta_1 x_1 + \beta_2 x_2 + \beta_3 x_1 x_2 + \beta_4 x_2^2$$

where

  y = Annual salary (in thousands of dollars)

  $x_1 = \begin{cases} 1 & \text{if female} \\ 0 & \text{if male} \end{cases}$

  x_2 = Experience (years)

  a. What hypothesis would you test to determine whether the *rate* of increase of mean salary with experience is different for males and females?
  b. What hypothesis would you test to determine whether there are differences in mean salaries that are attributable to sex?

7.23 Refer to Example 7.15 and find the equation relating $E(y)$ to x_1 and x_2 when A is at level A_2 and B is at level B_3.

7.24 During the 1970's, several industries, including the airline, trucking, natural gas, and cable television industries, were deregulated as a result of pressure from special interest groups and legislators. In the case of the airline industry, federal regulations on fares, schedules, and routes were thought to protect the airlines from price competition and generally prohibit new entry into the market. Thus, in theory, deregulation was expected to benefit consumers while having a negative impact on the airlines and their shareholders.

To test this theory, Davidson, Chandy, and Walker analyzed the impact of the airline Deregulation Act of 1978 on the security returns in the airline industry. Specifically, they examined the daily rates of returns of a sample of airline common stocks both prior to and following deregulation.

Thirty-two airlines engaged in air transportation for at least 1 year and listed on either the New York or the American Stock Exchange were selected for analysis. For each airline, daily stock returns were recorded for each of 120 days prior to deregulation and 120 days after deregulation. Data for the total of $n = 240$ observations (days) were then used to fit the model

*In practice, we would include other variables in the model. We include only two here to simplify the exercise.

$$E(y) = \beta_0 + \beta_1 x_1 + \beta_2 x_2 + \beta_3 x_1 x_2$$

where

y = Daily rate of return on the airline stock

x_1 = Average daily rate of return on the market

$x_2 = \begin{cases} 1 & \text{if after deregulation} \\ 0 & \text{if prior to deregulation} \end{cases}$

Thus, 32 regression analyses were conducted, one for each airline in the sample.

STOCK	$\hat{\beta}_0$	$\hat{\beta}_1$	$\hat{\beta}_2$	$\hat{\beta}_3$
AMR Corporation	.0005	2.8852*	−.0031	−.4106
Airborne Freight	.0024	.9748*	−.0012	−.6367*
Braniff International	.0007	1.4487*	−.0022	.5717
Canadian Pacific	.0011	.7687*	.0003	.3906
Continental Airlines	−.0021	2.1702*	.0004	−.7175*
Delta Airlines	−.0002	1.8557*	−.0005	−.6667*
Eastern Airlines	.0037	1.6353*	−.0069*	.9261*
Frontier	.0006	2.2188*	−.0033	−1.0355*
Greyhound	.0000	.3830*	.0020	.2163
Royal Dutch Airlines	.0004	1.3281	−.0024	−.5674
Lockheed Corporation	.0042	1.9337*	−.0053*	.6399
Northwest Airlines	−.0009	2.6056*	−.0005	−.1698
Ozark Airlines	.0032	1.1591*	−.0017	−.2921
PSA Airlines	.0033	.2555*	−.0036	1 4888*
Pan American	.0012	1.2136*	−.0025	−.0146
Piedmont	.0012	1.2138*	−.0025	−.0146
Puralotor, Inc.	.0009	.6829*	.0002	−.3787
Republic Airlines	.0023	.9524*	−.0031	1.3359*
Southwest Airlines	.0030	.8600*	.0016	−.2357
Tiger International	.0003	2.4979*	−.0009	−.7462*
Trans World Corp.	.0007	2.6399*	.0000	.1699
United Airlines	.0023	1.8137*	−.0039	.1444
US Air, Inc.	.0041	.8628	−.0029	1.4039*
W.A.F., Inc.	.0024	1.9271*	.0035	1.7067*
Western Airlines	.0012	1.5750*	−.0014	−.3987
World Airways	−.0014	3.2765*	.0015	−1.4218*

*Asterisk identifies β coefficients significant at $\alpha = .10$.
Source: Davidson, W. N., Chandy, P. R., and Walker, M. "The stock market effects of airline deregulation," *Quarterly Journal of Business and Economics,* Autumn 1984, Vol. 23, No. 4, pp. 31–45.

a. Write the equation of the line relating daily stock return y to average daily market return x_1 prior to deregulation (i.e., when $x_2 = 0$). Identify the y-intercept and slope of the line. [The slope is used to measure the stock's systematic risk and is often called the β **risk index** or β **value** for the stock. When the β value is greater than 1, the stock is classified as an *aggressive* or *risky* security since its daily rate of return is expected to move (upward or downward) faster than the market rate of return. In contrast, when the β value is less than 1, the stock is classified as a *defensive* or *stable* security since its daily rate of return moves slower than the market. A stock with a β value near 1 is called a *neutral* security, for its daily rate of return mirrors the market.]

b. Write the equation of the line relating daily stock return y to average daily market return x_1 after deregulation (i.e., when $x_2 = 1$). Identify the y-intercept and slope of the line.

c. What hypothesis would you test to determine whether the model is adequate for predicting y?

d. What hypothesis would you test to determine whether deregulation had an effect on the measure of systematic risk (i.e., β value) associated with the airline stock? [*Hint:* Use your answers to parts **a** and **b**.]

e. Of the 32 airline stocks analyzed, 26 had significant regressions (i.e., we could reject H_0: $\beta_1 = \beta_2 = \beta_3 = 0$) at $\alpha = .10$. The least squares prediction equations for these stocks are given in the table. Identify those stocks that have significant (at $\alpha = .10$) interaction between average market rate of return x_1 and deregulation x_2. For each of these stocks, how has deregulation affected the stock's β value?

SECTION 7.11

MODEL-BUILDING: AN EXAMPLE

We illustrate the modeling techniques outlined in this chapter with an example based, in part, on an actual trucking deregulation study. Consider the problem of modeling the price charged for motor transport service (such as trucking) in a particular state. In the early 1980's, several states removed regulatory constraints on the rate charged for intrastate trucking services. (Florida was the first state to embark on a deregulation policy, on July 1, 1980.) One of the goals of the regression analysis is to assess the impact of state deregulation on the supply price y charged per ton-mile.

Suppose that after a careful variable-screening process (for example, stepwise regression), the following independent variables were selected as the best predictors of supply price:

1. Distance shipped (hundreds of miles)
2. Weight of product shipped (thousands of pounds)
3. Deregulation in effect (yes or no)
4. Size of market destination (large or small)

Distance shipped and weight of product are quantitative variables since they each assume numerical values (miles and pounds, respectively) corresponding to the points on a line. Deregulation and market size are qualitative, or categorical, variables that we must describe with dummy (or coded) variables. The variable assignments are given as follows:

x_1 = Distance shipped

x_2 = Weight of product

$x_3 = \begin{cases} 1 & \text{if deregulation in effect} \\ 0 & \text{if not} \end{cases}$

$x_4 = \begin{cases} 1 & \text{if large market} \\ 0 & \text{if small market} \end{cases}$

Note that in defining the dummy variables, we have arbitrarily chosen "no" and "small" to be the base levels of deregulation and market size, respectively.

We will begin the model-building process by specifying the three models shown below. Notice that the first model specified is the complete second-order model. Recall from Section 7.10 that the complete second-order model contains quadratic (curvature) terms for quantitative variables and interactions among the quantitative and qualitative terms. For this example, the complete second-order model (Model 1) traces a parabolic surface for mean price $E(y)$ as a function of distance (x_1) and weight (x_2), and the response surfaces differ for the $2 \times 2 = 4$ combinations of the levels of deregulation and market size. Generally, the complete second-order model is a good place to start the model-building process since most real-world relationships are curvilinear. (Keep in mind, however, that you must have a sufficient number of data points to find estimates of all the parameters in the model.)

MODEL 1 $E(y) = \beta_0 + \beta_1 x_1 + \beta_2 x_2 + \beta_3 x_1 x_2 + \beta_4 x_1^2 + \beta_5 x_2^2$} Stage 1 terms

$\qquad\qquad + \beta_6 x_3 + \beta_7 x_4 + \beta_8 x_3 x_4$ } Stage 2 terms

$\qquad\qquad + \beta_9 x_1 x_3 + \beta_{10} x_1 x_4 + \beta_{11} x_1 x_3 x_4$

$\qquad\qquad + \beta_{12} x_2 x_3 + \beta_{13} x_2 x_4 + \beta_{14} x_2 x_3 x_4$

$\qquad\qquad + \beta_{15} x_1 x_2 x_3 + \beta_{16} x_1 x_2 x_4 + \beta_{17} x_1 x_2 x_3 x_4$ Stage 3 terms

$\qquad\qquad + \beta_{18} x_1^2 x_3 + \beta_{19} x_1^2 x_4 + \beta_{20} x_1^2 x_3 x_4$

$\qquad\qquad + \beta_{21} x_2^2 x_3 + \beta_{22} x_2^2 x_4 + \beta_{23} x_2^2 x_3 x_4$

MODEL 2 $E(y) = \beta_0 + \beta_1 x_1 + \beta_2 x_2 + \beta_3 x_1 x_2$ } Stage 1 terms

$\qquad\qquad + \beta_4 x_3 + \beta_5 x_4 + \beta_6 x_3 x_4$ } Stage 2 terms

$\qquad\qquad + \beta_7 x_1 x_3 + \beta_8 x_1 x_4 + \beta_9 x_1 x_3 x_4$

$\qquad\qquad + \beta_{10} x_2 x_3 + \beta_{11} x_2 x_4 + \beta_{12} x_2 x_3 x_4$ Stage 3 terms

$\qquad\qquad + \beta_{13} x_1 x_2 x_3 + \beta_{14} x_1 x_2 x_4 + \beta_{15} x_1 x_2 x_3 x_4$

Model 2 contains all the terms of Model 1 (the complete second-order model), except that the quadratic terms (terms involving x_1^2 and x_2^2) are dropped. This model also proposes four different response surfaces for the combinations of levels of deregulation and market size, but the surfaces are twisted planes (see Figure 7.9) rather than paraboloids. A direct comparison of Models 1 and 2 will allow us to test for the importance of the curvature terms.

MODEL 3 $E(y) = \beta_0 + \beta_1 x_1 + \beta_2 x_2 + \beta_3 x_1 x_2 + \beta_4 x_3 + \beta_5 x_4 + \beta_6 x_3 x_4$

Model 3 is a reduction of Model 2, with the quantitative–qualitative interaction terms omitted. This model assumes that the four response surfaces corresponding to the four deregulation–market size combinations, which are twisted planes, differ only with respect to the y-intercept in the space.

Data collected for $n = 132$ shipments were used to fit the models. The results are summarized in Table 7.7 (page 378).

TABLE 7.7
Trucking Deregulation
Regression Results

MODEL	SSE	R^2	df(ERROR)
1	203,570	.83	108
2	227,520	.81	116
3	395,165	.67	125

To determine whether the quadratic terms for distance (x_1) and weight (x_2), i.e., those involving x_1^2 and x_2^2, contribute information for predicting the mean supply price $E(y)$, we compare Models 1 and 2 using the partial F-test given in Section 4.11. If the quadratic terms are unimportant, then the β coefficients involving the x_1^2 and x_2^2 terms in Model 1 (i.e., $\beta_4, \beta_5, \beta_{18}, \beta_{19}, \ldots, \beta_{23}$) will all equal 0. Thus, we test

H_0: $\beta_4 = \beta_5 = \beta_{18} = \beta_{19} = \beta_{20} = \beta_{21} = \beta_{22} = \beta_{23} = 0$
H_a: At least one of the quadratic β's differs from 0

where the complete model is Model 1 and the reduced model is Model 2. The test statistic is

$$F = \frac{(SSE_2 - SSE_1)/\text{Number of } \beta \text{ parameters being tested}}{SSE_1/df(Error) \text{ for Model 1}}$$

$$= \frac{(227,520 - 203,570)/8}{203,570/108}$$

$$= \frac{2,993.75}{1,884.91} = 1.59$$

For $\alpha = .05$, $\nu_1 = $ number of β's tested (8), and $\nu_2 = df(Error)$ for Model 1 (108), the critical value (from Table 4 of Appendix D) is approximately $F_{.05} = 2.66$.

Since the calculated value $F = 1.59$ does not exceed this critical value, we have insufficient evidence (at $\alpha = .05$) to conclude that the curvature terms are important predictors of supply price y. We could collect more data and retest to determine whether x_1^2 and x_2^2 contribute information for the prediction of supply price. Lacking this additional information, we will simplify our model by dropping the terms involving x_1^2 and x_2^2.*

Can we simplify the model even further by dropping the terms involving quantitative–qualitative interactions? That is, do the response surfaces for the four combinations of deregulation and market size differ only with respect to the

*There is always a danger in dropping terms from the model. Essentially, we are accepting H_0: $\beta_4 = \beta_5 = \beta_{18} = \cdots = \beta_{23} = 0$ when $P(\text{Type II error}) = P(\text{accepting } H_0 \text{ when } H_0 \text{ is false}) = \beta$ is unknown. In practice, however, many researchers are willing to risk making a Type II error rather than use a more complex model for $E(y)$ when simpler models which are nearly as good as predictors (and which are easier to apply and interpret) are available. Note that we employed a relatively large amount of data $(n = 132)$ in fitting our models and that R^2 for Model 2 is only 2% less than R^2 for Model 1. If the quadratic terms are, in fact, important (i.e., we have made a Type II error), there is little lost in terms of explained variability in using Model 2.

y-intercept? If so, then the β coefficients involving the quantitative–qualitative interaction terms in Model 2 ($\beta_7, \beta_8, \ldots, \beta_{15}$) will all equal 0.

We compare Model 2 (now called the *complete* model) and Model 3 (the *reduced* model) with another partial F-test:

H_0: $\beta_7 = \beta_8 = \cdots = \beta_{15} = 0$

H_a: At least one of the quantitative–qualitative interaction β parameters differs from 0

Test statistic: $F = \dfrac{(SSE_3 - SSE_2)/\text{Number of } \beta \text{ parameters being tested}}{SSE_2/df(\text{Error})}$

$$= \frac{(395{,}165 - 227{,}520)/9}{227{,}520/116}$$

$$= \frac{18{,}627.22}{1{,}961.38} = 9.50$$

Rejection region: $F > F_{.05} = 1.96$ (from Table 4 of Appendix D) where $\nu_1 =$ number of β's tested (9) and $\nu_2 = df(\text{Error})$ for Model 2 (116).

Since the calculated value $F = 9.50$ exceeds the critical value $F_{.05} = 1.96$, there is sufficient evidence (at $\alpha = .05$) to indicate that the quantitative–qualitative interaction terms are important. We should retain these terms in the model.

The results of the tests above suggest that of the three models, Model 2 is the best for modeling the mean supply price $E(y)$. Further testing may lead to a simpler model, however. For example, we might want to test the importance of the quantitative–qualitative interaction ($x_1 x_2$) terms. If a partial F-test reveals no evidence of interaction between x_1 and x_2, we may wish to drop terms involving $x_1 x_2$ from the model.

A note of caution   Just as with t-tests on individual β parameters, you should avoid conducting too many partial F-tests. Regardless of the type of test (t-test or F-test), the more tests that are performed, the higher the overall Type I error rate will be. In practice, you should limit the number of models that you propose for $E(y)$ so that the overall Type I error rate α for conducting partial F-tests remains reasonably small.*

S E C T I O N 7.12

SUMMARY

Although this chapter on **model-building** covered many topics, only experience can make you competent in this fascinating area of statistics. Successful model-building requires a delicate blend of knowledge of the process being modeled, geometry, and formal statistical testing.

The first step is to **identify the response variable y and the set of independent variables**. Each independent variable is then classified as either **quantitative** or

---

*A technique, suggested by Bonferroni, is often applied to maintain control of the overall Type I error rate α. If p tests are to be performed, then conduct each individual test at significance level α/p. This will guarantee an overall Type I error rate less than or equal to α. For example, conducting each of $p = 5$ tests at the $.05/5 = .01$ level of significance guarantees an overall $\alpha \le .05$.

qualitative, and **dummy variables** are defined to represent the qualitative independent variables.

When the number of independent variables is manageable, the model-builder is ready to begin a systematic effort. At least **second-order models**, those containing **two-way interactions and quadratic terms** in the quantitative variables, should be considered. Remember that a model with no interaction terms implies that each of the independent variables affects the response independently of the other independent variables. **Quadratic terms add curvature** to the contour curves when $E(y)$ is plotted as a function of the independent variable. With higher-order models, you may decide to **code** the quantitative independent variables to reduce both rounding error and the built-in multicollinearity problem.

Many problems can arise in regression modeling, and the intermediate steps are often tedious and frustrating. However, the end result of a careful and determined modeling effort is very rewarding—you will have a better understanding of the process and will have a predictive model for the dependent variable y.

SUPPLEMENTARY EXERCISES 7.25–7.37

7.25 An appliance store is interested in modeling the total sales of television sets sold per month as a function of the following independent variables: (1) warranty period, (2) color or black-and-white, (3) solid-state or tube, (4) picture tube size, (5) brand (the store carries three brands).

a. Determine whether each of the independent variables is quantitative or qualitative. If it is quantitative, give the approximate range of levels you might expect to observe. If it is qualitative, give the number of levels and define dummy variables to describe each variable.

b. Write a main effects model to relate $E(y)$, the mean monthly sales for televisions sold, as a function of the five independent variables.

c. It is believed that the mean sales will be greater for larger picture tubes than for smaller picture tubes, but the rate of increase will depend on the brand. Why does the main effects model in part **b** not incorporate this belief? How can it be incorporated into the model?

7.26 To model the relationship between y, a dependent variable, and x, an independent variable, a researcher has taken one measurement on y at each of five different x-values. Drawing on his mathematical expertise, the researcher realizes that he can fit the fourth-order polynomial model

$$E(y) = \beta_0 + \beta_1 x + \beta_2 x^2 + \beta_3 x^3 + \beta_4 x^4$$

and it will pass exactly through all five points, yielding SSE = 0. The researcher, delighted with the "excellent" fit of the model, eagerly sets out to use it to make inferences. What problems will he encounter in attempting to make inferences?

7.27 Economic research has established evidence of a positive correlation between earnings and educational attainment (Miller and Volker, 1984). However, it is unclear whether higher wage rates for better educated workers reflect an individual's added value or merely the employer's use of higher education as a screening device in the recruiting process. One version of this "sheepskin screening" hypothesis supported by many economists is that wages will rise faster with extra years of education when the extra years culminate in a certificate (e.g., master's or Ph.D. degree, CPA certificate, or actuarial degree).

a. Write a first-order, main effects model for mean wage rate $E(y)$ of an employer as a function of employee's years of education and whether or not the employee is certified.

b. Write a first-order model for $E(y)$ that corresponds to the "sheepskin screening" hypothesis stated above.

c. Write the complete second-order model for $E(y)$ as a function of the two independent variables.

7.28 To make a product more appealing to the consumer, an automobile manufacturer is experimenting with a new type of paint that is supposed to help the car maintain its new-car look. The durability of this paint depends on the length of time the car body is in the oven after it has been painted. In the initial experiment, three groups of ten car bodies each were baked for three different lengths of time—12, 24, and 36 hours—at the standard temperature setting. Then, the paint finish of each of the 30 cars was analyzed to determine a durability rating, y.

a. Write a quadratic model relating the mean durability, $E(y)$, to the length of baking.

b. Could a cubic model be fit to the data? Explain.

7.29 Refer to Exercise 7.28. Suppose the Research and Development Department develops three new types of paint to be tested. Thus, 90 cars are to be tested—30 for each type of paint—in the manner described in Exercise 7.28. Write the complete second-order model for $E(y)$ as a function of the type of paint and bake time.

7.30 Use the coding system for observational data to fit a second-order model to the data on demand y and price p given in Example 5.2 (the data are reproduced in the accompanying table). Show that the inherent multicollinearity problem with fitting a polynomial model is reduced when the coded values of p are used.

DEMAND y, pounds	1,120	999	932	884	807	760	701	688
PRICE p, dollars	3.00	3.10	3.20	3.30	3.40	3.50	3.60	3.70

7.31 One factor that must be considered in developing a shipping system that is beneficial to both the customer and the seller is time of delivery. A manufacturer of farm equipment can ship its products by either rail or truck. Quadratic models are thought to be adequate in relating time of delivery to distance to be shipped for both modes of transportation. Consequently, it has been suggested that the following model be fit to begin the model-building process

$$E(y) = \beta_0 + \beta_1 x_1 + \beta_2 x_1^2 + \beta_3 x_2 + \beta_4 x_1 x_2 + \beta_5 x_1^2 x_2$$

where

y = Time of delivery

x_1 = Distance to be shipped

$x_2 = \begin{cases} 1 & \text{if rail} \\ 0 & \text{if truck} \end{cases}$

    a. Sketch the proposed relationships between delivery time y and distance x_1 for both modes of transportation.

    b. What hypothesis would you test to determine whether the data indicate that the quadratic distance terms are useful in the model, i.e., whether curvature is present in the relationship between mean delivery time and distance?

    c. What hypothesis would you test to determine whether there is a difference in mean delivery time by rail and by truck?

7.32 Refer to Exercise 7.31. Suppose the complete second-order model is fit to a total of 50 observations on time of delivery. The sum of squared errors is SSE = 226.12. Then, the reduced model

$$E(y) = \beta_0 + \beta_1 x_1 + \beta_2 x_1^2$$

is fit to the same data, and SSE = 279.34. Test to determine whether the data indicate that the mean delivery time differs for rail and truck deliveries.

7.33 A company wants to model the total weekly sales, y, of its product as a function of the variables packaging and location. Two types of packaging, P_1 and P_2, are used in each of four locations, L_1, L_2, L_3, and L_4.

    a. Write a main effects model to relate $E(y)$ to packaging and location. What implicit assumption are we making about the interrelationships between sales, packaging, and location when we use this model?

    b. Now write a model for $E(y)$ that includes interaction between packaging and location. How many parameters are in this model (remember to include β_0)? Compare this number to the number of packaging–location combinations being modeled.

7.34 Refer to Exercise 7.33. Suppose the main effects and interaction models are fit for 40 observations on weekly sales. The values of SSE are:

SSE for main effects model = 422.36

SSE for interaction model = 346.65

    a. Determine whether the data indicate that the interaction between location and packaging is important in estimating mean weekly sales. Use $\alpha = .05$.

    b. What implications does your conclusion in part **a** have for the company's marketing strategy?

7.35 Many companies must accurately estimate their costs before a job is begun in order to acquire a contract and make a profit. For example, a heating and plumbing contractor may base cost estimates for new homes on the total area of the house, the number of baths in the plans, and whether central air conditioning is to be installed.

    a. Write a main effects model relating the mean cost of material and labor, $E(y)$, to the area, number of baths, and central air conditioning variables.

    b. Write a complete second-order model for the mean cost as a function of the same three variables.

    c. How would you test the hypothesis that the second-order and interaction terms are useful for predicting mean cost?

7.36 Refer to Exercise 7.35. The contractor samples 25 recent jobs and fits both the complete second-order model (part **b**) and the reduced main effects model (part **a**), so that a test can be conducted to determine whether the additional complexity of the second-order model is necessary. The resulting values of SSE and R^2 are shown in the table.

	SSE	R^2
Main effects	8.548	.950
Second-order	6.133	.964

a. Is there sufficient evidence to conclude that the second-order terms are important for predicting the mean cost?

b. Suppose the contractor decides to use the main effects model to predict costs. Use the global F-test (Section 4.6) to determine whether the main effects model is useful for predicting costs.

7.37 An experiment was conducted to investigate the effect of financial incentive on the sales effort of a company's sales representatives. Five salespeople were selected from each of three divisions, A, B, and C, of the company. One person from each division was offered (confidentially) a commission equal to 1% of sales. Similarly, for each division, one person was selected to receive 1.5%, one 2.0%, one 2.5%, and one 3.0% of sales. The sales, y (in hundreds of thousands of dollars), for the $n = 15$ sales representatives are shown in the table.

		\multicolumn{5}{c}{COMMISSION (%)}
		1	1.5	2.0	2.5	3.0
	A	2.0	3.0	4.0	3.8	3.5
DIVISION	B	1.6	2.8	3.0	3.2	2.8
	C	1.0	2.0	2.5	2.7	2.5

a. Assume that the response curves for each division are second-order and that they may differ from division to division. Write an appropriate model for the mean sales, $E(y)$.

b. Fit the model from part a to the data using coded values for the quantitative factor, percent commission, and dummy variables for the qualitative factor, division.

c. Do the data provide sufficient evidence to indicate that the three sales curves differ?

REFERENCES

Blomstrom, M. "Multinationals and market structure in Mexico," *World Development*, Vol. 14, No. 4, 1986, pp. 523–530.

Draper, N. and Smith, H. *Applied Regression Analysis*, 2nd ed. New York: Wiley, 1981.

Esposito, F. F. and Esposito, L. "Excess capacity and market structure in U. S. manufacturing: New evidence," *Quarterly Journal of Business & Economics*, Vol. 25, No. 3, Summer 1986, pp. 3–13.

Graybill, F. A. *Theory and Application of the Linear Model*. North Scituate, Mass.: Duxbury, 1976.

Kaufman, R. L. "The impact of industrial and occupational structure on black–white employment allocation," *American Sociological Review*, Vol. 51, No. 6, June 1986, pp. 310–322.

Klassen, R. A. "The application of response surface methods to a tablet formulation problem." Paper presented at Joint Statistical Meetings, American Statistical Association and Biometric Society, August 1986, Chicago, Ill.

Mendenhall, W. *Introduction to Linear Models and the Design and Analysis of Experiments*. Belmont, Ca.: Wadsworth, 1968.

Miller, P. W. and Volker, P. A. "The screening hypothesis: An application of the Wiles test," *Economic Inquiry*, Vol. 22, January 1984, pp. 121–127.

Neter, J., Wasserman, W., and Kutner, M. H. *Applied Linear Statistical Models*. Homewood, Ill.: Richard D. Irwin, 1985.

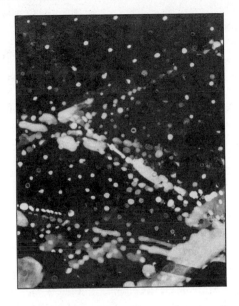

CHAPTER 8

OBJECTIVE

To present a method for analyzing data collected from multivariable designed experiments for comparing two or more population means; to identify its relation to regression analysis and their overlapping features

CONTENTS

THE ANALYSIS OF VARIANCE FOR DESIGNED EXPERIMENTS

SECTION 8.1

INTRODUCTION

As we have seen in the preceding chapters, the solutions to many business problems are based on inferences about population means. For example, a property appraiser interested in comparing the mean sale prices of homes in two adjacent neighborhoods might choose to apply the methods of Section 2.8. Or, the appraiser may decide to use regression analysis (Chapters 3–7) to estimate the mean sale price of a home. This chapter extends the methods of Chapters 2–7 to the comparison of more than two means.

When the data have been obtained according to certain specified sampling procedures, they are easy to analyze and also may contain more information pertinent to the population means than could be obtained using simple random sampling. The procedure for selecting sample data is called the **design of the experiment** and the statistical procedure for comparing the population means is called an **analysis of variance**. The objective of this chapter is to introduce some aspects of experimental design and the analysis of the data from such experiments using an analysis of variance. In optional Section 8.10, we illustrate the relationship between this method of analysis and multiple regression.

SECTION 8.2

EXPERIMENTAL DESIGN: TERMINOLOGY

The study of experimental design originated in England and, in its early years, was associated solely with agricultural experimentation. The need for experimental design in agriculture was very clear: It takes a full year to obtain a single observation on the yield of a new variety of wheat. Consequently, the need to save time and money led to a study of ways to obtain more information using smaller samples. Similar motivations led to its subsequent acceptance and wide use in all fields of scientific experimentation. Despite this fact, the terminology associated with experimental design clearly indicates its early association with the biological sciences.

We will call the process of collecting sample data an **experiment**. The planning of the sampling procedure is called the **design** of the experiment. The object upon which the response measurement y is taken is called an **experimental unit**.

DEFINITION 8.1

The process of collecting sample data is called an **experiment**.

DEFINITION 8.2

The plan for collecting the sample is called the **design** of the experiment.

DEFINTION 8.3

The object upon which the response y is measured is called an **experimental unit**.

Independent variables that may be related to a response variable y are called **factors**. The value—that is, the intensity setting—assumed by a factor in an experiment is called a **level**. For example, suppose a marketing study is conducted to investigate the effect on coffee sales of two factors, the brand and the shelf location. If coffee sales are recorded for each of two brands (brand A and brand B) and three shelf locations (bottom, middle, and top), then brand is at two levels and shelf location is at three levels. The combinations of levels of the factors for which y will be observed are called **treatments**. For example, if coffee sales are recorded for each of the six brand–shelf location combinations (brand A, bottom), (brand A, middle), (brand A, top), (brand B, bottom), (brand B, middle), and (brand B, top), then the experiment would involve six treatments. The term *treatments* is used to describe the factor-level combinations to be included in an experiment because many experiments involve "treating" or doing something to alter the nature of the experimental unit. Thus, we might view the six brand–shelf location combinations as treatments on the experimental units in the marketing study involving coffee sales.

DEFINTION 8.4

The independent variables that are related to a response variable y are called **factors**.

DEFINTION 8.5

The intensity setting of a factor is called a **level**.

DEFINTION 8.6

A **treatment** is a particular combination of levels of the factors involved in an experiment.

Now that you understand some of the terminology, it is helpful to think of the design of an experiment in four steps:

STEP 1 Select the factors to be included in the experiment and identify the parameters that are the object of the study. Usually, the target parameters are the population means associated with the factor-level combinations (i.e., treatments).

STEP 2 Choose the treatments (the factor-level combinations to be included in the experiment).

STEP 3 Determine the number of observations to be made for each treatment. (This will usually depend on the standard error(s) that you desire.)

STEP 4 Decide how the treatments will be assigned to the experimental units.

By following these steps, you can control the quantity of information in an experiment. We shall explain how this is done in Section 8.3.

SECTION 8.3

CONTROLLING THE INFORMATION IN AN EXPERIMENT

The problem of acquiring good experimental data is analogous to the problem faced by a communications engineer. The receipt of any signal, verbal or otherwise, depends on the volume of the signal and the amount of the background noise. The greater the volume of the signal, the greater will be the amount of information transmitted to the receiver. Conversely, the amount of information transmitted is reduced when the background noise is great. These intuitive thoughts about the factors that affect the information in an experiment are supported by the following fact: The standard errors of most estimators of the target parameters are proportional to σ (a measure of data variation or noise) and inversely proportional to the sample size (a measure of the volume of the signal). To illustrate, take the simple case where we wish to estimate a population mean μ by the sample mean $\bar{y}$. The standard error of the sampling distribution of $\bar{y}$ is

$$\sigma_{\bar{y}} = \frac{\sigma}{\sqrt{n}}$$

Note that, for a fixed sample size n, the smaller the value of σ, which measures the variability (noise) in the population of measurements, the smaller will be the standard error $\sigma_{\bar{y}}$. Similarly, by increasing the sample size n (the volume of the signal) in a given experiment, you decrease $\sigma_{\bar{y}}$.

The first three steps in the design of an experiment (see Section 8.2), the selection of the factors and treatments to be included in an experiment and the specification of the sample sizes, determine the volume of the signal. You must select the treatments so that the observed values of y provide information on the parameters of interest. Then the larger the treatment sample sizes, the greater will be the quantity of information in the experiment. Is it possible to observe y and obtain no information on a parameter of interest? The answer is "yes." To illustrate, suppose that you attempt to fit a first-order model

$$E(y) = \beta_0 + \beta_1 x$$

to a set of $n = 10$ data points, all of which were observed for a single value of x, say $x = 5$. The data points might appear as shown in Figure 8.1. Clearly, there is no possibility of fitting a line to these data points. The only way to obtain information on β_0 and β_1 is to observe y for *different* values of x. Consequently, the $n = 10$ data points in this example contain absolutely no information on the parameters β_0 and β_1.

FIGURE 8.1
Data Set with $n = 10$ Responses, All at $x = 5$

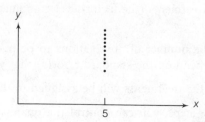

Step 4 in the design of an experiment provides an opportunity to reduce the noise (or experimental error) in an experiment. Known sources of data variation can be reduced or eliminated by **blocking**—that is, observing all treatments within relatively homogeneous **blocks** of experimental material. For example, suppose we want to compare the length of time required to assemble a digital watch using three different methods of assembly (treatments). One method for conducting the experiment would be to randomly assign five assemblers to each method of assembly and record the assembly time for each of the 15 assemblers included in the experiment. Using this method, we introduce an unwanted source of variation—namely, the portion contributed by the differences in dexterity of the 15 assemblers.

A second method of conducting the experiment allows us to block out the variation from assembler to assembler. We could use only five assemblers and require each to assemble three watches—one for each of the three methods of assembly. If a particular assembler is by nature slow or fast, then this characteristic will be reflected in the measurements obtained for all three assembly methods. In comparing the assembly times for two different assembly methods, we hope to cancel out the variation contributed by the assembler to each of the measurements.

SUMMARY OF STEPS IN EXPERIMENTAL DESIGN	
Volume-increasing:	1. Select the factors.
	2. Choose the treatments (factor–level combinations).
	3. Determine the sample size for each treatment.
Noise-reducing:	4. Assign the treatments to the experimental units.

In Sections 8.5–8.7 we consider three popular experimental designs and demonstrate how to execute the above steps for each. First, in Section 8.4 we present a short discussion of the logic behind the analysis of data collected from such experiments.

| | | | | | | | | | | | | |

SECTION 8.4

THE LOGIC BEHIND AN ANALYSIS OF VARIANCE

Once the data for a designed experiment have been collected, we will want to use the sample information to make inferences about the population means associated with the various treatments. The method used to compare the treatment means is known as **analysis of variance**, or ANOVA. The concept behind an analysis of variance can be explained using the following simple example.

Consider an experiment with a single factor at two levels (that is, two treatments). Suppose we want to decide whether the two treatment means differ based on the means of two independent random samples, each containing $n_1 = n_2 = 5$ measurements, and that the y-values appear as in Figure 8.2 (page 390). Note that the five circles on the left are plots of the y-values for sample 1 and the five

FIGURE 8.2

Plots of Data for
Two Samples

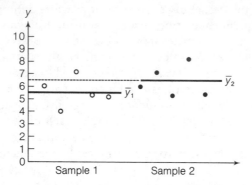

solid dots on the right are plots of the y-values for sample 2. Also, observe the horizontal lines that pass through the means for the two samples $\bar{y}_1$ and $\bar{y}_2$. Do you think the plots provide sufficient evidence to indicate a difference between the corresponding population means?

If you are uncertain whether the population means differ for the data in Figure 8.2, examine the situation for two different samples in Figure 8.3(a). We think that you will agree that for these data, it appears that the population means differ. Examine a third case in Figure 8.3(b). For these data, it appears that there is little or no difference between the population means.

FIGURE 8.3

Plots of Data for
Two Cases

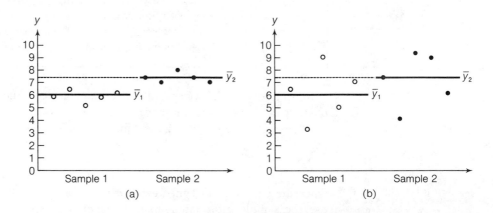

What elements of Figures 8.2 and 8.3 did we intuitively use to decide whether the data indicate a difference between the population means? The answer to the question is that we visually compared the distance (the variation) *between* the sample means to the variation *within* the y-values for each of the two samples. Since the difference between the sample means in Figure 8.3(a) is large relative to the within-sample variation, we inferred that the population means differ.

Conversely, in Figure 8.3(b), the variation between the sample means is small relative to the within-sample variation and therefore there is little evidence to infer that the means are significantly different.

The variation within samples is measured by the pooled s^2 that we computed for the independent random samples t-test of Section 2.8, namely

$$\textit{Within-sample variation:}\quad s^2 = \frac{\sum_{i=1}^{n_1}(y_{i1} - \bar{y}_1)^2 + \sum_{i=1}^{n_2}(y_{i2} - \bar{y}_2)^2}{n_1 + n_2 - 2}$$

$$= \frac{SSE}{n_1 + n_2 - 2}$$

where y_{i1} is the ith observation in sample 1 and y_{i2} is the ith observation in sample 2. The quantity in the numerator of s^2 is often denoted **SSE**, the **sum of squared errors**. As with regression analysis, SSE measures unexplained variability. But in this case, it measures variability *unexplained* by the differences between the sample means.

A measure of the between-sample variation is given by the weighted sum of squares of deviations of the individual sample means about the mean for all 10 observations, $\bar{y}$, divided by the number of samples minus 1, i.e.,

$$\textit{Between-sample variation:}\quad \frac{n_1(\bar{y}_1 - \bar{y})^2 + n_2(\bar{y}_2 - \bar{y})^2}{2 - 1} = \frac{SST}{1}$$

The quantity in the numerator is often denoted **SST**, the **sum of squares for treatments**, since it measures the variability *explained* by the differences between the sample means of the two treatments.

For this experimental design, SSE and SST sum to a known total, namely

$$SS(\text{Total}) = \sum (y_i - \bar{y})^2$$

Also, the ratio

$$F = \frac{\text{Between-sample variation}}{\text{Within-sample variation}}$$

$$= \frac{SST/1}{SSE/(n_1 + n_2 - 2)}$$

has an F-distribution with $\nu_1 = 1$ and $\nu_2 = n_1 + n_2 - 2$ degrees of freedom (df) and therefore can be used to test the null hypothesis of no difference between the treatment means. The additivity property of the sums of squares led early researchers to view this analysis as a **partitioning** of $SS(\text{Total}) = \Sigma(y_i - \bar{y})^2$ into sources corresponding to the factors included in the experiment and to SSE. The simple formulas for computing the sums of squares, the additivity property, and the form of the test statistic made it natural for this procedure to be called **analysis of variance**. We demonstrate the analysis of variance procedures for three special types of experimental designs in Sections 8.5, 8.6, and 8.7.

| | | | | | | | | | | | | |

SECTION 8.5

COMPLETELY RANDOMIZED DESIGNS

As noted in Section 8.2, the first two steps in designing an experiment are (1) deciding on the factors to be investigated and (2) selecting the factor-level combinations (treatments) to be included in the experiment. For example, suppose you wish to compare the length of time to assemble a device in a manufacturing operation for workers who have completed one of three training programs, A, B, and C. Then this experiment involves a single factor, training program, at three levels, A, B, and C. Since training program is the only factor, these levels (A, B, and C) represent the treatments. Now we must decide the sample size for each treatment (step 3) and how to assign the treatments to the experimental units, namely the specific workers (step 4).

The most common assignment of treatments to experimental units is called a **completely randomized design**. To illustrate, suppose we wish to obtain equal amounts of information on the mean assembly times for the three training procedures, i.e., we decide to assign equal numbers of workers to each of the three training programs. Also, suppose we use the procedure of Section 2.8 (Example 2.14) to select the sample size and we determine the number of workers in each of the three samples to be $n_1 = n_2 = n_3 = 10$. Then a completely randomized design is one in which the $n_1 + n_2 + n_3 = 30$ workers are **randomly assigned**, ten to each of the three treatments. *A random assignment is one in which any one assignment is as probable as any other*. This eliminates the possibility of bias that might occur if the workers were assigned in some systematic manner. For example, a systematic assignment might accidentally assign most of the manually dexterous workers to training program A, thus tending to underestimate the true mean assembly time corresponding to A.

Example 8.1 illustrates how a **random number table** can be used to assign the 30 workers to the three treatments.

EXAMPLE 8.1

Use the random number table, Table 7 in Appendix D, to assign $n = 30$ experimental units to three treatment groups.

SOLUTION

The first step is to number the 30 workers from 1 to 30. We will then use Table 7 in Appendix D to select two-digit numbers, discarding those that are larger than 30 or are identical, until we have a total of 20 two-digit numbers. We will then have 20 of the integers between 1 and 30 arranged in random order. The workers who have been assigned the first ten numbers in the sequence are assigned to training program A, the second group of ten workers are assigned to B, and the remaining workers are assigned to C.

To illustrate, suppose we start with the two-digit random number in row 5, column 6 of Table 7 and proceed down the column, selecting only two-digit numbers (the first two digits) less than or equal to 30 and deleting those that repeat. The first 20 are: 20, 18, 13, 16, 19, 04, 14, 06, 30, 25, 27, 17, 24, 21, 22, 02, 15, 05, 09, 08. The workers with the first ten numbers are assigned to program

A, the second ten to B, and the remaining ten to C. So the workers are assigned to the training program as follows:

A	B	C
20, 18, 13, 16, 19, 4, 14, 6, 30, 25	27, 17, 24, 21, 22, 2, 15, 5, 9, 8	1, 3, 7, 10, 11, 12, 23, 26, 28, 29

■

Example 8.2 will refresh your memory concerning the selection of sample sizes for comparing population means.

EXAMPLE 8.2

Refer to the experimental situation discussed in Example 8.1. Suppose you know from prior experience that the range of the assembly times for a given treatment will be roughly 8 minutes. Now, suppose you wish to estimate the difference in mean assembly times, corresponding to a pair of training programs, correct to within 1 minute. How many workers should be assigned to each program?

SOLUTION

We know from Section 2.3 that one-fourth of the range of a set of measurements provides a rough approximation to their standard deviation. Using this approximation, it follows that the standard deviation of the assembly times for a given treatment is

$$\sigma \approx \frac{\text{Range}}{4} = \frac{8}{4} = 2$$

Since the bound on the error of estimation will be less than

$$2\sigma_{\bar{y}_1 - \bar{y}_2} = 2\sigma \sqrt{\frac{1}{n_1} + \frac{1}{n_2}}$$

with probability near .95, we will set

$$2\sigma \sqrt{\frac{1}{n_1} + \frac{1}{n_2}} = \text{Bound on the error} = 1$$

Thus, we will substitute $\sigma \approx 2$ into this equation and solve for $n_1 = n_2$. [*Note:* We agreed earlier to use equal treatment group sizes. Therefore $n_1 = n_2 = n_3$.] Then,

$$2(2) \sqrt{\frac{1}{n_1} + \frac{1}{n_1}} = 1 \text{ minute}$$

$$\sqrt{n_1} = 4\sqrt{2}$$

$$n_1 = 32$$

Therefore, to estimate the difference between a pair of mean assembly times to within 1 minute with 95% confidence, we must have a total of 96 workers, and we must randomly assign 32 to each of the three training programs. ■

EXAMPLE 8.3

Suppose a beverage bottler wished to compare the effect of three different advertising displays on the sales of a beverage in supermarkets. Identify the experimental units you would use for the experiment, and explain how you would employ a completely randomized design to collect the sales data.

SOLUTION

Presumably, the bottler has a list of supermarkets in a number of different cities that market the beverage. If we decide to measure the sales increase (or decrease) as the monthly dollar increase in sales (over the previous month) for a given supermarket, then the experimental unit is a 1-month unit of time in a specific supermarket. Thus, we would randomly select a 1-month period of time for each of $n_1 + n_2 + n_3$ supermarkets and assign n_1 supermarkets to receive display D_1, n_2 to receive display D_2, and n_3 to receive display D_3.   ∎

In some experimental situations, we are unable to randomly assign the treatment to the experimental units due to the nature of the experimental units themselves. For example, suppose we want to compare the mean annual salaries of professors in two College of Business departments, accounting and economics. Then the treatments—accounting and economics—cannot be "assigned" to the professors (experimental units). A professor is a member of either the accounting department or the economics (or some other) department and cannot be arbitrarily assigned one of the treatments. Rather, we view the treatments (departments) as populations from which we will select independent random samples of experimental units (professors). A **completely randomized design**, then, involves a comparison of the means for a number, say p, of treatments, based on independent random samples of $n_1, n_2, \ldots, n_p$ observations, drawn from populations associated with treatments $1, 2, \ldots, p$, respectively.

DEFINITION 8.7

A **completely randomized design** to compare p treatment means is one in which the treatments are randomly assigned to the experimental units, or in which independent random samples are drawn from each of the p populations.

After collecting the data from a completely randomized design, we want to make inferences about p population means where μ_i is the mean of the population of measurements associated with treatment i, for $i = 1, 2, \ldots, p$. The null hypothesis to be tested is that the p treatment means are equal, i.e., $H_0: \mu_1 = \mu_2 = \cdots = \mu_p$, and the alternative hypothesis we wish to detect is that at least two of the treatment means differ.

An analysis of variance provides an easy way to analyze the data from a completely randomized design. This analysis partitions SS(Total) into two components, SSE and SST (see Figure 8.4). Recall that the quantity SST denotes the sum of squares for treatments and measures the variation explained by the differences between the treatment means. The sum of squares for error, SSE, is a measure of the unexplained variability, obtained by calculating a pooled measure

FIGURE 8.4

The Partitioning of SS(Total) for a Completely Randomized Design

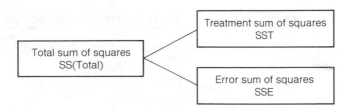

of the variability *within* the p samples. If the treatment means truly differ, then SSE should be substantially smaller than SST. We compare the two sources of variability by forming an F-statistic:

$$F = \frac{\text{SST}/(p - 1)}{\text{SSE}/(n - p)} = \frac{\text{MST}}{\text{MSE}}$$

where n is the total number of measurements. The numerator of the F-statistic, $\text{MST} = \text{SST}/(p - 1)$, denotes **mean square for treatments** and is based on $(p - 1)$ degrees of freedom—one for each of the p treatments minus one for the estimation of the overall mean. The denominator of the F-statistic, $\text{MSE} = \text{SSE}/(n - p)$, denotes **mean square for error** and is based on $(n - p)$ degrees of freedom—one for each of the n measurements minus one for each of the p treatment means being estimated. Thus, F is based on $\nu_1 = (p - 1)$ and $\nu_2 = (n - p)$ degrees of freedom. If the computed value of F exceeds the upper critical value, F_α, we reject H_0 and conclude that at least two of the treatment means differ.

The notation used in an analysis of variance of a completely randomized design is given in Table 8.1. The computing formulas and the analysis of variance F-test are given in the boxes on page 396. In this section we will illustrate the analysis of a completely randomized design using the analysis of variance formulas shown here. The regression approach to conducting an analysis of variance is illustrated in optional Section 8.10.

TABLE 8.1

Summary of Notation for a Completely Randomized Design

	POPULATION (TREATMENTS)				
	1	2	3	$\cdots$	p
Mean	μ_1	μ_2	μ_3	$\cdots$	μ_p
Variance	σ_1^2	σ_2^2	σ_3^2	$\cdots$	σ_p^2
	INDEPENDENT RANDOM SAMPLES				
	1	2	3	$\cdots$	p
Sample size	n_1	n_2	n_3	$\cdots$	n_p
Sample totals	T_1	T_2	T_3	$\cdots$	T_p
Sample means, $\dfrac{T_i}{n_i}$	$\overline{T}_1$	$\overline{T}_2$	$\overline{T}_3$	$\cdots$	$\overline{T}_p$

Total number of measurements:
$$n = n_1 + n_2 + n_3 + \cdots + n_p$$

ANALYSIS OF VARIANCE F-TEST FOR A COMPLETELY RANDOMIZED DESIGN

H_0: $\mu_1 = \mu_2 = \cdots = \mu_p$ (i.e., there is no difference in the treatment means)

H_a: At least two of the treatment means differ

Test statistic: $F = \dfrac{\text{MST}}{\text{MSE}} = \dfrac{\text{MST}}{s^2}$

Rejection region: $F > F_\alpha$, where F is based on $\nu_1 = (p - 1)$ and $\nu_2 = (n - p)$ degrees of freedom

---

COMPUTING FORMULAS FOR THE ANALYSIS OF VARIANCE FOR A COMPLETELY RANDOMIZED DESIGN

Sum of all n measurements $= \displaystyle\sum_{i=1}^{n} y_i$

Mean of all n measurements $= \bar{y}$

Sum of squares of all n measurements $= \displaystyle\sum_{i=1}^{n} y_i^2$

CM = Correction for mean

$\quad = \dfrac{(\text{Total of all observations})^2}{\text{Total number of observations}} = \dfrac{\left(\displaystyle\sum_{i=1}^{n} y_i\right)^2}{n}$

SS(Total) = Total sum of squares

$\quad$ = (Sum of squares of all observations) $-$ CM

$\quad = \displaystyle\sum_{i=1}^{n} y_i^2 - \text{CM}$

SST = Sum of squares for treatments

$\quad = \left(\begin{array}{c}\text{Sum of squares of treatment totals with}\\ \text{each square divided by the number of}\\ \text{observations for that treatment}\end{array}\right) - \text{CM}$

$\quad = \dfrac{T_1^2}{n_1} + \dfrac{T_2^2}{n_2} + \cdots + \dfrac{T_p^2}{n_p} - \text{CM}$

SSE = Sum of squares for error

$\quad$ = SS(Total) $-$ SST

MST = Mean square for treatments

$\quad = \dfrac{\text{SST}}{p - 1}$

MSE = Mean square for error

$\quad = \dfrac{\text{SSE}}{n - p}$

EXAMPLE 8.4

Suppose a large chain of department stores wants to compare the mean dollar amounts owed by its delinquent credit card customers in three different annual income groups: under $12,000, $12,000–$25,000, and over $25,000. Samples of ten customers with delinquent accounts are randomly selected from each group and the amount owed by each customer is recorded. The data are shown in Table 8.2. Do the data provide sufficient evidence to indicate a difference in the mean amount for the three income groups? Test using $\alpha = .05$.

TABLE 8.2
Income Class:
Dollars Owed

	UNDER $12,000	$12,000–$25,000	OVER $25,000
	$148	$513	$335
	76	264	643
	393	433	216
	520	94	536
	236	535	128
	134	327	723
	55	214	258
	166	135	380
	415	280	594
	153	304	465
Totals	$2,296	$3,099	$4,278

SOLUTION

This experiment involves a single factor, income class, which is at three levels. Thus, we have a completely randomized design with $p = 3$ treatments. Let μ_1, μ_2, μ_3 represent the mean amounts owed by customers with incomes under $12,000, between $12,000 and $25,000, and over $25,000, respectively. Then we want to test

$$H_0: \quad \mu_1 = \mu_2 = \mu_3$$

against

H_a: At least two of the three treatment means differ

From Table 8.2, the totals for the three samples are $T_1 = \$2,296$, $T_2 = \$3,099$, and $T_3 = \$4,278$.

$$\sum_{i=1}^{n} y_i = T_1 + T_2 + T_3 = 9,673$$

$$\sum_{i=1}^{n} y_i^2 = (148)^2 + (76)^2 + \cdots + (465)^2 = 4,088,341$$

Then, following the order of calculations listed in the previous box, we find

$$CM = \frac{\left(\sum_{i=1}^{n} y_i\right)^2}{n} = \frac{(9,673)^2}{30} = 3,118,897.633$$

$$SS(Total) = \sum_{i=1}^{n} y_i^2 - CM$$

$$= 4,088,341 - 3,118,897.633$$

$$= 969,443.367$$

$$SST = \frac{T_1^2}{n_1} + \frac{T_2^2}{n_2} + \frac{T_3^2}{n_3} - CM$$

$$= \frac{(2,296)^2}{10} + \frac{(3,099)^2}{10} + \frac{(4,278)^2}{10} - 3,118,897.633$$

$$= 3,317,670.1 - 3,118,897.633 = 198,772.467$$

$$SSE = SS(Total) - SST = 969,443.367 - 198,772.467$$

$$= 770,670.9$$

$$MST = \frac{SST}{p-1} = \frac{198,772.467}{2} = 99,386.233$$

$$MSE = \frac{SSE}{n-p} = \frac{770,670.9}{27} = 28,543.367$$

Then the value of the test statistic is

$$F = \frac{MST}{MSE} = \frac{99,386.233}{28,543.367} = 3.48$$

where the distribution of F is based on $\nu_1 = (p-1) = 3 - 1 = 2$ and $\nu_2 = n - p = 30 - 3 = 27$ df. For $\alpha = .05$, the critical value (obtained from Table 4 of Appendix D) is $F_{.05} = 3.35$ (see Figure 8.5).

FIGURE 8.5
Rejection Region
for Example 8.4;
Numerator df = 2,
Denominator df = 27,
$\alpha = .05$

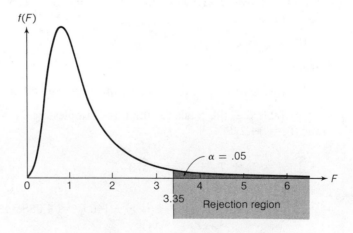

Since the computed value of F, 3.48, exceeds the critical value, $F_{.05} = 3.35$, we reject H_0 and conclude (at the $\alpha = .05$ level of significance) that the mean indebtedness of the delinquent credit card holders differs in at least two of the three income groups.   ■

The results of an analysis of variance (ANOVA) are often summarized in tabular form. The general form of an ANOVA table for a completely randomized design is shown in Table 8.3. The column head "Source" refers to the source of variation, and for each source, "df" refers to the degrees of freedom, "SS" to the sum of squares, "MS" to the mean square, and "F" to the F-statistic comparing the treatment mean square to the error mean square. Table 8.4 is the ANOVA summary table corresponding to the analysis of variance data for Example 8.4.

TABLE 8.3
ANOVA Summary Table for a Completely Randomized Design

SOURCE	df	SS	MS	F
Treatments	$p - 1$	SST	MST	MST/MSE
Error	$n - p$	SSE	MSE	
Total	$n - 1$	SS(Total)		

TABLE 8.4
ANOVA Summary Table for Example 8.4

SOURCE	df	SS	MS	F
Income group	2	198,772.467	99,386.233	3.48
Error	27	770,670.900	28,543.367	
Total	29	969,443.367		

Because the completely randomized design involves the selection of independent random samples, we can find a confidence interval for a single treatment mean using the method of Section 2.6 or for the difference between two treatment means using the methods of Section 2.8. The estimate of σ^2 will be based on the pooled sum of squares within all p samples; that is,

$$\text{MSE} = s^2 = \frac{\text{SSE}}{n - p}$$

This is the same quantity that is used as the denominator for the analysis of variance F-test. The formulas for the confidence intervals of Sections 2.6 and 2.8 are reproduced in the box.

CONFIDENCE INTERVALS FOR MEANS: COMPLETELY RANDOMIZED DESIGN

Single treatment mean (say, treatment i): $\overline{T}_i \pm t_{\alpha/2}\left(\dfrac{s}{\sqrt{n_i}}\right)$

Difference between two treatment means (say, treatments i and j):

$$(\overline{T}_i - \overline{T}_j) \pm t_{\alpha/2}s\sqrt{\frac{1}{n_i} + \frac{1}{n_j}}$$

where $s = \sqrt{\text{MSE}}$ and $t_{\alpha/2}$ is the tabulated value of t (Table 2 of Appendix D) that locates $\alpha/2$ in the upper tail of the t-distribution with $(n - p)$ df (the degrees of freedom associated with error in the ANOVA).

EXAMPLE 8.5

Refer to Example 8.4 and find a 95% confidence interval for the mean indebtedness of customers with incomes less than $12,000 per year.

SOLUTION

From Table 8.4, MSE = 28,543.367. Then

$$s = \sqrt{\text{MSE}} = \sqrt{28,543.367} = 168.9$$

The sample mean indebtedness for those with income levels under $12,000 is

$$\overline{T}_1 = \frac{T_1}{n_1} = \frac{2,296}{10}$$
$$= 229.6$$

The tabulated value of $t_{.025}$ for 27 df (the same as for MSE) is 2.052. Therefore, a 95% confidence interval for μ_1, the mean indebtedness of people with incomes less than $12,000, is

$$\overline{T}_1 \pm t_{\alpha/2}\left(\frac{s}{\sqrt{n_1}}\right) = 229.6 \pm 2.052\left(\frac{168.9}{\sqrt{10}}\right)$$
$$= 229.6 \pm 109.6$$

or ($120.0, $339.2).

Note that this confidence interval is quite wide—probably too wide to be of any practical value. The reason why the interval is so wide can be seen in the large amount of variation within each income class. For example, the indebtedness for customers with incomes under $12,000 varies from $55 to $520. The more variable the data, the larger will be the value of s that appears in the confidence interval and the wider will be the confidence interval. Consequently, if you want to obtain a more accurate estimate of treatment means with a narrower confidence interval, you will have to select larger samples of customers from within each income class. ∎

EXAMPLE 8.6

Find a 95% confidence interval for the difference in mean indebtedness between customers with incomes under $12,000 and those with incomes over $25,000.

SOLUTION

The mean of the sample for customers with incomes over $25,000 is

$$\overline{T}_3 = \frac{T_3}{n_3} = \frac{4,278}{10}$$
$$= 427.8$$

and, from Example 8.5, $\overline{T}_1 = 229.6$. The tabulated t-value, $t_{.025}$, is the same as for Example 8.5, namely 2.052. Then, the 95% confidence interval for $(\mu_3 - \mu_1)$, the difference in mean indebtedness between the two groups, is

$$(\overline{T}_3 - \overline{T}_1) \pm t_{.025}s\sqrt{\frac{1}{n_3} + \frac{1}{n_1}} = (427.8 - 229.6) \pm (2.052)(168.9)\sqrt{\frac{1}{10} + \frac{1}{10}}$$

or ($43.2, $353.2).

As with the confidence interval for a single mean, the confidence interval for the difference $(\mu_3 - \mu_1)$ is very wide. This is due to the excessive within-sample variation. To obtain a narrower confidence interval, the sample sizes for the three income groups must be increased. However, the fact that the interval contains only positive numbers means we can conclude, with 95% confidence, that the mean indebtedness of those with incomes over $25,000 exceeds the mean indebtedness of those with incomes under $12,000. ∎

CASE STUDY 8.1

In the late 1960's there was considerable discussion in business literature concerning college student attitudes toward the business community. Leslie M. Dawson (1969) states:

> It is not surprising that the possibility of hostile student attitudes toward business has become a source of great concern. A loss of faith in the viability of our business system and institutions on the part of college youth, leading to widespread aversion to business careers, could well sap the strength of our nation . . . . While we need the brightest and best of our young people in fields such as government, education, and science, we also need them in business if we are to endure as an industrial power of the first order.

Since many college graduates with negative attitudes toward business are eventually employed by the business community, Dawson decided to investigate the relationship between business employees' proficiency and their attitudes toward business. A total of 152 employees were divided into four *job performance rating categories* based on ratings by immediate superiors, using a uniform rating scale. The employees were then administered a questionnaire to obtain a quantitative measure of their attitudes toward business. This constitutes a completely randomized design, with the four different rating categories representing the *treatments*.

An analysis of variance of the data indicated a difference in the mean *attitude scores* of the four job rating categories at the $\alpha = .05$ level of significance. This points out a need to investigate further the nature of relationships that may exist between employees' attitudes and job performance. The information could then be used to institute programs for college students (such as part-time jobs) that may influence their attitudes toward business. The results might also be used to aid business in hiring graduating college students. As Dawson says, the study indicates a need "for greater efforts by businessmen and educators alike to encourage students to think more positively about business."

8.1 Researchers recently conducted an experiment to compare the mean job satisfaction rating $E(y)$ of workers using three types of work scheduling: flextime (which allows workers to set their individual work schedules), staggered starting hours, and fixed hours.

  a. Identify the treatments in the experiment.

  b. Suppose 60 workers are available for the study. Explain how you would employ a completely randomized design for this experiment.

  c. Use the random number table (Table 7 in Appendix D) to randomly assign the workers to the three work schedules.

8.2 A partially completed ANOVA table for a completely randomized design is shown here.

SOURCE	df	SS	MS	F
Treatments	4	24.7		
Error				
Total	34	62.4		

  a. Complete the ANOVA table.

  b. How many treatments are involved in the experiment?

  c. Do the data provide sufficient evidence to indicate a difference among the treatment means? Test using $\alpha = .10$.

8.3 The data for a completely randomized design with two treatments are shown in the accompanying table.

TREATMENT 1	TREATMENT 2
10	12
7	8
8	13
11	10
10	10
9	11
9	

  a. Calculate MST for the data. What type of variability is measured by this quantity?

  b. Calculate MSE for the data. What type of variability is measured by this quantity?

  c. How many degrees of freedom are associated with MST?

  d. How many degrees of freedom are associated with MSE?

  e. Compute the test statistic appropriate for testing $H_0: \mu_1 = \mu_2$ against the alternative that the two treatment means differ, using a significance level of $\alpha = .05$.

  f. Summarize the results from parts **a–e** in an ANOVA table.

  g. Specify the rejection region, using a significance level of $\alpha = .05$.

  h. State the proper conclusion.

8.4 Exercise 8.3 involves a test of the null hypothesis H_0: $\mu_1 = \mu_2$ based on independent random sampling (recall the definition of a completely randomized design). This test was conducted in Section 2.8 using a Student's t-statistic.

a. Use the Student's t-test to test H_0: $\mu_1 = \mu_2$ against the alternative hypothesis H_a: $\mu_1 \neq \mu_2$. Test using $\alpha = .05$.

b. It can be shown (proof omitted) that an F-statistic with $\nu_1 = 1$ numerator degree of freedom and ν_2 denominator degrees of freedom is equal to t^2, where t is a Student's t-statistic based on ν_2 degrees of freedom. Square the value of t calculated in part **a** and show that it is equal to the value of F calculated in Exercise 8.3.

c. Is the analysis of variance F-test for comparing two population means a one- or a two-tailed test of H_0: $\mu_1 = \mu_2$? [*Hint:* Although the t-test can be used to test for either H_a: $\mu_1 > \mu_2$ or H_a: $\mu_1 < \mu_2$, the alternative hypothesis for the F-test is H_a: The two means are different.]

8.5 The display consoles of modern computer-based systems use many abbreviated words in order to accommodate the large volume of information to be displayed. Therefore, operators must learn to quickly and accurately decode each abbreviation. An experiment was conducted to determine the optimal method for abbreviating any specific set of words on the sonar consoles used at the Naval Submarine Medical Research Laboratory in Groton, Connecticut (*Human Factors*, February 1984). Of the 20 Navy and civilian personnel who took part in the study, five were highly familiar with the sonar system. The 15 subjects unfamiliar with the system were randomly divided into three groups of five. Thus, the study consisted of a total of four groups (one experienced and three inexperienced groups), with five subjects per group. The experienced group and one inexperienced group (denoted TE and TI, respectively) were both assigned to learn the simple method of abbreviation. One of the remaining inexperienced groups was assigned the conventional single abbreviation method (denoted CS), while the other was assigned the conventional multiple abbreviation method (denoted (CM). Each subject was then given a list of 75 abbreviations to learn, one at a time, through the display console of a minicomputer. The number of trials until the subject accurately decoded at least 90% of the words on the list was recorded. [*Note:* The data in the accompanying table are simulated values based on the group means reported in the article.] Do the data provide sufficient evidence to indicate differences among the mean numbers of trials required for the four groups? Test using $\alpha = .05$.

CM	CS	TE	TI
4	6	5	8
7	9	5	4
5	5	7	8
6	7	8	10
8	6	7	3

8.6 Each year *Business Week* reports the total cash compensations (salary plus bonus) for the top corporate executives. The data (in thousands of dollars) in the table on page 404 were extracted from *Business Week*'s 1986 "Executive Compensation Scoreboard." Assume that the data represent independent random samples of 1986 total cash compensations for eight corporate executives in each of the three industries—banks, utilities, and office equipment/computers—and that the experiment is a completely randomized design.

a. Construct an ANOVA table for the data.

b. Is there evidence of a difference among the means of the 1986 total cash compensations for the three groups of corporate executives? Test using $\alpha = .01$.

c. Find a 90% confidence interval for the mean 1986 total cash compensation of executives in the banking industry.

BANKS AND BANK HOLDING COMPANIES	UTILITIES	OFFICE EQUIPMENT AND COMPUTERS
$ 755	$520	$438
712	295	828
845	553	622
985	950	453
1,300	930	562
1,143	428	348
733	510	405
1,189	864	938

Source: "Executive compensation scoreboard," *Business Week*, May 4, 1987, pp. 59–94.

8.7 How likely is a firm to hire a disabled (but rehabilitated) worker? Billions of dollars are spent annually in the United States on training disabled persons for the workplace, yet only 25% of all disabled served by vocational rehabilitation are gainfully employed. A study was conducted to investigate the low placement rate of disabled persons in the workforce. Each employer in a sample of 124 companies located in central Kentucky was asked to rate the employability of an individual with a physical handicap in his company, using a 5-point scale (1 = cannot accommodate, 5 = can accommodate quite easily). One goal of the study was to determine whether company size (measured by the number of employees) affects the employability ratings of disabled people. To accomplish this, the companies were divided into three groups according to size— small (1–15 employees), medium (16–55 employees), and large (over 55 employees)—and the mean employability ratings compared.

a. Give the appropriate null and alternative hypotheses for this study.

b. A partial ANOVA table for the data is shown here. Complete the table.

SOURCE	df	SS	MS	F
Company size				4.93
Error		190		
Total	123			

Source: Combs, I. H. and Omvig, C. P. "Accommodation of disabled people into employment: Perceptions of employers," *Journal of Rehabilitation*, Vol. 52, No. 2, April/May/June 1986, pp. 42–45.

c. Is there sufficient evidence to indicate that the mean employability ratings of physically handicapped persons differ among the three groups? Test using $\alpha = .05$.

8.8 One of the selling points of golf balls is their durability. An independent testing laboratory is commissioned to compare the durability of four different brands of golf balls. Balls of each type will be put into a machine that hits the balls with the same force that a golfer does on the course. The number of hits required until the outer covering is cracked is recorded for each ball, with the results given in the table. Ten balls from each manufacturer are randomly selected for testing.

BRAND			
A	B	C	D
310	261	233	275
235	219	289	290
279	263	301	265
306	247	264	284
237	288	273	239
284	197	208	257
259	207	245	232
273	221	271	251
219	244	298	266
301	228	276	287

a. Is there evidence that the mean durabilities of the four brands differ? Use $\alpha = .05$.

b. Estimate the difference between the mean durability of brands A and C using a 99% confidence interval.

8.9 R. H. Brockhaus (St. Louis University) conducted a study to determine whether entrepreneurs, newly hired (transferred) managers, and newly promoted managers differ in their risk-taking propensities. For the purposes of this study, entrepreneurs were defined as individuals who, within 3 months prior to the study, had ceased working for their employers to own and manage business ventures. Thirty-one individuals from each of the three groups were randomly selected to participate in the study. Each was administered a questionnaire which required the respondent to choose between a safe alternative and a more attractive but risky one. Test scores were designed to measure risk-taking propensity. (Lower scores are associated with greater conservatism in risk-taking situations.) The test scores for the three groups are summarized in the table. [*Note:* Although the individual scores are not provided, there is sufficient information in the table to conduct the analysis for this completely randomized design.]

GROUP	SAMPLE SIZE	SAMPLE MEAN	STANDARD DEVIATION	GROUP TOTALS
Entrepreneurs	31	71.00	11.94	2,201
Transferred managers	31	72.52	12.19	2,248
Promoted managers	31	66.97	10.84	2,076
	93			6,525

Source: Brockhaus, R. H. "Risk-taking propensity of entrepreneurs," *Academy of Management Journal*, September 1980, 23, pp. 509–520.

a. Use the sum of the group totals and the total sample size to compute CM.

b. Use the individual group totals and sample sizes to compute SST.

c. Compute SSE using the pooled sum of squares formula:

$$SSE = \sum (y_{i1} - \bar{y}_1)^2 + \sum (y_{i2} - \bar{y}_2)^2 + \sum (y_{i3} - \bar{y}_3)^2$$
$$= (n_1 - 1)s_1^2 + (n_2 - 1)s_2^2 + (n_3 - 1)s_3^2$$

d. Find SS(Total).

e. Construct an ANOVA table for the data.

f. Do the data provide sufficient evidence to indicate differences in the mean risk-taking propensities among the three groups? Test using $\alpha = .05$.

8.10 As oil drilling costs rise at unprecedented rates, the task of measuring drilling performance becomes essential to a successful oil company. One method of lowering drilling costs is to increase drilling speed. Researchers at Cities Service Co. have developed a drill bit, called the PD-1, which they believe penetrates rock at a faster rate than any other bit on the market. It is decided to compare the speed of the PD-1 with the two fastest drill bits known, the IADC 1-2-6 and the IADC 5-1-7, at 12 drilling locations in Texas. Four drilling sites were randomly assigned to each bit, and the rate of penetration (RoP) in feet per hour (fph) was recorded after drilling 3,000 feet at each site. The data are given in the table. Based on this information, can Cities Service Co. conclude that the mean RoP differs for at least two of the three drill bits? Test at the $\alpha = .05$ level of significance.

PD-1	IADC 1-2-6	IADC 5-1-7
35.2	25.8	14.7
30.1	29.7	28.9
37.6	26.6	23.3
34.3	30.1	16.2

8.11 A fast-food chain that specializes in Mexican food (tacos, burritos, etc.) is opening a new franchise in a university town. An important consideration in determining where the franchise will be located is traffic density. Five possible locations (each near a major intersection) are under consideration by the chain. To compare the density of traffic at the possible sites, company employees are placed at each of the five locations to count the number of cars passing each location daily for a period of 10 randomly selected days. (At location IV, the counter assigned to record traffic density became ill and could obtain data for only 8 of the days.) The results are listed in the accompanying table. Assuming the samples of days were independently selected, test for a difference in average daily traffic density among the five locations. Use a significance level of $\alpha = .10$.

		LOCATION		
I	II	III	IV	V
344	412	237	518	367
382	441	390	501	445
353	607	365	577	480
395	531	355	642	323
207	486	217	489	366
312	508	268	475	325
407	337	117	532	316
421	419	273	540	381
366	499	288		407
222	387	351		339

8.12 Due to the growing concern over the problems related to alcohol consumption, the Federal Trade Commission has considered banning alcohol advertising. It is unclear, however, whether alcohol advertising actually increases alcohol consumption. G. B. Wilcox conducted a study to examine the effect of price advertising on sales of beer in lower Michigan. The state of Michigan was selected since it has prohibited retailers from advertising the price of beer products since 1975, except for a brief period

(March 1982–May 1983) when the ban was temporarily lifted. The data in the table are the bimonthly total sales of brewed beverages (in thousands of 31-gallon barrels) over the period May 1981–April 1984. The data allow us to compare total sales of beer in three periods, before (period 1), during (period 2), and after (period 3) the lifting of the price advertising restrictions.

PERIOD 1: PRICE ADVERTISING RESTRICTED (May/June 1981–Jan./Feb. 1982)	PERIOD 2: NO RESTRICTIONS (Mar./Apr. 1982–May/June 1983)	PERIOD 3: PRICE ADVERTISING RESTRICTED (July/Aug. 1983–Mar./Apr. 1984)
462	522	433
417	508	470
516	427	609
605	477	442
654	603	446
	692	
	584	
	496	

Source: Wilcox, G. B. "The effect of price advertising on alcoholic beverage sales," *Journal of Advertising Research*, Vol. 25, No. 5, October/November 1985, pp. 33–37. Copyright © 1985 by the Advertising Research Foundation.

   a. Treat this as a completely randomized design, and construct an ANOVA summary table.
   b. Is there sufficient evidence to indicate differences in the average total sales of beer in the three periods? Test using $\alpha = .10$.

8.13 In the United States, two basic types of management attitudes prevail: Theory X bosses believe that workers are basically lazy and untrustworthy, and Theory Y managers hold that employees are hard-working, dependable individuals. Japanese firms take a third approach: Theory Z companies emphasize long-range planning, consensus decision-making, and strong, mutual worker-employer loyalty. Suppose we want to compare the hourly wage rates of workers at Theory X-, Y-, and Z-style corporations. Independent random samples of six engineering firms of each managerial philosophy were selected, and the starting hourly wage rates for laborers at each were recorded, as shown in the accompanying table.

MANAGERIAL ATTITUDE		
Theory X	Theory Y	Theory Z
5.20	6.25	5.50
5.20	6.80	5.75
6.10	6.87	4.60
6.00	7.10	5.36
5.75	6.30	5.85
5.60	6.35	5.90

   a. Is there evidence of a difference among the mean starting hourly wages of laborers at Theory X-, Y-, and Z-style firms? Test at a significance level of $\alpha = .01$.
   b. Calculate 99% confidence intervals for each of the three differences, $(\mu_X - \mu_Y)$, $(\mu_X - \mu_Z)$, and $(\mu_Y - \mu_Z)$.

c. Based on your intervals from part **b**, which type of engineering firm appears to pay its laborers a higher average starting hourly wage rate, Theory X-, Theory Y-, or Theory Z-style firms? [*Note:* We present a more formal method of multiple comparisons in Section 8.8.]

SECTION 8.6

RANDOMIZED BLOCK DESIGNS

Noise reduction in an experimental design, i.e., the removal of extraneous experimental variation, can be accomplished by an appropriate assignment of treatments to the experimental units. The idea is to compare treatments within blocks of relatively homogeneous experimental units. The most common design of this type—called a **randomized block design**—often provides more information per observation than the amount contained in a completely randomized design. For example, suppose you want to compare the length of time required to process a bank's daily receipts using three different computer programs, A, B, and C. A completely randomized design could be used by selecting 15 days (where the days are the experimental units) and randomly assigning the receipts for 5 days to be processed using each of the programs (treatments).

A better way to conduct the experiment—one that contains more information on the mean processing times—would be to utilize the receipts for only 5 days and process the data for each day using each of the three programs. This *randomized block* procedure acknowledges the fact that the length of time required to process a day's receipts varies substantially from day to day, depending on the level of the day's business, the complexity of the transactions, and so on. By comparing the three processing times for each day, we eliminate day-to-day variation from the comparison.

The randomized block design that we have just described is shown diagrammatically in Figure 8.6. The figure shows that there are five jobs. Each job can be viewed as a **block** of three experimental units—runs on the computer—one corresponding to the use of each of the programs, A, B, and C. The blocks are said to be **randomized** because the treatments (computer programs) are randomly assigned to the experimental units within a block. For our example, the programs employed to process a day's receipts would be run in a random order to avoid bias introduced by other unknown and unmeasured variables that may affect the processing time.

FIGURE 8.6

Diagram for a Randomized Block Design Containing $b = 5$ Blocks and $p = 3$ Treatments

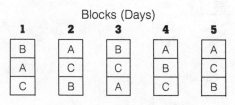

In general, a randomized block design to compare p treatments will contain b relatively homogeneous blocks, with each block containing p experimental units. Each treatment appears once in every block with the p treatments randomly assigned to the experimental units within each block.

> **DEFINITION 8.8**
>
> A **randomized block design** to compare p treatments involves b blocks, each containing p relatively homogeneous experimental units. The p treatments are randomly assigned to the experimental units within each block, with one experimental unit assigned per treatment.

EXAMPLE 8.7

Suppose you want to compare the abilities of four real estate appraisers, A, B, C, and D. One way to make the comparison would be to randomly allocate a number of pieces of real estate, say 40, ten to each of the four appraisers. Each appraiser would appraise the property and you would record y, the difference between the appraised and selling prices expressed as a percentage of the selling price. Thus, y measures the appraisal error expressed as a percentage of selling price, and the treatment allocation to experimental units that we have described is a completely randomized design.

a. Discuss the problems with using a completely randomized design for this experiment.

b. Explain how you could employ a randomized block design.

SOLUTION

a. The problem with using a completely randomized design for the appraisal experiment is that comparison of mean percentage errors will be influenced by the nature of the properties. Some properties will be easier to appraise than others, and the variation in percentage errors that can be attributed to this fact will make it more difficult to compare the treatment means.

b. To eliminate the effect of property-to-property variability in comparing appraiser means, you could select only ten properties and require each appraiser to appraise the value of each of the ten properties. Although in this case there is probably no need for randomization, it might be desirable to randomly assign the order (in time) of the appraisals. This randomized block design, consisting of $p = 4$ treatments and $b = 10$ blocks would appear as shown in Figure 8.7.

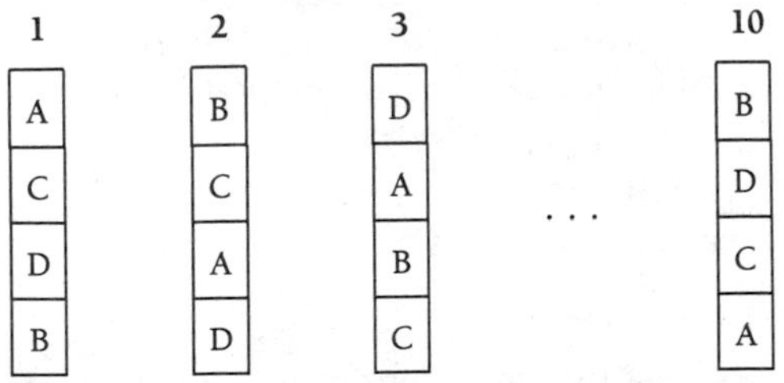

FIGURE 8.7
Diagram for a
Randomized Block
Design: Example 8.7

The analysis of variance for data collected from a randomized block design is similar to the completely randomized design. The partitioning of SS(Total) for the randomized block design is most easily seen by examining Figure 8.8 (page 410). Note that SS(Total) is now partitioned into *three* parts:

$$SS(Total) = SSB + SST + SSE$$

The notation that we will employ in the analysis of variance formulas is shown in Table 8.5. The formulas for calculating SST and SSB follow the same pattern as the formula for calculating SST for the completely randomized design. Once these quantities have been calculated, we find SSE by subtraction:

$$SSE = SS(Total) - SST - SSB$$

FIGURE 8.8

Partitioning of the Total Sum of Squares for the Randomized Block Design

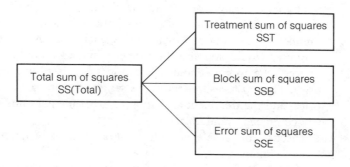

TABLE 8.5

Notation for the Analysis of Variance for a Randomized Block Design

	TREATMENT 1	TREATMENT 2	$\cdots$	TREATMENT p
Treatment totals	T_1	T_2	$\cdots$	T_p
Treatment means	$\overline{T}_1$	$\overline{T}_2$	$\cdots$	$\overline{T}_p$
	BLOCK 1	BLOCK 2	$\cdots$	BLOCK b
Block totals	B_1	B_2	$\cdots$	B_b
Block means	$\overline{B}_1$	$\overline{B}_2$	$\cdots$	$\overline{B}_b$

b = Number of measurements in a single treatment total
p = Number of measurements in a single block total
n = Total number of measurements in the complete experiment = $b \times p$

The formulas for calculating the pertinent sums of squares and mean squares are given in the next box.

We are interested in using the randomized block design to test the same null and alternative hypotheses we tested using the completely randomized design:

H_0: $\mu_1 = \mu_2 = \cdots = \mu_p$

H_a: At least two treatment means differ

The test statistic, also identical to that used for the completely randomized design, is

$$F = \frac{MST}{MSE}$$

COMPUTING FORMULAS FOR THE ANALYSIS OF VARIANCE
FOR A RANDOMIZED BLOCK DESIGN

$$\sum_{i=1}^{n} y_i = \text{Sum of all } n \text{ measurements}$$

$$\sum_{i=1}^{n} y_i^2 = \text{Sum of squares of all } n \text{ measurements}$$

$$CM = \text{Correction for mean}$$

$$= \frac{(\text{Total of all measurements})^2}{\text{Total number of measurements}} = \frac{\left(\sum_{i=1}^{n} y_i\right)^2}{n}$$

$$SS(\text{Total}) = \text{Total sum of squares}$$

$$= (\text{Sum of squares of all measurements}) - CM$$

$$= \sum_{i=1}^{n} y_i^2 - CM$$

$$SST = \text{Sum of squares for treatments}$$

$$= \left(\begin{array}{c} \text{Sum of squares of treatment totals with} \\ \text{each square divided by } b, \text{ the number of} \\ \text{measurements for that treatment} \end{array} \right) - CM$$

$$= \frac{T_1^2}{b} + \frac{T_2^2}{b} + \cdots + \frac{T_p^2}{b} - CM$$

$$SSB = \text{Sum of squares for blocks}$$

$$= \left(\begin{array}{c} \text{Sum of squares for block totals with} \\ \text{each square divided by } p, \text{ the number} \\ \text{of measurements in that block} \end{array} \right) - CM$$

$$= \frac{B_1^2}{p} + \frac{B_2^2}{p} + \cdots + \frac{B_b^2}{p} - CM$$

$$SSE = \text{Sum of squares for error}$$

$$= SS(\text{Total}) - SST - SSB$$

$$MST = \text{Mean square for treatments}$$

$$= \frac{SST}{p-1}$$

$$MSB = \text{Mean square for blocks}$$

$$= \frac{SSB}{b-1}$$

$$MSE = \text{Mean square for error}$$

$$= \frac{SSE}{n-p-b+1}$$

Since the sum of squares for blocks, SSB, measures the variation explained by differences among the block means, we can also test

H_0:   The b block means are equal

H_a: At least two block means differ

using the test statistic

$$F = \frac{\text{MSB}}{\text{MSE}}$$

Although a test of a hypothesis concerning a difference among treatment means is our primary objective, the latter test enables us to determine whether there is evidence of a difference among block means—that is, whether blocking is really effective. If there are no differences among block means, then blocking will not reduce the variability in the experiment and, consequently, will be ineffective. *In fact, if there are no differences among block means, you will lose information by blocking because blocking reduces the number of degrees of freedom associated with s^2.*

ANALYSIS OF VARIANCE *F*-TESTS FOR A RANDOMIZED BLOCK DESIGN

TEST FOR COMPARING TREATMENT MEANS

H_0: $\mu_1 = \mu_2 = \cdots = \mu_p$ (i.e., no differences among the p treatment means)

H_a: At least two of the treatment means differ

Test statistic: $F = \dfrac{\text{MST}}{\text{MSE}} = \dfrac{\text{MST}}{s^2}$

Rejection region: $F > F_\alpha$, where F is based on $\nu_1 = (p - 1)$ and $\nu_2 = (n - p - b + 1)$ degrees of freedom

TEST FOR COMPARING BLOCK MEANS

H_0:   There are no differences among the b block means

H_a: At least two of the block means differ

Test statistic: $F = \dfrac{\text{MSB}}{\text{MSE}} = \dfrac{\text{MSB}}{s^2}$

Rejection region: $F > F_\alpha$, where F is based on $\nu_1 = (b - 1)$ and $\nu_2 = (n - p - b + 1)$ degrees of freedom

---

The *F*-tests for comparing block and treatment means are shown in the previous box. The use of the analysis of variance computational formulas and the tests are demonstrated in Examples 8.8 and 8.9.

EXAMPLE 8.8

Prior to submitting a bid for a construction job, companies prepare a detailed analysis of the estimated labor and materials costs required to complete the job. This estimate will depend on the estimator who performs the analysis. An overly large estimate will reduce the chance of acceptance of a company's bid price and, if too low, will reduce the profit or even cause the company to lose money on the job. A company that employs three job cost estimators wanted to compare the mean level of the estimators' estimates. This was done by having each estimator estimate the cost of the same four jobs. The data (in hundreds of thousands of dollars) are shown in Table 8.6. Perform an analysis of variance on the data and test to determine whether there is sufficient evidence to indicate differences among treatment and block means. Test using $\alpha = .05$.

TABLE 8.6

Data for the Randomized Block Design of Example 8.8

		JOB				TOTALS	MEANS
		1	2	3	4		
	1	4.6	6.2	5.0	6.6	22.4	5.60
ESTIMATOR	2	4.9	6.3	5.4	6.8	23.4	5.85
	3	4.4	5.9	5.4	6.3	22.0	5.50
TOTALS		13.9	18.4	15.8	19.7	67.8	
MEANS		4.63	6.13	5.27	6.57		

SOLUTION

The data for this experiment were collected according to a randomized block design because we would expect estimates of the same job to be more nearly alike than estimates between jobs. Thus, the experiment involves three treatments (estimators) and four blocks (jobs).

The following calculations are required for the analysis of variance:

$$\sum_{i=1}^{n} y_i = 4.6 + 4.9 + \cdots + 6.3 = 67.8$$

$$\sum_{i=1}^{n} y_i^2 = (4.6)^2 + (4.9)^2 + \cdots + (6.3)^2 = 390.28$$

$$CM = \frac{\left(\sum_{i=1}^{n} y_i\right)^2}{n} = \frac{(67.8)^2}{12} = 383.07$$

$$SS(Total) = \sum_{i=1}^{n} y_i^2 - CM = 7.21$$

$$SST = \frac{\sum_{i=1}^{p} T_i^2}{b} - CM$$

$$= \frac{(22.4)^2 + (23.4)^2 + (22.0)^2}{4} - 383.07$$

$$= .26$$

$$SSB = \frac{\sum_{i=1}^{b} B_i^2}{p} - CM = \frac{(13.9)^2 + (18.4)^2 + (15.8)^2 + (19.7)^2}{3} - 383.07$$

$$= 6.763333$$

$$SSE = SS(Total) - SST - SSB$$

$$= 7.21 - .26 - 6.763333 = .186667$$

$$MST = \frac{SST}{p-1} = \frac{.26}{2} = .13$$

$$MSB = \frac{SSB}{b-1} = \frac{6.763333}{3} = 2.254$$

$$MSE = s^2 = \frac{SSE}{n-p-b+1} = \frac{.186667}{12-3-4+1} = \frac{.186667}{6} = .031111$$

The sources of variation and their respective degrees of freedom, sums of squares, and mean squares for a randomized block design are summarized in the analysis of variance table shown in Table 8.7. The analysis of variance table for this example is shown in Table 8.8. Note that the degrees of freedom for the three sources of variation, Treatments, Blocks, and Error, sum to the degrees of freedom for SS(Total). Similarly, the sums of squares for the three sources will always sum to SS(Total).

TABLE 8.7

Analysis of Variance Table for a Randomized Block Design

SOURCE	df	SS	MS	F
Treatments	$p-1$	SST	MST	MST/MSE
Blocks	$b-1$	SSB	MSB	MSB/MSE
Error	$n-p-b+1$	SSE	MSE	
Total	$n-1$	SS(Total)		

TABLE 8.8

Analysis of Variance Table for Example 8.8

SOURCE	df	SS	MS	F
Treatments (Estimators)	2	.260	.130	4.18
Blocks (Jobs)	3	6.763	2.254	72.46
Error	6	.187	.031	
Total	11	7.210		

The values of the F-statistics needed to test the hypotheses about differences among the treatment and block means are shown in the last column of the analysis of variance table. Since the F-statistic for testing treatment means is based on $\nu_1 = (p-1) = 2$ and $\nu_2 = (n-p-b+1) = 6$ degrees of freedom, we will reject the null hypothesis

H_0:   There are no differences among treatment means

if $F > F_{.05} = 5.14$. Since the computed value of F, 4.18, is less than $F_{.05}$, there is insufficient evidence, at the $\alpha = .05$ level of significance, to indicate differences among the mean level of estimates for the three estimators. The observed significance level for the test (obtained from a computer printout) is .07.

The F-statistic for testing

H_0: There are no differences among block means

is based on $\nu_1 = (b - 1) = 3$ and $\nu_2 = (n - b - p + 1) = 6$ degrees of freedom and we will reject H_0 if $F > F_{.05} = 4.76$. Since the computed value of F, 72.46, exceeds $F_{.05}$, there is sufficient evidence to indicate differences among the mean estimates for the four different jobs. ∎

Caution: The result of the test for the equality of block means must be interpreted with care, especially when the calculated value of the F-test statistic does not fall in the rejection region. This does not necessarily imply that the block means are the same, i.e., that blocking is unimportant. Reaching this conclusion would be equivalent to accepting the null hypothesis, a practice we have carefully avoided due to the unknown probability of committing a Type II error (that is, of accepting H_0 when H_a is true). In other words, even when a test for block differences is inconclusive, we may still want to use the randomized block design in similar future experiments. If the experimenter believes that the experimental units are more homogeneous within blocks than among blocks, he or she should use the randomized block design regardless of whether the test comparing the block means shows them to be different.

Confidence intervals for the difference between a pair of treatment means are shown in the accompanying box.

CONFIDENCE INTERVALS FOR THE DIFFERENCE, $(\mu_i - \mu_j)$, BETWEEN A PAIR OF TREATMENT OR BLOCK MEANS

TREATMENT MEANS

$$(\bar{T}_i - \bar{T}_j) \pm t_{\alpha/2}s\sqrt{\frac{2}{b}}$$

BLOCK MEANS

$$(\bar{B}_i - \bar{B}_j) \pm t_{\alpha/2}s\sqrt{\frac{2}{p}}$$

where

b = Number of blocks
p = Number of treatments
$s = \sqrt{MSE}$

and $t_{\alpha/2}$ is based on $(n - b - p + 1)$ degrees of freedom

EXAMPLE 8.9

Refer to Example 8.8. Find a 90% confidence interval for the difference between the mean level of estimates for estimators 1 and 2.

SOLUTION

From Example 8.8, we know that $b = 4$, $\overline{T}_1 = 5.60$, $\overline{T}_2 = 5.85$, and $s^2 = $ MSE $= .031111$. The degrees of freedom associated with s^2 (and, therefore, with $t_{\alpha/2}$) is 6. Therefore, $s = \sqrt{s^2} = \sqrt{.031111} = .176$ and $t_{\alpha/2} = t_{.05} = 1.943$. Substituting these values into the formula for the confidence interval for $(\mu_1 - \mu_2)$, we obtain

$$(\overline{T}_1 - \overline{T}_2) \pm t_{\alpha/2}s \sqrt{\frac{2}{b}}$$

$$(5.60 - 5.85) \pm 1.943(.176)\sqrt{\frac{2}{4}}$$

$$-.25 \pm .24$$

or, $-.49$ to $-.01$. Since each unit represents \$100,000, we estimate the difference between the mean level of job estimates for estimators 1 and 2 to be enclosed by the interval, $-\$49,000$ to $-\$1,000$. [*Note:* At first glance, this result may appear to contradict the result of the F-test for comparing treatment means. However, the observed significance level of the F-test (.07) implies that significant differences exist between the means at $\alpha = .10$, which is consistent with the fact that 0 is not within the 90% confidence interval.] ■

There is one very important point to note when you block the treatments in an experiment. When you take the difference between a pair of sample treatment means, it can be shown (proof omitted) that the block effects cancel. This fact enables us to calculate confidence intervals for the difference between treatment means using the formulas given in the box on page 415. But, if a sample treatment mean is used to estimate a *single treatment mean*, the block effects do not cancel. *Therefore, the only way that you can obtain an unbiased estimate of a single treatment mean (and corresponding confidence interval) in a blocked design is to randomly select the blocks from a large collection (population) of blocks and to treat the block effect as a second random component, in addition to random error.* Designs that contain two or more random components are called *nested designs* and are beyond the scope of this text. For more information on this topic, see the references at the end of this chapter.

EXERCISES 8.14–8.22

8.14 Retail store audits are periodic audits of a sample of retail sales to monitor inventory and purchases of a particular product. Such audits are often used by marketing researchers to estimate market share. A study was conducted to compare market shares of beer brands estimated by two different auditing methods.
a. Identify the treatments in the experiment.
b. Due to brand-to-brand variation in estimated market share, a randomized block design will be employed. Explain how the treatments might be assigned to the experimental units if ten beer brands are to be included in the study.

8.15 The analysis of variance for a randomized block design produced the ANOVA table entries shown here.

SOURCE	df	SS	MS	F
Treatments	3	27.1		
Blocks	5		14.90	
Error		33.4		
Total				

a. Complete the ANOVA table.

b. Do the data provide sufficient evidence to indicate a difference among the treatment means? Test using $\alpha = .01$.

c. Do the data provide sufficient evidence to indicate that blocking was a useful design strategy to employ for this experiment? Explain.

8.16 A supermarket advertisement asserted that "You'll save up to 21% with Albertson's lower prices" (*Gainesville Sun*, March 18, 1984). To substantiate the claim, Albertson's supermarket compared the prices of 49 grocery items at three competing supermarkets with the prices at its store on a given day. The survey results for seven items randomly selected from the 49 are shown in the accompanying table. This represents a randomized block design, where the seven grocery items represent the blocks.

GROCERY ITEM	ALBERTSON'S	KASH 'N KARRY	PUBLIX	FOOD 4 LESS
Cheerios cereal	$1.10	$1.18	$1.39	$1.18
Jell-o gelatin	.24	.24	.31	.26
Dial soap	.52	.60	.63	.55
Crisco oil	1.26	1.70	2.27	1.29
Kleenex	.67	.70	.79	.70
Star-Kist tuna	.63	.66	.79	.63
Del Monte peas	.43	.47	.65	.47

a. Construct an ANOVA table for the data.

b. Test to determine whether the mean prices of grocery items differ among the four supermarkets. Use $\alpha = .01$.

c. Food 4 Less is considered to be a "low cost, no frills" supermarket. Calculate a 99% confidence interval for the difference in mean prices per item between Albertson's supermarket and Food 4 Less. Interpret your result.

8.17 The popular Friday night Public Television show *Wall Street Week* features economic forecasters, money managers, and financial analysts, who discuss the current state of the economy and the stock market and frequently offer tips on which stocks they favor. The accompanying table (page 418), extracted from the *Orlando Sentinel* (February 27, 1982), gives the difference between the forecasted annual change in the Dow Jones Average (DJA) and the actual increase (or decrease) in the DJA, 1978–1981, for each of four *Wall Street Week* panelists. Assume that the data represent a randomized block design, with four treatments (panelists) and four blocks (years).

a. Construct an ANOVA table for the data.

b. Is there sufficient evidence to indicate that the mean forecast errors for the four panelists differ? Test using $\alpha = .05$.

PANELIST	1978	1979	1980	1981
Robert Nurock	21	35	65	7
Frank Cappiello	18	−1	3	22
Robert Stovall	12	24	24	−4
Carter Randall	5	8	12	15

c. Is there sufficient evidence to indicate that blocking on years was effective in reducing the year-to-year variations in differences between the forecasted and actual Dow Jones Average? Test using $\alpha = .05$.

8.18 A nuclear power plant, which uses water from the surrounding bay for cooling its condensers, is required by the Environmental Protection Agency (EPA) to determine whether discharging its heated water into the bay has a detrimental effect on the plant life in the water. The EPA requests that the power plant make its investigation at three strategically chosen locations, called *stations*. Stations 1 and 2 are located near the plant's discharge tubes, while station 3 is located farther out in the bay. During one randomly selected day in each of six months, a diver descends to each of the stations, randomly samples a square meter area of the bottom, and counts the number of blades of the different types of grasses present. The results for one important grass type are listed in the table.

MONTH	STATION		
	1	2	3
April	32	40	30
May	28	31	53
June	25	22	61
July	37	30	56
August	20	26	48
September	18	21	30

a. Is there sufficient evidence to indicate that the mean number of blades found per square meter per month differs for at least two of the three stations? Use $\alpha = .05$.
b. Is there sufficient evidence to indicate that the mean number of blades found per square meter differs among the six months? Use $\alpha = .05$.
c. Find a 99% confidence interval for the difference between the mean number of blades found in May and August.

8.19 Research into the human-engineering aspect of computing has grown tremendously due to the ever-increasing number of computer users. The end goal is to create "user-friendly" hardware and software that reduce as much as possible any form of human strain or stress resulting from computer usage. A recent topic of concern to computer terminal users is the color of the characters displayed on the video display screens. Early full-screen video display terminals presented the viewer with white characters on a black background. Initially, viewers found the high degree of contrast easy on the eyes. However, after an extended period of use, black and white displays were frequently found to cause temporary eye irritations. In the past few years, experimentation with other colors revealed that yellow/amber displays may be the easiest on the eyes. In one German study, video display terminals were produced with white/black and six different symbol colors. Thirty test subjects were asked to specify which color combination they preferred by ranking each of the seven color combinations on a scale from 0 (no preference) to 10. Although the raw data were not revealed, the mean preference scores for each color were provided by the researchers. Based on these means, we have simulated the individual preference scores for ten subjects in the accompanying table. Do the data provide sufficient evidence of a difference among the mean preference scores for the seven video display color combinations? Test using $\alpha = .05$.

SUBJECT	GREEN/ BLACK	WHITE/ BLACK	YELLOW/ WHITE	ORANGE/ WHITE	YELLOW	YELLOW/ AMBER	YELLOW/ ORANGE
1	7	6	7	2	8	9	3
2	8	6	9	4	9	8	1
3	5	5	7	1	6	8	2
4	3	4	2	0	2	6	0
5	9	8	8	3	9	9	2
6	7	5	6	2	7	7	1
7	6	7	8	4	6	9	5
8	6	5	8	1	8	9	1
9	9	9	8	2	9	8	0
10	9	8	8	3	9	10	1

Source: Adapted from Solomon, L. and Burawa, A. "Maximize your computing comfort & efficiency," *Computers & Electrònics*, April 1983, pp. 35–40.

8.20 The traditional retail store audit is one of the most widely used marketing research tools among consumer package goods companies. It involves periodic audits of a sample of retail stores to monitor inventory and purchases of a particular product. Researchers conducted a study to compare market data yielded by the traditional retail store audit with two alternative, less costly auditing procedures—weekend selldown audits and store purchases audits. The market shares of six major brands of beer distributed in eastern cities were estimated using each of the three store audit methods. The data are provided in the table.

BRAND	TRADITIONAL STORE AUDIT	WEEKEND SELLDOWN AUDIT	STORE PURCHASES AUDIT
1	18.0	19.0	20.7
2	15.3	17.3	14.0
3	8.9	8.5	10.1
4	6.5	4.9	6.1
5	5.3	6.1	4.6
6	3.4	3.0	3.1

Source: Prasad, V. K., Casper, W. R., and Schieffer, R. J. "Alternatives to the traditional retail store audit: A field study," *Journal of Marketing*, Winter 1984, *48*, pp. 54–61. Reprinted by permission of the American Marketing Association.

a. Construct an ANOVA summary table for the data.

b. Is there sufficient evidence to indicate a difference in the mean estimates of beer brand market shares produced by the three auditing methods? Test using $\alpha = .05$.

c. Estimate the difference between the mean estimates of beer brand market shares produced by the traditional store audit and the weekend selldown audit using a 95% confidence interval.

8.21 R. W. T. Gill considered the role of the "in-tray" exercise in assessing the management potential of future administrators and executives. The in-tray or in-basket exercise, developed 30 years ago as a training tool for officers in the U. S. Air Force, simulates the typical contents of an executive's in-tray with a variety of everyday problems in a written form—letters, memoranda, notes, reports, and telephone messages—requiring decisions and action. Trainees are provided with instructions, information on the company, its organization and the role to be played, and the in-tray contents, and are allotted a fixed amount of time to complete the tasks. After completing the tasks, the trainees' performances are assessed by one or more expert raters. However, the reliability of the assessors' ratings should be determined before using the in-tray measure of managerial effectiveness. "Inexperienced, overcautious, or overanxious raters," writes Gill, "can lead to unreliable evaluations and thus lower exercise validity." Ratings of overall in-tray performance are usually given on a scale of 1 (high performance) to 6 (low performance).

To investigate the phenomenon of rater reliability, Gill obtained data for seven subjects who were given the in-tray test. The subjects, all candidates for a general management position in a manufacturing company in the British motor industry, were from a variety of backgrounds—engineering, finance, production, and marketing. Overall in-tray performance of each candidate was assessed by three different raters. The results are given in the accompanying table. Note that each of the three raters judged the overall performance of all seven candidates. Thus, the candidates represent blocks and the sampling design used is a randomized block design.

CANDIDATE	RATER 1	RATER 2	RATER 3
A	4.5	4.5	5.0
B	2.5	4.5	4.5
C	5.0	3.0	4.0
D	4.0	4.5	4.5
E	1.5	2.0	4.5
F	3.5	4.5	4.5
G	4.0	4.0	4.0

Source: Gill, R. W. T. "The in-tray (in-basket) exercise as a measure of management potential," *Journal of Occupational Psychology*, 1979, 52, pp. 185–195.

a. Construct an ANOVA summary table for the data.

b. Is there evidence of a difference among the mean performance scores assessed by the three raters? Use a significance level of $\alpha = .05$.

c. Is there evidence of a difference among the mean performance scores of the candidates? That is, is there evidence that blocking by candidates was effective in removing an unwanted source of variability?

d. Find a 95% confidence interval for the difference between the mean performance scores assessed by raters 2 and 3. Interpret the interval.

8.22 Refer to the study on the effect of price advertising on beer sales in lower Michigan in Exercise 8.12. In that exercise you conducted the analysis as a completely randomized design. The fact that the experimental unit is a month (actually, a pair of successive months, since we are measuring total bimonthly beer sales) could introduce an unwanted source of variation into the analysis—namely, the month-to-month variation in beer sales.

a. Explain how to set up a randomized block design to reduce this unwanted source of variation.

b. The data of Exercise 8.12 are reorganized in the accompanying table. Notice that the months in which the beer sales are recorded are now identified for each period. Using only the data for months that appear across all three periods (why is this necessary?), construct an ANOVA table for the randomized block design described in part **a**.

MONTHS	PERIOD 1	PERIOD 2	PERIOD 3
January/February	654	692	442
March/April		522, 584	446
May/June	462	508, 496	
July/August	417	427	433
September/October	516	477	470
November/December	605	603	609

Source: Wilcox, G. B. "The effect of price advertising on alcoholic beverage sales," *Journal of Advertising Research,* Vol. 25, No. 5, October/November 1985, pp. 33–37.

c. Is there sufficient evidence to indicate that the mean total bimonthly beer sales differ among the three periods? Test using $\alpha = .10$.

d. Calculate the value of s for the experiment and compare this to the value for the completely randomized design in Exercise 8.12. Does it appear that blocking on months was effective in reducing the month-to-month variation in beer sales?

e. Conduct a test to determine whether blocking was effective in reducing month-to-month variation in beer sales. Use $\alpha = .10$. Does the result support your answer to part **d**?

S E C T I O N   8.7

TWO-WAY CLASSIFICATIONS OF DATA: FACTORIAL EXPERIMENTS

A randomized block design is often called a **two-way classification of data** because it has the following characteristics:

1. It involves two independent variables—one factor and one direction of blocking.
2. Each level of one independent variable occurs with every level of the other independent variable.

A two-way classification of data always permits the display of the data in a two-way table, one containing r rows and c columns. For example, the data for the randomized block design of Example 8.8 are displayed in a two-way table containing $r = 3$ rows and $c = 4$ columns in Table 8.6. Each of the $rc = (3)(4) = 12$ cells of the table contains one observation.

The treatment selection for a two-factor experiment may also yield a two-way classification of data. For example, suppose you want to relate the mean number of defects on a finished item—say, a new desk top—to two factors, type of nozzle for the varnish spray gun and length of spraying time. Suppose further that you want to investigate the mean number of defects per desk for three types (three levels) of nozzles (N_1, N_2, and N_3) and for two lengths (two levels) of spraying time (S_1 and S_2). If we choose the treatments for the experiment to include all combinations of the three levels of nozzle type with the two levels of spraying time, i.e., we observe number of defects for the factor–level combinations N_1S_1, N_1S_2, N_2S_1, N_2S_2, N_3S_1, N_3S_2, we will obtain a two-way classification of data. This selection of treatments is called a **complete 3 × 2 factorial experiment**. Note that the design, called a **factorial design**, will contain $3 \times 2 = 6$ treatments.

DEFINITION 8.9

A **factorial design** is a method for selecting the treatments (that is, the factor-level combinations) to be included in an experiment. A **complete factorial experiment** is one in which the treatments consist of all factor-level combinations.

If we were to include a third factor, say, paint type, at three levels, then a complete factorial experiment would include all $3 \times 2 \times 3 = 18$ combinations of nozzle type, spraying time, and paint type. The resulting collection of data would be called a **three-way classification of data**.

EXAMPLE 8.10

Suppose you plan to conduct a marketing study of coffee sales at a supermarket to investigate the effect of three factors: shelf display (one of three types, A_1, A_2, and A_3), shelf location (bottom, B_1, middle, B_2, and top, B_3) and brand (C_1 and C_2). Note that factor interactions could be very significant in this experiment. For example, brand C_1 might sell very well, regardless of the shelf display and shelf location, while the sales of C_2 may be greatly enhanced by using a particular display and locating it on a top shelf. Consequently, you will want to conduct a complete factorial experiment. Identify the 18 treatments for this $3 \times 3 \times 2$ factorial experiment.

SOLUTION

The complete factorial experiment includes all possible combinations of shelf display, shelf location, and brand. We therefore would include the following treatments: $A_1B_1C_1$, $A_1B_1C_2$, $A_1B_2C_1$, $A_1B_2C_2$, $A_1B_3C_1$, $A_1B_3C_2$, $A_2B_1C_1$, $A_2B_1C_2$, $A_2B_2C_1$, $A_2B_2C_2$, $A_2B_3C_1$, $A_2B_3C_2$, $A_3B_1C_1$, $A_3B_1C_2$, $A_3B_2C_1$, $A_3B_2C_2$, $A_3B_3C_1$, $A_3B_3C_2$. ∎

Factorial experiments are useful methods for selecting treatments because they permit us to make inferences about **factor interactions**. In this section we will learn how to perform an analysis of variance for a two-way classification of data. You will see that the computational procedure is the same for both the randomized block design and for the two-factor factorial experiment because each involves a two-way classification of data. You will also learn why the analysis of variance F-tests differ and how these tests can be used to interpret the results of the experiment.

Suppose a two-way classification represents a two-factor factorial experiment with factor A at a levels and factor B at b levels. Further, assume that the ab treatments of the factorial experiment are replicated r times so that there are r observations for each of the ab treatment combinations (i.e., there are r observations in each of the ab cells of the two-way table). Then the total number of observations is $n = abr$ and the total sum of squares, SS(Total), can be partitioned into four parts, SS(A), SS(B), SS(AB), and SSE (see Figure 8.9). The first two sums of squares, SS(A) and SS(B), are called **main effect sums of squares** to distinguish them from the **interaction sum of squares**, SS(AB).

FIGURE 8.9
Partitioning of the Total Sum of Squares for a Two-Factor Factorial Experiment

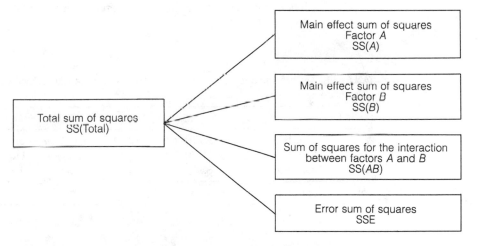

When the number of observations per cell for a two-way factorial experiment is the same for every cell (r observations per cell), the sums of squares and the degrees of freedom for the analysis of variance are additive:

$$SS(Total) = SS(A) + SS(B) + SS(AB) + SSE$$

and

$$abr - 1 = (a - 1) + (b - 1) + (a - 1)(b - 1) + ab(r - 1)$$

and the analysis of variance table would appear as shown in Table 8.9 (page 424). Note that for a factorial experiment, *the number r of observations per factor-level combination must always be 2 or more* (i.e., $r \geq 2$). Otherwise, you will not have any degrees of freedom for SSE.

TABLE 8.9 ANOVA Table for a Two-Way Classification of Data with r Observations per Cell: Neither Variable Is a Direction of Blocking

SOURCE	df	SS	MS	F
Main effects A	$(a-1)$	SS(A)	MS(A) = SS(A)/$(a-1)$	MS(A)/MSE
Main effects B	$(b-1)$	SS(B)	MS(B) = SS(B)/$(b-1)$	MS(B)/MSE
AB interaction	$(a-1)(b-1)$	SS(AB)	MS(AB) = SS(AB)/$[(a-1)(b-1)]$	MS(AB)/MSE
Error	$ab(r-1)$	SSE	MSE = SSE/$[ab(r-1)]$	
Total	$abr-1$	SS(Total)		

If either A or B represents a direction of blocking, then the degrees of freedom and sums of squares for AB interaction (lines 3 and 4 of Table 8.9) are combined to form a source of error variation. This is because the block–treatment interaction always represents experimental error.* The resulting analysis of variance would appear as shown in Table 8.10. Note that both the degrees of freedom and the sums of squares sum to those associated with the total sum of squares of deviations, SS(Total).

TABLE 8.10 ANOVA Table for a Two-Way Classification of Data with r Observations per Cell When B Is a Direction of Blocking

SOURCE	df	SS	MS	F
A	$a-1$	SS(A)	MS(A) = SS(A)/$(a-1)$	MS(A)/MSE
Blocks	$b-1$	SS(B)	MS(B) = SS(B)/$(b-1)$	MS(B)/MSE
Error	$abr-a-b+1$	SSE	MSE = SSE/$(abr-a-b+1)$	
Total	$abr-1$	SS(Total)		

The notation used in the formulas for the respective sums of squares and the formulas for the sums of squares are given in the boxes on pages 425–426.

*The failure of the difference between a pair of treatments to remain the same from block to block is experimental error. If we think of a randomized block design in the context of a factorial experiment with treatments as one factor and blocks as another, then the failure of the difference between two treatments to remain the same from block to block is, by Definition 8.8, block–treatment interaction. In other words, in a randomized block design, block–treatment interaction and experimental error are synonymous.

NOTATION FOR THE ANALYSIS OF VARIANCE FOR A TWO-WAY CLASSIFICATION OF DATA

a = Number of levels of independent variable 1

b = Number of levels of independent variable 2

r = Number of measurements for each pair of levels of independent variables 1 and 2

A_i = Total of all measurements of independent variable 1 at level i ($i = 1, 2, \ldots, a$)

$\bar{A}_i$ = Mean of all measurements of independent variable 1 at level i ($i = 1, 2, \ldots, a$)

$\quad = \dfrac{A_i}{br}$

B_j = Total of all measurements of independent variable 2 at level j ($j = 1, 2, \ldots, b$)

$\bar{B}_j$ = Mean of all measurements of independent variable 2 at level j ($j = 1, 2, \ldots, b$)

$\quad = \dfrac{B_j}{ar}$

AB_{ij} = Total of all measurements at the ith level of independent variable 1 and at the jth level of independent variable 2 ($i = 1, 2, \ldots, a$; $j = 1, 2, \ldots, b$)

n = Total number of measurements

$\quad = a \times b \times r$

We have already explained (Section 8.6) how to conduct the F-tests for determining whether a difference exists among treatment means (or block means) if the data are obtained from a randomized block design and the analysis of variance is as shown in Table 8.10. The confidence intervals for estimating the difference between two treatment means are also given in Section 8.6.

If the data are obtained from a two-factor factorial experiment, the analysis of variance will appear as shown in Table 8.9. To test a hypothesis for any one of the three sources of variation (the source corresponding to main effects factor A, the source corresponding to main effects factor B, or the source corresponding to the interaction between the two factors), you proceed in exactly the same manner as was done in earlier analyses—that is, you divide the appropriate mean square by MSE and use this F-ratio as a test statistic. The analysis of variance F-tests for a two-factor factorial experiment are summarized in the box at the top of page 427.

COMPUTING FORMULAS FOR THE ANALYSIS OF VARIANCE FOR A TWO-WAY CLASSIFICATION OF DATA

CM = Correction for the mean

$$= \frac{(\text{Total of all } n \text{ measurements})^2}{n}$$

$$= \frac{\left(\sum_{i=1}^{n} y_i\right)^2}{n}$$

$SS(\text{Total})$ = Total sum of squares

= Sum of squares of all n measurements $-$ CM

$$= \sum_{i=1}^{n} y_i^2 - CM$$

$SS(A)$ = Sum of squares for main effects, independent variable 1

$$= \left(\begin{array}{c}\text{Sum of squares of the totals } A_1, A_2, \ldots, A_a \\ \text{divided by the number of measurements} \\ \text{in a single total, namely } br\end{array}\right) - CM$$

$$= \frac{\sum_{i=1}^{a} A_i^2}{br} - CM$$

$SS(B)$ = Sum of squares for main effects, independent variable 2

$$= \left(\begin{array}{c}\text{Sum of squares of the totals } B_1, B_2, \ldots, B_b \\ \text{divided by the number of measurements} \\ \text{in a single total, namely } ar\end{array}\right) - CM$$

$$= \frac{\sum_{i=1}^{b} B_i^2}{ar} - CM$$

$SS(AB)$ = Sum of squares for AB interaction

$$= \left(\begin{array}{c}\text{Sum of squares of the cell totals} \\ AB_{11}, AB_{12}, \ldots, AB_{ab} \text{ divided by} \\ \text{the number of measurements} \\ \text{in a single total, namely } r\end{array}\right) - SS(A) - SS(B) - CM$$

$$= \frac{\sum_{i=1}^{b} \sum_{i=1}^{a} AB_{ij}^2}{r} - SS(A) - SS(B) - CM$$

ANALYSIS OF VARIANCE F-TESTS FOR A TWO-FACTOR FACTORIAL EXPERIMENT

TEST FOR FACTOR INTERACTION

H_0:   No interaction between factors A and B

H_a: Factors A and B interact

Test statistic: $F = \dfrac{MS(AB)}{MSE} = \dfrac{MS(AB)}{s^2}$

Rejection region: $F > F_\alpha$, where F is based on $\nu_1 = (a-1)(b-1)$ and
$\nu_2 = ab(r-1)$df

TEST FOR MAIN EFFECTS FOR FACTOR A

H_0:   There are no differences among the means for main effect A

H_a: At least two of the main effect A means differ

Test statistic: $F = \dfrac{MS(A)}{MSE} = \dfrac{MS(A)}{s^2}$

Rejection region: $F > F_\alpha$, where F is based on $\nu_1 = (a-1)$ and
$\nu_2 = ab(r-1)$df

TEST FOR MAIN EFFECTS FOR FACTOR B

H_0:   There are no differences among the means for main effect B

H_a: At least two of the main effect B means differ

Test statistic: $F = \dfrac{MS(B)}{MSE} = \dfrac{MS(B)}{s^2}$

Rejection region: $F > F_\alpha$, where F is based on $\nu_1 = (b-1)$ and
$\nu_2 = ab(r-1)$df

To estimate the mean for a single cell of the two-way table or the difference between the means for two cells (i.e., two different combinations of levels of the two independent variables), use the formulas in the next two boxes.

$100(1-\alpha)$% CONFIDENCE INTERVAL FOR THE MEAN OF A SINGLE CELL OF THE TWO-WAY TABLE: FACTORIAL EXPERIMENT

$$\bar{y}_{ij} \pm t_{\alpha/2}\frac{s}{\sqrt{r}}$$

where $\bar{y}_{ij}$ is the cell mean for the cell in the ith row, jth column,

r = Number of measurements per cell

$s = \sqrt{MSE}$

and $t_{\alpha/2}$ is based on $ab(r-1)$df.

> **100(1 − α)% CONFIDENCE INTERVAL FOR THE DIFFERENCE IN A PAIR OF CELL MEANS: FACTORIAL EXPERIMENT**
>
> Let
>
> $\bar{y}_1$ = Sample mean of the r measurements in the first cell
>
> $\bar{y}_2$ = Sample mean of the r measurements in the second cell
>
> Then, the $100(1 - \alpha)\%$ confidence interval for the difference between the cell means is
>
> $$(\bar{y}_1 - \bar{y}_2) \pm t_{\alpha/2}s\sqrt{\frac{2}{r}}$$
>
> where $s = \sqrt{\text{MSE}}$ and $t_{\alpha/2}$ is based on $ab(r - 1)$df.

Before we work through a numerical example of an analysis of variance for a factorial experiment, we need to understand the practical significance of the tests for factor interaction and factor main effects. We illustrate these concepts in Example 8.11.

EXAMPLE 8.11

A company that stamps gaskets out of sheets of rubber, plastic, and other materials, wants to compare the mean number of gaskets produced per hour for two different types of stamping machines. Practically, the manufacturer wants to determine whether one machine is more productive than the other and, even more important, whether one machine is more productive in producing rubber gaskets while the other is more productive in producing plastic gaskets. To answer these questions, the manufacturer decides to conduct a 2×3 factorial experiment using three types of gasket material, B_1, B_2, and B_3, with each of the two types of stamping machines, A_1 and A_2. Each machine is operated for three 1-hour time periods for each of the gasket materials, with the eighteen 1-hour time periods assigned to the six machine–material combinations in random order. (The purpose of the randomization is to eliminate the possibility that uncontrolled environmental factors might bias the results.) Suppose we have calculated and plotted the six treatment means. Two hypothetical plots of the six means are shown in Figures 8.10(a) and (b). The three means for stamping machine A_1 are connected by solid line segments and the corresponding three means for machine A_2 by dashed line segments. What do these plots imply about the productivity of the two stamping machines?

SOLUTION

Figure 8.10(a) suggests that machine A_1 produces a larger number of gaskets per hour, regardless of the gasket material, and is therefore superior to machine A_2. On the average, machine A_1 stamps more cork (B_1) gaskets per hour than rubber or plastic, but the *difference* in the mean numbers of gaskets produced by the two machines remains approximately the same, regardless of the gasket material. Thus, the difference in the mean number of gaskets produced by the two machines is *independent* of the gasket material used in the stamping process.

FIGURE 8.10

Hypothetical Plot of
the Means for the Six
Machine–Material
Combinations

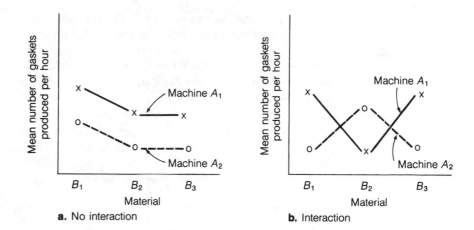

a. No interaction   **b.** Interaction

In contrast to Figure 8.10(a), Figure 8.10(b) shows the productivity for machine A_1 to be greater than for machine A_2, when the gasket material is cork (B_1) or plastic (B_3). But the means are reversed for rubber (B_2) gasket material. For this material, machine A_2 produces, on the average, more gaskets per hour than machine A_1. Thus, Figure 8.10(b) illustrates a situation where the mean value of the response variable *depends* on the combination of the factor levels. When this situation occurs, we say that the factors *interact*. Thus, one of the most important objectives of a factorial experiment is to detect factor interaction if it exists. ∎

DEFINITION 8.10

In a factorial experiment, when the difference in the mean levels of factor A depends on the different levels of factor B, we say that the factors A and B **interact**. If the difference is independent of the levels of B, then there is **no interaction** between A and B.

Tests for main effects are relevant only when no interaction exists between factors. Generally, the test for interaction is performed first. If there is evidence of factor interaction, then we will not perform the tests on the main effects. Rather, we will want to focus attention on the individual cell (treatment) means, perhaps locating one that is the largest or the smallest.

EXAMPLE 8.12

A manufacturer, whose daily supply of raw materials is variable and limited, can use the material to produce two different products in various proportions. The profit per unit of raw material obtained by producing each of the two products depends on the length of a product's manufacturing run and hence on the amount of raw material assigned to it. Other factors, such as worker productivity and machine breakdown, affect the profit per unit as well, but their net effect on profit is random and uncontrollable. The manufacturer has conducted an experiment to investigate the effect of the level of supply of raw materials, S, and the ratio of its assignment, R, to the two product manufacturing lines on the profit y per unit of raw material. The ultimate goal would be to be able to choose the best ratio R to match each day's supply of raw materials, S. The levels of supply of the raw material chosen for the experiment were 15, 18, and 21 tons; the levels of the ratio of allocation to the two product lines were $\frac{1}{2}$, 1, and 2. The response was the profit (in dollars) per unit of raw material supply obtained from a single day's production. Three replications of a complete 3×3 factorial experiment were conducted in a random sequence (i.e., a completely randomized design). The data for the 27 days are shown in Table 8.11.

TABLE 8.11

Data for Example 8.12

		RAW MATERIAL SUPPLY, TONS (S)		
		15	18	21
RATIO OF RAW MATERIAL ALLOCATION (R)	$\frac{1}{2}$	23, 20, 21	22, 19, 20	19, 18, 21
	1	22, 20, 19	24, 25, 22	20, 19, 22
	2	18, 18, 16	21, 23, 20	20, 22, 24

a. Calculate the appropriate sums of squares and construct an ANOVA table.
b. Do the data present sufficient evidence to indicate an interaction between supply S and ratio R?
c. Find a 95% confidence interval to estimate the mean profit per unit of raw materials when $S = 18$ tons and the ratio of production is $R = 1$.
d. Find a 95% confidence interval to estimate the difference in mean profit per unit of raw materials between $S = 18$, $R = \frac{1}{2}$, and $S = 18$, $R = 1$.

SOLUTION

a. The sums of squares for the ANOVA table are calculated as follows:

$$CM = \frac{(\text{Total of all } n \text{ measurements})^2}{n} = \frac{(558)^2}{27} = 11{,}532$$

$$SS(\text{Total}) = \sum_{i=1}^{n} y_i^2 - CM = 11{,}650 - 11{,}532 = 118$$

The next step is to construct a table showing the totals of y-values for each combination of levels of supply S and ratio R and then the totals for each level of S and for each level of R. These totals, computed from the raw data table, are shown in Table 8.12. Then,

$$SS(\text{Supply}) = \frac{\sum_{i=1}^{3} S_i^2}{9} - CM = \frac{(177)^2 + (196)^2 + (185)^2}{9} - CM$$

$$= \frac{103{,}970}{9} - 11{,}532 = 20.22$$

$$SS(Ratio) = \frac{\sum_{i=1}^{3} R_j^2}{9} - CM = \frac{(183)^2 + (193)^2 + (182)^2}{9} - CM$$

$$= \frac{103,862}{9} - 11,532 = 8.22$$

$$SS(SR) = \sum_{j=1}^{3} \sum_{i=1}^{3} \frac{SR_{ij}^2}{3} - SS(Supply) - SS(Ratio) - CM$$

$$= \frac{(64)^2 + (61)^2 + (58)^2 + \cdots + (66)^2}{3} - 20.22 - 8.22 - 11,532$$

$$= \frac{34,820}{3} - 20.22 - 8.22 - 11,532 = 46.22*$$

$$SSE = SS(Total) - SS(Supply) - SS(Ratio) - SS(SR)$$

$$= 118.00 - 20.22 - 8.22 - 46.22 = 43.33$$

The ANOVA table is given in Table 8.13.

TABLE 8.12
Totals for the Data in Table 8.11

		RAW MATERIAL SUPPLY, TONS (S)			
		15	18	21	
RATIO OF RAW MATERIAL ALLOCATION (R)	$\frac{1}{2}$	64	61	58	$R_1 = 183$
	1	61	71	61	$R_2 = 193$
	2	52	64	66	$R_3 = 182$
		$S_1 = 177$	$S_2 = 196$	$S_3 = 185$	Total = 558

TABLE 8.13
ANOVA Table for Example 8.12

SOURCE	df	SS	MS
Supply	2	20.22	10.11
Ratio	2	8.22	4.11
Supply–Ratio interaction	4	46.22	11.56
Error	18	43.33	2.41
Total	26	118.00	

b. To test the null hypothesis that supply and ratio do not interact, we use the test statistic

$$F = \frac{MS(SR)}{s^2} = \frac{11.56}{2.41} = 4.80$$

The degrees of freedom associated with MS(SR) and MSE are 4 and 18, respectively (given in Table 8.13). Therefore, we reject H_0 if $F > F_{.05}$, where $\nu_1 = 4$, $\nu_2 = 18$, and $F_{.05} = 2.93$. Since the computed value of $F(4.80)$ exceeds

*Use full accuracy in all sums of squares.

$F_{.05}$, we reject H_0 and conclude that supply and ratio interact. The presence of interaction tells you that the mean profit depends on the particular combination of levels of supply S and ratio R. Consequently, there is little point in checking to see whether the means differ for the three levels of supply or whether they differ for the three levels of ratio (i.e., we will not perform the tests for main effects). For example, the supply level that gave the highest mean profit (over all levels of R) might not be the same supply–ratio level combination that produces the largest mean profit per unit of raw material.

c. A 95% confidence interval for the mean $E(y)$ when supply $S = 18$ and ratio $R = 1$ is

$$\bar{y}_{18,1} \pm t_{.025}\left(\frac{s}{\sqrt{r}}\right)$$

where $\bar{y}_{18,1}$ is the mean of the $r = 3$ values of y obtained for $S = 18$, $R = 1$ and $t_{.025} = 2.101$ is based on 18 df. Substituting, we obtain

$$\frac{71}{3} \pm 2.101\left(\frac{1.55}{\sqrt{3}}\right)$$

$$23.67 \pm 1.88$$

Therefore, our interval estimate for the mean profit per unit of raw material where $S = 18$ and $R = 1$ is \$21.79 to \$25.55.

d. A 95% confidence interval for the difference in mean profit per unit of raw material for two different combinations of levels of S and R is

$$(\bar{y}_1 - \bar{y}_2) \pm t_{.025}s\sqrt{\frac{2}{r}}$$

where $\bar{y}_1$ and $\bar{y}_2$ represent the means of the $r = 3$ replications for the factor-level combinations $S = 18$, $R = \frac{1}{2}$ and $S = 18$, $R = 1$, respectively. Substituting, we obtain

$$\left(\frac{61}{3} - \frac{71}{3}\right) \pm (2.101)(1.55)\sqrt{\frac{2}{3}}$$

$$-3.33 \pm 2.66$$

Therefore, the interval estimate for the difference in mean profit per unit of raw material for the two factor-level combinations is $(-\$5.99, -\$.67)$. The negative values indicate that we estimate the mean for $S = 18$, $R = \frac{1}{2}$ to be less than the mean for $S = 18$, $R = 1$ by between \$.67 and \$5.99.   ∎

The techniques illustrated in the solution of Example 8.12 would answer most of the practical questions you might have about profit per unit of supply of raw materials if the two independent variables affecting the response were qualitative. But since both independent variables are quantitative, we can obtain much more information about their effect on response by performing a regression analysis.

For example, the analysis of variance in Example 8.12 enables us to estimate the mean profit per unit of supply for *only* the nine combinations of supply–ratio levels. It will not permit us to estimate the mean response for some other combination of levels of the independent variables not included among the nine used in the factorial experiment. For example, the prediction equation obtained from a regression analysis would enable us to estimate the mean profit per unit of supply when $S = 17$, $R = 1$. We could not obtain this estimate from the analysis of variance in Example 8.12.

The prediction equation (i.e., the estimated response surface) found by regression analysis also contributes other information not provided by an analysis of variance. For example, we might wish to estimate the rate of change in the mean profit, $E(y)$, for unit changes in S, R, or both for specific values of S and R. We illustrate some of these applications of regression analysis to an analysis of variance in optional Section 8.10.

EXERCISES 8.23–8.34

8.23 Suppose you plan to investigate the effect of hourly pay rate and length of workday on some measure y of worker productivity. Both pay rate and length of workday will be set at three levels and y will be observed for all combinations of these factors.
a. What type of experiment is this?
b. Identify the factors and state whether they are quantitative or qualitative.
c. Identify the treatments to be employed in the experiment.

8.24 A beverage company wants to investigate the effect of advertising expenditures and markets in two urban areas (M_1 and M_2) on the mean monthly sales of its best-selling diet cola. Advertising expenditure was set at $15,000, $18,000, $21,000, $24,000, and $27,000, respectively, for the next five months in each of the two markets and the monthly sales were observed.
a. List the factors involved in the experiment.
b. For each factor, state whether it is quantitative or qualitative.
c. How many treatments are involved in this experiment? List them.

8.25 Consider a factorial design with two factors, A and B, each at three levels. Suppose we select the following treatment (factor-level) combinations to be included in the experiment: A_1B_1, A_2B_1, A_3B_1, A_1B_2, and A_1B_3.
a. Is this a complete factorial experiment? Explain.
b. Explain why it is impossible to investigate AB interaction in this experiment.

8.26 The analysis of variance for a 3×2 factorial experiment, with four observations per treatment, produced the ANOVA table entries shown here.

SOURCE	df	SS	MS	F
A		100		
B	1			
AB	2		2.5	
Error			2.0	
Total		700		

a. Complete the ANOVA table.
b. Test for interaction between factor A and factor B. Use $\alpha = .05$.
c. Test for differences in main effect means for factor A. Use $\alpha = .05$.
d. Test for differences in main effect means for factor B. Use $\alpha = .05$.

8.27 Refer to Example 8.11 and the factorial experiment designed to measure the effect of two factors, gasket material and stamping machine, on productivity of a manufacturing process. The data for the 2×3 factorial experiment, number of gaskets produced per hour (in thousands), are shown in the accompanying table.

		GASKET MATERIAL			TOTAL
		Cork, B_1	Rubber, B_2	Plastic, B_3	
STAMPING MACHINE	A_1	4.31 4.27 4.40	3.36 3.42 3.48	4.01 3.94 3.89	35.08
	A_2	3.94 3.81 3.99	3.91 3.80 3.85	3.48 3.53 3.42	33.73
TOTAL		24.72	21.82	22.27	68.81

a. Construct an ANOVA summary table.
b. Is there evidence of interaction between gasket material and stamping machine? Test using $\alpha = .01$.
c. Explain the practical significance of the result obtained in part **b**.
d. Based on parts **b** and **c**, would you recommend that tests for main effects be conducted? Explain.
e. Find a 95% confidence interval for the difference in the mean number of gaskets produced by machines A_1 and A_2, when stamping cork (B_1) gaskets. Interpret the interval.

8.28 Many utility companies are experimenting with "time-of-day" pricing in order to defray the mounting costs of producing electricity during peak demand periods (e.g., 9 A.M. to 9 P.M. in the summer). *Time-of-day pricing* is a plan by which electric customers are charged a less expensive rate for using electricity during off-peak (less demanded) hours. One experiment (reported in *Journal of Consumer Research*, June 1982) was conducted to measure customer satisfaction with several time-of-day pricing schemes. The experiment consisted of two factors, price ratio (the ratio of peak to off-peak prices) and peak period length, each at three levels. The $3 \times 3 = 9$ combinations of price ratio and peak period length represent the nine time-of-day pricing schemes. For each pricing scheme, customers were randomly selected and asked to rate satisfaction with the plan using an index from 10 to 38, with 38 indicating extreme satisfaction. Suppose that four customers were sampled for each pricing scheme. The accompanying table gives the satisfaction scores for these customers. [*Note:* The data are based on mean scores provided in the *Journal of Consumer Research* article.]

a. Construct an ANOVA table for the data.
b. Compute the nine customer satisfaction index means.
c. Plot the nine means from part **b** on a graph similar to Figure 8.10. Does it appear that the two factors, price ratio and peak period length, interact? Explain.

| | | PRICING RATIO | | |
		2:1	4:1	8:1
PEAK PERIOD LENGTH	6 hours	25 28 26 27	31 29 26 27	24 28 25 26
	9 hours	26 27 29 30	25 24 30 26	33 28 25 27
	12 hours	22 20 25 21	33 27 25 27	30 31 26 27

d. Do the data provide sufficient evidence of interaction between price ratio and peak period length? Test using $\alpha = .05$.

e. Do the data provide sufficient evidence that mean customer satisfaction differs for the three peak period lengths? Test using $\alpha = .05$.

f. When is the test of part **e** appropriate?

g. Find a 90% confidence interval for the mean customer satisfaction rating of a pricing scheme with a peak period length of 9 hours and pricing ratio of 2:1.

8.29 Many television and radio commercials utilize high intensity stimulation (e.g., rapid changes in visual imagery, quick movements, bright lights, and loud sounds) to increase the general arousal of the viewer or listener. In theory, the more aroused the viewer, the more likely he or she is to remember the advertised product. To examine the effect of high intensity advertising on the audience, a study was conducted at a small midwestern liberal arts college. Two groups of psychology students (ten introverts and ten extroverts) took part in the experiment. (Scores on the Eysenck Personality Questionnaire were used to classify the students into one of the two groups.) Five subjects from each group were then randomly assigned to one of two experimental conditions: high volume commercial or normal volume commercial. Thus, the experiment consists of two factors, personality type and commercial volume, each at two levels. All students listened to a 5-minute tape-recorded "radio program" which included a commercial for a fictitious brand of chewing gum played at the assigned volume, and each subject's attitude toward the product was measured on a 14-point scale (1 = strongly dislike, 14 = strongly like). Since five responses were recorded for each of the $2 \times 2 = 4$ factor-level combinations, the factorial experiment includes five replicates of each treatment.

a. A partial ANOVA table for the experiment is shown below. Calculate as many of the missing entries as possible.

SOURCE	df	SS	MS	F
Personality (P)	1			
Volume (V)				4.61
VP				11.39
Error			15	
Total				

Source: Cetola, H. and Prinkley, K. "Introversion-extraversion and loud commercials," *Psychology & Marketing*, Vol. 3, No. 2, Summer 1986, pp. 123–132.

b. The means of the four treatments are given in the table. Plot the means in a graph similar to Figure 8.10. Does it appear that the two factors, personality type and commercial volume, interact?

	High Volume	Normal Volume
Extroverts	10.6	7.0
Introverts	7.2	6.0

c. Conduct a test to determine whether personality type interacts with commercial volume. Use $\alpha = .05$. Interpret the result.

8.30 An experiment was conducted to determine the effect of sintering time (two levels) on the compressive strength of two different metals. Five test specimens were sintered for each metal at each of the two sintering times. The data (in thousands of pounds per square inch) are shown in the accompanying table.

		SINTERING TIME	
		100 minutes	200 minutes
METAL	1	17.1 16.5 14.9 15.2 16.7	19.4 18.9 20.1 17.2 20.7
	2	12.3 13.8 10.8 11.6 12.1	15.6 17.2 16.7 16.1 18.3

a. Perform an analysis of variance for the data and construct an ANOVA table.
b. What is the practical significance of an interaction between sintering time and metal type?
c. Do the data provide sufficient evidence to indicate an interaction between sintering time and metal type? Test using $\alpha = .05$.

8.31 A company conducted an experiment to determine the effects of three types of incentive pay plans on worker productivity for both union and non-union workers. The company used plants in adjacent towns; one was unionized and the other was not. One-third of the production workers in each plant were assigned to each incentive plan. Then six workers were randomly selected from each group and their productivity (in number of items produced) was measured for a 1-week period. The six productivity measures for the 2×3 factor combinations are listed in the accompanying table.

		INCENTIVE PLAN		
		A	B	C
UNION AFFILIATION	Union	337 328 362 319 305 344	346 373 351 338 355 365	317 341 335 329 310 315
	Non-union	359 346 345 396 381 373	371 377 352 401 399 378	350 336 349 351 374 340

a. Perform an analysis of variance for the data and construct an ANOVA table.

b. If an interaction between union affiliation and incentive plan is present, what implication would it have on the selection of an incentive plan for a particular company plant?

c. Do the data present sufficient evidence to indicate an interaction between union affiliation and incentive plan? Test using $\alpha = .05$.

d. Find a 90% confidence interval for the mean productivity for a unionized worker on incentive plan B.

e. Find a 90% confidence interval for the difference in mean productivity between union and non-union workers on incentive plan B.

8.32 How does a worker's sex or rank in a company affect other people's evaluations of the worker's performance or qualifications? W. R. Morrow and G. Lowenberg of the University of Wisconsin–Parkside conducted a study to determine the effects of a writer's sex and organizational position on evaluations of business memos. In one portion of the study, each in a sample of approximately 100 subjects (all members of the Wisconsin Personnel and Industrial Relations Association) was asked to rate a poorly written memo on a 7-point scale (1 = very poor, 7 = very good). The attributed sex (male or female) of the supposed author of the memo and the attributed memo author's organizational position (executive or assistant) were varied from memo to memo. Thus, the experiment consists of two factors, author sex and author position, with each at two levels. Assume that 25 subjects are sampled for each of the four memo types. The accompanying table gives the totals of the rating scores for each memo type.

		AUTHOR SEX		TOTALS
		Male	Female	
AUTHOR POSITION	Executive	76	94	170
	Assistant	89	82	171
TOTALS		165	176	341

Source: Morrow, W. R. and Lowenberg, G. "Evaluation of business memos: Effects of writer sex on organizational position, memo quality, and rater sex," *Personnel Psychology*, Spring 1983, 36, pp. 73–85.

a. Compute the main effect and interaction sums of squares for the two factors, sex and position.

b. Assume that the value of s^2 obtained from the analysis of variance is 2.10. Calculate SSE and SS(Total).

c. Construct an ANOVA table for the data.

d. What do we mean when we say that the two factors, sex and position, interact? Illustrate with a graph.

e. Is there sufficient evidence to indicate an interaction between sex and position? Test using $\alpha = .01$.

8.33 A recent trend in advertising and marketing is the use of the chief executive officer (CEO) as the spokesperson for the company (e.g., Lee Iacocca for Chrysler, Frank Borman for Eastern, and Frank Sellinger for Schlitz). A study was conducted to determine the perceived effectiveness of CEOs as marketing spokespersons. Eleven magazine advertisements in which CEOs acted as spokespersons were selected for analysis. Each ad was rated by a sample of 58 MBA students on each of eight items designed to measure credibility. Thus, the experiment consists of two factors, CEO and credibility, with CEO at 11 levels and credibility at 8 levels (well known, believable, likeable, integrity, knowledgeable, expert, influential, and powerful). The 58 ratings for each of the $8 \times 11 = 88$ treatments (a total of 5,104 observations) were subjected to an analysis of variance, with the results shown in the accompanying ANOVA table. Interpret these results.

SOURCE	df	SS	MS	F
CEO	10	2,609.5	260.95	140.34
Credibility	7	1,012.5	144.64	77.79
CEO–Credibility	70	636.7	9.10	4.89
Error	5,016	9,326.2	1.86	
Total	5,103	13,584.9		

Source: Reidenbach, R. E. and Pitts, R. E. "Not all CEOs are created equal as advertising spokespersons: Evaluating the effective CEO spokesperson," *Journal of Advertising*, Vol. 15, No. 1, January 1986, pp. 30–36.

8.34 How do women compare with men in their ability to perform laborious tasks that require strength? Some information on this question is provided in a study, by M. D. Phillips and R. L. Pepper, of the firefighting ability of men and women. Phillips and Pepper conducted a 2×2 factorial experiment to investigate the effect of the factor sex (male or female) and the factor weight (light or heavy) on the length of time required for a person to perform a particular firefighting task. Eight persons were selected for each of the $2 \times 2 = 4$ sex–weight categories of the 2×2 factorial experiment and the length of time needed to complete the task was recorded for each of the 32 persons. The means and standard deviations of the four samples are shown in the table.

	LIGHT WEIGHT		HEAVY WEIGHT	
	Mean	Standard Deviation	Mean	Standard Deviation
FEMALE	18.30	6.81	14.50	2.93
MALE	13.00	5.04	12.25	5.70

Source: Phillips, M. D. and Pepper, R. L. "Shipboard fire-fighting performance of females and males," *Human Factors*, Vol. 24, No. 3, 1982. Copyright 1982 by the Human Factors Society, Inc. Reproduced by permission.

a. Calculate the total of the $n = 8$ time measurements for each of the four categories of the 2×2 factorial experiment.
b. Calculate CM.
c. Use the results of part **a** and **b** to calculate the sums of squares for sex, weight, and the sex–weight interaction.
d. Calculate each sample variance. Then calculate the sums of squares of deviations *within* each sample for each of the four samples.
e. Calculate SSE. [*Hint:*   SSE is the pooled sum of squares of the deviations calculated in part **d**.]

f. Now that you know SS(Sex), SS(Weight), SS(Sex–Weight), and SSE, find SS(Total).

g. Summarize the calculations in an analysis of variance table.

h. Explain the practical significance of the presence (or absence) of sex–weight interaction. Do the data provide evidence of an interaction between sex and weight?

i. Construct a 95% confidence interval for the difference in mean time to complete the task between light men and light women. Interpret the interval.

j. Construct a 95% confidence interval for the difference in mean time to complete the task between heavy men and heavy women. Interpret the interval.

| | | | | | | | | | | | |
SECTION 8.8

TUKEY'S PROCEDURE FOR MAKING MULTIPLE COMPARISONS

Many practical experiments are conducted to determine the largest (or the smallest) mean in a set. For example, suppose a drugstore is considering five floor displays for a new product. The drugstore would want to determine which display yields the greatest weekly sales of the product. Similarly, a production engineer might want to determine which among six machines or which among three foremen achieves the highest mean productivity per hour. A stockbroker might want to choose one stock, from among four, that yields the highest mean return, and so on.

Choosing the treatment with the largest mean from among five treatments might appear to be a simple matter. We could make, for example, $n_1 = n_2 = \cdots = n_5 = 10$ observations on each treatment, obtain the sample means $\bar{y}_1$, $\bar{y}_2, \ldots, \bar{y}_5$, and compare them using Student's t-tests to determine whether differences exist among the pairs of means. However, there is a problem associated with this procedure: *A Student's t-test, with its corresponding value of α, is valid only when the two treatments to be compared are selected prior to experimentation.* After you have looked at the data, you should not use a Student's t-statistic to compare the treatments for the largest and smallest sample means because they will always be farther apart, on the average, than any pair of treatments selected at random. Furthermore, if you conduct a series of t-tests, each with a chance α of indicating a difference between a pair of means if in fact no difference exists, then the risk of making *at least one* Type I error in a series of t-tests will be larger than the value of α specified for a single t-test.

There are a number of procedures for comparing and ranking a group of treatment means. The one that we present in this section, known as **Tukey's method for multiple comparisons**, utilizes the Studentized range

$$q = \frac{\bar{y}_{max} - \bar{y}_{min}}{s/\sqrt{n}}$$

(where $\bar{y}_{max}$ and $\bar{y}_{min}$ are the largest and smallest sample means, respectively) to determine whether the difference in any pair of sample means implies a difference in the corresponding treatment means. The logic behind this **multiple comparisons procedure** is that if we determine a critical value for the difference between the largest and smallest sample means, $|\bar{y}_{max} - \bar{y}_{min}|$, one that implies a difference in their respective treatment means, then any other pair of sample means that differ by as much as or more than this critical value would also imply a difference in the corresponding treatment means. Tukey's (1949) procedure selects this critical distance, ω, so that the probability of making one or more Type I errors (concluding that a difference exists between a pair of treatment means if, in fact, they are identical) is α. Therefore, the risk of making a Type I error applies to the whole procedure, i.e., to the comparisons of all pairs of means in the experiment, rather than to a single comparison. Consequently, the value of α selected by the researchers is called an **experimentwise error rate** (in contrast to a **comparisonwise error rate**).

Tukey's procedure relies on the assumption that the p sample means are based on independent random samples, *each containing an equal number n_t of observations*. Then if $s = \sqrt{\text{MSE}}$ is the computed standard deviation for the analysis, the distance ω is

$$\omega = q(p, \nu)\frac{s}{\sqrt{n_t}}$$

The tabulated statistic $q_\alpha(p, \nu)$ is the critical value of the Studentized range, the value that locates α in the upper tail of the q distribution. This critical value depends on α, the number of treatment means involved in the comparison, and ν, the number of degrees of freedom associated with MSE, as shown in the box. Values of $q(p, \nu)$ for $\alpha = .05$ and $\alpha = .01$ are given in Tables 11 and 12, respectively, of Appendix D.

TUKEY'S MULTIPLE COMPARISONS PROCEDURE: EQUAL SAMPLE SIZES

1. Calculate

$$\omega = q_\alpha(p, \nu)\frac{s}{\sqrt{n_t}}$$

where

p = Number of sample means
$s = \sqrt{\text{MSE}}$
ν = Number of degrees of freedom associated with MSE
n_t = Number of observations in each of the p samples

$q_\alpha(p, \nu)$ = Critical value of the Studentized range (Tables 11 and 12 of Appendix D)

2. Rank the p sample means and place a bar over those pairs that differ by less than ω. Any pair of sample means not connected by an overbar (i.e., differing by more than ω) implies a difference in the corresponding population means.

EXAMPLE 8.13

A transistor manufacturer conducted an experiment to investigate the effects of two factors on productivity (measured in thousands of dollars of items produced) per 40-hour week. The factors were:

Length of work week (two levels): five consecutive 8-hour days or four consecutive 10-hour days

Number of coffee breaks (three levels): 0, 1, or 2

The experiment was conducted over a 12-week period with the $2 \times 3 = 6$ treatments assigned in a random manner to the 12 weeks. The data for this two-factor factorial experiment are shown in Table 8.14.

TABLE 8.14
Data for Example 8.13

| | | COFFEE BREAKS | | |
		0	1	2
LENGTH OF	4 days	101	104	95
		102	107	92
WORK WEEK	5 days	95	109	83
		93	110	87

a. Perform an analysis of variance for the data.
b. Compare the six population means using Tukey's multiple comparisons procedure. Use $\alpha = .05$.

SOLUTION

a. The first step in conducting an analysis of variance is to calculate the cell, row, and column totals for the data in Table 8.14. These totals are shown in Table 8.15.

TABLE 8.15
Totals for the Data
in Table 8.14

| | | COFFEE BREAKS, B | | | |
		0	1	2	
LENGTH OF WORK WEEK, L	4 days	203	211	187	$L_1 = 601$
	5 days	188	219	170	$L_2 = 577$
		$B_1 = 391$	$B_2 = 430$	$B_3 = 357$	Total $= 1{,}178$

Then we compute

$$CM = \frac{(\text{Total of all } n \text{ measurements})^2}{n} = \frac{(1{,}178)^2}{12} = 115{,}640.33$$

$$SS(\text{Total}) = \sum_{i=1}^{n} y_i^2 - CM = (101)^2 + (102)^2 + \cdots + (87)^2 - 115{,}640.33$$

$$= 831.67$$

$$SS(\text{Length}) = \frac{\sum_{i=1}^{2} L_i^2}{6} - CM = \frac{(601)^2 + (577)^2}{6} - 115{,}640.33$$

$$= 48.00$$

$$SS(\text{Breaks}) = \frac{\sum_{j=1}^{3} B_j^2}{4} - CM = \frac{(391)^2 + (430)^2 + (357)^2}{4} - 115{,}640.33$$

$$= 667.17$$

$$SS(LB) = \frac{\sum_{i=1}^{2} \sum_{j=1}^{3} LB_{ij}^2}{2} - SS(\text{Length}) - SS(\text{Breaks}) - CM$$

$$= \frac{(203)^2 + (211)^2 + \cdots + (170)^2}{2} - 48.00 - 667.17 - 115{,}640.33$$

$$= 96.50$$

$$SSE = SS(\text{Total}) - SS(\text{Length}) - SS(\text{Breaks}) - SS(LB)$$

$$= 831.67 - 48.00 - 667.17 - 96.50$$

$$= 20.00$$

The ANOVA table for the data is given in Table 8.16.

TABLE 8.16
ANOVA Table for
Example 8.13

SOURCE	df	SS	MS	F
Length (L)	1	48.00	48.00	14.40
Breaks (B)	2	667.17	333.58	100.17
LB	2	96.5	48.25	14.49
Error	6	20.00	3.33	
Total	11	831.67		

To test for interaction between the two factors, length (L) and breaks (B), we compute

$$F = \frac{MS(LB)}{MSE} = \frac{48.25}{3.33} = 14.49 \quad (\text{see Table 8.16})$$

Since this value exceeds the critical value $F_{.05} = 5.14$ (based on $v_1 = 2$ and $v_2 = 6$ df) obtained from Table 4 of Appendix D, there is sufficient evidence of interaction between length and breaks. Since interaction implies that the level of length (L) that yields the highest mean productivity may differ across the levels of breaks (B), we ignore the tests for main effects and focus our investigation on the individual treatment means.

b. The sample means for the six factor-level combinations are shown in Table 8.17. Since the sample means in the table represent measures of productivity

TABLE 8.17
Sample Means for the
$p = 6$ Treatments of
Example 8.13

			COFFEE BREAKS, B		
			0	1	2
LENGTH OF	4 days		101.5	105.5	93.5
WORK WEEK, L	5 days		94.0	109.5	85.0

in the manufacture of transistors, we would want to find the length of work week and number of coffee breaks that yield the highest mean productivity.

The first step in the ranking procedure is to calculate ω for $p = 6$ (we are ranking six treatment means), $n_t = 2$ (two observations per treatment), $\alpha = .05$, and $s = \sqrt{\text{MSE}} = \sqrt{3.33} = 1.83$ (where MSE is given in Table 8.16). Since MSE is based on $\nu = 6$ degrees of freedom, we have

$$q_{.05}(6, 6) = 5.63$$

and

$$\omega = q_{.05}(6, 6)\left(\frac{s}{\sqrt{n_t}}\right) = (5.63)\left(\frac{1.83}{\sqrt{2}}\right) = 7.27$$

Therefore, population means corresponding to pairs of sample means that differ by more than $\omega = 7.27$ will be judged to be different. The six sample means are ranked as follows:

Sample means	85.0	93.5	94.0	101.5	105.5	109.5
Treatments	(5, 2)	(4, 2)	(5, 0)	(4, 0)	(4, 1)	(5, 1)
(Length, Breaks)						

Using $\omega = 7.27$ as a yardstick to determine differences between pairs of treatments, we have placed connecting bars over those means that *do not* significantly differ. The following conclusions can be drawn:

1. We see that there is evidence of a difference between the population mean of the treatment corresponding to a 5-day work week with two coffee breaks (with the smallest sample mean of 85.0) and every other treatment mean. Therefore, we can conclude that the 5-day, 2-break work week yields the lowest mean productivity among all length–break combinations.
2. The population mean of the treatment corresponding to a 5-day, 1-break work week (with the largest sample mean of 109.5) is significantly larger than the treatments corresponding to the four smallest sample means. However, there is no evidence of a difference between the 5-day, 1-break treatment mean and the 4-day, 1-break treatment mean (with a sample mean of 105.5).
3. There is no evidence of a difference between the 4-day, 1-break treatment mean (with a sample mean of 105.5) and the 4-day, 0-break treatment mean (with a sample mean of 101.5). Both these treatments, though, have significantly larger means than the treatments corresponding to the three smallest sample means.
4. There is no evidence of a difference among the treatments corresponding to the sample means 93.5 and 94.0. Further experimentation would be required to determine whether the observed differences in these means really imply a difference in the corresponding sample means.

In summary, the treatment means appear to fall into four groups, as shown below:

	TREATMENTS (LENGTH, BREAKS)
Group 1 (lowest mean productivity):	(5, 2)
Group 2:	(4, 2) and (5, 0)
Group 3:	(4, 0) and (4, 1)
Group 4 (highest mean productivity):	(4, 1) and (5, 1)

Notice that it is unclear where we should place the treatment corresponding to a 4-day, 1-break work week due to the overlapping bars above its sample mean, 105.5. That is, although there is sufficient evidence to indicate that treatments (4, 0) and (5, 1) differ, neither has been shown to differ significantly from treatment (4, 1). Tukey's method guarantees that the probability of making one or more Type I errors in the pairwise comparisons above is only $\alpha = .05$.

◼

Remember that Tukey's multiple comparisons procedure requires the sample sizes associated with the treatments to be equal. This, of course, will be satisfied

---

TUKEY'S APPROXIMATE MULTIPLE COMPARISONS PROCEDURE FOR UNEQUAL SAMPLE SIZES

1. Calculate for each treatment pair (i, j)

$$\omega_{ij} = q_\alpha(p, \nu)\frac{s}{\sqrt{2}} \sqrt{\frac{1}{n_i} + \frac{1}{n_j}}$$

where

p = Number of sample means

$s = \sqrt{\text{MSE}}$

ν = Number of degrees of freedom associated with MSE

n_i = Number of observations in sample for treatment i

n_j = Number of observations in sample for treatment j

$q_\alpha(p, \nu)$ = Critical value of the Studentized range (Tables 11 and 12 of Appendix D)

2. Rank the p sample means and place a bar over any treatment pair (i, j) that differs by less than ω_{ij}. Any pair of sample means not connected by an overbar (i.e., differing by more than ω) implies a difference in the corresponding population means.

Note: This procedure is approximate, i.e., the value of α selected by the researcher approximates the true probability of making at least one Type I error.

for the randomized block designs and factorial experiments described in Sections 8.6 and 8.7, respectively. The sample sizes, however, may not be equal in a completely randomized design (Section 8.5). In this case a modification of Tukey's method (sometimes called the Tukey–Kramer method) is necessary, as described in the box. The technique requires that the critical difference ω_{ij} be calculated for each pair of treatments (i, j) in the experiment and pairwise comparisons made based on the appropriate value of ω_{ij}. However, when Tukey's method is used with unequal sample sizes, the value of α selected a priori by the researcher only approximates the true experimentwise error rate. In fact, when applied to unequal sample sizes, the procedure has been found to be more conservative, i.e., less likely to detect differences between pairs of treatment means when they exist, than in the case of equal sample sizes. For this reason, researchers sometimes look to alternative methods of multiple comparisons when the sample sizes are unequal. Two of these methods are presented in optional Section 8.9.

| | | | | | | | | | | | | |

EXERCISES 8.35–8.43

8.35 Refer to the *Journal of Rehabilitation* (April/May/June 1986) study on the employability of physically handicapped persons in Exercise 8.7. A follow-up analysis was conducted using a technique similar to Tukey's multiple comparisons method. The ranking of the mean employability ratings for the three groups of companies is given below. Interpret the results. Assume $\alpha = .05$.

Sample mean rating	2.57	3.38	3.58
Company size	Small	Medium	Large

8.36 Refer to Exercise 8.10. Use Tukey's multiple comparisons procedure to compare the mean rates of penetration for the three types of drill bits. Identify the means that appear to differ. Use $\alpha = .05$.

8.37 Refer to Exercise 8.11. Use Tukey's multiple comparisons procedure to compare the daily traffic density averages for the five locations. Identify the means that appear to differ. Use $\alpha = .05$.

8.38 Refer to Exercise 8.16. Use Tukey's multiple comparisons procedure to compare the mean prices of grocery items for the four supermarkets. Identify the means that appear to differ. Use $\alpha = .01$.

8.39 Refer to Exercise 8.21. Use Tukey's multiple comparisons procedure to compare the mean in-tray performance scores assessed by the three raters. Identify the means that appear to differ. Use $\alpha = .05$.

8.40 Refer to Exercise 8.27. Use Tukey's multiple comparisons procedure to compare the productivity means for the $2 \times 3 = 6$ machine–material combinations. Identify the means that appear to differ. Use $\alpha = .05$.

8.41 Refer to Exercise 8.28. Use Tukey's multiple comparisons procedure to compare the mean satisfaction scores for the three peak period lengths under each of the three pricing ratios. Identify the means that appear to differ under each pricing ratio. Use $\alpha = .01$.

8.42 Refer to the 2×2 factorial experiment described in Exercise 8.29. Compare the four treatment means using Tukey's multiple comparisons method. Use $\alpha = .05$.

8.43 Refer to the 8×11 factorial experiment in Exercise 8.33. The mean ratings for the 11 CEOs are given in the table for the credibility item "well-known person." Compare the means using Tukey's multiple comparisons procedure. Use $\alpha = .05$. [*Note:* Higher ratings indicate a greater degree of credibility.]

CEO (COMPANY)	MEAN RATING
1. David Mahoney (Avis)	3.74
2. Franco Bolla (Bolla Wines)	4.57
3. Lee Iacocca (Chrysler)	6.47
4. L. Stanley Crane (Conrail)	2.21
5. Constantine Demmas (Demmas Financial Services)	2.16
6. W. H. Bricker (Diamond Shamrock)	2.67
7. Frank Borman (Eastern)	6.16
8. Herb Imhoff (General Employment Enterprises)	2.07
9. Guy Milner (Norrell Temporary Services)	2.88
10. Frank B. Hall (F. B. Hall & Company)	2.31
11. L. S. Shoen (U-Haul)	3.66

SECTION 8.9

OTHER MULTIPLE COMPARISONS METHODS (OPTIONAL)

In this optional section we present two alternatives to Tukey's method of multiple comparisons of treatment means. The choice of methods will depend on the type of experimental design employed and the particular error rate that the researcher wants to control, as described below.

SCHEFFÉ METHOD

Recall that Tukey's method of multiple comparisons is designed to control the experimentwise error rate, i.e., the probability of making at least one Type I error in the comparison of *all pairs* of treatment means in the experiment. Therefore, Tukey's method should be applied when you are interested in pairwise comparisons only.

Scheffé (1953) developed a more general procedure for comparing all possible linear combinations of the treatment means, called **contrasts**.

DEFINITION 8.11

A **contrast** L is a linear combination of the p treatment means in a designed experiment, i.e.,

$$L = \sum_{i=1}^{p} c_i \mu_i$$

where the constants $c_1, c_2, \ldots, c_p$ sum to 0, i.e., $\sum_{i=1}^{p} c_i = 0$.

For example, in an experiment with four treatments (A, B, C, D), you might want to compare the following contrasts, where μ_i represents the population mean for treatment i:

$$L_1 = \frac{\mu_A + \mu_B}{2} - \frac{\mu_C + \mu_D}{2}$$

$$L_2 = \mu_A - \mu_D$$

$$L_3 = \frac{\mu_B + \mu_C + \mu_D}{3} - \mu_A$$

The contrast L_2 involves a comparison of a pair of treatment means, while L_1 and L_3 are more complex comparisons of the treatments. Thus, pairwise comparisons are special cases of general contrasts.

As in Tukey's method, the value of α selected by the researcher using Scheffé's method applies to the procedure as a whole, i.e., to the comparisons of all possible contrasts (not just those considered by the researcher). Unlike Tukey's method, however, the probability of at least one Type I error, α, is exact regardless of whether the sample sizes are equal or unequal. For this reason, some researchers prefer Scheffé's method to Tukey's method in the unequal sample size case, even if only pairwise comparisons of treatment means are made.

The Scheffé method for general contrasts is outlined in the box.

SCHEFFÉ'S MULTIPLE COMPARISONS PROCEDURE FOR GENERAL CONTRASTS

1. For each contrast $L = \sum\limits_{i=1}^{p} c_i \mu_i$, calculate

$$\hat{L} = \sum_{i=1}^{p} c_i \bar{y}_i$$

and

$$S = \sqrt{(p-1)F_\alpha(p-1, \nu)\text{MSE} \sum_{i=1}^{p} \left(\frac{c_i^2}{n_i}\right)}$$

where

$$p = \text{Number of sample (treatment) means}$$
$$\text{MSE} = \text{Mean squared error}$$
$$n_i = \text{Number of observations in sample for treatment } i$$
$$\bar{y}_i = \text{Sample mean for treatment } i$$
$$F_\alpha(p-1, \nu) = \text{Critical value of } F\text{-distribution with } \nu_1 = p - 1$$
$$\text{numerator df and } \nu_2 = \nu \text{ denominator df}$$
$$\text{(Tables 3, 4, 5, and 6 of Appendix D)}$$
$$\nu = \text{Number of degrees of freedom associated with MSE}$$

2. Calculate the confidence interval $\hat{L} \pm S$ for each contrast. The confidence coefficient, $1 - \alpha$, applies to the procedure as a whole, i.e., to the entire set of confidence intervals for all possible contrasts.

In the special case of all pairwise comparisons in an experiment with four treatments, the relevant contrasts reduce to $L_1 = \mu_A - \mu_B$, $L_2 = \mu_A - \mu_C$, $L_3 = \mu_A - \mu_D$, and so forth. Notice that for each of these contrasts $\sum c_i^2/n_i$ reduces to $(1/n_i + 1/n_j)$, where n_i and n_j are the sizes of treatments i and j, respectively. [For example, for contrast L_1, $c_1 = 1$, $c_2 = -1$, $c_3 = c_4 = 0$, and $\sum c_i^2/n_i = (1/n_1 + 1/n_2)$.] Consequently, the formula for S in the box can be simplified and pairwise comparisons made using the technique of Section 8.8.

SCHEFFÉ'S MULTIPLE COMPARISONS PROCEDURE FOR PAIRWISE COMPARISONS OF TREATMENT MEANS

1. Calculate Scheffé's critical difference for each pair of treatments (i, j):

$$S_{ij} = \sqrt{(p-1)F_\alpha(p-1, \nu)\mathrm{MSE}\left(\frac{1}{n_i} + \frac{1}{n_j}\right)}$$

where

p = Number of sample (treatment) means

MSE = Mean squared error

n_i = Number of observations in sample for treatment i

n_j = Number of observations in sample for treatment j

$F_\alpha(p-1, \nu)$ = Critical value of F-distribution with $\nu_1 = p - 1$ numerator df and $\nu_2 = \nu$ denominator df (Tables 3, 4, 5, and 6 of Appendix D)

ν = Number of degrees of freedom associated with MSE

2. Rank the p sample means and place a bar over any treatment pair (i, j) that differs by less than S_{ij}. Any pair of sample means not connected by an overbar implies a difference in the corresponding population means.

EXAMPLE 8.14

Refer to the 2×3 factorial experiment to investigate the effects of length of work week and number of coffee breaks on mean productivity in Example 8.13. Compare the population means corresponding to the six length–break combinations (treatments) using the Scheffé method of multiple comparisons. Use $\alpha = .05$.

SOLUTION

From Table 8.16 in Example 8.13, we have MSE = 3.33, $p = 6$, and $\nu = $ df(Error) = 6. Also, there are two observations per treatment, $n_i = n_j = 2$, for all treatment pairs. Consequently, the critical difference S_{ij} will be the same for all treatment pairs (i, j). From Table 4 in Appendix D, we have $F_{.05} = 4.39$ (based on $p - 1 = 5$ numerator df and $\nu = 6$ denominator df). Then, Scheffé's critical difference S_{ij} is

$$S_{ij} = \sqrt{(p-1)F_{.05}\mathrm{MSE}\left(\frac{1}{n_i} + \frac{1}{n_j}\right)}$$

$$= \sqrt{(5)(4.39)(3.33)\left(\frac{1}{2} + \frac{1}{2}\right)} = 8.55$$

Treatment means differing by more than $S = 8.55$ will imply a significant difference between the corresponding population means. The rankings of the treatment means, with overbars indicating "no evidence of a difference," are shown here:

Sample means	85.0	93.5	94.0	101.5	105.5	109.5
Treatments	(5, 2)	(4, 2)	(5, 0)	(4, 0)	(4, 1)	(5, 1)
(Length, Breaks)						

Recall that the manufacturer's goal is to determine the treatment(s) yielding the highest mean productivity. Based on this result, the six means can be grouped as follows:

		TREATMENTS (LENGTH, BREAKS)
Group 1	(lowest mean productivity):	(5, 2) and (4, 2)
Group 2:		(4, 2), (5, 0), and (4, 0)
Group 3	(highest mean productivity):	(4, 0), (4, 1), and (5, 1)

You can see that the treatments corresponding to 5-day, 1-break and 4-day, 1-break work weeks produce the highest mean productivity. The Scheffé method did not detect a significant difference between these treatments and treatment (4, 0), or a significant difference between treatment (4, 0) and either treatment (4, 2) or (5, 0). Thus, we are not certain where to place treatment (4, 0)—in the group with the highest mean productivity (group 3) or in the middle group (group 2).   ∎

Note that in Example 8.14, the Scheffé method produced a critical difference of $S = 8.55$—a value larger than Tukey's critical difference of $\omega = 7.27$ (Example 8.13). This implies that Tukey's method produces narrower confidence intervals than Scheffé's method for differences in pairs of treatment means. Therefore, if only pairwise comparisons of treatments are to be made, Tukey's is the preferred method as long as the sample sizes are equal. The Scheffé method, on the other hand, yields narrower confidence intervals (i.e., smaller critical differences) for situations in which the goal of the researchers is to make comparisons of general contrasts.

BONFERRONI APPROACH   As noted earlier, Tukey's multiple comparisons procedure in the unequal sample size case is approximate. That is, the value of α selected a priori only approximates the true probability of making at least one Type I error. The Bonferroni approach is an exact method that is applicable in either the equal or the unequal sample size case. Further, Bonferroni's procedure covers all possible comparisons of treatments, including pairwise comparisons, general contrasts, or combinations of pairwise comparisons and more complex contrasts.

BONFERRONI MULTIPLE COMPARISONS PROCEDURE FOR GENERAL CONTRASTS

1. For each contrast $L = \sum_{i=1}^{p} c_i \mu_i$, calculate

$$\hat{L} = \sum_{i=1}^{p} c_i \bar{y}_i$$

and

$$B = t_{\alpha/(2g)} s \sqrt{\sum_{i=1}^{p} \left(\frac{c_i^2}{n_i} \right)}$$

where

p = Number of sample (treatment) means

g = Number of contrasts

$s = \sqrt{MSE}$

ν = Number of degrees of freedom associated with MSE

n_i = Number of observations in sample for treatment i

$\bar{y}_i$ = Sample mean for treatment i

$t_{\alpha/(2g)}$ = Critical value of t-distribution with ν df and tail area $\alpha/(2g)$ (Table 2 in Appendix D)

2. Calculate the confidence interval $\hat{L} \pm B$ for each contrast. The confidence coefficient for the procedure as a whole, i.e., for the entire set of confidence intervals, is *at least* $1 - \alpha$.

The Bonferroni approach is based on the following result (proof omitted): If g comparisons are to be made, each with confidence coefficient $1 - \alpha/g$, then the overall probability of making one or more Type I errors (i.e., the experimentwise error rate) is at most α. That is, the set of intervals constructed using the Bonferroni method yields an overall confidence level of at least $1 - \alpha$. For example, if you want to construct $g = 2$ confidence intervals with an experimentwise error rate of at most $\alpha = .05$, then each individual interval must be constructed using a confidence level of $1 - .05/2 = .975$.

When applied to only pairwise comparisons of treatments, the Bonferroni approach can be carried out as shown in the next box.

EXAMPLE 8.15   Refer to the two-factor factorial experiment in Example 8.13. Use Bonferroni's method to perform pairwise comparisons of the six treatment means. Use $\alpha = .05$.

BONFERRONI MULTIPLE COMPARISONS PROCEDURE FOR PAIRWISE COMPARISONS OF TREATMENT MEANS

1. Calculate for each treatment pair (i, j)

$$B_{ij} = t_{\alpha/(2g)}s \sqrt{\frac{1}{n_i} + \frac{1}{n_j}}$$

where

p = Number of sample (treatment) means in the experiment

g = Number of pairwise comparisons
 [*Note:* If all pairwise comparisons are to be made, then
 $g = p(p - 1)/2$]

$s = \sqrt{\text{MSE}}$

ν = Number of degrees of freedom associated with MSE

n_i = Number of observations in sample for treatment i

n_j = Number of observations in sample for treatment j

$t_{\alpha/(2g)}$ = Critical value of t-distribution with ν df and tail area $\alpha/(2g)$
 (Table 2 in Appendix D)

2. Rank the sample means and place a bar over any treatment pair (i, j) whose sample means differ by less than B_{ij}. Any pair of means not connected by an overbar implies a difference in the corresponding population means.

Note: The level of confidence associated with all inferences drawn from the analysis is at least $1 - \alpha$.

SOLUTION

From Examples 8.13 and 8.14, we have $p = 6$, $s = \sqrt{\text{MSE}} = 1.83$, $\nu = 6$, and $n_i = n_j = 2$ for all treatment pairs (i, j). For $p = 6$ means, the number of pairwise comparisons to be made is

$$g = \frac{p(p - 1)}{2} = \frac{6(5)}{2} = 15$$

Thus, we need to find the critical value, $t_{\alpha/(2g)} = t_{.05/[2(15)]} = t_{.0017}$, for the t-distribution based on $\nu = 6$ df. This value, although not shown in Table 2 in Appendix D, is approximately 4.7.* Substituting $t_{.0017} \approx 4.7$ into the equation for Bonferroni's critical difference B_{ij}, we have

$$B_{ij} \approx t_{.0017}s \sqrt{\frac{1}{n_i} + \frac{1}{n_j}} = (4.7)(1.83) \sqrt{\frac{1}{2} + \frac{1}{2}} = 8.60$$

for any treatment pair (i, j).

---

*We obtained the value using the SAS probability generating function for a Student's t-distribution.

Using the value $B = 8.60$ to detect significant differences between treatment means, we obtain the following results:

Sample means	85.0	93.5	94.0	101.5	105.5	109.5
Treatments	(5, 2)	(4, 2)	(5, 0)	(4, 0)	(4, 1)	(5, 1)
(Length, Breaks)						

You can see that the group of treatments with the highest mean productivity includes treatments (4, 0), (4, 1), and (5, 1). The bar over these three means, however, indicates that we are unable to detect differences between any pair of these treatments. All inferences derived from this analysis can be made at an overall confidence level of at least $1 - \alpha = .95$.   ∎

When applied to pairwise comparisons of treatments, the Bonferroni method, like the Scheffé procedure, produces wider confidence intervals (reflected by the magnitude of the critical difference) than Tukey's method. (In Example 8.15, Bonferroni's critical difference is $B \approx 8.60$ compared to Tukey's $\omega = 7.27$.) Therefore, if only pairwise comparisons are of interest, Tukey's procedure is again superior. However, if the sample sizes are unequal or more complex contrasts are to be compared, the Bonferroni technique may be preferred. Unlike the Tukey and Scheffé methods, however, you must know in advance how many contrasts are to be compared in order to properly use Bonferroni's procedure. Also, the value needed to calculate the critical difference, B, $t_{\alpha/(2g)}$, may not be available in the t-tables provided in most texts, and you will have to estimate its value.

In this section we have presented two alternatives to Tukey's multiple comparisons procedure. The technique you select will depend on several factors, including the sample sizes and the type of comparisons to be made. Keep in mind, however, that many other methods of making multiple comparisons are available, and one or more of these techniques may be more appropriate to use in your particular application. Consult the references given at the end of this chapter for details on other techniques.

EXERCISES 8.44–8.46

8.44 Refer to Exercises 8.10 and 8.36.
  a. Use the Scheffé method to perform all pairwise comparisons of the mean rates of penetration for the three types of drill bits. Use $\alpha = .05$.
  b. Use the Bonferroni approach to perform all pairwise comparisons of the mean rates of penetration for the three types of drill bits. Use $\alpha = .05$.
  c. Compare the results in parts a and b to the results of Tukey's method in Exercise 8.36.

8.45 Refer to Exercises 8.27 and 8.40.
  a. Use the Scheffé method to perform all pairwise comparisons of the productivity means for the $2 \times 3 = 6$ machine–materials combinations. Use $\alpha = .05$.
  b. Use the Bonferroni approach to perform all pairwise comparisons of the productivity means for the $2 \times 3 = 6$ machine–materials combinations. Use $\alpha = .05$.
  c. Compare the results in parts a and b to the results of Tukey's method in Exercise 8.40.

8.46 Refer to Exercises 8.29 and 8.42.
  a. Use the Scheffé method to perform all pairwise comparisons of the four treatment means. Use $\alpha = .05$.
  b. Use the Bonferroni approach to perform all pairwise comparisons of the four treatment means. Use $\alpha = .05$.
  c. Compare the results in parts **a** and **b** to the results of Tukey's method in Exercise 8.42.

SECTION 8.10

THE RELATIONSHIP BETWEEN ANALYSIS OF VARIANCE AND A REGRESSION ANALYSIS (OPTIONAL)

The preceding sections illustrated the analysis of variance approach to analyzing data collected from designed experiments. The analysis of variance sums of squares are easy to compute with the aid of a pocket or desk calculator. By forming a ratio of mean squares, we are able to test the hypothesis that a set of population means (treatment means, block means, or factor main effect means) are equal.

The same analysis can also be conducted using a multiple regression analysis. Each experimental design is associated with a linear model for the response y, called the **complete model**. The analysis of variance F-test for testing a set of means is equivalent to a partial F-test in regression in which the complete model is fit and compared to a **reduced model**. The difference between SSEs for the two models, called the **drop in SSE**, is equal to the sum of squares for the appropriate source of variation (e.g., treatments) that appears in the numerator of the F-statistic. Before you can apply regression analysis in an analysis of variance, you need to learn the appropriate complete and reduced models to fit for each type of experimental design.

Consider, first, the completely randomized design. Recall (from Section 8.5) that this design is one that involves a comparison of the means for p treatments based on independent random samples. Since we want to make inferences about the p population means, $\mu_1, \mu_2, \ldots, \mu_p$, the appropriate linear model for the response y is

$$E(y) = \beta_0 + \beta_1 x_1 + \beta_2 x_2 + \cdots + \beta_{p-1} x_{p-1}$$

where

$$x_1 = \begin{cases} 1 & \text{if treatment 2} \\ 0 & \text{if not} \end{cases} \qquad x_2 = \begin{cases} 1 & \text{if treatment 3} \\ 0 & \text{if not} \end{cases} \qquad \cdots \qquad x_{p-1} = \begin{cases} 1 & \text{if treatment } p \\ 0 & \text{if not} \end{cases}$$

and treatment 1 is the base level. Recall that this 0–1 system of coding implies that

$$\beta_0 = \mu_1$$
$$\beta_1 = \mu_2 - \mu_1$$
$$\beta_2 = \mu_3 - \mu_1$$
$$\vdots \qquad \vdots$$
$$\beta_{p-1} = \mu_p - \mu_1$$

The null hypothesis that the p population means are equal, i.e.,

$$H_0: \quad \mu_1 = \mu_2 = \cdots = \mu_p$$

is equivalent to the null hypothesis that all the treatment differences equal 0, i.e.,

$$H_0: \quad \beta_1 = \beta_2 = \cdots = \beta_{p-1} = 0$$

To test this hypothesis using regression, we compare the sum of squares for error, SSE_1, for the reduced model

$$E(y) = \beta_0$$

to the sum of squares for error, SSE_2, for the complete model

$$E(y) = \beta_0 + \beta_1 x_1 + \beta_2 x_2 + \cdots + \beta_{p-1} x_{p-1}$$

using the F-statistic

$$F = \frac{(SSE_1 - SSE_2)/\text{Number of } \beta \text{ parameters in } H_0}{SSE_2/[n - (\text{Number of } \beta \text{ parameters in the complete model})]}$$

$$= \frac{(SSE_1 - SSE_2)/(p - 1)}{SSE_2/(n - p)}$$

The quantity $SSE_1 - SSE_2$ in the numerator of the F-statistic is equal to the sum of squares for treatments, SST, in an analysis of variance. Also, the sum of squares for error for the complete model, SSE_2, is equal to SSE in an analysis of variance. Thus,

$$F = \frac{(SSE_1 - SSE_2)/(p - 1)}{SSE_2/(n - p)} = \frac{SST/(p - 1)}{SSE/(n - p)} = \frac{MST}{MSE}$$

and, therefore, the analysis of variance and regression test statistics are identical.

We illustrate the regression approach to conducting an analysis of variance for a completely randomized design in the following example.

EXAMPLE 8.16

Refer to Example 8.4, where the chain of department stores wants to compare the mean dollar amounts owed by its delinquent creditors in three income groups. The experiment is a completely randomized design with three treatments: under $12,000, $12,000–$25,000, and over $25,000. Analyze the data shown in Table 8.2 (page 397) using a regression analysis.

SOLUTION

The appropriate linear model for $p = 3$ treatments is

Complete model: $\quad y = \beta_0 + \beta_1 x_1 + \beta_2 x_2 + \varepsilon$

where

$$x_1 = \begin{cases} 1 & \text{if income class 2} \\ 0 & \text{if not} \end{cases}$$

$$x_2 = \begin{cases} 1 & \text{if income class 3} \\ 0 & \text{if not} \end{cases}$$

The Minitab regression analysis for the complete model is shown in Figure 8.11. Note that the SSE shown in the printout, $SSE_2 = 770{,}671$, agrees (except for rounding) with the value of SSE calculated in Example 8.4.

FIGURE 8.11

Minitab Computer Printout
for the Completely
Randomized Design,
Example 8.16

```
The regression equation is
Y = 230 + 80.3 X1 + 198 X2

Predictor        Coef        Stdev      t-ratio
Constant       229.60        53.43         4.30
X1              80.30        75.56         1.06
X2             198.20        75.56         2.62

s = 168.9      R-sq = 20.5%      R-sq(adj) = 14.6%

Analysis of Variance

SOURCE         DF          SS           MS
Regression      2      198772        99386
Error          27      770671        28543
Total          29      969443
```

The reduced model, obtained by omitting the two treatment parameters, is

Reduced model: $y = \beta_0 + \varepsilon$

Since we know that the least squares estimate of β_0 for this model is $\bar{y}$, it follows that the sum of squares for error for the reduced model is

$$SSE_1 = SS(Total) = \sum_{i=1}^{n} (y_i - \bar{y})^2 = \sum_{i=1}^{n} y_i^2 - \frac{\left(\sum_{i=1}^{n} y_i\right)^2}{n}$$

This quantity is shown in Figure 8.11 as

$SS(Total) = 969,443$

Then the drop in the sum of squares for error that is attributable to a difference in treatment means is

$SST = SSE_1 - SSE_2$

$= 969,443 - 770,671$

$= 198,772$

To test the null hypothesis $H_0: \beta_1 = \beta_2 = 0$ or, equivalently, $H_0: \mu_1 = \mu_2 = \mu_3$, we form the test statistic

$$F = \frac{(SSE_1 - SSE_2)/(p - 1)}{SSE_2/(n - p)}$$

$$= \frac{198,772/2}{770,671/27} = 3.48$$

This value is identical to the analysis of variance F-statistic computed in Example 8.4. For $\nu_1 = p - 1 = 3 - 1 = 2$ and $\nu_2 = n - p = 30 - 3 = 27$ df, $F_{.05} = 3.35$. Since the computed value of F, 3.48, exceeds the critical value, there is sufficient evidence (at $\alpha = .05$) to indicate that the means for the three groups of delinquent creditors differ.   ∎

The regression approach to analyzing data from a completely randomized design is summarized in the box.

ANOVA F-TEST FOR A COMPLETELY RANDOMIZED DESIGN WITH p TREATMENTS: REGRESSION APPROACH

H_0: $\beta_1 = \beta_2 = \cdots = \beta_{p-1} = 0$ (i.e., H_0: $\mu_1 = \mu_2 = \cdots = \mu_p$)

H_a: At least one of the β parameters listed in H_0 differs from 0
    (i.e., H_a: At least two means differ)

Complete model: $E(y) = \beta_0 + \beta_1 x_1 + \beta_2 x_2 + \cdots + \beta_{p-1} x_{p-1}$

where

$$x_1 = \begin{cases} 1 & \text{if treatment 2} \\ 0 & \text{if not} \end{cases} \qquad x_2 = \begin{cases} 1 & \text{if treatment 3} \\ 0 & \text{if not} \end{cases}$$

$$\ldots, \qquad x_{p-1} = \begin{cases} 1 & \text{if treatment } p \\ 0 & \text{if not} \end{cases}$$

Reduced model: $E(y) = \beta_0$

Test statistic: $F = \dfrac{(\text{SSE}_1 - \text{SSE}_2)/(p-1)}{\text{SSE}_2/(n-p)} = \dfrac{\text{MST}}{\text{MSE}}$

where

SSE$_1$ = SSE for reduced model

SSE$_2$ = SSE for complete model

Rejection region: $F > F_\alpha$, where the distribution of F is based on
    $\nu_1 = p - 1$ and $\nu_2 = n - p$ degrees of freedom.

EXAMPLE 8.17

To further relate regression analysis and analysis of variance, refer to Examples 8.4 and 8.16. Note that

$$\beta_1 = \mu_2 - \mu_1$$

Therefore, since the analysis of variance estimate of $(\mu_2 - \mu_1)$ is $(\overline{T}_2 - \overline{T}_1)$, it follows that $\hat{\beta}_1$ must equal $(\overline{T}_2 - \overline{T}_1)$. Further, the estimated standard error of

$$\hat{\beta}_1 = \overline{T}_2 - \overline{T}_1$$

is by ANOVA formulas

$$s\sqrt{\frac{1}{n_1} + \frac{1}{n_2}}$$

Find $\hat{\beta}_1$ and its standard error in the regression printout for Example 8.16 (Figure 8.11), and verify that, in fact,

$$\hat{\beta}_1 = \bar{T}_2 - \bar{T}_1 \quad \text{and} \quad s_{\hat{\beta}_1} = s\sqrt{\frac{1}{n_1} + \frac{1}{n_2}}$$

SOLUTION   From Table 8.2, $\bar{T}_2 - \bar{T}_1 = 309.9 - 229.6 = \$80.30$. This is exactly the same as the value given for $\hat{\beta}_1$ in the regression printout (Figure 8.11). Similarly, in Example 8.5, we found $s = \sqrt{\text{MSE}} = 168.9$. Therefore,

$$s\sqrt{\frac{1}{n_1} + \frac{1}{n_2}} = 168.9\sqrt{\frac{1}{10} + \frac{1}{10}} = 75.53$$

Consulting the regression printout (Figure 8.11), you will see that the standard error of $\hat{\beta}_1$ is 75.6. (Again, the different values are due to rounding errors.)

The point of this example is that the completely randomized design may be analyzed using either regression analysis or an analysis of variance. The conclusions will be identical. ∎

The regression approach to analyzing randomized block designs and factorial experiments is similar to that shown above. For each null hypothesis, complete and reduced models are fit. The drop in SSE, $SSE_1 - SSE_2$, always represents the analysis of variance SS for the appropriate source being tested. The ANOVA F-tests for randomized block designs and factorial experiments using regression analysis are shown in the boxes on pages 458 and 459, respectively.

EXAMPLE 8.18   Refer to the supply–ratio 3×3 factorial experiment of Example 8.12.

a. Write the complete model for the experiment.
b. What hypothesis would you test to determine whether supply and ratio interact?

SOLUTION   **a.** Both factors, supply and ratio, are quantitative. According to the relevant box, when the factors in a factorial experiment are quantitative, the main effects are represented by terms such as x, x^2, x^3, and so forth. Since each factor has three levels, we require two main effects, x and x^2, for each factor. (In general, the number of main effect terms will be one less than the number of levels for a factor.) Consequently, the complete factorial model for this 3×3 factorial experiment is

$$y = \beta_0 + \underbrace{\beta_1 x_1 + \beta_2 x_1^2}_{\text{Supply main effects}} + \underbrace{\beta_3 x_2 + \beta_4 x_2^2}_{\text{Ratio main effects}}$$
$$+ \underbrace{\beta_5 x_1 x_2 + \beta_6 x_1 x_2^2 + \beta_7 x_1^2 x_2 + \beta_8 x_1^2 x_2^2}_{\text{Supply–Ratio interaction}} + \varepsilon$$

ANOVA F-TEST FOR A RANDOMIZED DESIGN WITH p TREATMENTS AND b BLOCKS: REGRESSION APPROACH

TESTS FOR COMPARING TREATMENT MEANS

H_0: $\beta_1 = \beta_2 = \cdots = \beta_{p-1} = 0$
(i.e., H_0: The p treatment means are equal)

H_a: At least one of the β parameters listed in H_0 differs from 0
(i.e., H_a: At least two treatment means differ)

Complete model:

$$E(y) = \beta_0 + \overbrace{\beta_1 x_1 + \cdots + \beta_{p-1} x_{p-1}}^{(p-1) \text{ treatment terms}} + \overbrace{\beta_p x_p + \cdots + \beta_{p+b-2} x_{p+b-2}}^{(b-1) \text{ block terms}}$$

where

$$x_1 = \begin{cases} 1 & \text{if treatment 2} \\ 0 & \text{if not} \end{cases} \cdots \quad x_{p-1} = \begin{cases} 1 & \text{if treatment } p \\ 0 & \text{if not} \end{cases}$$

$$x_p = \begin{cases} 1 & \text{if block 2} \\ 0 & \text{if not} \end{cases} \cdots \quad x_{p+b-2} = \begin{cases} 1 & \text{if block } b \\ 0 & \text{if not} \end{cases}$$

Reduced model: $E(y) = \beta_0 + \beta_p x_p + \cdots + \beta_{p+b-2} x_{p+b-2}$

Test statistic: $F = \dfrac{(\text{SSE}_1 - \text{SSE}_2)/(p-1)}{\text{SSE}_2/(n-p-b+1)} = \dfrac{\text{MST}}{s^2}$

where

$\text{SSE}_1 = \text{SSE}$ for reduced model

$\text{SSE}_2 = \text{SSE}$ for complete model

Rejection region: $F > F_\alpha$ where F is based on $\nu_1 = (p-1)$ and $\nu_2 = (n-p-b+1)$ degrees of freedom.

TESTS FOR COMPARING BLOCK MEANS

H_0: $\beta_p = \beta_{p+1} = \cdots = \beta_{p+b-2} = 0$
(i.e., H_0: The b block means are equal)

H_a: At least one of the β parameters listed in H_0 differs from 0
(i.e., H_a: At least two block means differ)

Complete model: (See above)

Reduced model: $E(y) = \beta_0 + \beta_1 x_1 + \beta_2 x_2 + \cdots + \beta_{p-1} x_{p-1}$

Test statistic: $F = \dfrac{(\text{SSE}_1 - \text{SSE}_2)/(b-1)}{\text{SSE}_2/(n-p-b+1)} = \dfrac{\text{MSB}}{s^2}$

where

$\text{SSE}_1 = \text{SSE}$ for reduced model

$\text{SSE}_2 = \text{SSE}$ for complete model

Rejection region: $F > F_\alpha$, where F is based on $\nu_1 = (b-1)$ and $\nu_2 = (n-p-b+1)$ degrees of freedom.

ANOVA F-TEST FOR INTERACTION IN A TWO-FACTOR FACTORIAL EXPERIMENT WITH FACTOR A AT a LEVELS AND FACTOR B AT b LEVELS: REGRESSION APPROACH

H_0: $\beta_{a+b-1} = \beta_{a+b} = \cdots = \beta_{ab-1} = 0$
(i.e., H_0: No interaction between factors A and B

H_a: At least one of the β parameters listed in H_0 differs from 0
(i.e., H_a: Factors A and B interact)

Complete model:

$$E(y) = \beta_0 + \overbrace{\beta_1 x_1 + \cdots + \beta_{a-1} x_{a-1}}^{\text{Main effect } A \text{ terms}} + \overbrace{\beta_a x_a + \cdots + \beta_{a+b-2} x_{a+b-2}}^{\text{Main effect } B \text{ terms}}$$

$$+ \overbrace{\beta_{a+b-1} x_1 x_a + \beta_{a+b} x_1 x_{a+1} + \cdots + \beta_{ab-1} x_{a-1} x_{a+b-2}}^{AB \text{ interaction terms}}$$

where*

$$x_1 = \begin{cases} 1 & \text{if level 2 of factor } A \\ 0 & \text{if not} \end{cases} \quad \cdots \quad x_{a-1} = \begin{cases} 1 & \text{if level } a \text{ of factor } A \\ 0 & \text{if not} \end{cases}$$

$$x_a = \begin{cases} 1 & \text{if level 2 of factor } B \\ 0 & \text{if not} \end{cases} \quad \cdots \quad x_{a+b-2} = \begin{cases} 1 & \text{if level } b \text{ of factor } B \\ 0 & \text{if not} \end{cases}$$

Reduced model:

$$E(y) = \beta_0 + \overbrace{\beta_1 x_1 + \cdots + \beta_{a-1} x_{a-1}}^{\text{Main effect } A \text{ terms}} + \overbrace{\beta_a x_a + \cdots + \beta_{a+b-2} x_{a+b-2}}^{\text{Main effect } B \text{ terms}}$$

Test statistic: $\quad F = \dfrac{(SSE_1 - SSE_2)/[(a-1)(b-1)]}{SSE_2/[ab(r-1)]}$

where

$SSE_1 = $ SSE for reduced model

$SSE_2 = $ SSE for complete model

$r = $ Number of replications

Rejection region: $\quad F > F_\alpha$, where F is based on $\nu_1 = (a-1)(b-1)$ and $\nu_2 = ab(r-1)$ df.

*Note: The independent variables, $x_1, x_2, \ldots, x_{a+b-2}$, are defined for an experiment in which both factors represent *qualitative* variables. When a factor is *quantitative*, you may choose to represent the main effects with quantitative terms such as x, x^2, x^3, and so forth.

where

x_1 = Supply of raw material (in tons)

x_2 = Ratio of allocation

Note that the interaction terms for the model are constructed by taking the products of the various main effect terms, one from each factor. For example, we included terms involving the products of x_1 with x_2 and x_2^2. The remaining interaction terms were formed by multiplying x_1^2 by x_2 and by x_2^2.

b. To test the null hypothesis that supply and ratio do not interact, we must test the null hypothesis that the interaction terms are not needed in the linear model of part **a**:

$$H_0: \quad \beta_5 = \beta_6 = \beta_7 = \beta_8 = 0$$

This requires that we fit the reduced model

$$y = \beta_0 + \beta_1 x_1 + \beta_2 x_1^2 + \beta_3 x_2 + \beta_4 x_2^2$$

and perform the partial F-test outlined in Section 4.11. The test statistic is

$$F = \frac{(SSE_1 - SSE_2)/4}{s^2}$$

where

SSE_1 = SSE for reduced model

SSE_2 = SSE for complete model

s^2 = MSE for complete model   ■

EXAMPLE 8.19

Use the SAS computer software package to fit a second-order model to the data of the factorial experiment in Example 8.12.

SOLUTION

An analysis of variance for a complete two-variable factorial experiment *always* yields results, including an SSE, that would be obtained by fitting a complete factorial model of the type indicated in part **a** of Example 8.18. However, we would not expect the third- and fourth-order terms to contribute much to the model and would recommend a second-order model to fit the response surface of Example 8.12. The second-order model in two independent variables is

$$y = \beta_0 + \beta_1 x_1 + \beta_2 x_1^2 + \beta_3 x_2 + \beta_4 x_2^2 + \beta_5 x_1 x_2 + \varepsilon$$

where

x_1 = Supply of raw material (in tons)

x_2 = Ratio of allocation

The SAS regression analysis printout obtained by fitting this model to the data of Example 8.12 is shown in Figure 8.12. The estimates $\hat{\beta}_0 = -27.815$, $\hat{\beta}_1 =$

FIGURE 8.12   SAS Computer Printout for Fitting a Second-Order Model to the Data of Example 8.19

ANALYSIS OF VARIANCE

SOURCE	DF	SUM OF SQUARES	MEAN SQUARE	F VALUE	PROB>F
MODEL	5	63.50793651	12.70158730	4.895	0.0040
ERROR	21	54.49206349	2.59486017		
C TOTAL	26	118.00000			

ROOT MSE		1.610857	R-SQUARE	0.5382
DEP MEAN		20.66667	ADJ R-SQ	0.4283
C.V.		7.794469		

PARAMETER ESTIMATES

VARIABLE	DF	PARAMETER ESTIMATE	STANDARD ERROR	T FOR H0: PARAMETER=0	PROB > |T|
INTERCEP	1	-27.81481481	23.80152168	-1.169	0.2557
S	1	5.94444444	2.64418353	2.248	0.0354
SS	1	-0.18518519	0.07306996	-2.534	0.0193
R	1	-7.76190476	5.04522969	-1.538	0.1389
RR	1	-2.29629630	1.33939441	-1.714	0.1012
SR	1	0.74603175	0.20294890	3.676	0.0014

5.944, ... , $\hat{\beta}_5 = .746$ can be obtained from the printout. Hence, the second-order prediction equation relating supply of raw materials, S, and ratio of allocation, R, to profit per unit of raw material supply, y, is

$$\hat{y} = -27.815 + 5.944x_1 - .185x_1^2 - 7.762x_2 - 2.296x_2^2 + .746x_1x_2$$

Note that SSE = 54.492 given in Figure 8.12 differs from the value obtained in the analysis of variance of Example 8.12. This is because we omitted the third- and fourth-order terms from the linear model when performing the regression analysis so that now SSE is based on a larger number of degrees of freedom, namely $[n - (k + 1)] = 27 - 6 = 21$ df.   ∎

EXAMPLE 8.20

Do the data provide sufficient information to indicate that the complete factorial model given in Example 8.18 contributes more information for the prediction of y than the second-order model of Example 8.19?

SOLUTION

If the response to the question is "yes," then at least one of the parameters, β_6, β_7, or β_8, of the complete factorial model differs from 0 (i.e., they are needed in the model). Consequently, the null hypothesis, "The complete factorial model contributes no more information about y than that contributed by a second-order model" is equivalent to the hypothesis

$$H_0: \quad \beta_6 = \beta_7 = \beta_8 = 0$$

You will recall (from Section 4.11) that to test this hypothesis, we need to determine the drop in the sum of squares for error between the reduced model (i.e., the second-order model) and the complete model (i.e., the complete factorial model). From part a of Example 8.12, SSE for the complete model is $SSE_2 = 43.33$. From Figure 8.12, the reduced (second-order) model gives $SSE_1 = 54.492$.

Therefore, the drop in SSE associated with β_6, β_7, and β_8, based on 3 df, is

$$\text{Drop} = \text{SSE}_1 - \text{SSE}_2 = 54.492 - 43.33 = 11.16$$

The F-statistic needed for the test is

$$F = \frac{\text{MS(Drop)}}{\text{MSE(Complete model)}} = \frac{11.16/3}{2.41} = 1.54$$

[*Note:* MSE for the complete factorial model is given in Table 8.13.] Since MS(Drop) is based on 3 df and MSE(Complete model) is based on 18 df, we compare the computed value of F with $F_{.05} = 3.16$, where $F_{.05}$ is the tabulated value of F for 3 and 18 df. Since the computed value of F (1.54) is less than the critical value, $F_{.05} = 3.16$, we cannot reject the null hypothesis that $\beta_6 = \beta_7 = \beta_8 = 0$. That is, there is insufficient evidence to indicate that the third- and fourth-order terms associated with β_6, β_7, and β_8 contribute information for the prediction of y.

EXAMPLE 8.21

Use the second-order model of Example 8.19 and find a 95% confidence interval for the mean profit per unit supply of raw material when $S = 17$ and $R = 1$.

SOLUTION

The portion of the SAS printout for the second-order model with 95% confidence intervals for $E(y)$ is shown in Figure 8.13.

The confidence interval for $E(y)$ when $S = 17$ and $R = 1$ is given at the bottom of the printout in the row labeled OBSERVATION 28. You can see that the interval is (20.97, 23.72). Thus, we estimate (with confidence coefficient equal to .95) that the mean profit per unit of supply will lie between \$20.97 and \$23.72 when $S = 17$ tons and $R = 1$. Beyond this immediate result, you will note that this example illustrates the power and versatility of a regression analysis. In particular, there is no way to obtain this estimate from the analysis of variance in Example 8.12. However, a computerized regression package can be programmed to include the confidence interval automatically.   ∎

We conclude this section by pointing out another very powerful advantage of the regression approach to an analysis of variance. The ANOVA formulas for a factorial experiment and many other experimental designs can be used only when the sample sizes are equal. However, the regression approach applies to both equal and unequal sample sizes for the various factor-level combinations. A complete list of the advantages and disadvantages of the regression approach is provided in Section 8.13.

EXERCISES 8.47–8.55

8.47 Refer to Exercise 8.6.
  a. Give the complete and reduced models appropriate for conducting the ANOVA.
  b. If you have access to a computer package, fit the models specified in part **a** and compute the value of the F-statistic. This value should agree with the value calculated in Exercise 8.6.

FIGURE 8.13   SAS Printout with 95% Confidence Limits for $E(y)$ for the Seconc-Order Model in Example 8.21

OBSERVATION	R	S	OBSERVED VALUE	PREDICTED VALUE	RESIDUAL	LOWER 95% CL FOR MEAN	UPPER 95% CL FOR MEAN
1	0.5	15	23.00000000	20.82539683	2.17460317	19.15486668	22.49592698
2	0.5	15	20.00000000	20.82539683	-0.82539683	19.15486668	22.49592698
3	0.5	15	21.00000000	20.82539683	0.17460317	19.15486668	22.49592698
4	0.5	18	22.00000000	21.44444444	0.55555556	20.00286224	22.88602665
5	0.5	18	19.00000000	21.44444444	-2.44444444	20.00286224	22.88602665
6	0.5	18	20.00000000	21.44444444	-1.44444444	20.00286224	22.88602665
7	0.5	21	19.00000000	18.73015873	0.26984127	17.05962858	20.40068888
8	0.5	21	18.00000000	18.73015873	-0.73015873	17.05962858	20.40068888
9	0.5	21	21.00000000	18.73015873	2.26984127	17.05962858	20.40068888
10	1	15	22.00000000	20.81746032	1.18253968	19.36051446	22.27440617
11	1	15	20.00000000	20.81746032	-0.81746032	19.36051446	22.27440617
12	1	15	19.00000000	20.81746032	-1.81746032	19.36051446	22.27440617
13	1	18	24.00000000	22.55555556	1.44444444	21.11397335	23.99713776
14	1	18	25.00000000	22.55555556	2.44444444	21.11397335	23.99713776
15	1	18	22.00000000	22.55555556	-0.55555556	21.11397335	23.99713776
16	1	21	20.00000000	20.96031746	-0.96031746	19.50337160	22.41726332
17	1	21	19.00000000	20.96031746	-1.96031746	19.50337160	22.41726332
18	1	21	22.00000000	20.96031746	1.03968254	19.50337160	22.41726332
19	2	15	18.00000000	17.35714286	0.64285714	15.57067740	19.14360832
20	2	15	18.00000000	17.35714286	0.64285714	15.57067740	19.14360832
21	2	15	16.00000000	17.35714286	-1.35714286	15.57067740	19.14360832
22	2	18	23.00000000	21.33333333	1.66666667	19.89175113	22.77491553
23	2	18	20.00000000	21.33333333	-1.33333333	19.89175113	22.77491553
24	2	18	20.00000000	21.33333333	-1.33333333	19.89175113	22.77491553
25	2	21	20.00000000	21.97619048	-1.97619048	20.18972502	23.76265594
26	2	21	22.00000000	21.97619048	0.02380952	20.18972502	23.76265594
27	2	21	24.00000000	21.97619048	2.02380952	20.18972502	23.76265594
28 *	1	17	.	22.34656085	.	20.96874302	23.72437868

* OBSERVATION WAS NOT USED IN THIS ANALYSIS

8.48 Refer to Exercise 8.11.

  a. Give the complete and reduced models appropriate for conducting the ANOVA.

  b. If you have access to a computer package, fit the models specified in part **a** and compute the value of the F-statistic. This value should agree with the value calculated in Exercise 8.11.

8.49 Refer to Exercise 8.17.

  a. Give the complete and reduced models appropriate for conducting the ANOVA.

  b. If you have access to a computer package, fit the models specified in part **a** and compute the value of the F-statistic. This value should agree with the value calculated in Exercise 8.17.

8.50 Refer to Exercise 8.28.

  a. Give the complete and reduced models appropriate for conducting the ANOVA.

  b. If you have access to a computer package, fit the models specified in part **a** and compute the value of the F-statistic. This value should agree with the value calculated in Exercise 8.28.

8.51 An experiment was conducted to compare the abilities of four real estate appraisers. To eliminate the effect of property-to-property variation in comparing appraiser means, ten properties were randomly selected, and each appraiser was required to appraise each property. Thus, a randomized block experiment was conducted. The appraised values (in thousands of dollars) obtained from the four appraisers for each of the ten properties are shown in the table.

		PROPERTY									
		1	2	3	4	5	6	7	8	9	10
	A	46.3	63.0	43.1	84.6	51.0	49.2	60.0	31.5	47.2	55.0
	B	50.9	66.2	46.7	88.4	50.0	54.0	54.8	32.0	50.5	60.3
APPRAISER	C	48.1	64.7	46.0	83.0	48.8	51.4	62.1	30.5	46.0	57.1
	D	49.2	65.0	48.3	87.8	52.7	52.9	66.0	33.4	49.0	58.0

  a. Write the complete model for this randomized block design.

  b. Modify the model for y in part **a** by deleting the treatment (appraiser) parameters.

  c. Modify the model in part **a** by deleting the block (property) parameters.

  d. The values of SSE associated with the models in parts **a**, **b**, and **c** are shown in the accompanying table. Do the data provide sufficient evidence to indicate differences in the mean appraisal levels for the four appraisers? Test using $\alpha = .05$.

	MODEL		
	a	b	c
SSE	105.28175	167.61250	7,397.88700

  e. Intuitively, we know that blocking was necessary in this experiment because of the large amount of variation in the property values. As an exercise, use the information in part **d** to verify the contention that blocking increased the information in the experiment. Test the null hypothesis of "no difference in block means" using $\alpha = .05$.

8.52 Suppose you plan to investigate the effect of hourly pay rate and length of workday on some measure y of worker productivity. Both pay rate and length of workday will be set at three levels and y will be observed for all combinations of these factors. Thus, a 3×3 factorial experiment will be employed.

   a. Identify the factors and state whether they are quantitative or qualitative.

   b. Identify the treatments to be employed in the experiment.

   c. Write a complete factorial model for the experiment. [*Hint:* When the factors are quantitative, main effect terms include x and x^2 terms for each factor.]

   d. What is the order of the model specified in part c?

   e. Suppose you want to fit a second-order model to the data. Give the appropriate model.

   f. If you have only one observation for each combination and you fit a complete factorial model to the data, how many degrees of freedom will be available for estimating σ^2?

   g. Refer to part f. If you fit a second-order model to the data, how many degrees of freedom will be available for estimating σ^2?

   h. Suppose you replicated the experiment and hence obtained two observations for each treatment. How many degrees of freedom would be available for estimating σ^2 if you fit: (1) the complete factorial model? (2) a second-order model?

   i. Which model—complete factorial or second-order—do you think would be more appropriate for this experiment?

   j. Explain how to conduct a test for interaction between pay rate and length of workday using the complete factorial model.

   k. Explain how to conduct a test for interaction between pay rate and length of workday using the second-order model.

8.53 The data for the 3×3 factorial experiment described in Exercise 8.52 are shown in the accompanying table, and the SAS computer printout of the regression analysis for the second-order model is also shown (page 466).

		HOURLY PAY RATE, DOLLARS (x_1)		
		6.50	7.00	7.50
	8	350	375	402
		377	390	411
LENGTH OF WORKDAY, HOURS (x_2)	9	398	423	434
		386	424	429
	10	345	377	394
		351	381	389

   a. Do the data provide sufficient evidence to indicate that the model contributes information for the prediction of y?

   b. Find R^2 and interpret its meaning.

   c. Does a large value for R^2 imply that the second-order model is a good predictor of productivity?

   d. If the value for R^2 were small, would this imply that the model is inadequate, i.e., that we need to change the form of the model and/or add other independent variables to the model that may be related to worker productivity?

   e. Is there evidence of interaction between the two factors, pay rate and length of workday? Test using $\alpha = .05$.

SAS Computer Printout for Exercise 8.53

ANALYSIS OF VARIANCE

SOURCE	DF	SUM OF SQUARES	MEAN SQUARE	F VALUE	PROB>F
MODEL	5	11355.01389	2271.00278	33.863	0.0001
ERROR	12	804.76389	67.06365741		
C TOTAL	17	12159.77778			

ROOT MSE	8.18924	R-SQUARE	0.9338	
DEP MEAN	390.8889	ADJ R-SQ	0.9062	
C.V.	2.09503			

PARAMETER ESTIMATES

VARIABLE	DF	PARAMETER ESTIMATE	STANDARD ERROR	T FOR H0: PARAMETER=0	PROB > |T|
INTERCEP	1	-4026.63889	939.43751	-4.286	0.0011
X1	1	385.08333	235.19426	1.637	0.1275
X1X1	1	-24.66666667	16.37848069	-1.506	0.1579
X2	1	661.58333	84.14751627	7.862	0.0001
X2X2	1	-37.16666667	4.09462017	-9.077	0.0001
X1X2	1	0.25000000	5.79066738	0.043	0.9663

f. Extract the parameter estimates from the computer printout and give the prediction equation.

g. Graph the predicted productivity $\hat{y}$ as a function of pay rate x_1 for $x_2 = 8$ hours. Then obtain the predicted productivity curves for $x_2 = 9$ and $x_2 = 10$ hours.

8.54 Refer to Exercise 8.30. An appropriate linear model for the experiment is

$$E(y) = \beta_0 + \beta_1 x_1 + \beta_2 x_2 + \beta_3 x_1 x_2$$

where

$x_1 =$ Sintering time (in units of 100 minutes)

$x_2 = \begin{cases} 1 & \text{if metal 1} \\ 0 & \text{if metal 2} \end{cases}$

Fitting this model to the data using the SAS multiple regression package, we obtain the SAS printout shown here.

SAS Computer Printout for Exercise 8.54

ANALYSIS OF VARIANCE

SOURCE	DF	SUM OF SQUARES	MEAN SQUARE	F VALUE	PROB>F
MODEL	3	131.41200	43.80400000	34.772	0.0001
ERROR	16	20.15600000	1.25975000		
C TOTAL	19	151.56800			

ROOT MSE	1.122386	R-SQUARE	0.8670	
DEP MEAN	16.06	ADJ R-SQ	0.8421	
C.V.	6.988704			

PARAMETER ESTIMATES

VARIABLE	DF	PARAMETER ESTIMATE	STANDARD ERROR	T FOR H0: PARAMETER=0	PROB > |T|
INTERCEP	1	7.46000000	1.12238585	6.647	0.0001
X1	1	0.04660000	0.007098591	6.565	0.0001
X2	1	5.44000000	1.58729329	3.427	0.0035
X1X2	1	-0.01480000	0.01003892	-1.474	0.1598

a. Find the least squares prediction equation.

b. Give the prediction equation relating the mean compressive strength to sintering time for metal 1.

c. Give the prediction equation relating the mean compressive strength to sintering time for metal 2.

d. Graph the prediction lines in parts b and c.

e. What does the parameter β_3 measure?

f. Test H_0: $\beta_3 = 0$ against the alternative hypothesis H_a: $\beta_3 \neq 0$. Use $\alpha = .05$.

g. How does the test of part f relate to the test in part c of Exercise 8.30?

8.55 A $2 \times 2 \times 2 \times 2 = 2^4$ factorial experiment was conducted to investigate the effect of four factors on the light output, y, of flashbulbs. Two observations were taken for each of the factorial treatments. The factors are: amount of foil contained in a bulb (100 and 120 milligrams); speed of sealing machine (1.2 and 1.3 revolutions per minute); shift (day or night); machine operator (A or B). The data for the two replications of the 2^4 factorial experiment are shown here and the SAS computer printout for the regression analysis appears on page 468.

		AMOUNT OF FOIL			
		100 milligrams		120 milligrams	
		SPEED OF MACHINE			
		1.2 rpm	1.3 rpm	1.2 rpm	1.3 rpm
DAY SHIFT	Operator B	6; 5	5; 4	16; 14	13; 14
	Operator A	7; 5	6; 5	16; 17	16; 15
NIGHT SHIFT	Operator B	8; 6	7; 5	15; 14	17; 14
	Operator A	5; 4	4; 3	15; 13	13; 14

To simplify computations, we let

$$x_1 = \frac{\text{Amount of foil} - 110}{10} \qquad x_2 = \frac{\text{Speed of machine} - 1.25}{.05}$$

so that x_1 and x_2 will take values -1 and $+1$. Also,

$$x_3 = \begin{cases} -1 & \text{if night shift} \\ 1 & \text{if day shift} \end{cases} \qquad x_4 = \begin{cases} -1 & \text{if machine operator } B \\ 1 & \text{if machine operator } A \end{cases}$$

a. Do the data provide sufficient evidence to indicate that any of the factors contribute information for the prediction of y? Give the results of a statistical test to support your answer.

b. Identify the factors that appear to affect the amount of light y in the flashbulbs.

c. Give the complete factorial model for y. [*Hint:* For a factorial experiment with four factors, the complete model includes main effects for each factor, two-way cross product terms, three-way cross product terms, and four-way cross product terms.]

d. How many degrees of freedom will be available for estimating σ^2?

SAS Printout for Exercise 8.55

ANALYSIS OF VARIANCE

SOURCE	DF	SUM OF SQUARES	MEAN SQUARE	F VALUE	PROB>F
MODEL	15	745.46875	49.69791667	40.778	0.0001
ERROR	16	19.50000000	1.21875000		
C TOTAL	31	764.96875			

ROOT MSE	1.10397	R-SQUARE	0.9745	
DEP MEAN	10.03125	ADJ R-SQ	0.9506	
C.V.	11.00531			

PARAMETER ESTIMATES

VARIABLE	DF	PARAMETER ESTIMATE	STANDARD ERROR	T FOR HO: PARAMETER=0	PROB > |T|
INTERCEP	1	10.03125000	0.19515619	51.401	0.0001
X1	1	4.71875000	0.19515619	24.179	0.0001
X2	1	-0.34375000	0.19515619	-1.761	0.0973
X3	1	0.21875000	0.19515619	1.121	0.2789
X4	1	-0.15625000	0.19515619	-0.801	0.4351
X1X2	1	0.09375000	0.19515619	0.480	0.6375
X1X3	1	0.15625000	0.19515619	0.801	0.4351
X1X4	1	0.28125000	0.19515619	1.441	0.1688
X2X3	1	-0.15625000	0.19515619	-0.801	0.4351
X2X4	1	-0.03125000	0.19515619	-0.160	0.8748
X3X4	1	0.78125000	0.19515619	4.003	0.0010
X1X2X3	1	-0.21875000	0.19515619	-1.121	0.2789
X1X2X4	1	-0.09375000	0.19515619	-0.480	0.6375
X1X3X4	1	-0.03125000	0.19515619	-0.160	0.8748
X2X3X4	1	0.15625000	0.19515619	0.801	0.4351
X1X2X3X4	1	0.09375000	0.19515619	0.480	0.6375

SECTION 8.11

ASSUMPTIONS

For all the experiments and designs discussed in this chapter, the probabilistic model for the response variable y is the familiar regression model of Chapters 4 and 7. Therefore, the assumptions underlying the analyses are the same as those required for a regression analysis (see Section 4.2). The assumptions for each experimental design, stated in the terminology of an analysis of variance, are summarized in the boxes.

ASSUMPTIONS FOR A TEST TO COMPARE p POPULATION MEANS: COMPLETELY RANDOMIZED DESIGN

1. All p population probability distributions are normal.
2. The p population variances are equal.
3. The samples from each population are random and independent.

ASSUMPTIONS FOR A TEST TO COMPARE p POPULATION MEANS (TREATMENTS): RANDOMIZED BLOCK DESIGN

1. The population probability distribution of the difference between any pair of treatment observations within a block is approximately normal.
2. The variance of this difference is constant and the same for all pairs of observations.
3. The treatments are randomly assigned to the experimental units within each block.

ASSUMPTIONS FOR A TWO-WAY ANOVA: FACTORIAL EXPERIMENT

1. The population probability distribution of the observations for any factor-level combination is approximately normal.
2. The variance of the probability distribution is constant and the same for all factor-level combinations.
3. The treatments (factor-level combinations) are randomly assigned to the experimental units.
4. The observations for each factor-level combination represent independent random samples.

In most business applications the assumptions will not be satisfied exactly. These analysis of variance procedures are flexible, however, in the sense that slight departures from the assumptions will not significantly affect the analysis or the validity of the resulting inferences. On the other hand, gross violations of the assumptions (e.g., a nonconstant variance) will cast doubt on the validity of the inferences. Therefore, you should make it standard practice to conduct an analysis of the residuals from the ANOVA using the techniques of Chapter 6 in order to verify that the assumptions are (approximately) satisfied.

| | | | | | | | | | | | |

SECTION 8.12

COMPUTER PRINTOUTS FOR AN ANALYSIS OF VARIANCE

The calculations in an analysis of variance can become very tedious when the number of observations is large or when the observations contain a large number of digits. If you have access to a computer, this difficulty can be easily circumvented by using the regression approach, i.e., by fitting complete and reduced models (see Section 8.9). However, even the regression approach requires you to perform some calculations (for the partial F-tests) and you will need to fit at least two models to carry out the analysis. For these reasons, you may want to take advantage of the analysis of variance routines included in most statistical computer software packages.

All four packages discussed in this text have procedures for conducting an analysis of variance. The SAS, SPSS^x, BMDP, and Minitab packages are capable of analyzing experimental designs ranging from simple one-way classifications of data (completely randomized designs) to the more sophisticated factorial experiments.* In this section, we will examine the respective analysis of variance printouts for a factorial experiment. Since the printouts for completely randomized designs and randomized block designs appear similar to that of the factorial experiment, you should be able to understand the ANOVA results for any of these three experimental procedures.

Refer to the 3×3 factorial experiment of Example 8.12, where we investigated the effect of two factors, level of supply of raw materials (S) and ratio of its assignment (R) to the product manufacturing lines, on the profit (y) per unit of

---

*Currently, Minitab is limited to two-way classifications of data (i.e., two-factor factorial experiments); SAS, SPSS^x, and BMDP are capable of analyzing the more general k-way classifications of data, where k is any integer.

FIGURE 8.14 Printouts for the 3 × 3 Factorial Experiment on the Data in Table 8.11

(a) SAS Printout

DEPENDENT VARIABLE: Y

SOURCE	DF	SUM OF SQUARES	MEAN SQUARE	F VALUE	PR > F	R-SQUARE	C.V.
MODEL	8	74.66666667	9.33333333	3.88	0.0081	0.632768	7.5077
ERROR	18	43.33333333	2.40740741			ROOT MSE	Y MEAN
CORRECTED TOTAL	26	118.00000000				1.55158223	20.66666667

SOURCE	DF	ANOVA SS	F VALUE	PR > F
RATIO	2	8.22222222	1.71	0.2094
SUPPLY	2	20.22222222	4.20	0.0318
RATIO*SUPPLY	4	46.22222222	4.80	0.0082

(b) SPSSˣ Printout

SOURCE OF VARIATION	SUM OF SQUARES	DF	MEAN SQUARE	F	SIG. OF F
WITHIN CELLS	43.33333	18	2.40741		
CONSTANT	11532.00000	1	11532.00000	4790.21538	.000
SUPPLY	20.22222	2	10.11111	4.20000	.032
RATIO	8.22222	2	4.11111	1.70769	.209
SUPPLY BY RATIO	46.22222	4	11.55556	4.80000	.008

(c) BMDP Printout

SOURCE	SUM OF SQUARES	DEGREES OF FREEDOM	MEAN SQUARE	F	TAIL PROB.
MEAN	11532.00000	1	11532.00000	4790.21	0.0000
SUPPLY	20.22222	2	10.11111	4.20	0.0318
RATIO	8.22222	2	4.11111	1.71	0.2094
SR	46.22222	4	11.55556	4.80	0.0082
1 ERROR	43.33333	18	2.40741		

(d) Minitab Printout

ANALYSIS OF VARIANCE

DUE TO	DF	SS	MS=SS/DF
SUPPLY	2	20.22	10.11
RATIO	2	8.22	4.11
SUPPLY*RATIO	4	46.22	11.56
ERROR	18	43.33	2.41
TOTAL	26	118.00	

raw material. The SAS, SPSS[x], BMDP, and Minitab printouts for the data of Table 8.11 are shown in Figures 8.14(a)–(d), respectively.

You can see that all four printouts report the results in the form of an ANOVA summary table, giving the source of variation, degrees of freedom, and sum of squares. Note that SAS, SPSS[x], and BMDP [Figures 8.14(a)–(c), respectively] report the F-values and corresponding observed significance levels (i.e., p-values). Minitab is the only one of the four packages that does not give the value of the test statistic for the analysis of variance F-test. Consequently, you will have to compute the appropriate F-statistic by hand and find the corresponding rejection region if you use Minitab.

The SPSS[x] and BMDP printouts [Figures 8.14(b) and 8.14(c), respectively] include a source of variation that does not appear on either the SAS [Figure 8.14(a)] or Minitab [Figure 8.14(d)] printouts. This source, titled CONSTANT in the SPSS[x] printout and MEAN in the BMDP printout, represents the term β_0 in the complete model for the factorial experiment. Since the source of variation due to β_0 is associated with 1 degree of freedom, the sum of the entries in the DF column for both SPSS[x] and BMDP will equal the sample size n (in this example, $n = 12$), rather than the traditional degrees of freedom ($n - 1$). Note also that the source of variation due to error is titled WITHIN CELLS in the SPSS[x] printout [Figure 8.14(b)].

The SAS package is the only one of the four packages that reports the results of a test for the overall adequacy of the complete factorial model. Note that in the upper portion of the SAS printout [Figure 8.14(a)], SS(Total) is partitioned into two sources of variation, MODEL and ERROR. The source designated as MODEL represents the sum of the main effects for supply S, main effects for ratio R, and supply–ratio interaction effects. Therefore, the F-value given in the upper right of the printout is the value of the test statistic for testing overall model adequacy, i.e.,

H_0: All β coefficients (excluding β_0) in the complete factorial model equal 0

For instructions on how to access the ANOVA routines of SAS, SPSS[x], BMDP, and Minitab, consult Appendix C.

EXERCISES 8.56–8.58

8.56 Before purchasing a $10,000 life insurance policy, a young executive wants to determine if there are significant differences among the mean monthly premiums of four types of policies that he is considering: (1) individual convertible term, (2) individual renewable term, (3) group renewable term, and (4) government group insurance. For each of these policies he obtains price quotations from six randomly selected insurance companies (thus, 24 insurance companies—six for each type of policy—are included in the total sample). The data were analyzed for differences among the mean prices for the four policy types by using the ANOVA procedure of the SAS. The results appear on page 472.

SAS ANOVA Procedure for Exercise 8.56

DEPENDENT VARIABLE: PREMIUM

SOURCE	DF	SUM OF SQUARES	MEAN SQUARE	F VALUE	PR > F
MODEL	3	7.01500000	2.33833333	24.83	0.0001
ERROR	20	1.88333333	0.09416667		
CORRECTED TOTAL	23	8.89833333			

SOURCE	DF	ANOVA SS	F VALUE	PR > F
POLICIES	3	7.01500000	24.83	0.0001

a. Locate the following quantities on the SAS printout and interpret their values: SST, SSE, MST, MSE, F.

b. Are there significant differences among the mean monthly premiums of the four types of insurance policies?

8.57 A large clothing manufacturer conducted an experiment to study the effect of increases in its employees' hourly wages on their productivity. Four treatments were used in the experiment:

Treatment 1: No increase in hourly wage
Treatment 2: Increase hourly wage by $.50
Treatment 3: Increase hourly wage by $1.00
Treatment 4: Increase hourly wage by $1.50

Twelve employees were selected and grouped into three blocks of size four according to the length of time they had been with the company. The four treatments were randomly assigned to the four employees in each block. The employees were observed for 3 weeks and their productivity was measured as the average number of nondefective garments each produced per hour. The resulting productivity measures appear in the table.

	TREATMENT			
	1	2	3	4
Group 1 (less than 1 year)	2.4	3.0	3.1	3.2
Group 2 (1–5 years)	4.8	6.1	5.9	5.7
Group 3 (over 5 years)	5.1	7.0	7.2	7.3

To determine whether there is a difference among the mean productivity levels of the four pay programs, the data were subjected to an ANOVA. The SAS printout appears at the top of the next page.

a. Locate the key elements of the ANOVA on the printout and interpret their values.

b. Is there evidence of a difference among the mean productivity levels of the four pay programs? What is the observed significance level of the test?

c. Use a 95% confidence interval to estimate the difference in mean productivity for treatments 1 and 3.

SAS ANOVA Printout for Exercise 8.57

DEPENDENT VARIABLE: PRODUCT

SOURCE	DF	SUM OF SQUARES	MEAN SQUARE	F VALUE	PR > F	R-SQUARE	C.V.
MODEL	5	33.36166667	6.67233333	45.24	0.0001	0.974158	7.5801
ERROR	6	0.88500000	0.14750000		ROOT MSE		PRODUCT MEAN
CORRECTED TOTAL	11	34.24666667			0.38405729		5.06666667

SOURCE	DF	ANOVA SS	F VALUE	PR > F
TRTMENT	3	3.74000000	8.45	0.0142
GROUP	2	29.62166667	100.41	0.0001

8.58 In increasingly severe oil well environments, oil producers are interested in high-strength nickel alloys that are corrosion-resistant. Since nickel alloys are especially susceptible to hydrogen embrittlement, an experiment was conducted to compare the yield strengths of nickel alloy tensile specimens cathodically charged in a 4% sulfuric acid solution saturated with carbon disulfide, a hydrogen recombination poison. The alloys were tested under two material conditions (cold rolled and cold drawn), each at three different charging times (0, 25, and 50 days). Thus, a 2×3 factorial experiment was conducted, with material condition at two levels and charging time at three levels. Two hydrogen-charged tensile specimens were prepared for each of the $2 \times 3 = 6$ factor-level combinations. Their yield strengths (kilograms per square inch) are recorded in the accompanying table. The SAS ANOVA printout for the 2×3 factorial experiment is also reproduced here.

		ALLOY TYPE	
		Cold rolled	Cold drawn
CHARGING TIME	0 days	53.4; 52.6	47.1; 49.3
	25 days	55.2; 55.7	50.8, 51.4
	50 days	51.0; 50.5	45.2; 44.0

SAS ANOVA Printout for Exercise 8.58

DEPENDENT VARIABLE: YIELD

SOURCE	DF	SUM OF SQUARES	MEAN SQUARE	F VALUE	PR > F	R-SQUARE	C.V.
MODEL	5	142.54666667	28.50933333	43.97	0.0001	0.973436	1.5939
ERROR	6	3.89000000	0.64833333		ROOT MSE		YIELD MEAN
CORRECTED TOTAL	11	146.43666667			0.80519149		50.51666667

SOURCE	DF	ANOVA SS	F VALUE	PR > F
TIME	2	62.76166667	48.40	0.0002
MATERIAL	1	78.03000000	120.35	0.0001
TIME*MATERIAL	2	1.75500000	1.35	0.3272

a. Locate the key elements of the ANOVA on the printout and interpret their values.

b. Is there evidence of interaction between material condition and charging time? Interpret the observed significance level of the test.

S E C T I O N 8.13

SUMMARY

Experimental design is a plan (or strategy) to increase the amount of information in data pertinent to a parameter (or a set of parameters) by controlling two factors:

1. Volume of the signal contained in the data
2. Noise or random variation in the data that is measured by σ^2

The first three steps in designing an experiment are

STEP 1 Select the factors to be investigated.

STEP 2 Choose the factor-level combinations (treatments).

STEP 3 Determine the number of observations for each.

These steps affect the volume of the signal contained in the data because they enable us to shift the information in the experiment so that it focuses on the parameter(s) of interest.

The fourth step in designing an experiment is

STEP 4 Decide how to assign the treatments to the experimental units.

As you have seen, by appropriately assigning the treatments to relatively homogeneous blocks of experimental units, we can reduce the variation of treatment differences. The net effect of this action is to reduce the variance of the random error ε that appears in the linear model.

There are many different ways to select the treatments for an experiment and to assign them to experimental units. In this chapter we discussed **factorial experiments**, in which all possible treatments (factor-level combinations) are selected. We also considered two methods of assigning the treatments to the experimental units—the **completely randomized design** and the **randomized block design**. The particular choice will depend on your experimental objective and the sources of variation that you may be able to reduce by blocking.

After the data have been collected using a designed experiment, inferences about the treatment means are obtained by conducting an **analysis of variance**. An analysis of variance partitions the total sum of squares, SS(Total), into SSE and, depending on the design, sums of squares for treatments (SST), blocks (SSB), and factor main effects and interactions. Tests for treatment means, block means, factor interactions, and so forth are obtained by calculating the ratio of the appropriate mean square to MSE and conducting an F-test.

In optional Section 8.10 we showed that an analysis of variance and a regression analysis yield equivalent results. But an analysis of variance possesses both advantages and disadvantages compared with a regression analysis. One major advantage of an analysis of variance is that it is easy to perform on a pocket or desk calculator. Second, an analysis of variance is essential when y is affected by more than one source of random variation, and then the experimental design permits us to separate these sources and to estimate the variances of their respective random components. (A discussion of these models is not included in this text.) The disadvantages are its restrictions and limitations. They are as follows:

1. **The set of analysis of variance formulas appropriate for a particular experimental design applies only to that design.** If the data collected are observational (i.e., the independent variables are uncontrolled), an analysis of variance is inappropriate. No deviations from the design are permitted. Consequently, the method is of value only for special types of designed experiments.

2. **In contrast to a regression analysis, the analysis of variance formulas change from one design to another.** (There is a pattern, as indicated in Section 8.5, but the pattern is usually not apparent to a beginner.)

3. **An analysis of variance does not give you a prediction equation.** This is a great handicap when one (or more) of the independent variables is quantitative.

4. **Although a linear model is always implied in an analysis of variance, it is rarely presented or discussed when analyses of data have been performed.** Consequently, the thrust of an analysis of variance is often counter (although it need not be) to the notion of modeling, which is the modern quantitative way of analyzing business phenomena.

In conclusion, perhaps the most important point for you to note in this chapter is the following: **If your data can be modeled using a linear model that contains a single random error component (which is the model used throughout this text), then a regression analysis can do everything that an analysis of variance can do and it can do more!** But you will probably need a computer to do it. Regression analysis is simple, uses the same formulas for all analyses, is programmed for both mainframe and personal computers, and can be used to analyze data obtained from both designed and undesigned experiments. Consequently, a beginner may be advised to stick to regression analyses for analyzing the relationship between a set of independent variables and a response y if a computer is available to perform the computations.

8.59 How do you measure the quantity of information in a sample that is pertinent to a particular population parameter?

8.60 What steps in the design of an experiment affect the volume of the signal pertinent to a particular population parameter?

8.61 In what step in the design of an experiment can you possibly reduce the variation produced by extraneous and uncontrolled variables?

8.62 Explain the difference between a completely randomized design and a randomized block design. When is a randomized block design more advantageous?

8.63 Consider a two-factor factorial experiment where one factor is set at two levels and the other factor is set at four levels. How many treatments are included in the experiment? List them.

8.64 Complete the ANOVA summary table for a completely randomized design.

SOURCE	df	SS	MS	F
Treatments	9		15.2	
Error		200.7		
Total	39			

8.65 Complete the ANOVA summary table for a randomized block design.

SOURCE	df	SS	MS	F
Treatments	2			
Blocks			3.0	
Error	14	32		
Total	23	60		

8.66 Complete the ANOVA summary table for a two-factor factorial experiment.

SOURCE	df	SS	MS	F
A	1			
B		120		
AB	3		4.5	
Error			7.0	
Total	23	600		

	PLAN		
1	2	3	4
27	25	24	30
25	28	19	33
29	30	22	31
26	27	21	
	24	26	

8.67 In hopes of attracting more riders, a city transit company plans to have express bus service from a suburban terminal to the downtown business district. These buses will travel along a major city street where there are numerous traffic lights that will affect travel time. The city decides to study the effect of four different plans on the travel times for buses: (1) a special bus express lane; (2) traffic signal progression; (3) express lane plus traffic signal progression; and (4) control—no special travel arrangements. Travel times (in minutes) are measured for randomly selected weekdays during a morning rush-hour trip while each plan is in effect. The results are recorded in the table.

a. What type of experimental design does this represent?

b. Construct an ANOVA summary table for this experiment.

c. Is there evidence of a difference among the mean travel times for the four plans? Use $\alpha = .025$.

d. Form a 95% confidence interval for the difference between the mean travel times of buses under the express lane plan (plan 1) and the control plan which provides for no special travel arrrangements (plan 4).

8.68 Higher wholesale beef prices over the past few years have resulted in the sale of ground beef with higher fat content in an attempt to keep retail prices down. Four different supermarket chains were chosen, and four 1-pound packages of ground beef were randomly selected from each. The percentage of fat content was measured for each package, with the results shown in the accompanying table.

SUPERMARKET			
A	B	C	D
22	25	30	18
20	27	20	20
23	24	23	17
25	24	27	17

a. What type of experimental design does this represent?

b. Do the data provide evidence at the $\alpha = .05$ level that the mean percentage of fat content differs for the four supermarket chains?

c. Use a 90% confidence interval to estimate the mean percentage of fat content per pound at supermarket C.

d. Write the appropriate linear model for this experiment. Explain how to use this model to conduct the test of part **b**.

8.69 Methods of displaying goods can affect their sales. The manager of a supermarket would like to try three different display types for a certain diet beverage. The manager chooses the locations for the three displays in such a way that each display type will be equally accessible to the customers. Each display will be set up for eight 1-week periods. Between each of these experimental periods, a standard display will be used for 2 weeks. For each of the eight experimental weeks, the sales (in dollars) of diet beverage from each display is determined, with the results shown in the table.

PERIOD	DISPLAY		
	A	B	C
1	$225	$253	$208
2	237	235	213
3	210	222	205
4	219	233	212
5	241	244	236
6	207	199	231
7	252	260	244
8	214	216	201

a. What type of experimental design was employed?

b. Construct the ANOVA summary table for this experiment.

c. Is there evidence of a difference among the mean sales for the three types of display? Use $\alpha = .05$.

d. Do the data indicate that the use of weeks as blocks was necessary? Test at $\alpha = .05$.

e. What assumptions are necessary for the validity of the tests conducted in parts **c** and **d**?

f. Use Tukey's multiple comparisons procedure with $\alpha = .05$ to compare the mean sales for the three display types. Interpret the results.

8.70 The "major bottom indicator" of the New York Stock Exchange is a "buy" signal used by Gerald Appel (*Barron's*, February 27, 1984) to forecast upturns in the market. Appel recommends that the optimal time to buy stock is when the difference between the weekly close of the NYSE index and the 10-week moving average is -4.0 points or more. The percentage gains for stock held 4 weeks, 13 weeks, and 26 weeks after the buy signal for eight dates are shown in the table on page 478.

a. What type of experimental design does this represent?

b. Test to determine whether the mean percentage gains differ among the three holding periods. Use $\alpha = .05$.

c. Write the appropriate linear model for this experiment.

d. Explain how the model of part **c** can be used to conduct the test of part **b**.

DATE OF BUY SIGNAL	PERCENT GAIN IN NYSE INDEX		
	4 weeks	13 weeks	26 weeks
June 12, 1970	.0	11.1	21.1
December 21, 1973	3.0	4.7	−7.7
October 4, 1974	18.4	13.5·	30.6
November 17, 1978	1.5	5.4	7.1
November 16, 1979	5.5	11.9	3.6
April 11, 1980	7.0	22.1	32.7
October 16, 1981	2.5	−2.5	−2.7
July 9, 1982	−4.6	19.9	33.8

e. Construct a 95% confidence interval for the difference between the mean percentage gains for 4-week and 26-week holding periods. Interpret the interval.

8.71 The percentage of water removed from paper as it passes through a dryer depends on the temperature of the dryer and the speed of the paper passing through it. A laboratory experiment was conducted to investigate the relationship between dryer temperature T at three levels (100°, 120°, and 140°F) and exposure time E (which is related to speed) also at three levels (10, 20, and 30 seconds). Four paper specimens were prepared for each of the $3 \times 3 = 9$ conditions. The data (percentages of water removed) are shown in the accompanying table.

		TEMPERATURE (T)		
		100	120	140
	10	24 26 21 25	33 33 36 32	45 49 44 45
EXPOSURE TIME (E)	20	39 34 37 40	51 50 47 52	67 64 68 65
	30	58 55 56 53	75 71 70 73	89 87 86 83

a. What type of experimental design does this represent?
b. Perform an analysis of variance for the data and construct an analysis of variance table.
c. Do the data provide sufficient evidence to indicate that temperature and time interact? Test using $\alpha = .05$. What is the practical significance of this test?
d. Find a 90% confidence interval for the mean percentage of water removed when the temperature is 120°F and the exposure time is 20 seconds.
e. Find a 90% confidence interval for the difference in mean times for ($T = 120$, $E = 20$) and ($T = 120$, $E = 30$).
f. Use Tukey's multiple comparisons procedure to compare the mean percentages for the nine treatments. Use $\alpha = .05$.

8.72 Researchers at the University of Iowa conducted a study to evaluate the effectiveness of performance appraisal training in an organizational setting. Each member of a sample of 345 middle-level managers was randomly assigned to one of three training conditions: no training ($n_1 = 122$), computer assisted instruction ($n_2 = 135$), and computer assisted training plus a behavior modeling workshop ($n_3 = 88$). After formal training, the managers were administered a 25-question multiple choice test of managerial knowledge and the number of correct answers was recorded for each. The data are summarized in the table.

TRAINING GROUP	SAMPLE SIZE	SAMPLE MEAN	STANDARD DEVIATION
No training	122	16.75	1.37
Computer assisted training	135	18.35	1.33
Computer training plus workshop	88	18.88	1.20

Source: Davis, B. L. and Mount, M. K. "Effectiveness of performance appraisal training using computer assisted instruction and modeling behavior," *Personnel Psychology*, Autumn 1984, *37*, pp. 439–451.

a. What type of experimental design was employed?

b. Use the technique outlined in Exercise 8.9 to construct an ANOVA summary table for the data.

c. Is there sufficient evidence to indicate that the mean scores on the managerial knowledge test differ among managers trained under the three programs? Test using $\alpha = .01$.

d. Find a 99% confidence interval for the difference between the mean scores of managers with no training and managers with computer assisted training.

e. Construct a 99% confidence interval for the difference between the mean scores of the two groups of managers with computer assisted training. Is there evidence that the behavior modeling workshop increases mean score on the managerial knowledge exam?

f. Explain how you could conduct the analysis of part c using the regression approach.

8.73 From time to time, one branch office of a company must make shipments to a certain branch office in another state. There are three package delivery services between the two cities where the branch offices are located. Since the price structures for the three delivery services are quite similar, the company wants to compare the delivery times. The company plans to make several different types of shipments to its branch office. To compare the carriers, each shipment will be sent in triplicate, one with each carrier. The results listed in the table are the delivery times in hours.

SHIPMENT	CARRIER I	II	III
1	15.2	16.9	17.1
2	14.3	16.4	16.1
3	14.7	15.9	15.7
4	15.1	16.7	17.0
5	14.0	15.6	15.5

a. What type of experimental design does this represent?

b. Is there evidence that the mean delivery times differ for the three companies? Use $\alpha = .05$.

c. Use a 99% confidence interval to estimate the difference between the mean delivery time for carriers I and II.

d. What assumptions are necessary for the validity of the procedures you used in parts **a** and **b**?

e. Explain how to conduct the analysis of part **b** using the regression approach.

8.74 A real estate broker wanted to investigate the relationship between a response y, the per-square-foot increase in sales value per year of residential property, and two qualitative independent variables: region in the broker's community and salesperson who negotiated the sale. Properties included in the study were those that had been recently sold after being held for approximately 2 years. The broker divided the area of the community into four regions and then classified the properties according to salesperson (one of three) and region (one of four). Three properties were randomly selected from within each of the 12 salesperson–region categories, and the annual increase in value per square foot per year was calculated for each. The data are shown in the table.

		REGION				TOTALS
		R_1	R_2	R_3	R_4	
SALESPERSON	S_1	4.22, 3.90, 4.36	2.96, 3.31, 3.13	3.25, 3.73, 3.56	1.81, 2.22 2.21	38.66
	S_2	4.86, 4.42, 4.53	2.60, 2.84 2.49	3.01, 3.61 3.45	1.96, 1.96, 2.10	37.83
	S_3	3.69, 4.46, 4.08	2.50, 2.51 2.06	3.55, 3.10 3.34	2.49, 2.23 2.78	36.79
TOTALS		38.52	24.40	30.60	19.76	113.28

a. What type of experimental design was employed?

b. Compute the sums of squares and mean squares and construct an ANOVA table for the data.

c. What is the practical interpretation of an interaction between salesperson and region?

d. Do the data provide sufficient evidence to indicate that there is an interaction between salesperson and region? Test using $\alpha = .05$.

e. Explain how to conduct the test of part **d** using regression analysis.

f. Use Tukey's multiple comparisons procedure to compare the mean increases in sales value for the four regions separately for each of the three salespersons. Use $\alpha = .05$.

8.75 In purchasing negotiations, buyers and sellers tend to have conflicting bargaining goals. A buyer wants to obtain the best value at minimum cost, while the seller wants to maximize profit. In addition, the seller's commission often is directly contingent on his or her selling effectiveness. In contrast, buyers are rarely compensated directly for their purchasing performance in any one transaction. What effect do these different compensation systems have on the motivation of salespeople and professional buyers? Researchers conducted a study to determine what effects contingent and noncontingent pay systems might have on bargaining behavior and outcomes. They employed a 2×2 factorial experiment with two factors, buyer condition and seller condition, each at two levels (contingent and noncontingent reward). A sample of 160 industrial purchasers was randomly divided into two groups of buyers and sellers, with 80 subjects in each group. Each buyer was matched with a seller to form one of 80 different bargaining "dyads." The dyads were then divided equally among the $2 \times 2 = 4$ buyer–seller reward conditions. Thus, 20 bargaining dyads were assigned to each of the four "treatments" of the experiment.

All subjects participated in a game in which the sellers were told they represented an individual who wanted to sell an office building (for no less than $100,000) and buyers were informed they represented an individual interested in purchasing the building (for no more than $150,000). Sellers in the contingent reward group (SC) were told that they would be rewarded in cash in proportion to the final negotiated price, while sellers in the noncontingent reward group (SNC) were told they would be paid a fixed fee. Similarly, buyers in the contingent reward group (BC) were led to believe that they would be compensated according to the final agreed price, while buyers in the noncontingent reward group (BNC) received a fixed sum regardless of the outcome. One of the key bargaining variables recorded in the experiment was seller's initial offer. The accompanying table gives the mean of the seller's initial offer (in thousands of dollars) for each of the four buyer–seller reward conditions.

		BUYER CONDITION	
		Contingent (BC)	Noncontingent (BNC)
SELLER	Contingent (SC)	160.3	173.05
CONDITION	Noncontingent (SNC)	155.5	149.50

Source: McFillen, J. M., Reck, R. R., and Benton, W. C. "An experiment in purchasing negotiations," *Journal of Purchasing and Materials Management,* Summer 1983, Vol. 19, No. 1, pp. 2–8

a. Construct an ANOVA table for the 2×2 factorial experiment. (For this experiment, $s = \sqrt{\text{MSE}} \approx 21.00$.)

b. Plot the four cell means on the same graph. Does it appear that the two factors, buyer condition and seller condition, interact?

c. Test for interaction between the two factors, buyer condition and seller condition. Use $\alpha = .10$.

d. Give the complete and reduced models appropriate for testing for interaction using the regression approach.

e. Assuming that contingently rewarded sellers open with higher initial offers than noncontingently rewarded sellers, the researchers hypothesized that this difference is due to reward manipulation and therefore that the means for BC/SC and BNC/SC will not differ significantly. Construct a 90% confidence interval for the difference between mean seller's initial offer for the BC/SC and BNC/SC groups. Does the result refute or support the researchers' hypothesis?

8.76 To be able to provide its clients with comparative information on two large suburban residential communities, a realtor wants to know the average home value in each community. Eight homes are selected at random within each community and are appraised by the realtor. The appraisals (in thousands of dollars) are given in the table on page 482. Can you conclude that the average home value is different in the two communities? You have three ways of analyzing this problem, as described in parts a–c.

COMMUNITY	
A	B
43.5	73.5
49.5	62.0
38.0	47.5
66.5	36.5
57.5	44.5
32.0	56.0
67.5	68.0
71.5	63.5

a. Use the two-sample t-statistic (Section 2.8) to test H_0: $\mu_A = \mu_B$.

b. Consider the regression model

$$y = \beta_0 + \beta_1 x + \varepsilon$$

where

$$x = \begin{cases} 1 & \text{if community B} \\ 0 & \text{if community A} \end{cases}$$

$$y = \text{Appraised price}$$

Since $\beta_1 = \mu_B - \mu_A$, testing H_0: $\beta_1 = 0$ is equivalent to testing H_0: $\mu_A = \mu_B$. Use the partial reproduction of the SAS printout shown here to test H_0: $\beta_1 = 0$. Use $\alpha = .05$.

Portion of the SAS Computer Printout for Exercise 8.76

```
                         ANALYSIS OF VARIANCE

                        SUM OF           MEAN
        SOURCE    DF    SQUARES          SQUARE       F VALUE    PROB>F

        MODEL      1   40.64062500    40.64062500     0.215     0.6501
        ERROR     14   2648.71875     189.19420
        C TOTAL   15   2689.35938

        ROOT MSE      13.75479        R-SQUARE      0.0151
        DEP MEAN      54.84375        ADJ R-SQ     -0.0552
        C.V.          25.07996

                        PARAMETER ESTIMATES

                      PARAMETER       STANDARD      T FOR H0:
    VARIABLE    DF    ESTIMATE        ERROR         PARAMETER=0    PROB > |T|

    INTERCEP     1   53.25000000    4.86305198        10.950        0.0001
    X            1    3.18750000    6.87739406         0.463        0.6501
```

c. Use the ANOVA method to test H_0: $\mu_A = \mu_B$. Use $\alpha = .05$.

Using the results of the three tests above, verify that the tests are the equivalent (for this special case, $p = 2$) of the completely randomized design in terms of the test statistic value and rejection region. For the three methods used, what are the advantages and disadvantages (limitations) of using each in analyzing results for this type of experimental design?

8.77 A production manager who supervises an assembly operation wants to investigate the effect of the incoming rate (parts per minute) x_1 of components and room temperature x_2 on the productivity (number of items produced per minute) y. The component parts approach the worker on a belt and return to the worker if not selected on the first trip past the assembly point. It is thought that an increase in the arrival rate of components has a positive effect on the assembly rate, up to a point, after which increases may annoy the assembler and reduce productivity. Similarly, it is suspected that lowering the room temperature is beneficial to a point, after which reductions may reduce productivity. The experimenter used the same assembly position for each worker. Thirty-two workers were used for the experiment, two each assigned to the 16 independent variable level combinations of a 4 × 4 factorial experiment. The data, in parts per minute averaged over a 5-minute period, are shown in the table.

		RATE OF INCOMING COMPONENTS, PARTS PER MINUTE (x_1)			
		40	50	60	70
ROOM TEMPERATURE, °F (x_2)	65	24.0, 23.8	25.6, 25.4	29.2, 29.4	28.4, 27.6
	70	25.0, 26.0	28.8, 28.8	31.6, 32.0	30.2, 30.0
	75	25.6, 25.0	27.6, 28.0	29.8, 28.6	28.0, 27.0
	80	24.0, 24.6	27.6, 26.2	27.6, 28.6	26.0, 24.4

a. Perform an analysis of variance for the data. Display the computed quantities in an ANOVA table.

b. Write the linear model implied by the analysis of variance. [*Hint:* For a quantitative variable recorded at four levels, main effects include terms for x, x^2, and x^3.]

c. Do the data provide sufficient evidence to indicate differences among the mean responses for the 16 treatments of the 4×4 factorial experiment? Test using $\alpha = .05$.

d. Do the data provide sufficient evidence to indicate an interaction between arrival rate x_1, and room temperature x_2 on worker productivity? Test using $\alpha = .05$.

e. Find the value of R^2 that would be obtained if you were to fit the linear model in part **b** to the data.

f. Explain why a regression analysis would be a useful addition to the inferential methods used in parts **a**–**e**.

8.78 A second-order model would be a reasonable choice to model the data of Exercise 8.77. To simplify the analysis, we will code the arrival rate and temperature values as follows:

$$x_1 = \frac{\text{Arrival rate} - 55}{5} \qquad x_2 = \frac{\text{Temperature} - 72.5}{2.5}$$

A printout of the SAS regression analysis is shown here.

SAS Computer Printout for Exercise 8.78

ANALYSIS OF VARIANCE

SOURCE	DF	SUM OF SQUARES	MEAN SQUARE	F VALUE	PROB>F
MODEL	5	130.80680	26.16136000	27.726	0.0001
ERROR	26	24.53320000	0.94358462		
C TOTAL	31	155.34000			

ROOT MSE	0.9713828	R-SQUARE	0.8421	
DEP MEAN	27.325	ADJ R-SQ	0.8117	
C.V.	3.554923			

PARAMETER ESTIMATES

VARIABLE	DF	PARAMETER ESTIMATE	STANDARD ERROR	T FOR H0: PARAMETER=0	PROB > |T|
INTERCEP	1	29.85625000	0.34876060	85.607	0.0001
X1	1	0.56000000	0.07679456	7.292	0.0001
X2	1	-0.16250000	0.07679456	-2.116	0.0441
X1X2	1	-0.11350000	0.03434357	-3.305	0.0028
X1X1	1	-0.27500000	0.04292946	-6.406	0.0001
X2X2	1	-0.23125000	0.04292946	-5.387	0.0001

a. Write the second-order linear model for the response. Note the difference between this model and the ANOVA model in Exercise 8.77, part **a**.

b. Give the prediction relating the response y to the coded independent variables x_1 and x_2.

c. Why does the SSE given in the computer printout differ from the SSE obtained in Exercise 8.77?

d. Find the value of R^2 appropriate for your second-order model and interpret its value.

e. Do the data provide sufficient evidence to indicate that the complete factorial model provides more information for predicting y than a second-order model?

8.79 Suppose you want to investigate the effect of two factors—arrival rate of product components, A, and temperature of the room, T—on the length of time, y, required by individual workers to perform a product assembly operation. Each factor will be held at two levels: arrival rate at .5 and 1.0 component per second, and temperature at 70° and 80°F. Thus, a 2 × 2 factorial experiment will be employed. In order to block out worker-to-worker variability, each of ten randomly selected workers will be required to assemble the product under all four experimental conditions. Therefore, the four treatments (working conditions) will be assigned to the experimental units (workers) using a randomized block design, where the workers represent the blocks.

The appropriate complete model for the randomized block design is

$$\text{Complete model:} \quad E(y) = \beta_0 + \overbrace{\beta_1 x_1 + \beta_2 x_2 + \beta_3 x_1 x_2}^{\substack{\text{Treatment effects (main} \\ \text{effects and interaction terms} \\ \text{for arrival rate and temperature)}}}$$

$$+ \overbrace{\beta_4 x_3 + \beta_5 x_4 + \cdots + \beta_{12} x_{11}}^{\text{Block (worker) effects}}$$

where

$x_1 = $ Arrival rate $x_2 = $ Temperature

$$x_3 = \begin{cases} 1 & \text{if worker 1} \\ 0 & \text{if not} \end{cases} \qquad x_4 = \begin{cases} 1 & \text{if worker 2} \\ 0 & \text{if not} \end{cases} \quad \cdots \quad x_{11} = \begin{cases} 1 & \text{if worker 9} \\ 0 & \text{if not} \end{cases}$$

[Note that $x_3 = x_4 = \cdots = x_{11} = 0$ if worker (block) 10 is the assembler.]

The assembly time data for the 2 × 2 factorial experiment with a randomized block design are given in the table. The SAS printouts for the complete model and the reduced model

$$\text{Reduced model:} \quad E(y) = \beta_0 + \overbrace{\beta_4 x_3 + \beta_5 x_4 + \cdots + \beta_{12} x_{11}}^{\text{Workers}}$$

are also provided.

a. Do the data provide sufficient evidence to indicate a difference among the four treatment means?

b. Does the effect of a change in arrival rate on assembly time depend on temperature (i.e., do arrival rate and temperature interact)?

c. Estimate the mean loss (or gain) in assembly time as arrival rate is increased from .5 to 1.0 component per second and temperature is held at 70°F. What inference can you make based on this estimate?

				WORKER								
				1	2	3	4	5	6	7	8	9	10
ROOM TEMPERATURE	70°F	ARRIVAL RATE, COMPONENT PER SECOND	.5	1.7	1.3	1.7	2.0	2.0	2.3	2.0	2.8	1.5	1.6
			1.0	.8	.8	1.5	1.2	1.2	1.7	1.1	1.5	.5	1.0
	80°F		.5	1.3	1.5	2.3	1.6	2.2	2.1	1.8	2.4	1.3	1.8
			1.0	1.8	1.5	2.3	2.0	2.7	2.2	2.3	2.6	1.3	1.8

SAS Computer Printout: Complete Model for Exercise 8.79

```
                          ANALYSIS OF VARIANCE

                            SUM OF         MEAN
            SOURCE    DF    SQUARES       SQUARE     F VALUE      PROB>F

            MODEL     12  10.17400000   0.84783333   21.176      0.0001
            ERROR     27   1.08100000   0.04003704
            C TOTAL   39  11.25500000

            ROOT MSE      0.2000926     R-SQUARE     0.9040
            DEP MEAN         1.725      ADJ R-SQ     0.8613
            C.V.         11.59957

                          PARAMETER ESTIMATES

                     PARAMETER      STANDARD     T FOR H0:
        VARIABLE  DF  ESTIMATE        ERROR    PARAMETER=0    PROB > |T|

        INTERCEP   1    9.75500000   1.50701723     6.473       0.0001
        X1         1  -15.24000000   1.90245845    -8.011       0.0001
        X2         1   -0.10400000   0.02000926    -5.198       0.0001
        X1X2       1    0.19600000   0.02530993     7.744       0.0001
        X3         1   -0.15000000   0.14148681    -1.060       0.2985
        X4         1   -0.27500000   0.14148681    -1.944       0.0624
        X5         1    0.40000000   0.14148681     2.827       0.0087
        X6         1    0.15000000   0.14148681     1.060       0.2985
        X7         1    0.47500000   0.14148681     3.357       0.0024
        X8         1    0.52500000   0.14148681     3.711       0.0010
        X9         1    0.25000000   0.14148681     1.767       0.0885
        X10        1    0.77500000   0.14148681     5.478       0.0001
        X11        1   -0.40000000   0.14148681    -2.827       0.0087
```

SAS Computer Printout: Reduced Model for Exercise 8.79

```
                          ANALYSIS OF VARIANCE

                            SUM OF         MEAN
            SOURCE    DF    SQUARES       SQUARE     F VALUE      PROB>F

            MODEL      9   5.19500000   0.57722222    2.858       0.0147
            ERROR     30   6.06000000   0.20200000
            C TOTAL   39  11.25500000

            ROOT MSE      0.4494441     R-SQUARE     0.4616
            DEP MEAN         1.725      ADJ R-SQ     0.3000
            C.V.         26.05473

                          PARAMETER ESTIMATES

                     PARAMETER      STANDARD     T FOR H0:
        VARIABLE  DF  ESTIMATE        ERROR    PARAMETER=0    PROB > |T|

        INTERCEP   1    1.55000000   0.22472205     6.897       0.0001
        X3         1   -0.15000000   0.31780497    -0.472       0.6404
        X4         1   -0.27500000   0.31780497    -0.865       0.3937
        X5         1    0.40000000   0.31780497     1.259       0.2179
        X6         1    0.15000000   0.31780497     0.472       0.6404
        X7         1    0.47500000   0.31780497     1.495       0.1455
        X8         1    0.52500000   0.31780497     1.652       0.1090
        X9         1    0.25000000   0.31780497     0.787       0.4377
        X10        1    0.77500000   0.31780497     2.439       0.0209
        X11        1   -0.40000000   0.31780497    -1.259       0.2179
```

8.80 A company that employs a large number of salespeople is interested in learning which of the salespeople sell the most: those strictly on commission, those with a fixed salary, or those with a reduced fixed salary plus a commission. The previous month's records for a sample of salespeople are inspected and the amount of sales (in dollars) is recorded for each, as shown in the table.

COMMISSIONED	FIXED SALARY	COMMISSION PLUS SALARY
$425	$420	$430
507	448	492
450	437	470
483	432	501
466	444	
492		

a. Do the data provide sufficient evidence to indicate a difference in the mean sales for the three types of compensation? Use $\alpha = .05$.

b. Use a 90% confidence interval to estimate the mean sales for salespeople who receive a commission plus salary.

c. Use a 90% confidence interval to estimate the difference in mean sales between salespeople on commission plus salary versus those on fixed salary.

ON YOUR OWN . . .

Due to ever-increasing food costs, consumers are becoming more discerning in their choice of supermarkets. It usually is more convenient to shop at only one market, as opposed to buying different items at different markets. Thus, it would be useful to compare the mean food expenditure for a market basket of food items from store to store. Since there is a great deal of variability in the prices of products sold at any supermarket, we will consider an experiment that blocks on products.

Choose three (or more) supermarkets in your area that you want to compare; then choose approximately ten (or more) food products you typically purchase. For each food item, record the price each store charges in the following manner:

FOOD ITEM 1	FOOD ITEM 2 . . . FOOD ITEM 10
Price store 1	Price store 1 . . . Price store 1
Price store 2	Price store 2 . . . Price store 2
Price store 3	Price store 3 . . . Price store 3

Use the data you obtain to test

H_0: Mean expenditures at the stores are the same

H_a: Mean expenditures for at least two of the stores are different

Also, test to determine whether blocking on food items is advisable in this kind of experiment. Fully interpret the results of your analysis.

REFERENCES

Box, G. E. P., Hunter, W. G., and Hunter, J. S. *Statistics for Experimenters*, New York: Wiley, 1978.

Cochran, W. G. and Cox, G. M. *Experimental Designs*, 2nd ed. New York: Wiley, 1957.

Davies, O. L. *The Design and Analysis of Industrial Experiments*. New York: Hafner, 1956.

Dawson, L. M. "Campus attitudes toward business," *MSU Business Topics*, Summer 1969, Vol. 17, pp. 36–46. Published by Division of Research, Graduate School of Business Administration, Michigan State University.

Kirk, R. E. *Experimental Design: Procedures for the Behavioral Sciences*. Monterey, Calif.: Brooks/Cole, 1968.

Kramer, C. Y. "Extension of multiple range tests to group means with unequal number of replications," *Biometrics*, Vol. 12, 1956, pp. 307–310.

Mendenhall, W. *Introduction to Linear Models and the Design and Analysis of Experiments*. Belmont, Calif.: Wadsworth, 1968.

Miller, R. G. *Simultaneous Statistical Inference*, 2nd ed. New York: Springer-Verlag, 1981.

Neter, J., Wasserman, W., and Kutner, M. H. *Applied Linear Statistical Models*, 2nd ed. Homewood, Ill.: Richard D. Irwin, 1985.

Scheffé, H. "A method for judging all contrasts in the analysis of variance," *Biometrika*, Vol. 40, 1953, pp. 87–104.

Scheffé, H. *The Analysis of Variance*. New York: Wiley, 1959.

Tukey, J. W. "Comparing individual means in the analysis of variance," *Biometrics*, Vol. 5, 1949, pp. 99–114.

Winer, B. J. *Statistical Principles in Experimental Design*. New York: McGraw-Hill, 1962.

OBJECTIVE

To present models that allow for the correlation between observations taken sequentially over time; to show how these models can be used to forecast a future response.

CONTENTS

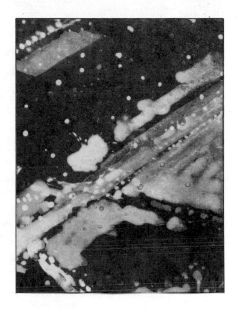

TIME SERIES MODELING AND FORECASTING

WHAT IS A TIME SERIES?

In many business and economic studies, the response variable y is measured sequentially in time. For example, we might record the number y of new housing starts for each month in a particular region. This collection of data is called a **time series**. Other examples of time series are data collected on the monthly production for a manufacturing company, the annual sales for a corporation, and the recorded month-end values of the prime interest rate.

DEFINITION 9.1

A **time series** is a collection of data obtained by observing a response variable at periodic points in time.

DEFINITION 9.2

If repeated observations on a variable produce a time series, the variable is called a **time series variable**. We use y_t to denote the value of the variable at time t.

If you were to develop a model relating the number of new housing starts to the prime interest rate over time, the model would be called a **time series model**, because both the dependent variable, new housing starts, and the independent variable, prime interest rate, are measured sequentially over time. Furthermore, time itself would probably play an important role in such a model, because the economic trends and seasonal cycles associated with different points in time would almost certainly affect both time series.

The construction of time series models is an important aspect of business and economic analyses, because many of the variables of most interest to business and economic researchers are time series. This chapter is an introduction to the very complex and voluminous body of material concerned with time series modeling and forecasting future values of a time series.

TIME SERIES COMPONENTS

Researchers often approach the problem of describing the nature of a time series y_t by identifying four kinds of change, or variation, in the time series values. These four components are commonly known as (1) secular trend; (2) cyclical effect; (3) seasonal variation; and (4) residual effect. The components of a time series are most easily identified and explained pictorially.

Figure 9.1(a) shows a **secular trend** in the time series values. The secular component describes the tendency of the value of the variable to increase or decrease over a long period of time. Thus, this type of change or variation is also known as the **long-term trend**. In Figure 9.1(a), the long-term trend is of an increasing nature. However, this does not imply that the time series has always moved upward from month to month and from year to year. You can see that the series fluctuates, but that the trend has been an increasing one over that period of time.

FIGURE 9.1 Illustrating the Components of a Time Series

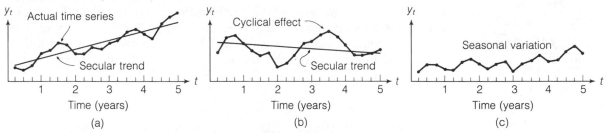

The **cyclical effect** in a time series, as shown in Figure 9.1(b), generally describes the fluctuation about the secular trend that is attributable to business and economic conditions at the time. These fluctuations are sometimes called **business cycles**. During a period of general economic expansion, the business cycle lies above the secular trend, while during a recession, when business activity is likely to slump, the cycle lies below the secular trend. You can see that the cyclical variation does not follow any definite trend, but moves rather unpredictably.

The **seasonal variation** in a time series describes the fluctuations that recur during specific portions of each year (e.g., monthly or seasonally). In Figure 9.1(c), you can see that the pattern of change in the time series within a year tends to be repeated from year to year, producing a wavelike or oscillating curve.

The final component, the **residual effect**, is what remains after the secular, cyclical, and seasonal components have been removed. This component is not systematic and may be attributed to unpredictable influences such as wars, hurricanes, presidential assassination, and randomness of human actions. Thus, the residual effect represents the random error component of a time series.

In many practical applications of time series to business, the objective is to *forecast* (predict) some *future value or values* of the series. To obtain forecasts, some type of model that can be projected into the future must be used to describe the time series. One of the most widely used models is the **additive model***

$$y_t = T_t + C_t + S_t + R_t$$

where T_t, C_t, S_t, and R_t represent the secular trend, cyclical effect, seasonal variation, and residual effect, respectively, of the time series variable y_t. Various methods exist for estimating the components of the model and forecasting the time series. These range from simple **descriptive techniques**, which rely on smoothing the pattern of the time series, to complex **inferential models** which combine regression analysis with specialized time series models. Several descriptive forecasting techniques are presented in optional Section 9.3, and forecasting using the general linear regression model of Chapter 4 is discussed in Section 9.4. The remainder of the chapter is devoted to the more complex and more powerful time series models.

---

*Another useful model is the **multiplicative model** $y_t = T_t C_t S_t R_t$. Note that this model can be written in the form of an additive model by taking natural logarithms:

$$\ln y_t = \ln T_t + \ln C_t + \ln S_t + \ln R_t$$

S E C T I O N 9.3

FORECASTING USING SMOOTHING TECHNIQUES (OPTIONAL)

Various descriptive methods are available for identifying and characterizing a time series. Generally, these methods attempt to remove the rapid fluctuations in a time series so that the secular trend can be seen. For this reason, they are sometimes called **smoothing techniques**. Once the secular trend is identified, forecasts for future values of the time series are easily obtained. In this section we present three of the more popular smoothing techniques.

MOVING AVERAGE METHOD

A widely used smoothing technique is the **moving average method**. A moving average, M_t, at time t is formed by averaging the time series values over adjacent time periods. Moving averages aid in identifying the secular trend of a time series because the averaging tends to modify the effect of short-term (cyclical or seasonal) variation. That is, a plot of the moving averages yields a "smooth" time series curve which clearly depicts the long-term trend.

For example, consider the 1985–1988 quarterly power loads for a utility company located in a southern part of the United States, given in Table 9.1.

TABLE 9.1
Quarterly Power Loads, 1985–1988

YEAR	QUARTER	TIME t	POWER LOAD y_t, megawatts
1985	I	1	103.5
	II	2	94.7
	III	3	118.6
	IV	4	109.3
1986	I	5	126.1
	II	6	116.0
	III	7	141.2
	IV	8	131.6
1987	I	9	144.5
	II	10	137.1
	III	11	159.0
	IV	12	149.5
1988	I	13	166.1
	II	14	152.5
	III	15	178.2
	IV	16	169.0

A graph of the quarterly time series, Figure 9.2, shows the pronounced seasonal variation, i.e., the fluctuation that recurs from year to year. The quarterly power loads tend to be highest in the summer months (quarter III) with another smaller peak in the winter months (quarter I), and lowest during the spring and fall (quarters II and IV). To clearly identify the long-term trend of the series, we need to average, or "smooth out," these seasonal fluctuations. We will apply the moving average method for this purpose.

FIGURE 9.2
Graph of Quarterly Power
Loads, Table 9.1

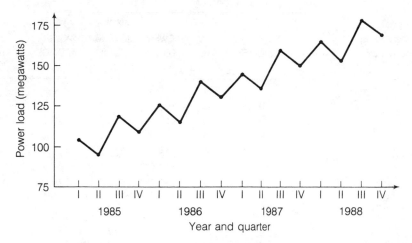

The first step in calculating a moving average for quarterly data is to sum the observed time values y_t—in this example, quarterly power loads—for the four quarters during the initial year 1985. Summing the values from Table 9.1, we have

$$y_1 + y_2 + y_3 + y_4 = 103.5 + 94.7 + 118.6 + 109.3$$
$$= 426.1$$

This sum is called a **4-point moving total**, which we denote by the symbol S_t. It is customary to use a subscript t to represent the time period at the midpoint of the four quarters in the total. Since for this sum, the midpoint is between $t = 2$ and $t = 3$, we will use the conventional procedure of "dropping it down one line" to $t = 3$. Thus, our first 4-point moving total is $S_3 = 426.1$.

We find the next moving total by eliminating the first quantity in the sum, $y_1 = 103.5$, and adding the next value in the time series sequence, $y_5 = 126.1$. This enables us to keep four quarters in the total of adjacent time periods. Thus, we have

$$S_4 = y_2 + y_3 + y_4 + y_5 = 94.7 + 118.6 + 109.3 + 126.1 = 448.7$$

Continuing this process of "moving" the 4-point total over the time series until we have included the last value, we find

$$S_5 = y_3 + y_4 + y_5 + y_6 \quad = 118.6 + 109.3 + 126.1 + 116.0 = 470.0$$
$$S_6 = y_4 + y_5 + y_6 + y_7 \quad = 109.3 + 126.1 + 116.0 + 141.2 = 492.6$$
$$\vdots \qquad\qquad\qquad \vdots \qquad\qquad\qquad\qquad \vdots \qquad\qquad\qquad \vdots$$
$$S_{15} = y_{13} + y_{14} + y_{15} + y_{16} = 166.1 + 152.5 + 178.2 + 169.0 = 665.8$$

The complete set of 4-point moving totals is given in the appropriate column of Table 9.2 (page 494). Notice that three data points will be "lost" in forming the moving totals.

YEAR	QUARTER	TIME t	POWER LOAD y_t	4-POINT MOVING TOTAL S_t	4-POINT MOVING AVERAGE M_t	RATIO y_t/M_t
1985	I	1	103.5	—	—	—
	II	2	94.7	—	—	—
	III	3	118.6	426.1	106.5	1.113
	IV	4	109.3	448.7	112.2	.974
1986	I	5	126.1	470.0	117.5	1.073
	II	6	116.0	492.6	123.2	.942
	III	7	141.2	514.9	128.7	1.097
	IV	8	131.6	533.3	133.3	.987
1987	I	9	144.5	554.4	138.6	1.043
	II	10	137.1	572.2	143.1	.958
	III	11	159.0	590.1	147.5	1.078
	IV	12	149.5	611.7	152.9	.978
1988	I	13	166.1	627.1	156.8	1.059
	II	14	152.5	646.3	161.6	.944
	III	15	178.2	665.8	166.5	1.071
	IV	16	169.0	—	—	—

After the 4-point moving totals are calculated, the second step is to determine the **4-point moving average**, denoted by M_t, by dividing each of the moving totals by 4. For example, the first three values of the 4-point moving average for the quarterly power load data are:

$$M_3 = \frac{y_1 + y_2 + y_3 + y_4}{4} = \frac{S_3}{4} = \frac{426.1}{4} = 106.5$$

$$M_4 = \frac{y_2 + y_3 + y_4 + y_5}{4} = \frac{S_4}{4} = \frac{448.7}{4} = 112.2$$

$$M_5 = \frac{y_3 + y_4 + y_5 + y_6}{4} = \frac{S_5}{4} = \frac{470.0}{4} = 117.5$$

All of the 4-point moving averages are given in the appropriate column of Table 9.2.

Both the original power load time series and the 4-point moving average are graphed in Figure 9.3. Notice that the moving average has "smoothed" the time series, i.e., the averaging has modified the effects of the short-term or seasonal variation. The plot of the 4-point moving average clearly depicts the secular (long-term) trend component of the time series.

In addition to identifying a long-term trend, moving averages provide us with a measure of the seasonal effects in a time series. The ratio between the observed power load y_t and the 4-point moving average M_t for each quarter measures the seasonal effect (primarily attributable to temperature differences) for that quarter.

FIGURE 9.3
Quarterly Power Loads
and 4-Point Moving
Average

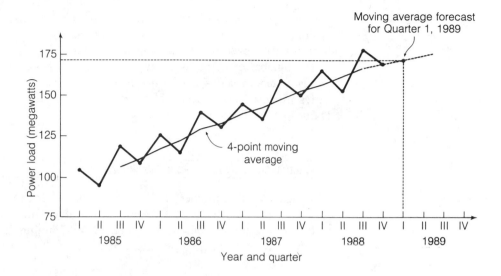

The ratios y_t/M_t are shown in the last column of Table 9.2. Note that the ratio is always greater than 1 in quarters I and III, and always less than 1 in quarters II and IV. The average of the ratios for a particular quarter, multiplied by 100, can be used to form a **seasonal index** for that quarter. For example, the seasonal index for quarter I is

$$(100)\frac{1.073 + 1.043 + 1.059}{3} = 105.8$$

implying that the time series value in quarter I is, on the average, 105.8% of the moving average value for that time period.

To forecast a future value of the time series, simply extend the moving average M_t on the graph to the future time period. For example, a graphical extension of the moving average for the quarterly power loads to quarter I of 1989 ($t = 17$) yields a moving average of approximately $M_{17} = 172$ (see Figure 9.3). Thus, if there were no seasonal variation in the time series, we would expect the power load for quarter I of 1989 to be approximately 172 megawatts. To adjust the forecast for seasonal variation, multiply the future moving average value $M_{17} = 172$ by the seasonal index for quarter I, then divide by 100:

$$F_{17} = M_{17}\left(\frac{\text{Seasonal index for quarter I}}{100}\right)$$

$$= 172\left(\frac{105.8}{100}\right)$$

$$\approx 182$$

where F_{17} is the forecast of y_{17}. Therefore, the moving average forecast for the power load in quarter I of 1989 is approximately 182 megawatts.

Moving averages are not restricted to 4 points. For example, you may wish to calculate a 7-point moving average for daily data, a 12-point moving average for monthly data, or a 5-point moving average for yearly data. Although the choice of the number of points is arbitrary, you should search for the number N that yields a smooth series, but is not so large that many points at the end of the series are "lost." The method of forecasting with a general N-point moving average is outlined in the box.

FORECASTING USING AN *N*-POINT MOVING AVERAGE

1. Select N, the number of consecutive time series values $y_1, y_2, \ldots, y_n$ that will be averaged. (The time series values must be equally spaced.)
2. Calculate the N-point moving total, S_t, by summing the time series values over N adjacent time periods, where

$$S_t = \begin{cases} y_{t-(N-1)/2} + \cdots + y_t + \cdots + y_{t+(N-1)/2} & \text{if } N \text{ is odd} \\ y_{t-N/2} + \cdots + y_t + \cdots + y_{t+N/2-1} & \text{if } N \text{ is even} \end{cases}$$

3. Compute the N-point moving average, M_t, by dividing the corresponding moving total by N:

$$M_t = \frac{S_t}{N}$$

4. Graph the moving average M_t on the vertical axis with time t on the horizontal axis. (This plot should reveal a "smooth" curve that identifies the long-term trend of the time series.*) Extend the graph to a future time period to obtain the forecasted value of M_t.
5. For a future time period t, the forecast of y_t is

$$F_t = \begin{cases} M_t & \text{if little or no seasonal variation exists in the time series} \\ M_t \cdot \left(\dfrac{\text{Seasonal index}}{100} \right) & \text{otherwise} \end{cases}$$

where the seasonal index for a particular quarter (or month) is the average of past values of the ratios

$$\frac{Y_t}{M_t}(100)$$

for that quarter (or month).

---

*When the number N of points is small, the plot may not yield a very smooth curve. However, the moving average will be smoother (or less variable) than the plot of the original time series values.

EXPONENTIAL
SMOOTHING

One problem with using a moving average to forecast future values of a time series is that values at the ends of the series are "lost," thereby requiring that we subjectively extend the graph of the moving average into the future. No exact calculation of a forecast is available since the moving average at a future time period t requires that we know one or more future values of the series. A technique which leads to forecasts that can be explicitly calculated is called **exponential smoothing**. Like the moving average method, exponential smoothing tends to deemphasize (or "smooth") most of the residual effects. However, exponential smoothing averages only past and current values of the time series.

To obtain an exponentially smoothed time series, we first need to choose a weight w, between 0 and 1, called the **exponential smoothing constant**. The exponentially smoothed series, denoted E_t, is then calculated as follows:

$$E_1 = y_1$$
$$E_2 = wy_2 + (1 - w)E_1$$
$$E_3 = wy_3 + (1 - w)E_2$$
$$\vdots \qquad \vdots$$
$$E_t = wy_t + (1 - w)E_{t-1}$$

You can see that the exponentially smoothed value at time t is simply a weighted average of the current time series value, y_t, and the exponentially smoothed value at the previous time period, E_{t-1}. Smaller values of w give less weight to the current value, y_t, while larger values give more weight to y_t.

For example, suppose we want to smooth the quarterly power loads given in Table 9.1 using an exponential smoothing constant of $w = .7$. Then we have

$$E_1 = y_1 = 103.5$$
$$E_2 = .7y_2 + (1 - .7)E_1$$
$$\quad = .7(94.7) + .3(103.5) = 97.3$$
$$E_3 = .7y_3 + (1 - .7)E_2$$
$$\quad = .7(118.6) + 3(97.3) = 112.2$$
$$\vdots$$

The exponentially smoothed values (using $w = .7$) for all the quarterly power loads are given in Table 9.3 (page 498). Both the actual and the smoothed time series values are graphed in Figure 9.4.

Exponential smoothing forecasts are obtained by taking a weighted average of the most recent value of the time series, y_t, and the most recent exponentially smoothed value, E_t. If n is the last time period in which y_t is observed, then the forecast for a future time period t is given by

$$F_t = wy_n + (1 - w)E_n \quad \text{(see the box on page 499)}$$

TABLE 9.3

Quarterly Power Load with Exponential Smoothing

YEAR	QUARTER	TIME t	POWER LOAD y_t	EXPONENTIALLY SMOOTHED POWER LOAD E_t
1985	I	1	103.5	103.5
	II	2	94.7	97.3
	III	3	118.6	112.2
	IV	4	109.3	110.2
1986	I	5	126.1	121.3
	II	6	116.0	117.6
	III	7	141.2	134.1
	IV	8	131.6	132.4
1987	I	9	144.5	140.9
	II	10	137.1	138.2
	III	11	159.0	152.8
	IV	12	149.5	150.5
1988	I	13	166.1	161.4
	II	14	152.5	155.2
	III	15	178.2	171.3
	IV	16	169.0	169.7

FIGURE 9.4

Plot of Exponentially Smoothed Power Loads

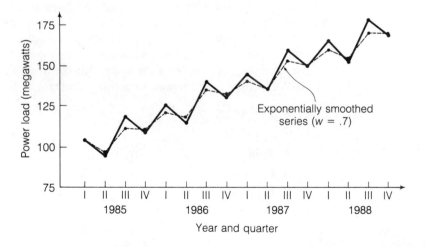

Exponentially smoothed series ($w = .7$)

Note that the right-hand side of the forecast equation does not depend on t; hence, F_t is used to forecast *all* future values of y_t.

For example, the forecast for the power load in quarter I of 1989 ($t = 17$) is calculated as follows:

$$F_{17} = wy_{16} + (1 - w)E_{16}$$
$$= .7(169.0) + .3(169.7)$$
$$= 169.2$$

The forecasts for quarter II of 1989 ($t = 18$), quarter III of 1989 ($t = 19$), and all other future time periods will be the same:

$$F_{18} = 169.2$$
$$F_{19} = 169.2$$
$$F_{20} = 169.2$$
$$\vdots$$

This points out one disadvantage of the exponential smoothing forecasting technique. Since the exponentially smoothed forecast is constant for all future values, any changes in trend and/or seasonality are not taken into account. Therefore, exponentially smoothed forecasts are appropriate only when the trend and seasonal components of the time series are relatively insignificant.

FORECASTING USING EXPONENTIAL SMOOTHING

1. The data consist of n equally-spaced time series values, $y_1, y_2, \ldots, y_n$.
2. Select a smoothing constant, w, between 0 and 1. (Smaller values of w give less weight to the current value of the series and yield a smoother series. Larger values of w give more weight to the current value of the series and yield a more variable series.)
3. Calculate the exponentially smoothed series, E_t, as follows.

$$E_1 = y_1$$
$$E_2 = wy_2 + (1 - w)E_1$$
$$E_3 = wy_3 + (1 - w)E_2$$
$$\vdots$$
$$E_n = wy_n + (1 - w)E_{n-1}$$

4. Calculate the forecast for any future time period t as follows:

$$F_t = wy_n + (1 - w)E_n, \quad t = n + 1, n + 2, \ldots$$

HOLT–WINTERS FORECASTING MODEL

One drawback to the exponential smoothing forecasting method is that the secular trend and seasonal components of a time series are not taken into account. The **Holt–Winters forecasting model** is an extension of the exponential smoothing method that explicitly recognizes the trend and seasonal variation in a time series.

Consider a time series with a trend component, but little or no seasonal variation. Then the Holt–Winters model for y_t is

$$E_t = wy_t + (1 - w)(E_{t-1} + T_{t-1})$$
$$T_t = v(E_t - E_{t-1}) + (1 - v)T_{t-1}$$

where E_t is the exponentially smoothed series, T_t is the trend component, and w and v are smoothing constants between 0 and 1. Note that the trend component T_t is a weighted average of the most recent change in the smoothed value (measured by the difference $E_t - E_{t-1}$) and the trend estimate of the previous time period (T_{t-1}). When seasonal variation is present in the time series, the Holt–Winters model takes the form

$$E_t = w(y_t/S_{t-P}) + (1 - w)(E_{t-1} + T_{t-1})$$
$$T_t = v(E_t - E_{t-1}) + (1 - v)T_{t-1}$$
$$S_t = u(y_t/E_t) + (1 - u)S_{t-P}$$

where S_t is the seasonal component, u is a constant between 0 and 1, and P is the number of time periods in a cycle (usually, a year). The seasonal component S_t is a weighted average of the ratio y_t/E_t (i.e., the ratio of the actual time series value to the smoothed value) and the seasonal component for the previous cycle. For example, for the quarterly power loads, $P = 4$ (four quarters in a year) and the seasonal component for, say, quarter III of 1986 ($t = 7$) is a weighted average of the ratio y_7/E_7 and the seasonal component for quarter III of 1985 ($t = 3$). That is,

$$S_7 = u(y_7/E_7) + (1 - u)S_3$$

Forecasts for future time periods, $t = n + 1, n + 2, \ldots$, using the Holt–Winters models are obtained by summing the most recent exponentially smoothed component with an estimate of the expected increase (or decrease) attributable to trend. For seasonal models, the forecast is multiplied by the most recent estimate of the seasonal component (similar to the moving average method).

The Holt–Winters forecasting methodology is summarized in the accompanying box.

EXAMPLE 9.1

Refer to the 1985–1988 quarterly power loads listed in Table 9.1. Use the Holt–Winters forecasting model with both trend and seasonal components to forecast the utility company's quarterly power loads in 1989. Use the smoothing constants $w = .7$, $v = .5$, and $u = .5$.

SOLUTION

First note that $P = 4$ for the quarterly time series. Following the formulas for E_t, T_t, and S_t given in the box, we calculate

$$E_2 = y_2 = 94.7$$
$$T_2 = y_2 - y_1 = 94.7 - 103.5 = -8.8$$
$$S_2 = y_2/E_2 = 94.7/94.7 = 1$$
$$E_3 = .7y_3 + (1 - .7)(E_2 + T_2)$$
$$\quad = .7(118.6) + .3(94.7 - 8.8) = 108.8$$
$$T_3 = .5(E_3 - E_2) + (1 - .5)T_2$$
$$\quad = .5(108.8 - 97.4) + .5(-8.8) = 2.6$$

FORECASTING USING THE HOLT–WINTERS MODEL

TREND COMPONENT ONLY

1. The data consist of n equally-spaced time series values, $y_1, y_2, \ldots, y_n$.

2. Select smoothing constants w and v, where $0 \leq w \leq 1$ and $0 \leq v \leq 1$.

3. Calculate the exponentially smoothed component, E_t, and the trend component, T_t, for $t = 2, 3, \ldots, n$ as follows:

$$E_t = \begin{cases} y_2, & t = 2 \\ wy_t + (1 - w)(E_{t-1} + T_{t-1}), & t > 2 \end{cases}$$

$$T_t = \begin{cases} y_2 - y_1, & t = 2 \\ v(E_t - E_{t-1}) + (1 - v)T_{t-1}, & t > 2 \end{cases}$$

[*Note:* E_1 and T_1 are not defined.]

4. The forecast for a future time period t is given by

$$F_t = \begin{cases} E_n + T_n, & t = n + 1 \\ E_n + 2T_n, & t = n + 2 \\ \vdots \\ E_n + kT_n, & t = n + k \end{cases}$$

TREND AND SEASONAL COMPONENTS

2. Select smoothing constants w, v, and u, where $0 \leq w \leq 1$, $0 \leq v \leq 1$, and $0 \leq u \leq 1$.

3. Determine P, the number of time periods in a cycle. Usually, $P = 4$ for quarterly data and $P = 12$ for monthly data.

4. Calculate the exponentially smoothed component, E_t, the trend component, T_t, and the seasonal component, S_t, for $t = 2, 3, \ldots, n$ as follows:

$$E_t = \begin{cases} y_2, & t = 2 \\ wy_t + (1 - w)(E_{t-1} + T_{t-1}), \\ \quad t = 3, 4, \ldots, P + 2 \\ w(y_t/S_{t-P}) + (1 - w)(E_{t-1} + T_{t-1}), \\ \quad t > P + 2 \end{cases}$$

$$T_t = \begin{cases} y_2 - y_1, & t = 2 \\ v(E_t - E_{t-1}) + (1 - v)T_{t-1}, & t > 2 \end{cases}$$

$$S_t = \begin{cases} y_t/E_t, & t = 2, 3, \ldots, P + 2 \\ u(y_t/E_t) + (1 - u)S_{t-P}, & t > P + 2 \end{cases}$$

[*Note:* E_1, T_1, and S_1 are not defined.]

5. The forecast for a future time period t is given by

$$F_t = \begin{cases} (E_n + T_n)S_{n+1-P}, & t = n + 1 \\ (E_n + 2T_n)S_{n+2-P}, & t = n + 2 \\ \vdots \\ (E_n + kT_n)S_{n+k-P}, & t = n + k \end{cases}$$

$$S_3 = y_3/E_3 = 118.6/108.8 = 1.090$$
$$E_4 = .7y_4 + (1 - .7)(E_3 + T_3)$$
$$\quad = .7(109.3) + .3(108.8 + 2.6) = 109.9$$
$$T_4 = .5(E_4 - E_3) + (1 - .5)T_3$$
$$\quad = .5(109.9 - 108.8) + .5(2.6) = 1.9$$
$$S_4 = y_4/E_4 = 109.3/109.9 = .994$$

$$\vdots$$

(Remember that beginning with $t = P + 3 = 7$, the formulas for E_t and S_t, shown in the box, are slightly different.) All the values of E_t, T_t, and S_t are given in Table 9.4.

TABLE 9.4
Holt–Winters Components for Quarterly Power Load Data

YEAR	QUARTER	TIME t	POWER LOAD y_t	E_t $(w = .7)$	T_t $(v = .5)$	S_t $(u = .5)$
1985	I	1	103.5	—	—	—
	II	2	94.7	94.7	−8.8	1.000
	III	3	118.6	108.8	2.6	1.090
	IV	4	109.3	109.9	1.9	.994
1986	I	5	126.1	121.8	6.9	1.035
	II	6	116.0	119.8	2.5	.968
	III	7	141.2	127.4	5.1	1.100
	IV	8	131.6	132.3	5.0	.995
1987	I	9	144.5	138.9	5.8	1.038
	II	10	137.1	142.6	4.8	.965
	III	11	159.0	145.4	3.8	1.097
	IV	12	149.5	149.9	4.2	.996
1988	I	13	166.1	158.2	6.3	1.044
	II	14	152.5	160.0	4.1	.959
	III	15	178.2	162.9	3.5	1.095
	IV	16	169.0	168.7	4.7	.999

The forecast for quarter I of 1989 (i.e., y_{17}) is given by

$$F_{17} = (E_{16} + T_{16})S_{17-4}$$
$$= (E_{16} + T_{16})S_{13} = (168.7 + 4.7)(1.044)$$
$$= 181.0$$

Similarly, the forecasts for y_{18}, y_{19}, and y_{20} (quarters II, III, and IV, respectively) are

$$F_{18} = (E_{16} + 2T_{16})S_{18-4}$$
$$= (E_{16} + 2T_{16})S_{14} = [168.7 + 2(4.7)](.959)$$
$$= 170.8$$

$$F_{19} = (E_{16} + 3T_{16})S_{19-4}$$
$$= (E_{16} + 3T_{16})S_{15} = [168.7 + 3(4.7)](1.095)$$
$$= 200.2$$

$$F_{20} = (E_{16} + 4T_{16})S_{20-4}$$
$$= (E_{16} + 4T_{16})S_{16} = [168.7 + 4(4.7)](.999)$$
$$= 187.3$$

We conclude this section with a comment. A major disadvantage of forecasting with smoothing techniques (the moving average method, exponential smoothing, or the Holt–Winters models) is that no measure of the forecast error (or reliability) is known. Although forecast errors can be calculated *after* the future values of the time series have been observed, we prefer to have some measure of the accuracy of the forecast *before* the actual values are observed. For this reason, smoothing techniques are generally regarded as descriptive rather than as inferential procedures. On the other hand, forecasts with inferential models (such as regression models) are accompanied by measures of the *standard error of the forecast*, which allow us to construct prediction intervals for the future time series value. We discuss inferential time series forecasting models in the remaining sections of this chapter.

| | | | | | | | | | | | | |

EXERCISES 9.1–9.7

9.1 The quarterly numbers of new privately owned housing starts (in thousands of dwellings) in the United States from Winter 1982 through Fall 1986 are recorded in the accompanying table.

YEAR	QUARTER	HOUSING STARTS	YEAR	QUARTER	HOUSING STARTS
1982	I	165.4	1985	I	326.1
	II	274.0		II	509.2
	III	284.3		III	469.1
	IV	286.2		IV	408.6
1983	I	296.5	1986	I	357.5
	II	443.7		II	550.0
	III	347.4		III	465.5
	IV	373.1		IV	384.6
1984	I	367.6			
	II	492.4			
	III	418.2			
	IV	349.4			

Source: Standard & Poor's Trade and Securities Statistics (annual). New York: Standard & Poor Corporation.

a. Plot the quarterly time series. Can you detect a long-term trend? Can you detect any seasonal variation?

b. Calculate the 4-point moving average for the quarterly housing starts.

c. Graph the 4-point moving average on the same set of axes you used for the graph in part **a**. Is the long-term trend more evident? What effects has the moving average method removed or "smoothed"?

d. Calculate the seasonal index for the number of housing starts in quarter I.

e. Use the moving average method to forecast the number of housing starts in quarter I of 1987.

9.2 Refer to the quarterly housing starts data given in Exercise 9.1.

a. Calculate the exponentially smoothed series for housing starts using a smoothing constant of $w = .2$.

b. Use the exponentially smoothed series from part **a** to forecast the quarterly number of new housing starts in 1987.

c. Use the Holt–Winters forecasting model with both trend and seasonal components to forecast the quarterly number of new housing starts in 1987. Use smoothing constants $w = .2$, $v = .5$, and $u = .7$.

9.3 The data in the table represent the annual oil production of the 13-nation Organization of Petroleum Exporting Countries (OPEC) cartel, expressed as a percentage of world total, for the years 1973–1986.

YEAR	OPEC PRODUCTION	YEAR	OPEC PRODUCTION
1973	67.8	1980	59.8
1974	67.9	1981	48.9
1975	65.6	1982	43.9
1976	68.0	1983	33.3
1977	66.5	1984	32.5
1978	64.3	1985	30.4
1979	63.2	1986	35.6

Source: International Energy Statistical Review, 1986. Central Intelligence Agency.

a. Plot the yearly time series. Can you detect a long-term trend?

b. Calculate and plot a 3-point moving average for annual OPEC oil production.

c. Calculate and plot the exponentially smoothed series for annual OPEC oil production using a smoothing constant of $w = .3$.

d. Forecast OPEC oil production in 1987 using the moving average method.

e. Forecast OPEC oil production in 1987 using exponential smoothing with $w = .3$.

f. Forecast OPEC oil production in 1987 using the Holt–Winters forecasting model with trend. Use smoothing constants $w = .3$ and $v = .8$.

9.4 The closing IBM common stock price from January 1984 to December 1986 is shown in the accompanying table.

1984	IBM STOCK PRICE	1985	IBM STOCK PRICE	1986	IBM STOCK PRICE
January	114.1	January	136.3	January	151.4
February	110.2	February	134.0	February	150.7
March	114.0	March	127.0	March	149.1
April	113.6	April	126.4	April	156.2
May	107.6	May	128.5	May	152.3
June	105.6	June	123.6	June	146.4
July	110.7	July	131.3	July	132.4
August	123.7	August	126.5	August	138.6
September	124.3	September	123.7	September	134.4
October	124.6	October	129.7	October	123.5
November	121.7	November	139.6	November	127.1
December	123.1	December	155.4	December	120.0

Source: Daily Stock Price Record, New York Stock Exchange.

a. Construct and plot a 12-point moving average for the stock price. Use the moving average method to forecast the price in January 1987.

b. Construct and plot the exponentially smoothed series for the stock price using a smoothing constant of $w = .5$. Use the exponential smoothing technique to forecast the price in January 1987.

c. Use the Holt–Winters forecasting model with trend and seasonal components to forecast the price in January 1987. Use smoothing constants $w = .5$, $v = .5$, and $u = .5$.

d. The actual IBM stock price in January 1987 was 133.0. Calculate the forecast errors for the forecasts obtained in parts a–c. (The forecast error is measured by the difference, $y_t - F_t$.)

9.5 The Consumer Price Index (CPI) measures the increase (or decrease) in the prices of goods and services relative to a base year. The CPI for the years 1970–1987 (using 1967 as a base period) is shown in the table.

YEAR	CPI	YEAR	CPI	YEAR	CPI
1970	116.3	1976	170.5	1982	289.1
1971	121.3	1977	181.5	1983	298.4
1972	125.3	1978	195.4	1984	311.1
1973	133.3	1979	217.4	1985	322.2
1974	147.7	1980	246.8	1986	328.4
1975	161.2	1981	272.4	1987	330.5

Source: *Survey of Current Business*, United States Department of Commerce.

a. Graph the time series. Do you detect a long-term trend?

b. Calculate and plot a 5-point moving average for the CPI. Use the moving average to forecast the CPI in 1988.

c. Calculate and plot the exponentially smoothed series for the CPI using a smoothing constant of $w = .4$. Use the exponentially smoothed values to forecast the CPI in 1988.

d. Use the Holt–Winters forecasting model with trend to forecast the CPI in 1988. Use smoothing constants $w = .4$ and $v = .5$.

9.6 Standard & Poor's 500 Stock Composite Average (S&P 500) is a stock market index. Like the Dow Jones Industrial Average (DJA), it is an indicator of stock market activity. The table at the top of page 506 contains end-of-quarter values of the S&P 500 for the years 1975–1986.

a. Calculate a 4-point moving average for the quarterly stock market index.

b. Plot the quarterly index and the 4-point moving average on the same graph. Can you identify the secular trend of the time series? Can you identify any seasonal variations about the secular trend?

c. Use the moving average method to forecast the quarterly S&P 500 for 1987.

d. Calculate and plot the exponentially smoothed series for the quarterly S&P 500 using a smoothing constant of $w = .3$.

e. Use the exponential smoothing technique with $w = .3$ to forecast the quarterly S&P 500 for 1987.

YEAR	QUARTER	S&P 500	YEAR	QUARTER	S&P 500	YEAR	QUARTER	S&P 500
1975	I	83.36	1979	I	101.59	1983	I	151.07
	II	95.19		II	102.91		II	165.11
	III	83.87		III	109.32		III	168.66
	IV	98.19		IV	107.94		IV	164.12
1976	I	102.77	1980	I	105.36	1984	I	159.18
	II	104.28		II	113.72		II	153.18
	III	105.24		III	127.14		III	166.10
	IV	107.46		IV	131.44		IV	167.24
1977	I	98.42	1981	I	134.94	1985	I	180.66
	II	100.48		II	129.06		II	191.85
	III	96.53		III	118.77		III	182.08
	IV	95.10		IV	119.13		IV	211.28
1978	I	89.21	1982	I	117.09	1986	I	232.33
	II	95.53		II	109.82		II	245.30
	III	102.54		III	121.72		III	238.27
	IV	96.11		IV	138.79		IV	248.61

Source: Standard & Poor's Trade and Securities Statistics (annual). New York: Standard & Poor Corporation.

f. Use the Holt–Winters forecasting model with trend and seasonal components to forecast the quarterly S&P 500 for 1987. Use smoothing constants $w = .3$, $v = .8$, and $u = .5$.

9.7 Consider the gold price time series recorded in the table. (Gold prices are given in dollars per troy ounce.)

YEAR	PRICE OF GOLD	YEAR	PRICE OF GOLD	YEAR	PRICE OF GOLD
1970	36.41	1976	124.80	1982	374.18
1971	41.25	1977	148.30	1983	449.03
1972	58.61	1978	193.50	1984	360.29
1973	97.81	1979	307.80	1985	317.30
1974	159.70	1980	606.01	1986	367.87
1975	161.40	1981	450.63	1987	408.91

Sources: Standard & Poor's Trade and Securities Statistics (annual). New York: Standard & Poor Corporation. *Survey of Current Business*, United States Department of Commerce.

a. Calculate a 3-point moving average for the gold price time series. Plot the gold prices and the 3-point moving average on the same graph. Can you detect the long-term trend and any cyclical patterns in the time series?

b. Use the moving averages to forecast the price of gold in 1988.

c. Calculate and plot the exponentially smoothed gold price series using a smoothing constant of $w = .8$.

d. Use the exponentially smoothed series to forecast the price of gold in 1988.

e. Use the Holt–Winters forecasting model with trend to forecast the price of gold in 1988. Use smoothing constants $w = .8$ and $v = .4$.

S E C T I O N 9.4

FORECASTING: THE REGRESSION APPROACH

Many firms use past sales to forecast future sales. Suppose a wholesale distributor of sporting goods is interested in forecasting its sales revenue for each of the next 5 years. Since an inaccurate forecast may have dire consequences to the distributor, some measure of the forecast's reliability is required. To make such forecasts and assess their reliability, an **inferential time series forecasting model** must be constructed. The familiar general linear regression model of Chapter 4 represents one type of inferential model since it allows us to calculate prediction intervals for the forecasts.

To illustrate the technique of forecasting with regression, consider the data in Table 9.5. The data are annual sales (in thousands of dollars) for a firm (say, the sporting goods distributor) in each of its 35 years of operation. A plot of the data (Figure 9.5) reveals a linearly increasing trend, so the first-order (straight-line) model

$$E(y_t) = \beta_0 + \beta_1 t$$

TABLE 9.5

A Firm's Yearly Sales Revenue (thousands of dollars)

t	y_t	t	y_t	t	y_t
1	4.8	13	48.4	25	100.3
2	4.0	14	61.6	26	111.7
3	5.5	15	65.6	27	108.2
4	15.6	16	71.4	28	115.5
5	23.1	17	83.4	29	119.2
6	23.3	18	93.6	30	125.2
7	31.4	19	94.2	31	136.3
8	46.0	20	85.4	32	146.8
9	46.1	21	86.2	33	146.1
10	41.9	22	89.9	34	151.4
11	45.5	23	89.2	35	150.9
12	53.5	24	99.1		

FIGURE 9.5

Plot of Sales Data

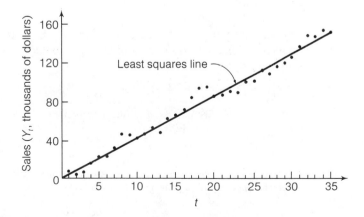

seems plausible for describing the secular trend. The SAS computer printout for the model is shown in Figure 9.6. Note that the model apparently provides an excellent fit to the data, with $R^2 = .98$, $F = 1,615.72$ (p-value $< .0001$), and $s = 6.39$. The least squares prediction equation, whose coefficients are shaded in Figure 9.6, is

$$\hat{y}_t = \hat{\beta}_0 + \hat{\beta}_1 t = .4015 + 4.2956t$$

We can obtain sales forecasts and corresponding 95% prediction intervals for years 36–40 by employing the formulas given in Section 3.9. However, these values are given in the bottom portion of the SAS printout shown in Figure 9.6. For example, for $t = 36$, we have $\hat{y}_{36} = 155.0$ with the 95% prediction interval (141.3, 168.8). That is, we predict that sales revenue in year $t = 36$ will fall between \$141,300 and \$168,800 with 95% confidence.

Note that the prediction intervals for $t = 36, 37, \ldots, 40$ widen as we attempt to forecast further into the future. Intuitively, we know that the farther into the future we forecast, the less certain we are of the accuracy of the forecast since some unexpected change in business and economic conditions may make the model inappropriate. Since we have less confidence in the forecast for, say, $t = 40$ than for $t = 36$, it follows that the prediction interval for $t = 40$ must be wider in order to attain a 95% level of confidence. For this reason, time series forecasting (regardless of the forecasting method) is generally confined to the short term.

Multiple regression models can also be used to forecast future values of a time series with seasonal variation. We illustrate with an example.

EXAMPLE 9.2

Refer to the 1985–1988 quarterly power loads listed in Table 9.1.

a. Propose a model for quarterly power load, y_t, that will account for both the secular trend and seasonal variation present in the series.

b. Fit the model to the data, and use the least squares prediction equation to forecast the utility company's quarterly power loads in 1989. Construct 95% prediction intervals for the forecasts.

SOLUTION

a. A common way to describe seasonal differences in a time series is with dummy variables.* For quarterly data, a model that includes both trend and seasonal components is

$$E(y_t) = \beta_0 + \underbrace{\beta_1 t}_{\substack{\text{Secular} \\ \text{trend}}} + \underbrace{\beta_2 Q_1 + \beta_3 Q_2 + \beta_4 Q_3}_{\text{Seasonal component}}$$

---

*Another way to account for seasonal variation is with trigonometric (sine and cosine) terms. We discuss seasonal models with trigonometric terms in Section 9.6.

FIGURE 9.6   SAS Printout for the Straight-Line Model

ANALYSIS OF VARIANCE

SOURCE	DF	SUM OF SQUARES	MEAN SQUARE	F VALUE	PROB>F
MODEL	1	65875.20817	65875.20817	1615.724	0.0001
ERROR	33	1345.45355	40.77131958		
C TOTAL	34	67220.66171			

ROOT MSE	6.385242	R-SQUARE	0.9800
DEP MEAN	77.72286	ADJ R-SQ	0.9794
C.V.	8.215398		

PARAMETER ESTIMATES

VARIABLE	DF	PARAMETER ESTIMATE	STANDARD ERROR	T FOR H0: PARAMETER=0	PROB > |T|
INTERCEP	1	0.40151261	2.20570829	0.182	0.8567
T	1	4.29563025	0.10686692	40.196	0.0001

OBS	ACTUAL	PREDICT VALUE	STD ERR PREDICT	LOWER95% PREDICT	UPPER95% PREDICT	RESIDUAL
1	4.8000	4.6971	2.1132	-8.9866	18.3809	0.1029
2	4.0000	8.9928	2.0220	-4.6338	22.6194	-4.9928
3	5.5000	13.2884	1.9325	-0.2843	26.8611	-7.7884
4	15.6000	17.5840	1.8448	4.0619	31.1061	-1.9840
5	23.1000	21.8797	1.7593	8.4048	35.3545	1.2203
6	23.3000	26.1753	1.6761	12.7444	39.6062	-2.8753
7	31.4000	30.4709	1.5959	17.0805	43.8613	0.9291
8	46.0000	34.7666	1.5189	21.4133	48.1198	11.2334
9	46.1000	39.0622	1.4457	25.7428	52.3818	7.0378
10	41.9000	43.3578	1.3769	30.0684	56.6472	-1.4578
11	45.5000	47.6534	1.3132	34.3908	60.9161	-2.1534
12	53.5000	51.9491	1.2554	38.7096	65.1886	1.5509
13	48.4000	56.2447	1.2043	43.0249	69.4645	-7.8447
14	61.6000	60.5403	1.1609	47.3366	73.7441	1.0597
15	65.6000	64.8360	1.1259	51.6448	78.0272	0.7640
16	71.4000	69.1316	1.1003	55.9494	82.3138	2.2684
17	83.4000	73.4272	1.0846	60.2504	86.6041	9.9728
18	93.6000	77.7229	1.0793	64.5478	90.8979	15.8771
19	94.2000	82.0185	1.0846	68.8416	95.1953	12.1815
20	85.4000	86.3141	1.1003	73.1319	99.4964	-0.9141
21	86.2000	90.6097	1.1259	77.4185	103.8	-4.4097
22	89.9000	94.9054	1.1609	81.7016	108.1	-5.0054
23	89.2000	99.2010	1.2043	85.9812	112.4	-10.0010
24	99.1000	103.5	1.2554	90.2571	116.7	-4.3966
25	100.3	107.8	1.3132	94.5296	121.1	-7.4923
26	111.7	112.1	1.3769	98.7985	125.4	-0.3879
27	108.2	116.4	1.4457	103.1	129.7	-8.1835
28	115.5	120.7	1.5189	107.3	134.0	-5.1792
29	119.2	125.0	1.5959	111.6	138.4	-5.7748
30	125.2	129.3	1.6761	115.8	142.7	-4.0704
31	136.3	133.6	1.7593	120.1	147.0	2.7339
32	146.8	137.9	1.8448	124.3	151.4	8.9383
33	146.1	142.2	1.9325	128.6	155.7	3.9427
34	151.4	146.5	2.0220	132.8	160.1	4.9471
35	150.9	150.7	2.1132	137.1	164.4	0.1514
36	.	155.0	2.2057	141.3	168.8	.
37	.	159.3	2.2995	145.5	173.1	.
38	.	163.6	2.3944	149.8	177.5	.
39	.	167.9	2.4903	154.0	181.9	.
40	.	172.2	2.5870	158.2	186.2	.

SUM OF RESIDUALS   -7.10543E-14
SUM OF SQUARED RESIDUALS   1345.454
PREDICTED RESID SS (PRESS)   1484.211

where

t = Time period, ranging from $t = 1$ for quarter I of 1985 to $t = 16$ for quarter IV of 1988

y_t = Power load (megawatts) in time t

$$Q_1 = \begin{cases} 1 & \text{if quarter I} \\ 0 & \text{if not} \end{cases} \qquad Q_2 = \begin{cases} 1 & \text{if quarter II} \\ 0 & \text{if not} \end{cases}$$

$$Q_3 = \begin{cases} 1 & \text{if quarter III} \\ 0 & \text{if not} \end{cases} \qquad \text{Base level = quarter IV}$$

The β coefficients associated with the seasonal dummy variables determine the mean increase (or decrease) in power load for each quarter, relative to the base level quarter, quarter IV.

b. The model is fit to the data from Table 9.1 using the SAS multiple regression routine. The resulting SAS printout is shown in Figure 9.7. Note that the model appears to fit the data quite well: $R^2 = .997$, indicating that the model accounts for 99.7% of the sample variation in power loads over the 4-year period; $F = 968.96$ strongly supports the hypothesis that the model has predictive utility (p-value = .0001); and the standard deviation, ROOT MSE = 1.53, implies that the model predictions will usually be accurate to within approximately $\pm 2(1.53)$, or about ± 3.06 megawatts.

Forecasts and corresponding 95% prediction intervals for the 1989 power loads are reported in the bottom portion of the printout in Figure 9.7. For example, the forecast for power load in quarter I of 1989 is 184.7 megawatts with the 95% prediction interval (180.5, 188.9). Therefore, using a 95% prediction interval, we expect the power load in quarter I of 1989 to fall between 180.5 and 188.9 megawatts. ∎

Many descriptive forecasting techniques have proved their merit by providing good forecasts for particular applications. Nevertheless, the advantage of forecasting using the regression approach is clear: Regression analysis provides us with a measure of reliability for each forecast through prediction intervals. However, there are two problems associated with forecasting time series using a multiple regression model.

PROBLEM 1

We are using the least squares prediction equation to forecast values outside the region of observation of the independent variable, t. For example, in Example 9.2, we are forecasting for values of t between 17 and 20 (the four quarters of 1989), even though the observed power loads are for t-values between 1 and 16. As noted in Chapter 5, it is risky to use a least squares regression model for prediction outside the range of the observed data because some unusual change, economic or political, may make the model inappropriate for predicting future

FIGURE 9.7 SAS Printout of Least Squares Fit to Quarterly Power Loads

ANALYSIS OF VARIANCE

SOURCE	DF	SUM OF SQUARES	MEAN SQUARE	F VALUE	PROB>F
MODEL	4	9101.67800	2275.41950	968.962	0.0001
ERROR	11	25.83137500	2.34830682		
C TOTAL	15	9127.50938			

ROOT MSE		1.532419	R-SQUARE	0.9972	
DEP MEAN		137.3062	ADJ R-SQ	0.9961	
C.V.		1.116059			

PARAMETER ESTIMATES

VARIABLE	DF	PARAMETER ESTIMATE	STANDARD ERROR	T FOR H0: PARAMETER=0	PROB > |T|
INTERCEP	1	90.20625000	1.14931396	78.487	0.0001
T	1	4.96437500	0.08566480	57.951	0.0001
Q1	1	10.09312500	1.11364246	9.063	0.0001
Q2	1	-4.84625000	1.09704478	-4.418	0.0010
Q3	1	14.36437500	1.08696452	13.215	0.0001

OBS	ID	ACTUAL	PREDICT VALUE	STD ERR PREDICT	LOWER95% PREDICT	UPPER95% PREDICT	RESIDUAL
1	1985_1	103.5	105.3	0.9226	101.3	109.2	-1.7637
2	1985_2	94.7000	95.2888	0.9226	91.3518	99.2257	-0.5888
3	1985_3	118.6	119.5	0.9226	115.5	123.4	-0.8638
4	1985_4	109.3	110.1	0.9226	106.1	114.0	-0.7637
5	1986_1	126.1	125.1	0.7851	121.3	128.9	0.9788
6	1986_2	116.0	115.1	0.7851	111.4	118.9	0.8538
7	1986_3	141.2	139.3	0.7851	135.5	143.1	1.8788
8	1986_4	131.6	129.9	0.7851	126.1	133.7	1.6788
9	1987_1	144.5	145.0	0.7851	141.2	148.8	-0.4787
10	1987_2	137.1	135.0	0.7851	131.2	138.8	2.0963
11	1987_3	159.0	159.2	0.7851	155.4	163.0	-0.1787
12	1987_4	149.5	149.8	0.7851	146.0	153.6	-0.2787
13	1988_1	166.1	164.8	0.9226	160.9	168.8	1.2638
14	1988_2	152.5	154.9	0.9226	150.9	158.8	-2.3612
15	1988_3	178.2	179.0	0.9226	175.1	183.0	-0.8362
16	1988_4	169.0	169.6	0.9226	165.7	173.6	-0.6362
17	1989_1	.	184.7	1.1493	180.5	188.9	.
18	1989_2	.	174.7	1.1493	170.5	178.9	.
19	1989_3	.	198.9	1.1493	194.7	203.1	.
20	1989_4	.	189.5	1.1493	185.3	193.7	.

SUM OF RESIDUALS   3.55271E-13
SUM OF SQUARED RESIDUALS   25.83137
PREDICTED RESID SS (PRESS)   55.61836

events. Because forecasting always involves predictions about future values of a time series this problem obviously cannot be avoided. However, it is important that the forecaster recognize the dangers of this type of prediction.

PROBLEM 2

Recall the standard assumptions made about the random error component of a multiple regression model (Section 4.2). We assume that the errors have mean 0, constant variance, normal probability distributions, and are *independent*. The latter assumption is often violated in time series that exhibit short-term trends. As an illustration, refer to the plot of the sales revenue data shown in Figure 9.5.

Notice that the observed sales tend to deviate about the least squares line in positive and negative runs. That is, if the difference between the observed sales and predicted sales in year t is positive (or negative), the difference in year $t +$ 1 tends to be positive (or negative). Since the variation in the yearly sales is systematic, the implication is that the errors are correlated. (We gave a formal statistical test for correlated errors in Section 6.6.) Violation of this standard regression assumption could lead to unreliable forecasts.

Time series models have been developed specifically for the purpose of making forecasts when the errors are known to be correlated. These models include an **autoregressive term** for the correlated errors that result from cyclical, seasonal, or other short-term effects. Time series autoregressive models are the subject of Sections 9.6–9.10.

EXERCISES 9.8–9.15

9.8 The accompanying table records the volume of wheat (in thousands of bushels) harvested by members of a farmers' marketing cooperative for the period 1975–1988. The cooperative is interested in detecting the long-term linear trend of the wheat harvest.

YEAR	TIME	WHEAT HARVESTED	YEAR	TIME	WHEAT HARVESTED
1975	1	75	1982	8	91
1976	2	78	1983	9	92
1977	3	82	1984	10	92
1978	4	82	1985	11	93
1979	5	84	1986	12	96
1980	6	85	1987	13	101
1981	7	87	1988	14	102

a. Graph the wheat harvest time series.

b. Propose a model for the long-term linear trend of the time series.

c. Fit the model, using the method of least squares. Plot the least squares line on the graph of part **a**. Can you identify the long-term trend?

d. How well does the linear model describe the long-term trend? [*Hint:* Check the value of r^2.]

e. Use the least squares model to forecast the volume of wheat harvested in 1989. Construct a 95% prediction interval for the forecast.

9.9 A realtor working in a large city wants to identify the secular trend in the weekly number of one-family houses sold by her firm. For the past 15 weeks she has collected data on her firm's home sales, as shown in the table.

WEEK	HOMES SOLD	WEEK	HOMES SOLD	WEEK	HOMES SOLD
t	y_t	t	y_t	t	y_t
1	59	6	137	11	88
2	73	7	106	12	75
3	70	8	122	13	62
4	82	9	93	14	44
5	115	10	86	15	45

a. Plot the time series. Is there visual evidence of a quadratic trend?

b. The realtor hypothesizes the model $E(y_t) = \beta_0 + \beta_1 t + \beta_2 t^2$ for the secular trend of the weekly time series. Fit the model to the data, using the method of least squares. (You will need access to a statistical computer program package.)

c. Plot the least squares model on the graph of part a. How well does the quadratic model describe the secular trend?

d. Use the model to forecast home sales in week 16 with a 95% prediction interval.

9.10 Refer to the quarterly S&P 500 values given in Exercise 9.6.

a. Hypothesize a time series model to account for trend and seasonal variation.

b. Fit the model in part a to the data.

c. Use the least squares model from part b to forecast the S&P 500 for all four quarters of 1987. Obtain 95% prediction intervals for the forecasts.

9.11 Information on intercity passenger traffic (excluding travel by private automobiles) since 1940 is given in the table. The data are recorded as percentages of total passenger miles traveled.

YEAR	TIME	RAILROADS	BUSES	AIR CARRIERS
1940	1	67.1	26.5	2.8
1945	2	74.3	21.4	2.7
1950	3	46.3	37.7	14.3
1955	4	36.5	32.4	28.9
1960	5	28.6	25.7	42.1
1965	6	17.9	24.2	54.7
1970	7	7.3	16.9	73.1
1975	8	5.8	14.2	77.7
1980	9	4.7	11.4	83.9
1985	10	3.6	7.9	88.4

Source: Statistical Abstract of the United States, 1987. Interstate Commerce Commission, Civil Aeronautics Board.

a. Let y_t be the percentage of total passenger-miles at time t for a particular mode of transportation. Consider the linear model $E(y_t) = \beta_0 + \beta_1 t$. Which modes of transportation do you think have a secular trend adequately represented by this model?

b. Fit the model in part a to the data for each mode of transportation, using the method of least squares.

c. Plot the data and the least squares line for each mode of transportation. Which models adequately describe the secular trend of percentage of total passenger-miles traveled? Does this agree with your answer to part **a**?

d. Refer to your answer for part **c**. Use the least squares prediction equations to forecast the percentage of total passenger-miles to be traveled for the respective modes of transportation in 1990. If you have access to a linear regression computer routine, obtain 95% prediction intervals. What are the risks associated with this forecast procedure?

9.12 The annual price (in cents per pound) of galvanized steel from 1971 to 1984 is shown in the table.

YEAR	t	y_t	YEAR	t	y_t
1971	1	9.61	1978	8	20.47
1972	2	10.88	1979	9	22.32
1973	3	10.59	1980	10	23.88
1974	4	12.39	1981	11	26.88
1975	5	14.80	1982	12	26.75
1976	6	16.07	1983	13	28.43
1977	7	18.10	1984	14	30.30

Source: Standard & Poor's Trade and Securities Statistics (Annual). New York: Standard & Poor Corporation.

a. Plot the time series. Is there visual evidence of a linear trend?

b. Fit the model $E(y_t) = \beta_0 + \beta_1 t$ to the data, using the method of least squares.

c. Plot the least squares line on the graph of part **a**. How well does the linear model describe the time series?

d. Use the fitted least squares model to forecast the price of galvanized steel for the years 1985–1989. Obtain 95% prediction intervals for the forecasts and verify that the width of the interval increases the farther you forecast into the future.

9.13 Refer to the monthly IBM common stock prices given in Exercise 9.4.

a. Fit the model $E(y_t) = \beta_0 + \beta_1 t$ to the data.

b. Plot the least squares line and the actual time series on the same graph. Is there visual evidence that the model provides a reasonable fit to the data?

c. Test to determine whether the model is useful for predicting IBM monthly stock price. Use $\alpha = .05$.

d. Use the fitted model to forecast the IBM stock price in January 1987 with a 95% prediction interval. Check to see that the actual value falls within the prediction interval. (Recall that the actual price in January 1987 was 133.0.).

9.14 Refer to Exercise 9.13.

a. Propose a model for $E(y_t)$ that accounts for possible seasonal variation in the monthly series. [*Hint:* Consider a model with dummy variables for the 12 months, January, February, etc.]

b. Fit the model of part **a**. Compare the regression results to the model with trend only, obtained in Exercise 9.13.

c. Test the hypothesis that the monthly dummy variables are useful predictors of IBM stock price. [*Hint:* Conduct a partial F-test.]

d. Use the fitted least squares model from part **b** to forecast the IBM stock price in January 1987 with a 95% prediction interval. Compare the interval with the interval obtained in part **d** of Exercise 9.13.

9.15 The Employee Retirement Income Security Act (ERISA) of 1974 was originally established to enhance retirement security income. J. Ledolter (University of Iowa) and M. L. Power (Iowa State University) investigated the effects of ERISA on the growth in the number of private retirement plans (*Journal of Risk and Insurance*, December 1983). Using quarterly data from 1956 through the third quarter of 1982 ($n = 107$ quarters), Ledolter and Power fit quarterly time series models for the number of pension qualifications and the number of profit-sharing plan qualifications. One of the various models investigated was the quadratic model $E(y_t) = \beta_0 + \beta_1 t + \beta_2 t^2$, where y_t is the logarithm of the dependent variable (number of pension or number of profit-sharing qualifications) in quarter t. The results (modified for the purpose of this exercise) are summarized below:

Pension plan qualifications:

$$\hat{y}_t = 6.19 + .039t - .00024t^2$$
$$t \text{ (for } H_0: \beta_2 = 0) = -1.39$$

Profit-sharing plan qualifications:

$$\hat{y}_t = 6.22 + .035t - .00021t^2$$
$$t \text{ (for } H_0: \beta_2 = 0) = -1.61$$

a. Is there evidence that the quarterly number of pension plan qualifications increases at a decreasing rate over time? Test using $\alpha = .05$. [*Hint.* Test $H_0: \beta_2 = 0$ against $H_a: \beta_2 < 0$.]
b. Forecast the number of pension plan qualifications for the fourth quarter of 1982 (i.e., $t = 108$). [*Hint:* Since y_t is the logarithm of the number of pension plan qualifications, to obtain the forecast you must take the antilogarithm of $\hat{y}_{108}$, i.e., $e^{\hat{y}_{108}}$.]
c. Is there evidence that the quarterly number of profit-sharing plan qualifications increases at a decreasing rate over time? Test using $\alpha = .05$. [*Hint:* Test $H_0: \beta_2 = 0$ against $H_a: \beta_2 < 0$.]
d. Forecast the number of profit-sharing plan qualifications for the fourth quarter of 1982 (i.e., $t = 108$). [*Hint:* Since y_t is the logarithm of the number of profit-sharing plan qualifications, to obtain the forecast you must take the antilogarithm of $\hat{y}_{108}$, i.e., $e^{\hat{y}_{108}}$.]

SECTION 9.5

AUTOCORRELATION AND AUTOREGRESSIVE MODELS

In Chapter 6 we presented the Durbin–Watson test for detecting correlated residuals in a regression analysis. Correlated residuals are quite common when the response is a *time series* variable. Correlation of residuals for a regression model with a time series response is called **autocorrelation**, because it refers to correlation between residuals from the *same* time series model at different points in time.

A special case of autocorrelation that has many applications to business and economic phenomena is the case in which neighboring residuals one time period apart (say, at times t and $t + 1$) are correlated. This type of correlation is called **first-order autocorrelation**. In general, correlation between time series residuals m time periods apart is mth-order autocorrelation.

DEFINITION 9.3

Autocorrelation is the correlation between time series residuals at different points in time. The special case in which neighboring residuals one time period apart (at times t and $t + 1$) are correlated is called **first-order autocorrelation**.

To see how autocorrelated residuals affect the regression model, we will assume a model similar to the general linear model of Chapter 4,

$$y_t = E(y_t) + R_t$$

where $E(y_t)$ is the regression model

$$E(y_t) = \beta_0 + \beta_1 x_1 + \cdots + \beta_k x_k$$

and R_t represents the random residual. We assume that the residual R_t has mean 0 and constant variance σ^2, but that it is autocorrelated. The effect of autocorrelation on the general linear model depends on the pattern of the autocorrelation. One of the most common patterns is that the autocorrelation between residuals at consecutive time points is positive. Thus, when the residual at time t, R_t, is indicating that the observed value y_t is more than the mean value $E(y_t)$, then the residual at time $(t + 1)$ will have a tendency (probability greater than .5) to be positive. This would occur, for example, if you were to model a monthly economic index (e.g., the Consumer Price Index) with a straight-line model. In times of recession, the observed values of the index will tend to be less than the predictions of a straight line for most or all of the months during the period. Similarly, in extremely inflationary periods, the residuals are likely to be positive because the observed value of the index will lie above the straight-line model. In either case, the fact that residuals at consecutive time points tend to have the same sign implies that they are **positively correlated**.

A second property commonly observed for autocorrelated residuals is that the size of the autocorrelation between values of the residual R at two different points in time diminishes rapidly as the distance between the time points increases. Thus, the autocorrelation between R_t and R_{t+m} becomes smaller (i.e., weaker) as the distance m between the time points becomes larger.

A residual model that possesses this property—positive autocorrelation diminishing rapidly as distance between time points increases—is the **first-order autoregressive model**:

$$R_t = \phi R_{t-1} + \varepsilon_t, \quad -1 < \phi < 1$$

where ε_t, a residual called **white noise**, is uncorrelated with any and all other residual components. Thus, the value of the residual R_t is equal to a constant multiple, ϕ (Greek letter "phi"), of the previous residual, R_{t-1}, plus random error. In general, the constant ϕ is between -1 and $+1$, and the numerical value of ϕ determines the sign (positive or negative) and strength of the autocorrelation. In fact, it can be shown (proof omitted) that the autocorrelation between two residuals that are m time units apart, R_t and R_{t+m}, is

$$\text{Autocorrelation}(R_t, R_{t+m}) = \phi^m$$

Since the absolute value of ϕ will be less than 1, the autocorrelation between R_t and R_{t+m}, ϕ^m, will decrease as m increases. This means that neighboring values of R_t, i.e., $m = 1$, will have the highest correlation, and the correlation diminishes rapidly as the distance m between time points is increased. This points to an interesting property of the autoregressive time series model. The autocorrelation function depends only on the distance m between R-values, and not on the time t. Time series models that possess this property are said to be **stationary**.

DEFINITION 9.4

A **stationary time series model** for regression residuals is one that has mean 0, constant variance, and autocorrelations that depend only on the distance between time points.

The autocorrelation function of first-order autoregressive models is shown for several values of ϕ in Figure 9.8 (page 518). Note that positive values of ϕ yield positive autocorrelation for all residuals, while negative values of ϕ imply negative correlation for neighboring residuals, positive correlation between residuals two time points apart, negative correlation for residuals three time points apart, and so forth. The appropriate pattern will, of course, depend on the particular application, but the occurrence of a positive autocorrelation pattern is more common.

Although the first-order autoregressive model provides a good representation for many autocorrelation patterns, more complex patterns can be described by higher-order autoregressive models. The general form of a pth-order autoregressive model is

$$R_t = \phi_1 R_{t-1} + \phi_2 R_{t-2} + \cdots + \phi_p R_{t-p} + \varepsilon_t$$

The inclusion of p parameters, $\phi_1, \phi_2, \ldots, \phi_p$ permits more flexibility in the pattern of autocorrelations exhibited by a residual time series. When an autoregressive model is used to describe residual autocorrelations, the observed autocorrelations are used to estimate these parameters. Methods for estimating these parameters will be presented in Section 9.7.

FIGURE 9.8

Autocorrelation Functions
for Several First-Order
Autoregressive Models:
$R_t = \phi R_{t-1} + \varepsilon_t$

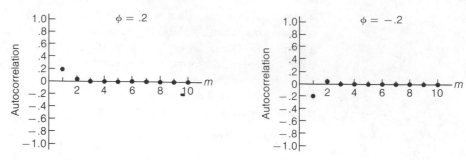

(a) Weak autocorrelation

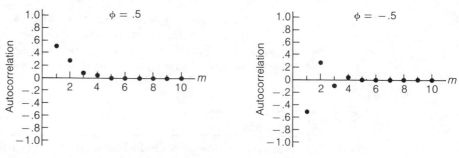

(b) Moderate autocorrelation

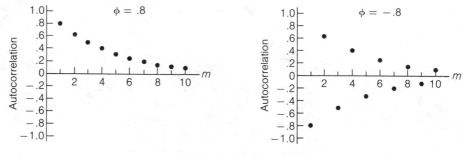

(c) Strong autocorrelation

‖ ‖ ‖ ‖ ‖ ‖ ‖ ‖ ‖ ‖ ‖ ‖ ‖ ‖ ‖

EXERCISES 9.16–9.19

9.16 Suppose that the random component of a time series model follows the first-order
autoregressive model $R_t = \phi R_{t-1} + \varepsilon_t$, where ε_t is a white noise process. Consider
four versions of this model: $\phi = .9$, $\phi = -.9$, $\phi = .2$, and $\phi = -.2$.

a. Calculate the first ten autocorrelations, Autocorrelation (R_t, R_{t+m}), $m = 1, 2, 3,$
. . . , 10, for each of the four models.

b. Plot the autocorrelations against the distance in time separating the R-values (m)
for each case.

c. Examine the rate at which the correlation diminishes in each plot. What does this
imply?

9.17 When using time series to analyze quarterly data (data in which seasonal effects are present) it is highly possible that the random component of the model R_t, also exhibits the same seasonal variation as the dependent variable. In these cases, the following non–first-order autoregressive model is sometimes postulated for the correlated error term, R_t:

$$R_t = \phi R_{t-4} + \varepsilon_t$$

where $|\phi| < 1$ and ε_t is a white noise process. The autocorrelation function for this model is given by

$$\text{Autocorrelation } (R_t, R_{t+m}) = \begin{cases} \phi^{m/4} & \text{if } m = 4, 8, 12, 16, 20, \dots \\ 0 & \text{if otherwise} \end{cases}$$

a. Calculate the first 20 autocorrelations ($m = 1, 2, \dots, 20$) for the model with constant coefficient $\phi = .5$.

b. Plot the autocorrelations against m, the distance in time separating the R-values. Compare the rate at which the correlation diminishes with the first-order model $R_t = .5R_{t-1} + \varepsilon_t$.

9.18 Consider the autocorrelation pattern shown in the figure. Write a first-order autoregressive model that exhibits this pattern.

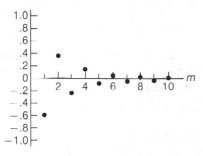

9.19 Write the general form for a fourth-order autoregressive model.

SECTION 9.6

CONSTRUCTING TIME SERIES MODELS

Recall that the general form of the times series model is

$$y_t = E(y_t) + R_t$$

We are assuming that the expected value of y_t is

$$E(y_t) = \beta_0 + \beta_1 x_1 + \beta_2 x_2 + \cdots + \beta_k x_k$$

where $x_1, x_2, \dots, x_k$ are independent variables, which themselves may be time series, and the residual component is

$$R_t = \phi_1 R_{t-1} + \phi_2 R_{t-2} + \cdots + \phi_p R_{t-p} + \varepsilon_t$$

where ε_t is white noise (uncorrelated error). Thus, a time series model consists of a pair of models: one model for the deterministic component $E(y_t)$ and one model for the autocorrelated residuals R_t.

The deterministic portion of the model is chosen in exactly the same manner as for the regression models of the preceding chapters except that some of the independent variables might be time series variables or might be trigonometric functions of time (such as sin t or cos t). It is helpful to think of the deterministic component as consisting of the trend (T_t), cyclical (C_t), and seasonal (S_t) effects described in Section 9.2.

For example, we may want to model the number of new housing starts, y_t, as a function of the prime interest rate, x_t. Then, one model for the mean of y_t is

$$E(y_t) = \beta_0 + \beta_1 x_t$$

for which the mean number of new housing starts is a multiple β_1 of the prime interest rate, plus a constant β_0. Another possibility is a second-order relationship.

$$E(y_t) = \beta_0 + \beta_1 x_t + \beta_2 x_t^2$$

which permits the *rate* of increase in the mean number of housing starts to increase or decrease with the prime interest rate.

Yet another possibility is to model the mean number of new housing starts as a function of both the prime interest rate and the year, t. Thus, the model

$$E(y_t) = \beta_0 + \beta_1 x_t + \beta_2 t + \beta_3 x_t t$$

implies that the mean number of housing starts increases linearly in x_t, the prime interest rate, but the rate of increase depends on the year t. If we wanted to adjust for seasonal (cyclical) effects due to t, we might introduce time into the model using trigonometric functions of t. This topic will be explained in greater detail below.

Another important type of model for $E(y_t)$ is the **lagged independent variable model**. *Lagging* means that we are pairing observations on a dependent variable and independent variable at two different points in time, with the time corresponding to the independent variable lagging behind the time for the dependent variable. Suppose, for example, we believe that the monthly mean number of new housing starts is a function of the *previous* month's prime interest rate. Thus, we model y_t as a linear function of the lagged independent variable, prime interest rate, x_{t-1}:

$$E(y_t) = \beta_0 + \beta_1 x_{t-1}$$

or, alternatively, as the second-order function,

$$E(y_t) = \beta_0 + \beta_1 x_{t-1} + \beta_2 x_{t-1}^2$$

For this example, the independent variable, prime interest rate x_t, is lagged 1 month behind the response y_t.

Many time series have distinct seasonal patterns. Retail sales are usually highest around Christmas, spring, and fall, with relative lulls in the winter and summer periods. Energy usage is highest in summer and winter, and lowest in in spring and fall. Teenage unemployment rises in the summer months when schools are not in session, and falls near Christmas when many businesses hire part-time help.

When a time series' seasonality is exhibited in a relatively consistent pattern from year to year, we can model the pattern using trigonometric terms in the model for $E(y_t)$. For example, the model of a monthly series with mean $E(y_t)$ might be

$$E(y_t) = \beta_0 + \beta_1\left(\cos\frac{2\pi}{12}t\right) + \beta_2\left(\sin\frac{2\pi}{12}t\right)$$

This model would appear as shown in Figure 9.9. Note that the model is **cyclic**, with a **period** of 12 months. That is, the mean $E(y_t)$ completes a cycle every 12 months and then repeats the same cycle over the next 12 months. Thus, the **expected peaks and valleys** of the series remain the same from year to year. The coefficients β_1 and β_2 determine the **amplitude and phase shift** of the model. The amplitude is the magnitude of the seasonal effect, while the phase shift locates the peaks and valleys in time. For example, if we assume month 1 is January, the mean of the time series depicted in Figure 9.9 has a peak each April and a valley each October.

FIGURE 9.9

A Seasonal Time Series Model

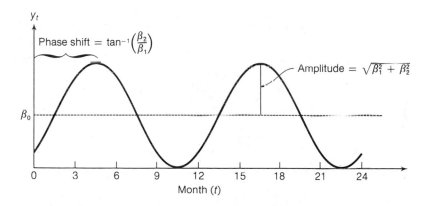

If the data are monthly or quarterly, we can treat the season as a qualitative independent variable (see Example 9.2), and write the model

$$E(y_t) = \beta_0 + \beta_1 S_1 + \beta_2 S_2 + \beta_3 S_3$$

where

$$S_1 = \begin{cases} 1 & \text{if season is spring (II)} \\ 0 & \text{otherwise} \end{cases} \qquad S_2 = \begin{cases} 1 & \text{if season is summer (III)} \\ 0 & \text{otherwise} \end{cases}$$

$$S_3 = \begin{cases} 1 & \text{if season is fall (IV)} \\ 0 & \text{otherwise} \end{cases}$$

Thus, S_1, S_2, and S_3 are dummy variables that describe the four levels of season, letting winter (I) be the base level. The β coefficients determine the mean value of y_t for each season, as shown in Figure 9.10. Note that for the dummy variable model and the trigonometric model we assume the seasonal effects are approximately the same from year to year. If they tend to increase or decrease with time, an interaction of the seasonal effect with time may be necessary. (An example will be given in Section 9.9.)

FIGURE 9.10
Seasonal Model for
Quarterly Data Using
Dummy Variables

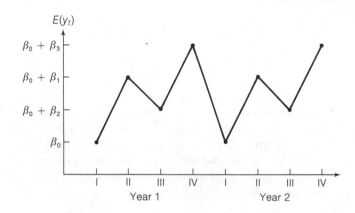

The appropriate form of the deterministic time series model will depend on both theory and data. Economic theory often provides several plausible models relating the mean response to one or more independent variables. The data can then be used to determine which, if any, of the models is best supported. The process is often an iterative one, beginning with preliminary models based on theoretical notions, using data to refine and modify these notions, collecting additional data to test the modified theories, and so forth.

CHOOSING THE RESIDUAL COMPONENT

The appropriate form of the residual component R_t will depend on the pattern of autocorrelation in the residuals (see Section 9.5). The autoregressive model is very useful for this aspect of time series modeling, with the general form

$$R_t = \phi_1 R_{t-1} + \phi_2 R_{t-2} + \cdots + \phi_p R_{t-p} + \varepsilon_t$$

which is called an autoregressive model of order p. Recall that the name *autoregressive* comes from the fact that R_t is regressed on its own past values. As the order p is increased, more complex autocorrelation functions can be modeled. There are several other types of models that can be used for the random component, but the autoregressive model is very flexible and receives more application in business forecasting than the others.

The simplest autoregressive model is the **first-order autoregressive model**

$$R_t = \phi R_{t-1} + \varepsilon_t$$

Recall that the autocorrelation between residuals at two different points in time diminishes as the distance between the time points increases. Since many business and economic time series exhibit this property, the first-order autoregressive model is a popular choice for the residual component.

To summarize, we describe a general approach for constructing a time series:

1. Construct a regression model for the trend, seasonal, and cyclical components of $E(y_t)$. This model may be a polynomial in t for the trend (usually a straight-line or quadratic model) with trigonometric terms or dummy variables for the seasonal (cyclical) effects. The model may also include other time series variables as independent variables. For example, last year's rate of inflation may be used as a predictor of this year's GNP.

2. Next, construct a model for the random component (residual effect) of the model. A model that is widely used in practice is the first-order autoregressive model

$$R_t = \phi R_{t-1} + \varepsilon_t$$

When the pattern of autocorrelation is more complex, use the general pth-order autoregressive model

$$R_t = \phi_1 R_{t-1} + \phi_2 R_{t-2} + \cdots + \phi_p R_{t-p} + \varepsilon_t$$

3. The two components are then combined so that the model can be used for forecasting:

$$y_t = E(y_t) + R_t$$

Prediction intervals are calculated to measure the reliability of the forecasts. In the following two sections we will demonstrate how time series models are fit to data and used for forecasting. In Section 9.9 we will present an example in which we fit a seasonal time series model to a set of data.

EXERCISES 9.20–9.24

9.20 Suppose you are interested in buying gold on the commodities market. Your broker has advised you that your best strategy is to sell back the gold at the first substantial jump in price. Hence, you are interested in a short-term investment. Before buying, you would like to model the closing price of gold, y_t, over time (in days), t.
a. Write a first-order model for the deterministic portion of the model, $E(y_t)$.
b. If a plot of the daily closing prices for the past month reveals a quadratic trend, write a plausible model for $E(y_t)$.
c. Since the closing price of gold on day $(t + 1)$ is very highly correlated with the closing price on day t, your broker suggests that the random error components of the model are not white noise. Given this information, postulate a model for the error term, R_t.

9.21 An economist wishes to model the GNP over time (in years) and also as a function of certain personal consumption expenditures. Let t = Time in years and let

y_t = GNP at time t
x_{1t} = Durable goods at time t
x_{2t} = Nondurable goods at time t
x_{3t} = Services at time t

a. The economist believes that y_t is linearly related to the independent variables x_{1t}, x_{2t}, x_{3t}, and t. Write the first-order model for $E(y_t)$.

b. Rewrite the model if interaction between the independent variables and time is present.

c. Postulate a model for the random error component, R_t. Explain why this model is appropriate.

9.22 Airlines sometimes overbook flights because of "no-show" passengers, i.e., passengers who have purchased a ticket but fail to board the flight. An airline supervisor wishes to be able to predict, for a Miami-to-New York flight, the monthly accumulation of no-show passengers during the upcoming year, using data from the past 3 years. Let y_t = Number of no-shows during month t.

a. Using dummy variables, propose a model for $E(y_t)$ that will take into account the seasonal (fall, winter, spring, summer) variation that may be present in the data.

b. Postulate a model for the error term R_t.

c. Write the full time series model for y_t (include random error terms).

d. Suppose the airline supervisor believes that the seasonal variation in the data is not constant from year to year; in other words, that there exists interaction between time and season. Rewrite the full model with the interaction terms added.

9.23 A farmer is interested in modeling the daily price of hogs at a livestock market. The farmer knows that the price varies over time (days) and also is reasonably confident that a seasonal effect is present.

a. Write a seasonal time series model with trigonometric terms for $E(y_t)$, where y_t = Selling price (in dollars) of hogs on day t.

b. Interpret the β parameters.

c. Include in the model an interaction between time and the trigonometric components. What does the presence of interaction signify?

d. Is it reasonable to assume that the random error component of the model, R_t, is white noise? Explain. Postulate a more appropriate model for R_t.

9.24 Suppose a CPA firm wants to model its monthly income, y_t. The firm is growing at an increasing rate, so that the mean income will be modeled as a second-order function of t. In addition, the mean monthly income increases significantly each year from January through April due to processing of tax returns.

a. Write a model for $E(y_t)$ to reflect both the second-order function of time, t, and the January–April jump in mean income.

b. Suppose the size of the January–April jump grows each year. How could this information be included in the model? Assume that 5 years of monthly data are available.

SECTION 9.7

FITTING TIME SERIES REGRESSION MODELS

We have proposed a general form for a time series model:

$$y_t = E(y_t) + R_t$$

where

$$E(y_t) = \beta_0 + \beta_1 x_1 + \cdots + \beta_k x_k$$

and, using an autoregressive model for R_t,

$$R_t = \phi R_{t-1} + \phi_2 R_{t-2} + \cdots + \phi_p R_{t-p} + \varepsilon_t$$

We now want to develop estimators for the parameters β_0, β_1, . . . , β_k of the regression model, and for the parameters ϕ_1, ϕ_2, . . . , ϕ_p of the autoregressive model. The ultimate objective is to use the model to obtain forecasts (predictions) of future values of y_t, as well as to make inferences about the structure of the model itself.

We will introduce the techniques of fitting a time series model with a simple example. Refer to the data in Table 9.5 (page 507), the annual sales for a firm in each of its 35 years of operation. Recall that the objective is to forecast future sales in years 36–40. In Section 9.4 we used a simple straight-line model for the mean sales

$$E(y_t) = \beta_0 + \beta_1 t$$

to make the forecasts.

The computer printout showing the least squares estimates of β_0 and β_1 is reproduced in Figure 9.11. Although the model is useful for predicting annual sales (p-value for H_0: $\beta_1 = 0$ is less than .0001), the Durbin–Watson statistic is $d = .821$, which is less than the tabled value, $d_L = 1.40$ (Table 9 of Appendix D), for $\alpha = .05$, $n = 35$, and $k = 1$ independent variable. Thus, there is evidence that the residuals are positively correlated. The plot of the least squares residuals over time, in Figure 9.12 (page 526), shows the pattern of positive autocorrelation. The residuals tend to cluster in positive and negative runs; if the residual at time t is positive, the residual at time $(t + 1)$ tends to be positive.

FIGURE 9.11 SAS Computer Printout: Regression Analysis for Least Squares Sales Data

```
DEP VARIABLE: Y
                              ANALYSIS OF VARIANCE

                         SUM OF           MEAN
          SOURCE    DF   SQUARES          SQUARE       F VALUE      PROB>F

          MODEL      1   65875.20817   65875.20817    1615.724     0.0001
          ERROR     33    1345.45355     40.77131958
          C TOTAL   34   67220.66171

               ROOT MSE      6.385242    R-SQUARE      0.9800
               DEP MEAN     77.72286     ADJ R-SQ      0.9794
               C.V.          8.215398

                          PARAMETER ESTIMATES

                       PARAMETER        STANDARD       T FOR H0:
          VARIABLE  DF  ESTIMATE        ERROR          PARAMETER=0   PROB > |T|

          INTERCEP   1  0.40151261      2.20570829       0.182       0.8567
          T          1  4.29563025      0.10686692      40.196       0.0001


          DURBIN-WATSON D                  0.821
          (FOR NUMBER OF OBS.)                35
          1ST ORDER AUTOCORRELATION       0.590
```

What are the consequences of fitting the least squares model when autocorrelated residuals are present? Although *the least squares estimators of β_0 and β_1 remain unbiased* even if the residuals are autocorrelated, i.e., $E(\hat{\beta}_0) = \beta_0$ and $E(\hat{\beta}_1) = \beta_1$, the *standard errors given by least squares theory are usually smaller than the true standard errors* when the residuals are positively autocorrelated.

FIGURE 9.12

Annual Sales Time Series
Example: Least Squares
Residual Plot

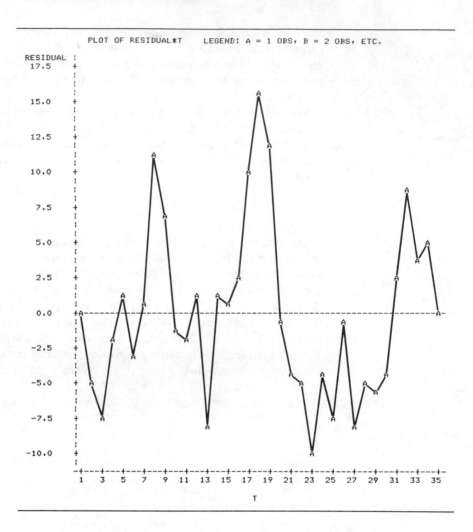

Consequently, t-values computed by the methods of Chapter 4 (which apply when the errors are uncorrelated) will usually be inflated and will lead to a higher Type I error rate (α) than the value of α selected for a test. Thus, the application of standard least squares techniques to time series often produces statistical test results that are misleading and leads to overoptimistic evaluations of a model's predictive ability. And there is a second reason for seeking methods that specifically take into account the autocorrelated residuals. If we can successfully model the residual autocorrelation, we should achieve a smaller MSE and correspondingly narrower prediction intervals than those given by the least squares model.

To account for the autocorrelated residual, we will postulate a first-order autoregressive model,

$$R_t = \phi R_{t-1} + \varepsilon_t$$

Thus, we use the pair of models

$$y_t = \beta_0 + \beta_1 t + R_t$$
$$R_t = \phi R_{t-1} + \varepsilon_t$$

to describe the yearly sales of the firm. In order to estimate the parameters of the time series model (β_0, β_1, and ϕ), a modification of the least squares method is required. To do this, we use a *transformation* that is much like the variance-stabilizing transformations discussed in Chapter 6.

First, we multiply the model

$$y_t = \beta_0 + \beta_1 t + R_t \tag{1}$$

by ϕ at time $(t - 1)$ to obtain

$$\phi y_{t-1} = \phi \beta_0 + \phi \beta_1 (t - 1) + \phi R_{t-1} \tag{2}$$

Taking the difference between equations (1) and (2), we have

$$y_t - \phi y_{t-1} = \beta_0 (1 - \phi) + \beta_1 [t - \phi(t - 1)] + (R_t - \phi R_{t-1})$$

or, since $R_t = \phi R_{t-1} + \varepsilon_t$, then

$$y_t^* = \beta_0^* + \beta_1 t^* + \varepsilon_t$$

where $y_t^* = y_t - \phi y_{t-1}$, $t^* = t - \phi(t - 1)$ and $\beta_0^* = \beta_0(1 - \phi)$. Thus, we can use the transformed dependent variable y_t^* and transformed independent variable t^* and obtain least squares estimates of β_0^* and β_1. The residual ε_t is uncorrelated, so that the assumptions necessary for the least squares estimators are all satisfied. The estimator of the original intercept, β_0, can be calculated by

$$\hat{\beta}_0 = \frac{\hat{\beta}_0^*}{1 - \phi}$$

This transformed model appears to solve the problem of first-order autoregressive residuals. However, making the transformation requires knowing the value of the parameter ϕ. Also, we lose the initial observation, since the values of y_t^* and t^* can be calculated only for $t \geq 2$. The methods for estimating ϕ and adjustments for the values at $t = 1$ will not be detailed here. Instead, we will present output from the SAS computer package, which both performs the transformation and estimates the model parameters, β_0, β_1, and ϕ.

The SAS printout of the straight-line, autoregressive time series model fit to the sales data is shown in Figure 9.13 (page 528). Note that the format of the time series printout is different from that of the standard regression printout. The estimates of β_0 and β_1 in the deterministic component $E(y_t)$ appear at the bottom of the printout under the column heading B VALUE. The estimate of the first-order autoregressive parameter ϕ is given in the middle portion of the printout titled ESTIMATES OF THE AUTOREGRESSIVE PARAMETERS beneath the column heading COEFFICIENT.

FIGURE 9.13 SAS Computer Printout for the Combined Straight-Line Autoregressive Residual Fit to the Sales Data

```
A U T O R E G   P R O C E D U R E


              ORDINARY LEAST SQUARES ESTIMATES

      SSE           1345.454    DFE              33
      MSE           40.77132    ROOT MSE    6.385242
      SBC           234.1562    AIC         231.0455
      REG RSQ         0.9800    TOTAL RSQ     0.9800
      DURBIN-WATSON   0.8207


  VARIABLE DF      B VALUE      STD ERROR    T RATIO APPROX PROB

  INTERCPT  1   0.40151261    2.20570829     0.182    0.8567
  T         1   4.29563025    0.10686692    40.196    0.0001

           ESTIMATES OF AUTOCORRELATIONS

LAG  COVARIANCE  CORRELATION  -1 9 8 7 6 5 4 3 2 1 0 1 2 3 4 5 6 7 8 9 1
  0    38.4415    1.000000    |                    |********************|
  1    22.6661    0.589624    |                    |************        |


            PRELIMINARY MSE=    25.07708

      ESTIMATES OF THE AUTOREGRESSIVE PARAMETERS
   LAG   COEFFICIENT        STD ERROR        T RATIO
    1    -0.58962415       0.14277861      -4.129639


            YULE-WALKER ESTIMATES

      SSE            877.6854   DFE              32
      MSE           27.42767    ROOT MSE    5.237143
      SBC           223.1868    AIC         218.5208
      REG RSQ         0.9412    TOTAL RSQ     0.9869


  VARIABLE DF      B VALUE      STD ERROR    T RATIO APPROX PROB

  INTERCPT  1   0.40575699    3.99697517     0.102    0.9198
  T         1   4.29593038    0.18983105    22.630    0.0001
```

The interpretations of two quantities shown on the SAS printout for the time series procedure differ from those described in the preceding sections. First, the quantity printed as REG RSQ (in the lower portion of the printout) is not the value of R^2 based on the values of y_t. Instead, it is based on the values of the transformed variable, y_t^*. When we refer to R^2 in this chapter, we will always mean the value R^2 based on the original time series variable y_t. This value, which will usually be larger than the REG RSQ value, is given on the printout as TOTAL RSQ (shaded). Second, the SAS time series model is defined so that ϕ takes the *opposite* sign from the value contained in our model. Consequently, you must multiply the estimate of ϕ shown in the printout by (-1) to obtain the estimate of ϕ for our model.

Therefore, the fitted models are

$$\hat{y}_t = .4058 + 4.2959t + \hat{R}_t$$
$$\hat{R}_t = .5896\hat{R}_{t-1}$$

with

$$MSE = 27.43$$

and

$$R^2 = .9869 \quad \text{(TOTAL RSQ on the printout, Figure 9.13)}$$

A comparison of the least squares(Figure 9.6) and autoregressive (Figure 9.13) computer printouts is given in Table 9.6. Note that the autoregressive model reduces MSE and increases R^2. The values of the estimators β_0 and β_1 change very little, but the estimated standard errors are considerably increased, thereby decreasing the t-value for testing $H_0: \beta_1 = 0$. Note that the implication that the linear relationship between sales y_t and year t is of significant predictive value is the same using either method. However, you can see that the underestimation of standard errors by using least squares in the presence of residual autocorrelation could result in the inclusion of unimportant independent variables in the model, since the t-values will usually be inflated.

TABLE 9.6

	LEAST SQUARES	AUTOREGRESSIVE
R^2	.98	.99
MSE	40.77	27.43
$\hat{\beta}_0$	.4015	.4058
$\hat{\beta}_1$	4.2956	4.2959
Standard error $(\hat{\beta}_0)$	2.2057	3.9970
Standard error $(\hat{\beta}_1)$	.1069	.1898
t-statistic for $H_0: \beta_1 = 0$	40.20	22.63
	$(p < .0001)$	$(p < .0001)$
$\hat{\phi}$	—	.5896
t-statistic for $H_0: \phi = 0$	—	4.13

The estimated value of ϕ is .5896, and an approximate t-test* of the hypothesis $H_0: \phi = 0$ yields a t-value of 4.13. With 32 df, this value is significant at less than $\alpha = .01$. Thus, the result of the Durbin–Watson d-test is confirmed: There is adequate evidence of positive residual autocorrelation.[†] Furthermore, the first-order autoregressive model appears to describe this residual correlation well.

---

*An explanation of this t-test has been omitted. Consult the references at the end of the chapter for details of this test.

[†]This result is to be expected since it can be shown (proof omitted) that $\hat{\phi} \approx 1 - d/2$, where d is the value of the Durbin–Watson statistic.

The steps for fitting a time series model to a set of data are summarized in the box. Once the model is estimated, the model can be used to forecast future values of the time series y_t.

STEPS FOR FITTING TIME SERIES MODELS

1. Use the least squares approach to obtain initial estimates of the β parameters. Do *not* use the t- or F-tests to assess the importance of the parameters, since the estimates of their standard errors may be biased (often underestimated).
2. Analyze the residuals to determine whether they are autocorrelated. The Durbin–Watson test is one technique for making this determination.
3. If there is evidence of autocorrelation, construct a model for the residuals. The autoregressive model is one useful model. Consult the references at the end of the chapter for more types of residual models and for methods of identifying the most suitable model.
4. Reestimate the β parameters, taking the residual model into account. This involves a simple transformation if an autoregressive model is used, and several statistical packages have computer routines to accomplish this.

EXERCISES 9.25–9.29

9.25 The Gross National Product (GNP) is a measure of total U.S. output, and is therefore an important indicator of the U.S. economy. The quarterly GNP values (in billions of dollars) from 1975–1985 are given in the table. Let y_t be the GNP in quarter t, $t = 1, 2, 3, \ldots , 44$.

YEAR	QUARTER			
	1	2	3	4
1975	1,479.8	1,516.7	1,578.5	1,621.8
1976	1,672.0	1,698.6	1,729.0	1,772.5
1977	1,834.8	1,895.1	1,954.4	1,988.9
1978	2,031.7	2,139.5	2,202.5	2,281.6
1979	2,335.5	2,377.9	2,454.8	2,502.9
1980	2,572.9	2,578.8	2,639.1	2,736.0
1981	2,866.6	2,912.5	3,004.9	3,032.2
1982	3,021.4	3,070.2	3,090.7	3,109.6
1983	3,171.5	3,272.0	3,362.2	3,432.0
1984	3,676.5	3,757.5	3,812.2	3,852.5
1985	3,909.3	3,965.0	4,030.5	4,087.7

Source: Standard & Poor's Trade and Securities Statistics (Annual). New York: Standard & Poor's Corporation.

a. Hypothesize a time series model for quarterly GNP that includes a straight-line long-term trend and autocorrelated residuals.

b. The SAS printout for the pair of models

$$y_t = \beta_0 + \beta_1 t + R_t$$
$$R_t = \phi R_{t-1} + \varepsilon_t$$

is shown here. Write the least squares prediction equation.

SAS Printout for Exercise 9.25

A U T O R E G   P R O C E D U R E

ORDINARY LEAST SQUARES ESTIMATES

SSE	246375.5	DFE	42
MSE	5866.083	ROOT MSE	76.59036
SBC	512.1736	AIC	508.6052
REG RSQ	0.9909	TOTAL RSQ	0.9909
DURBIN-WATSON	0.2649		

VARIABLE	DF	B VALUE	STD ERROR	T RATIO	APPROX PROB
INTERCPT	1	1302.76522	23.4921909	55.455	0.0001
T	1	61.32387	0.9092805	67.442	0.0001

ESTIMATES OF AUTOCORRELATIONS

LAG	COVARIANCE	CORRELATION	-1 9 8 7 6 5 4 3 2 1 0 1 2 3 4 5 6 7 8 9 1
0	5599.44	1.000000	|***********************|
1	4620.35	0.825145	|********************|

PRELIMINARY MSE=   1786.978

ESTIMATES OF THE AUTOREGRESSIVE PARAMETERS

LAG	COEFFICIENT	STD ERROR	T RATIO
1	-0.82514548	0.08822573	-9.352663

YULE-WALKER ESTIMATES

SSE	63839.13	DFE	41
MSE	1557.052	ROOT MSE	39.4595
SBC	457.6783	AIC	452.3257
REG RSQ	0.9575	TOTAL RSQ	0.9976

VARIABLE	DF	B VALUE	STD ERROR	T RATIO	APPROX PROB
INTERCPT	1	1327.28665	54.7378551	24.248	0.0001
T	1	61.02839	2.0089776	30.378	0.0001

---

c. Interpret the estimates of the model parameters, β_0, β_1, and ϕ.

d. Interpret the value of R^2.

9.26 Refer to Exercise 9.8.

a. Hypothesize a time series model for annual volume of wheat harvested, y_t, that takes into account the residual autocorrelation.

b. If you have access to a computer package that uses the modified least squares method, fit the autoregressive time series model. Interpret the estimates of the model parameters.

9.27 Refer to Exercises 9.4 and 9.13.

    a. Hypothesize a time series model for monthly price of IBM common stock, y_t, that takes into account the residual autocorrelation.

    b. If you have access to a computer package that uses the modified least squares method, fit the autoregressive time series model. Interpret the estimates of the model parameters.

9.28 Refer to Exercise 9.15 and the study on the long-term effects of the Employment Retirement Income Security Act (ERISA). Ledolter and Power also fit quarterly time series models for the number of pension plan terminations and the number of profit-sharing plan terminations from the first quarter of 1956 through the third quarter of 1982 ($n = 107$ quarters). To account for residual correlation, they fit straight-line autoregressive models of the form

$$y_t = \beta_0 + \beta_1 t + \phi R_{t-1} + \varepsilon_t$$

The results were as follows:

    Pension plan terminations: $\hat{y}_t = 3.54 + .039t + .40\hat{R}_{t-1}$

    Profit-sharing plan terminations: $\hat{y}_t = 3.45 + .038t + .22\hat{R}_{t-1}$

    a. Interpret the estimates of the model parameters for pension plan terminations.

    b. Interpret the estimates of the model parameters for profit-sharing plan terminations.

9.29 The Dow Jones Industrial Average (DJA) is a widely followed stock market indicator. The values of the DJA from 1964 to 1983 are given in the accompanying table. Suppose we want to model the yearly DJA, y_t, as a function of t, where t is the number of years since 1963 (i.e., $t = 1$ for 1964, $t = 2$ for 1965, ... , $t = 20$ for 1983). A time series model that includes a long-term trend and autocorrelated residuals is the regression–autoregression pair

$$y_t = \beta_0 + \beta_1 t + R_t$$
$$R_t = \phi R_{t-1} + \varepsilon_t$$

The SAS printout for the time series model is also reproduced here.

YEAR	DJA	YEAR	DJA
1964	874.13	1974	759.37
1965	969.26	1975	802.49
1966	785.69	1976	974.92
1967	905.11	1977	835.15
1968	943.75	1978	805.01
1969	800.36	1979	838.74
1970	838.92	1980	963.99
1971	884.76	1981	899.01
1972	950.71	1982	1,046.54
1973	923.88	1983	1,258.64

Source; Standard & Poor's Trade and Securities Statistics (Annual). New York: Standard & Poor's Corporation.

Printout for Exercise 9.29

```
                    ORDINARY LEAST SQUARES ESTIMATES

          SSE            207388.1    DFE                  18
          MSE            11521.56    ROOT MSE        107.3385
          SBC            247.6813    AIC             245.6898
          REG RSQ          0.1479    TOTAL RSQ         0.1479
          DURBIN-WATSON    1.2054


   VARIABLE DF        B VALUE        STD ERROR    T RATIO APPROX PROB

   INTERCPT  1      825.758316     49.8621265     16.561      0.0001
   T         1        7.358398      4.1624099      1.768      0.0940


                    ESTIMATES OF AUTOCORRELATIONS

 LAG  COVARIANCE  CORRELATION  -1 9 8 7 6 5 4 3 2 1 0 1 2 3 4 5 6 7 8 9 1
   0    10369.4    1.000000    |                    |********************|
   1    2036.84    0.196428    |                    |****                |


                 PRELIMINARY MSE=     9969.313

          ESTIMATES OF THE AUTOREGRESSIVE PARAMETERS
          LAG    COEFFICIENT       STD ERROR       T RATIO
           1    -0.19642777      0.23781062     -0.825984


                     YULE-WALKER ESTIMATES

          SSE            195597.6    DFE                  17
          MSE            11505.74    ROOT MSE        107.2648
          SBC            249.5457    AIC             246.5585
          REG RSQ          0.1386    TOTAL RSQ         0.1964


   VARIABLE DF        B VALUE        STD ERROR    T RATIO APPROX PROB

   INTERCPT  1      820.217046     60.1458941     13.637      0.0001
   T         1        8.257379      4.9923977      1.654      0.1165
```

a. Identify and interpret estimates of the model parameters.

b. Interpret the value of R^2.

FORECASTING WITH TIME SERIES MODELS

The ultimate objective of fitting a time series model is often to forecast future values of the series. We will demonstrate the techniques for the simple model

$$y_t = \beta_0 + \beta_1 x_t + R_t$$

with the first-order autoregressive residual

$$R_t = \phi R_{t-1} + \varepsilon_t$$

Suppose we use the data (y_1, x_1), (y_2, x_2), . . . , (y_n, x_n) to obtain estimates of β_0, β_1, and ϕ, using the method presented in Section 9.7. We now want to forecast the value of y_{n+1}. From the model,

$$y_{n+1} = \beta_0 + \beta_1 x_{n+1} + R_{n+1}$$

where

$$R_{n+1} = \phi R_n + \varepsilon_{n+1}$$

Combining these, we obtain

$$y_{n+1} = \beta_0 + \beta_1 x_{n+1} + \phi R_n + \varepsilon_{n+1}$$

The forecast of y_{n+1} is obtained by estimating each of the unknown quantities in this equation:*

$$\hat{y}_{n+1} = \hat{\beta}_0 + \hat{\beta}_1 x_{n+1} + \hat{\phi}\hat{R}_n$$

where $\hat{\beta}_0$, $\hat{\beta}_1$, and $\hat{\phi}$ are the estimates based on the time series model-fitting approach presented in Section 9.7 and ε_{n+1} is estimated by its expected value 0. The estimate $\hat{R}_n$ of the residual R_n is obtained by noting that

$$R_n = y_n - (\beta_0 + \beta_1 x_n)$$

so that

$$\hat{R}_n = y_n - (\hat{\beta}_0 + \hat{\beta}_1 x_n)$$

The two-step-ahead forecast of y_{n+2} is similarly obtained. The true value of y_{n+2} is

$$\begin{aligned} y_{n+2} &= \beta_0 + \beta_1 x_{n+2} + R_{n+2} \\ &= \beta_0 + \beta_1 x_{n+2} + \phi R_{n+1} + \varepsilon_{n+2} \end{aligned}$$

and the forecast at $t = n + 2$ is

$$\hat{y}_{n+2} = \hat{\beta}_0 + \hat{\beta}_1 x_{n+2} + \hat{\phi}\hat{R}_{n+1}$$

The residual R_{n+1} (and all future residuals) can now be obtained from the recursive relation

$$R_{n+1} = \phi R_n + \varepsilon_{n+1}$$

so that

$$\hat{R}_{n+1} = \hat{\phi}\hat{R}_n$$

Thus, the forecasting of future y-values is an iterative process, with each new forecast making use of the previous residual to obtain the estimated residual for the future time period. The general forecasting procedure using time series models with first-order autoregressive residuals is outlined in the next box.

EXAMPLE 9.3

Suppose we want to forecast the sales of the company for the data analyzed in Section 9.7. Recall that we fit the regression–autoregression pair of models

$$y_t = \beta_0 + \beta_1 t + R_t \qquad R_t = \phi R_{t-1} + \varepsilon_t$$

Using 35 years of sales data, we obtained the estimated models

$$\hat{y}_t = .4058 + 4.2959t + \hat{R}_t \qquad \hat{R}_t = .5896\hat{R}_{t-1}$$

---

*Note that the forecast requires the value of x_{n+1}. When x_t is itself a time series, the future value x_{n+1} will generally be unknown and must also be estimated. Often, $x_t = t$ (as in Example 9.3). In this case, the future time period (e.g., $t = n + 1$) is known and no estimate is required.

FORECASTING USING TIME SERIES MODELS WITH FIRST-ORDER AUTOREGRESSIVE RESIDUALS

$$y_t = \beta_0 + \beta_1 x_{1t} + \beta_2 x_{2t} + \cdots + \beta_k x_{kt} + R_t$$
$$R_t = \phi R_{t-1} + \varepsilon_t$$

STEP 1 Use a statistical software package to obtain the estimated model

$$\hat{y}_t = \hat{\beta}_0 + \hat{\beta}_1 x_{1t} + \hat{\beta}_2 x_{2t} + \cdots + \hat{\beta}_k x_{kt} + \hat{R}_t, \quad t = 1, 2, \ldots, n$$
$$\hat{R}_t = \hat{\phi}\hat{R}_{t-1}$$

STEP 2 Compute the estimated residual for the last time period in the data (i.e., $t = n$) as follows:

$$\hat{R}_n = y_n - \hat{y}_n$$
$$= y_n - (\hat{\beta}_0 + \hat{\beta}_1 x_{1n} + \hat{\beta}_2 x_{2n} + \cdots + \hat{\beta}_k x_{kn})$$

STEP 3 To forecast the value y_{n+1}, compute

$$\hat{R}_{n+1} = \hat{\phi}\hat{R}_n \quad \text{(where } \hat{R}_n \text{ is obtained from step 2)}$$
$$\hat{y}_{n+1} = \hat{\beta}_0 + \hat{\beta}_1 x_{1,n+1} + \hat{\beta}_2 x_{2,n+1} + \cdots + \hat{\beta}_k x_{k,n+1} + \hat{R}_{n+1}$$

STEP 4 To forecast the value y_{n+2}, compute

$$\hat{R}_{n+2} = \hat{\phi}\hat{R}_{n+1} \quad \text{(where } \hat{R}_{n+1} \text{ is obtained from step 3)}$$
$$\hat{y}_{n+2} = \hat{\beta}_0 + \hat{\beta}_1 x_{1,n+2} + \hat{\beta}_2 x_{2,n+2} + \cdots + \hat{\beta}_k x_{k,n+2} + \hat{R}_{n+2}$$

Future forecasts are obtained in a similar manner.

Combining these, we have

$$\hat{y}_t = .4058 + 4.2959t + .5896\hat{R}_{t-1}$$

a. Use the fitted model to forecast sales in years $t = 36$, 37, and 38.
b. Calculate approximate 95% prediction intervals for the forecasts.

SOLUTION

a. The forecast for the 36th year requires an estimate of the residual R_{35},

$$\hat{R}_{35} = y_{35} - [\hat{\beta}_0 + \hat{\beta}_1(35)]$$
$$= 150.9 - [.4058 + 4.2959(35)]$$
$$= .1377$$

Then

$$\hat{R}_{36} = \hat{\phi}\hat{R}_{35} = (.5896)(.1377) = .0812$$

and

$$\hat{y}_{36} = \hat{\beta}_0 + \hat{\beta}_1(36) + \hat{R}_{36}$$
$$= .4058 + 4.2959(36) + .0812$$
$$= 155.14$$

To calculate the sales forecast for year 37, we first calculate

$$\hat{R}_{37} = \hat{\phi}\hat{R}_{36}$$
$$= (.5896)(.0812)$$
$$= .0479$$

The sales forecast is then

$$\hat{y}_{37} = \hat{\beta}_0 + \hat{\beta}_1(37) + \hat{R}_{37}$$
$$= .4058 + 4.2959(37) + .0479$$
$$= 159.40$$

Similarly, for $t = 38$,

$$\hat{R}_{38} = \hat{\phi}\hat{R}_{37}$$
$$= (.5896)(.0479)$$
$$= .0282$$

and

$$\hat{y}_{38} = \hat{\beta}_0 + \hat{\beta}_1(38) + \hat{R}_{38}$$
$$= .4058 + 4.2959(38) + .0282$$
$$= 163.68$$

We can proceed in this manner to generate sales forecasts as far into the future as desired. However, the potential for error increases as the distance into the future increases. Forecast errors are traceable to three primary causes:

1. The form of the model may change at some future time. This is an especially difficult source of error to quantify, since we will not usually know when or if the model changes, nor the extent of the change. The possibility of a change in the model structure is the primary reason we have consistently urged you to avoid predictions outside the observed range of the independent variables. However, time series forecasting leaves us little choice—by definition, the forecast will be a prediction at a future time.

2. A second source of forecast error is the uncorrelated residual, ε_t, with variance σ^2. For a first-order autoregressive residual, the forecast variance of the one-step-ahead prediction is σ^2, while that for the two-step-ahead prediction is $\sigma^2(1 + \phi^2)$, and, in general, for m steps ahead, the forecast variance* is $\sigma^2(1 + \phi^2 + \phi^4 + \cdots + \phi^{2(m-1)})$. Thus, the forecast variance increases as the distance is increased. These variances allow us to form approximate 95% prediction intervals for the forecasts (see the box).

3. A third source of variability is that attributable to the error of estimating the model parameters. This is generally of less consequence than the others, and is usually ignored in forming prediction intervals.

*See Fuller (1976).

APPROXIMATE 95% FORECASTING LIMITS USING TIME SERIES MODELS WITH FIRST-ORDER AUTOREGRESSIVE RESIDUALS

One-Step-Ahead Forecast

$$\hat{y}_{n+1} \pm 2\sqrt{MSE}$$

Two-Step-Ahead Forecast

$$\hat{y}_{n+2} \pm 2\sqrt{MSE(1 + \hat{\phi}^2)}$$

Three-Step-Ahead Forecast

$$\hat{y}_{n+3} \pm 2\sqrt{MSE(1 + \hat{\phi}^2 + \hat{\phi}^4)}$$

$$\vdots$$

m-Step-Ahead Forecast

$$\hat{y}_{n+m} \pm 2\sqrt{MSE(1 + \hat{\phi}^2 + \hat{\phi}^4 + \cdots + \hat{\phi}^{2(m-1)})}$$

[*Note:* MSE estimates σ^2, the variance of the uncorrelated residual ε_t.]

---

b. To obtain a prediction interval, we first estimate σ^2 by the MSE, the mean square for error from the time series regression analysis. For the sales data, we form an approximate 95% prediction interval for the sales in year 36:

$$\hat{y}_{36} \pm z_{.025}\sqrt{MSE}$$
$$155.1 \pm 1.96\sqrt{27.42767}$$
$$155.1 \pm 10.1$$

or (144.8, 165.4). Thus, we forecast that the sales in year 36 will be between $145,000 and $165,000.

The approximate 95% prediction interval for year 37 is

$$\hat{y}_{37} \pm z_{.025}\sqrt{MSE(1 + \hat{\phi}^2)}$$
$$159.4 \pm 196\sqrt{27.42767[1 + (.5896)^2]}$$
$$159.4 \pm 11.9$$

or (147.5, 171.3). Note that this interval is wider than that for the one-step-ahead forecast. The intervals will continue to widen as we attempt to forecast farther ahead.

The forecasts and prediction intervals for years 36–40 are shown in Figure 9.14 (page 538). We again stress that the accuracy of these forecasts and intervals depends on the assumption that the model structure does not change during the forecasting period. If, for example, the company merges with another company during the year 37, the structure of the sales model will almost surely change, and therefore, prediction intervals past year 37 are probably useless.

FIGURE 9.14

Forecasts and Prediction
Intervals for Years 36–40:
Straight-Line Model with
Autoregressive Residual

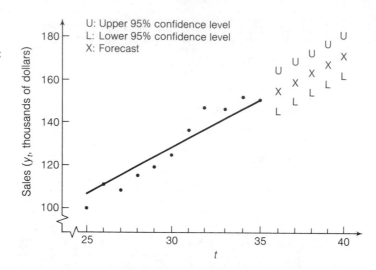

FIGURE 9.14

Forecasts and Prediction
Intervals for Years 36–40:
Straight-Line Model with
Autoregressive Residual

It is important to note that the forecasting procedure makes explicit use of the residual autocorrelation. The result is a better forecast than would be obtained using the standard least squares procedure of Chapter 4 (which ignores residual correlation). Generally, this is reflected by narrower prediction intervals for the time series forecasts than for the least squares prediction.* The end result, then, of using a time series model when autocorrelation is present is that you obtain more reliable estimates of the β coefficients, smaller residual variance, and more accurate prediction intervals for future values of the time series.

EXERCISES 9.30–9.37

9.30 The annual time series model $y_t = \beta_0 + \beta_1 t + \phi R_{t-1} + \varepsilon_t$ was fit to data collected for $n = 30$ years with the following results:

$$\hat{y}_t = 10 + 2.5t + .64\hat{R}_{t-1}$$

$$y_{30} = 82 \qquad \text{MSE} = 4.3$$

a. Calculate forecasts for y_t for $t = 31$, $t = 32$, and $t = 33$.
b. Construct approximate 95% prediction intervals for the forecasts obtained in part **a**.

9.31 The quarterly time series model $y_t = \beta_0 + \beta_1 t + \beta_2 t^2 + \phi R_{t-1} + \varepsilon_t$ was fit to data collected for $n = 48$ quarters, with the following results:

$$\hat{y}_t = 220 + 17t - .3t^2 + .82\hat{R}_{t-1}$$

$$y_{48} = 350 \qquad \text{MSE} = 10.5$$

a. Calculate forecasts for y_t for $t = 49$, $t = 50$, and $t = 51$.
b. Construct approximate 95% prediction intervals for the forecasts obtained in part **a**.

*When n is large, approximate 95% prediction intervals obtained from the standard least squares procedure reduce to $\hat{y}_t \pm 2\sqrt{\text{MSE}}$ for *all* future values of the time series. These intervals may actually be narrower than the more accurate prediction intervals produced from the time series analysis.

9.32 Use the fitted times series model of Exercise 9.25 to forecast GNP for the four quarters of 1986 and calculate approximate 95% forecast limits. Do these bounds contain the actual 1986 GNP values shown in the accompanying table?

QUARTER	1986 GNP
1	4,149.2
2	4,175.6
3	4,240.7
4	4,268.4

9.33 Use the fitted time series model of Exercise 9.26 to forecast annual volume of wheat harvest for 1989. Place approximate 95% confidence bounds on the forecast.

9.34 Use the fitted time series model of Exercise 9.27 to forecast the price of IBM common stock in February 1987. Place approximate 95% confidence bounds on the forecast.

9.35 Refer to Exercise 9.28. The values of MSE for the quarterly time series models of retirement plan terminations are as follows:

Pension plan termination: MSE = .0440

Profit-sharing plan termination: MSE = .0402

a. Forecast the number of pension plan terminations for the fourth quarter of 1982 (i.e., $t = 108$). Assume that $y_{107} = 7.5$. [*Hint:* Remember that the forecasted number of pension plan terminations is $e^{\hat{y}_{108}}$.]

b. Place approximate 95% confidence bounds on the forecast obtained in part **a**. [*Hint:* First, calculate upper and lower confidence limits for y_{108}, then take antilogarithms.]

c. Repeat parts **a** and **b** for the number of profit-sharing plan terminations in the fourth quarter of 1982. Assume that $y_{107} = 7.6$.

YEAR	DJA
1984	1,187.35
1985	1,500.60
1986	1,926.12

9.36 Use the fitted time series model of Exercise 9.29 to forecast the DJA for the years 1984–1986 (i.e., $t = 21, 22,$ and 23). Calculate approximate 95% prediction intervals for the forecasts. Do these intervals contain the actual 1984–1986 DJA values shown in the accompanying table? If not, give a plausible explanation.

9.37 Taxes, a major source of income for the state of Florida, have grown steadily since 1962. The data in the table on page 540 give the total state tax collections for the years 1962–1985. A tax economist fit the model $y_t = \beta_0 + \beta_1 x + \beta_2 x^2 + \varepsilon$ to the data. Results from the SAS printout gave the least squares model

$$\hat{y} = 1,400.904 + 204.626t + 13.9949t^2$$

A plot of the residuals revealed that the usual least squares assumption of independent error terms may have been violated. Since the data were recorded over time (years), it seems reasonable to assume that the tax collections at time $(t + 1)$ are highly correlated with the tax collections at time t, and hence that the random errors in the model follow the same pattern.

YEAR	YEAR – 1970 t	TOTAL TAX COLLECTIONS (MILLION DOLLARS) y	YEAR	YEAR – 1970 t	TOTAL TAX COLLECTIONS (MILLION DOLLARS) y
1962	–8	564	1974	4	2,794
1963	–7	592	1975	5	2,791
1964	–6	709	1976	6	2,936
1965	–5	762	1977	7	3,275
1966	–4	819	1978	8	3,764
1967	–3	877	1979	9	4,291
1968	–2	973	1980	10	4,804
1969	–1	1,269	1981	11	5,314
1970	0	1,421	1982	12	5,556
1971	1	1,587	1983	13	6,225
1972	2	1,996	1984	14	7,329
1973	3	2,488	1985	15	7,883

Source: Statistical Abstracts of the United States 1963–1987, United States Bureau of the Census.

a. Hypothesize a model for the correlated error term, R_t.

b. Rewrite the model fit by the tax economist, using the correlated error term.

c. Fit the time series model in part **b** to the data given in the table using an available statistical computer program package.

d. Use the fitted time series model from part **c** to forecast total tax collection dollars in 1986 and 1987. Place approximate 95% prediction intervals around the forecasts.

SECTION 9.9

SEASONAL TIME SERIES MODELS: AN EXAMPLE

We have used a simple regression model to illustrate the methods of model estimation and forecasting when the residuals are autocorrelated. In this section we present a more realistic example that requires a seasonal model for $E(y_t)$, as well as an autoregressive model for the residual.

Critical water shortages have dire consequences for both business and private sectors of communities. Forecasting water usage for months in advance is essential to avoid such shortages. Suppose a community has monthly water usage records over the past 15 years. A plot of the last 6 years of the time series, y_t, is shown in Figure 9.15. Note that both an increasing trend and a seasonal pattern appear prominent in the data. The water usage seems to peak during the summer months and decline during the winter months. Thus, we might propose the following model:

$$E(y_t) = \beta_0 + \beta_1 t + \beta_2\left(\cos\frac{2\pi}{12}t\right) + \beta_3\left(\sin\frac{2\pi}{12}t\right)$$

Since the amplitude of the seasonal effect (that is, the magnitude of the peaks and valleys) appears to increase with time, we include in the model an interaction between time and trigonometric components, to obtain

$$E(y_t) = \beta_0 + \beta_1 t + \beta_2\left(\cos\frac{2\pi}{12}t\right) + \beta_3\left(\sin\frac{2\pi}{12}t\right) + \beta_4 t\left(\cos\frac{2\pi}{12}t\right) + \beta_5 t\left(\sin\frac{2\pi}{12}t\right)$$

FIGURE 9.15

Water Usage Time Series

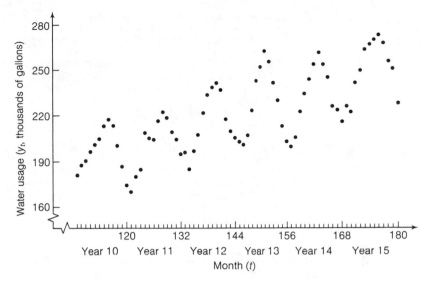

The model for the random component R_t must allow for short-term cyclic effects. For example, in an especially hot summer, if the water usage, y_t, exceeds the expected usage, $E(y_t)$, for July, we would expect the same thing to happen in August. Thus, we propose a first-order autoregressive model for the random component:*

$$R_t = \phi R_{t-1} + \varepsilon_t$$

We now fit the models to the time series y_t, where y_t is expressed in thousands of gallons. The SAS output is shown in Figure 9.16 (page 542). The estimated models are given by

$$\hat{y}_t = 100.083 + .826t - 10.801\left(\cos\frac{2\pi}{12}t\right) - 7.086\left(\sin\frac{2\pi}{12}t\right) - .0556t\left(\cos\frac{2\pi}{12}t\right) - .0296t\left(\sin\frac{2\pi}{12}t\right) + \hat{R}_t$$

$$\hat{R}_t = .6617\hat{R}_{t-1}$$

with MSE = 23.135. The R^2-value of .99 (TOTAL RSQ) indicates that the model provides a good fit to the data.

We now use the models to forecast water usage for the next 12 months. The forecast for the first month is obtained as follows. The last residual value is $\hat{R}_{180} = -1.3247$, so that

$$\hat{R}_{181} = \hat{\phi}\hat{R}_{180} = (.6617)(-1.3247) = -.8766$$

Then,

$$\hat{y}_{181} = \hat{\beta}_0 + \hat{\beta}_1(181) + \hat{\beta}_2\left(\cos\frac{2\pi}{12}181\right) + \hat{\beta}_3\left(\sin\frac{2\pi}{12}181\right)$$

$$+ \hat{\beta}_4(181)\left(\cos\frac{2\pi}{12}181\right) + \hat{\beta}_5(181)\left(\sin\frac{2\pi}{12}181\right) + \hat{R}_{181} = 238.0$$

*A more complex time series model may be more appropriate. We use the simple first-order autoregressive model so you can follow the modeling process more easily.

FIGURE 9.16 SAS Computer Printout for Water Usage Model

ESTIMATES OF THE AUTOREGRESSIVE PARAMETERS

LAG	COEFFICIENT	STD ERROR	T RATIO
1	-0.66167894	0.055886	-11.839831

YULE-WALKER ESTIMATES

SSE	4025.513	DFE	174
MSE	23.135	ROOT MSE	4.810
REG RSQ	0.9431	TOTAL RSQ	0.9900

VARIABLE	DF	B VALUE	STD ERROR	T RATIO	APPROX PROB
INTERCEPT	1	100.083218977	2.07617706007	48.206	0.0001
T	1	0.826274293	0.01979498750	41.742	0.0001
CS	1	-10.801144	1.85586558083	-5.820	0.0001
SN	1	-7.0857642	1.89574083666	-3.738	0.0003
CST	1	-0.055634923	0.01771077652	-3.141	0.0020
SNT	1	-0.029630055	0.01820045673	-1.628	0.1053

Approximate 95% prediction bounds on this forecast are given by $\pm 2\sqrt{\text{MSE}} = \pm 2\sqrt{23.135} = \pm 9.6$* That is, we expect our forecast for 1 month ahead to be within 9,600 gallons of the actual water usage. This forecasting process is then repeated for the next 11 months. The forecasts and their bounds are shown in Figure 9.17. Also shown are the actual values of water usage during year 16. Note that the forecast prediction intervals widen as we attempt to forecast farther into the future. This property of the prediction intervals makes long-term forecasts very unreliable.

FIGURE 9.17
Forecasts of Water Usage

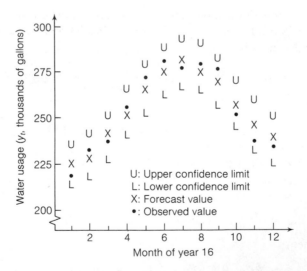

We have barely scratched the surface of time series modeling. The variety and complexity of available techniques are overwhelming. However, if we have convinced you that time series modeling is a useful and powerful tool for business

*We are ignoring the errors in the parameter estimates in calculating the forecast reliability. These errrors should be small for a series of this length.

forecasting, we have accomplished our purpose. The successful construction of times series models requires much experience, and, like regression modeling, entire texts are devoted to the subject (see the references at the end of the chapter).

We conclude with a warning: Many oversimplified forecasting methods have been proposed. They usually consist of graphical extensions of a trend or seasonal pattern to future time periods. Although such pictorial techniques are easy to understand and therefore are intuitively appealing, they should be avoided. There is no measure of reliability for these forecasts, and thus the risk associated with making decisions based on them is very high.

| | | | | | | | | | | | |

SECTION 9.10

OTHER TIME SERIES MODELS

There are many models for autocorrelated residuals in addition to the autoregressive model, but the autoregressive model provides a good approximation for the autocorrelation pattern in many applications. Recall that the autocorrelations for autoregressive models diminish rapidly as the time distance m between the residuals increases. Occasionally, residual autocorrelations appear to change abruptly from nonzero for small values of m to 0 for larger values of m. For example, neighboring residuals ($m = 1$) may be correlated, while residuals that are further apart ($m > 1$) are uncorrelated. This pattern can be described by a **first-order moving average model**:

$$R_t = \varepsilon_t + \theta\varepsilon_{t-1}$$

Note that the residual R_t is a linear combination of the current and previous *uncorrelated* (white noise) residuals. It can be shown that the autocorrelations for this model are

$$\text{Autocorrelation}(R_t, R_{t+m}) = \begin{cases} \dfrac{\theta}{1 + \theta^2} & \text{if } m = 1 \\ 0 & \text{if } m > 1 \end{cases}$$

This pattern is shown in Figure 9.18.

FIGURE 9.18

Autocorrelations for the First-Order Moving Average Model: $R_t = \varepsilon_t + \theta\varepsilon_{t-1}$

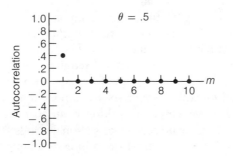
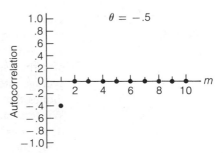

More generally, a qth-order moving average model is given by

$$R_t = \varepsilon_t + \theta\varepsilon_{t-1} + \theta_2\varepsilon_{t-2} + \cdots + \theta_q\varepsilon_{t-q}$$

Residuals within q time points are correlated, while those farther than q time points apart are uncorrelated. For example, a regression model for the quarterly earnings per share for a company may have residuals that are autocorrelated when within 1 year ($m = 4$ quarters) of one another, but uncorrelated when farther apart. An example of this pattern is shown in Figure 9.19.

Some autocorrelation patterns require even more complex residual models. A more general model is a combination of the **autoregressive–moving average (ARMA) models**,

$$R_t = \phi_1 R_{t-1} + \cdots + \phi_p R_{t-p} + \varepsilon_t + \theta_1 \varepsilon_{t-1} + \cdots + \theta_q \varepsilon_{t-q}$$

FIGURE 9.19

Autocorrelations for a Fourth-Order Moving Average Model

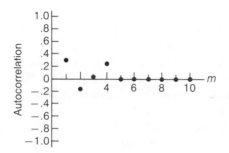

Like the autoregressive model, the ARMA model has autocorrelations that diminish as the distance m between residuals increases. However, the patterns that can be described by ARMA models are more general than those of either autoregressive or moving average models.

The methods for fitting time series models when the residual is either moving average or ARMA are more complicated than those given for the autoregressive residual model in Section 9.7. Consult the references at the end of the chapter for details of these methods.

S E C T I O N   9.11

SUMMARY

Time series are often modeled as a combination of four components: **secular, seasonal, cyclical,** and **residual.** Both descriptive and inferential techniques are available for **estimating** the time series components and **forecasting** future values of the time series. The **moving average method** is a smoothing technique that uses estimates of the secular and seasonal components to forecast future values of a time series. However, the method requires you to extrapolate the moving average into the future to obtain the forecasts. Two alternative smoothing techniques that lead to explicit forecasts are **exponential smoothing** and the **Holt–Winters model.** Exponential smoothing is an adaptive forecasting method for time series with little or no secular or seasonal trends. The Holt–Winters model is an extension of the exponential smoothing technique that allows for trend and seasonal components.

One type of **inferential time series model** employs a combination of the **deterministic component** $E(y_t)$ of the typical multiple regression model with an autoregressive model for the **autocorrelated residual.** The deterministic portion of the model accounts for the trend and seasonal components, and the autocorrelated residual deals with the problem of correlated errors.

The forecaster should be very careful to **distinguish between descriptive and inferential time series models**. If descriptive models (e.g., smoothing techniques) are used to predict future values of the series, no assessment of forecast reliability is possible. Only when a probabilistic model (e.g., a time series autoregressive model) is constructed can a prediction interval be used to evaluate the reliability of the forecast. Even then, if the structure of the model changes at some future time, forecasts beyond that point are probably useless. Careful application of time series modeling and forecasting will usually be rewarded with a better understanding of the phenomenon and with useful forecasts that assist in planning future strategy.

| | | | | | | | | | | | | |

SUPPLEMENTARY EXERCISES 9.38–9.46

9.38 The level at which lending institutions set mortgage interest rates has a significant effect on the volume of buying, selling, and construction of residential and commercial real estate. The data in the table are the annual average interest rates on 30-year, fixed-rate mortgages for the period 1980–1987. Forecast the 1988 average mortgage interest rate using each of the methods listed below the table.

YEAR	MORTGAGE INTEREST RATE (%)	YEAR	MORTGAGE INTEREST RATE (%)
1980	13.77	1984	13.87
1981	16.63	1985	12.43
1982	16.09	1986	10.18
1983	13.23	1987	9.55

Source: Mortgage Bankers Association of America.

a. A 3-point moving average
b. The exponential smoothing technique ($w = .2$)
c. The Holt–Winters model with trend ($w = .2$ and $v = .5$)
d. Simple linear regression   (Obtain a 95% prediction interval.)
e. A straight-line, first-order autoregressive model   (Obtain an approximate 95% prediction interval.)

9.39 The table at the top of page 546 lists the monthly retail sales in the United States for the period January 1982 through December 1986.
a. Calculate and plot a 12-point moving average for the total retail sales time series. Can you detect the secular trend? Does there appear to be a seasonal pattern?
b. Use the moving average from part a to forecast retail sales in January 1987.
c. Calculate and plot the exponentially smoothed series using $w = .6$.
d. Obtain the forecast for January 1987 using the exponential smoothing technique.
e. Obtain the forecast for January 1987 using the Holt–Winters model with trend and seasonal components and smoothing constants $w = .6$, $v = .7$, and $u = .5$.
f. Propose a time series model for total retail sales that accounts for secular trend, seasonal variation, and residual autocorrelation.
g. Fit the time series model specified in part f, using an available computer package.

	1982	1983	1984	1985	1986
January	77.34	81.34	92.63	98.47	105.60
February	76.21	78.88	93.46	95.34	99.66
March	86.57	93.75	104.20	109.90	114.20
April	87.96	93.97	104.40	113.00	115.70
May	90.81	97.84	111.60	120.20	125.40
June	88.97	100.61	112.00	114.80	120.40
July	91.21	99.56	106.00	115.20	120.70
August	89.64	100.23	110.80	120.80	124.10
September	88.16	97.97	103.60	113.80	124.60
October	91.41	100.38	109.10	115.80	123.10
November	94.19	104.32	113.00	118.10	120.80
December	113.10	125.15	131.50	138.70	151.30

Source: Standard & Poor's Trade and Securities Statistics (annual). New York: Standard & Poor Corporation.

h. Use the time series model to forecast retail sales in January 1987. Obtain an approximate 95% prediction interval for the forecast.

9.40 A traditional pulse rate of the economic health of the accommodations (hotel–motel) industry is the trend in room occupancy. Average monthly occupancies for the years 1983–1984 are given in the table for hotels and motels in the cities of Atlanta, Georgia, and Phoenix, Arizona.

MONTH 1983	PERCENTAGE OF ROOMS OCCUPIED		MONTH 1984	PERCENTAGE OF ROOMS OCCUPIED	
	Atlanta	Phoenix		Atlanta	Phoenix
January	59	67	January	64	72
February	63	85	February	69	91
March	68	83	March	73	87
April	70	69	April	67	75
May	63	63	May	68	70
June	59	52	June	71	61
July	68	49	July	67	46
August	64	49	August	71	44
September	62	56	September	65	63
October	73	69	October	72	73
November	62	63	November	63	71
December	47	48	December	47	51

Source: Trends in the Hotel Industry, 1985.

a. Graph the time series values for both cities on the same set of axes. Use two different colors so that you will be able to distinguish between the two time series.

b. Do you detect any differences in the seasonal and long-term trends of the two time series?

c. Postulate a time series model that will be useful in forecasting occupancy rates for Atlanta and Phoenix.

d. Fit the model in part **c** to each set of data using an available statistical computer program package.

e. Interpret the differences in the modified least squares parameter estimates for the two models. Does this support your answer to part **b**?

f. Would you recommend using the model to forecast monthly occupancy rates in 1989? Explain.

9.41 In May 1978, the first casino (Resorts International Hotel and Casino) opened in Atlantic City, New Jersey. In the first few years following the casino opening, employment in hotels and other lodging places accelerated along Atlantic City's boardwalk, as shown in the table.

YEAR	QUARTER	EMPLOYMENT IN ATLANTIC CITY HOTELS
1978	I	1,711
	II	4,065
	III	5,787
	IV	5,019
1979	I	5,459
	II	9,184
	III	12,168
	IV	11,842
1980	I	13,730
	II	14,964
	III	18,058
	IV	21,393

Source: *Business Review*, January–February 1982.

a. Use a smoothing technique to forecast employment in Atlantic City hotels in Quarter I, 1981. Comment on the reliability of this forecast.

b. Propose a time series model for the quarterly series that will account for secular trend, seasonal variation, and residual autocorrelation. If you have access to a computer package with a modified least squares routine, fit the model.

c. Use the fitted time series model of part **b** to forecast employment in Quarter I, 1981. Place an approximate 95% prediction interval about the forecast.

d. Actual employment in Quarter I of 1981 was 22,772. Check to see whether the forecasting technique of part **c** has captured this value.

e. Would you recommend using the fitted model in part **b** for forecasting quarterly employment in 1988? Explain.

9.42 The table at the top of page 548 lists the yearly outlays for the Federal Space Program (in millions of dollars) from 1966 through 1984.

a. Plot the time series.

b. Calculate moving averages for 3 points, 5 points, and 7 points. Plot each moving average series on the same graph. Which moving average best characterizes the long-term trend of outlays for the Space Program?

c. Graphically extend the moving average you chose in part **b** to forecast the outlay for the Space Program in 1988. Comment on the reliability of this forecast. Is there a better forecasting method available? Explain.

YEAR	OUTLAY	YEAR	OUTLAY
1966	6,956	1976	5,320
1967	6,970	1977	5,983
1968	6,529	1978	6,509
1969	5,976	1979	7,419
1970	5,341	1980	8,689
1971	4,741	1981	9,978
1972	4,575	1982	12,091
1973	4,825	1983	15,564
1974	4,640	1984	17,449
1975	4,914		

Source: *Statistical Abstract of the United States*, United States Department of Commerce, Bureau of the Census, 1987.

9.43 Beer production in the United States for the years 1973–1985 is given in the accompanying table. Suppose you are interested in forecasting U.S. beer production in 1986. Since a plot of the time series y_t reveals a linearly increasing trend, you hypothesize the model

$$E(y_t) = \beta_0 + \beta_1 t$$

for the secular trend.

YEAR	TIME t	U.S. BEER PRODUCTION y_t, millions of barrels	YEAR	TIME t	U.S. BEER PRODUCTION y_t, millions of barrels
1973	1	148.6	1980	8	194.1
1974	2	156.2	1981	9	193.7
1975	3	160.6	1982	10	196.2
1976	4	163.7	1983	11	196.0
1977	5	170.5	1984	12	193.0
1978	6	179.1	1985	13	194.2
1979	7	184.2			

Source: *Standard & Poor's Trade and Securities Statistics* (annual). New York: Standard & Poor Corporation.

a. Fit the model to the data using the method of least squares.

b. Plot the least squares model from part **a** and extend the line to forecast y_{1986}, the U.S. beer production (in millions of barrels) in 1986. How reliable do you think this forecast is?

c. Calculate and plot the residuals for the model from part **a**. Is there visual evidence of residual autocorrelation?

d. How could you test to determine whether residual autocorrelation exists? If you have access to a computer package, carry out the test. Use $\alpha = .05$.

e. Hypothesize a time series model that accounts for the residual autocorrelation. If you have access to a computer package with a modified least squares routine, fit the model.

f. Compute a 95% prediction interval for y_{1986}, the U.S. beer production in 1986. Why is this forecast preferred to that of part **b**?

9.44 The accompanying table records the monthly number of mortgage applications (in thousands) for new home construction processed by the Federal Housing Administration (FHA) for the period 1984–1986.

	1984	1985	1986
January	9.7	11.1	24.1
February	10.3	11.5	24.8
March	12.9	12.9	39.1
April	11.4	15.8	51.0
May	11.1	15.2	41.0
June	8.2	16.6	26.9
July	8.0	17.6	24.7
August	7.8	17.1	20.2
September	7.4	16.3	21.7
October	9.7	17.1	18.9
November	9.8	14.8	16.2
December	9.2	14.8	16.7

Source: Survey of Current Business, United States Department of Commerce, Bureau of Economic Analysis.

a. Graph the monthly time series.

b. Calculate a 12-point moving average for the monthly time series. Plot the 12-point moving average on the graph constructed in part a. Can you identify the long-term trend of the series? Is any seasonal component evident?

c. Use the moving average technique to forecast FHA mortgage applications for the first three months of 1987. Comment on the reliability of these forecasts.

d. Obtain the forecasts for the first three months of 1987 using the Holt–Winters model with smoothing constants $w = .4$, $v = .5$, and $u = .2$. Comment on the reliability of these forecasts.

e. Hypothesize a time series model for the monthly series that accounts for trend, seasonal variation, and residual autocorrelation. If you have access to a computer package with a modified least squares routine, fit the model.

f. Use the fitted time series model to forecast the number of mortgage applications in May 1987. Calculate an approximate 95% prediction interval for the forecast.

9.45 The number of motor vehicle deaths in the United States for each month of 1982 and 1983 is recorded in the table at the top of page 550. The National Safety Council would like to model y_t, total deaths per month, as a function of time, t, for forecasting purposes. Since the number of motor vehicle deaths seems to peak during the summer months and decline during the winter months, the following seasonal model (with trigonometric terms) is proposed:

$$y_t = \beta_0 + \beta_1 t + \beta_2\left(\cos\frac{2\pi}{12}t\right) + \beta_3\left(\sin\frac{2\pi}{12}t\right) + \phi R_{t-1} + \varepsilon_t$$

a. If you have access to a computer package with a modified least squares routine, fit the proposed time series model to the data.

b. From the output, write the least squares prediction equation for y_t.

MONTH 1982	TIME t	TOTAL DEATHS y_t	MONTH 1983	TIME t	TOTAL DEATHS y_t
January	1	3,070	January	13	3,050
February	2	2,930	February	14	2,880
March	3	3,430	March	15	3,210
April	4	3,700	April	16	3,430
May	5	4,120	May	17	3,840
June	6	4,000	June	18	3,900
July	7	4,500	July	19	4,370
August	8	4,340	August	20	4,260
September	9	4,110	September	21	4,110
October	10	4,300	October	22	4,200
November	11	3,620	November	23	3,620
December	12	3,880	December	24	3,730

Source: Accident Facts, 1984. National Safety Council.

c. Forecast y_{25}, the number of motor vehicle deaths during January 1984, with an approximate 95% prediction interval.

9.46 Since the energy shortage, the price of foreign crude oil has skyrocketed. Consequently, crude oil imports into the United States have declined. The data in the table are the amounts of crude oil (in millions of barrels) imported into the United States from the Organization of Petroleum-Exporting Countries (OPEC) for the years 1973–1985.

YEAR	t	IMPORTS, y_t	YEAR	t	IMPORTS, y_t
1973	1	767	1980	8	1,414
1974	2	926	1981	9	1,067
1975	3	1,171	1982	10	633
1976	4	1,663	1983	11	540
1977	5	2,058	1984	12	553
1978	6	1,892	1985	13	479
1979	7	1,866			

Source: Statistical Abstracts of the United States, 1975–1987, U.S. Bureau of the Census.

a. Plot the time series.
b. Hypothesize a straight-line autoregressive time series model for annual amount of imported crude oil, y_t.
c. If you have access to a computer package, fit the proposed model to the data.
d. From the output, write the modified least squares prediction equation for y_t.
e. Forecast the amount of foreign crude oil imported into the United States from OPEC in 1988. Place approximate 95% bounds on the forecast value.

REFERENCES

Anderson, T. W. *The Statistical Analysis of Time Series.* New York: Wiley, 1971.

Box, G. E. P. and Jenkins, G. M. *Time Series Analysis: Forecasting and Control*, 2nd ed. San Francisco: Holden-Day, 1977.

Fuller, W. A. *Introduction to Statistical Time Series.* New York: Wiley, 1976.

Granger, C. W. J. *Spectral Analysis of Economic Time Series.* Princeton: Princeton University Press, 1964.

Granger, C. W. J. and Newbold, P. *Forecasting Economic Time Series.* New York: Academic Press, 1977.

Ledolter, J. and Power, M. L. "A study of ERISA's impact on private retirement plan growth." *Journal of Risk and Insurance*, December 1983, pp. 225–241.

Makridakis, S. et al. *The Forecasting Accuracy of Major Time Series Methods.* New York: Wiley, 1984.

Nelson, C. R. *Applied Time Series Analysis for Managerial Forecasting.* San Francisco: Holden-Day, 1973.

Seitz, N. *Business Forecasting: Concepts and Microcomputer Applications.* Reston, Virginia: Reston Publishing Company, 1984.

OBJECTIVE

To introduce a number of special regression techniques for problems that require more advanced methods of analysis

CONTENTS

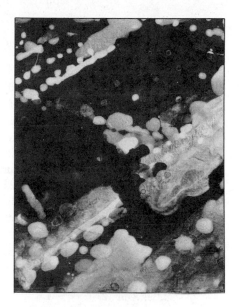

SPECIAL TOPICS
IN REGRESSION
(OPTIONAL)

SECTION 10.1

INTRODUCTION

The procedures presented in Chapters 3–9 provide the tools basic to a regression analysis. An understanding of these techniques will enable you to successfully apply regression analysis to a variety of problems encountered in practice. For some studies, however, you may require more sophisticated techniques. In this chapter we introduce several special topics in regression for the advanced student.

SECTION 10.2

PIECEWISE LINEAR REGRESSION

Occasionally, the linear relationship between a dependent variable y and an independent variable x may differ for different intervals over the range of x. For example, it is known that the compressive strength y of concrete depends on the proportion x of water mixed with the cement. A certain type of concrete, when mixed in batches with varying water/cement ratios (measured as a percentage), may yield compressive strengths (measured in pounds per square inch) that follow the pattern shown in Figure 10.1. Note that the compressive strength decreases at a much faster rate for batches with water/cement ratios greater than 70%. That is, the slope of the relationship between compressive strength (y) and water/cement ratio (x) changes when $x = 70$.

FIGURE 10.1

Relationship Between Compressive Strength (y) and Water/Cement Ratio (x)

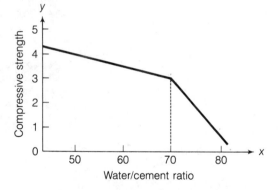

A model that proposes different straight-line relationships for different intervals over the range of x is called a **piecewise linear regression model**. As its name suggests, the linear regression model is fit in pieces. For the concrete example, the piecewise model would consist of two pieces, $x \leq 70$ and $x > 70$. The model can be expressed as follows:

$$y = \beta_0 + \beta_1 x_1 + \beta_2(x_1 - 70)x_2 + \varepsilon$$

where

$$x_1 = \text{Water/cement ratio } (x)$$

$$x_2 = \begin{cases} 1 & \text{if } x_1 > 70 \\ 0 & \text{if } x_1 \leq 70 \end{cases}$$

The value of the dummy variable x_2 controls the values of the slope and y-intercept for each piece. For example, when $x_1 \leq 70$, then $x_2 = 0$ and the equation is given by

$$y = \beta_0 + \beta_1 x_1 + \beta_2(x_1 - 70)(0) + \varepsilon$$
$$ = \underbrace{\beta_0}_{y\text{-intercept}} + \underbrace{\beta_1 x_1}_{\text{Slope}} + \varepsilon$$

Conversely, if $x_1 > 70$, then $x_2 = 1$ and we have

$$y = \beta_0 + \beta_1 x_1 + \beta_2(x_1 - 70)(1) + \varepsilon$$
$$ = \beta_0 + \beta_1 x_1 + \beta_2 x_1 - 70\beta_2 + \varepsilon$$

or

$$y = \underbrace{(\beta_0 - 70\beta_2)}_{y\text{-intercept}} + \underbrace{(\beta_1 + \beta_2)x_1}_{\text{Slope}} + \varepsilon$$

Thus, β_1 and $(\beta_1 + \beta_2)$ represent the slopes of the lines for the two intervals of x, $x \leq 70$ and $x > 70$, respectively. Similarly, β_0 and $(\beta_0 - 70\beta_2)$ represent the respective y-intercepts. The slopes and y-intercepts of the two lines are illustrated graphically in Figure 10.2. [*Note:* The value at which the slope changes, 70 in this example, is often referred to as a **knot value**. Usually, the knot values of a piecewise regression are unknown and must be estimated from the sample data. This is often accomplished by visually inspecting the scattergram for the data and locating the points on the x-axis at which the slope appears to change.]

FIGURE 10.2

Slopes and y intercepts for Piecewise Linear Regression

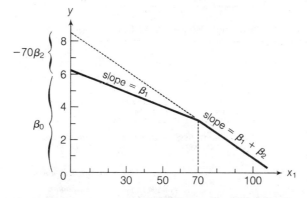

Piecewise regression models can be fit using the standard multiple regression routines of most computer packages by making the appropriate transformations on the independent variables. For example, consider the data on compressive strength (y) and water/cement ratio (x) for 18 batches of concrete recorded in Table 10.1 on page 556. (The water/cement ratio is computed by dividing the weight of water used in the mix by the weight of the cement.) To obtain the least squares fit of the piecewise linear regression model for the data of Table 10.1, we specify the model

$$E(y) = \beta_0 + \beta_1 x_1 + \beta_2 x_2^*$$

where

$$x_2^* = (x_1 - 70)x_2 \quad \text{and} \quad x_2 = \begin{cases} 1 & \text{if } x_1 > 70 \\ 0 & \text{if } x_1 \le 70 \end{cases}$$

The SAS printout for the piecewise linear regression is shown in Figure 10.3. From the printout we obtain the least squares prediction equation:

$$\hat{y} = 7.792 - .06633x_1 - .10119x_2^*$$

Note that the estimated mean change in compressive strength for a 1% increase in water/cement ratio $\hat{\beta}_1 = -.06633$ for ratios less than or equal to 70% and $\hat{\beta}_1 + \hat{\beta}_2 = -.06633 + (-.10119) = -.16752$ for ratios greater than 70%.

TABLE 10.1

Data on Compressive Strength and Water/Cement Ratios for 18 Batches of Cement

BATCH	COMPRESSIVE STRENGTH y, pounds per square inch	WATER/CEMENT RATIO x, percent	BATCH	COMPRESSIVE STRENGTH y, pounds per square inch	WATER/CEMENT RATIO x, percent
1	4.67	47	10	2.21	73
2	3.54	68	11	4.10	60
3	2.25	75	12	1.13	85
4	3.82	65	13	1.67	80
5	4.50	50	14	1.59	75
6	4.07	55	15	3.91	63
7	.76	82	16	3.15	70
8	3.01	72	17	4.37	50
9	4.29	52	18	3.75	57

FIGURE 10.3 SAS Printout for Piecewise Linear Regression of Data in Table 10.1

```
DEP VARIABLE: Y
                                       ANALYSIS OF VARIANCE

                                   SUM OF          MEAN
                SOURCE     DF      SQUARES        SQUARE      F VALUE      PROB>F

                MODEL       2    24.71775224   12.35887612   114.441      0.0001
                ERROR      15     1.61989776    0.10799318
                C TOTAL    17    26.33765000

                   ROOT MSE     0.3286232      R-SQUARE      0.9385
                   DEP MEAN        3.155       ADJ R-SQ      0.9303
                   C.V.         10.41595

                                   PARAMETER ESTIMATES

                          PARAMETER     STANDARD     T FOR HO:
                VARIABLE  DF   ESTIMATE      ERROR    PARAMETER=0    PROB > :T:

                INTERCEP   1   7.79198302   0.67696058    11.510       0.0001
                X1         1  -0.06633080   0.01123476    -5.904       0.0001
                X2STAR     1  -0.10118610   0.02812449    -3.598       0.0026
```

Piecewise regression is not limited to two pieces, nor is it limited to straight lines. One or more of the pieces may require a quadratic or higher-order fit. Also, piecewise regression models can be proposed to allow for discontinuities or jumps in the regression function. Such models require additional dummy variables to be introduced. Several different piecewise linear regression models relating y to an independent variable x are shown in the box.

PIECEWISE LINEAR REGRESSION MODELS RELATING y TO AN INDEPENDENT VARIABLE x_1

TWO STRAIGHT LINES (CONTINUOUS)

$$E(y) = \beta_0 + \beta_1 x_1 + \beta_2(x_1 - k)x_2$$

where

k = Knot value (i.e., the value of the independent variable x_1 at which the slope changes)

$$x_2 = \begin{cases} 1 & \text{if } x_1 > k \\ 0 & \text{if not} \end{cases}$$

	$x_1 \leq k$	$x_1 > k$
y-intercept	β_0	$\beta_0 - k\beta_2$
Slope	β_1	$\beta_1 + \beta_2$

THREE STRAIGHT LINES (CONTINUOUS)

$$E(y) = \beta_0 + \beta_1 x_1 + \beta_2(x_1 - k_1)x_2 + \beta_3(x_1 - k_2)x_3$$

where k_1 and k_2 are knot values of the independent variable x_1, $k_1 < k_2$, and

$$x_2 = \begin{cases} 1 & \text{if } x_1 > k_1 \\ 0 & \text{if not} \end{cases} \qquad x_3 = \begin{cases} 1 & \text{if } x_1 > k_2 \\ 0 & \text{if not} \end{cases}$$

	$x_1 \leq k_1$	$k_1 < x_1 \leq k_2$	$x_1 > k_2$
y-intercept	β_0	$\beta_0 - k_1\beta_2$	$\beta_0 - k_1\beta_2 - k_2\beta_3$
Slope	β_1	$\beta_1 + \beta_2$	$\beta_1 + \beta_2 + \beta_3$

TWO STRAIGHT LINES (DISCONTINUOUS)

$$E(y) = \beta_0 + \beta_1 x_1 + \beta_2(x_1 - k)x_2 + \beta_3 x_2$$

where

k = Knot value (i.e., the value of the independent variable x_1 at which the slope changes—also the point of discontinuity)

$$x_2 = \begin{cases} 1 & \text{if } x_1 > k \\ 0 & \text{if not} \end{cases}$$

	$x_1 \leq k$	$x_1 > k$
y-intercept	β_0	$\beta_0 - k\beta_2 + \beta_3$
Slope	β_1	$\beta_1 + \beta_2$

Tests of model adequacy, tests and confidence intervals on individual β parameters, confidence intervals for $E(y)$, and prediction intervals for y for piecewise regression models are conducted in the usual manner.

| | | | | | | | | | | |

EXERCISES 10.1–10.5

10.1 Consider a two-piece linear relationship between y and x with no discontinuity and a slope change at $x = 15$.
   a. Specify the appropriate piecewise linear regression model for y.
   b. In terms of the β coefficients, give the y-intercept and slope for observations with $x \le 15$; for observations with $x > 15$.
   c. Explain how you could determine whether the two slopes proposed by the model are, in fact, different.

10.2 Consider a three-piece linear relationship between y and x with no discontinuity and slope changes at $x = 1.45$ and $x = 5.20$.
   a. Specify the appropriate piecewise linear regression model for y.
   b. In terms of the β coefficients, give the y-intercept and slope for each of the three intervals, $x \le 1.45$, $1.45 < x \le 5.20$, and $x > 5.20$.
   c. Explain how you could determine whether at least two of the three slopes proposed by the model are, in fact, different.

10.3 Consider a two-piece linear relationship between y and x with discontinuity and slope change at $x = 320$.
   a. Specify the appropriate piecewise linear regression model for y.
   b. In terms of the β coefficients, give the y-intercept and slope for observations with $x \le 320$; for observations with $x > 320$.
   c. Explain how you could determine whether the two straight lines proposed by the model are, in fact, different.

10.4 The total amount y of corn produced in the United States is recorded (in millions of bushels) for the years 1950–1986 in the accompanying table.

YEAR x	PRODUCTION y	YEAR x	PRODUCTION y	YEAR x	PRODUCTION y
1950	3,075	1963	4,019	1976	6,289
1951	2,926	1964	3,484	1977	6,505
1952	3,292	1965	4,084	1978	7,268
1953	3,210	1966	4,117	1979	7,928
1954	3,058	1967	4,760	1980	6,639
1955	3,220	1968	4,450	1981	8,119
1956	3,445	1969	4,687	1982	8,235
1957	3,400	1970	4,152	1983	4,175
1958	3,725	1971	5,646	1984	7,674
1959	4,197	1972	5,580	1985	8,865
1960	4,314	1973	5,671	1986	8,253
1961	3,598	1974	4,701		
1962	3,606	1975	5,841		

Sources: Agricultural Statistics, 1984, U.S. Department of Agriculture (1968–1982); Historical Statistics of the U.S., Colonial Times to 1970, U.S. Department of Commerce, Bureau of the Census (1950–1967); Commodity Year Book 1986, Commodity Research Bureau (1983–1985); Survey of Current Business, U.S. Department of Commerce.

a. Plot the annual corn production y against year x. Note that corn production appears to increase at a much faster rate over the period 1971–1986.

b. Propose a piecewise linear model for annual corn production y with a knot at $x = 1970$.

c. If you have access to a computer program package, fit the model proposed in part b. Give the least squares prediction equation.

d. Is the model adequate for predicting annual corn production y? Test using $\alpha = .05$.

e. Give the estimates of the expected annual increases in corn production over the two periods, 1950–1970 and 1971–1986.

f. Is there sufficient evidence to indicate that annual corn production increases at a faster rate after 1970? Test using $\alpha = .05$.

10.5 The manager of a packaging plant wants to model the unit cost y of shipping lots of a semifragile product as a linear function of lot size x. Due to economies of scale, the manager believes that the cost per unit will decrease at a faster rate for lot sizes of more than 1,000. Data collected on unit cost and lot size for 15 recent shipments are given in the table.

SHIPPING COST y, $ per unit	LOT SIZE x	SHIPPING COST y, $ per unit	LOT SIZE x
1.29	1,150	2.90	520
2.20	840	2.63	670
2.26	900	.55	1,420
2.38	800	2.31	850
1.77	1,070	1.90	1,000
1.25	1,220	2.15	910
1.87	980	1.20	1,230
.71	1,300		

a. Specify the appropriate piecewise linear model for y.

b. If you have access to a computer program package, fit the model to the data. Give the least squares prediction equation.

c. Is the model adequate for predicting unit cost y? Test using $\alpha = .10$.

d. Give a 90% confidence interval for the mean increase in shipping cost per unit for every unit increase in lot size for lots with 1,000 or fewer units.

SECTION 10.3

INVERSE PREDICTION

Often, the goal of regression is to predict the value of one variable when another variable takes on a specified value. For most simple linear regression problems, we are interested in predicting y for a given x. We provided a formula for a prediction interval for y when $x = x_p$ in Section 3.9. In this section, we discuss **inverse prediction**—that is, predicting x for a given value of the dependent variable y.

Inverse prediction has many applications in the engineering and physical sciences, in medical research, and in business. For example, when calibrating a new instrument, scientists often search for approximate measurements y, which are easy and inexpensive to obtain, and which are related to the precise, but more expensive and time-consuming measurements x. If a regression analysis reveals that x and y are highly correlated, then the scientist could choose to use the quick and inexpensive approximate measurement value, say $y = y_p$, to estimate the unknown precise measurement x. (In this context, the problem of inverse prediction is sometimes referred to as a **linear calibration** problem.) Physicians often use inverse prediction to determine the required level of dosage of a drug. Suppose a regression analysis conducted on patients with high blood pressure showed that a linear relationship exists between decrease in blood pressure y and dosage x of a new drug. Then a physician treating a new patient may want to determine what dosage x to administer in order to reduce the patient's blood pressure by an amount $y = y_p$. To illustrate inverse prediction in a business setting, consider a firm that sells a particular product. Suppose the firm's monthly market share y is linearly related to its monthly television advertising expenditure x. For a particular month, the firm may want to know how much it must spend on advertising x in order to attain a specified market share $y = y_p$.

The classical approach to inverse prediction is first to fit the familiar straight-line model

$$y = \beta_0 + \beta_1 x + \varepsilon$$

to a sample of n data points and obtain the least squares prediction equation

$$\hat{y} = \hat{\beta}_0 + \hat{\beta}_1 x$$

Solving the least squares prediction equation for x, we have

$$x = \frac{\hat{y} - \hat{\beta}_0}{\hat{\beta}_1}$$

Now let y_p be an observed value of y in the future with unknown x. Then a point estimate of x is given by

$$\hat{x} = \frac{y_p - \hat{\beta}_0}{\hat{\beta}_1}$$

Although no exact expression for the standard error of $\hat{x}$ (denoted $s_{\hat{x}}$) is known, we can algebraically manipulate the formula for a prediction interval for y given x (see Section 3.9) to form a prediction interval for x given y. It can be shown (proof omitted) that an approximate $(1 - \alpha)100\%$ prediction interval for x when $y = y_p$ is

$$\hat{x} \pm t_{\alpha/2} s_{\hat{x}} \approx \hat{x} \pm t_{\alpha/2}\left(\frac{s}{\hat{\beta}_1}\right)\sqrt{1 + \frac{1}{n} + \frac{(\hat{x} - \bar{x})^2}{SS_{xx}}}$$

where the distribution of t is based on $(n - 2)$ degrees of freedom, $s = \sqrt{MSE}$, and

$$SS_{xx} = \sum x^2 - \frac{\left(\sum x\right)^2}{n}$$

This approximation is appropriate as long as the quantity

$$D = \left(\frac{t_{\alpha/2}s}{\hat{\beta}_1}\right)^2 \cdot \frac{1}{SS_{xx}}$$

is small. The procedure for constructing an approximate confidence interval for x in inverse prediction is summarized in the box.

INVERSE PREDICTION: APPROXIMATE $(1 - \alpha)100\%$ PREDICTION INTERVAL FOR x WHEN $y = y_p$ IN SIMPLE LINEAR REGRESSION

$$\hat{x} \pm t_{\alpha/2}\left(\frac{s}{\hat{\beta}_1}\right)\sqrt{1 + \frac{1}{n} + \frac{(\hat{x} - \bar{x})^2}{SS_{xx}}}$$

where

$$\hat{x} = \frac{y_p - \hat{\beta}_0}{\hat{\beta}_1}$$

$\hat{\beta}_0$ and $\hat{\beta}_1$ are the y-intercept and slope, respectively, of the least squares line

$n = $ Sample size

$$\bar{x} = \frac{\sum x}{n}$$

$$SS_{xx} = \sum x^2 - \frac{\left(\sum x\right)^2}{n}$$

$$s = \sqrt{MSE}$$

and the distribution of t is based on $(n - 2)$ degrees of freedom.
The approximation is appropriate when the quantity

$$D = \left(\frac{t_{\alpha/2}s}{\hat{\beta}_1}\right)^2 \cdot \frac{1}{SS_{xx}}$$

is small.*

EXAMPLE 10.1

A firm that sells copiers advertises regularly on television. One goal of the firm is to determine the amount it must spend on television advertising in a single month in order to gain a market share of 10%. For one year, the firm varied its monthly television advertising expenditures (x) and at the end of each month determined its market share (y). The data for the 12 months are recorded in Table 10.2 (page 562).

*Neter, Wasserman, and Kutner (1985) and others suggest using the approximation when D is less than .1.

TABLE 10.2

A Firm's Market Share and Television Advertising Expenditure for 12 Months, Example 10.1

MONTH	MARKET SHARE y, percent	TELEVISION ADVERTISING EXPENDITURE x, $ thousands
January	7.5	23
February	8.5	25
March	6.5	21
April	7.0	24
May	8.0	26
June	6.5	22
July	9.5	27
August	10.0	31
September	8.5	28
October	11.0	32
November	10.5	30
December	9.0	29

a. Fit the straight-line model $y = \beta_0 + \beta_1 x + \varepsilon$ to the data.

b. Is there evidence that television advertising expenditure x is linearly related to market share y? Test using $\alpha = .05$.

c. Use inverse prediction to estimate the amount that must be spent on television advertising in a particular month in order for the firm to gain a market share of $y = 10\%$. Construct an approximate 95% prediction interval for monthly television advertising expenditure x.

SOLUTION

a. The SAS printout for the simple linear regression is shown in Figure 10.4. From the printout, we obtain the least squares line

$$\hat{y} = -1.975 + .397x$$

The least squares line is plotted along with 12 data points in Figure 10.5.

b. To determine whether television advertising expenditure x is linearly related to market share y, we test the hypothesis

$$H_0: \quad \beta_1 = 0$$

against

$$H_a: \quad \beta_1 \neq 0$$

The value of the test statistic, shaded on the printout, is $t = 9.12$, and the associated p-value of the test is $p = .0001$ (also shaded). Thus, there is sufficient evidence, at any value of α greater than .0001, to indicate that television advertising expenditure x and market share y are linearly related.

Caution: One should avoid using inverse prediction when there is insufficient evidence to reject the null hypothesis $H_0: \beta_1 = 0$. Inverse predictions made when x and y are *not* linearly related may lead to nonsensical results. Therefore, you should always conduct a test of model adequacy to be sure that x and y are linearly related before you carry out an inverse prediction.

FIGURE 10.4 Least Squares Fit of Straight-Line Model, Example 10.1

ANALYSIS OF VARIANCE

SOURCE	DF	SUM OF SQUARES	MEAN SQUARE	F VALUE	PROB>F
MODEL	1	22.52141608	22.52141608	83.174	0.0001
ERROR	10	2.70775058	0.27077506		
C TOTAL	11	25.22916667			

ROOT MSE	0.5203605	R-SQUARE	0.8927	
DEP MEAN	8.541667	ADJ R-SQ	0.8819	
C.V.	6.092025			

PARAMETER ESTIMATES

VARIABLE	DF	PARAMETER ESTIMATE	STANDARD ERROR	T FOR H0: PARAMETER=0	PROB > |T|
INTERCEP	1	-1.97494172	1.16288320	-1.698	0.1203
X	1	0.39685315	0.04351473	9.120	0.0001

FIGURE 10.5 Scattergram of Data and Least Squares Line, Example 10.1

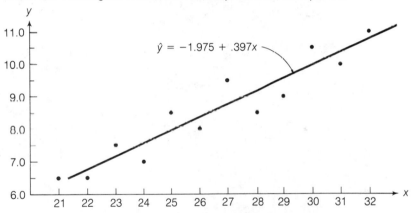

$$\hat{y} = -1.975 + .397x$$

c. Since the model is found to be adequate, we can use the model to predict x from y. For this example, we want to estimate the television advertising expenditure x that yields a market share of $y_p = 10\%$.

Substituting $y_p = 10$, $\hat{\beta}_0 = -1.975$, and $\hat{\beta}_1 = .397$ into the formula for $\hat{x}$ given in the box, we have

$$\hat{x} = \frac{y_p - \hat{\beta}_0}{\hat{\beta}_1}$$

$$= \frac{10 - (-1.975)}{.397}$$

$$= 30.16$$

Thus, we estimate that the firm must spend \$30,160 on television advertising in a particular month in order to gain a market share of 10%.

Before we construct an approximate 95% prediction interval for x, we check to determine whether the approximation is appropriate, that is, whether the quantity

$$D = \left(\frac{t_{\alpha/2}s}{\hat{\beta}_1}\right)^2 \cdot \frac{1}{SS_{xx}}$$

is small. For $\alpha = .05$, $t_{\alpha/2} = t_{.025} = 2.228$ for $n - 2 = 12 - 2 = 10$ degrees of freedom. From the printout (Figure 10.4), $s = $ ROOT MSE $= .5204$ and $\hat{\beta}_1 = .397$. The value of SS_{xx} is not shown on the printout and must be calculated:

$$SS_{xx} = \sum x^2 - \frac{\left(\sum x\right)^2}{n}$$

$$= 8,570 - \frac{(318)^2}{12}$$

$$= 8,570 - 8,427$$

$$= 143$$

Substituting these values into the formula for D, we have

$$D = \left(\frac{t_{\alpha/2}s}{\hat{\beta}_1}\right)^2 \cdot \frac{1}{SS_{xx}}$$

$$= \left[\frac{(2.228)(.5204)}{.397}\right]^2 \cdot \frac{1}{143}$$

$$= \frac{8.5295}{143}$$

$$= .0596$$

Since the value of D is small (i.e., less than .1), we may use the formula for the approximate 95% prediction interval given in the box:

$$\hat{x} \pm t_{\alpha/2}\left(\frac{s}{\hat{\beta}_1}\right)\sqrt{1 + \frac{1}{n} + \frac{(\hat{x} - \bar{x})^2}{SS_{xx}}}$$

$$30.16 \pm (2.228)\frac{(.5204)}{.397}\sqrt{1 + \frac{1}{12} + \frac{\left(30.16 - \frac{318}{12}\right)^2}{143}}$$

$$30.16 \pm (2.9205)(1.0849)$$

$$30.16 \pm 3.17$$

or (26.99, 33.33). Therefore, using the 95% prediction interval, we estimate that the amount of monthly television advertising expenditure required to gain a market share of 10% falls between $26,999 and $33,330. ∎

Another approach to the inverse prediction problem is to regress x on y, i.e., fit the model (called the **inverse estimator model**)

$$x = \beta_0 + \beta_1 y + \varepsilon$$

and then use the standard formula for a prediction interval given in Section 3.9. However, in theory this method requires that x be a random variable. In many business applications, the value of x is set in advance (i.e., controlled) and therefore is *not* a random variable. (For example, the firm in Example 10.1 selected the amount x spent on advertising *prior* to each month.) Thus, the inverse model above may violate the standard least squares assumptions given in Chapter 4. Some researchers advocate the use of the inverse model despite this caution, while others have developed different estimators of x using a modification of the classical approach. Consult the references given at the end of this chapter for details on the various alternative methods of inverse prediction.

EXERCISES 10.6–10.9

10.6 The data for Exercise 3.22 are reproduced in the accompanying table. Use inverse prediction to estimate the subsidy rate x for a particular industry that laid off workers at a rate of $y = 6.50$ workers per 1,000. Construct an approximate 99% prediction interval for x.

INDUSTRY	SUBSIDY RATE x	LAYOFF RATE y
Apparel	57%	12.54
Chemicals	32	1.78
Construction	31	7.10
Electrical machinery	29	8.38
Fabricated metals	27	11.72
Food	36	5.10
Machinery	32	4.44
Miscellaneous manufacturing	61	9.82
Primary metals	23	7.34
Retail	27	1.98
Wholesale trade	33	1.86

Source: Tropel, R. H. "On layoffs and unemployment insurance," *American Economic Review*, 1983. Vol. 83, pp. 541–559.

10.7 The data for Exercise 3.24 are reproduced in the table at the top of page 566.
   a. Use inverse prediction to estimate the GMAT score x of a former University of Florida MBA student with a starting salary of $y = \$35,000$. Construct an approximate 95% prediction interval for x.
   b. Explain why the interval of part **a** is so wide.

STARTING SALARY y	GMAT x
$40,000	510
33,000	510
40,000	550
35,000	600
28,000	600
28,000	520
34,000	560
30,000	530
32,500	530
31,000	590

Source: Graduate College of Business Administration, University of Florida.

10.8 The data in Table 3.7 are reproduced here. Use inverse prediction to estimate the distance from nearest fire station, x, for a residential fire that caused $18,200 in damages. Construct a 90% prediction interval for x.

DISTANCE FROM FIRE STATION x, miles	FIRE DAMAGE y, thousands of dollars
3.4	26.2
1.8	17.8
4.6	31.3
2.3	23.1
3.1	27.5
5.5	36.0
.7	14.1
3.0	22.3
2.6	19.6
4.3	31.3
2.1	24.0
1.1	17.3
6.1	43.2
4.8	36.4
3.8	26.1

10.9 A pharmaceutical company has developed a new drug designed to reduce a smoker's reliance on tobacco. Since certain dosages of the drug may reduce one's pulse rate to dangerously low levels, the product testing division of the pharmaceutical company wants to model the relationship between decrease in pulse rate y (beats/minute) and dosage x (cubic centimeters). Different dosages of the drug were administered to eight randomly selected patients, and 30 minutes later the decrease in each patient's pulse rate was recorded, with the results given in the table.

PATIENT	DOSAGE x, cubic centimeters	DECREASE IN PULSE RATE y, beats/minute
1	2.0	12
2	4.0	20
3	1.5	6
4	1.0	3
5	3.0	16
6	3.5	20
7	2.5	13
8	3.0	18

a. Fit the straight-line model $E(y) = \beta_0 + \beta_1 x$ to the data.

b. Conduct a test for model adequacy. Use $\alpha = .05$.

c. Use inverse prediction to estimate the appropriate dosage x to administer in order to reduce a patient's pulse rate $y = 10$ beats per minute. Construct an approximate 95% prediction interval for x.

SECTION 10.4

WEIGHTED LEAST SQUARES

Consider the general linear model

$$y = \beta_0 + \beta_1 x_1 + \beta_2 x_2 + \cdots + \beta_k x_k + \varepsilon$$

To obtain the least squares estimates of the unknown β parameters, recall (from Section 4.3) that we minimize the quantity

$$\text{SSE} = \sum_{i=1}^{n} (y_i - \hat{y}_i)^2 = \sum_{i=1}^{n} [y_i - (\hat{\beta}_0 + \hat{\beta}_1 x_{1i} + \hat{\beta}_2 x_{2i} + \cdots + \hat{\beta}_k x_{ki})]^2$$

with respect to $\hat{\beta}_0, \hat{\beta}_1, \ldots, \hat{\beta}_k$.

The least squares criterion weighs each observation equally in determining the estimates of the β's. Sometimes we will want to weigh some observations more heavily than others. To do this we minimize

$$\text{WSSE} = \sum_{i=1}^{n} w_i (y_i - \hat{y}_i)^2$$

$$= \sum_{i=1}^{n} w_i [y_i - (\hat{\beta}_0 + \hat{\beta}_1 x_{1i} + \hat{\beta}_2 x_{2i} + \cdots + \hat{\beta}_k x_{ki})]^2$$

where w_i is the weight assigned to the ith observation. This procedure is known as **weighted least squares** and the resulting parameter estimates are called **weighted least squares estimates**. [Note that the ordinary least squares procedure assigns a weight of $w_i = 1$ to each observation.]

Weighted least squares has applications in the following two areas:

1. Stabilizing the variance of ε in order to satisfy the standard regression assumption of homoscedasticity

2. Dampening the influence of outlying observations on the regression analysis

Although the two applications are related, our discussion of weighted least squares in this section is geared toward the first application.

The regression routines of most statistical computer program packages have options for conducting a weighted least squares analysis. However, the weights w_i must be specified. When using weighted least squares as a variance-stabilizing technique, the weight for the ith observation should be the reciprocal of the variance of that observation's error term, σ_i^2, i.e.,

$$w_i = \frac{1}{\sigma_i^2}$$

In this manner, observations with larger error variances will receive less weight (and hence have less influence on the analysis) than observations with smaller error variances.

In practice, the actual variances σ_i^2 will usually be unknown. Fortunately, in many business applications, the error variance σ_i^2 is proportional to one or more of the levels of the independent variables. This fact will allow us to determine the appropriate weights to use. For example, in a simple linear regression problem suppose we know that the error variance σ_i^2 increases proportionally with the value of the independent variable x_i, i.e.,

$$\sigma_i^2 = kx_i$$

where k is some unknown constant. Then the appropriate (albeit unknown) weight to use is

$$w_i = \frac{1}{kx_i}$$

Fortunately, it can be shown (proof omitted) that k can be ignored and the weights can be assigned as follows:

$$w_i = \frac{1}{x_i}$$

If the functional relationship between σ_i^2 and x_i is not known prior to conducting the analysis, the weights can be estimated based on the results of an ordinary (unweighted) least squares fit. For example, in simple linear regression, one approach is to divide the regression residuals into several groups of approximately equal size based on the value of the independent variable x and calculate the variance of the observed residuals in each group. An examination of the relationship between the residual variances and several different functions of x (such as x, x^2, and $\sqrt{x}$) may reveal the appropriate weights to use.

EXAMPLE 10.2

A Department of Transportation (DOT) official is investigating the possibility of collusive bidding among the state's road construction contractors. One aspect of the investigation involves a comparison of the winning (lowest) bid price y on a job with the length x of new road construction, a measure of job size. The data listed in Table 10.3 were supplied by the DOT for a sample of 11 new road construction jobs with approximately the same number of bidders.

	LENGTH			LENGTH	
TABLE 10.3					
Sample Data for New	JOB	OF ROAD x, miles	WINNING BID PRICE y, $ thousands	JOB	OF ROAD x, miles	WINNING BID PRICE y, $ thousands

TABLE 10.3
Sample Data for New
Road Construction Jobs,
Example 10.2

JOB	LENGTH OF ROAD x, miles	WINNING BID PRICE y, $ thousands	JOB	LENGTH OF ROAD x, miles	WINNING BID PRICE y, $ thousands
1	2.0	10.1	7	7.0	71.1
2	2.4	11.4	8	11.5	132.7
3	3.1	24.2	9	10.9	108.0
4	3.5	26.5	10	12.2	126.2
5	6.4	66.8	11	12.6	140.7
6	6.1	53.8			

a. Use the method of least squares to fit the straight-line model

$$E(y) = \beta_0 + \beta_1 x$$

b. Calculate and plot the regression residuals against x. Do you detect any evidence of heteroscedasticity?

c. Use the method described in the preceding paragraph to find the approximate weights necessary to stabilize the error variances with weighted least squares.

d. Carry out the weighted least squares analysis using the weights determined in part c.

e. The **weighted least squares residuals** are defined as

$$\sqrt{w_i}(y_i - \hat{y}_i)$$

where $\hat{y}_i$ is the predicted value from the weighted least squares fit and w_i is the weight. Plot the weighted least squares residuals against x to determine whether the variances have stabilized.

SOLUTION

a. The simple linear regression analysis was conducted using the SAS regression package. The SAS printout appears in Figure 10.6. The least squares line, obtained from the printout, is

$$\hat{y} = -15.11 + 12.07x$$

Note that the model is statistically useful (reject H_0: $\beta_1 = 0$) at $p = .0001$.

b. The regression residuals are calculated and reported in the bottom portion of the SAS printout shown in Figure 10.6 (page 570). A plot of the residuals against the predictor variable x is shown in Figure 10.7. The residual plot clearly shows that the residual variance increases as length of road x increases, strongly suggesting the presence of heteroscedasticity. A procedure such as weighted least squares is needed to stabilize the variances.

FIGURE 10.6 SAS Printout for Least Squares Fit of Straight-Line Model, Example 10.2

```
DEP VARIABLE: Y
                                         ANALYSIS OF VARIANCE

                              SUM OF          MEAN
            SOURCE    DF      SQUARES        SQUARE       F VALUE     PROB>F

            MODEL      1    24557.87884   24557.87884    850.451     0.0001
            ERROR      9      259.88662     28.87629072
            C TOTAL   10    24817.76545

                    ROOT MSE      5.373666      R-SQUARE     0.9895
                    DEP MEAN     70.13636       ADJ R-SQ     0.9884
                    C.V.          7.661741

                                      PARAMETER ESTIMATES

                            PARAMETER       STANDARD      T FOR H0:
            VARIABLE   DF     ESTIMATE         ERROR      PARAMETER=0    PROB > :T:

            INTERCEP    1   -15.11237262    3.34221489      -4.522       0.0014
            X           1    12.06867566    0.41384231      29.162       0.0001


                                                     PREDICT
                OBS            ID       ACTUAL        VALUE       RESIDUAL

                 1             2       10.1000       9.0250       1.0750
                 2            2.4      11.4000      13.8524      -2.4524
                 3            3.1      24.2000      22.3005       1.8995
                 4            3.5      26.5000      27.1280      -0.6280
                 5            6.4      66.8000      62.1272       4.6728
                 6            6.1      53.8000      58.5065      -4.7065
                 7             7       71.1000      69.3684       1.7316
                 8           11.5     132.7         123.7         9.0226
                 9           10.9     108.0         116.4        -8.4362
                10           12.2     126.2         132.1        -5.9255
                11           12.6     140.7         137.0         3.7471

    SUM OF RESIDUALS         -2.86438E-14
    SUM OF SQUARED RESIDUALS    259.8866
```

c. In order to apply weighted least squares, we first need to determine the weights. Since it is not clear what function of x the error variance is proportional to, we will apply the procedure described previously to estimate the weights.

First, we need to divide the data into several groups according to the value of the independent variable x. Ideally, we want to form one group of data points for each different value of x. However, unless each value of x is replicated, not all of the group residual variances can be calculated. Therefore, we resort to grouping the data according to "nearest neighbors." One choice would be to use three groups, $2 \leq x \leq 4$, $6 \leq x \leq 7$, and $10 \leq x \leq 13$. These groups have approximately the same numbers of observations (namely, 4, 3, and 4 observations, respectively).

Next, we calculate the sample variance s_i^2 of the residuals included in each group. The three residual variances are given in Table 10.4. These variances are compared to three different functions of $\bar{x}$ ($\bar{x}$, $\bar{x}^2$, and $\sqrt{\bar{x}}$), as shown in Table 10.4 (page 572), where $\bar{x}$ is the mean road length x for each group.

Note that the ratio $s_i^2/\bar{x}^2$ yields a value near .5 for each of the three groups. This result suggests that the residual variance of each group is proportional to $\bar{x}^2$, i.e.,

$$\sigma_i^2 = k\bar{x}_i^2, \quad i = 1, 2, 3$$

FIGURE 10.7 Plot of Least Squares Residuals Versus x, Example 10.2

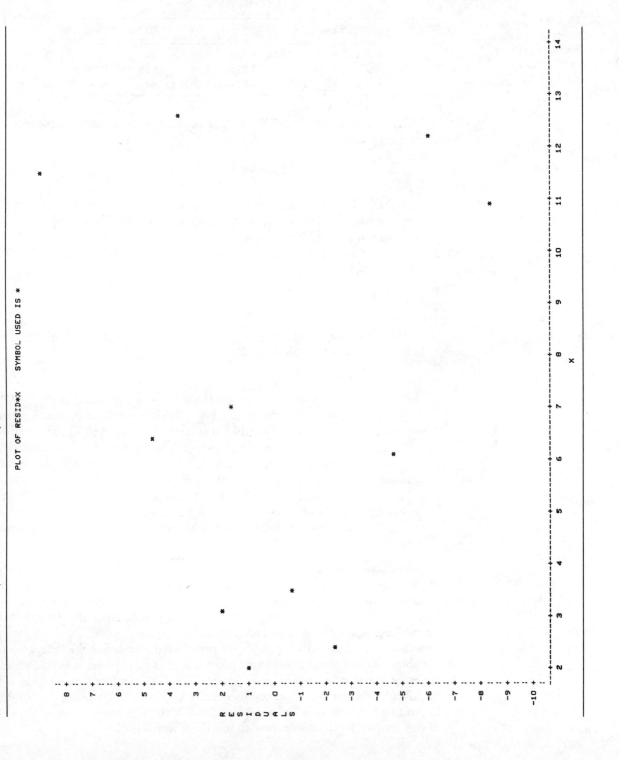

TABLE 10.4

Comparison of Residual Variances to Three Functions of $\bar{x}$, Example 10.2

GROUP	RANGE OF x	$\bar{x}_i$	s_i^2	$s_i^2/\bar{x}_i$	$s_i^2/\bar{x}_i^2$	$s_i^2/\sqrt{\bar{x}_i}$
1	$2 \leq x \leq 4$	2.75	3.722	1.353	.492	2.244
2	$6 \leq x \leq 7$	6.5	23.016	3.541	.545	9.028
3	$10 \leq x \leq 13$	11.8	67.031	5.681	.481	19.514

where k is approximately .5. Thus, a reasonable approximation to the weight for each group is

$$w_i = \frac{1}{\bar{x}_i^2}$$

With this weighting scheme, observations associated with large values of length of road x will have less influence on the regression residuals than observations associated with smaller values of x.

d. A weighted least squares analysis was conducted on the data in Table 10.3 using the weights

$$w_i = \frac{1}{\bar{x}_i^2}$$

The weighted least squares estimates are shown in the SAS printout reproduced in Figure 10.8. The prediction equation is

$$\hat{y} = -15.274 + 12.120x$$

Note that the test of model adequacy, H_0: $\beta_1 = 0$, is significant at $p = .0001$. Also, the standard error of the model, $s = \sqrt{\text{MSE}}$, is significantly smaller than the value of s for the unweighted least squares analysis (.669 as compared to 5.37). This last result is expected because, in the presence of heteroscedasticity, the unweighted least squares estimates are subject to greater sampling error than the weighted least squares estimates.

e. A plot of the weighted least squares residuals against x is shown in Figure 10.9 (page 574). The lack of a discernible pattern in the residual plot suggests that the weighted least squares procedure has corrected the problem of unequal variances.   ■

Before concluding this section, we mention that the "nearest neighbor" technique, illustrated in Example 10.2, will not always be successful in finding the optimal or near-optimal weights in weighted least squares. First, it may not be easy to identify the appropriate groupings of data points, especially if more than one independent variable is included in the regression. Second, the relationship between the residual variance and some preselected function of the independent variables may not reveal a consistent pattern over groups. In other words, unless the right function (or approximate function) of x is examined, the weights will be difficult to determine. More sophisticated techniques for choosing the weights in weighted least squares are available. Consult the references given at the end of this chapter for details on how to use these techniques.

FIGURE 10.8 SAS Printout of Weighted Least Squares Fit for Straight-Line Model, Example 10.2

DEP VARIABLE: Y

ANALYSIS OF VARIANCE

SOURCE	DF	SUM OF SQUARES	MEAN SQUARE	F VALUE	PROB>F
MODEL	1	457.47787	457.47787	1021.773	0.0001
ERROR	9	4.02956477	0.44772942		
C TOTAL	10	461.50743			

ROOT MSE	0.6691259	R-SQUARE	0.9913	
DEP MEAN	28.20272	ADJ R-SQ	0.9903	
C.V.	2.372558			

PARAMETER ESTIMATES

VARIABLE	DF	PARAMETER ESTIMATE	STANDARD ERROR	T FOR H0: PARAMETER=0	PROB > :T:
INTERCEP	1	-15.27436091	1.60067934	-9.542	0.0001
X	1	12.12037303	0.37917418	31.965	0.0001

OBS	ID	ACTUAL	PREDICT VALUE	RESIDUAL
1	2	10.1000	8.9664	1.1336
2	2.4	11.4000	13.8145	-2.4145
3	3.1	24.2000	22.2988	1.9012
4	3.5	26.5000	27.1469	-0.6469
5	6.4	66.8000	62.2960	4.5040
6	6.1	53.8000	58.6599	-4.8599
7	7	71.1000	69.5683	1.5317
8	11.5	132.7	124.1	8.5901
9	10.9	108.0	116.8	-8.8377
10	12.2	126.2	132.6	-6.3942
11	12.6	140.7	137.4	3.2577

SUM OF RESIDUALS   -421.735
SUM OF SQUARED RESIDUALS   30358.76

| | | | | | | | | | | | | | |

EXERCISES 10.10–10.12

10.10 Consider the straight-line model $y_i = \beta_0 + \beta_1 x_i + \varepsilon_i$. Give the appropriate weights w_i to use in a weighted least squares regression if the variance of the random error ε_i, i.e., σ_i^2, is proportional to:
a. x_i^2 **b.** $\sqrt{x_i}$ **c.** x_i

d. $\dfrac{1}{n_i}$,   where n_i is the number of observations at level x_i **e.** $\dfrac{1}{x_i}$

10.11 A machine that mass produces rubber gaskets can be set at one of three different speeds: 100, 150, or 200 gaskets per minute. As part of a quality control study, the machine was monitored several different times at each of the three speeds and the number of defectives produced per hour was recorded. The data are provided in the table on page 575. Since the number of defectives (y) is thought to be linearly related to speed (x), the following straight-line model is proposed:

$$y = \beta_0 + \beta_1 x + \varepsilon$$

a. Fit the model using the method of least squares. Is there evidence that the model is useful for predicting y? Test using $\alpha = .05$.
b. Plot the residuals from the least squares model against x. What does the plot reveal about the standard least squares assumption of homoscedasticity?

FIGURE 10.9 Plot of Weighted Residuals Versus x, Example 10.2

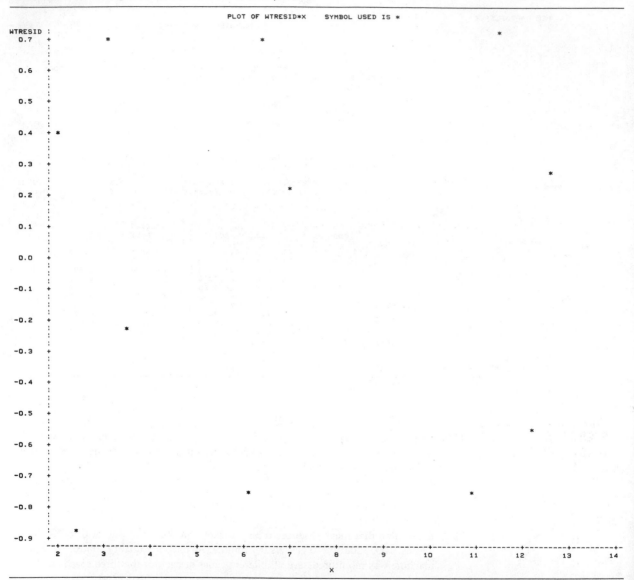

c. Estimate the appropriate weights to use in a weighted least squares regression. [*Hint:*   Calculate the variance of the least squares residuals at each level of x.]

d. Refit the model using weighted least squares. Compare the standard deviation of the weighted least squares slope to the standard deviation of the unweighted least squares slope.

e. Plot the weighted residuals against x to determine whether weighted least squares has corrected the problem of unequal variances.

MACHINE SPEED x, gaskets per minute	NUMBER OF DEFECTIVES y	MACHINE SPEED x, gaskets per minute	NUMBER OF DEFECTIVES y
100	15	150	35
100	23	150	24
100	11	200	26
100	14	200	48
100	18	200	27
150	19	200	38
150	29	200	30
150	20		

10.12 Refer to the data on salary (y) and years of experience (x) for 50 executives, given in Table 6.7. (The data are reproduced here for convenience.) Recall that the least squares fit of the quadratic model $E(y) = \beta_0 + \beta_1 x + \beta_2 x^2$ yielded regression residuals with unequal variances (see Figure 6.13) Apply the method of weighted least squares to correct this problem. [*Hint*: Estimate the weights using the "nearest neighbor" technique outlined in this section.]

YEARS OF EXPERIENCE x	SALARY y	YEARS OF EXPERIENCE x	SALARY y
7	$26,075	28	$64,785
28	79,370	26	61,581
23	65,726	27	70,678
18	41,983	20	51,301
19	62,309	18	39,346
15	41,154	1	24,833
24	53,610	26	65,929
13	33,697	20	41,721
2	22,444	26	82,641
8	32,562	28	99,139
20	43,076	23	52,624
21	56,000	17	50,594
18	58,667	25	53,272
7	22,210	26	65,343
2	20,521	19	46,216
18	49,727	16	54,288
11	33,233	3	20,844
21	43,628	12	32,586
4	16,105	23	71,235
24	65,644	20	36,530
20	63,022	19	52,745
20	47,780	27	67,282
15	38,853	25	80,931
25	66,537	12	32,303
25	67,447	11	38,371

SECTION 10.5

**MODELING
QUALITATIVE
DEPENDENT
VARIABLES**

For all models discussed in the previous sections of this text, the response (dependent) variable y is a *quantitative* variable. In this section we consider models for which the response y is a **qualitative variable at two levels**, or, as it is sometimes called, a **binary variable**.

For example, an entrepreneur may want to relate the success or failure of a new business to the characteristics (such as age, years of experience, and years of education) of the owner. The value of the response of interest to the entrepreneur is either *yes*, the new business is a success, or *no*, the new business is a failure. (A success implies the business did not fail.) Similarly, a state attorney general investigating collusive practices among bidders for road construction contracts may want to determine what contract-related variables (such as number of bidders, bid amount, and cost of materials) are useful indicators of whether a bid is fixed (i.e., whether the bid price is intentionally set higher than the fair market value). Here, the value of the response variable is either *fixed* bid or *competitive* bid.

Just as with qualitative independent variables, we use **dummy** (i.e., coded 0–1) **variables** to represent the qualitative response variable. For example, the response of interest to the entrepreneur is recorded as

$$y = \begin{cases} 1 & \text{if new business a success} \\ 0 & \text{if new business a failure} \end{cases}$$

where the assignment of 0 and 1 to the two levels is arbitrary. The general linear model takes the usual form

$$y = \beta_0 + \beta_1 x_1 + \beta_2 x_2 + \cdots + \beta_k x_k + \varepsilon$$

However, when the response is binary, the expected response

$$E(y) = \beta_0 + \beta_1 x_1 + \beta_2 x_2 + \cdots + \beta_k x_k$$

has a special meaning. It can be shown* that $E(y) = p$, where p is the probability that $y = 1$ for given values of $x_1, x_2, \ldots, x_k$. Thus, for the entrepreneur, the mean response $E(y)$ represents the probability that a new business with certain owner-related characteristics will be a success.

When the ordinary least squares approach is used to fit models with a binary response, several well-known problems are encountered. A discussion of these problems and their solutions follows.

PROBLEM 1

Nonnormal errors   The standard least squares assumption of normal errors is violated since the response y and hence, the random error ε, can take on only two values. To see this, consider the simple model $y = \beta_0 + \beta_1 x + \varepsilon$. Then we can write

$$\varepsilon = y - (\beta_0 + \beta_1 x)$$

Thus, when $y = 1$, $\varepsilon = 1 - (\beta_0 + \beta_1 x)$ and when $y = 0$, $\varepsilon = -\beta_0 - \beta_1 x$.

---

*The result is a straightforward application of the expectation theorem for a random variable. Let $p = P(y = 1)$ and $1 - p = P(y = 0)$, $0 \le p \le 1$. Then, by definition, $E(y) = \Sigma_y \, y_i \cdot p(y) = (1)P(y = 1) + (0)P(y = 0) = P(y = 1) = p$. Students familiar with discrete random variables will recognize y as the **Bernoulli random variable**, i.e., a binomial random variable with $n = 1$.

When the sample size n is large, however, any inferences derived from the least squares prediction equation remain valid in most practical situations even though the errors are nonnormal.*

PROBLEM 2

Unequal variances It can be shown[†] that the variance σ^2 of the random error is a function of p, the probability that the response y equals 1. Specifically,

$$\sigma^2 = V(\varepsilon) = p(1 - p)$$

Since, for the general linear model,

$$p = E(y) = \beta_0 + \beta_1 x_1 + \beta_2 x_2 + \cdots + \beta_k x_k$$

this implies that σ^2 is not constant and, in fact, depends on the values of the independent variables; hence, the standard least squares assumption of equal variances is also violated. One solution to this problem is to use weighted least squares (see Section 10.4), where the weights are inversely proportional to σ^2, i.e.,

$$w_i = \frac{1}{\sigma_i^2} = \frac{1}{p_i(1 - p_i)}$$

Unfortunately, the true proportion

$$p_i = E(y_i) = \beta_0 + \beta_1 x_{1i} + \beta_2 x_{2i} + \cdots + \beta_k x_{ki}$$

is unknown since $\beta_0, \beta_1, \ldots, \beta_k$ are unknown population parameters. However, a technique called **two-stage least squares** can be applied to circumvent this difficulty. Two-stage least squares, as its name implies, involves conducting an analysis in two steps:

STAGE 1 Fit the regression model using the *ordinary least squares* procedure and obtain the predicted values $\hat{y}_i$, $i = 1, 2, \ldots, n$. Recall that $\hat{y}_i$ estimates p_i for the binary model.

STAGE 2 Refit the regression model using *weighted least squares*, where the estimated weights are calculated as follows:

$$w_i = \frac{1}{\hat{y}_i(1 - \hat{y}_i)}$$

Further iterations—revising the weights at each step—can be performed if desired. In most practical problems, however, the estimates of p_i obtained in stage 1 are adequate for use in weighted least squares.

---

*This property is due to the asymptotic normality of the least squares estimates of the model parameters under very general conditions.

[†]Using the properties of expected values with the Bernoulli random variable, we have $V(y) = E(y^2) - [E(y)]^2 = \Sigma y^2 \cdot p(y) - (p)^2 = (1)^2 P(y = 1) + (0)^2 P(y = 0) - p^2 = p - p^2 = p(1 - p)$. Since in regression, $V(\varepsilon) = V(y)$, the result follows.

PROBLEM 3

Restricting the predicted response to be between 0 and 1   Since the predicted value $\hat{y}$ estimates $E(y) = p$, the probability that the response y equals 1, we would like $\hat{y}$ to have the property that $0 \le \hat{y} \le 1$. There is no guarantee, however, that the regression analysis will always yield predicted values in this range. Thus, the regression may lead to nonsensical results, i.e., negative estimated probabilities or predicted probabilities greater than 1. To avoid this problem, you may want to fit a model with a mean response function $E(y)$ that automatically falls between 0 and 1. (We consider one such model in the next section.)

In summary, the purpose of this section has been to identify some of the problems resulting from fitting a linear model with a binary response and to suggest ways in which to circumvent these problems. Another approach is to fit a model specially designed for a binary response, called a **logistic model**. Logistic models are the subject of Section 10.6.

EXERCISES
10.13–10.15

10.13 Discuss the problems associated with fitting a model where the response y is recorded as 0 or 1.

10.14 A retailer of home personal computers (PCs) conducted a study to relate PC ownership with annual income of heads of households. Data collected for a random sample of 20 households were used to fit the straight-line model

$$E(y) = \beta_0 + \beta_1 x$$

where

$$y = \begin{cases} 1 & \text{if own PC} \\ 0 & \text{if not} \end{cases}$$

x = Annual income (in dollars)

The data are shown in the accompanying table. Fit the model using two-stage least squares. Is the model useful for predicting y? Test using $\alpha = .05$.

HOUSEHOLD	y	x	HOUSEHOLD	y	x
1	0	$16,300	11	1	$22,400
2	0	11,200	12	0	10,600
3	0	36,500	13	0	21,400
4	1	21,700	14	0	8,300
5	1	40,200	15	1	27,500
6	0	12,400	16	0	15,700
7	0	15,000	17	0	12,100
8	0	9,200	18	1	59,600
9	1	36,700	19	1	20,200
10	0	62,000	20	0	33,100

10.15 Suppose you are investigating allegations of sex discrimination in the hiring practices of a particular firm. An equal-rights group claims that females are less likely to be hired than males with the same background, experience, and other qualifications.

Data collected on 28 former applicants (shown in the table) will be used to fit the model:

$$E(y) = \beta_0 + \beta_1 x_1 + \beta_2 x_2 + \beta_3 x_3$$

where

$$y = \begin{cases} 1 & \text{if hired} \\ 0 & \text{if not} \end{cases}$$

x_1 = years of higher education   (4, 6, or 8)

x_2 = years of experience

$$x_3 = \begin{cases} 1 & \text{if male applicant} \\ 0 & \text{if female applicant} \end{cases}$$

HIRING STATUS y	EDUCATION x_1, years	EXPERIENCE x_2, years	SEX x_3	HIRING STATUS y	EDUCATION x_1, years	EXPERIENCE x_2, years	SEX x_3
0	6	2	0	1	4	5	1
0	4	0	1	0	6	4	0
1	6	6	1	0	8	0	1
1	6	3	1	1	6	1	1
0	4	1	0	0	4	7	0
1	8	3	0	0	4	1	1
0	4	2	1	0	4	5	0
0	4	4	0	0	6	0	1
0	6	1	0	1	8	5	1
1	8	10	0	0	4	9	0
0	4	2	1	0	8	1	0
0	8	5	0	0	6	1	1
0	4	2	0	1	4	10	1
0	6	7	0	1	6	12	0

a. Interpret each of the β's in the multiple regression model.

b. If you have access to a computer program package, fit the multiple regression model using two-stage least squares.

c. Conduct a test of model adequacy. Use $\alpha = .05$.

d. Is there sufficient evidence to indicate that sex is an important predictor of hiring status? Test using $\alpha = .05$.

e. Calculate a 95% confidence interval for the mean response $E(y)$ when $x_1 = 4$, $x_2 = 3$, and $x_3 = 0$. Interpret the interval.

SECTION 10.6

LOGISTIC REGRESSION

Often, the relationship between a qualitative binary response y and a single predictor variable x is curvilinear. One particular curvilinear pattern frequently encountered in practice is the S-shaped curve shown in Figure 10.10 (page 580). A model which accounts for this type of curvature is the **logistic** (or **logit**) **model,**

$$E(y) = \frac{\exp(\beta_0 + \beta_1 x)}{1 + \exp(\beta_0 + \beta_1 x)}$$

FIGURE 10.10

Graph of $E(y)$ for the Logistic Model

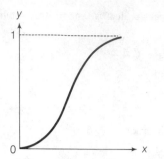

The logistic model was originally developed for use in **survival analysis**, where the response y is typically measured as 0 or 1, depending on whether or not the experimental unit (for example, a patient) "survives." Note that the curve shown in Figure 10.10 has asymptotes at 0 and 1—that is, the mean response $E(y)$ can never fall below 0 or above 1. Thus, the logistic model assures that the estimated response $\hat{y}$ (i.e., the estimated probability that $y = 1$) lies between 0 and 1.

In general, the logistic model can be written

$$E(y) = \frac{\exp(\beta_0 + \beta_1 x_1 + \beta_2 x_2 + \cdots + \beta_k x_k)}{1 + \exp(\beta_0 + \beta_1 x_1 + \beta_2 x_2 + \cdots + \beta_k x_k)}$$

Note that the general logistic model is not a linear function of the β parameters (see Section 4.1). Obtaining the parameter estimates of a **nonlinear regression model**, such as the logistic model, is a numerically tedious process and often requires sophisticated computer programs. In this section we briefly discuss two approaches to the problem, and give an example of a computer printout for the second.

1. *Least squares estimation using a transformation*   One method of fitting the model involves a transformation on the mean response $E(y)$. Recall (from Section 10.5) that for a binary response, $E(y) = p$, where p denotes the probability that $y = 1$. Then the logistic model

$$p = \frac{\exp(\beta_0 + \beta_1 x_1 + \cdots + \beta_k x_k)}{1 + \exp(\beta_0 + \beta_1 x_1 + \cdots + \beta_k x_k)}$$

implies (proof omitted) that

$$\ln\left(\frac{p}{1-p}\right) = \beta_0 + \beta_1 x_1 + \cdots + \beta_k x_k$$

Set

$$p^* = \ln\left(\frac{p}{1-p}\right)$$

The transformed logistic model

$$p^* = \beta_0 + \beta_1 x_1 + \cdots + \beta_k x_k$$

is now linear in the β's and the method of least squares can be applied.

Although the transformation succeeds in linearizing the response function, two other problems remain. First, since the true probability p is unknown, the values of the response p^*, necessary for input into the regression, are also unknown. To carry out the least squares analysis, we must obtain estimates of p^* for each combination of the independent variables. A good choice is the estimator

$$p^* = \ln\left(\frac{\hat{p}}{1 - \hat{p}}\right)$$

where $\hat{p}$ is the sample proportion of 1's for the particular combination of x's. To obtain these estimates, however, *we must have replicated observations of the response y at each combination of the levels of the independent variables.* Thus, the least squares transformation approach is limited to replicated experiments, which occur infrequently in a practical business setting.

The second problem concerns the problem of unequal variances. The transformed logistic model yields error variances that are inversely proportional to $p(1 - p)$. Since p, or $E(y)$, is a function of the independent variables, the regression errors are heteroscedastic. To stabilize the variances, weighted least squares should be used. This technique also requires that replicated observations be available for each combination of the x's and, in addition, that the number of observations at each combination be relatively large. If the experiment is replicated, with n_j (large) observations at each combination of the levels of the independent variables, then the appropriate weights to use are

$$w_j = n_j \hat{p}_j (1 - \hat{p}_j)$$

where

$$\hat{p}_j = \frac{\text{Number of 1's for combination } j \text{ of the } x\text{'s}}{n_j}$$

2. *Maximum likelihood estimation* Estimates of the β parameters in the logistic model also can be obtained by applying a common statistical technique, called **maximum likelihood estimation**. Like the least squares estimators, the maximum likelihood estimators have certain desirable properties.* (In fact, when the errors of a linear regression model are normally distributed, the least squares estimates and maximum likelihood estimates are equivalent.) Many of the available statistical computer program packages employ maximum likelihood estimation to fit logistic regression models. Therefore, one practical advantage of using the maximum likelihood method (rather than the transformation approach) to fit logistic regression models is that computer programs are readily available. Another advantage is that the data need not be replicated to apply maximum likelihood estimation.

*For details on how to obtain maximum likelihood estimators and their distributional properties, consult the references given at the end of the chapter.

The maximum likelihood estimates of the parameters of a logistic model have distributional properties that are different from the standard F- and t-distributions of least squares regression. Under certain conditions, the test statistics for testing individual parameters and overall model adequacy have approximate **chi-square** (χ^2) **distributions**. The χ^2-distribution is similar to the F-distribution in that it depends on degrees of freedom and is nonnegative, as shown in Figure 10.11. We illustrate the application of maximum likelihood estimation for logistic regression with an example.

FIGURE 10.11

Several Chi-Square Probability Distributions

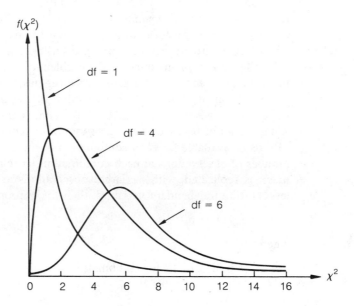

Consider the problem of collusive (i.e., noncompetitive) bidding among road construction contractors. Recall (from Section 10.5) that contractors sometimes scheme to set bid prices higher than the fair market (or competitive) price. Suppose an investigator has obtained information on the bid status (fixed or competitive) for a sample of 31 contracts. In addition, two variables thought to be related to bid status are also recorded for each contract: number of bidders x_1 and the difference between the winning (lowest) bid and the estimated competitive bid (called the engineer's estimate) where the difference x_2 is measured as a percentage of the estimate. The data appear in Table 10.5, with the response y recorded as follows:

$$y = \begin{cases} 1 & \text{if fixed bid} \\ 0 & \text{if competitive bid} \end{cases}$$

TABLE 10.5 Data for a Sample of 31 Road Construction Bids

CONTRACT	BID STATUS y	NUMBER OF BIDDERS x_1	DIFFERENCE BETWEEN WINNING BID AND ENGINEER'S ESTIMATE x_2, %	CONTRACT	BID STATUS y	NUMBER OF BIDDERS x_1	DIFFERENCE BETWEEN WINNING BID AND ENGINEER'S ESTIMATE x_2, %
1	1	4	19.2	17	0	10	6.6
2	1	2	24.1	18	1	5	−2.5
3	0	4	−7.1	19	0	13	24.2
4	1	3	3.9	20	0	7	2.3
5	0	9	4.5	21	1	3	36.9
6	0	6	10.6	22	0	4	−11.7
7	0	2	−3.0	23	1	2	22.1
8	0	11	16.2	24	1	3	10.4
9	1	6	72.8	25	0	2	9.1
10	0	7	28.7	26	0	5	2.0
11	1	3	11.5	27	0	6	12.6
12	1	2	56.3	28	1	5	18.0
13	0	5	−.5	29	0	3	1.5
14	0	3	−1.3	30	1	4	27.3
15	0	3	12.9	31	0	10	−8.4
16	0	8	34.1				

An appropriate model for $E(y)$ is the logistic model

$$E(y) = \frac{\exp(\beta_0 + \beta_1 x_1 + \beta_2 x_2)}{1 + \exp(\beta_0 + \beta_1 x_1 + \beta_2 x_2)}$$

The model was fit to the data using the logistic regression option of the SAS System package. The resulting printout is shown in Figure 10.12 (page 584).

The maximum likelihood estimates of β_0, β_1, and β_2 are given in Figure 10.12 under the column heading BETA. These estimates (shaded in the printout) are $\hat{\beta}_0 = 1.4212$, $\hat{\beta}_1 = -.7553$, and $\hat{\beta}_2 = .1122$. Therefore, the prediction equation for the probability of a fixed bid (i.e., the probability that $y = 1$) is

$$\hat{y} = \frac{\exp(1.4212 - .7553 x_1 + .1122 x_2)}{1 + \exp(1.4212 - .7553 x_1 + .1122 x_2)}$$

The standard errors of the β estimates are given under the column STD. ERROR and the ratios of the β estimates to their respective standard errors are given under the column CHI-SQUARE. As in regression with a linear model, this ratio provides a test statistic for testing the contribution of each variable to the model (i.e., for testing H_0: $\beta_i = 0$).* The observed significance

*In the logistic regression model, the ratio $\hat{\beta}_i/s_{\hat{\beta}_i}$ has an approximate χ^2-distribution with 1 degree of freedom. Consult the references for more details of the χ^2-distribution and its use in logistic regression.

FIGURE 10.12
SAS Printout for Logistic
Regression on Bid Status

```
                    LOGISTIC REGRESSION PROCEDURE

                      DEPENDENT VARIABLE: Y

           31 OBSERVATIONS
           19  Y      = 0
           12  Y      = 1
            0 OBSERVATIONS DELETED DUE TO MISSING VALUES

       -2 LOG LIKELIHOOD FOR MODEL CONTAINING INTERCEPT ONLY=  41.38

     MODEL CHI-SQUARE=   13.47 WITH   2 D.F.   (SCORE STAT.) P=0.0012.
     CONVERGENCE IN  7 ITERATIONS WITH  0 STEP HALVINGS   R= 0.593.
     MAX ABSOLUTE DERIVATIVE=0.4939D-07.              -2 LOG L=   22.84.
     MODEL CHI-SQUARE=   18.54 WITH   2 D.F.   (-2 LOG L.R.) P=0.0001.
```

VARIABLE	BETA	STD. ERROR	CHI-SQUARE	P	R
INTERCEPT	1.42119760	1.28676876	1.22	0.2694	
X1	-0.75533943	0.33880242	4.97	0.0258	-0.268
X2	0.11220495	0.05139319	4.77	0.0290	0.259

```
C=0.904       SOMER DYX=0.807        GAMMA=0.807        TAU-A=0.396
```

```
           PREDICTED PROBABILITIES AND 95% CONFIDENCE LIMITS
```

OBS	Y	X1	X2	_LOWER_	_P_	_UPPER_
1	1	4	19.2	0.3298	0.6351	0.8603
2	1	2	24.1	0.5363	0.9318	0.9938
3	0	4	-7.1	0.0104	0.0834	0.4402
4	1	3	3.9	0.1586	0.3996	0.7014
5	0	9	4.5	0.0002	0.0076	0.2684
6	0	6	10.6	0.0258	0.1277	0.4472
7	0	2	-3.0	0.1027	0.3951	0.7884
8	0	11	16.2	0.0001	0.0062	0.3684
9	1	6	72.8	0.3517	0.9937	1.0000
10	0	7	28.7	0.0614	0.3439	0.8078
11	1	3	11.5	0.3157	0.6096	0.8409
12	1	2	56.3	0.6967	0.9980	1.0000
13	0	5	-0.5	0.0125	0.0823	0.3879
14	0	3	-1.3	0.0745	0.2708	0.6314
15	0	3	12.9	0.3407	0.6463	0.8659
16	0	8	34.1	0.0317	0.3110	0.8618
17	0	10	6.6	0.0001	0.0045	0.2662
18	1	5	-2.5	0.0085	0.0669	0.3742
19	0	13	24.2	0.0000	0.0034	0.4575
20	0	7	2.3	0.0017	0.0264	0.3066
21	1	3	36.9	0.5473	0.9643	0.9983
22	0	4	-11.7	0.0041	0.0515	0.4162
23	1	2	22.1	0.5187	0.9161	0.9910
24	1	3	10.4	0.2946	0.5798	0.8201
25	0	2	9.1	0.3390	0.7174	0.9263
26	0	5	2.0	0.0200	0.1061	0.4079
27	0	6	12.6	0.0348	0.1549	0.4819
28	1	5	18.0	0.1787	0.4168	0.7013
29	0	3	1.5	0.1148	0.3370	0.6659
30	1	4	27.3	0.4005	0.8120	0.9654
31	0	10	-8.4	0.0000	0.0008	0.1586

levels of the tests (i.e., the p-values) are given under the column headed P. Note that both independent variables, number of bidders (x_1) and percentage difference (x_2), have p-values less than .03 (implying that we would reject H_0: $\beta_1 = 0$ and H_0: $\beta_2 = 0$ for $\alpha = .03$).

The test statistic for testing the overall adequacy of the logistic model, i.e., for testing H_0: $\beta_1 = \beta_2 = 0$, is given in the upper portion of the printout (shaded) as MODEL CHI-SQUARE $= 18.54$, with observed significance level (shaded) $P = 0.0001$.[*] Based on the p-value of the test, we can reject H_0 and conclude that at least one of the β coefficients is nonzero. Thus, the model is adequate for predicting bid status, y.

The statistic $R = 0.593$ (shaded in the upper portion of the printout), when squared, is similar to the multiple coefficient of determination R^2 of a standard regression analysis.[†] A loose interpretation is that the fitted logistic model explains approximately $R^2 = (.593)^2 = .352$ of the variation in bid status y.

Finally, the bottom portion of the printout gives predicted values and lower and upper 95% prediction limits for each observation used in the analysis in the columns titled __P__, __LOWER__, and __UPPER__, respectively. The 95% prediction interval for y for a contract with $x_1 = 3$ bidders and winning bid amount $x_2 = 11.5\%$ above the engineer's estimate is shaded on the printout. We estimate the probability of this particular contract being fixed to fall between .3157 and .8409. Note that all the predicted values and limits lie between 0 and 1, a property of the logistic model.

In summary, we have presented two approaches to fitting logistic regression models. If the data are replicated, you may want to apply the transformation approach. The maximum likelihood estimation approach can be applied to any data set, but you need access to a statistical computer program package (such as SAS, SPSS^x, or BMDP) with logistic regression procedures.

This section should be viewed only as an overview of logistic regression. Many of the details of fitting logistic regression models using either technique have been omitted. Before conducting a logistic regression analysis, we strongly recommend that you consult the references given at the end of this chapter.

| | | | | | | | | | | | | |

EXERCISES
10.16–10.17

10.16 If you have access to a computer with a logistic regression routine, fit the logistic model to the data given in Exercise 10.14. Request lower and upper 95% prediction limits for each observation used in the analysis.

10.17 If you have access to a computer with a logistic regression routine, fit the logistic model to the data given in Exercise 10.15. Request lower and upper 95% prediction limits for each observation used in the analysis.

[*]The test statistic has an approximate χ^2-distribution with $k = 2$ degrees of freedom, where k is the number of β parameters in the model (excluding β_0).

[†]The value of R^2 for logistic regression is not necessarily bounded by 0 and 1 and does not necessarily increase when variables are added to the model. Consult the references for more on the derivation and use of R in logistic regression.

RIDGE REGRESSION

When the sample data for regression exhibit multicollinearity, the least squares estimates of the β coefficients may be subject to extreme roundoff error as well as inflated standard errors (see Section 5.5). Since their magnitudes and signs may change considerably from sample to sample, the least squares estimates are said to be *unstable*. A technique developed for stabilizing the regression coefficients in the presence of multicollinearity is **ridge regression**.

Ridge regression is a modification of the method of least squares to allow *biased* estimators of the regression coefficients. At first glance, the idea of biased estimation may not seem very appealing. But consider the sampling distributions of two different estimators of a regression coefficient β, one unbiased and the other biased, portrayed in Figure 10.13.

FIGURE 10.13

Sampling Distributions of Two Estimators of a Regression Coefficient β

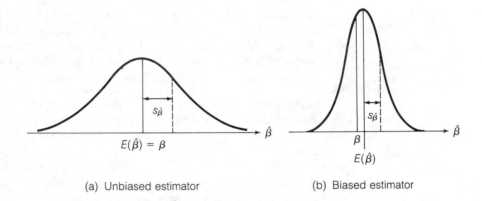

(a) Unbiased estimator   (b) Biased estimator

Figure 10.13(a) shows an unbiased estimator of β with a fairly large variance. In contrast, the estimator shown in Figure 10.13(b) has a slight bias but is much less variable. In this case, we would prefer the biased estimator over the unbiased estimator since it will lead to more precise estimates of the true β (i.e., narrower confidence intervals for β). One way to measure the "goodness" of an estimator of β is to calculate the **mean square error** of $\hat{\beta}$, denoted by $\text{MSE}(\hat{\beta})$, where $\text{MSE}(\hat{\beta})$ is defined as

$$\text{MSE}(\hat{\beta}) = E[(\hat{\beta} - \beta)^2]$$
$$= V(\hat{\beta}) + [E(\hat{\beta}) - \beta]^2$$

The difference $E(\hat{\beta}) - \beta$ is called the **bias** of $\hat{\beta}$. Therefore, $\text{MSE}(\hat{\beta})$ is just the sum of the variance of $\hat{\beta}$ and the squared bias:

$$\text{MSE}(\hat{\beta}) = V(\hat{\beta}) + (\text{Bias in } \hat{\beta})^2$$

Let $\hat{\beta}_{LS}$ denote the least squares estimate of β. Then, since $E(\hat{\beta}_{LS}) = \beta$, the bias is 0 and

$$\text{MSE}(\hat{\beta}_{LS}) = V(\hat{\beta}_{LS})$$

We have previously stated that the variance of the least squares regression coefficient, and hence $\text{MSE}(\hat{\beta}_{LS})$, will be quite large in the presence of multicollinearity. The idea behind ridge regression is to introduce a small amount of bias in the ridge estimator of β, denoted by $\hat{\beta}_R$, so that its mean square error is considerably smaller than the corresponding mean square error for least squares, i.e.,

$$\text{MSE}(\hat{\beta}_R) < \text{MSE}(\hat{\beta}_{LS})$$

In this manner, ridge regression will lead to narrower confidence intervals for the β coefficients, and hence, more stable estimates.

Although the mechanics of a ridge regression are beyond the scope of this text, we point out that some of the more sophisticated computer packages (including the SAS System) are now capable of conducting this type of analysis. To obtain the ridge regression coefficients, the user must specify the value of a biasing constant c, where $c \geq 0$.* Researchers have shown that as the value of c increases, the bias in the ridge estimates increases while the variance decreases. The idea is to choose c so that the total mean square error for the ridge estimators is smaller than the total mean square error for the least squares estimates. Although such a c exists, the optimal value, unfortunately, is unknown.

Various methods for choosing the value of c have been proposed. One commonly used graphical technique employs a **ridge trace**. Values of the estimated ridge regression coefficients are calculated for different values of c ranging from 0 to 1 and are plotted. The plots for each of the independent variables in the model are overlayed to form the ridge trace. An example of ridge trace for a model with three independent variables is shown in Figure 10.14 (page 588). Initially, the estimated coefficients may fluctuate dramatically as c is increased from 0 (especially if severe multicollinearity is present). Eventually, however, the ridge estimates will stabilize. After careful examination of the ridge trace, the analyst chooses the smallest value of c for which it appears that all the ridge estimates are stable. The choice of c, therefore, is subjective.

Once the value of c has been determined (using the ridge trace or some other analytical technique), the corresponding ridge estimates may be used in place of the least squares estimates. If the optimal (or near-optimal) value of c has been selected, the new estimates will have reduced variances (which lead to narrower confidence intervals for the β's). Also, some of the other problems associated with multicollinearity (e.g., incorrect signs on the β's) should have been corrected.

*In matrix notation, the ridge estimator $\hat{\beta}_R$ is calculated as follows:

$$\hat{\beta}_R = (\mathbf{X'X}) + c\mathbf{I})^{-1}\mathbf{X'Y}$$

When $c = 0$, the least squares estimator

$$\hat{\beta}_{LS} = (\mathbf{X'X})^{-1}\mathbf{X'Y}$$

is obtained. See Appendix A for details on the matrix mechanics of a regression analysis.

FIGURE 10.14

Ridge Trace of β Coefficients of a Model With Three Independent Variables

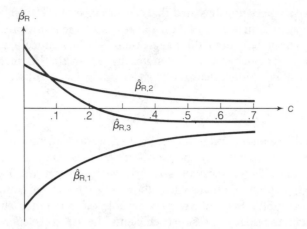

In conclusion, we caution that one should not assume that ridge regression is a panacea for multicollinearity or poor data. Although there are probably ridge regression estimates that are better than the least squares estimates when multicollinearity is present, the choice of the biasing constant c is crucial. Unfortunately, much of the controversy in ridge regression centers on how to find the optimal value of c. In addition, the exact distributional properties of the ridge estimators are unknown when c is estimated from the data. For these reasons, some statisticians recommend that ridge regression be used only as an exploratory data analysis tool for identifying unstable regression coefficients, and not for estimating parameters and testing hypotheses in a linear regression model.

SECTION 10.8

ROBUST REGRESSION

Consider the problem of fitting the linear regression model

$$y = \beta_0 + \beta_1 x_1 + \beta_2 x_2 + \cdots + \beta_k x_k + \varepsilon$$

by the method of least squares when the errors ε are nonnormal. In practice, moderate departures from the assumption of normality tend to have minimal effect on the validity of the least squares results (see Section 6.4). However, when the distribution of ε is **heavy-tailed** (longer-tailed) in comparison to the normal distribution, however, the method of least squares may not be appropriate. For example, the heavy-tailed error distribution shown in Figure 10.15 will most

FIGURE 10.15

Probability Distribution of ε: Normal Versus Heavy-Tailed

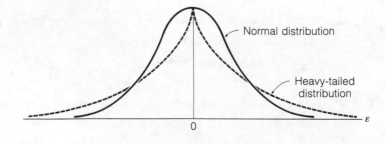

likely produce outliers with strong influence on the regression analysis. Furthermore, since they tend to "pull" the least squares fit too much in their direction, these outliers will have smaller than expected residuals and, consequently, are more difficult to detect.

Robust regression procedures are available for errors that follow a nonnormal distribution. In the context of regression, the term *robust* describes a technique that yields estimates for the β's that are nearly as good as the least squares estimates when the assumption of normality is satisfied, and significantly better for a heavy-tailed distribution. Robust regression is designed to dampen the effect of outlying observations that otherwise would exhibit strong influence on the analysis. This has the effect of leaving the residuals of influential observations large so that they may be more easily identified.

A number of different robust regression procedures exist and they fall into one of three general classes: **M-estimators**, **R-estimators**, and **L-estimators**. Of the three, robust techniques that produce *M-estimates* of the β coefficients receive the most attention in the literature.

M-estimates of the β coefficients are obtained by minimizing the quantity

$$\sum_{i=1}^{n} f(\hat{\varepsilon}_i)$$

where

$$\hat{\varepsilon}_i = y_i - (\hat{\beta}_0 + \hat{\beta}_1 x_{1i} + \hat{\beta}_2 x_{2i} + \cdots + \hat{\beta}_k x_{ki})$$

are the unobservable residuals and $f(\hat{\varepsilon}_i)$ is some function of the residuals. Note that the function $f(\hat{\varepsilon}_i) = \hat{\varepsilon}_i^2$ yields the ordinary least squares estimates since we are minimizing

$$\sum_{i=1}^{n} f(\hat{\varepsilon}_i) = \sum_{i=1}^{n} \hat{\varepsilon}_i^2$$

$$= \sum_{i=1}^{n} [y_i - (\hat{\beta}_0 + \hat{\beta}_1 x_{1i} + \hat{\beta}_2 x_{2i} + \cdots + \hat{\beta}_k x_{ki})]^2$$

$$= \text{SSE}$$

and, therefore, is appropriate when the errors are normal. For errors with heavier-tailed distributions, the analyst chooses some other function $f(\hat{\varepsilon}_i)$ that places less weight on the errors in the tails of the distribution. For example, the function $f(\hat{\varepsilon}_i) = |\hat{\varepsilon}_i|$ is appropriate when the errors follow the heavy-tailed distribution pictured in Figure 10.15. Since we are minimizing

$$\sum_{i=1}^{n} f(\hat{\varepsilon}_i) = \sum_{i=1}^{n} |\hat{\varepsilon}_i|$$

$$= \sum_{i=1}^{n} |y_i - (\hat{\beta}_0 + \hat{\beta}_1 x_{1i} + \hat{\beta}_2 x_{2i} + \cdots + \hat{\beta}_k x_{ki})|$$

the *M-estimators* of robust regression yield the estimates obtained from the **method of absolute deviations** (see Section 6.5).

The other types of robust estimation, R-estimation and L-estimation, take a different approach. R-estimators are obtained by minimizing the quantity

$$\sum_{i=1}^{n} [y_i - (\hat{\beta}_0 + \hat{\beta}_1 x_{1i} + \hat{\beta}_2 x_{2i} + \cdots + \hat{\beta}_k x_{ki})] R_i$$

where R_i is the rank of the ith residual when the residuals are placed in ascending order. L-estimation is similar to R-estimation in that it involves ordering of the data, but uses measures of location (such as the sample median) to estimate the regression coefficients.

The numerical techniques for obtaining robust estimates (M-, R-, or L-estimates) are quite complex and require sophisticated computer programs. At present, statistical computer program packages for robust regression are not widely available. However, the growing demand for packaged robust regression programs, especially M-estimation procedures, leads us to believe that these programs will be available in the near future.

Much of the current research in the area of robust regression is focused on the distributional properties of the robust estimators of the β coefficients. At present, there is little information available on robust confidence intervals, prediction intervals, and hypothesis testing procedures. For this reason, some researchers* recommend that robust regression be used in conjunction with and as a check on the method of least squares. If the results of the two procedures are substantially the same, use the least squares fit since confidence intervals and tests on the regression coefficients can be made. On the other hand, if the two analyses yield quite different results, use the robust fit to identify any influential observations. A careful examination of these data points may reveal the problem with the least squares fit.

SECTION 10.9

MODEL VALIDATION

Regression analysis is one of the most widely used statistical tools for estimation and prediction in the business world. All too frequently, however, a regression model deemed to be an adequate predictor of some response y performs poorly when applied in practice. For example, a model developed for forecasting new housing starts, although found to be statistically useful based on a test for overall model adequacy, may fail to take into account any extreme changes in future home mortgage rates due to new government policy. This points out an important problem. **There is no assurance that a model that fits the sample data well will be a successful predictor of y when applied to new data.** For this reason, it is important to assess the **validity** of the regression model in addition to its **adequacy** before using it in practice.

In the preceding chapters, we have presented several techniques for checking *model adequacy* (for example, tests of overall model adequacy, partial F-tests, residual analysis, influence diagnostics). In short, checking model adequacy involves determining whether the regression model adequately fits the *sample*

*See Montgomery and Peck (1982).

data. **Model validation**, however, involves an assessment of how the fitted regression model will perform in practice—that is, how successful it will be when applied to new or future data. A number of different model validation techniques have been proposed, several of which are briefly discussed in this section. You will need to consult the references for more details on how to apply some of these techniques.

1. *Examining the predicted values* Sometimes, the predicted values $\hat{y}$ of the fitted regression model can help to identify an invalid model. Nonsensical or unreasonable predicted values may indicate that the form of the model is incorrect or that the β coefficients are poorly estimated. For example, a binary response model may yield predicted probabilities that are negative or greater than 1. In this case, the user may want to consider a model that produces predicted values between 0 and 1 (such as a logistic model) in practice. On the other hand, if the predicted values of the fitted model all seem reasonable, the user should refrain from using the model in practice until further checks of model validity are carried out.

2. *Examining the estimated model parameters* Typically, the user of a regression model has some knowledge of the relative size and sign (positive or negative) of the model parameters. This information should be used as a check on the estimated β coefficients. Coefficients with signs opposite to what is expected or with unusually small or large values, or unstable coefficients (i.e., coefficients with large standard errors) forewarn that the final model may perform poorly when applied to new or different data.

3. *Collecting new data for prediction* One of the most effective ways of validating a regression model is to use the model to predict y for a new sample. By directly comparing the predicted values to the observed values of the new data, we can determine the accuracy of the predictions and use this information to assess how well the model performs "in practice."

   Several measures of model validity have been proposed for this purpose. One simple technique is to calculate the percentage of variability in the new data explained by the model, denoted $R^2_{prediction}$, and compare it to the coefficient of determination R^2 for the least squares fit. Let $y_1, y_2, \ldots, y_{n_1}$ represent the n_1 observations used to fit the final regression model and $y_{n_1+1}, y_{n_1+2}, \ldots, y_{n_1+n_2}$ represent the n_2 observations in the new data set. Then

$$R^2_{prediction} = 1 - \frac{\sum_{i=n_1+1}^{n_1+n_2} (y_i - \hat{y}_i)^2}{\sum_{i=n_1+1}^{n_1+n_2} (y_i - \bar{y})^2}$$

where $\hat{y}_i$ is the predicted value for the ith observation using the fitted model and $\bar{y}$ is the sample mean of the original data.* If $R^2_{prediction}$ compares favorably to R^2 from the least squares fit, we will have increased confidence in the

*Alternatively, the sample mean of the new data may be used.

usefulness of the model. However, if a significant drop in R^2 is observed, we should be cautious about using the model for prediction in practice.

A similar type of comparison can be made between the mean square error, MSE, for the least squares fit and the mean squared prediction error

$$\text{MSE}_{\text{prediction}} = \frac{\sum\limits_{i=n_1+1}^{n_1+n_2} (y_i - \hat{y}_i)^2}{n_2 - k},$$

where k is the number of β coefficients (excluding β_0) in the model. Whichever measure of model validity you decide to use, the number of observations in the new data set should be large enough to reliably assess the model's prediction performance. Montgomery and Peck (1982) recommend 15–20 new observations.

4. *Data-splitting (cross validation)*   For those applications where it is impossible or impractical to collect new data, the original sample data can be split into two parts, with one part used to estimate the model parameters and the other part used to assess the fitted model's predictive ability. **Data-splitting** (or **cross-validation**, as it is sometimes known) can be accomplished in a variety of ways. A common technique is to randomly assign half the observations to the estimation data set and the other half to the prediction data set.* Measures of model validity, such as $R^2_{\text{prediction}}$ or $\text{MSE}_{\text{prediction}}$ can then be calculated. Of course, a sufficient number of observations must be available in order for data-splitting to be effective. For the estimation and prediction data sets of equal size, Snee (1977) recommends that the entire sample consist of at least $n = 2k + 25$ observations, where k is the number of β parameters in the model.

The appropriate model validation technique(s) will vary from application to application. Keep in mind that a favorable result is still no guarantee that the model will always perform successfully in practice. However, we have much greater confidence in a validated model than in one that simply fits the sample data well.

| | | | | | | | | | | | |

S E C T I O N 10.10

SUMMARY

A number of special topics in regression have been introduced in this chapter. **Piecewise linear regression** can be employed when the theoretical relationship between the dependent variable y and a single independent variable x differs for different intervals over the range of x. **Inverse prediction** is a technique used for predicting a value of x for a given value of y. When the response y is **binary** (i.e., takes on only two values, 0 or 1), a **logistic regression model** may be more appropriate than a linear regression model.

Several methods are available for situations when the standard least squares assumptions about the random errors are violated. **Weighted least squares** is a variance-stabilizing technique which can also be used to dampen the influence of outlying observations. When the distribution of the errors is nonnormal, **robust**

---

*Random splits are usually applied in cases where there is no logical basis for dividing the data. Consult the references for other, more formal, data-splitting techniques.

regression yields estimates of the β's with certain optimal properties. **Ridge regression** was developed in order to stabilize the estimated β coefficients in the presence of multicollinearity.

Finally, it is often important to assess a model's predictive performance before releasing it for use in the real world. Several **model validation techniques** are available for this purpose.

REFERENCES

Andrews, D. F. "A robust method for multiple linear regression," *Technometrics*, 16, 1974, pp. 523–531.

Cox, D. R. *The Analysis of Binary Data.* London: Methuen, 1970.

Draper, N. R. and Van Nostrand, R. C. "Ridge regression and James–Stein estimation: Review and comments," *Technometrics*, 21, 1979, p. 451.

Geisser, S. " The predictive sample reuse method with applications," *Journal of the American Statistical Association*, 70, 1975, pp. 320–328.

Graybill, F. A. *Theory and Application of the Linear Model.* North Scituate, Mass.: Duxbury Press, 1976.

Halperin, M., Blackwelder, W. C., and Verter, J. I. "Estimation of the multivariate logistic risk function: A comparison of the discriminant function and maximum likelihood approaches," *Journal of Chronic Diseases*, 24, 1971, pp. 125–158.

Hauck, W. W. and Donner, A. "Wald's test as applied to hypotheses in logit analysis," *Journal of the American Statistical Association*, 72, 1977, pp. 851–853.

Hill, R. W. and Holland, P. W. "Two robust alternatives to least squares regression," *Journal of the American Statistical Association*, 72, 1977, pp. 828–833.

Hoerl, A. E. and Kennard, R. W. "Ridge regression: Biased estimation for nonorthogonal problems," *Technometrics*, 12, 1970, pp 55–67.

Hoerl, A. E., Kennard, R. W., and Baldwin, K. F. "Ridge regression: Some simulations," *Communications in Statistics*, A5, 1976, pp. 77–88.

Hogg, R. V. "Statistical robustness: One view of its use in applications today," *The American Statistician*, 33, 1979, pp. 108–115.

Montgomery, D. C. and Peck, E. A. *Introduction to Linear Regression Analysis.* New York: Wiley, 1982.

Mosteller, F. and Tukey, J. W. *Data Analysis and Regression: A Second Course in Statistics.* Reading, Mass.: Addison-Wesley, 1977.

Neter, J., Wasserman, W., and Kutner, M. H. *Applied Linear Statistical Models*, 2nd ed. Homewood, Ill.: Richard D. Irwin, 1985.

Obenchain, R. L. "Classical F-tests and confidence intervals for ridge regression," *Technometrics*, 19, 1977, pp. 429–439.

Snee, R. D. "Validation of regression models: Methods and examples," *Technometrics*, 19, 1977, pp. 415–428.

Tsiatis, A. A. "A note on the goodness-of-fit test for the logistic regression model," *Biometrika*, 67, 1980, pp. 250–251.

Walker, S. H. and Duncan, D. B. "Estimation of the probability of an event as a function of several independent variables," *Biometrika*, 54, 1967, pp. 167–179.

OBJECTIVE

To demonstrate how regression analysis can be used to model and compare the relationship between the sale prices and assessed values of residential properties for six different city neighborhoods

CONTENTS

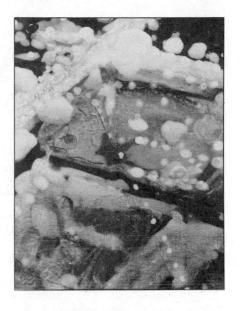

AN EXAMINATION OF THE RELATIONSHIP BETWEEN THE SALE PRICES AND ASSESSED VALUES OF PROPERTIES FOR SIX NEIGHBORHOODS

SECTION 11.1

THE PROBLEM

This case study concerns a problem of interest to real estate appraisers, tax assessors, real estate investors, and home buyers—namely, the relationship between the appraised value of a property and its sale price. The sale price for any given property will vary depending on the price set by the seller, the strength of appeal of the property to a specific buyer, and the state of the money and real estate markets. Therefore, we can think of the sale price of a specific property as possessing a relative frequency distribution. The mean of this distribution might be regarded as a measure of the fair value of the property. Presumably, this is the value that a property appraiser or a tax assessor would like to attach to a given property.

The purpose of this case study is to examine the relationship between the mean sale price $E(y)$ of a property and the following independent variables:

1. Appraised land value of the property
2. Appraised value of the improvements on the property
3. Neighborhood in which the property is listed

The objectives of the study are twofold:

1. To determine whether the data indicate that appraised values of land and improvements are related to sale prices. That is, do the data supply sufficient evidence to indicate that these variables contribute information for the prediction of sale price?
2. To acquire the prediction equation relating appraised value of land and improvements to sale price and to determine whether this relationship is the same for a variety of neighborhoods. In other words, do the appraisers use the same appraisal criteria for various types of neighborhoods?

SECTION 11.2

THE DATA

The data for the study were supplied by the property appraiser's office of a midsized Florida city and consist of the appraised land and improvement values and sale prices for residential properties sold in the city during 1986. Six neighborhoods, each relatively homogeneous but differing sociologically and in property types and values, were identified within the city and surrounding area. The subset of sales and appraisal data pertinent to these six neighborhoods were used to develop a prediction equation relating sale prices to appraised land and improvement values. The data are given in Appendix E.

SECTION 11.3

THE MODELS

If the mean sale price $E(y)$ of a property were, in fact, equal to its appraised value x, the relationship between $E(y)$ and x would be a straight line with slope equal to 1, as shown in Figure 11.1. But it is unlikely that this ideal situation will exist. The property appraiser's data could be several years old and consequently may represent (because of inflation) only a percentage of the actual mean sale price. Also, it is not a foregone conclusion that the relationship between $E(y)$ and x is linear (a straight-line relationship). The appraiser might have a tendency to overappraise or underappraise properties in specific price ranges, say very low-priced or very high-priced properties. If this were true, a second-order curve would provide a better model. Consequently, we will fit both first-order and second-order models to sale price y to see whether the second-order terms improve the prediction equation.

FIGURE 11.1
The Theoretical
Relationship Between
Mean Sale Price and
Appraised Value x

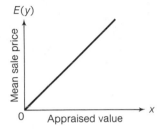

You will recall that we want to relate three independent variables to the sale price y: the qualitative factor, neighborhood (six levels), and the two quantitative factors, appraised land value and appraised improvement value. We consider the following three models as candidates for this relationship:

MODEL 1

Model 1 is a first-order model that will trace a response plane for mean sale price $E(y)$ as a function of x_1 = appraised land value and x_2 = appraised improvement value. This model will assume that the response planes are identical for all six neighborhoods, i.e., that a first-order model is appropriate for relating sale price to x_1 and x_2 and that the relationship between the sale price and the appraised value of a property is the same for all neighborhoods. This model is:

MODEL 1
First-Order Model,
Identical for All
Neighborhoods

$$E(y) = \beta_0 \quad + \quad \overbrace{\beta_1 x_1}^{\text{Appraised land value}} \quad + \quad \overbrace{\beta_2 x_2}^{\text{Appraised improvement value}}$$

MODEL 2

Model 2 will assume that the relationship between mean sale price $E(y)$ and x_1 and x_2 is first-order (a planar response surface), but that the planes differ depending on the neighborhood. This model would be appropriate if the appraiser's procedure for establishing appraised values produced a relationship between mean sale price $E(y)$ and x_1 and x_2 that differed in at least two neighborhoods. Consequently, to test the null hypothesis that the relationship between $E(y)$ and appraised values x_1 and x_2 is identical for all neighborhoods, we would utilize the drop in SSE from Model 1 to Model 2. Model 2 is

MODEL 2

First-Order Model in x_1 and x_2 That Differs from One Neighborhood to Another

$$E(y) = \beta_0 + \overbrace{\beta_1 x_1}^{\substack{\text{Appraised land} \\ \text{value}}} + \overbrace{\beta_2 x_2}^{\substack{\text{Appraised} \\ \text{improvement value}}}$$

$$+ \overbrace{\beta_3 x_3 + \beta_4 x_4 + \beta_5 x_5 + \beta_6 x_6 + \beta_7 x_7}^{\text{Main effect terms for neighborhoods}}$$

$$+ \overbrace{\beta_8 x_1 x_3 + \beta_9 x_1 x_4 + \cdots + \beta_{12} x_1 x_7}^{\text{Interaction, appraised land by neighborhood}}$$

$$+ \overbrace{\beta_{13} x_2 x_3 + \beta_{14} x_2 x_4 + \cdots + \beta_{17} x_2 x_7}^{\text{Interaction, appraised improvement by neighborhood}}$$

where

x_1 = Appraised land value

x_2 = Appraised improvement value

$$x_3 = \begin{cases} 1 & \text{if neighborhood A} \\ 0 & \text{if not} \end{cases} \qquad x_4 = \begin{cases} 1 & \text{if neighborhood B} \\ 0 & \text{if not} \end{cases}$$

$$\vdots \qquad\qquad \vdots$$

$$x_7 = \begin{cases} 1 & \text{if neighborhood E} \\ 0 & \text{if not} \end{cases}$$

The sixth neighborhood, neighborhood F, was chosen as the baseline. Consequently, the model will predict $E(y)$ for neighborhood F when $x_3 = x_4 = \cdots = x_7 = 0$.

MODEL 3

Model 3 is a second-order model that is similar to first-order Model 2 except that we will add terms corresponding to x_1^2, x_2^2, $x_1 x_2$, and all the interactions between these terms and the five dummy variables that define the neighborhoods. Consequently, the second-order model will trace (geometrically) a second-order response surface, one for each neighborhood. The second-order model is given as follows:

MODEL 3 Second-Order Model in x_1 and x_2 That Differs from One Neighborhood to Another

$$E(y) = \beta_0 + \overbrace{\beta_1 x_1 + \beta_2 x_2 + \beta_3 x_1 x_2 + \beta_4 x_1^2 + \beta_5 x_2^2}^{\text{Second-order model in } x_1 \text{ and } x_2}$$

$$+ \overbrace{\beta_6 x_3 + \beta_7 x_4 + \beta_8 x_5 + \beta_9 x_6 + \beta_{10} x_7}^{\text{Main effect terms for neighborhoods}}$$

$$+ \beta_{11} x_3 x_1 + \beta_{12} x_3 x_2 + \beta_{13} x_3 x_1 x_2 + \beta_{14} x_3 x_1^2 + \beta_{15} x_3 x_2^2$$

$$+ \beta_{16} x_4 x_1 + \beta_{17} x_4 x_2 + \beta_{18} x_4 x_1 x_2 + \beta_{19} x_4 x_1^2 + \beta_{20} x_4 x_2^2 + \cdots$$

$$+ \beta_{31} x_7 x_1 + \beta_{32} x_7 x_2 + \beta_{33} x_7 x_1 x_2 + \beta_{34} x_7 x_1^2 + \beta_{35} x_7 x_2^2$$

$\left. \right\}$ Interaction terms: second-order model terms by neighborhood

We will fit Models 1, 2, and 3 to the data. If the drop in SSE between Models 1 and 2 is statistically significant, we will know that there is evidence to indicate a different linear relationship between mean sales price $E(y)$ and appraised values x_1 and x_2 for at least two of the neighborhoods. If the drop in SSE between Models 2 and 3 is statistically significant, we will know that a second-order model will significantly improve the prediction equation.

| | | | | | | | | | | | | |

SECTION 11.4

THE REGRESSION ANALYSES

TABLE 11.1

The SAS computer printouts for Models 1, 2, and 3 are shown in Figures 11.2, 11.3, and 11.4 (pages 600—601), respectively. Checking these printouts, you will note that the sums of squares for error and their respective degrees of freedom for the three models are as listed in Table 11.1.

MODEL	SSE	df(ERROR)
1	$SSE_1 = 93{,}233{,}186{,}667$	648
2	$SSE_2 = 71{,}997{,}720{,}827$	633
3	$SSE_3 = 54{,}177{,}821{,}240$	615

To test the hypothesis that a single first-order model is appropriate for all neighborhoods, we wish to test the null hypothesis that the parameters $\beta_3 = \beta_4 = \cdots = \beta_{17} = 0$ in Model 2, i.e., that Model 1, $E(y) = \beta_0 + \beta_1 x_1 + \beta_2 x_2$, is the appropriate model. To test

$$H_0: \quad \beta_3 = \beta_4 = \cdots = \beta_{17} = 0$$

we use the test statistic

$$F = \frac{\dfrac{\text{Drop in SSE}}{\text{Number of } \beta \text{ parameters in } H_0}}{s_2^2} = \frac{\dfrac{SSE_1 - SSE_2}{15}}{s_2^2}$$

where the mean square for the numerator is based on $\nu_1 = 15$ df, and s_2^2, the estimate of σ^2 using Model 2, is based on $\nu_2 = 633$ df. Substituting SSE_1, SSE_2, and s_2^2 (i.e., MSE_2) into the formula for F, we obtain

$$F = \frac{\dfrac{SSE_1 - SSE_2}{15}}{s_2^2} = \frac{\dfrac{93{,}233{,}186{,}667 - 71{,}997{,}720{,}827}{15}}{113{,}740{,}475}$$

$$= 12.45$$

The tabulated critical value of F for $\alpha = .05$ and $\nu_1 = 15$ and $\nu_2 = 633$ df is not shown in the table, but you can see that it is approximately equal to 1.72. Since the computed value of F, 12.45, exceeds this value, we have evidence to indicate that the first-order response planes differ for at least two neighborhoods. The practical implication of this result is that the appraiser is not assigning appraised values to properties in such a way that the first-order relationship between mean sales $E(y)$ and appraised values x_1 and x_2 is the same for all neighborhoods.

FIGURE 11.2   Model 1: First-Order Model in Land and Improvements, Appraised Values Identical for All Neighborhoods

ANALYSIS OF VARIANCE

SOURCE	DF	SUM OF SQUARES	MEAN SQUARE	F VALUE	PROB>F
MODEL	2	827669866237	413834933118	2876.283	0.0001
ERROR	648	93233186667	143878374		
C TOTAL	650	920903052903			

ROOT MSE	11994.93	R-SQUARE	0.8988
DEP MEAN	72306.45	ADJ R-SQ	0.8984
C.V.	16.58902		

PARAMETER ESTIMATES

VARIABLE	DF	PARAMETER ESTIMATE	STANDARD ERROR	T FOR H0: PARAMETER=0	PROB > |T|
INTERCEP	1	-4747.88287	1139.23391	-4.168	0.0001
X1	1	2.00688149	0.09516215	21.089	0.0001
X2	1	1.09856123	0.03602482	30.495	0.0001

FIGURE 11.3   Model 2: First-Order Model, Different for Each Neighborhood

ANALYSIS OF VARIANCE

SOURCE	DF	SUM OF SQUARES	MEAN SQUARE	F VALUE	PROB>F
MODEL	17	848905332076	49935607769	439.031	0.0001
ERROR	633	71997720827	113740475		
C TOTAL	650	920903052903			

ROOT MSE	10664.92	R-SQUARE	0.9218
DEP MEAN	72306.45	ADJ R-SQ	0.9197
C.V.	14.74961		

PARAMETER ESTIMATES

VARIABLE	DF	PARAMETER ESTIMATE	STANDARD ERROR	T FOR H0: PARAMETER=0	PROB > |T|
INTERCEP	1	-2533.26136	3304.92377	-0.767	0.4437
X1	1	2.46408707	0.39540178	6.232	0.0001
X2	1	0.81249869	0.09764477	8.321	0.0001
X3	1	-30696.62785	4792.73263	-6.405	0.0001
X4	1	-9278.39887	4792.92113	-1.936	0.0533
X5	1	-6130.17189	6636.14518	-0.924	0.3560
X6	1	5469.57016	4475.29296	1.222	0.2221
X7	1	20992.11396	6127.76901	3.426	0.0007
X1X3	1	-0.36549298	0.42590097	-0.858	0.3911
X2X3	1	0.66246593	0.11501968	5.760	0.0001
X1X4	1	0.17863972	0.45297759	0.394	0.6934
X2X4	1	0.27526749	0.13388240	2.056	0.0402
X1X5	1	-1.15513312	0.49573717	-2.330	0.0201
X2X5	1	0.55978295	0.17246101	3.246	0.0012
X1X6	1	-0.55600423	0.51782573	-1.074	0.2834
X2X6	1	0.13585357	0.14826493	0.916	0.3599
X1X7	1	-0.98259014	0.77153132	-1.274	0.2033
X2X7	1	-0.12955381	0.16334377	-0.793	0.4280

Can the prediction equation be improved by the addition of second-order terms? That is, do the data provide sufficient evidence to indicate that the second-order model is a better predictor of sale price than the first-order model? To answer this question, we will test the null hypothesis that the parameters associated with all second-order terms in Model 3 equal 0. Checking the equation

FIGURE 11.4 Model 3: Second-Order Model, Different for Each Neighborhood

ANALYSIS OF VARIANCE

SOURCE	DF	SUM OF SQUARES	MEAN SQUARE	F VALUE	PROB>F
MODEL	35	866725231663	24763578048	281.104	0.0001
ERROR	615	54177821240	88094018.28		
C TOTAL	650	920903052903			

ROOT MSE	9385.841	R-SQUARE	0.9412	
DEP MEAN	72306.45	ADJ R-SQ	0.9378	
C.V.	12.98064			

PARAMETER ESTIMATES

VARIABLE	DF	PARAMETER ESTIMATE	STANDARD ERROR	T FOR H0: PARAMETER=0	PROB > |T|
INTERCEP	1	-12344.47126	9942.98331	-1.242	0.2149
X1	1	6.97567066	1.63757577	4.260	0.0001
X2	1	-0.15944712	0.28299662	-0.563	0.5734
X1X2	1	0.000210252	0.000044306	4.745	0.0001
X1SQ	1	-0.000477487	0.000101298	-4.714	0.0001
X2SQ	1	-0.000022715	.00000560042	-4.056	0.0001
X3	1	42777.18647	14157.59086	3.022	0.0026
X4	1	-25166.06351	13625.74313	-1.847	0.0652
X5	1	71141.78997	33051.55712	2.152	0.0318
X6	1	42267.14205	12170.91963	3.473	0.0006
X7	1	55290.89367	15911.12706	3.475	0.0005
X1X3	1	-5.63269104	1.71074028	-3.293	0.0010
X2X3	1	0.15161101	0.43528802	0.348	0.7277
X1X2X3	1	-0.000172004	0.000044636	-3.853	0.0001
X1SQX3	1	0.000439761	0.000101788	4.320	0.0001
X2SQX3	1	0.000024847	.00000602505	4.124	0.0001
X1X4	1	-6.37177746	1.79213408	-3.555	0.0004
X2X4	1	2.66153633	0.46596877	5.712	0.0001
X1X2X4	1	-0.000246023	0.000046171	-5.328	0.0001
X1SQX4	1	0.000579204	0.000103997	5.569	0.0001
X2SQX4	1	0.000017561	.00000679018	2.586	0.0099
X1X5	1	-9.05967716	3.66907687	-2.469	0.0138
X2X5	1	-0.006856490	0.70296033	-0.010	0.9922
X1X2X5	1	-0.000191231	0.000061011	-3.134	0.0018
X1SQX5	1	0.000554053	0.000137901	4.018	0.0001
X2SQX5	1	0.000032410	0.000011011	2.943	0.0034
X1X6	1	-9.56188518	2.01847810	-4.737	0.0001
X2X6	1	0.83904018	0.44583045	1.882	0.0603
X1X2X6	1	-0.000162315	0.000052718	-3.079	0.0022
X1SQX6	1	0.000606352	0.000114142	5.312	0.0001
X2SQX6	1	0.000018101	.00000980195	1.847	0.0653
X1X7	1	-2.88971386	2.94120679	-0.982	0.3262
X2X7	1	-1.07562502	0.61286334	-1.755	0.0797
X1X2X7	1	-0.000220311	0.000096025	-2.294	0.0221
X1SQX7	1	0.000374275	0.000222536	1.682	0.0931
X2SQX7	1	0.000047020	0.000011429	4.114	0.0001

of the second-order model, you will see that there are 18 second-order terms and that the parameters included in H_0 will be

$$H_0: \quad \beta_3 = \beta_4 = \beta_5 = \beta_{13} = \beta_{14} = \beta_{15} = \beta_{18} = \beta_{19} = \beta_{20} = \cdots = \beta_{35} = 0$$

To test H_0, we form the test statistic

$$F = \frac{\dfrac{\text{Drop in SSE}}{\text{Number of } \beta \text{ parameters in } H_0}}{s_3^2} = \frac{\dfrac{\text{SSE}_2 - \text{SSE}_3}{18}}{s_3^2}$$

where s_3^2 is the estimate of σ^2 obtained from the complete second-order model (Model 3). Substituting the values of SSE_2, SSE_3, and s_3^2 (i.e., MSE_3) into the formula for F, we obtain

$$F = \frac{\dfrac{71,997,720,827 - 54,177,821,240}{18}}{88,094,018.28}$$

$$= 11.24$$

The degrees of freedom for F are $\nu_1 = 18$ and $\nu_2 = 615$ (the degrees of freedom for SSE_3 and s_3^2 are given on the computer printout, Figure 11.4) and the tabulated F-value for $\alpha = .05$ is approximately 1.68. Since the computed value of F, 11.24, exceeds the critical value, $F_{.05} = 1.68$, there is sufficient evidence to indicate that the second-order terms contribute information for the prediction of y.

We have already shown that the first-order prediction equations vary among neighborhoods, and we would expect the same result for the second-order model, i.e., the second-order response surfaces that predict sale prices as a function of the appraised values x_1 and x_2 vary in shape from neighborhood to neighborhood. To test this theory, we would need to fit a second-order model that would imply the same second-order surface for all neighborhoods. This reduced model is identical to Model 3 except that all terms involving the neighborhood dummy variables, $x_3, x_4, \ldots, x_7$, have been omitted:

MODEL 4

A Single Second-Order Surface Appropriate for All Neighborhoods

$$E(y) = \beta_0 + \beta_1 x_1 + \beta_2 x_2 + \beta_3 x_1 x_2 + \beta_4 x_1^2 + \beta_5 x_2^2$$

To determine whether the second-order surfaces differ from neighborhood to neighborhood, we check the reduction in SSE from the reduced model (Model 4) to the complete model (Model 3). Then the null hypothesis that a single surface is adequate for all neighborhoods is equivalent to the hypothesis that the parameters $\beta_6, \beta_7, \ldots, \beta_{35}$ in Model 3 all equal 0, i.e., we are hypothesizing that Model 4 is the appropriate model. Then the null hypothesis is

$$H_0: \beta_6 = \beta_7 = \cdots = \beta_{35} = 0$$

and the alternative hypothesis is that at least one of these parameters does not equal 0. The F-statistic is

$$F = \frac{\dfrac{\text{Drop in SSE}}{\text{Number of } \beta \text{ parameters in } H_0}}{s_3^2} = \frac{\dfrac{SSE_4 - SSE_3}{30}}{s_3^2}$$

The regression analysis computer printout for Model 4, shown in Figure 11.5, gives $SSE_4 = 76,356,480,863$. Similarly, the values of the SSE_3 and s_3^2 are given on the computer printout for the complete model in Figure 11.4. Substituting these values into the formula for the F-statistic, we obtain

FIGURE 11.5 Model 4: Second-Order Model, Same for All Neighborhoods

```
                          ANALYSIS OF VARIANCE

                          SUM OF          MEAN
          SOURCE    DF    SQUARES         SQUARE       F VALUE      PROB>F

          MODEL      5  844546572040  168909314408    1426.814     0.0001
          ERROR    645   76356480863    118382141
          C TOTAL  650  920903052903

               ROOT MSE       10880.36      R-SQUARE      0.9171
               DEP MEAN       72306.45      ADJ R-SQ      0.9164
               C.V.           15.04756

                         PARAMETER ESTIMATES

                          PARAMETER       STANDARD     T FOR H0:
          VARIABLE  DF    ESTIMATE        ERROR        PARAMETER=0   PROB > |T|

          INTERCEP   1    9914.64794     1915.51144     5.176        0.0001
          X1         1    0.81611885     0.18370421     4.443        0.0001
          X2         1    0.93263893     0.08883570    10.498        0.0001
          X1X2       1    0.000019606    .00000449054   4.366        0.0001
          X1SQ       1   -.0000056195    .00000583723  -0.963        0.3361
          X2SQ       1   -.0000015889    .00000122741  -1.295        0.1960
```

$$F = \dfrac{\dfrac{\text{SSE}_4 - \text{SSE}_3}{30}}{s_3^2} = \dfrac{\dfrac{76{,}356{,}480{,}863 - 54{,}177{,}821{,}240}{30}}{88{,}094{,}018.28}$$

$$= 8.39$$

Again, the tabulated value of $F_{.05}$ for $\nu_1 = 30$ and $\nu_2 = 615$ df is not given in the $F_{.05}$ table, but you can see that it is approximately equal to 1.53. Since the computed F-value, 8.39, exceeds this critical value, it supports the alternative hypothesis that the prediction equations differ from one neighborhood to another. This result might have been expected, since we found a difference in first-order prediction equations from one neighborhood to another.

The results of the preceding tests suggest that Model 3 is the best of the four models for modeling mean sale price $E(y)$. The global F-value for testing H_0: $\beta_1 = \beta_2 = \cdots = \beta_{35} = 0$ is highly significant ($F = 281.10$, p-value $< .0001$) and the R^2-value indicates that the model is explaining over 94% of the variability in sale prices. Although a few of the t-tests involving the individual β parameters in Model 3 are nonsignificant, this does not imply that these terms should be dropped from the model.

Whenever a model includes a large number of interactions and squared terms, the t-tests will often be nonsignificant even if the global F-test is highly significant. This result is due partly to the unavoidable intercorrelations between the main effects for a variable, its interactions, and its squared terms (see the discussion on multicollinearity in Section 5.3). We warned in Chapter 4 of the dangers of conducting a series of t-tests to determine model adequacy. For a model with a large number of β's, such as Model 3, you should avoid conducting any t-tests at all and rely on the global F-test and partial F-tests to determine the important terms for predicting y.

A GRAPHICAL EXAMINATION OF THE PREDICTION EQUATION

Substituting the estimates of the Model 3 parameters (Figure 11.4) into the prediction equation, we have

$$\hat{y} = -12{,}344.47 + 6.9757x_1 - .1594x_2 + .00021x_1x_2 - .00048x_1^2$$
$$- .000023x_2^2 + 42{,}777.19x_3 - 25{,}166.06x_4 + \cdots + .000374x_7x_1^2$$
$$+ .000047x_7x_2^2$$

We have noted that $\hat{y}$ graphs as a set of six second-order surfaces, one for each neighborhood. The equation of the second-order surface for a particular neighborhood is obtained by substituting the appropriate values of the dummy variables into the prediction equation. For example, the prediction equation for neighborhood A is obtained by setting $x_3 = 1$ and $x_4 = x_5 = x_6 = x_7 = 0$. This surface, which would be difficult to sketch in its exact form, might appear as shown in Figure 11.6.

FIGURE 11.6
A Sketch of the Second-Order Surface for a Single Neighborhood

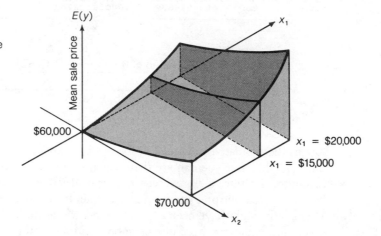

One way to obtain a more precise graphical description of the response surface for a single neighborhood would be to graph $\hat{y}$ as a function of appraised improvement value x_2 for different levels of appraised land value x_1. These curves would be obtained if you sliced the surface shown in Figure 11.6 with planes parallel to the y,x_2-plane, as shown in Figure 11.6. The curves are shown in Figure 11.7 for appraised land value $x_1 = \$10{,}000$. Similar sets of curves are shown in Figures 11.8, 11.9, and 11.10 (pages 606–608) for $x_1 = \$14{,}000$, $\$18{,}000$, and $\$22{,}000$, respectively. Note that not all figures have curves corresponding to all neighborhoods. This is because the appraised land values for sales in some neighborhoods might not have been as low as $\$10{,}000$ or as high as $\$22{,}000$. Some curves are shortened for similar reasons, i.e., the appraised improvement values for some neighborhoods might not span the range shown on the x_2-axis.

Before we examine the curves in Figures 11.7–11.10, you will want some information concerning the nature of the neighborhoods. Figure 11.11 (page 609) gives the means, standard deviations, and other descriptive statistics for the sale price y and the appraised values x_1 and x_2 for each of the six neighborhoods.

The mean sale prices confirm what we (the authors) know to be true, i.e., that neighborhoods A and B are two of the relatively expensive residential areas in the city. Most of the inhabitants are older, established professional or business people. In contrast, neighborhoods C, E, and F are less expensive residential areas inhabited primarily by young married couples who are starting their careers. Neighborhood D is considered to be one of the least expensive residential areas in the city and is inhabited by mostly low-income families.

Turning to Figures 11.7–11.10, you will note that the estimated mean sale price $\hat{y}$ increases as the appraised improvement value x_2 increases, but the increase is not always linear. The relationship appears to be nearly a straight line for some neighborhoods (for example, neighborhood D), but is curvilinear for other neighborhoods.

FIGURE 11.7 Graph of Predicted Sale Price Versus Appraised Value of Improvements When Appraised Land Value Is $10,000

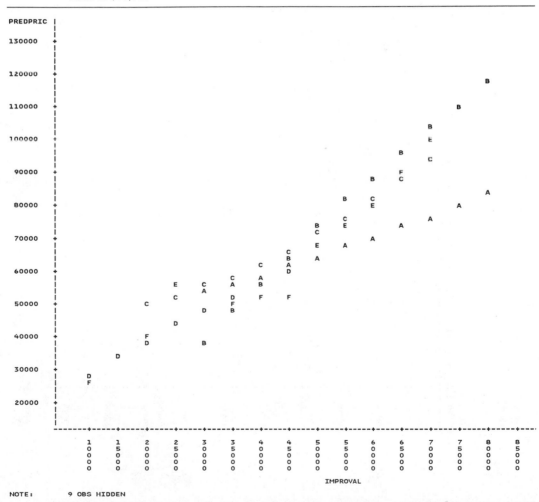

NOTE: 9 OBS HIDDEN

FIGURE 11.8   Graph of Predicted Sale Price Versus Appraised Value of Improvements When Appraised Land Value Is $14,000

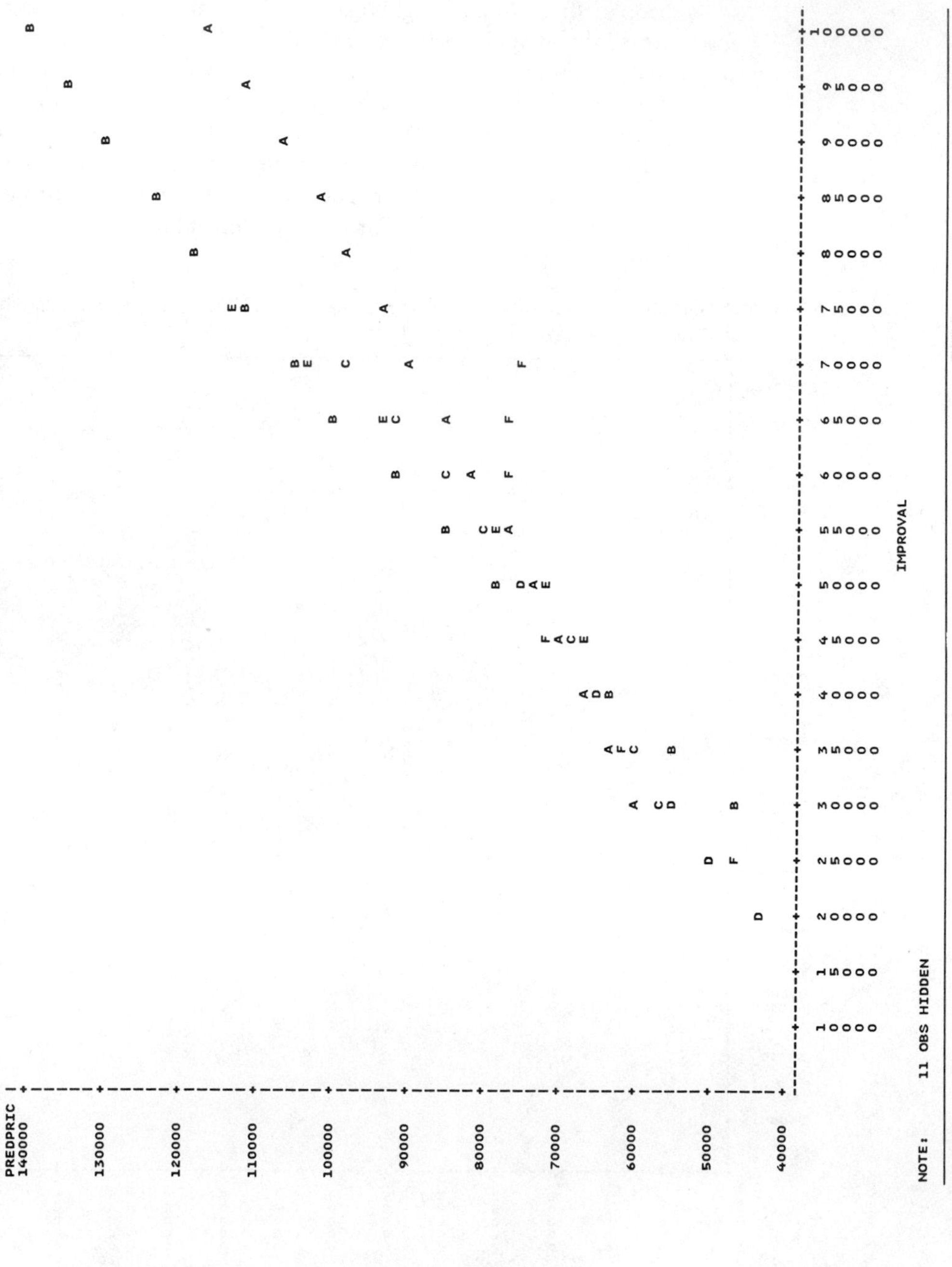

FIGURE 11.9 Graph of Predicted Sale Price Versus Appraised Land Improvements When Appraised Land Value Is $18,000

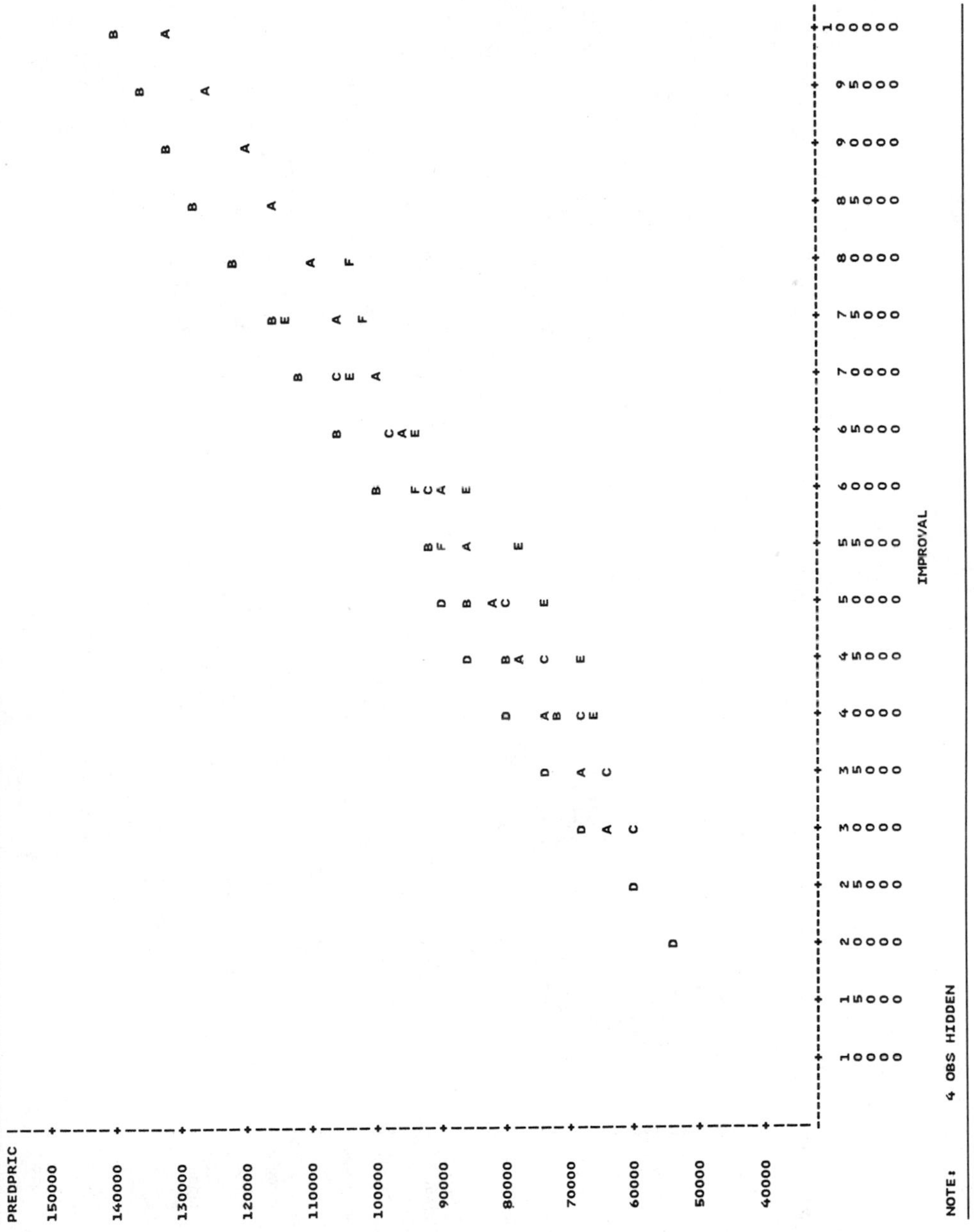

FIGURE 11.10   Graph of Predicted Sale Price Versus Appraised Value of Improvements When Appraised Land Value Is $22,000

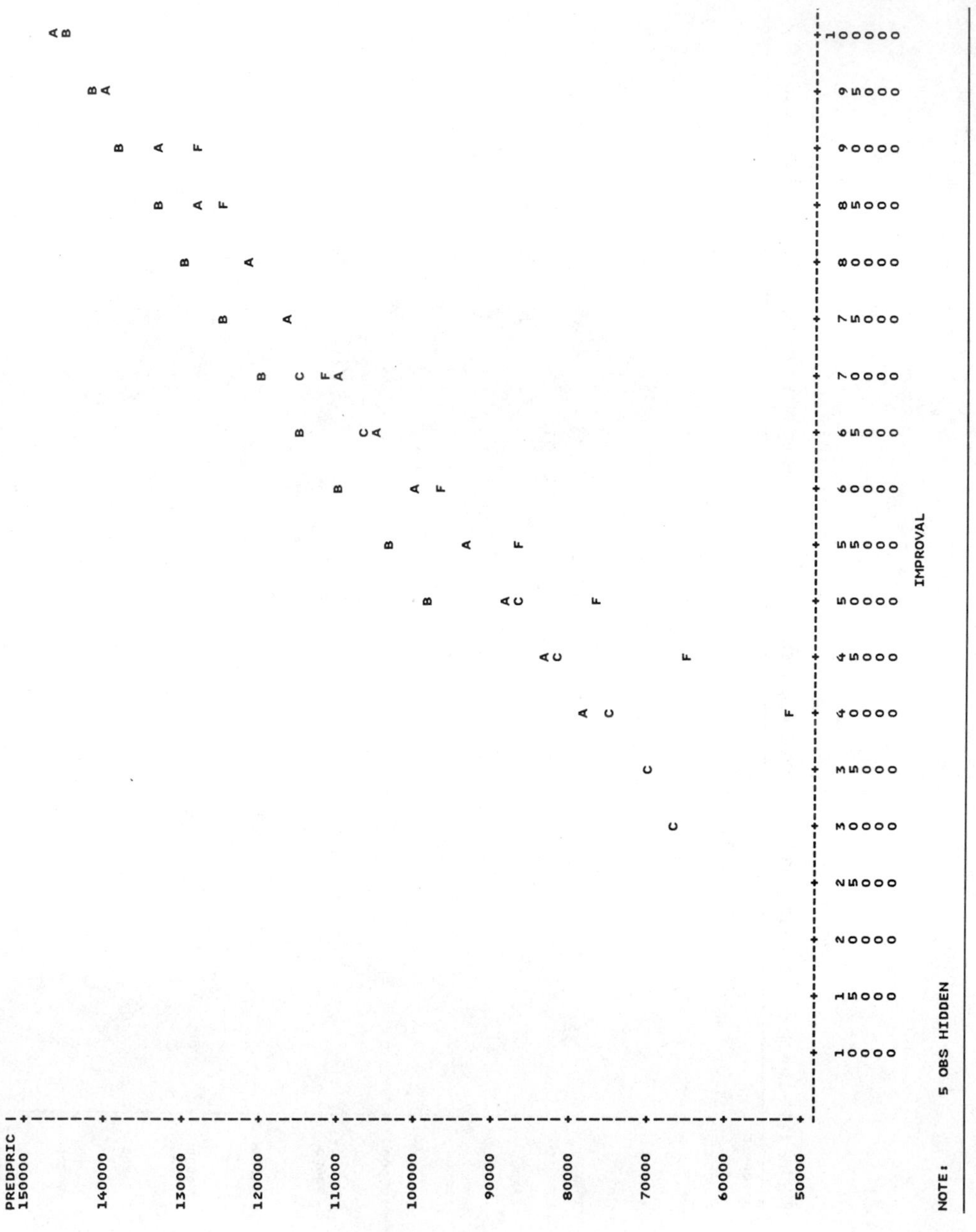

FIGURE 11.11 Real Estate Analyses: 1986 Sales in Six Neighborhoods.

DATA FROM APPENDIX F --- SALES IN SIX NEIGHBORHOODS
FIGURE 11.11: MEANS IN SIX NEIGHBORHOODS

VARIABLE	N	MEAN	STANDARD DEVIATION	MINIMUM VALUE	MAXIMUM VALUE	STD ERROR OF MEAN	SUM	VARIANCE	C.V.
--- NBRHOOD=A ---									
SALEPRIC	90	112496.666667	53271.7085570	57000.000000	400000.00000	5615.3311296	10124700.000	2837874932.6	47.354
LANDVAL	90	25100.8888889	9364.4740633	15000.000000	82080.00000	987.1023377	2259080.000	87693374.5	37.307
IMPROVAL	90	63086.244444	24382.0645015	30572.000000	150391.00000	2570.0952627	5677762.000	594485069.4	38.649
--- NBRHOOD=B ---									
SALEPRIC	80	90802.5000000	47006.8214145	44200.000000	365000.00000	5255.5224044	7264200.0000	2209641259.5	51.768
LANDVAL	80	15209.3125000	9065.3546120	8000.000000	65625.00000	1013.5409235	1216745.0000	82181216.3	59.604
IMPROVAL	80	57383.7500000	21873.6132120	32877.000000	170108.00000	2445.5443028	4590700.0000	478454954.9	38.118
--- NBRHOOD=C ---									
SALEPRIC	89	72505.6179775	17742.4590867	51500.000000	175000.00000	1880.6969018	6453000.0000	314794854.44	24.470
LANDVAL	89	12689.6629213	4509.1211725	9400.000000	23000.00000	477.9658884	1129380.0000	20332173.75	35.534
IMPROVAL	89	47044.9101124	9484.7701326	34606.000000	97182.00000	1005.3836233	4186997.0000	89960864.47	20.161
--- NBRHOOD=D ---									
SALEPRIC	140	43875.7142857	14823.4301773	13800.000000	119000.00000	1252.8085084	6142600.0000	219734082.22	33.785
LANDVAL	140	8644.9142857	3126.3541676	2000.000000	22000.00000	264.2251527	1210288.0000	9774090.38	36.164
IMPROVAL	140	25775.4357143	9369.3899675	8591.000000	66388.00000	791.8579795	3608561.0000	87785468.36	36.350
--- NBRHOOD=E ---									
SALEPRIC	136	61026.4705882	12074.1059652	23200.000000	111500.00000	1035.3460449	8299600.0000	145784034.86	19.785
LANDVAL	136	10117.3529412	1853.8221719	2500.000000	20000.00000	158.9639398	1375960.0000	3436656.64	18.323
IMPROVAL	136	40382.1617647	9379.2837642	9321.000000	76735.00000	804.2669476	5491974.0000	87970963.93	23.226
--- NBRHOOD=F ---									
SALEPRIC	116	75753.4482759	31103.5365836	29700.000000	221200.00000	2887.8908722	8787400.0000	967429988.01	41.059
LANDVAL	116	15788.7931034	5107.6033953	9000.000000	34000.00000	474.2290699	1831500.0000	26087612.44	32.350
IMPROVAL	116	48469.9224138	20682.6787594	14713.000000	127940.00000	1920.3385133	5622511.0000	427773200.66	42.671

There are some other interesting features to note. For example, in Figure 11.7 observe the different predicted sales prices for an appraised improvement value of $45,000. You can see that properties in neighborhood C (a lower-priced neighborhood) sell at a higher predicted price than those in the more expensive neighborhoods A and B when the appraised land value is low, i.e., $x_1 = \$10,000$. This would suggest that the appraised values of properties in this price range in neighborhoods A and B are too high in comparison to similar properties in neighborhood C. One reason for this could be that a low appraised property value in either neighborhood A or B might correspond to a very small-sized lot and this might have a strong depressive effect on sale prices. In contrast, you can see in Figure 11.9 ($x_1 = \$18,000$) that properties in neighborhoods A and B bring higher prices for an appraised home value of $45,000 than for similarly appraised homes in neighborhood C. Thus, neighborhood A and B homes are appraised at a lower value than homes in neighborhood C that sell for the same price.

The two major points to be derived from an analysis of the sale price–appraised value curves are:

1. The fact that the curves for neighborhoods A and B frequently lie above those for neighborhoods C, D, E, and F indicates that properties in the higher-priced neighborhoods are being underappraised relative to sale price in comparison with properties in the lower-priced neighborhoods.
2. The fact that, for most neighborhoods, the graphs tend to curve upward as the appraised improvement value increases indicates that more expensive properties within most neighborhoods are underappraised in comparison with the appraised values of less expensive properties.

SECTION 11.6

PREDICTING THE SALE PRICE OF A PROPERTY

How well do appraised land value x_1 and appraised improvement value x_2 predict property sale price? Recall that from Model 3 (Figure 11.4), we obtained $R^2 = .9412$, indicating that the model accounts for approximately 94% of the variability of the sale price values y about their mean $\bar{y}$. This seems to indicate that the model provides a fairly good fit to the data, but note that $s = 9,385.841$. If the model is used to predict sale price for particular values of x_1 and x_2, we can direct the SAS program to print the desired prediction interval. However, when n is large, an approximate formula* for the prediction interval is

$$\hat{y} \pm t_{\alpha/2}s$$

Consequently, an 80% prediction interval will be approximately

$$\hat{y} \pm z_{.10}s \qquad \text{where} \quad z_{.10} = 1.28 \quad \text{and} \quad s = 9,386$$
$$\hat{y} \pm (1.28)(9,386)$$
$$\hat{y} \pm \$12,014$$

*The exact formula is given in Section A.9 of Appendix A.

Therefore, approximately 80% of the predicted prices will lie within $12,014 of the actual corresponding property sale prices. A 90% prediction interval would be approximately

$$\hat{y} \pm 1.645s$$
$$\hat{y} \pm \$15,440$$

We leave it to you to answer our initial question and to decide whether these errors of prediction are small enough (and the corresponding confidence coefficients are large enough) to indicate that the prediction equation could be of practical value in predicting property sale prices. For our part, we feel certain that a much more accurate predictor of sale price could be developed by relating y to the variables that describe the property (such as location, square footage, and number of bedrooms) and those that describe the market (mortgage interest rates, availability of money, and so forth).

SECTION 11.7

CONCLUSIONS

The results of the regression analyses described in Section 11.4 indicate that the relationships between property sale prices and appraised values are not consistent from one neighborhood to another. Further, the bound on the prediction error for the prediction equation is rather sizable, thus indicating that there is room for improvement in the methods used to determine appraised property values.

EXERCISES 11.1–11.2

11.1 Explain why the tests of model adequacy conducted in Section 11.4 give no assurances that Model 3 will be a successful predictor of sale price in the future.

11.2 If you have access to a statistical computer package, use the data-splitting technique of Section 10.9 to assess the validity of Model 3.

ON YOUR OWN . . .

Figures 11.12 and 11.13 (page 612) are sketches of the city and surrounding area and an enlargement showing the location of the business district, neighborhoods A, B, . . . , F, and shopping malls. The complete set of data for 1986 residential sales (of which Appendix E is a subset) is presented in Appendix F. This data set includes the following:

1. Location of the property: Two numbers are used to identify each of the large squares in the sketch. One number is the range and the other is the township. Each range–township combination is associated with one of the large squares shown on the sketch. A third number gives the section (a small area) within the range–township square. Each range–township square is divided into 36 sections. The location of a particular section within a typical range–township square is shown in the upper left corner of Figure 11.12.

FIGURE 11.12
Location of the City and
Surrounding Areas

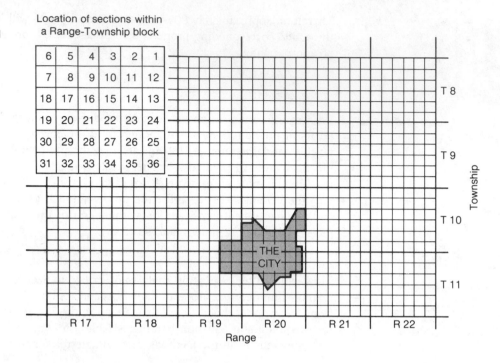

FIGURE 11.13
The City and Adjacent
Areas

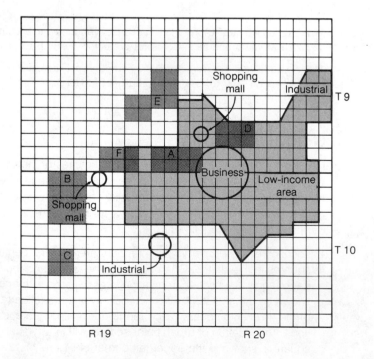

2. Appraised land value
3. Appraised improvement value
4. Sale price
5. Ratio of sale price to total appraised value of the property
6. Home size (area) in square feet

This data set provides many opportunities for analysis. We leave these projects to you and your instructor.*

*This data set, as well as all appendix data sets, can be obtained on magnetic tape or floppy diskette by contacting the publisher.

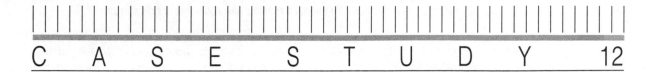

C A S E   S T U D Y   12

OBJECTIVE

To show how a data transformation can be used to improve the predictive ability of a regression model

CONTENTS

RELATING CONSUMPTION OF A PRODUCT TO PRICE AND ADVERTISING EXPENDITURE

INTRODUCTION

What is the optimal amount of money that a firm should spend for advertising a product? A theory that proposes to answer this question was set forth in the 1950's by Dorfman and Steiner (1954). Essentially, their theory states that "a firm which can influence the demand for its product by advertising will, in order to maximize its profits, choose the advertising budget and price such that the increase in gross revenue resulting from a one dollar increase in advertising expenditures is equal to the ordinary elasticity of demand for the firm's product."

Putting the Dorfman–Steiner theory into mathematical terms, let q be the demand for the product, let p be its price, and let a represent the amount spent on advertising. Then demand q for the product is some function of price p and advertising expenditure a, say

$$q = D(p, a)$$

The essence of the Dorfman–Steiner theory is that the firm's profits for any given demand q are maximized when the elasticity of demand is equal to the marginal value product of advertising.*

R. W. Ward (1975) successfully applied the Dorfman–Steiner theory, not to a firm, but to a commodity industry. In particular, Ward noted that the industry that produces processed grapefruit tends to operate with a near uniform system of pricing and that advertising is generic in form and is under direct control of the industry through the Florida Citrus Commission. The commission regulates the amount of advertising and the nature of the commercials, but it does not regulate price. Consequently, Ward reasoned, the processed grapefruit industry bears many similarities to a firm producing a single product and hence the Dorfman–Steiner theory applies. The price is not fixed across firms, but it tends to be fairly uniform, and advertising expenditure a is controlled by the industry.

In order to apply the Dorfman–Steiner theory to a particular firm or commodity industry, a good approximation must be found to the function relating demand q to price of the commodity p and advertising expenditure a, i.e.,

$$q = D(p, a)$$

This case study considers only this portion of Ward's research—the acquisition of a model relating q to p and a. It provides an interesting departure from the previous case study because one of the independent variables, advertising expenditure a, enters into the model as a reciprocal $(1/a)$. The net effect is to produce a response surface with a shape that differs from those we have previously encountered.

---

*If you are familiar with calculus, the marginal elasticity of demand is equal to $(\partial q/\partial p)(p/q)$, where the symbol $(\partial q/\partial p)$ represents the rate of change in demand q with respect to price p when advertising expenditure a is held constant. Similarly, the marginal value product is equal to $(\partial q/\partial a)(p)$ where $\partial q/\partial a$ represents the rate of change in demand q with respect to advertising a when price p is held constant. From calculus, you will recognize $\partial q/\partial p$ and $\partial q/\partial a$ as the partial derivatives of the demand function q with respect to p and a, respectively.

Ward (1975) presents data on the annual demand q, annual average price p, and total annual advertising expenditure a for the processed grapefruit industry for the producing seasons 1966–1967 through 1972–1973 to acquire a model relating q to p and a. The data are shown in Table 12.1 and will be used in Section 12.4 to find a reasonable approximation to the function relating q to p and a.

TABLE 12.1
Annual Demand, Price, and Advertising Expenditures for Processed Grapefruit

OBSERVATION	q, million gallons	p, dollars per gallon	a, million dollars
1 1972–1973	53.52	1.294	1.837
2 1971–1972	51.34	1.344	1.053
3 1970–1971	49.31	1.332	.905
4 1969–1970	45.93	1.274	.462
5 1968–1969	51.65	1.056	.576
6 1967–1968	38.26	1.102	.260
7 1966–1967	44.29	.930	.363

The limited number of data points ($n = 7$) implies that the model relating demand q to price p and advertising expenditure a must contain at most two or three parameters. Otherwise, we will not have a sufficient number of degrees of freedom to test the model. Based on theoretical considerations and experience, Ward constructed several different models for the demand function q and fit all to the data using regression analyses. The most promising model seemed to be

$$E(q) = \beta_0 + \beta_1 p + \beta_2\left(\frac{1}{a}\right)$$

for $p \geq p_0$, $a \geq a_0$, where $a_0 > 0$ and $p_0 > 0$, and q is annual demand in millions of gallons, p is price in dollars per gallon, and a is advertising expenditure in millions of dollars. Note that $E(q)$ traces a response surface over the plane defined by the two independent variables, p and a. To understand the nature of the surface, we will write the model in our standard notation:

$$E(y) = \beta_0 + \beta_1 x_1 + \beta_2 x_2$$

where

$$y = q \qquad x_1 = p \qquad x_2 = \frac{1}{a}$$

You can see that $E(y)$ is a first-order model in x_1 and x_2 and, consequently, that the model does not allow for interaction between x_1 and x_2. The implication is that price p and the reciprocal $(1/a)$ of advertising expenditure affect the mean demand $E(q)$ independently of one another. Therefore, we can deduce the nature of the response surface by separately examining the effect of p and $(1/a)$ on $E(q)$.

Suppose we fix advertising expenditure at some value, say $a = 1$. Substituting $a = 1$ into the model $E(q) = \beta_0 + \beta_1 p + \beta_2(1/a)$, we obtain an equation relating p to $E(q)$:

$$E(q) = (\beta_0 + \beta_2) + \beta_1 p \qquad \text{when} \quad a = 1$$
$$= (\text{Constant}) + \beta_1 p$$

You can see that this is the equation of a straight line with $E(q)$ intercept equal to $(\beta_0 + \beta_2)$ and slope equal to β_1. Since we would expect demand to decrease as price increases, β_1 should be negative and the graph of

$$E(q) = (\text{Constant}) + \beta_1 p$$

should appear as shown in Figure 12.1

FIGURE 12.1

The Relation Between
Mean Demand and Price
When $a = 1$

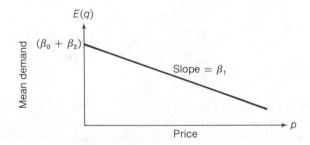

Similarly, we can investigate the effect of increasing advertising expenditure a on mean demand, $E(q)$, by substituting some value of price p into the model, say $p = 1.0$, and examining the equation

$$E(q) = (\beta_0 + \beta_1) + \beta_2\left(\frac{1}{a}\right) \qquad \text{when } p = 1.0$$
$$= (\text{Constant}) + \beta_2\left(\frac{1}{a}\right)$$

If β_2 is positive, $E(q)$ decreases as advertising expenditure a increases. In contrast, $E(q)$ increases as a increases when β_2 is negative. In this particular case, we will want the mean demand to increase as advertising expenditure increases. Consequently, we assume that β_2 should be negative and that the function

$$E(q) = (\text{Constant}) + \beta_2\left(\frac{1}{a}\right)$$

should appear as shown in Figure 12.2.

The form of the response surface is given by Figures 12.1 and 12.2. If the surface is sliced by a plane parallel to the $E(q),a$-plane, we obtain a curve similar to Figure 12.2. If the surface is sliced by a plane parallel to the $E(q),p$-plane, we obtain a straight line similar to that shown in Figure 12.1. The resulting surface would appear as shown in Figure 12.3.

FIGURE 12.2
The Relation Between
Mean Demand and
Advertising Expenditure
When $p = 1.0$

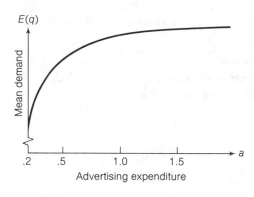

FIGURE 12.3
The Response Surface
Corresponding to $E(q) =$
$\beta_0 + \beta_1 p + \beta_2(1/a)$

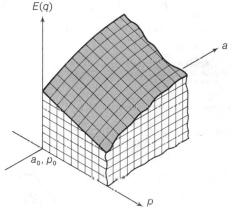

I I I I I I I I I I I I I

S E C T I O N 12.4

THE REGRESSION ANALYSIS

The regression analysis for the model

$$E(q) = \beta_0 + \beta_1 p + \beta_2\left(\frac{1}{a}\right)$$

was performed using the SPSS[x] regression analysis computer program. As an option, we requested that the program print out the residuals. The computer printout is shown in Figure 12.4 (page 620). This printout suggests that the model provides a fairly good approximation to the mean demand $E(q)$. The F-statistic for testing whether p and $(1/a)$ contribute information for the prediction of demand, $F = 46.98566$, is highly significant ($p = .0017$) and the value of R^2, .95917, is large. When viewing the value of R^2, keep in mind that we fit a model containing three parameters to only seven data points. Consequently, we would expect a good fit to the data if good judgment was employed in selecting the form of the model.

The best criterion for evaluating the adequacy of the model is the magnitude of the residuals. Are the errors of prediction small enough to indicate that the prediction equation provides a good approximation to $E(q)$? Remember, Ward used an approximation to the demand function, $q = D(p, a)$, to determine the

FIGURE 12.4 SPSS^x Computer Printout for Dorfman–Steiner Case Study: Model Linear in p and $1/a$; No Interaction

```
LISTWISE DELETION OF MISSING DATA

EQUATION NUMBER 1    DEPENDENT VARIABLE..    Q

BEGINNING BLOCK NUMBER  1.  METHOD:  ENTER      P       X2


VARIABLE(S) ENTERED ON STEP NUMBER   1..    X2
                                     2..    P


MULTIPLE R          .97937      ANALYSIS OF VARIANCE
R SQUARE            .95917                      DF    SUM OF SQUARES    MEAN SQUARE
ADJUSTED R SQUARE   .93876      REGRESSION       2         162.26149       81.13075
STANDARD ERROR     1.31404      RESIDUAL         4           6.90685        1.72671

                                F =     46.98566     SIGNIF F =   .0017


------------------ VARIABLES IN THE EQUATION ------------------

VARIABLE          B          SE B        BETA         T    SIG T

X2          -5.334763      .629369   -1.158831    -8.476   .0011
P          -10.091949     4.515770    -.305530    -2.235   .0891
(CONSTANT)  69.753526     6.249481                11.161   .0004


END BLOCK NUMBER   1   ALL REQUESTED VARIABLES ENTERED.

CASEWISE PLOT OF STANDARDIZED RESIDUAL

*: SELECTED   M: MISSING

               -3.0          0.0          3.0
     CASE #   O:.............:.............:O        Q      *PRED      *RESID
       1      .            *.           .        53.52    53.7905      -.2705
       2      .             .*          .        51.34    51.1237       .2163
       3      .        *    .           .        49.31    50.4163     -1.1063
       4      .             . *         .        45.93    45.3493       .5807
       5      .             .      *    .        51.65    49.8347      1.8153
       6      .             .*          .        38.26    38.1139       .1461
       7      .        *    .           .        44.29    45.6717     -1.3817
     CASE #   O:.............:.............:O        Q      *PRED      *RESID
               -3.0          0.0          3.0
```

Dorfman–Steiner solution for optimal advertising expenditure. Consequently, we would want to know how greatly the deviations between observed and predicted values of demand disturb the Dorfman–Steiner solution.

Since there is so little data, the model

$$E(q) = \beta_0 + \beta_1 p + \beta_2\left(\frac{1}{a}\right)$$

is probably a good choice for $E(q)$. You will notice that the computed value of the t-statistic ($t = -8.476$) for testing the null hypothesis H_0: $\beta_2 = 0$ is highly significant ($p = .0011$). And despite the fact that we would expect demand to be related to price, the computed value of the t-statistic ($t = -2.235$) corresponding to the coefficient β_1 is only modestly significant ($p = .0891$). This could be caused by the inability of a t-test, based on only 4 df, to detect small values of β_1 (values that differ from 0). It could also be possible that price p is not entered properly into the model. Perhaps the effect of increasing price on demand is not linear, or it may be that the effect of price on demand depends on the level of advertising expenditure (i.e., there may exist a price–advertising interaction).*

We tested this latter theory by adding an interaction term,

$$\beta_3 p\left(\frac{1}{a}\right)$$

to obtain the model

$$E(q) = \beta_0 + \beta_1 p + \beta_2\left(\frac{1}{a}\right) + \beta_3 p\left(\frac{1}{a}\right)$$

A t-test of H_0: $\beta_3 = 0$ was statistically significant at the $\alpha \approx .12$ level. This significance level is small enough to suggest that an interaction term might improve the model. More data are needed to test this and other model modifications.

SECTION 12.5

SUMMARY

This case study demonstrates that you can acquire many unusual and useful response surface shapes by varying the manner in which the independent variables enter a model. For example, if you expect $E(y)$ to be curvilinearly related to an independent variable x, you might be inclined to use the second-order model

$$E(y) = \beta_0 + \beta_1 x + \beta_2 x^2$$

We have shown in this case study that the two-parameter model,

$$E(y) = \beta_0 + \beta_1\left(\frac{1}{x}\right)$$

may provide a good fit to the data. The independent variable could be entered into the model in many other ways, say as $\log x$, e^x, or e^{-x}, and you could even add a second-order term involving these variables. For example, you might try

$$E(y) = \beta_0 + \beta_1(\log x) \quad \text{or} \quad E(y) = \beta_0 + \beta_1(\log x) + \beta_2(\log x)^2$$

When model-building, you will likely fit many different types of models to the same set of data. If you add enough terms to your model or try enough different functional forms for $E(y)$, you will eventually obtain a good fit to your data. At this juncture, the results of any statistical tests that you conduct will be suspect. You have used the data to formulate your theory (i.e., choose your model); therefore, it will not be surprising if the data agree with your model. A good procedure to avoid this self-delusion is to split your data set (if you have enough data) into two parts. Use one portion of the data to choose a model and the other portion to test it. Of course, we could not have applied this procedure to Ward's data because the entire data set contains only seven observations.

*Ward considered these alternatives in his analyses.

We do not intend to suggest that you blindly try all types of models in order to obtain one that adequately models a mean response. There is no substitute for a theoretical knowledge of the phenomenon you are attempting to model. This knowledge will aid you in postulating reasonable choices for the functional form of your model.

ON YOUR OWN . . .

If you have access to a computer, try other transformations on the independent variables and fit these models to Ward's data.

REFERENCES

Dorfman, R. and Steiner, P. O. "Optimal advertising and optimal quality," *American Economic Review*, 44, 1954, pp. 826–836.

Ward, R. W. "Revising the Dorfman–Steiner static advertising theorem: An application to the processed grapefruit industry," *American Journal of Agricultural Economics*, 57 (3), 1975.

C A S E   S T U D Y   13

OBJECTIVE

To present an example of regression modeling that can be used to detect the possibility of collusive bidding

CONTENTS

AN ANALYSIS OF BIDDING COMPETITION

THE PROBLEM

Many products and services are purchased by governments and businesses on the basis of competitive bids, and frequently contracts are awarded to the lowest bidders. This process works extremely well in competitive markets, but it has the potential to increase the cost of purchasing if the markets are noncompetitive or if collusive practices are present.

Numerous methods exist for detecting the possibility of collusive practices among bidders. In general, these procedures involve the detection of significant departures from normal market conditions such as (1) systematic rotation of the winning bid, (2) stable market shares over time, (3) geographic market divisions, (4) lack of relationship between delivery costs and bid levels, (5) high degree of uniformity and stability in bid levels over time, and (6) presence of a baseline point pricing scheme (Rothrock and McClave, 1979). This chapter demonstrates two applications of regression analysis to the detection of collusive bidding. In particular, we will show you how to use regression analysis to detect statistically significant changes in the annual pattern of mean low-bid prices for different markets. Then we will examine the residuals, the deviations between the predicted and the actual low-bid prices, to detect unusually large positive residuals.

| | | | | | | | | | | | | |

S E C T I O N   13.2

THE DATA

Each county in the state of Florida negotiates an annual contract for bread to supply the county's public schools. Sealed bids are submitted by the vendors and the lowest bid (price per pound of bread) is selected as the bid winner. The data of interest in this study are the annual low-bid prices for each of the three major bread purchases—white bread, hamburger buns, and hot dog buns—for most of the 67 county school boards in the state of Florida for the years 1970–1975. In order to reduce the size of this analysis, we will confine our attention to only one of these products—white bread.

The state was divided into eight geographic market areas. Each region contained counties that were adjacent to one another, possessed similar economic characteristics, and were served by the same group of vendors. There were 13 vendors that baked and supplied the bread, but only three of these possessed multiplant operations and sold bread throughout most of the state. The other vendors confined their selling efforts to a much smaller number of the eight geographic market areas. The total number of winning bids included in the analysis is 303. The following data were recorded for each of the 303 low bids and are provided in Appendix G.

1. LBPRICE: Low-bid price (y)
2. MKT: Geographic region containing a group of buyers with essentially the same set of vendors
3. YEAR: Year in which contract was let

SECTION 13.3

MODELS FOR STABLE AND UNSTABLE MARKET CONDITIONS

The low-bid price to a particular school system should be proportional to the cost of producing and delivering a pound of bread. This cost will vary from year to year and from market to market. The annual variation in the cost per pound of bread would generally be inflationary due to increasing costs of fuel, labor, and raw materials over the period 1970–1975, and would tend to rise in unequal jumps from year to year. The year-to-year effect of cost increases on low-bid price for a single market might appear as shown in Figure 13.1.

FIGURE 13.1

Hypothetical Relationship Between Low-Bid Price and Year for a Single Market

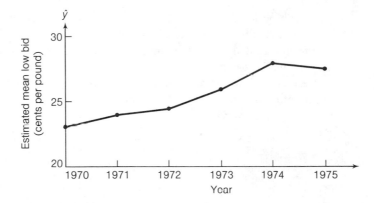

A second source of variation in cost per pound of bread (and hence bid prices) would be the different characteristics of the eight markets. The distances required to transport the bread would vary from market to market as would the costs of labor, rent, and other forms of overhead. The net effect of these costs would be to raise or lower the intercept of the low-bid price–year curve in Figure 13.1, depending on the particular market.

Annual changes in the cost of fuel, labor, and other purchases would not produce identical changes in the cost per pound of bread in all eight markets, but we would expect the overall changes in cost to be roughly of the same magnitude. Consequently, assuming that no unusual disturbance affects the cost in any of the markets (we will call this a **stable market condition**), the cost–year curves for the eight markets would be identical except for a vertical shift and might appear as shown in Figure 13.2 (page 626). Thus, Figure 13.2 shows the effect of the two sources of variation, the variation between markets and the inflationary variation due to time, on low-bid price in a stable market. The fact that the pattern of low-bid prices from market to market remains the same from year to year implies that the two factors, market and time, do not interact. The linear model that characterizes this situation is

MODEL 1
A Stable Market Condition

$$E(y) = \beta_0 + \overbrace{\beta_1 x_1 + \beta_2 x_2 + \cdots + \beta_7 x_7}^{\text{Main effects, market}}$$

$$+ \overbrace{\beta_8 x_8 + \beta_9 x_9 + \cdots + \beta_{12} x_{12}}^{\text{Main effects, time}}$$

$$x_1 = \begin{cases} 1 & \text{if market 1} \\ 0 & \text{if not} \end{cases} \quad x_2 = \begin{cases} 1 & \text{if market 2} \\ 0 & \text{if not} \end{cases} \cdots \quad x_7 = \begin{cases} 1 & \text{if market 7} \\ 0 & \text{if not} \end{cases}$$

$$x_8 = \begin{cases} 1 & \text{if 1970} \\ 0 & \text{if not} \end{cases} \quad x_9 = \begin{cases} 1 & \text{if 1971} \\ 0 & \text{if not} \end{cases} \cdots \quad x_{12} = \begin{cases} 1 & \text{if 1974} \\ 0 & \text{if not} \end{cases}$$

You will notice that time, a quantitative factor, was entered into the model as if it were a qualitative factor. The reason for this is that we expected an irregular pattern of annual inflationary increases in the mean low-bid prices. If we expected $E(y)$ to increase over time in a smooth and regular pattern, we would have entered time into the model with $x_8 = \text{Year} - c$ (where c is a constant, say 1972) and with terms involving x_8, x_8^2, and so forth.

FIGURE 13.2

Hypothetical Pattern of Annual Low-Bid Prices for a Stable Market Condition

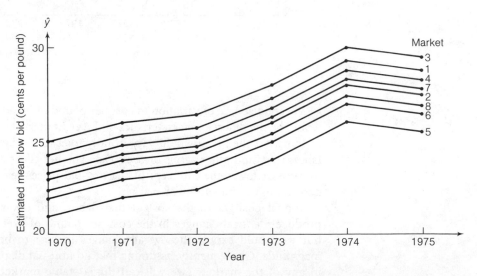

Suppose the markets are not stable over time—that is, suppose the low-bid price in a particular market increases substantially in a given year in relation to the low-bid prices for the other markets. This situation, which represents a market–time interaction, signals some very unusual increase in costs in that market, or it may be a warning of the possibility of collusive bidding. A graph of this situation might appear as shown in Figure 13.3.

FIGURE 13.3

Hypothetical Pattern of
Annual Low-Bid Prices for
an Unstable Market
Condition

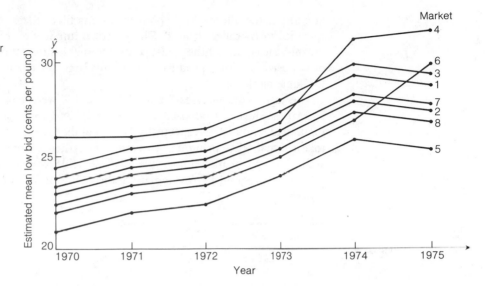

An interactive model that would characterize the situation described in Figure
13.3 is

MODEL 2

Unstable Market Condition

$$
\begin{aligned}
E(y) = \beta_0 + &\overbrace{\beta_1 x_1 + \beta_2 x_2 + \cdots + \beta_7 x_7}^{\text{Main effects, market}} \\
&+ \overbrace{\beta_8 x_8 + \beta_9 x_9 + \cdots + \beta_{12} x_{12}}^{\text{Main effects, year}} \\
&\left. \begin{aligned}
&+ \beta_{13} x_1 x_8 + \beta_{14} x_2 x_8 + \cdots + \beta_{19} x_7 x_8 \\
&+ \beta_{20} x_1 x_9 + \beta_{21} x_2 x_9 + \cdots + \beta_{26} x_7 x_9 \\
&+ \beta_{27} x_1 x_{10} + \beta_{28} x_2 x_{10} + \cdots + \beta_{33} x_7 x_{10} \\
&+ \beta_{34} x_1 x_{11} + \beta_{35} x_2 x_{11} + \cdots + \beta_{40} x_7 x_{11} \\
&+ \beta_{41} x_1 x_{12} + \beta_{42} x_2 x_{12} + \cdots + \beta_{47} x_7 x_{12}
\end{aligned} \right\} \begin{aligned} \text{Year–market} \\ \text{interaction} \end{aligned}
\end{aligned}
$$

| | | | | | | | | | | | |

S E C T I O N   13.4

**THE REGRESSION
ANALYSES: TESTING
TO DETECT UNSTABLE
MARKET CONDITIONS**

We have presented two models in Section 13.3. Model 1 represents a stable
market condition, the situation where the differences in mean low-bid prices
between any pair of markets remain the same from year to year. Model 2
represents an unstable market condition, a situation where differences in the
mean low-bid prices between any pair of markets is allowed to change from year
to year. Model 2 contains all the terms contained in Model 1 and, in addition,
contains 35 year–market interaction terms. Testing the null hypothesis that "the
market is stable" is equivalent to testing the null hypothesis that the 35 interaction
parameters of Model 2 equal 0, i.e.,

$$H_0: \beta_{13} = \beta_{14} = \beta_{15} = \cdots = \beta_{47} = 0$$

Support of the alternative hypothesis means that *at least one* of the parameters specified in H_0 differs from 0. The practical implication of this situation is that during at least one of the years, some external factors (such as economic factors or collusive bidding) produced unusually large (or small) mean low-bid prices for some markets.

The regression analyses for Models 1 and 2 were conducted using the SAS GLM procedure (as opposed to SAS REG), and their printouts are shown in Figures 13.4 and 13.5, respectively. Note that the GLM printout looks very similar to the usual REG printouts presented in earlier chapters, but includes some additional information under the ANOVA table which we will discuss in the following paragraphs.* A summary of the values of SSE and R^2 for Models 1 and 2 is shown in Table 13.1.

TABLE 13.1

A Summary of the Values (Rounded) of SSE, s^2, and R^2 for Models 1 and 2

MODEL	SSE	s^2	df	R^2
1	.216155	.000745	290	.732
2	.177537	.000696	255	.780

To test the null hypothesis that stable market conditions exist, i.e.,

$$H_0: \quad \beta_{13} = \beta_{14} = \cdots = \beta_{47} = 0$$

we obtain the drop in SSE from the reduced model (Model 1) to the complete model (Model 2), and calculate the value of the F-statistic:

$$F = \frac{\dfrac{\text{Drop in SSE}}{\text{Number of } \beta \text{ parameters in } H_0}}{s_2^2} = \frac{\dfrac{.216155 - .177537}{35}}{.000696}$$

$$= 1.58$$

The critical value of F for $\alpha = .05$, and $\nu_1 = 35$ and $\nu_2 = 255$ df is approximately equal to 1.50 (see Table 4 of Appendix D). Since the observed value of F, 1.58, exceeds this tabulated value, we reject the null hypothesis and conclude that there is evidence of a year–market interaction.

In conducting this test for factor interaction, we employed the standard procedure of fitting complete and reduced models (Chapter 4) so that our test would not be tied to a single computer program package. However, as explained in Chapter 4, some computer programs (e.g., the SAS GLM procedure) automatically compute the drop in SSE associated with sets of parameters in the complete model. In particular, the SAS computer printout in Figure 13.5 gives these quantities under the column titled TYPE III SS. Reading in that column in the MKT*YEAR row, you obtain the drop in SSE associated with the 35 interaction terms as .03861794.† The next column gives the computed value of the F-statistic

*The appropriate SAS GLM and SAS REG commands are given in Appendix C.

†In this particular case, the TYPE III SS and TYPE I SS are equal. This will not usually be true. Consequently, be certain that you read the TYPE III SS column of the printout.

FIGURE 13.4 Model 1: Low-Bid Price Versus Market and Year

DEPENDENT VARIABLE: LBPRICE

SOURCE	DF	SUM OF SQUARES	MEAN SQUARE	F VALUE	PR > F	R-SQUARE	C.V.
MODEL	12	0.59109360	0.04925780	66.09	0.0001	0.732232	11.2455
ERROR	290	0.21615516	0.00074536			ROOT MSE	LBPRICE MEAN
CORRECTED TOTAL	302	0.80724876				0.02730133	0.24277662

SOURCE	DF	TYPE I SS	F VALUE	PR > F	DF	TYPE III SS	F VALUE	PR > F
MKT	7	0.04055373	7.77	0.0001	7	0.03899047	7.47	0.0001
YEAR	5	0.55053987	147.72	0.0001	5	0.55053987	147.72	0.0001

PARAMETER		ESTIMATE	T FOR H0: PARAMETER=0	PR > |T|	STD ERROR OF ESTIMATE
INTERCEPT		0.28273797 B	38.07	0.0001	0.00742624
MKT	1	0.01646959 B	2.16	0.0317	0.00762948
	2	0.01487183 B	1.88	0.0611	0.00790900
	3	-0.01322826 B	-1.82	0.0702	0.00728017
	4	0.00831731 B	1.07	0.2874	0.00780301
	5	-0.00114983 B	-0.13	0.8945	0.00865911
	6	0.00537925 B	0.64	0.5224	0.00839998
	7	-0.00198967 B	-0.21	0.8347	0.00952599
	8	0.00000000 B	.	.	.
YEAR	70	-0.09001053 B	-16.05	0.0001	0.00560670
	71	-0.07795952 B	-14.43	0.0001	0.00540136
	72	-0.07607686 B	-14.40	0.0001	0.00528409
	73	-0.05303058 B	-9.99	0.0001	0.00531023
	74	0.02274620 B	4.37	0.0001	0.00520704
	75	0.00000000 B	.	.	.

FIGURE 13.5 Model 2: Low-Bid Price Versus Market and Year

SOURCE	DF	SUM OF SQUARES	MEAN SQUARE	F VALUE	PR > F	R-SQUARE	C.V.
MODEL	47	0.62971154	0.01339812	19.24	0.0001	0.780071	10.8685
ERROR	255	0.17753722	0.00069622			ROOT MSE	LBPRICE MEAN
CORRECTED TOTAL	302	0.80724876				0.02638606	0.24277662

SOURCE	DF	TYPE I SS	F VALUE	PR > F	DF	TYPE III SS	F VALUE	PR > F
MKT	7	0.04055373	8.32	0.0001	7	0.03951572	8.11	0.0001
YEAR	5	0.55053987	158.15	0.0001	5	0.38679865	111.11	0.0001
MKT*YEAR	35	0.03861794	1.58	0.0243	35	0.03861794	1.58	0.0243

PARAMETER		ESTIMATE	T FOR H0: PARAMETER=0	PR > |T|	STD ERROR OF ESTIMATE
INTERCEPT		0.28377037 B	18.63	0.0001	0.01523400
MKT	1	0.05439012 B	3.09	0.0022	0.01759071
	2	-0.01800741 B	-0.97	0.3354	0.01865776
	3	-0.00991037 B	-0.59	0.5551	0.01668801
	4	-0.01054815 B	-0.60	0.5493	0.01759071
	5	-0.02950370 B	-1.53	0.1270	0.01926966
	6	-0.01657037 B	-0.86	0.3906	0.01926966
	7	-0.01710370 B	-0.79	0.4280	0.02154413
	8	0.00000000 B	.	.	.
YEAR	70	-0.09780741 B	-4.54	0.0001	0.02154413
	71	-0.07643703 B	-3.17	0.0017	0.02408707
	72	-0.07087041 B	-3.29	0.0012	0.02154413
	73	-0.04896296 B	-2.27	0.0239	0.02154413
	74	0.01434674 B	0.67	0.5062	0.02154413
	75	0.00000000 B	.	.	.
MKT*YEAR	1 70	-0.03593945 B	-1.43	0.1529	0.02507062
	1 71	-0.05763596 B	-1.75	0.0821	0.02728687
	1 72	-0.05305258 B	-2.13	0.0339	0.02487702
	1 73	-0.04908642 B	-1.97	0.0496	0.02487702
	1 74	-0.03908754 B	-1.77	0.0782	0.02487702
	1 75	0.00000000 B	.	.	.
	2 70	-0.00539269 B	-0.20	0.8382	0.02638606
	2 71	-0.00980255 B	-0.35	0.7285	0.02820789
	2 72	-0.01738942 B	-0.67	0.5054	0.02607005
	2 73	0.00289339 B	0.11	0.9117	0.02607005
	2 74	0.01511957 B	0.58	0.5625	0.02607005
	2 75	0.00000000 B	.	.	.
	3 70	0.00063630 B	0.03	0.9787	0.02384498
	3 71	-0.00658963 B	-0.25	0.8002	0.02600639
	3 72	-0.01213712 B	-0.51	0.6080	0.02375055
	3 73	-0.00630119 B	-0.27	0.7904	0.02375122
	3 74	-0.00307354 B	-0.13	0.8968	0.02367055
	3 75	0.00000000 B	.	.	.
	4 70	0.04467407 B	1.71	0.0881	0.02609124
	4 71	0.02844444 B	0.82	0.4122	0.02781335
	4 72	0.02548994 B	1.01	0.3150	0.02551736
	4 73	0.00990741 B	0.40	0.6930	0.02507062
	4 74	0.02251111 B	0.90	0.3664	0.02487702
	4 75	0.00000000 B	.	.	.
	5 70	0.05234843 B	1.81	0.0713	0.02890448
	5 71	0.03831794 B	1.24	0.2151	0.03084650
	5 72	0.02093741 B	0.75	0.4516	0.02788280
	5 73	0.01136296 B	0.41	0.6840	0.02788280
	5 74	0.05552592 B	2.04	0.0426	0.02725141
	5 75	0.00000000 B	.	.	.
	6 70	0.02827407 B	1.01	0.3115	0.02788280
	6 71	0.02523703 B	0.86	0.3899	0.02930319
	6 72	0.01960741 B	0.72	0.4725	0.02725141
	6 73	0.02976296 B	1.07	0.2868	0.02788280
	6 74	0.03045926 B	1.12	0.2647	0.02788280
	6 75	0.00000000 B	.	.	.
	7 70	0.02314074 B	0.62	0.5557	0.03731553
	7 71	0.00754814 B	0.23	0.8155	0.03231620
	7 72	0.02391851 B	0.79	0.4332	0.03046800
	7 73	0.01040740 B	0.34	0.7329	0.03046800
	7 74	0.02565925 B	0.84	0.4005	0.03046800
	7 75	0.00000000 B	.	.	.
	8 70	0.00000000 B	.	.	.
	8 71	0.00000000 B	.	.	.
	8 72	0.00000000 B	.	.	.
	8 73	0.00000000 B	.	.	.
	8 74	0.00000000 B	.	.	.
	8 75	0.00000000 B	.	.	.

based on this drop in SSE, namely 1.58. The last column gives the significance level for the F-statistic, .0243. In other words, the probability of observing a value of F as large as or larger than 1.58 is only .0243, assuming the null hypothesis is true. Consequently, there is ample evidence to indicate that something is causing the mean low-bid price in some markets to move in an unstable pattern from one year to another.

Now that we have evidence to indicate a year–market interaction, we will examine computer plots of the curves tracing estimated mean low-bid price versus year for the eight markets (see Figure 13.6, page 632). The 1's trace the estimated mean low-bid prices for market 1 over the years 1970–1975. Similarly, the 2's correspond to the estimated mean low-bid prices for market 2, the 3's correspond to market 3, . . . , and the 8's correspond to market 8. The first thing to notice about the estimated mean low-bid plots is the grouping of estimated low bids for any given year. The range of the estimated mean low-bid prices for the years 1970–1974, approximately 4¢ per pound, may seem small, but in this high-volume business, a 1¢-per-pound increase in contract price can produce a substantial increase in income. Observe the changes in the relative position of the estimated mean low-bid prices for the various markets as you move from year to year. We have already shown that some of these changes cannot be explained solely by random error. Rather, they represent a combination of random variation and year–market interaction.

The second interesting aspect of Figure 13.6 is the inflationary year-to-year shift of the groups—relatively gradual through 1973, rising sharply in 1974, and then dropping substantially during the latter part of the 1974–1975 recession.

When you attempt to locate specific deviations from stable market conditions, you are led immediately to market 1. Notice the large positive deviation of the estimated mean low-bid price for market 1 in 1975. In fact, the estimated mean low-bid price for market 1 is only slightly larger than the mean for all markets for each of the years 1970–1974, but it is much larger than the mean of all markets for 1975. In fact, as the estimated mean low-bid price decreases substantially from 1974 to 1975 for all other markets, the estimated mean low-bid price for market 1 increases. If you examine the computer printout for Figure 13.5, you will find that our speculation of a market 1 by year interaction is not a figment of our imagination. Most of the individual t-tests associated with the market 1 by year interaction parameters are statistically significant.

The high estimated mean low-bid price for market 1 in 1975 may not be the only deviation from stable market conditions that is important, but it is certainly the most obvious. What does it mean? The answer to this question would have to come from some economists and accountants after a thorough review of the suppliers' financial accounts and an examination of the various economic factors at play in the markets. Perhaps the seemingly excessive estimated mean low-bid price in market 1 in 1975 can be explained by unusual economic factors that affected only market 1 in 1975. Or, perhaps the large deviation in bid price from the market means for 1975 is a sign of collusive bidding.

FIGURE 13.6 Plot of Predicted Values of Price Versus Market for Each Year: Numbers Identify the Market Associated with a Plotted Predicted Value

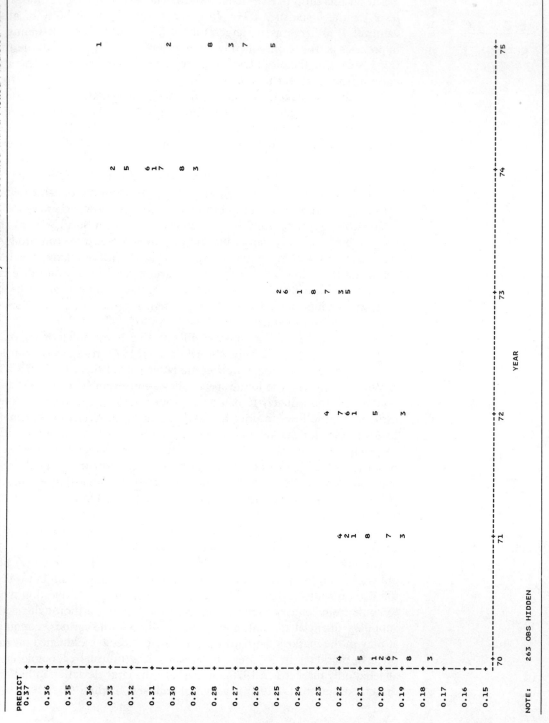

NOTE: 263 OBS HIDDEN

A RESIDUAL ANALYSIS OF THE DATA

An analysis of residuals provides a second method for detecting possible collusive bidding by examining the pattern of low bids *within* given markets. Low bids in a particular year and within a given market should lie, with high probability, within 2 estimated standard deviations of the estimated market mean for that year, i.e., within 2s of $\hat{y}$. Large positive deviations exceeding this value ($\hat{y}$ + 2s) would be suspect and should be examined for source of cause. Notice the difference between this technique and the technique of Section 13.4, where we employed a regression analysis to detect unusual year-to-year changes in the relative values of the *mean* low-bid prices for a set of markets. In contrast, a residual analysis examines the behavior of low bids about the means within market–years.

What type of residual pattern within market–years might provide a clue to the possibility of collusive bidding? This is a question for a specialist to answer, someone who has studied bidding patterns within market–years for proven cases of collusive bidding. Variation in low bids within a market–year would probably be due to varying unit costs of transportation or differing cost-capacity considerations within each firm. Consequently, we would question unusually large deviations from the estimated mean for a predicted market–year, in other words, unusually large residuals.

A computer plot of the relative frequency distribution of the 303 Model 2 regression analysis residuals is shown in Figure 13.7 (page 634), and a computer printout of the complete set of residuals is shown in Figure 13.8 (pages 635–639). The value of s given in the Model 2 regression analysis (Figure 13.5) is approximately .026. We have used this value to construct intervals of $\pm 2s$ and $\pm 3s$ about the mean of the residuals, which is 0, in the relative frequency distribution in Figure 13.7. You can see that a few of the low bids lie more than 3 standard deviations below the mean 0, and several lie more than 3 standard deviations above the mean. A person checking for clues to collusive bidding can identify the large residuals by locating them on the computer printout in Figure 13.8.

The residual patterns for markets and years can be seen in Figures 13.9 and 13.10 (pages 640–641). Both these figures are plots of the 303 residuals against the predicted estimated low-bid price $\hat{y}$. The points locating the individual residual values are identified by numbers. The number that identifies a residual on Figure 13.9 is the last digit of the year in which the low bid occurred; the number that identifies a residual in Figure 13.10 indicates the market in which the low bid occurred. Since both Figures 13.9 and 13.10 are plots of the residuals versus $\hat{y}$, they are identical (except for the identifying numbers). Consequently, you can determine the market–year associated with any residual by locating its position on both Figures 13.9 and 13.10 and reading the year and market identification numbers. For example, the large positive residual located in the upper right-hand corner of the distribution is identified by a 4 in Figure 13.9 and a 2 in Figure 13.10. Consequently, this is the residual for a low bid obtained in 1974 in market 2.

Three horizontal lines are shown on Figures 13.9 and 13.10. The middle line locates the mean of the residuals, which is 0. The upper and lower lines locate

FIGURE 13.7

The Relative Frequency
Distribution of Residuals
for Model 2

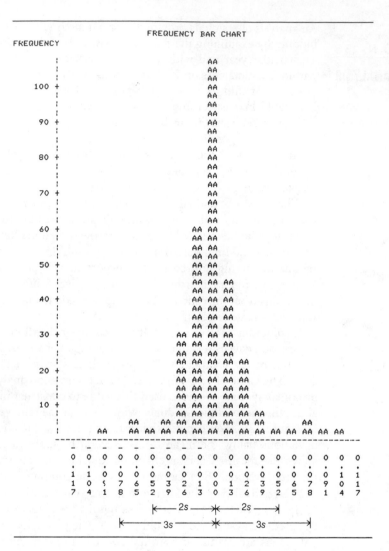

the mean of the residuals ±2s, respectively. You can tell whether individual residuals are outliers by observing their location relative to the ±2s band.

The residual patterns shown in Figures 13.9 and 13.10 are interesting, but they may or may not be relevant. For example, note that most of the residuals lie well within the ±2s band. Notice the tight vertical line of 3's (corresponding to market 3) plotted above $\hat{y} = .27$ in Figure 13.10. Checking Figure 13.9, you can see that these residuals correspond to low bids made in 1975. Notice how closely the low bids group about the predicted value of the mean for that market–year. An investigation of the supplier associated with these low bids might show that most of the low bids were submitted by the same supplier, or it might indicate a very small variation in the low bids among a group of different suppliers. Unusual regularities or patterns of this type indicate unusual market conditions which may warrant further investigation.

FIGURE 13.8 SAS Computer Printout of the Residuals for Model 2

```
                                PREDICT    STD ERR                 STD ERR   STUDENT                         COOK'S
OBS          ID      ACTUAL     VALUE      PREDICT     RESIDUAL    RESIDUAL  RESIDUAL    -2-1-0 1 2               D

  1   70_3           0.1503    0.1767    0.007617     -0.0264     0.0253    -1.0459    |    **|     |        0.002
  2   71_3           0.1667    0.1908    0.007052     -0.0242     0.0254    -0.9505    |     *|     |        0.001
  3   72_3           0.1833    0.1909    0.007052     -.007582    0.0254    -0.2982    |      |     |        0.000
  4   73_3           0.2133    0.2186    .0073182     -.005261    0.0254    -0.2075    |      |     |        0.000
  5   74_3           0.3067    0.2913    0.007052      0.0154     0.0254     0.6054    |      |*    |        0.001
  6   75_3           0.2800    0.2739    .0068129      0.00614    0.0255     0.2409    |      |     |        0.000
  7   70_3           0.2000    0.1767    0.007617      0.0233     0.0253     0.9227    |      |*    |        0.002
  8   71_3           0.2100    0.1908    0.007052      0.0192     0.0254     0.7538    |      |*    |        0.001
  9   72_3           0.1799    0.1909    0.007052     -0.0110     0.0254    -0.4345    |      |     |        0.000
 10   73_3           0.1533    0.2186    .0073182     -0.0653     0.0254    -2.5743    | *****|     |        0.012
 11   74_3           0.2000    0.2913    0.007052     -0.0913     0.0254    -3.5898    |******|     |        0.021
 12   75_3           0.2800    0.2739    .0068129      0.00614    0.0255     0.2409    |      |     |        0.000
 13   70_1           0.2068    0.2044    .0093289     .0023864    0.0247     0.0967    |      |     |        0.000
 14   71_1           0.2209    0.2141    .0093289     .0068125    0.0247     0.2760    |      |     |        0.000
 15   72_1           0.2209    0.2143    .0087954     .0065995    0.0249     0.2653    |      |     |        0.000
 16   73_1           0.2550    0.2401    .0087954      0.0149     0.0249     0.5985    |      |*    |        0.001
 17   75_1           0.3640    0.3382    .0087954      0.0258     0.0249     1.0387    |      |**   |        0.003
 18   75_3           0.2800    0.2739    .0068129      0.00614    0.0255     0.2409    |      |     |        0.000
 19   75_5           0.2480    0.2543      0.0118     -.006267    0.0236    -0.2655    |      |     |        0.000
 20   70_8           0.2090    0.1860      0.0152      0.0230     0.0215     1.0693    |      |**   |        0.012
 21   71_8           0.2280    0.2073      0.0187      0.0207     0.0187     1.1077    |      |**   |        0.026
 22   72_8           0.2300    0.2130      0.0152      0.0170     0.0215     0.7908    |      |*    |        0.007
 23   73_8           0.2307    0.2348      0.0152     -.004074    0.0215    -0.1891    |      |     |        0.000
 24   74_8           0.2767    0.2981      0.0152     -0.0214     0.0215    -0.9954    |     *|     |        0.010
 25   75_8           0.2425    0.2838      0.0152     -0.0412     0.0215    -1.9141    |   ***|     |        0.038
 26   70_1           0.2029    0.2044    .0093289     -.001505    0.0247    -0.0610    |      |     |        0.000
 27   71_1           0.2280    0.2141    .0093289      0.0139     0.0247     0.5637    |      |*    |        0.001
 28   72_1           0.2255    0.2143    .0087954      0.0112     0.0249     0.4484    |      |     |        0.001
 29   73_1           0.2400    0.2401    .0087954     -1.1E-04    0.0249    -.004466    |      |     |        0.000
 30   74_1           0.3345    0.3085    .0087954      0.0260     0.0249     1.0464    |      |**   |        0.003
 31   75_1           0.3400    0.3382    .0087954     .0018395    0.0249     0.0739    |      |     |        0.000
 32   71_7           0.2033    0.1978      0.0152     .0055556    0.0215     0.2579    |      |     |        0.001
 33   72_7           0.2560    0.2198      0.0152      0.0362     0.0215     1.6813    |      |***  |        0.029
 34   73_7           0.2560    0.2281      0.0152      0.0279     0.0215     1.2945    |      |**   |        0.017
 35   74_7           0.3600    0.3067      0.0152      0.0533     0.0215     2.4755    |      |**** |        0.064
 36   75_7           0.2667    0.2667      0.0152      6.0E-16    0.0215     2.8E-14    |      |     |        0.000
 37   70_3           0.1913    0.1767    0.007617      0.0146     0.0253     0.5797    |      |*    |        0.001
 38   71_3           0.2067    0.1908    0.007052      0.0158     0.0254     0.6227    |      |*    |        0.001
 39   72_3           0.2133    0.1909    0.007052      0.0224     0.0254     0.8817    |      |*    |        0.001
 40   73_3           0.2267    0.2186    .0073182     .0080728    0.0254     0.3184    |      |     |        0.000
 41   74_3           0.3040    0.2913    0.007052      0.0127     0.0254     0.5005    |      |*    |        0.000
 42   75_3           0.2667    0.2739    .0068129     -.007193    0.0255    -0.2822    |      |     |        0.000
 43   70_3           0.1900    0.1767    0.007617      0.0133     0.0253     0.5269    |      |*    |        0.001
 44   71_3           0.2150    0.1908    0.007052      0.0242     0.0254     0.9505    |      |*    |        0.001
 45   72_3           0.2200    0.1909    0.007052      0.0291     0.0254     1.1439    |      |**   |        0.002
 46   74_3           0.3534    0.2913    0.007052      0.0622     0.0254     2.4449    |      |**** |        0.010
 47   75_3           0.2600    0.2739    .0068129     -0.0139     0.0255    -0.5437    |     *|     |        0.000
 48   71_7           0.2000    0.1978      0.0152     .0022222    0.0215     0.1031    |      |     |        0.000
 49   72_7           0.2133    0.2198      0.0152     -.006444    0.0215    -0.2991    |      |     |        0.001
 50   73_7           0.2333    0.2281      0.0152     .0052222    0.0215     0.2424    |      |     |        0.001
 51   74_7           0.2700    0.3067      0.0152     -0.0367     0.0215    -1.7019    |   ***|     |        0.030
 52   75_7           0.2667    0.2667      0.0152      6.0E-16    0.0215     2.8E-14    |      |     |        0.000
 53   70_3           0.1600    0.1767    0.007617     -0.0167     0.0253    -0.6606    |     *|     |        0.001
 54   71_3           0.1733    0.1908    0.007052     -0.0175     0.0254    -0.6883    |     *|     |        0.001
 55   72_3           0.1867    0.1909    0.007052     -.004249    0.0254    -0.1671    |      |     |        0.000
 56   73_3           0.2407    0.2186    .0073182      0.0221     0.0254     0.8707    |      |*    |        0.001
 57   74_3           0.2977    0.2913    0.007052     .0063924    0.0254     0.2514    |      |     |        0.000
 58   75_3           0.2760    0.2739    .0068129      0.00214    0.0255     0.0840    |      |     |        0.000
 59   70_8           0.1956    0.1860      0.0152     .0095926    0.0215     0.4453    |      |     |        0.002
 60   72_8           0.2222    0.2130      0.0152     .0092593    0.0215     0.4298    |      |     |        0.002
 61   73_8           0.2356    0.2348      0.0152      7.5E-04    0.0215     0.0347    |      |     |        0.000
 62   74_8           0.3200    0.2981      0.0152      0.0219     0.0215     1.0160    |      |**   |        0.011
 63   75_8           0.3111    0.2838      0.0152      0.0273     0.0215     1.2691    |      |**   |        0.017
 64   73_4           0.2300    0.2342    .0093289     -.004167    0.0247    -0.1688    |      |     |        0.000
 65   74_4           0.3133    0.3101    .0087954     .0032593    0.0249     0.1310    |      |     |        0.000
 66   75_4           0.2667    0.2732    .0087954     -.006556    0.0249    -0.2635    |      |     |        0.000
```

(continued)

FIGURE 13.8   (continued)

							plot	
67	70_3	0.1757	0.1767	0.007617	-.001022	0.0253	-0.0405	0.000
68	71_3	0.2167	0.1908	0.007052	-0.0258	0.0254	-1.0160	0.002
69	72_3	0.1700	0.1909	0.007052	-0.0209	0.0254	-0.8226	0.001
70	73_3	0.1533	0.2186	0.0073182	-0.0653	0.0254	-2.5743	0.012
71	74_3	0.3133	0.2913	0.007052	-0.0221	0.0254	0.8676	0.001
73	70_1	0.2660	0.2739	0.0068129	-0.00786	0.0255	-0.3083	0.000
74	71_1	0.2000	0.2044	0.0093289	-0.004414	0.0247	-0.1788	0.000
75	72_1	0.2000	0.2141	0.0087954	-0.0141	0.0247	-0.5708	0.001
76	73_1	0.2160	0.2143	0.0087954	-0.0143	0.0249	-0.5748	0.001
77	74_1	0.2160	0.2401	0.0087954	-0.0241	0.0249	-0.9692	0.002
78	75_1	0.3289	0.3085	0.0087954	0.0204	0.0249	0.8190	0.002
79	73_1	0.3289	0.3382	0.0087954	-0.0272	0.0249	-0.3727	0.000
80	75_1	0.2050	0.1909	0.0087052	0.0141	0.0254	0.5539	0.000
81	73_3	0.2979	0.2186	0.0073182	0.0793	0.0254	3.1291	0.017
82	74_3	0.3133	0.2913	0.007052	0.0221	0.0254	0.8676	0.001
83	71_2	0.2533	0.2739	0.0068129	-0.0205	0.0255	-0.8052	0.001
84	75_3	0.2088	0.2155	0.009973	-0.00673	0.0244	-0.2758	0.000
85	73_2	0.2088	0.2136	0.009973	-0.004781	0.0244	-0.1957	0.000
86	72_2	0.2550	0.2499	0.009973	0.0050786	0.0244	0.2079	0.000
87	75_2	0.3200	0.3312	0.009973	-0.0112	0.0244	-0.4600	0.001
88	71_2	0.2560	0.3018	0.009973	-0.0458	0.0241	-1.9005	0.015
89	72_2	0.1848	0.1986	0.0108	-0.0118	0.0241	-0.5720	0.000
90	73_2	0.2068	0.2136	0.009973	0.008738	0.0244	0.3577	0.000
91	74_2	0.2068	0.2499	0.009973	0.0050786	0.0244	0.2079	0.000
92	71_2	0.2550	0.3312	0.009973	-0.001238	0.0244	-0.0507	0.000
93	70_1	0.3300	0.3018	0.0108	0.0182	0.0241	0.7565	0.002
94	72_1	0.3200	0.2044	0.0093289	0.0156	0.0247	-0.1048	0.001
95	74_1	0.2200	0.2141	0.0087954	-0.002587	0.0249	-0.1126	0.000
96	73_1	0.2141	0.2143	0.0087954	-0.002801	0.0249	-0.3422	0.000
97	75_1	0.2115	0.3085	0.0087954	-0.00115	0.0249	-0.4620	0.000
98	71_1	0.3000	0.1767	0.007617	-0.003356	0.0253	-0.1328	0.000
99	72_1	0.3267	0.1908	0.007052	-0.005833	0.0254	-0.2294	0.000
100	74_1	0.1733	0.1909	0.007052	-0.005915	0.0254	-0.2327	0.000
101	75_1	0.1850	0.2186	0.0073182	0.0064062	0.0254	0.2527	0.000
102	70_3	0.2250	0.2186	0.007052	-0.001524	0.0254	-0.0599	0.001
103	71_3	0.2897	0.2739	0.007052	-0.0171	0.0255	-0.6724	0.006
104	72_3	0.2910	0.2201	0.0068129	-0.004089	0.0236	-0.1733	0.000
105	73_3	0.2160	0.2196	0.0108	-0.0296	0.0241	-1.2301	0.000
106	74_3	0.1900	0.2279	0.0108	-0.007905	0.0244	-0.3236	0.000
107	75_3	0.2200	0.2279	0.009973	5.0E-04	0.0247	0.0203	0.000
108	70_4	0.2347	0.2342	0.0093289	-0.0032593	0.0249	-0.3037	0.002
109	71_4	0.3133	0.3101	0.0087954	-0.007556	0.0249	-0.5687	0.001
110	72_4	0.2657	0.2732	0.0087954	1.4E-17	0.0236	-0.2614	0.001
111	73_4	0.1920	0.1920	0.0264	-0.0134	0.0241	-0.5965	0.008
112	75_4	0.2067	0.2201	0.0118	-0.06296	0.0244	-1.6544	0.000
114	70_7	0.2133	0.2196	0.0108	-0.0146	0.0247	-0.1370	0.000
115	71_4	0.1933	0.2279	0.0093289	-0.0408	0.0249	-0.2635	0.000
116	73_4	0.2667	0.3101	0.0087954	-0.003407	0.0249	.0038986	0.002
117	72_4	0.2280	0.2279	0.009973	-0.006556	0.0247	-0.2498	0.000
118	75_4	0.2867	0.2342	0.0093289	9.5E-05	0.0244	-0.9409	0.000
119	73_4	0.2600	0.3101	0.0087954	-0.006167	0.0247	-0.6315	0.000
120	74_4	0.2134	0.2342	0.0093289	-0.0234	0.0249	0.5641	0.002
121	71_4	0.2279	0.2732	0.0087954	-0.0132	0.0249	0.5616	0.001
122	70_1	0.2232	0.2044	0.0093289	0.0089864	0.0247	0.3597	0.001
123	72_1	0.2550	0.2141	0.0087954	0.0089495	0.0249	0.5985	0.000
124	73_1	0.3067	0.2401	0.0087954	0.0149	0.0249	-0.0742	0.001
125	74_1	0.3400	0.3085	0.0087954	-0.001847	0.0249	-1.3566	0.000
126	75_1	0.1667	0.1977	0.0132	0.0018395	0.0236	-0.3390	0.013
127	70_6	0.2080	0.2160	0.0118	-0.0310	0.0236	1.2E-14	0.001
128	72_6	0.2240	0.2160	0.0118	-0.008	0.0236	1.3E-14	0.000
129	73_6	0.2480	0.2480	0.0132	2.8E-16	0.0229		
130	74_6	0.3120	0.3120	0.0118	3.1E-16	0.0236		

FIGURE 13.8 (continued)

OBS	ID	ACTUAL	PREDICT VALUE	STD ERR PREDICT	RESIDUAL	STD ERR RESIDUAL	STUDENT RESIDUAL	-2-1-0 1 2	COOK'S D
133	75_6	0.2640	0.2672	0.0118	-0.0032	0.0236	-0.1356		0.000
134	70_1	0.1922	0.2044	0.093289	-0.0122	0.0247	-0.4948		0.001
135	71_1	0.2185	0.2141	0.093289	.0044625	0.0247	0.1808		0.000
136	72_1	0.2185	0.2143	0.087954	.0042495	0.0249	0.1708		0.000
137	73_1	0.2400	0.2401	0.087954	-1.1E-04	0.0249	-.004466		0.000
138	74_1	0.3067	0.3085	0.087954	-.001847	0.0249	-0.0742		0.000
139	70_2	0.2052	0.1986	0.0108	.0066222	0.0241	0.2749		0.000
140	71_2	0.2200	0.2155	0.009973	.0044619	0.0244	0.1826		0.000
141	72_2	0.2200	0.2136	0.009973	.0064419	0.0244	0.2628		0.001
142	73_2	0.2600	0.2499	0.009973	0.0101	0.0244	0.4126		0.001
143	74_2	0.2640	0.3312	0.009973	-0.0672	0.0244	-2.7524	*****	0.026
144	75_2	0.3067	0.3018	0.0108	.0048889	0.0241	0.2030		0.000
145	70_3	0.1450	0.1767	0.007617	-.0317	0.0253	-1.2544	**	0.003
146	71_3	0.1650	0.1908	0.007052	-.0258	0.0254	-1.0160	**	0.002
147	72_3	0.1750	0.1909	0.007052	-.0159	0.0254	-0.6259	*	0.001
148	70_5	0.1969	0.2088	0.007052	-0.0119	0.0215	-0.5516	*	0.003
149	71_5	0.2000	0.2162	0.0152	-0.0162	0.0215	-0.7504	*	0.006
150	72_5	0.2000	0.2045	0.0132	-.004479	0.0229	-0.1960		0.000
151	73_5	0.2400	0.2167	0.0132	0.0233	0.0229	1.0211	**	0.007
152	74_5	0.3000	0.3241	0.0118	-0.0241	0.0236	-1.0226	**	0.005
153	75_5	0.2667	0.2543	0.0118	0.0124	0.0236	0.5254	*	0.001
154	71_7	0.1900	0.1978	0.0152	-.007778	0.0215	-0.3610	*	0.001
155	73_7	0.1900	0.2198	0.0152	-0.0298	0.0215	-1.3822	***	0.025
156	74_7	0.1950	0.2281	0.0152	-0.0331	0.0215	-1.5369	***	0.006
157	75_7	0.2900	0.2667	0.0152	-0.0167	0.0215	-0.7736	*	0.001
158	70_2	0.2667	0.2667	0.0152	6.0E-16	0.0215	2.8E-14		0.000
159	71_2	0.1848	0.1986	0.0108	-0.0138	0.0241	-0.5720	*	0.000
160	72_2	0.2115	0.2155	0.009973	-.004038	0.0244	-0.1653		0.000
161	73_2	0.2068	0.2136	0.009973	-.006781	0.0244	-0.2776		0.000
162	74_2	0.2524	0.2499	0.009973	-.0025286	0.0244	-0.1035		0.000
163	70_3	0.3300	0.3312	0.009973	-.001238	0.0253	-0.0507		0.000
164	71_3	0.1867	0.1767	0.007617	-.0099778	0.0254	-0.3950		0.000
165	72_3	0.2000	0.1908	0.007052	.0091667	0.0254	0.3605		0.000
166	73_3	0.2000	0.1908	0.007052	.0090845	0.0254	0.3605		0.000
167	74_3	0.2407	0.2186	0.007052	0.0221	0.0254	0.5573		0.000
168	75_3	0.2977	0.2913	0.0108	.063924	0.0254	0.8707	*	0.001
169	70_2	0.1867	0.1986	0.007052	.00214	0.0254	0.2514		0.000
170	71_2	0.2067	0.2155	0.0108	-0.0119	0.0255	0.0840		0.000
171	72_2	0.2067	0.1986	0.009973	-.008871	0.0241	-0.4945	*	0.000
172	73_2	0.2040	0.2136	0.009973	-.006914	0.0244	-0.3632	*	0.000
173	74_2	0.3067	0.2499	0.0108	-0.0459	0.0244	-1.8798	***	0.012
174	75_2	0.2560	0.3312	0.009973	-0.0246	0.0244	-1.0058	**	0.004
175	70_4	0.2000	0.3018	0.0108	-0.0458	0.0241	-1.9905	***	0.015
176	71_4	0.2067	0.2196	0.0108	-0.0201	0.0236	-0.8512	*	0.001
177	72_4	0.2267	0.2279	0.0108	-.0279	0.0241	-1.1423	**	0.005
178	73_4	0.3000	0.2342	0.093289	-0.0130	0.0247	-0.3039		0.000
179	74_4	0.2667	0.3101	0.087954	-0.0075	0.0249	-0.4050		0.000
180	75_4	0.1842	0.2732	0.087954	-0.0101	0.0249	-0.2635		0.000
181	71_3	0.1859	0.1908	0.007052	-.006556	0.0254	-0.0622		0.000
182	72_3	0.2208	0.1909	0.007052	-.006667	0.0254	-0.1986		0.000
183	73_3	0.3040	0.2186	0.007182	-.005049	0.0254	0.5005	*	0.001
184	75_3	0.2944	0.2913	0.007052	0.0127	0.0254	0.8058	*	0.001
185	70_6	0.2080	0.1986	0.0068129	.0022062	0.0255	0.4522		0.000
186	71_6	0.1977	0.1977	0.0132	0.0205	0.0236	5.9E-15		0.001
187	72_6	0.2160	0.2160	0.0118	0.0132	0.0229	0.3590	*	0.001
188	73_6	0.2160	0.2160	0.0132	1.4E-16	0.0236	1.2E-14		0.000
189	74_6	0.2480	0.2480	0.0118	-.0C08	0.0229	1.3E-14		0.000
190	75_6	0.3120	0.3120	0.0118	2.8E-16	0.0236	-0.5424	*	0.000
191	70_8	0.2800	0.2800	0.0118	3.1E-16	0.0236	-1.5145	***	0.002
192	71_8	0.1533	0.1860	0.0187	-0.0128	0.0215	-1.1077	**	0.024
193	72_8	0.1867	0.2073	0.0152	-0.0326	0.0187	-1.1206	**	0.026
194	73_8	0.1867	0.2130	0.0152	-0.0207	0.0215	0.1544		0.016
195	74_8	0.2130	0.2348	0.0152	-0.0263	0.0215	-0.0206		0.000
196	73_8	0.2381	0.2130	0.0152	.0033258	0.0215	-1.1206		0.000
197	74_8	0.2977	0.2981	0.0152	-4.4E-04	0.0215	-0.0206		0.000
198	75_8	0.2977	0.2838	0.0152	0.0139	0.0215	0.6450	*	0.004

(continued)

```
199  71_3   0.1725   0.1908   0.007052   -0.0183     0.0254   -0.7210   *        0.001
200  72_3   0.1887   0.1909   0.007052   -.002165    0.0254   -0.0852            0.000
201  73_3   0.2100   0.2186   .0073182   -.008594    0.0254   -0.3390            0.000
202  74_3   0.2900   0.2913   .0073182   -.001174    0.0254   -0.0501            0.000
203  75_3   0.2700   0.2739   .0068129   -0.00386    0.0255   -0.1514            0.000
204  70_1   0.2000   0.2044   .0093289   -0.004414   0.0247   -0.1788            0.000
205  71_1   0.2058   0.2141   .0093289   -.008287    0.0247   -0.3358            0.000
206  72_1   0.2058   0.2143   .0087954   -.008501    0.0249   -0.3417            0.000
207  73_1   0.2350   0.2401   .0087954   -.005111    0.0249   -0.2055            0.000
208  74_1   0.2793   0.3085   .0087954   -0.0292     0.0249   -1.1743   ****     0.004
209  75_1   0.3111   0.3382   .0087954   -0.0270     0.0249   -1.0873   ***      0.003
210  70_6   0.2080   0.1977   0.0132      0.0103     0.0229    0.4522            0.001
211  71_6   0.2240   0.2160   0.0118      0.008      0.0236    0.3390            0.000
212  72_6   0.2080   0.2160   0.0118     -0.008      0.0236   -0.3390            0.001
213  73_6   0.2480   0.2480   0.0132      2.8E-16    0.0229    1.2E-14           0.000
214  74_6   0.3120   0.3120   0.0118      3.1E-16    0.0236    1.3E-14           0.000
215  75_6   0.2640   0.2672   0.0118     -0.0032     0.0236   -0.1356            0.000
216  70_5   0.2067   0.2045   0.0132      .0021875   0.0229    0.0957            0.000
217  71_5   0.2333   0.2167   0.0132      0.0167     0.0229    0.7294   *        0.004
218  73_5   0.3040   0.3241   0.0118     -0.0201     0.0236   -0.8531            0.004
219  74_5   0.2500   0.2543   0.0118     -.004267    0.0236   -0.1808            0.029
220  75_5   0.2450   0.2088   0.0152      0.0362     0.0215    1.6799   ***      0.034
221  70_5   0.2550   0.2162   0.0152      0.0388     0.0215    1.8025   ***      0.007
222  71_5   0.2100   0.2045   0.0132      .005208    0.0229    1.0211   **       0.069
223  72_5   0.2400   0.2167   0.0132      0.0233     0.0229    1.0211            0.000
224  73_5   0.4100   0.3241   0.0118      0.0859     0.0236    3.6384   *****    0.069
225  74_5   0.2160   0.2160   0.0118      1.4E-16    0.0236    5.9E-15           0.000
226  71_6   0.2120   0.2160   0.0118     -0.004      0.0236    1.3E-14           0.000
227  72_6   0.3120   0.3120   0.0118     -0.0032     0.0236   -0.1356            0.000
228  74_6   0.2640   0.2672   0.0118      0.0781     0.0236    2.7269   ******   0.039
229  75_6   0.2844   0.2201   0.0118      0.0761     0.0241    3.2444   ******   0.044
230  70_4   0.2978   0.2196   .009973     0.0761     0.0244    3.1150   *******  0.034
231  71_4   0.3040   0.2342   .0093289    0.0698     0.0247    2.8293   *******  0.024
232  72_4   0.3040   0.3101   .0087954    0.0953     0.0249    3.8292   *******  0.038
233  73_4   0.4053   0.3101   .0087954    0.0734     0.0249    2.9523   *******  0.023
234  74_4   0.3467   0.3101   .0087954   -.006074    0.0249   -0.2442            0.000
235  75_4   0.3040   0.2732   .0087954   -0.0199     0.0249   -0.7995   *        0.002
236  70_3   0.2533   0.1767   0.007617   -0.0167     0.0253   -0.6606   *        0.001
237  72_3   0.1600   0.1908   .0007052   -0.0208     0.0254   -0.8194   *        0.001
238  73_3   0.1700   0.1909   0.007052   -0.0109     0.0254   -0.4293            0.000
239  75_3   0.2080   0.2186   .0073182    0.0347     0.0254    1.3703   **       0.000
240  70_1   0.2533   0.2913   .0068129   -.007624    0.0254   -0.2999            0.003
241  71_1   0.2836   0.2745   .0068129    6.4E-04    0.0255   -0.0251            0.000
242  73_1   0.2745   0.2044   .0093289   -.004414    0.0247   -0.1788            0.007
243  75_1   0.2000   0.2141   .0093289   -0.0141     0.0247   -0.5708            0.000
244  70_1   0.2000   0.2143   .0087954   -0.0143     0.0249   -0.5748            0.001
245  71_1   0.2400   0.2401   .0087954   -1.1E-04    0.0249   -.004466           0.001
246  73_1   0.3289   0.3085   .0087954    0.0204     0.0249    0.8190            0.000
247  74_1   0.3382   0.3382   .0087954   -.009272    0.0249   -0.3727            0.002
248  75_1   0.1933   0.2201   0.0118     -0.0268     0.0236   -1.1337   **       0.000
249  70_4   0.2167   0.2196   0.0108     -.002963    0.0236   -1.1230   **       0.007
250  71_4   0.2133   0.2279   .009973    -0.0146     0.0241   -0.5965   *        0.001
251  72_4   0.2267   0.2342   .0093289   -0.0075     0.0247   -0.3039            0.001
252  73_4   0.2342   0.3101   .0087954   -0.0221     0.0249   -0.8873   *        0.000
253  74_4   0.3101   0.2732   .0087954   -.006556    0.0249   -0.2635            0.002
254  75_4   0.2667   0.3241   0.0118     -0.0208     0.0236   -0.8813   *        0.004
255  70_3   0.3033   0.2543   .007617    -9.3E-04    0.0236   -0.0395            0.000
256  71_3   0.3241   0.1767   .007617     0.0313     0.0253    1.2394   *        0.000
257  72_3   0.2543   0.1908   .007052     0.0158     0.0254    0.6227   *        0.001
258  73_3   0.1767   0.2186   .0073182    0.0347     0.0254    1.3703   **       0.003
259  74_3   0.2080   0.2913   .0073182   -0.0103     0.0254   -0.1514            0.003
260  75_3   0.2067   0.2739   .0068129   .0063924    0.0255   -0.4522            0.000
261  70_6   0.2080   0.2186   .0073182   -0.00386    0.0254    0.03347           0.001
262  71_6   0.2977   0.2739   .0068129   -0.0103     0.0236   -0.1514            0.000
263  72_6   0.2700   0.2913   0.0132      1.4E-16    0.0229    5.9E-15           0.001
264  72_6   0.2120   0.2160   0.0118     -0.004      0.0236   -0.1695            0.000
```

FIGURE 13.8 (continued)

FIGURE 13.8 (continued)

OBS	ID	ACTUAL	PREDICT VALUE	STD ERR PREDICT	RESIDUAL	STD ERR RESIDUAL	STUDENT RESIDUAL	-2-1-0 1 2	COOK'S D.
265	73_6	0.2480	0.2480	0.0132	2.8E-16	0.0229	1.2E-14		0.000
266	74_6	0.3120	0.3120	0.0118	3.1E-16	0.0236	1.3E-14		0.000
267	75_6	0.2640	0.2672	0.0118	-0.0032	0.0236	-0.1356		0.005
268	71_4	0.1933	0.2196	0.0108	-0.0263	0.0241	-1.0917	**	0.005
269	72_4	0.2167	0.2279	0.009973	-0.0112	0.0244	-0.4600		0.001
270	73_4	0.2300	0.2342	.0093289	-.004167	0.0247	-0.1688		0.000
271	74_4	0.2667	0.3101	.0087954	-.0367	0.0249	-1.4769	**	0.006
272	75_4	0.2667	0.2732	.0087954	-.006556	0.0249	-0.2635		0.008
273	70_2	0.2320	0.1986	0.0108	0.0334	0.0241	1.3876	***	0.006
274	71_2	0.2480	0.2155	0.009973	0.0325	0.0244	1.3288	***	0.007
275	72_2	0.2480	0.2136	0.009973	0.0344	0.0244	1.4090	***	0.002
276	73_2	0.2680	0.2499	0.009973	0.0181	0.0244	0.7401	*	0.000
277	74_2	0.3280	0.3312	0.009973	-.003238	0.0244	-0.1326		0.018
278	75_2	0.3520	0.3018	0.0108	0.0502	0.0241	2.0850	***	0.000
279	70_3	0.1800	0.1767	0.007617	.0033111	0.0253	0.1311		0.000
280	71_3	0.2000	0.1908	0.007052	.0091667	0.0254	0.3605		0.000
281	72_3	0.2000	0.1909	0.007052	.0090845	0.0254	0.3573		0.012
282	73_3	0.1533	0.2186	0.0073182	-0.0653	0.0254	-2.5743	*****	0.010
283	74_3	0.2267	0.2913	0.007052	-0.0646	0.0254	-2.5410	*****	0.000
284	75_3	0.2700	0.2739	0.0068129	-.00386	0.0255	-0.1514		0.013
285	70_5	0.1845	0.2088	0.0152	-0.0243	0.0215	-1.1283	***	0.012
286	71_5	0.1935	0.2162	0.0152	-0.0227	0.0215	-1.0521	***	0.000
287	72_5	0.2012	0.2045	0.0132	-.003229	0.0229	-0.1413		0.053
288	73_5	0.1533	0.2167	0.0132	-0.0633	0.0229	-2.7716	*****	0.004
289	74_5	0.2533	0.2543	0.0118	-0.0208	0.0236	-0.8813	*	0.000
290	75_5	0.1980	0.1986	0.0108	-9.3E-04	0.0241	-0.0395		0.000
291	70_2	0.2070	0.2155	0.009973	5.3E-04	0.0244	-0.0240		0.000
292	71_2	0.1980	0.2136	0.009973	-.008538	0.0244	-0.3495		0.001
293	72_2	0.2550	0.2499	0.009973	-0.0156	0.0244	-0.6378	*	0.001
294	73_2	0.2550	0.3312	0.009973	.0050786	0.0244	0.2079		0.000
295	74_2	0.4400	0.3312	0.009973	0.1088	0.0244	4.4522	*****	0.069
296	75_2	0.3200	0.3018	0.0108	0.0182	0.0241	0.7565	**	0.002
297	73_1	0.2400	0.2401	.0087954	-1.1E-04	0.0249	-.004466		0.000
298	74_1	0.2850	0.3085	.0087954	-0.0235	0.0249	-0.9452	*	0.002
299	75_1	0.3289	0.3382	.0087954	-.009272	0.0249	-0.3727		0.000
300	72_1	0.2232	0.2143	.0087954	-.0089495	0.0249	0.3597		0.000
301	73_1	0.2400	0.2401	.0087954	-1.1E-04	0.0249	-.004466		0.000
302	74_1	0.3067	0.3085	.0087954	-.001847	0.0249	-0.0742		0.000
303	75_1	0.3750	0.3382	.0087954	0.0368	0.0249	1.4809	***	0.006

SUM OF RESIDUALS   5.10980E-13
SUM OF SQUARED RESIDUALS   0.1775372
PREDICTED RESID SS (PRESS)   0.2549934

FIGURE 13.9   A Plot of Residuals Versus Predicted Value $\hat{y}$: Numbers Identify the Year Associated with the Residual

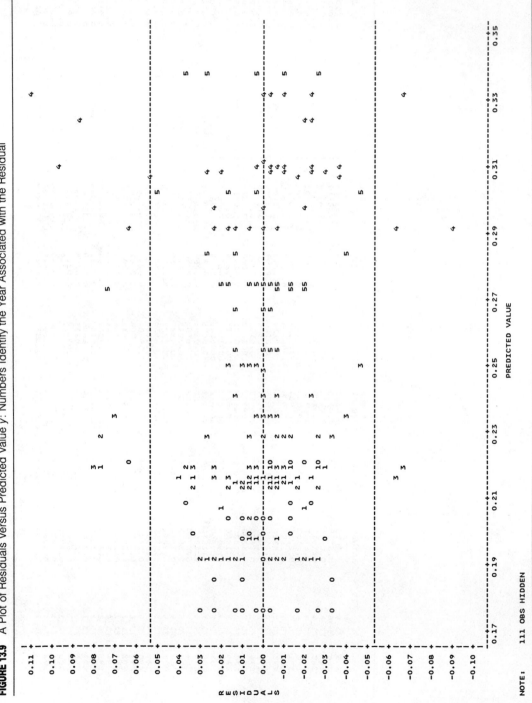

FIGURE 13.10   A Plot of Residuals Versus Predicted Value $\hat{y}$; Numbers Identify the Market Associated with the Residual

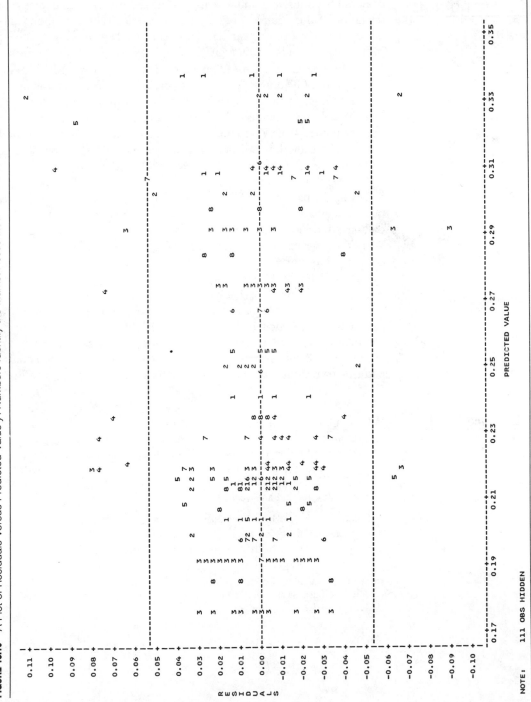

We have demonstrated two applications of regression analysis for the detection of possible collusive bidding. The first procedure seeks to detect an overly large deviation of the mean low bid for a particular market from the mean low bid for all markets for a given year. The second procedure investigates the deviation of individual low bids *within* a given market–year.

The first procedure is based on a model for a stable market. This model, which assumes that inflationary year-to-year increases in costs should apply uniformly (more or less) to all markets and affect the mean low-bid prices for all markets in a similar way, contains only main effect terms for market and year. An unstable market model—one that allows for market–year interactions—models the situation where the low-bid price for one or more markets deviates substantially from the mean of the low-bid prices for all markets in one or more years. It is our contention that forces operating in one or more market–years, and that these forces might be explained by unusual market conditions or, in the case of large positive deviations, by collusive bidding. Note that we do not mean to imply that collusive bidding could not occur when market conditions appear to be stable. We are saying that unstable market conditions may be a clue to the presence of collusive bidding, and the procedure that we have described will detect this condition if it exists.

The second procedure that we employed required an examination of the regression residuals, the deviations of the observed low bids about the estimated mean low bid for a particular market in a particular year. We would expect these residuals to vary in a random manner and would expect most of them to lie within $2s$ or $3s$ of their mean (0). Unusually large positive residuals or unusual patterns of residuals might provide a clue to the presence of collusive bidding.

| | | | | | | | | | | | |

EXERCISES 13.1–13.2

13.1 In addition to the predicted values and residuals, Figure 13.8 lists the value of Cook's distance, D_i, for each of the 303 observations in the analysis. Recall that D_i measures the overall influence for observation i on the analysis (see Section 6.5). Check the printout in Figure 13.8 and identify those observations with the largest Cook's D-values. These observations might be the result of collusive bidding in the corresponding market and year.

13.2 Refer to observations selected in Exercise 13.1. Do any of these observations have *unusual* influence on the regression results? Use the fact that the 50th percentile of an F-distribution with $\nu_1 = 48$ and $\nu_2 = 255$ degrees of freedom is .9887.

REFERENCE

Rothrock, T. P. and McClave, J. T. "An analysis of bidding competition in the Florida school bidding competition using a statistical model." Paper presented at the TIMS/ORSA Joint National Meeting, Chicago, 1979.

OBJECTIVE

To show how regression modeling can aid in establishing equitable rates of compensation

CONTENTS

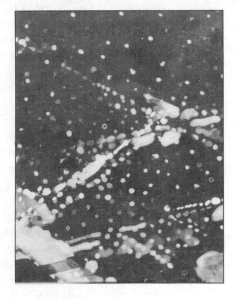

USING REGRESSION ANALYSIS TO DETERMINE EQUITABLE RATES OF COMPENSATION

BACKGROUND

Statistics can be used to assist in the determination of equitable rates of compensation to public utilities, health facilities, and others, for services supplied to federal or state governments or, as in the case of utilities, to the public. This chapter considers the problem of determining equitable Medicaid compensation to nursing homes in the state of Florida.

Nursing homes must apply to the state government and be approved for the care of Medicaid patients. Reimbursement is based on a specific number of beds per nursing home, a specific type of care, and the cost of such services. The reimbursement for a given nursing home is based on the number of beds approved for the Medicaid program times the number of days each bed is occupied. The appropriate maximum dollar amount of compensation per bed-day is set by each state.

The study we report arose as a result of a dispute between the Florida Health Care Association, Inc., and the Department of Health and Rehabilitative Services (DHRS) of the state of Florida over the department's procedure for establishing an equitable maximum level of compensation for each nursing home. Putting the issue in a nutshell, the state's procedure for determining equitable compensation per bed-day for a particular type of nursing care implies that the costs per bed-day differ from one nursing home to another due only to random variations in capital costs, operating costs, and inefficiencies of operation. Consequently, a single maximum rate of compensation should be appropriate for all nursing homes. In contrast, the Florida Health Care Association, Inc., contends that the cost per bed-day depends on the location of the nursing home within the state, the size of the nursing home, and its type (proprietary, public owned, etc.), and hence that these factors should be considered in setting the rate of compensation for a particular nursing home.

We will examine the DHRS procedure for establishing an equitable maximum rate of compensation in Section 14.2. In Section 14.3 we will set up a model, and in Section 14.4 we will subject the DHRS data to a regression analysis and present strong evidence that the cost per bed-day is a function of location of a nursing home, its size, and its type. The regression analysis was performed for the Florida Health Care Association, Inc., by a private statistical consulting firm.*

THE DHRS PROCEDURE FOR ESTABLISHING MAXIMUM BED-DAY COMPENSATION

Each nursing home in the state of Florida with an approved Medicaid program is required to calculate (according to a specified procedure) its projected cost per bed-day for each type of approved nursing care. This cost per bed-day, along with the number of bed-days of care provided during the year, must be reported to the DHRS. The data are then used to establish a maximum bed-day rate of compensation for each type of care, and these rates are used to establish the maximum amount of compensation per bed-day that will be paid to any nursing home. To simplify our discussion, we will explain how the DHRS calculates the rate of compensation for a specific type of nursing care, "intermediate care." The rates for other types of nursing care would be determined in a similar manner.

*The data, analyses, and figures are presented in this chapter with the permission of Info Tech, Inc., Gainesville, Florida.

The maximum rate of bed-day compensation for intermediate nursing care is based on the distribution of the costs of all reported bed-days. For example, if nursing home 1 reported 50,000 bed-days with a cost of $29 per bed-day, then each of the 50,000 bed-days would be assigned a cost of $29 and the data obtained from nursing home 1 would be the set of fifty thousand 29's. Similarly, if nursing home 2 had 60,000 bed-days with a cost of $32 per bed-day, the data obtained from nursing home 2 would be a set of sixty thousand 32's. Consequently, if there were a total of 50,000,000 bed-days reported to the DHRS, then the bed-day distribution would contain 50,000,000 bed-day cost measurements.

The DHRS maximum rate of bed-day compensation is computed from the distribution of all reported bed-day costs by arranging the bed-day costs in order of magnitude and locating the 60th percentile for this distribution. This number is the maximum DHRS compensation for each bed-day of intermediate care. Therefore, 60% of all bed-days will be fully compensated at cost and 40% will receive undercompensation. (The 60th percentile for the distribution of projected bed-day costs for nursing homes in Florida was $23.46 in 1979, the year the study was performed.)

Why the arbitrary choice of the 60th percentile in the bed-day cost distribution as a fair rate for bed-day compensation? Why not the 50th percentile? And was the 60th percentile a choice that provided an equitable rate of compensation? Not so, say the nursing home managers. They note that it is a nursing home that is operated efficiently or inefficiently, not a bed-day, and it is a nursing home that will make a profit or show a deficit. Consequently, they argue that the distribution of bed-day costs for a given nursing home should be used to determine the rate of bed-day compensation for that particular nursing home. They note that the DHRS procedure gives more weight to the bed-day costs of larger nursing homes and that these homes might tend to have lower costs due to economies of scale. But most importantly, they contend (in opposition to the belief of the DHRS) that other factors in addition to efficiency of operation affect the cost per bed-day for a particular nursing home and that these factors should be considered when establishing an equitable maximum rate of compensation for a particular nursing home. We will now examine these contentions.

| | | | | | | | | | | |

SECTION 14.3

A MODEL RELATING COST PER BED-DAY TO LOCATION, TYPE, AND SIZE OF NURSING HOME

The Florida Health Care Association, Inc., believes that many variables, in addition to efficiency of operation, affect the bed-day costs for a particular type of nursing care. This study examines three: location of a nursing home, its type, and its size. If a regression analysis indicates that one or more of these factors is important in predicting bed-day costs, it supports the contention of the Florida Health Care Association, Inc., that additional variables, such as location, type, and size, should be considered in establishing an equitable rate for bed-day compensation. Further, it would indicate that a more extensive study should be conducted to develop a nursing care bed-day rate of compensation that takes into account other important variables that affect the bed-day costs.

The response y and the factors included in the study were

y = Projected 1979 bed-day cost for intermediate care at a specific nursing home

FACTORS

1. *Location* The state was divided into three regions which possessed similar economic and social characteristics *within* regions and different economic and social characteristics *between* regions: (a) south Florida, (b) central Florida, and (c) north Florida.

2. *Type of home* A, church-operated; B, government-owned; C, nonproprietary (charity-owned, etc.); D, proprietary (privately owned)

3. *Size of home* The maximum of bed-days available for one year, MXBD

Note that location (three levels) and type of home (four levels) are both qualitative factors, and their factor-level combinations identify $3 \times 4 = 12$ qualitative categories of nursing homes. If we imagine that for each of these categories, size is related to cost in a linear (first-order) manner, then the model relating y to the factors would graph as a straight line, one line for each location–type combination. The linear model for this sitaution is

$$
\begin{aligned}
E(y) = \beta_0 + &\overbrace{\beta_1 x_1 + \beta_2 x_2}^{\substack{\text{Main effect,}\\ \text{location}}} + \overbrace{\beta_3 x_3 + \beta_4 x_4 + \beta_5 x_5}^{\substack{\text{Main effect,}\\ \text{type}}} \\
& \left. \begin{aligned} &+ \beta_6 x_1 x_3 + \beta_7 x_1 x_4 + \beta_8 x_1 x_5 \\ &+ \beta_9 x_2 x_3 + \beta_{10} x_2 x_4 + \beta_{11} x_2 x_5 \end{aligned} \right\} \text{Location–type interaction} \\
&+ \overbrace{\beta_{12} x_6}^{\text{Size}} \\
& \left. \begin{aligned} &+ \beta_{13} x_6 x_1 + \beta_{14} x_6 x_2 + \beta_{15} x_6 x_3 + \beta_{16} x_6 x_4 \\ &+ \beta_{17} x_6 x_5 + \beta_{18} x_6 x_1 x_3 + \beta_{19} x_6 x_1 x_4 + \beta_{20} x_6 x_1 x_5 \\ &+ \beta_{21} x_6 x_2 x_3 + \beta_{22} x_6 x_2 x_4 + \beta_{23} x_6 x_2 x_5 \end{aligned} \right\} \begin{aligned} &\text{Size–location,} \\ &\text{size–type, and} \\ &\text{size–location–type} \\ &\text{interactions} \end{aligned}
\end{aligned}
$$

where

$$x_1 = \begin{cases} 1 & \text{if central location} \\ 0 & \text{otherwise} \end{cases} \qquad x_2 = \begin{cases} 1 & \text{if north location} \\ 0 & \text{otherwise} \end{cases}$$

$$x_3 = \begin{cases} 1 & \text{if church-operated} \\ 0 & \text{otherwise} \end{cases} \qquad x_4 = \begin{cases} 1 & \text{if government-operated} \\ 0 & \text{otherwise} \end{cases}$$

$$x_5 = \begin{cases} 1 & \text{if nonproprietary} \\ 0 & \text{otherwise} \end{cases}$$

x_6 = Maximum number of bed-days (in units of 10,000)

The first 12 terms, β_0 and those involving only the dummy variables x_1, $x_2, \ldots, x_5$, provide 12 different constants (y-intercepts)—one for each of the 12 location–type combinations. Addition of the term $\beta_{12} x_6$ introduces a first-order contribution to $E(y)$ that is dependent on the size x_6 of the nursing home. This term, along with the first 12, produces a model that graphs as a set of six parallel lines, each with a slope equal to β_{12} and with a y-intercept determined by the location–type combination.

Since we would not expect the rate of increase in cost for a unit increase in size to be the same for every location–type combination, we add the last 11 terms, the interactions of size x_6 with the dummy variable parts of the terms involving β_1, β_2, . . . , β_{11}. These terms permit a different slope for every location–size combination and thus complete our model.*

| | | | | | | | | | | | |

SECTION 14.4

THE REGRESSION ANALYSIS

The SAS GLM computer printout for the $n = 254$ nursing homes included in the study is shown in Figure 14.1 (page 648). The first question that we would ask is whether the factors location, type, and size of nursing home contribute information for the prediction of y. If they do not, then $\beta_1 = \beta_2 = \cdots = \beta_{23} = 0$. To test this hypothesis, we turn to the analysis of variance and compare the mean square for MODEL (i.e., for the set of parameters β_1, β_2, . . . , β_{23}), 147.77884165, with MSE = 17.28870730. Comparing this F-value, 8.55, with the tabulated $F_{.05} \approx 1.53$ (based on $\nu_1 = 23$ and $\nu_2 = 230$ df), we reject the null hypothesis that $\beta_1 = \beta_2 = \cdots = \beta_{23} = 0$. The data provide sufficient evidence to indicate that at least one of the factors, location, type, or size, contributes information for the prediction of bed-day costs. This contradicts the contention of the DHRS that bed-day costs are independent of location, type, and size of the nursing home.

The next step in the analysis is to examine the various sources of variation (in the ANOVA table), the F-values for the corresponding F-tests, and the significance level of the tests. Checking the TYPE III sums of squares, F-values and p-values for size–location (MXBDPIOG*REGION), size–type (MXBDPIOG* HOMETYPE), and size–location–type (MXBDPI*REGION*HOMETY) interactions with 2, 3, and 6 df, respectively, we note that two of the three (size–location and size–location–type) are highly significant ($p \approx .01$). This tells us that there is evidence of a difference in the slopes of the lines that depict the bed-day cost as a function of size for the 12 location–type combinations. Further, it supports the contention that size, location, and type of nursing home contribute information for the prediction of bed-day costs. To see this more clearly, we will examine the graphs of the cost–size lines for the 12 location–type combinations.

The parameter estimates are given in the ESTIMATE column of the computer printout in Figure 14.1. Substituting these values into the prediction equation, we obtain

$$\hat{y} = 32.360 - 4.149x_1 - 6.170x_2 + 11.564x_3 - 8.651x_4 - 2.157x_5$$
$$+ .075x_1x_3 + 17.457x_1x_4 + .394x_1x_5 - 8.020x_2x_3 + 8.240x_2x_4$$
$$+ 5.237x_2x_5 - .598x_6 + .294x_6x_1 + .590x_6x_2 - .604x_6x_3$$
$$+ 6.466x_6x_4 + 1.805x_6x_5 - .006x_6x_1x_3 - 8.025x_6x_1x_4$$
$$- 1.346x_6x_1x_5 + 1.094x_6x_2x_3 - 5.717x_6x_2x_4 - 2.941x_6x_2x_5$$

*It is conceivable that one or more of the curves relating cost per bed-day to size is curvilinear, a situation for which a second-order model might be appropriate. If we believe this to be true, we should add a term involving x_6^2 along with interaction terms involving x_6^2 with the dummy variable portions of the terms involving β_1, β_2, . . . , β_{11}. We discuss this possibility in Section 14.4.

FIGURE 14.1 First-Order Model Relating Nursing Home Costs to Location, Type, and Size

DEPENDENT VARIABLE: INTRAD2   1979-80 INTERM. RATE; HEW INFL. RATES

SOURCE	DF	SUM OF SQUARES	MEAN SQUARE	F VALUE	PR > F	R-SQUARE	C.V.
MODEL	23	3398.91335784	147.77884165	8.55	0.0001	0.460850	14.6259
ERROR	230	3976.40267826	17.28870730			ROOT MSE	INTRAD2 MEAN
CORRECTED TOTAL	253	7375.31603610				4.15796913	28.42880175

SOURCE	DF	TYPE I SS	F VALUE	PR > F	DF	TYPE III SS	F VALUE	PR > F
REGION	2	900.99940564	26.06	0.0001	2	99.16786174	2.87	0.0588
HOMETYPE	3	1120.59490066	21.61	0.0001	3	295.29382850	5.69	0.0010
MXBDP1OG	1	110.73855193	6.41	0.0120	1	8.92760344	0.52	0.4731
REGION*HOMETYPE	6	691.39983445	6.67	0.0001	6	150.75215834	1.45	0.1953
MXBDP1OG*REGION	2	30.48582586	0.88	0.4155	2	150.63311671	4.36	0.0139
MXBDP1OG*HOMETYPE	3	245.14257884	4.73	0.0034	3	86.84574205	1.67	0.1715
MXBDP1*REGION*HOMETY	6	299.55226026	2.89	0.0099	6	299.55226026	2.89	0.0099

PARAMETER		ESTIMATE		T FOR H0: PARAMETER=0	PR > |T|	STD ERROR OF ESTIMATE
INTERCEPT		32.35960088 B		28.18	0.0001	1.14842781
REGION	X1	-4.14933097 B		-2.83	0.0051	1.46630853
	X2	-6.17040838 B		-3.58	0.0004	1.72387687
HOMETYPE	X3	11.56385643 B		2.51	0.0127	4.60223414
	X4	-8.65090634 B		-1.04	0.3010	8.34575108
	X5	-2.15654512 B		-0.90	0.3700	2.40105918
REGION*HOMETYPE	X1*X3	0.07497234 B		0.01	0.9892	5.52963455
	X1*X4	17.45727580 B		1.77	0.0775	9.84601733
	X1*X5	-0.39403229 B		0.10	0.9207	3.95545452
	X2*X3	-8.01965929 B		-1.39	0.1665	5.77870349
	X2*X4	8.24031980 B		0.90	0.3679	9.13404134
	X2*X5	5.23716883 B		1.08	0.2806	4.84214181
MXBDP1OG	X6	-0.59780840 B		-2.61	0.0095	0.22872237
MXBDP1OG*REGION	X1*X6	0.29436696 B		0.92	0.3573	0.31917296
	X2*X6	0.58959911 B		1.41	0.1585	0.41773363
MXBDP1OG*HOMETYPE	X3*X6	-0.60379133 B		-1.03	0.3061	0.58863754
	X4*X6	6.46593501 B		3.28	0.0012	1.97102201
	X5*X6	1.80486095 B		3.60	0.0004	0.50164686
MXBDP1*REGION*HOMETY	X1*X3*X6	-0.00587825 B		-0.01	0.9935	0.72039857
	X1*X4*X6	-8.02472147 B		-3.06	0.0025	2.62099253
	X1*X5*X6	-1.34641188 B		-1.76	0.0803	0.76651534
	X2*X3*X6	1.09354291 B		0.99	0.3246	1.10774097
	X2*X4*X6	-5.71728716 B		-2.50	0.0130	2.28353471
	X2*X5*X6	-2.94051534 B		-2.15	0.0326	1.36780880

The equation of the line depicting predicted bed-day cost $\hat{y}$ as a function of nursing home size x_6 for a church-operated nursing home in south Florida can be obtained by substituting the appropriate values for the dummy variables, namely $x_1 = x_2 = 0$, $x_3 = 1$, $x_4 = x_5 = 0$, into the prediction equation. The resulting equation is

$$\hat{y} = 43.924 - 1.202x_6$$

The equations of the corresponding lines for the other 11 location–type combinations can be obtained in a similar manner. The graphs of these lines are shown in Figures 14.2, 14.3, and 14.4 (pages 650–652). Figure 14.2 gives the cost–size lines for north Florida. Note that there is very little change (the slope of the line is close to 0) in cost as size increases for privately owned nursing homes (symbol D). In contrast, cost decreases with size for nonproprietary homes (symbol C) and rises for government- (symbol B) and church-supported (symbol A) nursing homes.

Figure 14.3 gives the corresponding cost–size lines for central Florida. Cost is fairly stable for proprietary and nonproprietary nursing homes, and cost decreases as size increases for church-owned and government nursing homes. The cost–size lines for south Florida, shown in Figure 14.4, are very similar to those in Figures 14.2 and 14.3, except for a noticeable change in the intercepts of the lines. For example, although the cost–size lines for proprietary homes have slopes near 0, the cost line for south Florida is well above the line for north Florida. This difference in y-intercepts for the two lines is represented in the model by β_2.* You will note that the t-value associated with this term, $t = -3.58$, is highly significant, thus indicating that the difference observed on the graphs is not imaginary. There appears to be a substantial difference in cost per bed-day between south and north Florida. This is not too surprising, because we would expect labor and capital costs (taxes, mortgages, and so forth) to be higher in the more heavily populated and tourist-oriented southern area of the state.

Do the variables location, type, and size account for most of the variability in nursing home bed-day costs? Note that R^2 equals only .460850, and this implies that there is a relatively large amount of unexplained variation in the residuals. Why? One explanation is that there are undoubtedly many important variables related to bed-day costs that were not included in the study (because the data were not available). Two possibly important variables are the age of the nursing home and whether the location is rural or urban.

And you might ask whether a straight-line relation between bed-day cost and nursing home size is appropriate. Would second-order response curves, one corresponding to each of the 12 location–type categories, provide a better model for the bed-day costs? The model depicting this situation is

---

*If you substitute the appropriate values for the dummy variables into the model, you will see that the intercept for the south Florida proprietary line is β_0; the intercept for the north Florida proprietary line is $(\beta_0 + \beta_2)$.

FIGURE 14.2 Predicted Cost–Size Relationships in North Florida

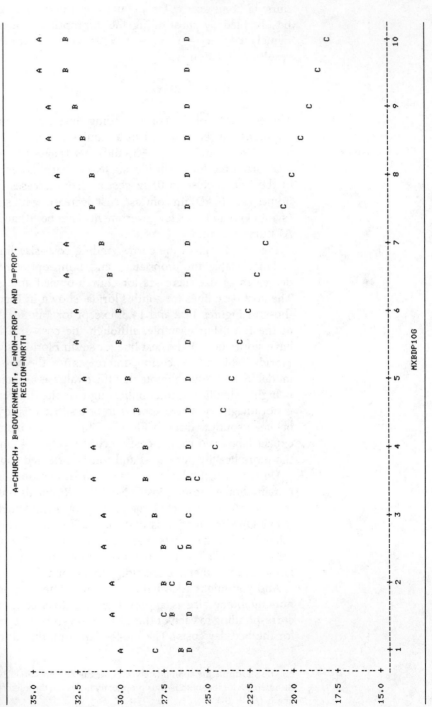

A=CHURCH, B=GOVERNMENT, C=NON-PROP, AND D=PROP.
REGION=NORTH

FIGURE 14.3   Predicted Cost–Size Relationships in Central Florida

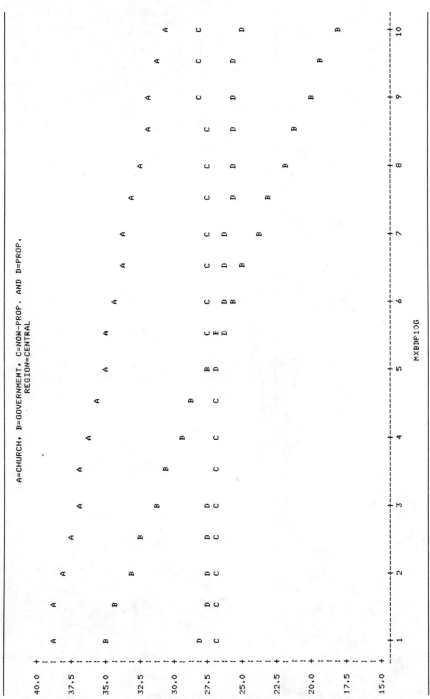

A=CHURCH, B=GOVERNMENT, C=NON-PROP., AND D=PROP.
REGION=CENTRAL

FIGURE 14.4   Predicted Cost–Size Relationships in South Florida

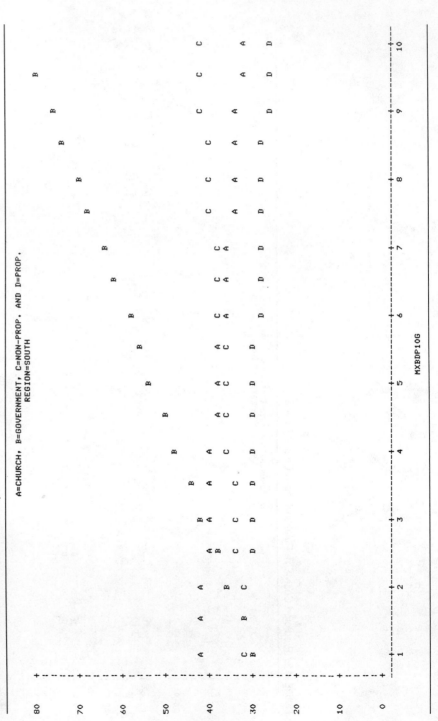

A=CHURCH, B=GOVERNMENT, C=NON-PROP, AND D=PROP, REGION=SOUTH

$$
\begin{aligned}
E(y) = \beta_0 + \overbrace{\beta_1 x_1 + \beta_2 x_2}^{\substack{\text{Main effect,} \\ \text{location}}} + \overbrace{\beta_3 x_3 + \beta_4 x_4 + \beta_5 x_5}^{\substack{\text{Main effect,} \\ \text{type}}} \\
\left. \begin{aligned}
+\ \beta_6 x_1 x_3 + \beta_7 x_1 x_4 + \beta_8 x_1 x_5 \\
+\ \beta_9 x_2 x_3 + \beta_{10} x_2 x_4 + \beta_{11} x_2 x_5
\end{aligned} \right\} \text{Location–type interaction}
\end{aligned}
$$

$$
\begin{aligned}
+\ \overbrace{\beta_{12} x_6 + \beta_{13} x_6^2}^{\text{Size}} & \\
\left. \begin{aligned}
+\ \beta_{14} x_6 x_1 + \beta_{15} x_6 x_2 + \beta_{16} x_6 x_3 + \beta_{17} x_6 x_4 \\
+\ \beta_{18} x_6 x_5 + \beta_{19} x_6 x_1 x_3 + \beta_{20} x_6 x_1 x_4 + \beta_{21} x_6 x_1 x_5 \\
+\ \beta_{22} x_6 x_2 x_3 + \beta_{23} x_6 x_2 x_4 + \beta_{24} x_6 x_2 x_5 \\
+\ \beta_{25} x_6^2 x_1 + \beta_{26} x_6^2 x_2 + \beta_{27} x_6^2 x_3 + \beta_{28} x_6^2 x_4 \\
+\ \beta_{29} x_6^2 x_5 + \beta_{30} x_6^2 x_1 x_3 + \beta_{31} x_6^2 x_1 x_4 \\
+\ \beta_{32} x_6^2 x_1 x_5 + \beta_{33} x_6^2 x_2 x_3 + \beta_{34} x_6^2 x_2 x_4 \\
+\ \beta_{35} x_6^2 x_2 x_5
\end{aligned} \right\} \begin{aligned} & \text{Size–location,} \\ & \text{size–type, and} \\ & \text{size–location–type} \\ & \text{interactions} \end{aligned}
\end{aligned}
$$

A least squares fit of this model to the data did not produce a reduction in the sum of squares for error that was sufficient to indicate that the second-order terms (those involving x_6^2) contribute information for the prediction of bed-day costs (i.e., that at least one of the parameters $\beta_{13}, \beta_{25}, \ldots, \beta_{35}$ differs from 0). This conflicts with our intuition, because we would expect some curvature in the cost–size curves, i.e., we would expect that bed-day costs would not continue to rise in a linear manner as the size of the nursing home increases. There are at least two possible reasons for this result. The relation between cost and size may be approximately linear over the range of size covered in this analysis, or it may be that curvature is present but that we were unable to detect it because of the large value of σ^2 associated with the model (remember, $R^2 = .46$ is rather small). The inclusion of several other important variables in the model (perhaps age of the nursing home or its rural–urban classification) would produce smaller errors of prediction and possibly enable us to detect curvature in the cost–size curves.

| | | | | | | | | | | | |

SECTION 14.5

SUMMARY

The regression analysis of the nursing home bed-day costs provides three important results. It demonstrates that location, type, and size of nursing home contribute information for the prediction of bed-day costs, i.e., that bed-day costs do vary with the size, location, and type of nursing home. It indicates a substantial difference in the level of costs between nursing homes located in north and south Florida, and it shows that cost remains fairly stable as size increases for proprietary homes but increases substantially for government homes located in the north and south. In general, these results support the contention that the DHRS should take these and other variables into consideration when determining a maximum bed-day rate of compensation.

The second result of the study is the acquisition of a prediction equation that can be used to predict the bed-day costs for a nursing home. This predictor will take into account the location, type, and size of the nursing home. It certainly is not the best predictor of bed-day costs, but it will provide a more equitable rate of compensation than that provided by the current DHRS method. Other variables thought to be related to bed-day costs can be monitored in the future and can be used to develop a better predictor and procedure for establishing equitable rates of Medicaid compensation.

The third important result of the study is that it shows that regression analysis can be used to assist in establishing equitable rates of compensation and suggests that this technique might be applied in many different areas. Hospitals encounter the same problem of recovering the cost of care for Medicaid patients, and similar variations in costs would occur in connection with the compensation for many government-sponsored social programs. These are just a few of the many rate-setting problems that might be susceptible to a solution by regression analysis.

OBJECTIVE

To show how regression analysis can be used to develop a model relating sale price of condominium units to a set of independent variables and, particularly, to show how the model can be used to reveal some interesting relationships among these variables

CONTENTS

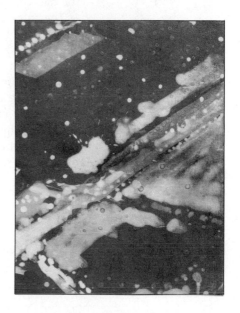

AN INVESTIGATION OF FACTORS AFFECTING THE SALE PRICE OF CONDOMINIUM UNITS SOLD AT PUBLIC AUCTION

This chapter contains a partial investigation of the factors that affect the sale price of oceanside condominium units. It represents an extension of an analysis of the same data by Herman Kelting (1979). And, because there are many different theories (and models) that might be proposed in this type of study, we present the data for your personal analysis in Appendix H.

The sales data were obtained for a new oceanside condominium complex consisting of two adjacent and connected eight-floor buildings. The complex contains 200 equal-sized (approximately 500 square foot) units. The locations of the buildings relative to the ocean, the swimming pool, the parking lot, etc., are shown in Figure 15.1. There are several features of the complex that you should note. The units facing south, called *ocean-view*, face the beach and ocean. In addition, units in building 1 have a good view of the pool. Units to the rear of the building, called *bay-view*, face the parking lot and an area of land that ultimately borders a bay. The view from the upper floors of these units is primarily of wooded, sandy terrain. The bay is very distant and barely visible.

FIGURE 15.1   Layout of Condominium Complex

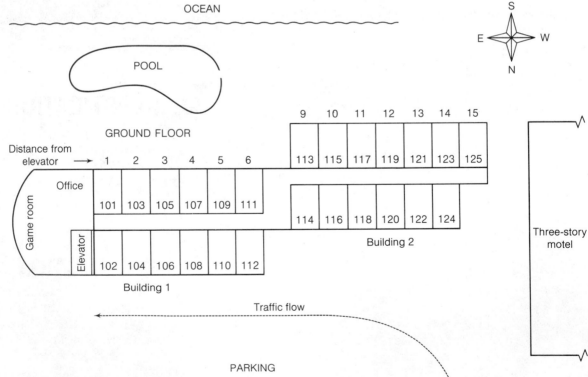

The only elevator in the complex is located at the east end of building 1, as are the office and the game room. People moving to or from the higher floor units in building 2 would likely use the elevator and move through the passages to their units. Thus, units on the higher floors and at a greater distance from the elevator would be less convenient; they would require greater effort in moving baggage, groceries, etc., and would be farther away from the game room, the office, and the swimming pool. These units also possess an advantage: There would be the least amount of traffic through the hallways in the area and hence they are the most private.

Lower-floor oceanside units are most suited to active people; they open onto the beach, ocean, and pool. They are within easy reach of the game room and they are easily reached from the parking area.

Checking Figure 15.1, you will see that some of the units in the center of the complex, units numbered __11 and __14, have part of their view blocked. We would expect this to be a disadvantage. We will show you later that this expectation is true for the ocean-view units and that these units sold at a lower price than adjacent ocean-view units.

The condominium complex was completed at the time of the 1975 recession; sales were slow and the developer was forced to sell most of the units at auction approximately 18 months after opening. Many unsold units were furnished by the developer and rented prior to the auction.

This condominium complex was particularly suited to our study. The single elevator located at one end of the complex produces a remarkably high level of both inconvenience and privacy for the people occupying units on the top floors in building 2. Consequently, the data provide a good opportunity to investigate the relationship between sale price, height of the unit (floor number), distance of the unit from the elevator, and presence or absence of an ocean view. The presence or absence of furniture in each of the units also enables us to investigate the effect of the availability of furniture on sale price. Finally, the auction data are completely buyer-specified and hence consumer-oriented in contrast to most other real estate sales data which are, to a high degree, seller- and broker-specified.

SECTION 15.2

THE DATA

In addition to the sale price, the following data were recorded for each of the 106 units sold at the auction:

1. *Floor height* The floor location of the unit; this variable, x_1, could take values 1, 2, . . . , 8.
2. *Distance from elevator* This distance, measured along the length of the complex, was expressed in number of condominium units. An additional two units of distance was added to the units in building 2 to account for the walking distance in the connecting area between the two buildings. Thus, the distance of unit 105 from the elevator would be 3, and the distance between unit 113 and the elevator would be 9. This variable, x_2, could take values 1, 2, . . . , 15.

3. *View of ocean*   The presence or absence of an ocean view was recorded for each unit and entered into the model with a dummy variable, x_3, where $x_3 = 1$ if the unit possessed an ocean view and $x_3 = 0$ if not. Note that units not possessing an ocean view would face the parking lot.

4. *End unit*   We expected the partial reduction of view of end units on the ocean side (numbers ending in 11) to reduce their sale price. The ocean view of these end units is partially blocked by building 2. This qualitative variable was entered into the model with a dummy variable, x_4, where $x_4 = 1$ if the unit has a unit number ending in 11 and $x_4 = 0$ if not.

5. *Furniture*   The presence or absence of furniture was recorded for each unit. This qualitative variable was entered into the model using a single dummy variable, x_5, where $x_5 = 1$ if the unit was furnished and $x_5 = 0$ if not.

The raw data used in this analysis are presented in Appendix H.

| | | | | | | | | | | | |

SECTION 15.3

THE MODELS

We noted above that we recorded data on five independent variables, two quantitative (floor height x_1 and distance from elevator x_2) and three qualitative (view of ocean, end unit, and furniture). We postulated four models relating mean sale price to these five factors. The models, numbered 1–4, are developed in sequence, Model 1 being the simplest and Model 4 the most complex. Each of Models 2 and 3 contains all the terms of the preceding models along with new terms that we think will improve their predictive ability. Thus, Model 2 contains all the terms contained in Model 1 plus some new terms, and hence it should predict mean sale price as well as or better than Model 1. Similarly, Model 3 should predict as well as or better than Model 2. Model 4 does not contain all the terms contained in Model 3, but that is only because we have entered floor height into Model 4 as a qualitative independent variable. Consequently, Model 4 contains all the predictive ability of Model 3 and it could be an improvement over Model 3 if our theory is correct. The logic employed in this sequential model building procedure will be explained in the following discussion.

The simplest theory that we might postulate is that the five factors affect the price in an independent manner and that the effect of the two quantitative factors on sale price is linear. Thus, we envision a set of planes, each identical except for their y-intercepts. We would expect sale price planes for ocean-view units to be higher than those with a bay view, those corresponding to end units (_11) would be lower than for non-end units, and those with furniture would be higher than those without.

MODEL 1

$$E(y) = \beta_0 + \beta_1 x_1 + \beta_2 x_2 + \beta_3 x_3 + \beta_4 x_4 + \beta_5 x_5$$

where

$x_1 =$ Floor height $(x_1 = 1, 2, \ldots, 8)$

$x_2 =$ Distance from elevator $(x_2 = 1, 2, \ldots, 15)$

$x_3 = \begin{cases} 1 & \text{if an ocean view} \\ 0 & \text{if not} \end{cases}$ $x_4 = \begin{cases} 1 & \text{if an end unit} \\ 0 & \text{if not} \end{cases}$ $x_5 = \begin{cases} 1 & \text{if furnished} \\ 0 & \text{if not} \end{cases}$

The second theory that we considered was that the effects on sale price of floor height and distance from elevator might not be linear. Consequently, we constructed Model 2, which is similar to Model 1 except that second-order terms are included for x_1 and x_2. This model envisions a single second-order response surface for $E(y)$ in x_1 and x_2 that possesses identically the same shape, regardless of the view, whether the unit is an end unit, and whether the unit is furnished. Expressed in other terminology, Model 2 assumes that there is no interaction between any of the qualitative factors (view of ocean, end unit, and furniture) and the quantitative factors (floor height and distance from elevator).

MODEL 2

$$E(y) = \beta_0 + \overbrace{\beta_1 x_1 + \beta_2 x_2 + \beta_3 x_1 x_2 + \beta_4 x_1^2 + \beta_5 x_2^2}^{\text{Second-order model in } x_1 \text{ and } x_2}$$

$$+ \quad \overbrace{\beta_6 x_3}^{\text{View of ocean}} \quad + \quad \overbrace{\beta_7 x_4}^{\text{End unit}} \quad + \quad \overbrace{\beta_8 x_5}^{\text{Furniture}}$$

Model 2 may possess a serious shortcoming. It assumes that the shape of the second-order response surface, relating mean sale price $E(y)$ to x_1 and x_2 is identical for ocean-view and bay-view units. Since we think that there is a strong possibility that completely different preference patterns may govern the purchase of these two groups of units, we will construct a model that provides for two completely different second-order response surfaces—one for ocean-view units and one for bay-view units. Further, we will assume that the effects of the two qualitative factors, end unit and furniture, are additive; i.e., their presence or absence will simply shift the mean sale price response surface up or down by a fixed amount. Thus, Model 3 is given as follows:

MODEL 3

$$E(y) = \beta_0 + \overbrace{\beta_1 x_1 + \beta_2 x_2 + \beta_3 x_1 x_2 + \beta_4 x_1^2 + \beta_5 x_2^2}^{\text{Second-order model in } x_1 \text{ and } x_2}$$

$$\overbrace{\beta_6 x_3}^{\text{View of ocean}} \quad + \quad \overbrace{\beta_7 x_4}^{\text{End unit}} \quad + \quad \overbrace{\beta_8 x_5}^{\text{Furniture}}$$
$$+ \quad \beta_6 x_3 \quad + \quad \beta_7 x_4 \quad + \quad \beta_8 x_5$$

$$+ \overbrace{\beta_9 x_1 x_3 + \beta_{10} x_2 x_3 + \beta_{11} x_1 x_2 x_3 + \beta_{12} x_1^2 x_3 + \beta_{13} x_2^2 x_3}^{\substack{\text{Interaction of the second-order model} \\ \text{with view of ocean}}}$$

As a fourth possibility, we constructed a model similar to Model 3 but entered floor height as a qualitative factor at eight levels. The reasons for doing this are:

1. Higher-floor units have better views but less accessibility to the outdoors. This latter characteristic could be a particularly undesirable feature for these units.
2. The views of some lower-floor bayside units were blocked by a nearby three-story motel.

If our supposition is correct and if these features would have a depressive effect on the sale price of these units, then the relationship between floor height and mean sale price would not be second-order (a smooth curvilinear relationship). Entering floor height as a qualitative factor would permit a better fit to this irregular relationship and improve the prediction equation. Thus, Model 4 is identical to Model 3 except that Model 4 contains seven main effect terms for floor height (in contrast to two for Model 3) and it also contains the corresponding interactions of these variables with the other variables included in Model 3. We will subsequently show that there was no evidence to indicate that Model 4 contributes more information for the prediction of y than Model 3. For this reason and also because Model 4 contains so many terms (49), we will not give the formula for this model.*

SECTION 15.4

THE REGRESSION ANALYSES

This section gives the regression analyses for the four models described in Section 15.3. You will see that our approach is to build the model in a sequential manner. In each case we use an F-test to see whether a particular model contributes more information for the prediction of sale price than its predecessor.

This procedure is more conservative than if we were to employ a step-down procedure, i.e., assume model 4 to be the appropriate model and then test and possibly delete terms. Deleting terms can be particularly risky because, in doing so, you are tacitly accepting the null hypothesis. Thus, you risk deleting important terms from the model and do so with an unknown probability of committing a Type II error.

Do not be unduly influenced by the individual t-tests associated with an analysis. As you will see, it is possible for a set of terms to contribute information for the prediction of y when none of their respective t-values are statistically significant. This is because the t-test tests the contribution of a single term, given that all the other terms are retained in the model. Therefore, if a set of terms contributes overlapping information, it is possible that none of the terms individually would be statistically significant when the set, as a whole, contributes information for the prediction of y.

The SAS regression analysis computer printouts for fitting Models 1, 2, 3, and 4 to the data are shown in Figures 15.2, 15.3, 15.4, and 15.5, respectively. A summary containing SSE values for these models, their respective degrees of freedom, and R^2-values is provided in Table 15.1.

Examining the computer printout for the first-order model (Model 1) in Figure 15.2, you can see that the value of the F-statistic for testing the null hypothesis

$$H_0: \quad \beta_1 = \beta_2 = \cdots = \beta_5 = 0$$

is 48.42. This is statistically significant at a level of $\alpha = .0001$. Consequently, there is ample evidence to indicate that the model contributes information for the prediction of y. At least one of the five factors contributes information for the prediction of sale price.

*Some of these terms were not estimable because sales were not consummated for some combinations of the independent variables. This is why the computer printout in Figure 15.5 shows only 41 df for the model.

FIGURE 15.2

SAS Regression Analysis: Model 1

ANALYSIS OF VARIANCE

SOURCE	DF	SUM OF SQUARES	MEAN SQUARE	F VALUE	PROB>F
MODEL	5	236620761	47324152.11	48.419	0.0001
ERROR	100	97737975.28	977379.75		
C TOTAL	105	334358736			

ROOT MSE	988.6252	R-SQUARE	0.7077	
DEP MEAN	19176.04	ADJ R-SQ	0.6931	
C.V.	5.155524			

PARAMETER ESTIMATES

VARIABLE	DF	PARAMETER ESTIMATE	STANDARD ERROR	T FOR H0: PARAMETER=0	PROB > |T|
INTERCEP	1	17770.00813	416.07044	42.709	0.0001
X1	1	-73.68311471	52.97836340	-1.391	0.1674
X2	1	-86.41176445	24.44923524	-3.534	0.0006
X3	1	3136.59235	222.70788	14.084	0.0001
X4	1	-1781.05526	397.45816	-4.481	0.0001
X5	1	986.99033	204.77016	4.820	0.0001

FIGURE 15.3

SAS Regression Analysis: Model 2

ANALYSIS OF VARIANCE

SOURCE	DF	SUM OF SQUARES	MEAN SQUARE	F VALUE	PROB>F
MODEL	8	244325938	30540742.24	32.904	0.0001
ERROR	97	90032797.92	928173.17		
C TOTAL	105	334358736			

ROOT MSE	963.4174	R-SQUARE	0.7307	
DEP MEAN	19176.04	ADJ R-SQ	0.7085	
C.V.	5.024069			

PARAMETER ESTIMATES

VARIABLE	DF	PARAMETER ESTIMATE	STANDARD ERROR	T FOR H0: PARAMETER=0	PROB > |T|
INTERCEP	1	19461.81838	764.78661	25.447	0.0001
X1	1	-683.94788	245.08261	-2.791	0.0063
X2	1	-264.54977	122.58866	-2.158	0.0334
X1X2	1	4.81613443	13.54099144	0.356	0.7229
X1X1	1	57.93750913	23.74605653	2.440	0.0165
X2X2	1	11.33982388	7.70152793	1.472	0.1441
X3	1	3051.55998	219.25992	13.918	0.0001
X4	1	-1680.93326	409.72999	-4.103	0.0001
X5	1	1114.78991	205.04676	5.437	0.0001

TABLE 15.1

A Summary of the Values (Rounded) of SSE and R^2 for Models 1, 2, 3, and 4

MODEL	SSE	df(Error)	R^2
1	97,737,975	100	.7077
2	90,032,798	97	.7307
3	78,428,545	92	.7654
4	51,647,128	64	.8455

If you examine the t-tests for the individual parameters, you will see that all are statistically significant except the test for β_1, the parameter associated with floor height x_1 (p-value $= .1674$). The failure of floor height x_1 to reveal itself as an important information contributor goes against our intuition and it demonstrates the pitfalls that can occur when you attempt to interpret the results

FIGURE 15.4

SAS Regression Analysis:
Model 3

ANALYSIS OF VARIANCE

SOURCE	DF	SUM OF SQUARES	MEAN SQUARE	F VALUE	PROB>F
MODEL	13	255930191	19686937.74	23.094	0.0001
ERROR	92	78428545.24	852484.19		
C TOTAL	105	334358736			

	ROOT MSE	923.3007	R-SQUARE	0.7654
	DEP MEAN	19176.04	ADJ R-SQ	0.7323
	C.V.	4.814867		

PARAMETER ESTIMATES

VARIABLE	DF	PARAMETER ESTIMATE	STANDARD ERROR	T FOR H0: PARAMETER=0	PROB > |T|
INTERCEP	1	14412.04197	2795.75178	5.155	0.0001
X1	1	819.81446	941.47799	0.871	0.3861
X2	1	-413.28219	303.20570	-1.363	0.1762
X1X2	1	33.53465808	34.14413805	0.982	0.3286
X1X1	1	-54.41406576	80.65633152	-0.675	0.5016
X2X2	1	11.87166612	16.59695367	0.715	0.4762
X3	1	8247.71451	2885.53387	2.858	0.0053
X4	1	-1632.24166	400.11602	-4.079	0.0001
X5	1	1242.52585	204.36346	6.080	0.0001
X1X3	1	-1350.92966	974.75335	-1.386	0.1691
X2X3	1	50.74158830	334.14998	0.152	0.8796
X1X2X3	1	-29.98215571	37.25553940	-0.805	0.4230
X1X1X3	1	91.37334655	84.49650785	1.081	0.2824
X2X2X3	1	5.55440578	19.06625192	0.291	0.7715

FIGURE 15.5

SAS Regression Analysis:
Model 4

ANALYSIS OF VARIANCE

SOURCE	DF	SUM OF SQUARES	MEAN SQUARE	F VALUE	PROB>F
MODEL	41	282711608	6895405.07	8.545	0.0001
ERROR	64	51647128.00	806986.37		
C TOTAL	105	334358736			

	ROOT MSE	898.3242	R-SQUARE	0.8455
	DEP MEAN	19176.04	ADJ R-SQ	0.7466
	C.V.	4.684618		

of t-tests in a regression analysis. Intuitively, we would expect floor height to be an important factor. You might argue that units on the higher floors possess a better view and hence should command a higher mean sale price. Or, you might argue that units on the lower floors have greater accessibility to the pool and ocean and, consequently, should be in greater demand. Why, then, is the t-test for floor height not statistically significant? The answer is that both of the preceding arguments are correct, one for the oceanside and one for the bayside. Thus, you will subsequently see that there is an interaction between floor height and view of ocean. Ocean-view units on the lower floors sell at higher prices than ocean-view units on the higher floors. In contrast, bay-view units on the higher floors command higher prices than bay-view units on the lower floors. These two contrasting effects tend to cancel (because we have not included interaction terms in the model) and thereby give the false impression that floor height is not an important variable for predicting mean sale price.

But, of course, we are looking ahead. Our next step is to determine whether Model 2 is better than Model 1.

Are floor height x_1 and distance from elevator x_2 related to sale price in a curvilinear manner, i.e., should we be using a second-order response surface instead of a first-order surface to relate $E(y)$ to x_1 and x_2? To answer this question, we will examine the drop in SSE from Model 1 to Model 2. The null hypothesis "Model 2 contributes no more information for the prediction of y than Model 1" is equivalent to testing

$$H_0: \quad \beta_3 = \beta_4 = \beta_5 = 0$$

where β_3, β_4, and β_5 appear in Model 2. The F-statistic for the test, based on 3 and 97 df, is

$$F = \frac{\dfrac{\text{Drop in SSE}}{\text{Number of } \beta \text{ parameters in } H_0}}{s_2^2} = \frac{\dfrac{97{,}737{,}975 - 90{,}032{,}798}{3}}{928{,}173.17}$$

$$= 2.77$$

An approximate value for $F_{.05}$ based on 3 and 97 df, obtained from Table 4 in Appendix D, is 2.70. Since the computed F-value exceeds the tabulated value, $F_{.05} = 2.70$, we reject H_0. There is evidence to indicate that Model 2 contributes more information for the prediction of y than Model 1. This tells us that there is evidence of curvature in the response surfaces relating mean sale price, $E(y)$, to floor height x_1 and distance from elevator x_2.

You will recall that the difference between Models 2 and 3 is that Model 3 allows for two differently shaped second-order surfaces, one for ocean-view units and another for bay-view units; Model 2 employs a single surface to represent both types of units. Consequently, we wish to test the null hypothesis that "a single second-order surface adequately characterizes the relationship between $E(y)$, floor height x_1, and distance from elevator x_2 for both ocean-view and bay-view units" [i.e., Model 2 adequately models $E(y)$] against the alternative hypothesis that you need two different second-order surfaces, i.e, you need Model 3. Thus,

$$H_0: \quad \beta_9 = \beta_{10} = \cdots = \beta_{13} = 0$$

where β_9, β_{10}, ... , β_{13} are parameters in Model 3. The F-statistic for this test, based on 5 and 92 df, is

$$F = \frac{\dfrac{\text{Drop in SSE}}{\text{Number of } \beta \text{ parameters in } H_0}}{s_3^2} = \frac{\dfrac{\text{SSE}_2 - \text{SSE}_3}{5}}{s_3^2}$$

$$= \frac{\dfrac{90{,}032{,}798 - 78{,}428{,}545}{5}}{852{,}484.19}$$

$$= 2.72$$

An approximate value for $F_{.05}$ based on 5 and 92 df, obtained from Table 4 in Appendix D, is 2.33. Since the computed F-value exceeds this tabulated value, we reject H_0 and conclude that there is evidence to indicate that we need two different second-order surfaces to relate $E(y)$ to x_1 and x_2, one each for ocean-view and bay-view units.

Finally, we question whether Model 4 will provide an improvement over Model 3, i.e., will we gain information for predicting y by entering floor height into the model as a qualitative factor at eight levels? The F-statistic to test the null hypothesis "Model 4 contributes no more information for predicting y than does Model 3" compares the drop in SSE from Model 3 to Model 4 with s_4^2. This F-statistic, based on 28 df (the difference in the numbers of parameters in Models 4 and 3) and 64 df, is

$$F = \frac{\dfrac{\text{Drop in SSE}}{\text{Number of } \beta \text{ parameters in } H_0}}{s_4^2} = \frac{\dfrac{SSE_3 - SSE_4}{28}}{s_4^2}$$

$$= \frac{\dfrac{78{,}428{,}545 - 51{,}647{,}128}{28}}{806{,}986}$$

$$= 1.19$$

Checking Table 4 in Appendix D, you will find that the value for $F_{.05}$, based on 28 and 64 df, is approximately 1.65. Since the computed F is less than this value, there is not sufficient evidence to indicate that Model 4 is a significant improvement over Model 3.

Having checked the four theories of Section 15.3, we have evidence to indicate that Model 2 is significantly better than Model 1 and that Model 3 is better than Model 2. We will examine the prediction equation for Model 3 and see what it tells us about the relationship between the mean sale price $E(y)$ and the five factors employed in our study; but first, it is important that we examine the residuals for Model 3 to determine whether the standard least squares assumptions about the random error term are satisfied.

SECTION 15.5

AN ANALYSIS OF THE RESIDUALS FROM MODEL 3

The four standard least squares assumptions about the random error term ε are (from Chapter 4) the following:

1. The mean is 0.
2. The variance (σ^2) is constant for all settings of the independent variables.
3. The errors follow a normal distribution.
4. The errors are independent.

If one or more of these assumptions are violated, any inferences derived from the Model 3 regression analysis are suspect. It is unlikely that assumption 1 (0 mean) is violated because the method of least squares guarantees that the mean of the residuals is 0. The same can be said for assumption 4 (independent errors) since the sale price data are not a time series. However, verifying assumptions 2 and 3 requires a thorough examination of the residuals from Model 3.

Recall that we can check for heteroscedastic errors (i.e., errors with unequal variances) by plotting the residuals against the predicted values. This residual plot is shown in Figure 15.6. If the variances were not constant, we would expect to see a cone-shaped pattern (since the response is sale price) in Figure 15.6 (page 666), with the spread of the residuals increasing as $\hat{y}$ increases. Note, however, that the residuals appear to be randomly scattered around 0. Therefore, assumption 2 (constant variance) appears to be satisfied.

To check the normality assumption (assumption 3), we have generated a histogram of the residuals in Figure 15.7 (page 667). It is very evident that the distribution of residuals is not normal, but skewed to the right. At this point, we could opt to use a transformation on the response (similar to the variance-stabilizing transformations discussed in Section 6.3) to normalize the residuals. However, a nonnormal error distribution is often due to the presence of a single outlier. If this outlier is eliminated (or corrected) the normality assumption may then be satisfied.

We can use the residual plot in Figure 15.6 to detect outliers in the analysis. In Section 6.4, we defined outliers to be residuals that exceed $3s$ (in absolute value) and suspect outliers as residuals that fall between $2s$ and $3s$ away from 0. The $\pm 2s$ and $\pm 3s$ lines are marked on Figure 15.6 (where $s = 923$ from Figure 15.3). Note that there is one outlier and one suspect outlier, both with large *positive* residuals (approximately 5,500 and 2,500, respectively). Should we automatically eliminate these two observations from the analysis and refit Model 3? Although many analysts adopt such an approach, we should carefully examine the observations before deciding to eliminate them. We may discover a correctable recording (or coding) error, or we may find that the outliers are very influential and are due to an inadequate model (in which case, it is the model that needs fixing, not the data).

An examination of the SAS printout of the Model 3 residuals (not shown) reveals that the two observations in question are identified by observation numbers 35 and 49 (where the observations are numbered from 1 to 106). The sale prices, floor heights, and so forth, for these two data points were found to be recorded and coded correctly. To determine how influential these outliers are on the analysis, influence diagnostics (i.e., Cook's D and leverage) were generated using SAS. The results are summarized in Table 15.2.

TABLE 15.2

Influence Diagnostics for Two Outliers in Model 3

OBSERVATION	RESPONSE y	PREDICTED VALUE $\hat{y}$	RESIDUAL $y - \hat{y}$	LEVERAGE h	COOK'S DISTANCE D
35	26,500	21,070.4	5,429.6	.0605	.169
49	21,000	18,414.3	2,585.7	.1607	.128

Based on the "rules of thumb" given in Section 6.5, neither observation has strong influence on the analysis. Both leverage (h) values fall below $2(k + 1)/n = 2(14)/106 = .264$, indicating that the observations are not influential with

FIGURE 15.6   Plot of Residuals Versus $\hat{y}$ for Model 3

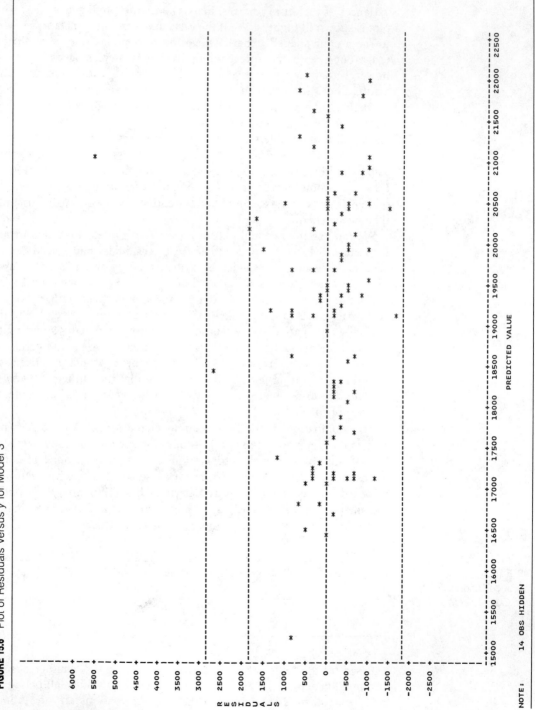

NOTE: 14 OBS HIDDEN

FIGURE 15.7

Histogram of Residuals from Model 3

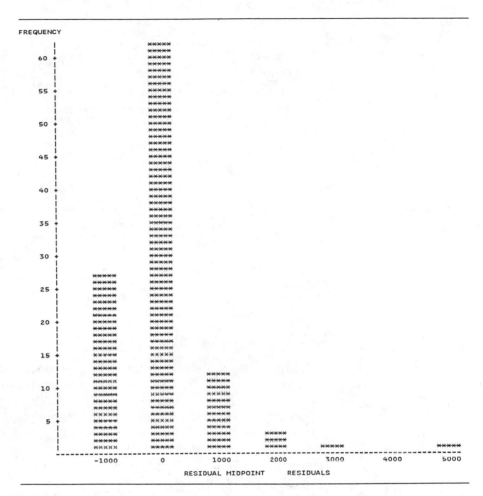

respect to their x-values; and both Cook's D-values fall below .96 (the 50th percentile of an F-distribution with $\nu_1 = k + 1 = 14$ and $\nu_2 = n - (k + 1) = 106 - 14 = 92$ degrees of freedom), implying that they do not exhibit strong overall influence on the regression results (e.g., the β estimates). Consequently, if we remove these outliers from the data and refit Model 3, the least squares prediction equation will not be greatly affected and the normality assumption will probably be more nearly satisfied.

The SAS printout for the refitted model is shown in Figure 15.8 (page 668). Note that df(error) is reduced from 92 to 90 (since we eliminated the two outlying observations) and the β estimates remain relatively unchanged. The model standard deviation, however, is decreased significantly from 923 to 658, implying that the refitted model will yield more accurate predictions of sale price. A residual plot for the refitted model is shown in Figure 15.9 and a histogram of the residuals in Figure 15.10 (pages 669–670). The residual plot (Figure 15.9) reveals no evidence of outliers, and the histogram of the residuals (Figure 15.10) is now approximately normal.

FIGURE 15.8
SAS Printout for Model 3
with Outliers Deleted

ANALYSIS OF VARIANCE

SOURCE	DF	SUM OF SQUARES	MEAN SQUARE	F VALUE	PROB>F
MODEL	13	237649572	18280736.29	42.254	0.0001
ERROR	90	38937243.63	432636.04		
C TOTAL	103	276586815			

ROOT MSE	657.7507	R-SQUARE	0.8592	
DEP MEAN	19088.08	ADJ R-SQ	0.8389	
C.V.	3.445872			

PARAMETER ESTIMATES

VARIABLE	DF	PARAMETER ESTIMATE	STANDARD ERROR	T FOR H0: PARAMETER=0	PROB > |T|
INTERCEP	1	15390.19258	2002.11309	7.687	0.0001
X1	1	310.30678	682.48668	0.455	0.6504
X2	1	-203.35236	222.11810	-0.916	0.3624
X1X2	1	26.08674126	24.39118628	1.070	0.2877
X1X1	1	-10.85790288	58.55677607	-0.185	0.8533
X2X2	1	1.09206281	12.11658757	0.090	0.9284
X3	1	7537.23222	2065.74801	3.649	0.0004
X4	1	-1500.60982	285.47441	-5.257	0.0001
X5	1	1075.38112	146.64993	7.333	0.0001
X1X3	1	-1024.61067	706.31066	-1.451	0.1504
X2X3	1	-167.24636	243.44976	-0.687	0.4939
X1X2X3	1	-25.43791062	26.60757746	-0.956	0.3416
X1X1X3	1	67.67866117	61.32668383	1.104	0.2727
X2X2X3	1	19.21642228	13.82949599	1.390	0.1681

| | | | | | | | | | | |

SECTION 15.6

WHAT THE MODEL 3 REGRESSION ANALYSIS TELLS US

We have settled on Model 3 (with two observations deleted) as our choice to relate mean sale price $E(y)$ to five factors: the two quantitative factors, floor height x_1 and distance from elevator x_2; and the three qualitative factors, view of ocean, end unit, and furniture. This model postulates two different second-order surfaces relating mean sale price $E(y)$ to x_1 and x_2, one for ocean-view units and one for bay-view units. The effect of each of the two qualitative factors, end unit (numbered __11) and furniture, is to produce a change in mean sale price that is identical for all combinations of values of x_1 and x_2. In other words, assigning a value of 1 to one of the dummy variables increases (or decreases) the estimated mean sale price by a fixed amount. The net effect is to push the second-order surface upward or downward, with the direction depending on the level of the specific qualitative factor. The estimated increase (or decrease) in mean sale price due to a given qualitative factor is given by the estimated value of the β parameter associated with its dummy variable.

For example, the prediction equation (with rounded values given for the parameter estimates) obtained from Figure 15.8 is

$$\hat{y} = 15,390.2 + 310.3x_1 - 203.4x_2$$
$$+ 26.1x_1x_2 - 10.9x_1^2 + 1.1x_2^2$$
$$+ 7,537.2x_3 - 1,500.6x_4 + 1,075.4x_5$$
$$- 1,024.6x_1x_3 - 167.2x_2x_3 - 25.4x_1x_2x_3$$
$$+ 67.7x_1^2x_3 + 19.2x_2^2x_3$$

FIGURE 15.9 Plot of Residuals Versus ŷ for Model 3 with Outliers Deleted

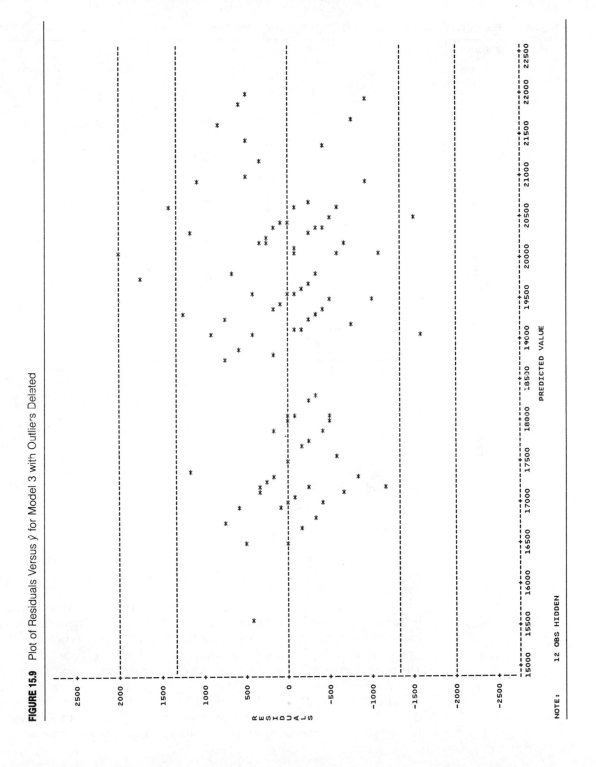

FIGURE 15.10   Histogram of Residuals from Model 3 with Outliers Deleted

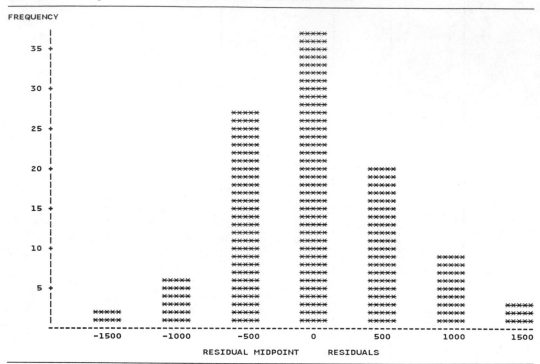

Since the dummy variables for end unit and furniture are, respectively, x_4 and x_5, the estimated changes in mean sale price for these qualitative factors are:

End unit ($x_4 = 1$): $\hat{\beta}_7 = -\$1,500.6$

Furnished ($x_5 = 1$): $\hat{\beta}_8 = +\$1,075.4$

Thus, if you substitute $x_4 = 1$ into the prediction equation, the estimated mean decrease in sale price for an end unit is $1,500.6, regardless of the view, floor, and whether it is furnished.

The effect of floor height x_1 and distance from elevator x_2 can be determined by plotting $\hat{y}$ as a function of one of the variables for given values of the other. For example, suppose we wish to examine the relationship between $\hat{y}$, x_1, and x_2 for bay-view ($x_3 = 0$), non-end units ($x_4 = 0$) with no furniture ($x_5 = 0$). The prediction curve relating $\hat{y}$ to distance from elevator x_2 can be graphed for each floor by first setting $x_1 = 1$; then $x_1 = 2, \ldots, x_1 = 8$. The graphs of these curves are shown in Figure 15.11; the 1's indicate the curve for the first floor, the 2's the curve for the second floor, and so forth. In Figure 15.11 (page 672), we can see some interesting patterns in the estimated mean sale prices:

1. The higher the floor of a bay-view unit, the higher will be the mean sale price. Low floors look out onto the parking lot and, all other variables held constant, are least preferred.
2. The relationship is curvilinear and is not the same for each floor.
3. Units on the first floor near the office have a higher estimated mean sale price than second- or third-floor units located in the west end of the complex. Perhaps the reason for this is that these units are close to the pool and the game room, and these advantages outweigh the disadvantage of a poor view.
4. The mean sale price decreases as the distance from the elevator and center of activity increases for the lower floors, but the decrease is less as you move upward, floor to floor. Finally, note that the estimated mean sale price increases substantially for units on the highest floor that are farthest away from the elevator. These units are subjected to the least human traffic and are, therefore, the most private. Consequently, a possible explanation for their high price is that buyers place a higher value on the privacy provided by the units than the negative value that they assign to their inconvenience. One additional explanation for the generally higher estimated sale price for units at the ends of the complex may be that they possess more windows.

A similar set of curves is shown in Figure 15.12 (page 673) for ocean-view units ($x_3 = 1$). You will note some amazing differences between these curves and those for the bay-view units in Figure 15.11 (these differences explain why we needed two separate second-order surfaces to describe these two sets of units). The preference for floors is completely reversed on the ocean side of the complex: the lower the floor, the higher the estimated mean sale price. Apparently, people selecting the ocean-view units are primarily concerned with accessibility to the ocean, pool, beach, and game room. Note that the estimated mean sale price is highest near the elevator. It drops and then rises as you reach the units farthest from the elevator. An explanation for this phenomenon is similar to the one that we used for the bayside units. Units near the elevator are more accessible and nearer to the recreational facilities. Those farthest from the elevators afford the greatest privacy. Units near the center of the complex offer reduced amounts of both accessibility *and* privacy. Notice that units adjacent to the elevator command a higher estimated mean sale price than those near the west end of the complex, thus suggesting that accessibility has a greater influence on price than privacy.

Rather than examine the graphs of $\hat{y}$ as a function of distance from elevator x_2, you may want to see how $\hat{y}$ behaves as a function of floor height x_1 for units located at various distances from the elevator. These estimated mean sale price curves are shown for bay-view units in Figure 15.13 and for ocean-view units in Figure 15.14 (pages 674–675). To avoid congestion in the graphs, we have shown only the curves for distances $x_2 = 1, 3, 5, \ldots, 13, 15$. A curve traced by a sequence of 1's indicates that it applies to units located a distance of 1 from the elevator. Similarly, a curve traced by a sequence of 5's applies to units located a distance of 5 from the elevator. We leave it to you and to the real estate experts to deduce the practical implications of these curves.

FIGURE 15.11 Plot of Predicted Price Versus Distance from Elevator: Bay-View Units

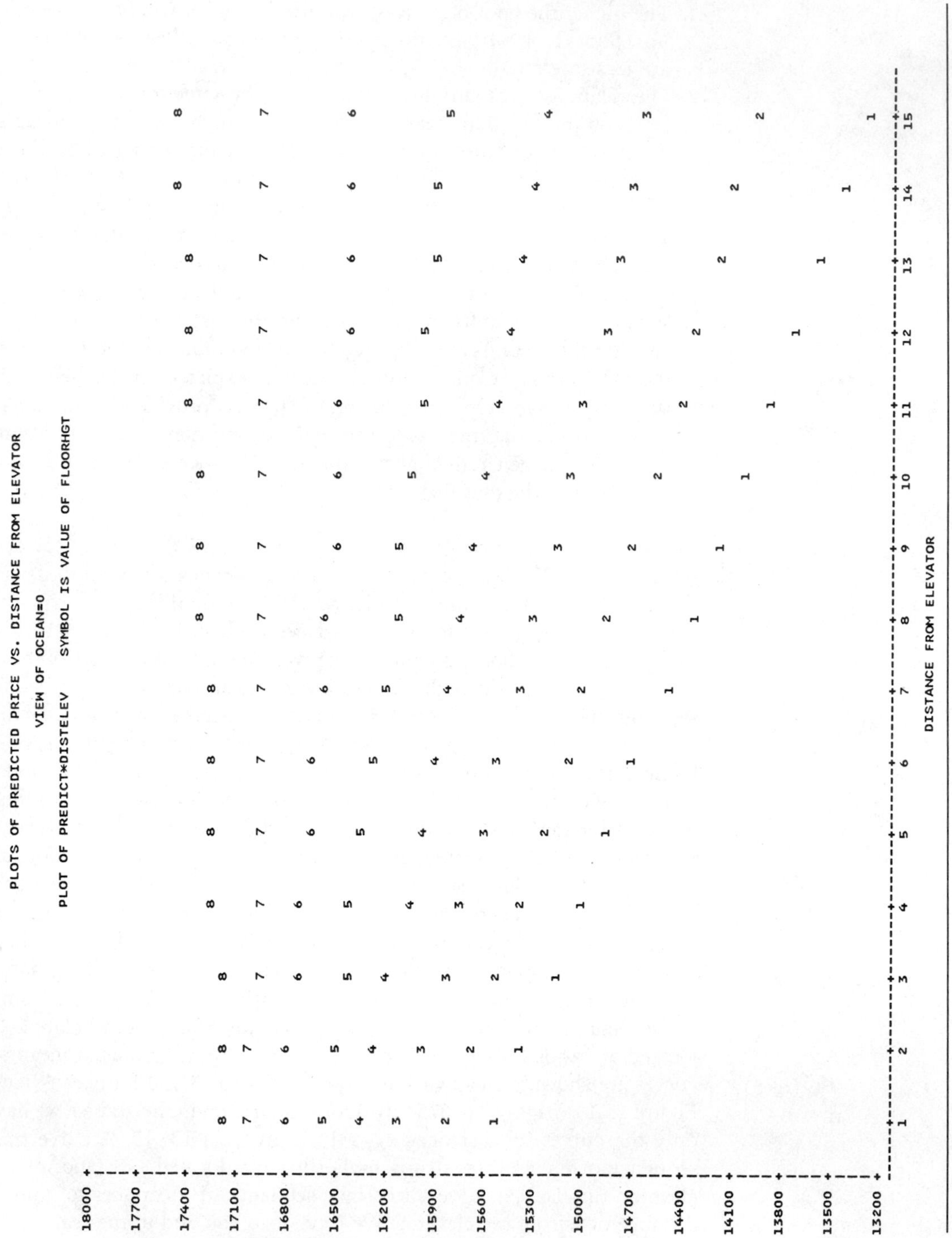

PLOTS OF PREDICTED PRICE VS. DISTANCE FROM ELEVATOR

VIEW OF OCEAN=0

PLOT OF PREDICT*DISTELEV   SYMBOL IS VALUE OF FLOORHGT

FIGURE 15.12 Plot of Predicted Price Versus Distance from Elevator

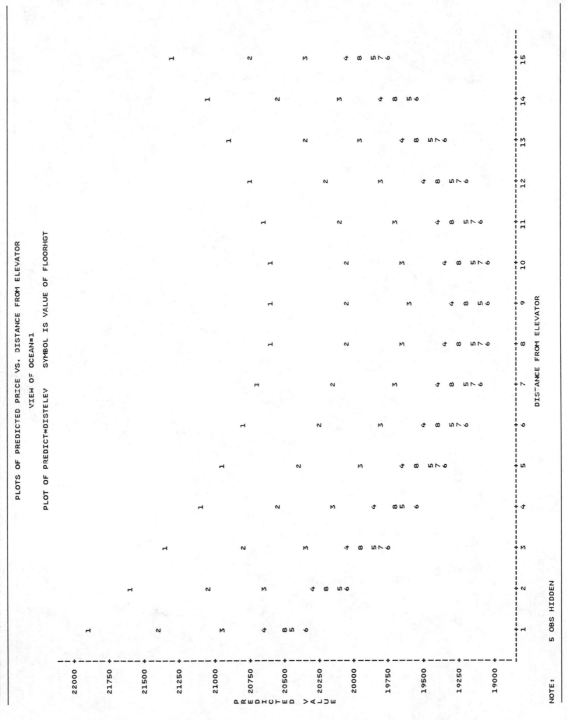

Plot of Predicted Price Versus Distance from Elevator: Ocean-View Units

FIGURE 15.13   Plot of Predicted Price Versus Floor Height: Bay-View Units

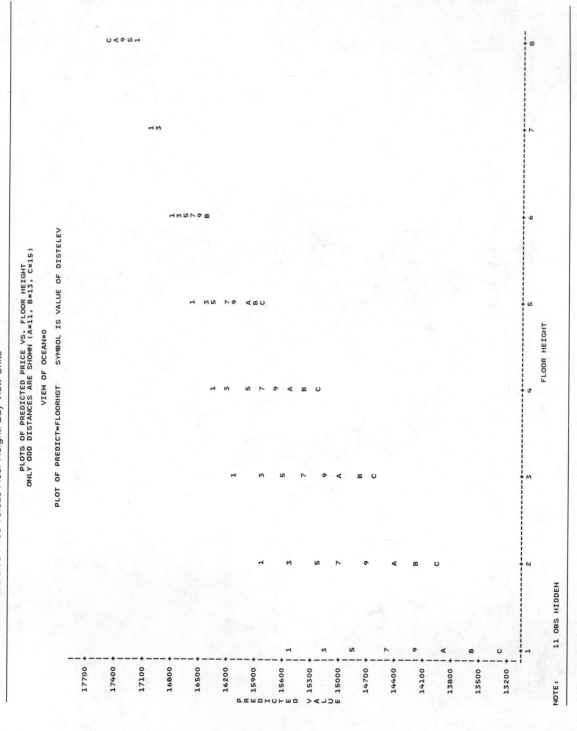

FIGURE 15.14   Plot of Predicted Price Versus Floor Height: Ocean-View Units

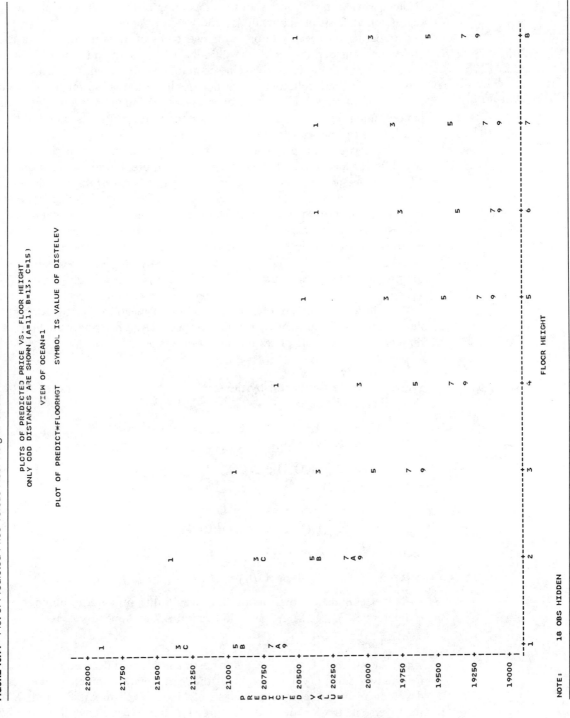

PLOTS OF PREDICTED PRICE VS. FLOOR HEIGHT
ONLY ODD DISTANCES ARE SHOWN (A=11, B=13, C=15)

VIEW OF OCEAN=1

PLOT OF PREDICT*FLOORHGT   SYMBOL IS VALUE OF DISTELEV

FLOOR HEIGHT

NOTE: 18 OBS HIDDEN

| | | | | | | | | | | | | | |

S E C T I O N   15.7

**COMPARING THE
MEAN SALE PRICE
FOR TWO TYPES OF
UNITS (OPTIONAL)**

[*Note:* This section requires an understanding of the mechanics of a multiple regression analysis presented in Appendix A.]

Comparing the mean sale price for two types of units might seem a useless endeavor, considering that all the units have been sold and that we will never be able to sell the units in the same economic environment that existed at the time the data were collected. Nevertheless, this information might be useful to a real estate appraiser or to a developer who is pricing units in a new and similar condominium complex. We will assume that a comparison is useful and show you how it can be accomplished.

Suppose you want to estimate the difference in mean sale price for units in two different locations and with or without furniture. For example, suppose you wish to estimate the difference in mean sale price between the first-floor ocean-view and bay-view units located at the east end of building 1 (units 101 and 102 in Figure 15.1). Both these units afford a maximum of accessibility, but they possess different views. We assume that both are furnished. The estimate of the mean sale price $E(y)$ for the first-floor bay-view unit will be the value of $\hat{y}$ when $x_1 = 1$, $x_2 = 1$, $x_3 = 0$, $x_4 = 0$, $x_5 = 1$. Similarly, the estimated value of $E(y)$ for the first-floor ocean-view unit is obtained by substituting $x_1 = 1$, $x_2 = 1$, $x_3 = 1$, $x_4 = 0$, and $x_5 = 1$ into the prediction equation.

We will let $\hat{y}_o$ and $\hat{y}_b$ represent the estimated mean sale prices for the first-floor ocean-view and bay-view units, respectively. Then the estimator of the difference in mean sale prices for the two units is

$$\ell = \hat{y}_o - \hat{y}_b$$

We represent this estimator by the symbol ℓ, because it is a linear function of the parameter estimators $\hat{\beta}_0, \hat{\beta}_1, \ldots, \hat{\beta}_{13}$; i.e.,

$$\begin{aligned}
\hat{y}_o = {}& \hat{\beta}_0 + \hat{\beta}_1(1) + \hat{\beta}_2(1) + \hat{\beta}_3(1)(1) + \hat{\beta}_4(1)^2 + \hat{\beta}_5(1)^2 \\
& + \hat{\beta}_6(1) + \hat{\beta}_7(0) + \hat{\beta}_8(1) + \hat{\beta}_9(1)(1) + \hat{\beta}_{10}(1)(1) \\
& + \hat{\beta}_{11}(1)(1)(1) + \hat{\beta}_{12}(1)^2(1) + \hat{\beta}_{13}(1)^2(1) \\
\hat{y}_b = {}& \hat{\beta}_0 + \hat{\beta}_1(1) + \hat{\beta}_2(1) + \hat{\beta}_3(1)(1) + \hat{\beta}_4(1)^2 + \hat{\beta}_5(1)^2 \\
& + \hat{\beta}_6(0) + \hat{\beta}_7(0) + \hat{\beta}_8(1) + \hat{\beta}_9(1)(0) + \hat{\beta}_{10}(1)(0) \\
& + \hat{\beta}_{11}(1)(1)(0) + \hat{\beta}_{12}(1)^2(0) + \hat{\beta}_{13}(1)^2(0)
\end{aligned}$$

then

$$\ell = \hat{y}_o - \hat{y}_b = \hat{\beta}_6 + \hat{\beta}_9 + \hat{\beta}_{10} + \hat{\beta}_{11} + \hat{\beta}_{12} + \hat{\beta}_{13}$$

A 95% confidence interval for the mean value of a linear function of the estimators $\hat{\beta}_0, \hat{\beta}_1, \ldots, \hat{\beta}_k$, given in Section A.7 of Appendix A, is

$$\ell \pm t_{.025}s\sqrt{\mathbf{a}'(\mathbf{X}'\mathbf{X})^{-1}\mathbf{a}}$$

where in our case, $\ell = \hat{y}_o - \hat{y}_b$ is the estimate of the difference in mean values for the two units, $E(y_o) - E(y_b)$; s is the least squares estimate of the standard deviation from the regression analysis of Model 3 (Figure 15.8); and $(\mathbf{X}'\mathbf{X})^{-1}$, the inverse matrix for the Model 3 regression analysis, is shown in Figure 15.15.

FIGURE 15.15 SAS Printout of the $(X'X)^{-1}$ Matrix for Model 3

INVERSE	INTERCEP	X1	X2	X1X2	X1X1	X2X2	X3	X4
INTERCEP	9.265194	-2.99228	-0.25651	0.01708831	0.2368571	0.01335886	-9.26105	-0.00350909
X1	-2.99228	1.076628	-0.00585467	0.00232133	-0.090368	-0.00156845	2.993569	-0.00109312
X2	-0.25651	-0.00585467	0.002321133	-0.01140368	-0.004048291	-0.00445162	0.2560235	0.000412168
X1X2	0.01708831	0.00232133	-0.01140368	0.00896821	-0.00175128	0.0000312222	-0.0170757	-0.0001107253
X1X1	0.2368571	-0.090368	-0.004048291	-0.00175128	0.00755996	0.000170988	-0.23705	-0.0001635619
X2X2	0.01335886	-0.00156845	-0.00445162	0.0000312222	0.000170988	0.000270988	-0.0133355	-0.00019801
X3	-9.26105	2.993569	0.2560235	-0.0170757	-0.23705	-0.0133355	9.863521	-0.00346016
X4	-0.00350909	-0.00109312	0.000412168	-0.0001107253	-0.0001635619	-0.00019801	-0.00346016	0.18837
X5	-0.0362161	-0.0112817	0.00425385	-0.000110692	0.001688068	-0.00020436	-0.03053109	0.004816517
X1X3	2.99394	-1.07611	0.005659513	-0.00251605	0.09029058	0.001577825	-3.16263	0.01236905
X2X3	0.2609271	-0.00225471	-0.114556	0.008981714	-0.00425418	0.004476544	-0.325309	-0.020075
X1X2X3	-0.0168751	0.009022947	-0.00399605	-0.000137448	0.00790486	-0.000129609	0.02213208	-0.000318277
X1X1X3	-0.237302	0.001163105	-0.00425868	-0.000090486	0.000836949	-0.000109564	0.2497183	-0.000886821
X2X2X3	-0.01373351	-0.00145126	0.004495806	-0.000032272	-0.000109564	0.000341465	0.01594939	-0.001543755
PRICPAID	15390.19	310.3068	-203.352	26.08674	-10.8579	1.092063	7537.232	-1500.61

INVERSE	X5	X1X3	X2X3	X1X2X3	X1X1X3	X2X2X3	PRICPAID
INTERCEP	-0.0362161	2.99394	0.2609271	-0.0168751	-0.237302	-0.01373351	15390.19
X1	-0.0112817	-1.07611	-0.00225471	0.009022947	0.001163105	-0.00145126	310.3068
X2	0.00425385	0.005659513	-0.114556	-0.00399605	-0.00425868	0.004495806	-203.352
X1X2	-0.000110692	-0.00251605	0.008981714	-0.000137448	-0.000090486	-0.000032272	26.08674
X1X1	0.001688068	0.09029058	-0.00425418	0.00790486	0.000836949	-0.000109564	-10.8579
X2X2	-0.00020436	0.001577825	0.004476544	-0.000129609	-0.000109564	0.000341465	1.092063
X3	-0.03053109	-3.16263	-0.325309	0.02213208	0.2497183	0.01594939	7537.232
X4	0.004816517	0.01236905	-0.020075	-0.000318277	-0.000886821	-0.001543755	-1500.61
X5	0.04970968	0.009001128	-0.0103167	-0.00425868	0.01369923	-0.00972204	1075.381
X1X3	0.009001128	1.163105	-0.0145736	-0.0973915	0.008693132	0.001387393	-1024.61
X2X3	-0.0103167	-0.0145736	0.1369923	-0.00567736	-0.000172506	0.000420689	-167.246
X1X2X3	-0.00425868	-0.0973915	-0.00567736	0.000836949	-0.000836949	0.000172506	-25.4379
X1X1X3	0.01369923	0.008693132	-0.000172506	-0.000836949	0.008693132	0.000420689	67.67866
X2X2X3	-0.00972204	0.001387393	0.000420689	0.000172506	0.000420689	0.000420689	19.21642
PRICPAID	1075.381	-1024.61	-167.246	-25.4379	67.67866	19.21642	38937244

The **a** matrix is a column matrix containing elements $a_0, a_1, a_2, \ldots, a_{13}$, where $a_0, a_1, a_2, \ldots, a_{13}$ are the coefficients of $\hat{\beta}_0, \hat{\beta}_1, \ldots, \hat{\beta}_{13}$ in the linear function ℓ, i.e.,

$$\ell = a_0\hat{\beta}_0 + a_1\hat{\beta}_1 + \cdots + a_{13}\hat{\beta}_{13}$$

Since our linear function is

$$\ell = \hat{\beta}_6 + \hat{\beta}_9 + \hat{\beta}_{10} + \hat{\beta}_{11} + \hat{\beta}_{12} + \hat{\beta}_{13}$$

it follows that $a_6 = a_9 = a_{10} = a_{11} = a_{12} = a_{13} = 1$ and $a_0 = a_1 = a_2 = a_3 = a_4 = a_5 = a_7 = a_8 = 0$.

Substituting the values of $\hat{\beta}_6, \hat{\beta}_9, \hat{\beta}_{10}, \ldots, \hat{\beta}_{13}$ (given in Figure 15.8) into ℓ, we have

$$\begin{aligned}
\ell = \hat{y}_o - \hat{y}_b &= \hat{\beta}_6 + \hat{\beta}_9 + \hat{\beta}_{10} + \hat{\beta}_{11} + \hat{\beta}_{12} + \hat{\beta}_{13} \\
&= 7{,}537.23 - 1{,}024.61 - 167.25 - 25.44 + 67.68 + 19.22 \\
&= \$6{,}406.83
\end{aligned}$$

The value of $t_{.025}$ needed for the confidence interval is approximately equal to 1.96 (because of the large number of degrees of freedom) and the value of s, given in Figure 15.8, is $s = 657.75$. Finally, the matrix product $\mathbf{a}'(\mathbf{X}'\mathbf{X})^{-1}\mathbf{a}$ can be obtained by multiplying the **a** matrix (described in the preceding paragraph) and the $(\mathbf{X}'\mathbf{X})^{-1}$ matrix given in Figure 15.15. Substituting these values into the formula for the confidence interval, we obtain (using a computer):

$$\overbrace{\hat{y}_o - \hat{y}_b}^{\ell} \pm t_{.025}s\sqrt{\mathbf{a}'(\mathbf{X}'\mathbf{X})^{-1}\mathbf{a}}$$
$$\$6{,}406.83 \pm \$2{,}733.46$$

Therefore, we estimate the difference in the mean sale prices of first-floor ocean-view and bay-view units (units 101 and 102) to lie within the interval $3,673.37 to $9,140.29.

You can use the technique described above to compare the mean sale prices for any pair of units.

S E C T I O N   15.8

CONCLUSIONS

You may be able to propose a better model for mean sale price than Model 3, but we think that Model 3 provides a good fit to the data. Further, it reveals some interesting information on the preferences of buyers of oceanside condominium units.

Lower floors are preferred on the ocean side; the closer they lie to the elevator and pool, the higher the estimated price. Some preference is given to the privacy of units located in the upper floor west-end.

Higher floors are preferred on the bay-view side (the side facing away from the ocean) with maximum preference given to units near the elevator (convenient and close to activities) and, to a lesser degree, to the privacy afforded by the west-end units.

| | | | | | | | | | | | |

EXERCISE 15.1

15.1 The data used in this study are presented in Appendix H.* If you have access to a computer, fit Models 1, 2, and 3 to the data set. Note that a computer program package other than the one used in the study may produce slightly different answers due to rounding.

ON YOUR OWN...

Postulate some models that you think might be an improvement over Model 3. Fit these models to the data set in Appendix H. Test to see whether they do, in fact, contribute more information for predicting sale price than Model 3.

REFERENCE

Kelting, H. "Investigation of condominium sale prices in three market scenarios: Utility of stepwise, interactive, multiple regression analysis and implications for design and appraisal methodology." Unpublished paper, University of Florida, Gainesville, 1979.

*Of the 200 units in the condominium complex, 106 were sold at auction and the remainder were sold (some more than once) at the developer's fixed price. The 106 units analyzed in this study are identified by observation (OBS) numbers 77–182 in Appendix H.

OBJECTIVE

To present a time series approach to modeling daily peak electricity demands on a power corporation and to show how to use the time series model for short-term forecasting

CONTENTS

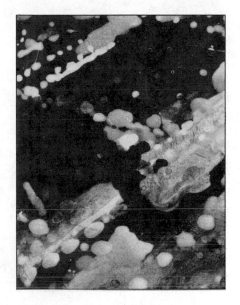

MODELING DAILY PEAK ELECTRICITY DEMANDS

SECTION 16.1

THE PROBLEM

In order to operate effectively, power companies must be able to predict daily peak demand for electricity. *Demand* (or *load*) is defined as the rate (measured in megawatts) at which electric energy is delivered to customers. Since demand is normally recorded on an hourly basis, daily peak demand refers to the maximum hourly demand in a 24-hour period. Power companies are continually developing and refining statistical models of daily peak demand.

Models of daily peak demand serve a twofold purpose. First, the models provide short-term *forecasts* that will assist in the economic planning and dispatching of electric energy. Second, models that relate peak demand to one or more weather variables provide estimates of historical peak demands under a set of alternative weather conditions. That is, since changing weather conditions represent the primary source of variation in peak demand, the model can be used to answer the often asked question, "What would the peak daily demand have been had normal weather prevailed?" This second application, commonly referred to as *weather normalization*, is mainly an exercise in *backcasting* (i.e., adjusting historical data) rather than forecasting (Jacob, 1985).

Since peak demand is recorded over time (days), the dependent variable is a time series and one approach to modeling daily peak demand is to use a time series model. This chapter presents some key results of a study designed to compare several alternative methods of modeling 1983 daily peak demand for the Florida Power Corporation (FPC). For this case study, we focus on two time series models and a multiple regression model proposed in the original FPC study. Then we demonstrate how to forecast daily peak demand using one of the time series models. (We leave the problem of backcasting as an exercise.)

SECTION 16.2

THE DATA

The data for the study consist of daily observations on peak demand recorded by the FPC for the period beginning November 1, 1982, and ending October 31, 1983, and several factors that are known to influence demand. It is typically assumed that demand consists of two components, a non–weather-sensitive "base" demand which is not influenced by temperature changes and a weather-sensitive demand component which is highly responsive to changes in temperature.

The principal factor that affects the usage of non–weather-sensitive appliances (such as refrigerators, generators, lights, and computers) is the *day of the week*. Typically, Saturdays have lower peak demands than weekdays due to decreased commercial and industrial activity, while Sundays and holidays exhibit even lower peak demand levels as commercial and industrial activity declines even further.

The single most important factor affecting the usage of weather-sensitive appliances (such as heaters and air conditioners) is *temperature*. During the winter months, as temperatures drop below comfortable levels, customers begin to operate their electric heating units, thereby increasing the level of demand placed on the system. Similarly, during the summer months, as temperatures climb above comfortable levels, the use of air conditioning drives demand upward. Since the

FPC serves 32 counties along west-central and northern Florida, it was necessary to capture temperature conditions from multiple weather stations. This was accomplished by identifying three primary weather stations within the FPC service area and recording the temperature value at the hour of peak demand each day at each station. A weighted average of these three daily temperatures was used to represent coincident temperature (i.e., temperature at the hour of peak demand) for the entire FPC service area, where the weights were proportional to the percentage of total electricity sales attributable to the weather zones surrounding each of the three weather stations.

To summarize, the independent variable (y_t) and the independent variables recorded for each of the 365 days of the November 1982–October 1983 year were

Dependent Variable:

y_t = Peak demand (in megawatts) observed on day t

Independent Variables:

Day of the week: Weekday, Saturday, or Sunday/holiday

Temperature: Coincident temperature (in degrees), i.e., the temperature recorded at the hour of the peak demand on day t, calculated as a weighted average of three daily temperatures.

| | | | | | | | | | | | | |

SECTION 16.3

THE MODELS

In any modeling procedure, it is often helpful to plot the data in a scattergram. Figure 16.1 (page 684) shows a graph of the daily peak demand (y_t) from November 1982 through October 1983. The seasonal weather impacts on peak demand are readily apparent from the figure. One way to account for this seasonal variation is to include dummy variables for months or trigonometric terms in the model (refer to Section 9.6). However, since temperature is such a strong indicator of the weather, the FPC chose a simpler model with temperature as the sole seasonal weather variable.

Figure 16.2 (page 685) presents a scatterplot of daily peak demands versus coincident temperature. Note the nonlinear relationship that exists between the two variables. During the cool winter months, peak demand is inversely related to temperature; lower temperatures cause increased usage of heating equipment, which in turn cause higher peak demands. In contrast, the summer months reveal a positive relationship between peak demand and temperature; higher temperatures yield higher peak demands due to greater usage of air conditioners. You might think that a second-order (quadratic) model would be a good choice to account for the U-shaped distribution of peak demands shown in Figure 16.2. The FPC, however, rejected such a model for two reasons:

1. A quadratic model yields a symmetrical shape (i.e., a parabola) and would, therefore, not allow independent estimates of the winter and summer peak demand–temperature relationship.

FIGURE 16.1 Daily Peak Megawatt Demands, November 1982–October 1983

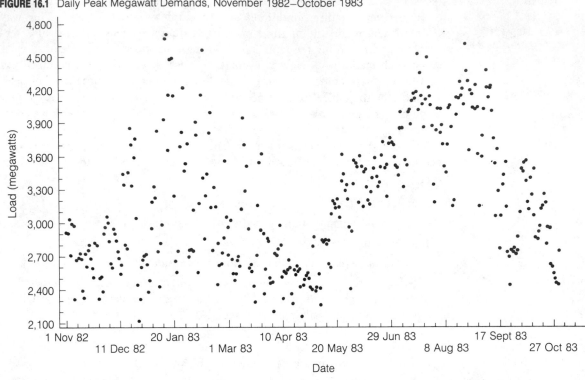

Source: Florida Power Corporation.

2. In theory, there exists a mild temperature range where peak demand is assumed to consist solely of the non–weather-sensitive base demand component. For this range, a temperature change will not spur any additional heating or cooling and, consequently, has no impact on demand. The lack of linearity in the bottom portion of the U-shaped parabola fit by the quadratic model would tend to yield overestimates of peak demand at the extremes of the mild temperature range and underestimates for temperatures in the middle of this range (see Figure 16.3).

The solution was to model daily peak demand with a piecewise linear regression model (see Section 10.2). This approach has the advantage of allowing the peak demand–temperature relationship to vary between some prespecified temperature ranges, as well as providing a mechanism for joining the separate pieces.

Using the piecewise linear specification as the basic model structure, the following model of daily peak demand was proposed:

MODEL 1

$$y_t = \beta_0 + \underbrace{\beta_1(x_{1t} - 59)x_{2t} + \beta_2(x_{1t} - 78)x_{3t}}_{\text{Temperature}} + \underbrace{\beta_3 x_{4t} + \beta_4 x_{5t}}_{\text{Day of the week}} + \varepsilon_t$$

FIGURE 16.2 Daily Peak Demand Versus Temperature, November 1982–October 1983

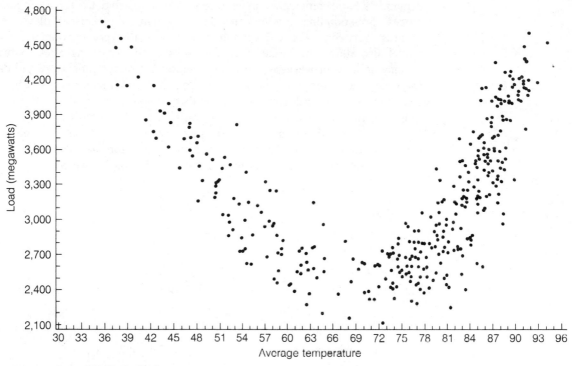

Source: Florida Power Corporation.

FIGURE 16.3

Theoretical Relationship Between Daily Peak Demand and Temperature

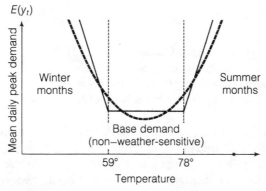

where

x_{1t} = Coincident temperature on day t

$x_{2t} = \begin{cases} 1 & \text{if } x_{1t} < 59 \\ 0 & \text{if not} \end{cases}$ $x_{3t} = \begin{cases} 1 & \text{if } x_{1t} > 78 \\ 0 & \text{if not} \end{cases}$

$x_{4t} = \begin{cases} 1 & \text{if Saturday} \\ 0 & \text{if not} \end{cases}$ $x_{5t} = \begin{cases} 1 & \text{if Sunday or holiday} \\ 0 & \text{if not} \end{cases}$ (Base level = Weekday)

ε_t = Uncorrelated error term

Model 1 proposes three different straight-line relationships between peak demand (y_t) and coincident temperature x_{1t}, one for each of the three temperature ranges corresponding to winter months (less than 59°), non–weather-sensitive months (between 59° and 78°), and summer months (greater than 78°).* The model also allows for variations in demand due to day of the week (Saturday, Sunday/holiday, or weekday). Since interaction between temperature and day of the week is omitted, the model is assuming that the differences between mean peak demand for weekdays and weekends/holidays is constant for the winter-sensitive, summer-sensitive, and non–weather-sensitive months.

We will illustrate the mechanics of the piecewise linear terms by finding the equations of the three demand–temperature lines for weekdays (i.e., $x_{4t} = x_{5t} = 0$). Substituting $x_{4t} = 0$ and $x_{5t} = 0$ into the model, we have

Winter-sensitive months ($x_{1t} < 59°$, $x_{2t} = 1$, $x_{3t} = 0$):
$$E(y_t) = \beta_0 + \beta_1(x_{1t} - 59)(0) + \beta_2(x_{1t} - 78)(1) + \beta_3(0) + \beta_4(0)$$
$$= \beta_0 + \beta_1(x_{1t} - 59)$$
$$= (\beta_0 - 59\beta_1) + \beta_1 x_{1t}$$

Summer-sensitive months ($x_{1t} > 78°$, $x_{2t} = 0$, $x_{3t} = 1$):
$$E(y_t) = \beta_0 + \beta_1(x_{1t} - 59)(0) + \beta_2(x_{1t} - 78)(1) + \beta_3(0) + \beta_4(0)$$
$$= \beta_0 + \beta_2(x_{1t} - 78)$$
$$= (\beta_0 - 78\beta_2) + \beta_2 x_{1t}$$

Non–weather-sensitive months ($59° \leq x_{1t} \leq 78°$, $x_{2t} = x_{3t} = 0$):
$$E(y_t) = \beta_0 + \beta_1(x_{1t} - 59)(0) + \beta_2(x_{1t} - 78)(0) + \beta_3(0) + \beta_4(0)$$
$$= \beta_0$$

Note that the slope of the demand–temperature line for winter-sensitive months (when $x_{1t} < 59$) is β_1 (which we expect to be negative), while the slope for summer-sensitive months (when $x_{1t} > 78$) is β_2 (which we expect to be positive). The intercept term β_0 represents the mean daily peak demand observed in the non–weather-sensitive period (when $59 \leq x_{1t} \leq 78$). Notice also that the peak demand during non–weather-sensitive days does not depend on temperature (x_{1t}).

Model 1 is a multiple regression model that relies on the standard regression assumptions of independent errors (ε_t uncorrelated). This may be a serious shortcoming in view of the fact that the data are in the form of a time series. To account for possible autocorrelated residuals, two time series models were proposed:

MODEL 2

$$y_t = \beta_0 + \beta_1(x_{1t} - 59)x_{2t} + \beta_2(x_{1t} - 78)x_{3t} + \beta_3 x_{4t} + \beta_4 x_{5t} + R_t$$
$$R_t = \phi_1 R_{t-1} + \varepsilon_t$$

*The temperature values, 59° and 78°, identify where the winter- and summer-sensitive portions of demand join the base demand component. These "knot values" were determined from visual inspection of the graph in Figure 16.2.

Model 2 proposes a regression–autoregression pair of models for daily peak demand (y_t). The deterministic component, $E(y_t)$, is identical to the deterministic component of Model 1; however, a first-order autoregressive model is chosen for the random error component. Recall (from Section 9.5) that a first-order autoregressive model is appropriate when the correlation between residuals diminishes as the distance between time periods (in this case, days) increases.

MODEL 3

$$y_t = \beta_0 + \beta_1(x_{1t} - 59)x_{2t} + \beta_2(x_{1t} - 78)x_{3t} + \beta_3 x_{4t} + \beta_4 x_{5t} + R_t$$
$$R_t = \phi_1 R_{t-1} + \phi_2 R_{t-2} + \phi_5 R_{t-5} + \phi_7 R_{t-7} + \varepsilon_t$$

Model 3 extends the first-order autoregressive error model of Model 2 to a seventh-order autoregressive model with lags at 1, 2, 5, and 7. In theory, the peak demand on day t will be highly correlated with the peak demand on day $t + 1$. However, there also may be significant correlation between demand 2 days, 5 days, and/or 1 week (7 days) apart. This more general error model is proposed to account for any residual correlation that may occur as a result of the week-to-week variation in peak demand, in addition to the day-to-day variation.

SECTION 16.4

THE REGRESSION AND AUTOREGRESSION ANALYSES

The multiple regression computer printout for Model 1 is shown in Figure 16.4, and a plot of the least squares fit is shown in Figure 16.5 (page 688). The model appears to provide a good fit to the data, with $R^2 = .8307$ and $F = 441.729$ (significant at $p = .0001$). The value of ROOT MSE, $s = 245.585$, implies that we can expect to predict daily peak demand accurate to within $2s \approx 491$ megawatts of its true value. However, we must be careful not to conclude at this point

FIGURE 16.4 SAS Printout for Multiple Regression Model of Daily Peak Demand, Model 1

```
DEP VARIABLE: LOAD

                    SUM OF        MEAN
SOURCE     DF      SQUARES       SQUARE      F VALUE    PROB > F
MODEL       4    106565982     26641496     441.729     0.0001
ERROR     360     21712247     60311.797
C TOTAL   364    128278229

        ROOT MSE      245.585    R-SQUARE
        DEP MEAN     3191.863    ADJ R-SQ     0.8289
        C.V.         7.694083

                   PARAMETER    STANDARD    T FOR H0:
VARIABLE DF        ESTIMATE       ERROR    PARAMETER = 0   PROB > !T!
INTERCEP  1        2670.171    21.251829      125.644       0.0001
AVTW      1       -82.039853    2.941928      -27.886       0.0001
AVTS      1        114.443      3.050468       37.516       0.0001
SAT       1       -164.932     37.990216       -4.341       0.0001
SUN       1       -285.114     35.328293       -8.070       0.0001

DURBIN-WATSON D              0.705
(NUMBER OF OBS)                365
1ST ORDER AUTOCORRELATION    0.648
```

FIGURE 16.5 Daily Peak Demand Versus Temperature: Actual Versus Fitted Piecewise Linear Model

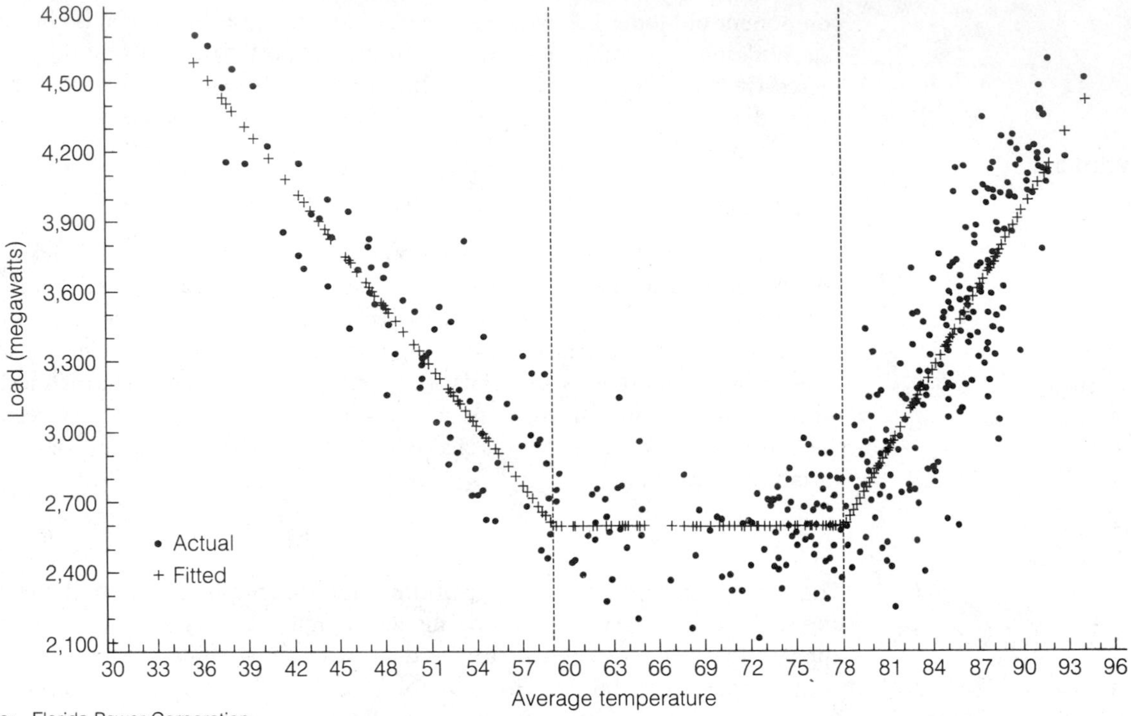

Source: Florida Power Corporation.

that the model is useful for predicting peak demand. Recall that in the presence of autocorrelated residuals, the standard errors of the regression coefficients are underestimated, thereby inflating the corresponding t-statistics for testing H_0: $\beta_i = 0$. At worst, this could lead to the false conclusion that a β parameter is significantly different from 0; at best, the results, although significant, give an overoptimistic view of the predictive ability of the model.

To determine whether the residuals of the multiple regression model are positively autocorrelated, we conduct the Durbin–Watson test:

H_0: Uncorrelated residuals

H_a: Positive residual correlation

Recall that the Durbin–Watson test is designed specifically for detecting first-order autocorrelation in the residuals, R_t. Thus, we can write the null and alternative hypotheses as

H_0: $\phi_1 = 0$

H_a: $\phi_1 > 0$

where $R_t = \phi_1 R_{t-1} + \varepsilon_t$, and ε_t = uncorrelated error (white noise).

The test statistic, given at the bottom of the printout in Figure 16.4, is $d = .705$. Recall that small values of d lead us to reject H_0: $\phi_1 = 0$ in favor of the alternative H_a: $\phi_1 > 0$. For $\alpha = .01$, $n = 365$, and $k = 4$ (the number of β parameters in the model, excluding β_0), the lower bound on the critical value (obtained from Table 9 of Appendix D) is approximately $d_L = 1.46$. Since the value of the test statistic, $d = .705$, falls well below the lower bound, there is strong evidence at $\alpha = .01$ of positive (first-order) autocorrelated residuals. Thus, we need to incorporate terms for residual autocorrelation into the model.

The time series printouts for Models 2 and 3 are shown in Figures 16.6 and 16.7 (page 690), respectively. A summary of the results for all three models is given in Table 16.1.

TABLE 16.1
Summary of Results for Models 1, 2, and 3

	MODEL 1	MODEL 2	MODEL 3
R^2	.8307	.9225	.9351
MSE	60,311.797	27,687.44	23,398.43
s	245.585	166.394	152.966

FIGURE 16.6 SAS Printout for First-Order Autoregressive Time Series Model of Daily Peak Demand, Model 2

```
                    DEPENDENT VARIABLE = LOAD
                 ORDINARY LEAST SQUARES ESTIMATES
                     VARIABLE DF      B VALUE
                      INTERCPT 1      2670.171
                      AVTW     1       -82.0399
                      AVTS     1       114.4427
                      SAT      1      -164.932
                      SUN      1      -285.114


                  ESTIMATES OF AUTOCORRELATIONS
  LAG    COVARIANCE    CORRELATION    -1 9 8 7 6 5 4 3 2 1 0 1 2 3 4 5 6 7 8 9 1
   0      59485.6       1.000000      !                    !********************!
   1      38519.4       0.647541      !                    !************        !

                   PRELIMINARY MSE =      34542.75
             ESTIMATES OF THE AUTOREGRESSIVE PARAMETERS
             LAG    COEFFICIENT      STD ERROR     T RATIO
              1     -0.64754083       0.039887   -16.234581
                    YULE-WALKER ESTIMATES
            SSE        9939789         DFE           359
            MSE        27687.44        ROOT MSE   166.3943
            REG RSQ    0.7626          TOTAL RSQ    0.9225


  VARIABLE DF       B VALUE        STD ERROR     T RATIO   APPROX PROB
   INTERCPT 1    2812.96710162   29.8790879359    94.145     0.0001
   AVTW     1     -65.337453      2.6639248330   -24.527     0.0001
   AVTS     1      83.45523858    3.8531993234    21.659     0.0001
   SAT      1    -130.82831      22.4136023517    -5.837     0.0001
   SUN      1    -275.55071      21.3736779578   -12.892     0.0001
```

FIGURE 16.7 SAS Printout for Seventh-Order Autoregressive Time Series Model of Daily Peak Demand, Model 3

```
                      DEPENDENT VARIABLE = LOAD
                    ORDINARY LEAST SQUARES ESTIMATES

                      VARIABLE  DF      B VALUE
                      INTERCPT   1     2670.171
                      AVTW       1      -82.0399
                      AVTS       1      114.4427
                      SAT        1     -164.932
                      SUN        1     -285.114

                    ESTIMATES OF AUTOCORRELATIONS

LAG   COVARIANCE   CORRELATION  -1 9 8 7 6 5 4 3 2 1 0 1 2 3 4 5 6 7 8 9 1
 0     59485.6     1.000000      !                     !********************!
 1     38519.4     0.647541      !                     !*************       !
 2     35741       0.600834      !                     !************        !
 3     32868.2     0.552540      !                     !***********         !
 4     29917.9     0.502943      !                     !**********          !
 5     31340.9     0.526865      !                     !***********         !
 6     30061.9     0.505364      !                     !**********          !
 7     31508       0.529674      !                     !***********         !

                    PRELIMINARY MSE =    28841.4
                ESTIMATES OF THE AUTOREGRESSIVE PARAMETERS

        LAG   COEFFICIENT      STD ERROR      T RATIO
         1    -0.36793644      0.049902      -7.373164
         2    -0.20702784      0.051722      -4.002705
         3     0.00000000      0.000000
         4     0.00000000      0.000000
         5    -0.13526445      0.049072      -2.756459
         6     0.00000000      0.000000
         7    -0.15338478      0.048430      -3.167144

                    EXPECTED AUTOCORRELATIONS

                       LAG  AUTOCORR
                        0    0.9668
                        1    0.6117
                        2    0.5640
                        3    0.4794
                        4    0.4494
                        5    0.4819
                        6    0.4469
                        7    0.4888

                      YULE-WALKER ESTIMATES
        SSE          8329842     DFE              356
        MSE         23398.43     ROOT MSE    152.9655
        REG RSQ       0.8112     TOTAL RSQ     0.9351


VARIABLE  DF       B VALUE      STD ERROR      T RATIO   APPROX PROB
INTERCPT   1   2809.95021301  58.2345577940    48.252      0.0001
AVTW       1     -71.28248      2.2620998226   -31.512      0.0001
AVTS       1      79.12014515   4.1805721109    18.926      0.0001
SAT        1    -150.52375     23.4727721121    -6.413      0.0001
SUN        1    -262.27342     21.6832940758   -12.096      0.0001
```

The addition of the first-order autoregressive error term in Model 2 yielded a drastic improvement to the fit of the model. The value of R^2 (i.e., TOTAL RSQ) increased from .83 to .92, and the standard deviation s (i.e., ROOT MSE) decreased from 245.6 to 166.4. These results support the conclusion reached by the Durbin–Watson test—namely, that the first-order autoregressive lag parameter ϕ_1 is significantly different from 0.

Does the more general autoregressive error model (Model 3) provide a better approximation to the pattern of correlation in the residuals than the first-order autoregressive model (Model 2)? To test this hypothesis, we would need to test H_0: $\phi_2 = \phi_5 = \phi_7 = 0$. Although we omit discussion of tests on autoregressive parameters in this text,* we can arrive at a decision from a pragmatic point of view by again comparing the values of R^2 and s for the two models. The more complex autoregressive model proposed by Model 3 yields a slight increase in R^2 (.935 as compared to .923 for Model 2) and a slight decrease in the value of s (153.00 as compared to 166.4 for Model 2). The additional lag parameters, although they may be statistically significant, may not be practically significant. The practical analyst may decide that the first-order autoregressive process proposed by Model 2 is the more desirable option since it is easier to use to forecast peak daily demand (and therefore more explainable to managers) while yielding approximate prediction errors (as measured by $2s$) that are only slightly larger than those for Model 3.

For the purposes of illustration, we use Model 2 to forecast daily peak demand in the following section.

SECTION 16.5

FORECASTING DAILY PEAK ELECTRICITY DEMAND

Suppose the FPC decided to use Model 2 to forecast daily peak demand for the first seven days of November 1983. The estimated model,[†] obtained from Figure 16.6, is given by

$$\hat{y}_t = 2{,}812.967 - 65.337(x_{1t} - 59)x_{2t} + 83.455(x_{1t} - 78)x_{3t}$$
$$- 130.828x_{4t} - 275.551x_{5t} + \hat{R}_t$$
$$\hat{R}_t = .6475\hat{R}_{t-1}$$

The forecast for November 1, 1983 ($t = 366$) requires an estimate of the residual R_{365}, where $\hat{R}_{365} = y_{365} - \hat{y}_{365}$. The last day of the November 1982–October 1983 time period ($t = 365$) was October 31, 1983, a Monday. On this day the peak demand was recorded as $y_{365} = 2{,}752$ megawatts and the coincident temperature as $x_{1,365} = 77°$. Substituting the appropriate values of the dummy variables into the equation for $\hat{y}_t$ (i.e., $x_{2t} = 0$, $x_{3t} = 0$, $x_{4t} = 0$, and $x_{5t} = 0$), we have

$$\hat{R}_{365} = y_{365} - \hat{y}_{365}$$
$$= 2{,}752 - [2{,}812.967 - 65.337(77 - 59)(0) + 83.455(77 - 78)(0)$$
$$- 130.828(0) - 275.551(0)]$$
$$= 2{,}752 - 2{,}812.967 = -60.967$$

*For details of tests on autoregressive parameters, see Fuller (1976).
†Remember that the estimate of ϕ_1 is obtained by multiplying the value reported on the SAS printout by (-1).

Then the formula for calculating the forecast for Tuesday, November 1, 1983, is

$$\hat{y}_{366} = 2{,}812.967 - 65.337(x_{1,366} - 59)x_{2,366} + 83.455(x_{1,366} - 78)x_{3,366}$$
$$- 130.828x_{4,366} - 275.551x_{5,366} + \hat{R}_{366}$$

where

$$\hat{R}_{366} = \hat{\phi}_1\hat{R}_{365} = (.6475)(-60.967) = -39.476$$

Note that the forecast requires an estimate of coincident temperature on that day, $\hat{x}_{1,366}$. If the FPC wants to forecast demand under normal weather conditions, then this estimate can be obtained from historical data for that day. Or, the FPC may choose to rely on a meteorologist's weather forecast for that day. For this example, assume that $\hat{x}_{1,366} = 76°$ (the actual temperature recorded by the FPC). Then $x_{2,366} = x_{3,366} = 0$ (since $59 \leq \hat{x}_{1,366} \leq 78$) and $x_{4,366} = x_{5,366} = 0$ (since the target day is a Tuesday). Substituting these values and the value of $\hat{R}_{366}$ into the equation, we have

$$\hat{y}_{366} = 2{,}812.967 - 65.337(76 - 59)(0) + 83.455(76 - 78)(0)$$
$$- 130.828(0) - 275.551(0) - 39.476$$
$$= 2{,}773.49$$

Similarly, a forecast for Wednesday, November 2, 1983 (i.e., $t = 367$) can be obtained:

$$\hat{y}_{367} = 2{,}812.967 - 65.337(x_{1,367} - 59)x_{2,367} + 83.455(x_{1,367} - 78)x_{3,367}$$
$$- 130.828x_{3,367} - 275.551x_{4,367} + \hat{R}_{367}$$

where $\hat{R}_{367} = \hat{\phi}_1\hat{R}_{366} = (.6475)(-39.476) = -25.561$, and $x_{3,367} = x_{4,367} = 0$. For an estimated coincident temperature of $\hat{x}_{1,367} = 77°$ (again, this is the actual temperature recorded on that day), we have $x_{2,367} = 0$ and $x_{3,367} = 0$. Substituting these values into the prediction equation, we obtain

$$\hat{y}_{367} = 2{,}812.967 - 65.337(77 - 59)(0) + 83.455(77 - 78)(0)$$
$$- 130.828(0) - 275.551(0) - 25.561$$
$$= 2{,}812.967 - 25.561$$
$$= 2{,}787.41$$

Approximate 95% prediction intervals for the two forecasts are calculated as follows:

Tuesday, Nov. 1, 1983:
$$\hat{y}_{366} \pm 1.96 \sqrt{\text{MSE}}$$
$$= 2{,}773.49 \pm 1.96\sqrt{27.687.44}$$
$$= 2{,}773.49 \pm 326.14 \quad \text{or} \quad (2{,}447.35, 3{,}099.63)$$

Wednesday, Nov. 2, 1983:

$$\hat{y}_{367} \pm 1.96\sqrt{\text{MSE}(1 + \hat{\phi}_1^2)}$$
$$= 2{,}787.41 \pm 1.96\sqrt{(27{,}687.44)[1 + (.6475)^2]}$$
$$= 2{,}787.41 \pm 388.53 \quad \text{or} \quad (2{,}398.88,\ 3{,}175.94)$$

The forecasts, approximate 95% prediction intervals, and actual daily peak demands (recorded by the FPC) for the first seven days of November 1983 are given in Table 16.2. Note that actual peak demand y_t falls within the corresponding prediction interval for all seven days. Thus, the model appears to be useful for making short-term forecasts of daily peak demand. Of course, if the prediction intervals were extremely wide this result would be of no practical value. For example, the forecast error $y_t - \hat{y}_t$, measured as a percentage of the actual value y_t, may be large even when y_t falls within the prediction interval. Various techniques, such as the percent forecast error, are available for evaluating the accuracy of forecasts. Consult the references given at the end of Chapter 9 for details on these techniques.

TABLE 16.2 Forecasts and Actual Peak Demands for the First Seven Days of November 1983

DATE	t	FORECAST $\hat{y}_t$	APPROXIMATE 95% PREDICTION INTERVAL	ACTUAL DEMAND y_t	ACTUAL TEMPERATURE x_{1t}
Tues., Nov. 1	366	2,773.49	(2,447.35, 3,099.63)	2,799	76
Wed., Nov. 2	367	2,787.41	(2,398.88, 3,175.94)	2,784	77
Thurs., Nov. 3	368	2,796.42	(2,384.53, 3,208.31)	2,845	77
Fri., Nov. 4	369	2,802.25	(2,380.92, 3,223.58)	2,701	76
Sat., Nov. 5	370	2,675.20	(2,249.97, 3,100.43)	2,512	72
Sun., Nov. 6	371	2,532.92	(2,106.07, 2,959.77)	2,419	71
Mon., Nov. 7	372	2,810.06	(2,382.59, 3,237.53)	2,749	68

SECTION 16.6

SUMMARY

This case study presents a time series approach to modeling and forecasting daily peak demand observed at Florida Power Corporation. A graphical analysis of the data provided the means of identifying and formulating a piecewise linear regression model relating peak demand to temperature and day of the week. The multiple regression model, although providing a good fit to the data, exhibited strong signs of positive residual autocorrelation.

Two autoregressive time series models were proposed to account for the autocorrelated errors. Both models were shown to provide a drastic improvement in model adequacy. Either could be used to provide reliable short-term forecasts of daily peak demand or for weather normalization (i.e., estimating the peak demand if normal weather conditions had prevailed).

| | | | | | | | | | | |

EXERCISES 16.1–16.2

16.1 All three models discussed in this case study make the underlying assumption that the peak demand–temperature relationship is independent of day of the week. Write a model that includes interaction between temperature and day of the week. Show the effect the interaction has on the straight-line relationships between peak demand and temperature. Explain how you could test the significance of the interaction terms.

16.2 Consider the problem of using Model 2 for weather normalization. Suppose the temperature on Saturday, March 5, 1983 (i.e., $t = 125$) was $x_{1,121} = 25°$, unusually cold for that day. Normally, temperatures range from $40°$ to $50°$ on March 5 in the FPC service area. Substitute $x_{1,121} = 45°$ into the prediction equation to obtain an estimate of the peak demand expected if normal weather conditions had prevailed on March 5, 1983. Calculate an approximate 95% prediction interval for the estimate. [*Hint*:   Use $\hat{y}_{121} \pm 1.96 \sqrt{\text{MSE}}$.]

REFERENCES

Fuller, W. A. *Introduction to Statistical Time Series*. New York: Wiley, 1976.

Jacob, M. F. "A time series approach to modeling daily peak electricity demands." Paper presented at the SAS Users Group International Annual Conference, Reno, Nevada, 1985.

THE MECHANICS OF A MULTIPLE REGRESSION ANALYSIS

CONTENTS

INTRODUCTION

The rationale behind a multiple regression analysis and the types of inferences it permits you to make were the subjects of Chapter 4. We noted that the method of least squares most often leads to a very difficult computational problem, namely the solution of a set of $(k + 1)$ simultaneous linear equations in the unknown values of the estimates $\hat{\beta}_0, \hat{\beta}_1, \ldots, \hat{\beta}_k$, and that the formulas for the estimated standard errors $s_{\hat{\beta}_0}, s_{\hat{\beta}_1}, \ldots, s_{\hat{\beta}_k}$ were too complicated to express as ordinary algebraic formulas. We circumvented both these problems in a very easy way— we relied on the least squares estimates, confidence intervals, tests, etc., provided by a standard regression analysis computer package. Thus, Chapter 4 provides a basic working knowledge of the types of inferences you might wish to make from a multiple regression analysis and explains how to interpret the results. If we can do this, why would we wish to know the actual process performed by the computer?

There are several answers to this question:

1. Many multiple regression computer packages do not print all the information you may want. As one illustration, we noted in Chapter 4 that very often the objective of a regression analysis is to develop a prediction equation that can be used to estimate the mean value of y (say, mean profit or mean yield) for given values of the predictor variables $x_1, x_2, \ldots, x_k$. Or we may wish to predict some future value of y for given values of $x_1, x_2, \ldots, x_k$. Many computer packages do not give the confidence interval for $E(y)$ or a prediction interval for y. Thus, you might need to know how to find the necessary quantities from the analysis and perform the computations yourself.

2. A multiple regression computer package may possess the capability of computing some specific quantity that you desire, but you may find the instructions on how to "call" for this special calculation difficult to understand. It may be easier to identify the components required for your computation and do it yourself.

3. For some designed experiments, finding the least squares equations and solving them is a trivial operation. Understanding the process by which the least squares equations are generated and understanding how they are solved will help you understand how experimental design affects the results of a regression analysis. Thus, a knowledge of the computations involved in performing a regression analysis will help you to better understand the contents of Chapters 8 and 10.

To summarize, "knowing how it is done" is not essential for performing an ordinary regression analysis or interpreting its results. But "knowing how" helps, and it is essential for an understanding of many of the finer points associated with a multiple regression analysis. This appendix explains "how it is done" without getting into the unpleasant task of performing the computations for solving the least squares equations. This mechanical and tedious procedure can be left to a computer (the solutions are verifiable). We illustrate the procedure in Appendix B.

SECTION A.2

**MATRICES
AND MATRIX
MULTIPLICATION**

While it is very difficult to give the formulas for the multiple regression least squares estimators and for their estimated standard errors in ordinary algebra, it is easy to do so using **matrix algebra**. Thus, by arranging the data in particular rectangular patterns called **matrices** and performing various operations with them, we can obtain the least squares estimates and their estimated standard errors. In this section and Sections A.3 and A.4 we will define what we mean by a matrix and explain various operations that can be performed with them. We will explain how to use this information to conduct a regression analysis in Section A.5.

Three matrices, **A**, **B**, and **C**, are shown below. Note that each matrix is a rectangular arrangement of numbers with one number in every row–column position.

$$A = \begin{bmatrix} 2 & 3 \\ 0 & 1 \\ -1 & 6 \end{bmatrix} \quad B = \begin{bmatrix} 3 & 0 & 1 \\ -1 & 0 & 1 \\ 4 & 2 & 0 \end{bmatrix} \quad C = \begin{bmatrix} 1 \\ 2 \\ 1 \end{bmatrix}$$

DEFINITION A.1

A **matrix** is a rectangular array of numbers.*

The numbers that appear in a matrix are called **elements** of the matrix. If a matrix contains r rows and c columns, there will be an element in each of the row–column positions of the matrix, and the matrix will have $r \times c$ elements. For example, the matrix **A** shown above contains $r = 3$ rows, $c = 2$ columns, and $rc = (3)(2) = 6$ elements, one in each of the 6 row–column positions.

DEFINITION A.2

A number in a particular row–column position is called an **element** of the matrix.

Notice that the matrices **A**, **B**, and **C** contain different numbers of rows and columns. The numbers of rows and columns give the **dimensions** of a matrix.

When we give a formula in matrix notation, the elements of a matrix will be represented by symbols. For example, if we have a matrix

$$A = \begin{bmatrix} a_{11} & a_{12} & a_{13} \\ a_{21} & a_{22} & a_{23} \end{bmatrix}$$

the symbol a_{ij} will denote the element in the ith row and jth column of the matrix. The first subscript always identifies the row and the second identifies the

*For our purpose, we assume that the numbers are real.

column in which the element is located. For example, the element a_{12} is in the first row and second column of the matrix **A**. The rows are always numbered from top to bottom, and the columns are always numbered from left to right.

DEFINITION A.3

A matrix containing r rows and c columns is said to be an *r × c matrix* where r and c are the **dimensions** of the matrix.

DEFINITION A.4

If $r = c$, a matrix is said to be a **square matrix**.

Matrices are usually identified by capital letters, such as **A**, **B**, **C**, corresponding to the letters of the alphabet employed in ordinary algebra. The difference is that in ordinary algebra, a letter is used to denote a single real number while in matrix algebra, *a letter denotes a rectangular array of numbers*. The operations of matrix algebra are very similar to those of ordinary algebra—you can add matrices, subtract them, multiply them, and so on. But since we are concerned only with the applications of matrix algebra to the solution of the least squares equations, we will define only the operations and types of matrices that are pertinent to that subject.

The most important operation for us is matrix multiplication, which requires **row–column multiplication**. To illustrate this process, suppose we wish to find the product **AB**, where

$$\mathbf{A} = \begin{bmatrix} 2 & 1 \\ 4 & -1 \end{bmatrix} \qquad \mathbf{B} = \begin{bmatrix} 2 & 0 & 3 \\ -1 & 4 & 0 \end{bmatrix}$$

We will always multiply the rows of **A** (the matrix on the left) by the columns of **B** (the matrix on the right). The product formed by the first row of **A** times the first column of **B** is obtained by multiplying the elements in corresponding positions and summing these products. Thus, the first row, first column product, shown diagrammatically below, is

$$(2)(2) + (1)(-1) = 4 - 1 = 3$$

$$\mathbf{AB} = \begin{bmatrix} 2 & 1 \\ 4 & -1 \end{bmatrix}\begin{bmatrix} 2 & 0 & 3 \\ -1 & 4 & 0 \end{bmatrix} = \begin{bmatrix} 3 & \\ & \end{bmatrix}$$

Similarly, the first row, second column product is

$$(2)(0) + (1)(4) = 0 + 4 = 4$$

So far we have

$$\mathbf{AB} = \begin{bmatrix} 3 & 4 \\ & \end{bmatrix}$$

To find the complete matrix product **AB**, all we need to do is find each element in the **AB** matrix. Thus, we will define an element in the ith row, jth column of **AB** as the product of the ith row of **A** and the jth column of **B**. We complete the process in Example A.1.

EXAMPLE A.1

Find the product **AB**, where

$$A = \begin{bmatrix} 2 & 1 \\ 4 & -1 \end{bmatrix} \qquad B = \begin{bmatrix} 2 & 0 & 3 \\ -1 & 4 & 0 \end{bmatrix}$$

SOLUTION

If we represent the product **AB** as

$$C = \begin{bmatrix} c_{11} & c_{12} & c_{13} \\ c_{21} & c_{22} & c_{23} \end{bmatrix}$$

we have already found $c_{11} = 3$ and $c_{12} = 4$. Similarly, the element c_{21}, the element in the second row, first column of **AB**, is the product of the second row of **A** and the first column of **B**:

$$(4)(2) + (-1)(-1) = 8 + 1 = 9$$

Proceeding in a similar manner to find the remaining elements of **AB**, we have

$$AB = \begin{bmatrix} 2 & 1 \\ 4 & -1 \end{bmatrix} \begin{bmatrix} 2 & 0 & 3 \\ -1 & 4 & 0 \end{bmatrix} = \begin{bmatrix} 3 & 4 & 6 \\ 9 & -4 & 12 \end{bmatrix}$$ ∎

Now, try to find the product **BA**, using matrices **A** and **B** from Example A.1. You will observe two very important differences between multiplication in matrix algebra and multiplication in ordinary algebra:

1. You cannot find the product **BA** because you cannot perform row–column multiplication. You can see that the dimensions do not match by placing the matrices side-by-side.

$$\underset{2 \times 3 \quad 2 \times 2}{BA} \qquad \text{does not exist}$$

   The number of elements (3) in a row of **B** (the matrix on the left) does not match the number of elements (2) in a column of **A** (the matrix on the right). Therefore, you cannot perform row–column multiplication, and the matrix product **BA** does not exist. The point is, not all matrices can be multiplied. You can find products for matrices **A** and **B** only when **A** is $r \times d$ and **B** is $d \times c$. That is:

REQUIREMENT FOR MULTIPLICATION

$$\underset{r \times d \quad d \times c}{AB}$$

The two inner dimension numbers must be equal. The dimensions of the product will always be given by the outer dimension numbers:

DIMENSIONS OF AB ARE $r \times c$

---

2. The second difference between ordinary and matrix multiplication is that in ordinary algebra, $ab = ba$. In matrix algebra, **AB** usually does not equal **BA**. In fact, as noted in item 1 above, it may not even exist.

DEFINITION A.5

The product **AB** of an $r \times d$ matrix **A** and a $d \times c$ matrix **B** is an $r \times c$ matrix **C**, where the element c_{ij} $(i = 1, 2, \ldots, r; j = 1, 2, \ldots, c)$ of **C** is the product of the ith row of **A** and the jth column of **B**.

EXAMPLE A.2

Given the matrices below, find **IA** and **IB**.

$$A = \begin{bmatrix} 2 \\ 1 \\ 3 \end{bmatrix} \qquad B = \begin{bmatrix} 3 & 0 \\ 1 & 2 \\ 4 & -1 \end{bmatrix} \qquad I = \begin{bmatrix} 1 & 0 & 0 \\ 0 & 1 & 0 \\ 0 & 0 & 1 \end{bmatrix}$$

SOLUTION

Notice that the product

exists and that it will be of dimensions 3×1. Performing the row–column multiplication yields

$$IA = \begin{bmatrix} 1 & 0 & 0 \\ 0 & 1 & 0 \\ 0 & 0 & 1 \end{bmatrix} \begin{bmatrix} 2 \\ 1 \\ 3 \end{bmatrix} = \begin{bmatrix} 2 \\ 1 \\ 3 \end{bmatrix}$$

Similarly,

exists and is of dimensions 3×2. Performing the row–column multiplications, we obtain

$$\text{IB} = \begin{bmatrix} 1 & 0 & 0 \\ 0 & 1 & 0 \\ 0 & 0 & 1 \end{bmatrix} \begin{bmatrix} 3 & 0 \\ 1 & 2 \\ 4 & -1 \end{bmatrix} = \begin{bmatrix} 3 & 0 \\ 1 & 2 \\ 4 & -1 \end{bmatrix}$$

Notice that the **I** matrix possesses a special property. We have $\text{IA} = \text{A}$ and $\text{IB} = \text{B}$. We will comment further on this property in Section A.3.   ∎

EXERCISES A.1–A.6

A.1 Given the matrices A, B, and C:

$$A = \begin{bmatrix} 3 & 0 \\ -1 & 4 \end{bmatrix} \qquad B = \begin{bmatrix} 2 & 1 \\ 0 & -1 \end{bmatrix} \qquad C = \begin{bmatrix} 1 & 0 & 3 \\ -2 & 1 & 2 \end{bmatrix}$$

a. Find AB.   b. Find AC.   c. Find BA.

A.2 Given the matrices A, B, and C:

$$A = \begin{bmatrix} 3 & 1 & 3 \\ 2 & 0 & 4 \\ -4 & 1 & 2 \end{bmatrix} \qquad B = [1 \quad 0 \quad 2] \qquad C = \begin{bmatrix} 3 \\ 0 \\ 2 \end{bmatrix}$$

a. Find AC.   b. Find BC.
c. Is it possible to find AB? Explain

A.3 If A is a 3×2 matrix and B is a 2×4 matrix:
a. What are the dimensions of AB?
b. Is it possible to find the product BA? Explain.

A.4 If matrices B and C are of dimensions 1×3 and 3×1, respectively:
a. What are the dimensions of the product BC?
b. What are the dimensions of CB?
c. If B and C are the matrices shown in Exercise A.2, find CB.

A.5 Given the matrices A, B, and C:

$$A = \begin{bmatrix} 1 & 0 & 0 \\ 0 & 3 & 0 \\ 0 & 0 & 2 \end{bmatrix} \qquad B = \begin{bmatrix} 2 & 3 \\ -3 & 0 \\ 4 & -1 \end{bmatrix} \qquad C = [3 \quad 0 \quad 2]$$

a. Find AB.   b. Find CA.   c. Find CB.

A.6 Given the matrices:

$$A = [3 \quad 0 \quad -1 \quad 2] \qquad B = \begin{bmatrix} 2 \\ -1 \\ 0 \\ 3 \end{bmatrix}$$

a. Find AB.   b. Find BA.

SECTION A.3

IDENTITY MATRICES AND MATRIX INVERSION

In ordinary algebra, the number 1 is the identity element for the multiplication operation. That is, 1 is the element such that any other number, say c, multiplied by the identity element is equal to c. Thus, $4(1) = 4$, $(-5)(1) = -5$, and so forth.

The corresponding identity element for multiplication in matrix algebra, identified by the symbol $\mathbf{I}$, is a matrix such that

$$\mathbf{AI} = \mathbf{IA} = \mathbf{A} \quad \text{for any matrix } \mathbf{A}$$

The difference between identity elements in ordinary algebra and matrix algebra is that in ordinary algebra there is only one identity element, the number 1. In matrix algebra, the identity matrix must possess the correct dimensions in order for the product $\mathbf{IA}$ to exist. Consequently, there is an infinitely large number of identity matrices—all square and possessing the same pattern. The 1×1, 2×2, and 3×3 identity matrices are

$$\mathbf{I}_{1 \times 1} = [1] \qquad \mathbf{I}_{2 \times 2} = \begin{bmatrix} 1 & 0 \\ 0 & 1 \end{bmatrix} \qquad \mathbf{I}_{3 \times 3} = \begin{bmatrix} 1 & 0 & 0 \\ 0 & 1 & 0 \\ 0 & 0 & 1 \end{bmatrix}$$

In Example A.2, we demonstrated the fact that this matrix satisfies the property

$$\mathbf{IA} = \mathbf{A}$$

DEFINITION A.6

If $\mathbf{A}$ is any matrix, then a matrix $\mathbf{I}$ is defined to be an identity matrix if $\mathbf{AI} = \mathbf{IA} = \mathbf{A}$. The matrices that satisfy this definition possess the pattern

$$\mathbf{I} = \begin{bmatrix} 1 & 0 & 0 & \cdots & 0 \\ 0 & 1 & 0 & \cdots & 0 \\ 0 & 0 & 1 & \cdots & 0 \\ \cdot & \cdot & \cdot & \cdots & \cdot \\ \cdot & \cdot & \cdot & \cdots & \cdot \\ \cdot & \cdot & \cdot & \cdots & \cdot \\ 0 & 0 & 0 & \cdots & 1 \end{bmatrix}$$

EXAMPLE A.3

If $\mathbf{A}$ is the matrix shown below, find $\mathbf{IA}$ and $\mathbf{AI}$.

$$\mathbf{A} = \begin{bmatrix} 3 & 4 & -1 \\ 1 & 0 & 2 \end{bmatrix}$$

SOLUTION

$$\underset{\substack{2 \times 2 \quad 2 \times 3}}{\mathbf{IA}} = \begin{bmatrix} 1 & 0 \\ 0 & 1 \end{bmatrix} \begin{bmatrix} 3 & 4 & -1 \\ 1 & 0 & 2 \end{bmatrix} = \begin{bmatrix} 3 & 4 & -1 \\ 1 & 0 & 2 \end{bmatrix} = \mathbf{A}$$

$$\underset{\substack{2 \times 3 \quad 3 \times 3}}{\mathbf{AI}} = \begin{bmatrix} 3 & 4 & -1 \\ 1 & 0 & 2 \end{bmatrix} \begin{bmatrix} 1 & 0 & 0 \\ 0 & 1 & 0 \\ 0 & 0 & 1 \end{bmatrix} = \begin{bmatrix} 3 & 4 & -1 \\ 1 & 0 & 2 \end{bmatrix} = \mathbf{A}$$

Notice that the identity matrices used to find the products $\mathbf{IA}$ and $\mathbf{AI}$ were of different dimensions. This was necessary in order for the products to exist.   ∎

The identity element assumes importance when we consider the process of division and its role in the solution of equations. In ordinary algebra, division is essentially multiplication using the reciprocals of elements. For example, the equation

$$2X = 6$$

can be solved by dividing both sides of the equation by 2, *or* it can be solved by *multiplying* both sides of the equation by $\frac{1}{2}$, which is the reciprocal of 2. Thus,

$$\left(\frac{1}{2}\right)2X = \frac{1}{2}(6)$$
$$X = 3$$

What is the reciprocal of an element? It is the element such that the reciprocal times the element is equal to the identity element. Thus, the reciprocal of 3 is $\frac{1}{3}$ because

$$3\left(\frac{1}{3}\right) = 1$$

The identity matrix plays the same role in matrix algebra. Thus, the reciprocal of a matrix $\mathbf{A}$, called $\mathbf{A}$-*inverse* and denoted by the symbol $\mathbf{A}^{-1}$, is a matrix such that $\mathbf{AA}^{-1} = \mathbf{A}^{-1}\mathbf{A} = \mathbf{I}$.

Inverses are defined only for square matrices, but not all square matrices possess inverses. Those that do have inverses play an important role in solving the least squares equations and in other aspects of a regression analysis. We will show you one important application of the inverse matrix in Section A.4. The procedure for finding the inverse of a matrix is demonstrated in Appendix B.

DEFINITION A.7

The square matrix $\mathbf{A}^{-1}$ is said to be the **inverse** of the square matrix $\mathbf{A}$ if

$$\mathbf{A}^{-1}\mathbf{A} = \mathbf{AA}^{-1} = \mathbf{I}$$

The procedure for finding an inverse matrix is computationally quite tedious and is performed most often using a computer. There is one exception. Finding the inverse of one type of matrix, called a **diagonal matrix**, is easy. A diagonal matrix is one that has nonzero elements down the **main diagonal** (running from top left of the matrix to bottom right) and 0 elements elsewhere. For example, the identity matrix is a diagonal matrix (with 1's along the main diagonal), as are the following matrices:

$$A = \begin{bmatrix} 3 & 0 & 0 \\ 0 & 1 & 0 \\ 0 & 0 & 2 \end{bmatrix} \qquad B = \begin{bmatrix} 5 & 0 & 0 & 0 \\ 0 & 2 & 0 & 0 \\ 0 & 0 & 1 & 0 \\ 0 & 0 & 0 & 5 \end{bmatrix}$$

DEFINITION A.8

A **diagonal matrix** is one that contains nonzero elements on the main diagonal and 0 elements elsewhere.

You can verify that the inverse of

$$A = \begin{bmatrix} 3 & 0 & 0 \\ 0 & 1 & 0 \\ 0 & 0 & 2 \end{bmatrix} \quad \text{is} \quad A^{-1} = \begin{bmatrix} \frac{1}{3} & 0 & 0 \\ 0 & 1 & 0 \\ 0 & 0 & \frac{1}{2} \end{bmatrix}$$

i.e., $AA^{-1} = I$. In general, the inverse of a diagonal matrix is given by the following theorem which is stated without proof:

THEOREM A.1

The **inverse of a diagonal matrix**

$$D = \begin{bmatrix} d_{11} & 0 & 0 & \cdots & 0 \\ 0 & d_{22} & 0 & \cdots & 0 \\ 0 & 0 & d_{33} & \cdots & 0 \\ \cdot & \cdot & \cdot & \cdots & \cdot \\ \cdot & \cdot & \cdot & \cdots & \cdot \\ \cdot & \cdot & \cdot & \cdots & \cdot \\ 0 & 0 & 0 & \cdots & d_{nn} \end{bmatrix} \quad \text{is} \quad D^{-1} = \begin{bmatrix} 1/d_{11} & 0 & 0 & \cdots & 0 \\ 0 & 1/d_{22} & 0 & \cdots & 0 \\ 0 & 0 & 1/d_{33} & \cdots & 0 \\ \cdot & \cdot & \cdot & \cdots & \cdot \\ \cdot & \cdot & \cdot & \cdots & \cdot \\ \cdot & \cdot & \cdot & \cdots & \cdot \\ 0 & 0 & 0 & \cdots & 1/d_{nn} \end{bmatrix}$$

EXERCISES A.7–A.11

A.7 Let $A = \begin{bmatrix} 3 & 0 & 2 \\ -1 & 1 & 4 \end{bmatrix}$.

a. Give the identity matrix that will be used to obtain the product IA.
b. Show that $IA = A$.
c. Give the identity matrix that will be used to find the product AI.
d. Show that $AI = A$.

A.8 Given the matrices A and B below, show that $AB = I$, that $BA = I$, and consequently, verify that $B = A^{-1}$

$$A = \begin{bmatrix} 1 & 0 & 0 \\ 0 & 2 & 0 \\ 0 & 0 & 3 \end{bmatrix} \qquad B = \begin{bmatrix} 1 & 0 & 0 \\ 0 & \frac{1}{2} & 0 \\ 0 & 0 & \frac{1}{3} \end{bmatrix}$$

A.9 If

$$A = \begin{bmatrix} 12 & 0 & 0 & 8 \\ 0 & 12 & 0 & 0 \\ 0 & 0 & 8 & 0 \\ 8 & 0 & 0 & 8 \end{bmatrix}$$

verify that

$$A^{-1} = \begin{bmatrix} \frac{1}{4} & 0 & 0 & -\frac{1}{4} \\ 0 & \frac{1}{12} & 0 & 0 \\ 0 & 0 & \frac{1}{8} & 0 \\ -\frac{1}{4} & 0 & 0 & \frac{3}{8} \end{bmatrix}$$

A.10 If

$$A = \begin{bmatrix} 3 & 0 & 0 \\ 0 & 5 & 0 \\ 0 & 0 & 7 \end{bmatrix}$$

show that

$$A^{-1} = \begin{bmatrix} \frac{1}{3} & 0 & 0 \\ 0 & \frac{1}{5} & 0 \\ 0 & 0 & \frac{1}{7} \end{bmatrix}$$

A.11 Verify Theorem A.1.

SECTION A.4

SOLVING SYSTEMS OF SIMULTANEOUS LINEAR EQUATIONS

Consider the following set of simultaneous linear equations in two unknowns:

$$2v_1 + v_2 = 7$$

$$v_1 - v_2 = 2$$

Note that the solution for these equations is $v_1 = 3$, $v_2 = 1$.

Now define the matrices

$$A = \begin{bmatrix} 2 & 1 \\ 1 & -1 \end{bmatrix} \qquad V = \begin{bmatrix} v_1 \\ v_2 \end{bmatrix} \qquad G = \begin{bmatrix} 7 \\ 2 \end{bmatrix}$$

Thus, A is the matrix of coefficients of v_1 and v_2, V is a column matrix containing the unknowns (written in order, top to bottom), and G is a column matrix containing the numbers on the right-hand side of the equal signs.

Now, the given system of simultaneous equations can be rewritten as a **matrix equation**:

$$AV = G$$

By a matrix equation, we mean that the product matrix **AV**, is equal to the matrix **G**. *Equality of matrices means that corresponding elements are equal.* You can see that this is true for the expression $AV = G$, since

$$\underset{2 \times 2 \quad 2 \times 1}{\underset{AV}{}} = \begin{bmatrix} 2 & 1 \\ 1 & -1 \end{bmatrix} \begin{bmatrix} v_1 \\ v_2 \end{bmatrix} = \begin{bmatrix} (2v_1 + v_2) \\ (v_1 - v_2) \end{bmatrix} = \underset{2 \times 1}{G}$$

The matrix procedure for expressing a system of two simultaneous linear equations in two unknowns can be extended to express a set of k simultaneous equations in k unknowns. If the equations are written in the orderly pattern

$$a_{11}v_1 + a_{12}v_2 + \cdots + a_{1k}v_k = g_1$$
$$a_{21}v_1 + a_{22}v_2 + \cdots + a_{2k}v_k = g_2$$
$$\vdots$$
$$a_{k1}v_1 + a_{k2}v_2 + \cdots + a_{kk}v_k = g_k$$

then the set of simultaneous linear equations can be expressed as the matrix equation $AV = G$, where

$$A = \begin{bmatrix} a_{11} & a_{12} & \cdots & a_{1k} \\ a_{21} & & \cdots & a_{2k} \\ \vdots & & & \vdots \\ a_{k1} & & \cdots & a_{kk} \end{bmatrix} \qquad V = \begin{bmatrix} v_1 \\ v_2 \\ \vdots \\ v_k \end{bmatrix} \qquad G = \begin{bmatrix} g_1 \\ g_2 \\ \vdots \\ g_k \end{bmatrix}$$

Now let us solve this system of simultaneous equations. (If they are uniquely solvable, it can be shown that A^{-1} exists.) Multiplying both sides of the matrix equation by A^{-1}, we have

$$(A^{-1})AV = (A^{-1})G$$

But since $A^{-1}A = I$, we have

$$(I)V = A^{-1}G$$
$$V = A^{-1}G$$

In other words, if we know A^{-1}, we can find the solution to the set of simultaneous linear equations by obtaining the product $A^{-1}G$.

MATRIX SOLUTION TO A SET OF SIMULTANEOUS LINEAR EQUATIONS, AV = G
Solution: $V = A^{-1}G$

EXAMPLE A.4

Apply the boxed result to find the solution to the set of simultaneous linear equations

$$2v_1 + v_2 = 7$$
$$v_1 - v_2 = 2$$

SOLUTION

The first step is to obtain the inverse of the coefficient matrix,

$$A = \begin{bmatrix} 2 & 1 \\ 1 & -1 \end{bmatrix}$$

namely,

$$A^{-1} = \begin{bmatrix} \frac{1}{3} & \frac{1}{3} \\ \frac{1}{3} & -\frac{2}{3} \end{bmatrix}$$

(This matrix can be found using a packaged computer program for matrix inversion or, for this simple case, you could use the procedure explained in Appendix B.) As a check, note that

$$A^{-1}A = \begin{bmatrix} \frac{1}{3} & \frac{1}{3} \\ \frac{1}{3} & -\frac{2}{3} \end{bmatrix} \begin{bmatrix} 2 & 1 \\ 1 & -1 \end{bmatrix} = \begin{bmatrix} 1 & 0 \\ 0 & 1 \end{bmatrix} = I$$

The second step is to obtain the product $A^{-1}G$. Thus,

$$V = A^{-1}G = \begin{bmatrix} \frac{1}{3} & \frac{1}{3} \\ \frac{1}{3} & -\frac{2}{3} \end{bmatrix} \begin{bmatrix} 7 \\ 2 \end{bmatrix} = \begin{bmatrix} 3 \\ 1 \end{bmatrix}$$

Since

$$V = \begin{bmatrix} v_1 \\ v_2 \end{bmatrix} = \begin{bmatrix} 3 \\ 1 \end{bmatrix}$$

it follows that $v_1 = 3$ and $v_2 = 1$. You can see that these values of v_1 and v_2 satisfy the simultaneous linear equations and are the values that we specified as a solution at the beginning of this section. ∎

EXERCISES A.12–A.13

A.12 Suppose the simultaneous linear equations

$$3v_1 + v_2 = 5$$
$$v_1 - v_2 = 3$$

are expressed as a matrix equation,

$$AV = G$$

a. Find the matrices A, V, and G.
b. Verify that

$$A^{-1} = \begin{bmatrix} \frac{1}{4} & \frac{1}{4} \\ \frac{1}{4} & -\frac{3}{4} \end{bmatrix}$$

[*Note:* A procedure for finding A^{-1} is given in Appendix B.]
c. Solve the equations by finding $V = A^{-1}G$.

A.13 For the simultaneous linear equations

$$10v_1 + 20v_3 - 60 = 0$$

$$20v_2 - 60 = 0$$

$$20v_1 + 68v_3 - 176 = 0$$

a. Find the matrices **A**, **V**, and **G**.

b. Verify that

$$\mathbf{A}^{-1} = \begin{bmatrix} \frac{17}{70} & 0 & -\frac{1}{14} \\ 0 & \frac{1}{20} & 0 \\ -\frac{1}{14} & 0 & \frac{1}{28} \end{bmatrix}$$

c. Solve the equations by finding $\mathbf{V} = \mathbf{A}^{-1}\mathbf{G}$.

S E C T I O N A.5

THE LEAST SQUARES EQUATIONS AND THEIR SOLUTIONS

In order to apply matrix algebra to a regression analysis, we must place the data in matrices in a particular pattern. We will suppose that the linear model is

$$y = \beta_0 + \beta_1 x_1 + \beta_2 x_2 + \cdots + \beta_k x_k + \varepsilon$$

where (from Chapter 4) $x_1, x_2, \ldots, x_k$ could actually represent the squares, cubes, cross products, or other functions of predictor variables, and ε is a random error. We will assume that we have collected n data points, i.e., n values of y and corresponding values of $x_1, x_2, \ldots, x_k$, and that these are denoted as shown in the table:

DATA POINT	y-Value	x_1	x_2	$\cdots$	x_k
1	y_1	x_{11}	x_{21}		x_{k1}
2	y_2	x_{12}	x_{22}		x_{k2}
.	.	.	.		.
.	.	.	.		.
n	y_n	x_{1n}	x_{2n}		x_{kn}

Then the two data matrices **Y** and **X** are as shown in the next box.

THE DATA MATRICES Y AND X AND THE $\hat{\beta}$ MATRIX

$$\mathbf{Y} = \begin{bmatrix} y_1 \\ y_2 \\ y_3 \\ \cdot \\ \cdot \\ \cdot \\ y_n \end{bmatrix} \qquad \mathbf{X} = \begin{bmatrix} 1 & x_{11} & x_{21} & \cdots & x_{k1} \\ 1 & x_{12} & x_{22} & \cdots & x_{k2} \\ 1 & x_{13} & x_{23} & \cdots & x_{k3} \\ \cdot & \cdot & \cdot & & \cdot \\ \cdot & \cdot & \cdot & & \cdot \\ \cdot & \cdot & \cdot & & \cdot \\ 1 & x_{1n} & x_{2n} & \cdots & x_{kn} \end{bmatrix} \qquad \hat{\beta} = \begin{bmatrix} \hat{\beta}_0 \\ \hat{\beta}_1 \\ \hat{\beta}_2 \\ \cdot \\ \cdot \\ \cdot \\ \hat{\beta}_k \end{bmatrix}$$

Notice that the first column in the X matrix is a column of 1's. Thus, we are inserting a value of x, namely x_0, as the coefficient of β_0, where x_0 is a variable always equal to 1. Therefore, there is one column in the X matrix for each β parameter. Also, remember that a particular data point is identified by specific rows of the Y and X matrices. For example, the y-value y_3 for data point 3 is in the third row of the Y matrix, and the corresponding values of $x_1, x_2, \ldots, x_k$ appear in the third row of the X matrix.

The $\hat{\boldsymbol{\beta}}$ matrix shown in the box contains the least squares estimates (which we are attempting to obtain) of the coefficients $\beta_0, \beta_1, \ldots, \beta_k$ of the linear model

$$y = \beta_0 + \beta_1 x_1 + \beta_2 x_2 + \cdots + \beta_k x_k + \varepsilon$$

In order to write the least squares equation, we need to define what we mean by the **transpose of a matrix**. If

$$\mathbf{Y} = \begin{bmatrix} 5 \\ 1 \\ 0 \\ 4 \\ 2 \end{bmatrix} \qquad \mathbf{X} = \begin{bmatrix} 1 & 0 \\ 1 & 1 \\ 1 & 4 \\ 1 & 2 \\ 1 & 6 \end{bmatrix}$$

then the transpose matrices of the Y and X matrices, denoted as $\mathbf{Y}'$ and $\mathbf{X}'$, respectively, are

$$\mathbf{Y}' = \begin{bmatrix} 5 & 1 & 0 & 4 & 2 \end{bmatrix} \qquad \mathbf{X}' = \begin{bmatrix} 1 & 1 & 1 & 1 & 1 \\ 0 & 1 & 4 & 2 & 6 \end{bmatrix}$$

DEFINITION A.9

The **transpose of a matrix A**, denoted as $\mathbf{A}'$, is obtained by interchanging corresponding rows and columns of the A matrix. That is, the ith row of the A matrix becomes the ith column of the $\mathbf{A}'$ matrix.

Using the Y and X data matrices, their transposes, and the $\hat{\boldsymbol{\beta}}$ matrix, we can write the least squares equations (proof omitted) as:

LEAST SQUARES MATRIX EQUATION

$$(\mathbf{X}'\mathbf{X})\hat{\boldsymbol{\beta}} = \mathbf{X}'\mathbf{Y}$$

Thus, $(\mathbf{X}'\mathbf{X})$ is the coefficient matrix of the least squares estimates $\hat{\beta}_0$, $\hat{\beta}_1, \ldots, \hat{\beta}_k$, and $\mathbf{X}'\mathbf{Y}$ gives the matrix of constants that appear on the right-hand side of the equality signs. In the notation of Section A.4,

$$\mathbf{A} = \mathbf{X}'\mathbf{X} \qquad \mathbf{V} = \hat{\boldsymbol{\beta}} \qquad \mathbf{G} = \mathbf{X}'\mathbf{Y}$$

The solution, which follows from Section A.4, is:

LEAST SQUARES SOLUTION

$$\hat{\boldsymbol{\beta}} = (\mathbf{X}'\mathbf{X})^{-1}\mathbf{X}'\mathbf{Y}$$

Thus, to solve the least squares matrix equation, the computer calculates $(\mathbf{X}'\mathbf{X})$, $(\mathbf{X}'\mathbf{X})^{-1}$, $\mathbf{X}'\mathbf{Y}$, and, finally, the product $(\mathbf{X}'\mathbf{X})^{-1}\mathbf{X}'\mathbf{Y}$. We will illustrate this process using the data for the advertising example from Section 3.3.

EXAMPLE A.5

Find the least squares line for the data given in Table A.1.

TABLE A.1

MONTH	ADVERTISING EXPENDITURE x, hundreds of dollars	SALES REVENUE y, thousands of dollars
1	1	1
2	2	1
3	3	2
4	4	2
5	5	4

SOLUTION

The model is

$$y = \beta_0 + \beta_1 x_1 + \varepsilon$$

and the $\mathbf{Y}$, $\mathbf{X}$, and $\hat{\boldsymbol{\beta}}$ matrices are

$$
\mathbf{Y} = \begin{bmatrix} 1 \\ 1 \\ 2 \\ 2 \\ 4 \end{bmatrix}
\qquad
\mathbf{X} = \begin{matrix} x_0 \; x_1 \\ \begin{bmatrix} 1 & 1 \\ 1 & 2 \\ 1 & 3 \\ 1 & 4 \\ 1 & 5 \end{bmatrix} \end{matrix}
\qquad
\hat{\boldsymbol{\beta}} = \begin{bmatrix} \hat{\beta}_0 \\ \hat{\beta}_1 \end{bmatrix}
$$

Then,

$$
\mathbf{X}'\mathbf{X} = \begin{bmatrix} 1 & 1 & 1 & 1 & 1 \\ 1 & 2 & 3 & 4 & 5 \end{bmatrix}
\begin{bmatrix} 1 & 1 \\ 1 & 2 \\ 1 & 3 \\ 1 & 4 \\ 1 & 5 \end{bmatrix}
= \begin{bmatrix} 5 & 15 \\ 15 & 55 \end{bmatrix}
$$

$$
\mathbf{X}'\mathbf{Y} = \begin{bmatrix} 1 & 1 & 1 & 1 & 1 \\ 1 & 2 & 3 & 4 & 5 \end{bmatrix}
\begin{bmatrix} 1 \\ 1 \\ 2 \\ 2 \\ 4 \end{bmatrix}
= \begin{bmatrix} 10 \\ 37 \end{bmatrix}
$$

The last matrix that we need is $(X'X)^{-1}$. This matrix, which can be found by using a packaged computer program or by using the method of Appendix B, is

$$(X'X)^{-1} = \begin{bmatrix} 1.1 & -.3 \\ -.3 & .1 \end{bmatrix}$$

Then the solution to the least squares equation is

$$\hat{\beta} = (X'X)^{-1}X'Y = \begin{bmatrix} 1.1 & -.3 \\ -.3 & .1 \end{bmatrix}\begin{bmatrix} 10 \\ 37 \end{bmatrix} = \begin{bmatrix} -.1 \\ .7 \end{bmatrix}$$

Thus, $\hat{\beta}_0 = -.1$, $\hat{\beta}_1 = .7$, and the prediction equation is

$$\hat{y} = -.1 + .7x$$

You can verify that this is the same answer as obtained in Section 3.3.   ∎

EXAMPLE A.6

Find the least squares solution for fitting the monthly kilowatt-hour usage y to size of home x for the model

$$y = \beta_0 + \beta_1 x + \beta_2 x^2 + \varepsilon$$

(The computer printout for this regression analysis was presented and discussed in Section 4.3.) The data are shown in Table A.2.

TABLE A.2
Data for Power Usage Study

SIZE OF HOME x, square feet	MONTHLY USAGE y, kilowatt-hours	SIZE OF HOME x, square feet	MONTHLY USAGE y, kilowatt-hours
1,290	1,182	1,840	1,711
1,350	1,172	1,980	1,804
1,470	1,264	2,230	1,840
1,600	1,493	2,400	1,956
1,710	1,571	2,930	1,954

SOLUTION

The Y, X, and $\hat{\beta}$ matrices are shown below:

$$Y = \begin{bmatrix} 1,182 \\ 1,172 \\ 1,264 \\ 1,493 \\ 1,571 \\ 1,711 \\ 1,804 \\ 1,840 \\ 1,956 \\ 1,954 \end{bmatrix} \qquad X = \begin{array}{ccc} x_0 & x & x^2 \\ \begin{bmatrix} 1 & 1,290 & 1,664,100 \\ 1 & 1,350 & 1,822,500 \\ 1 & 1,470 & 2,160,900 \\ 1 & 1,600 & 2,560,000 \\ 1 & 1,710 & 2,924,100 \\ 1 & 1,840 & 3,385,600 \\ 1 & 1,980 & 3,920,400 \\ 1 & 2,230 & 4,972,900 \\ 1 & 2,400 & 5,760,000 \\ 1 & 2,930 & 8,584,900 \end{bmatrix} \end{array}$$

Then:

$$\mathbf{X'X} = \begin{bmatrix} 10 & 18,800 & 37,755,400 \\ 18,800 & 37,755,400 & 8,093.9 \times 10^7 \\ 37,755,400 & 8,093.9 \times 10^7 & 1.843 \times 10^{14} \end{bmatrix}$$

$$\mathbf{X'Y} = \begin{bmatrix} 15,947 \\ 31,283,250 \\ 6.53069 \times 10^{10} \end{bmatrix}$$

And (obtained using a standard computer package):

$$(\mathbf{X'X})^{-1} = \begin{bmatrix} 26.9156 & -.027027 & 6.3554 \times 10^{-6} \\ -.027027 & 2.75914 \times 10^{-5} & -6.5804 \times 10^{-9} \\ 6.3554 \times 10^{-6} & -6.5804 \times 10^{-9} & 1.5934 \times 10^{-12} \end{bmatrix}$$

Finally, performing the multiplication, we obtain

$$\hat{\boldsymbol{\beta}} = (\mathbf{X'X})^{-1}\mathbf{X'Y}$$

$$= \begin{bmatrix} 26.9156 & -.027027 & 6.3554 \times 10^{-6} \\ -.027027 & 2.75914 \times 10^{-5} & -6.5804 \times 10^{-9} \\ 6.3554 \times 10^{-6} & -6.5804 \times 10^{-9} & 1.5934 \times 10^{-12} \end{bmatrix} \begin{bmatrix} 15,947 \\ 31,283,250 \\ 6.53069 \times 10^{10} \end{bmatrix}$$

$$= \begin{bmatrix} -1,216.14389 \\ 2.39893 \\ -.00045 \end{bmatrix}$$

Thus, $\hat{\beta}_0 = -1,216.14389$, $\hat{\beta}_1 = 2.39893$, $\hat{\beta}_2 = -.00045$, and the prediction equation is

$$\hat{y} = -1,216.14389 + 2.39893x - .00045x^2$$

These are the values shown in the computer regression analysis printout, Figure 4.2.   ∎

EXERCISES A.14–A.16

A.14 Use the method of least squares to fit a straight line to the five data points:

x	−2	−1	0	1	2
y	4	3	3	1	−1

a. Construct $\mathbf{Y}$ and $\mathbf{X}$ matrices for the data.
b. Find $\mathbf{X'X}$ and $\mathbf{X'Y}$.
c. Find the least squares estimates $\hat{\boldsymbol{\beta}} = (\mathbf{X'X})^{-1}\mathbf{X'Y}$. [*Note:* See Theorem A.1 for information on finding $(\mathbf{X'X})^{-1}$.]
d. Give the prediction equation.

Note that the matrix procedure gives the same solution as obtained in Exercise 3.7.

A.15 Use the method of least squares to fit the model $E(y) = \beta_0 + \beta_1 x$ to the six data points:

x	1	2	3	4	5	6
y	1	2	2	3	5	5

a. Construct **Y** and **X** matrices for the data.
b. Find **X'X** and **X'Y**.
c. Verify that

$$(\mathbf{X'X})^{-1} = \begin{bmatrix} \frac{13}{15} & -\frac{7}{35} \\ -\frac{7}{35} & \frac{2}{35} \end{bmatrix}$$

d. Find the $\hat{\boldsymbol{\beta}}$ matrix.
e. Give the prediction equation.

Note that the matrix procedure gives the same solution as obtained in Exercise 3.6.

A.16 An experiment was conducted in which two y observations were collected for each of five values of x:

x	\multicolumn{2}{c}{-2}	-1		0		1		2	
y	1.1	1.3	2.0	2.1	2.7	2.8	3.4	3.6	4.1	4.0

Use the method of least squares to fit the second-order model, $E(y) = \beta_0 + \beta_1 x + \beta_2 x^2$, to the ten data points.
a. Give the dimensions of the **Y** and **X** matrices.
b. Verify that

$$(\mathbf{X'X})^{-1} = \begin{bmatrix} \frac{17}{70} & 0 & -\frac{1}{14} \\ 0 & \frac{1}{20} & 0 \\ -\frac{1}{14} & 0 & \frac{1}{28} \end{bmatrix}$$

c. Both **X'X** and $(\mathbf{X'X})^{-1}$ are symmetric matrices. What is a symmetric matrix?
d. Find the $\hat{\boldsymbol{\beta}}$ matrix and the least squares prediction equation.
e. Plot the data points and graph the prediction equation.

SECTION A.6

CALCULATING SSE AND s^2

You will recall that the variances of the estimators of all the β parameters and of $\hat{y}$ will depend on the value of σ^2, the variance of the random error ε that appears in the linear model. Since σ^2 will rarely be known in advance, we must use the sample data to estimate its value.

FORMULAS FOR SSE AND s^2
$$SSE = \mathbf{Y'Y} - \hat{\boldsymbol{\beta}}'\mathbf{X'Y}$$ $$s^2 = \frac{SSE}{n - \text{Number of } \beta \text{ parameters in model}}$$

We will demonstrate the use of these formulas with the advertising–sales data of Example A.5.

EXAMPLE A.7

Find the SSE for the advertising–sales data of Example A.5.

SOLUTION

From Example A.5,

$$\hat{\boldsymbol{\beta}} = \begin{bmatrix} -.1 \\ .7 \end{bmatrix} \quad \text{and} \quad \mathbf{X'Y} = \begin{bmatrix} 10 \\ 37 \end{bmatrix}$$

Then,

$$\mathbf{Y'Y} = \begin{bmatrix} 1 & 1 & 2 & 2 & 4 \end{bmatrix} \begin{bmatrix} 1 \\ 1 \\ 2 \\ 2 \\ 4 \end{bmatrix} = 26$$

and

$$\hat{\boldsymbol{\beta}}' \mathbf{X'Y} = \begin{bmatrix} -.1 & .7 \end{bmatrix} \begin{bmatrix} 10 \\ 37 \end{bmatrix} = 24.9$$

So

$$\text{SSE} = \mathbf{Y'Y} - \hat{\boldsymbol{\beta}}' \mathbf{X'Y} = 26 - 24.9 = 1.1$$

(Note that this is the same answer as was obtained in Section 3.3.)
Finally,

$$s^2 = \frac{\text{SSE}}{n - \text{Number of } \beta \text{ parameters in model}} = \frac{1.1}{5 - 2} = .367$$

This estimate is needed to construct a confidence interval for β_1, to test a hypothesis concerning its value, or to construct a confidence interval for the mean sales for a given advertising expenditure.   ∎

SECTION A.7

STANDARD ERRORS OF ESTIMATORS, TEST STATISTICS, AND CONFIDENCE INTERVALS FOR $\beta_0, \beta_1, \ldots, \beta_k$

The importance of this Appendix is evidenced by the fact that all the relevant information pertaining to the standard errors of the sampling distributions of $\hat{\beta}_0, \hat{\beta}_1, \ldots, \hat{\beta}_k$ (and hence of $\hat{Y}$) is contained in $(\mathbf{X'X})^{-1}$. Thus, if we denote the $(\mathbf{X'X})^{-1}$ matrix as

$$(\mathbf{X'X})^{-1} = \begin{bmatrix} c_{00} & c_{01} & \cdots & c_{0k} \\ c_{10} & c_{11} & \cdots & c_{1k} \\ c_{20} & c_{21} & \cdots & c_{2k} \\ \cdot & \cdot & \cdot & \cdot \\ \cdot & \cdot & \cdot & \cdot \\ \cdot & \cdot & \cdot & \cdot \\ c_{k0} & c_{k1} & \cdots & c_{kk} \end{bmatrix}$$

then it can be shown (proof omitted) that the standard errors of the sampling distributions of $\hat{\beta}_0, \hat{\beta}_1, \ldots, \hat{\beta}_k$ are

$$\sigma_{\hat{\beta}_0} = \sigma\sqrt{c_{00}}$$
$$\sigma_{\hat{\beta}_1} = \sigma\sqrt{c_{11}}$$
$$\sigma_{\hat{\beta}_2} = \sigma\sqrt{c_{22}}$$

.

.

.

$$\sigma_{\hat{\beta}_k} = \sigma\sqrt{c_{kk}}$$

where σ is the standard deviation of the random error ε. In other words, the diagonal elements of $(\mathbf{X'X})^{-1}$ give the values of $c_{00}, c_{11}, \ldots, c_{kk}$ that are required for finding the standard errors of the estimators $\hat{\beta}_0, \hat{\beta}_1, \ldots, \hat{\beta}_k$. The estimated values of the standard errors are obtained by replacing σ by s in the formulas for the standard errors. Thus, for example, the estimated standard error of $\hat{\beta}_1$ is $s_{\hat{\beta}_1} = s\sqrt{c_{11}}$.

The confidence interval for a single β parameter, β_i, is given in the box.

CONFIDENCE INTERVAL FOR β_i

$$\hat{\beta}_i \pm t_{\alpha/2}(\text{Estimated standard error of } \hat{\beta}_i)$$

or

$$\hat{\beta}_i \pm t_{\alpha/2}s\sqrt{c_{ii}}$$

where $t_{\alpha/2}$ is based on the number of degrees of freedom associated with s.

Similarly, the test statistic for testing the null hypothesis H_0: $\beta_i = 0$ is as shown in the next box.

TEST STATISTIC FOR H_0: $\beta_i = 0$

$$t = \frac{\hat{\beta}_i}{s\sqrt{c_{ii}}}$$

EXAMPLE A.8

Refer to Example A.5 and find the estimated standard error for the sampling distribution of $\hat{\beta}_1$, the estimator of the slope of the line β_1. Then give a 95% confidence interval for β_1.

SOLUTION

The $(X'X)^{-1}$ matrix for the least squares solution of Example A.5 was

$$(X'X)^{-1} = \begin{bmatrix} 1.1 & -.3 \\ -.3 & .1 \end{bmatrix}$$

Therefore, $c_{00} = 1.1$, $c_{11} = .1$, and the estimated standard error for $\hat{\beta}_1$ is

$$s_{\hat{\beta}_1} = s\sqrt{c_{11}} = \sqrt{.367}(\sqrt{.1}) = .192$$

The value for s, $\sqrt{.367}$, was obtained from Example A.7.

A 95% confidence interval for β_1 is

$$\hat{\beta}_1 \pm t_{\alpha/2}s\sqrt{c_{11}}$$
$$.7 \pm (3.182)(.192) = (.09, 1.31)$$

The t-value, $t_{.025}$, is based on $(n - 2) = 3$ df. Observe that this is the same confidence interval as the one obtained in Section 3.6.   ∎

EXAMPLE A.9

Refer to Example A.6 and the least squares solution for fitting power usage y to the size of a home x using the model

$$y = \beta_0 + \beta_1 x + \beta_2 x^2 + \varepsilon$$

a. Compute the estimated standard error for $\hat{\beta}_1$ and compare this result with the regression analysis printout shown in Figure A.1.

b. Compute the value of the test statistic for testing H_0: $\beta_2 = 0$. Compare this with the value given in the computer printout shown in Figure A.1.

FIGURE A.1 SAS Computer Printout for the Power Usage Data, Examples A.6 and A.9

ANALYSIS OF VARIANCE

SOURCE	DF	SUM OF SQUARES	MEAN SQUARE	F VALUE	PROB>F
MODEL	2	831069.55	415534.77	189.710	0.0001
ERROR	7	15332.55363	2190.36480		
C TOTAL	9	846402.10			

ROOT MSE	46.80133	R-SQUARE	0.9819	
DEP MEAN	1594.7	ADJ R-SQ	0.9767	
C.V.	2.934805			

PARAMETER ESTIMATES

VARIABLE	DF	PARAMETER ESTIMATE	STANDARD ERROR	T FOR H0: PARAMETER=0	PROB > |T|
INTERCEP	1	-1216.14389	242.80637	-5.009	0.0016
X	1	2.39893018	0.24583560	9.758	0.0001
XX	1	-0.000450040	0.000059077	-7.618	0.0001

SOLUTION

The fitted model is

$$\hat{y} = -1{,}216.14389 + 2.39893x - .00045x^2$$

The $(\mathbf{X}'\mathbf{X})^{-1}$ matrix, obtained in Example A.6, is

$$(\mathbf{X}'\mathbf{X})^{-1} = \begin{bmatrix} 26.9156 & -.027027 & 6.3554 \times 10^{-6} \\ -.027027 & 2.75914 \times 10^{-5} & -6.5804 \times 10^{-9} \\ 6.3554 \times 10^{-6} & -6.5804 \times 10^{-9} & 1.5934 \times 10^{-12} \end{bmatrix}$$

and the computer printout for the complete regression analysis is shown in Figure A.1. Note from $(\mathbf{X}'\mathbf{X})^{-1}$ that

$$c_{00} = 26.9156$$
$$c_{11} = 2.75914 \times 10^{-5}$$
$$c_{22} = 1.5934 \times 10^{-12}$$

and that from the printout, $s = 46.801$.

a. The estimated standard error of $\hat{\beta}_1$ is

$$s_{\hat{\beta}_1} = s\sqrt{c_{11}}$$
$$= (46.801)\sqrt{2.75914 \times 10^{-1}} = .24583$$

Notice that this agrees with the value of $s_{\hat{\beta}_1}$ shown in the computer printout (Figure A.1).

b. The value of the test statistic for testing $H_0: \beta_2 = 0$ is

$$t = \frac{\hat{\beta}_2}{s\sqrt{c_{22}}} = \frac{-.00045}{(46.801)\sqrt{1.5934 \times 10^{-12}}} = -7.62$$

Notice that this value of the t-statistic agrees with the value given in the column headed T FOR H0: PARAMETER = 0 shown in the printout (Figure A.1). ∎

EXERCISES A.17–A.19

A.17 Do the data given in Exercise A.14 provide sufficient evidence to indicate that x contributes information for the prediction of y? Test $H_0: \beta_1 = 0$ against $H_a: \beta_1 \neq 0$ using $\alpha = .05$.

A.18 Find a 90% confidence interval for the slope of the line in Exercise A.17.

A.19 The term in the second-order model $E(y) = \beta_0 + \beta_1 x + \beta_2 x^2$ that controls the curvature in its graph is $\beta_2 x^2$. If $\beta_2 = 0$, $E(y)$ graphs as a straight line. Do the data given in Exercise A.16 provide sufficient evidence to indicate curvature in the model for $E(y)$? Test $H_0: \beta_2 = 0$ against $H_a: \beta_2 \neq 0$ using $\alpha = .10$.

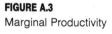

SECTION A.8

A CONFIDENCE INTERVAL FOR A LINEAR FUNCTION OF THE β PARAMETERS; A CONFIDENCE INTERVAL FOR $E(y)$

Suppose we were to postulate that the mean value of the productivity y of a company is related to the size of the company, x, and that the relationship could be modeled by the expression

$$E(y) = \beta_0 + \beta_1 x + \beta_2 x^2$$

A graph of $E(y)$ might appear as shown in Figure A.2.

We might have several reasons for collecting data on the productivity and size of a set of n companies and finding the least squares prediction equation,

$$\hat{y} = \hat{\beta}_0 + \hat{\beta}_1 x + \hat{\beta}_2 x^2$$

FIGURE A.2
Graph of Mean
Productivity $E(y)$

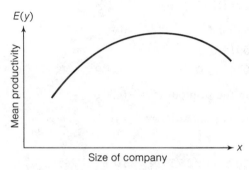

For example, we might wish to estimate the mean productivity for a company of a given size (say, $x = 2$). That is, we might wish to estimate

$$
\begin{aligned}
E(y) &= \beta_0 + \beta_1 x + \beta_2 x^2 \\
&= \beta_0 + 2\beta_1 + 4\beta_2 \quad \text{where} \quad x = 2
\end{aligned}
$$

Or we might wish to estimate the marginal increase in productivity, the slope of a tangent to the curve, when $x = 2$ (see Figure A.3). The marginal productivity for y when $x = 2$ is the rate of change of $E(y)$ with respect to x, evaluated at $x = 2$.* The marginal productivity for a value of x, denoted by the symbol $dE(y)/dx$, can be shown (proof omitted) to be

$$\frac{dE(y)}{dx} = \beta_1 + 2\beta_2 x$$

FIGURE A.3
Marginal Productivity

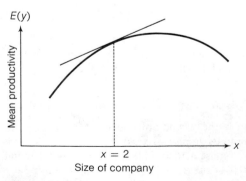

*If you have had calculus, note that the marginal productivity for y given x is the first derivative of $E(y) = \beta_0 + \beta_1 x + \beta_2 x^2$ with respect to x.

Therefore, the marginal productivity at $x = 2$ is

$$\frac{dE(y)}{dx} = \beta_1 + 2\beta_2(2) = \beta_1 + 4\beta_2$$

Note that for $x = 2$, both $E(y)$ and the marginal productivity are *linear* functions of the unknown parameters β_0, β_1, β_2 in the model. The problem we pose in this section is that of finding confidence intervals for linear functions of β parameters or testing hypotheses concerning their values. The information necessary to solve this problem is rarely given in a standard multiple regression analysis computer printout, but we can find these confidence intervals or values of the appropriate test statistics from knowledge of $(\mathbf{X}'\mathbf{X})^{-1}$.

We will suppose that we have a model,

$$y = \beta_0 + \beta_1 x_1 + \cdots + \beta_k x_k + \varepsilon$$

and that we are interested in making an inference about a linear function of the β parameters, say

$$a_0\beta_0 + a_1\beta_1 + \cdots + a_k\beta_k$$

where $a_0, a_1, \ldots, a_k$ are known constants. Further, we will use the corresponding linear function of least squares estimates,

$$\ell = a_0\hat{\beta}_0 + a_1\hat{\beta}_1 + \cdots + a_k\hat{\beta}_k$$

as our best estimate of $a_0\beta_0 + a_1\beta_1 + \cdots + a_k\beta_k$.

Then, for the assumptions on the random error ε (stated in Section 4.2), the sampling distribution for the estimator ℓ will be normal, with mean and standard error as given in the next box.

MEAN AND STANDARD ERROR OF ℓ

$$E(\ell) = a_0\beta_0 + a_1\beta_1 + \cdots + a_k\beta_k$$
$$\sigma_\ell = \sigma\sqrt{\mathbf{a}'(\mathbf{X}'\mathbf{X})^{-1}\mathbf{a}}$$

where σ is the standard deviation of ε, $(\mathbf{X}'\mathbf{X})^{-1}$ is the inverse matrix obtained in fitting the least squares model to the set of data, and

$$\mathbf{a} = \begin{bmatrix} a_0 \\ a_1 \\ a_2 \\ \cdot \\ \cdot \\ \cdot \\ a_k \end{bmatrix}$$

This indicates that ℓ is an unbiased estimator of

$$E(\ell) = a_0\beta_0 + a_1\beta_1 + \cdots + a_k\beta_k$$

and that its sampling distribution would appear as shown in Figure A.4.

FIGURE A.4
Sampling Distribution for ℓ

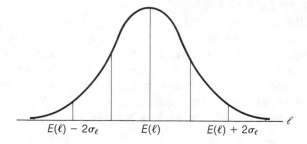

It can be shown that a $100(1 - \alpha)\%$ confidence interval for $E(\ell)$ is:

A 100(1 − α)% CONFIDENCE INTERVAL FOR $E(\ell)$

$$\ell \pm t_{\alpha/2}s\sqrt{\mathbf{a'(X'X)}^{-1}\mathbf{a}}$$

where

$$E(\ell) = a_0\beta_0 + a_1\beta_1 + \cdots + a_k\beta_k$$

$$\ell = a_0\hat{\beta}_0 + a_1\hat{\beta}_1 + \cdots + a_k\hat{\beta}_k \qquad \mathbf{a} = \begin{bmatrix} a_0 \\ a_1 \\ a_2 \\ . \\ . \\ . \\ a_k \end{bmatrix}$$

s and $\mathbf{(X'X)}^{-1}$ are obtained from the least squares procedure, and $t_{\alpha/2}$ is based on the number of degrees of freedom associated with s.

The linear function of the β parameters that is most often the focus of our attention is

$$E(y) = \beta_0 + \beta_1x_1 + \cdots + \beta_kx_k$$

That is, we want to find a confidence interval for $E(y)$ for specific values of x_1, $x_2, \ldots, x_k$. For this special case,

$$\ell = \hat{y}$$

and the **a** matrix is

$$\mathbf{a} = \begin{bmatrix} 1 \\ x_1 \\ x_2 \\ \cdot \\ \cdot \\ \cdot \\ x_k \end{bmatrix}$$

where the symbols $x_1, x_2, \ldots, x_k$ in the $\mathbf{a}$ matrix indicate the specific numerical values assumed by these variables. Thus, the procedure for forming a confidence interval for $E(y)$ is as shown in the box.

A 100$(1 - \alpha)$% CONFIDENCE INTERVAL FOR $E(y)$

$$\ell \pm t_{\alpha/2} s \sqrt{\mathbf{a}'(\mathbf{X}'\mathbf{X})^{-1}\mathbf{a}}$$

where

$$E(y) = \beta_0 + \beta_1 x_1 + \beta_2 x_2 + \cdots + \beta_k x_k$$

$$\ell = \hat{y} = \hat{\beta}_0 + \hat{\beta}_1 x_1 + \cdots + \hat{\beta}_k x_k \qquad \mathbf{a} = \begin{bmatrix} 1 \\ x_1 \\ x_2 \\ \cdot \\ \cdot \\ \cdot \\ x_k \end{bmatrix}$$

s and $(\mathbf{X}'\mathbf{X})^{-1}$ are obtained from the least squares analysis, and $t_{\alpha/2}$ is based on the number of degrees of freedom associated with s, namely $n - (k + 1)$.

EXAMPLE A.10

Refer to the data of Example A.5 for sales revenue y and advertising expenditure x. Find a 95% confidence interval for the mean sales revenue $E(y)$ when advertising expenditure is $x = 4$.

SOLUTION

The confidence interval for $E(y)$ for a given value of x is

$$\hat{y} \pm t_{\alpha/2} s \sqrt{\mathbf{a}'(\mathbf{X}'\mathbf{X})^{-1}\mathbf{a}}$$

Consequently, we need to find and substitute the values of $\mathbf{a}'(\mathbf{X}'\mathbf{X})^{-1}\mathbf{a}$, $t_{\alpha/2}$, and $\hat{y}$ into this formula. Since we wish to estimate

$$\begin{aligned} E(y) &= \beta_0 + \beta_1 x \\ &= \beta_0 + \beta_1(4) \qquad \text{when} \quad x = 4 \\ &= \beta_0 + 4\beta_1 \end{aligned}$$

it follows that the coefficients of β_0 and β_1 are $a_0 = 1$ and $a_1 = 4$, and thus,

$$a = \begin{bmatrix} 1 \\ 4 \end{bmatrix}$$

From Examples A.5 and A.7, $\hat{y} = -.1 + .7x$,

$$(X'X)^{-1} = \begin{bmatrix} 1.1 & -.3 \\ -.3 & .1 \end{bmatrix}$$

$s^2 = .367$, and $s = .61$. Then,

$$a'(X'X)^{-1}a = [1 \quad 4] \begin{bmatrix} 1.1 & -.3 \\ -.3 & .1 \end{bmatrix} \begin{bmatrix} 1 \\ 4 \end{bmatrix}$$

We first calculate

$$a'(X'X)^{-1} = [1 \quad 4] \begin{bmatrix} 1.1 & -.3 \\ -.3 & .1 \end{bmatrix} = [-.1 \quad .1]$$

Then,

$$a'(X'X)^{-1}a = [-.1 \quad .1] \begin{bmatrix} 1 \\ 4 \end{bmatrix} = .3$$

The t-value, $t_{.025}$, based on 3 df is 3.182. So, a 95% confidence interval for the mean sales revenue with an advertising expenditure of 4 is

$$\hat{y} \pm t_{\alpha/2} s \sqrt{a'(X'X)^{-1}a}$$

Since $\hat{y} = -.1 + .7x = -.1 + (.7)(4) = 2.7$, the 95% confidence interval for $E(y)$ when $x = 4$ is

$$2.7 \pm (3.182)(.61)\sqrt{.3}$$
$$2.7 \pm 1.1$$

Notice that this is exactly the same result as obtained in Example 3.3.   ■

EXAMPLE A.11   An economist recorded a measure of productivity y and the size x for each of 100 companies producing cement. A regression model,

$$y = \beta_0 + \beta_1 x + \beta_2 x^2 + \varepsilon$$

fit to the $n = 100$ data points produced the following results:

$$\hat{y} = 2.6 + .7x - .2x^2$$

where x is coded to take values in the interval $-2 < x < 2$,* and

$$(X'X)^{-1} = \begin{bmatrix} .0025 & .0005 & -.0070 \\ .0005 & .0055 & 0 \\ -.0070 & 0 & .0050 \end{bmatrix} \qquad s = .14$$

*We give a formula for *coding* observational data in Section 7.6.

Find a 95% confidence interval for the marginal increase in productivity given that the coded size of a plant is $x = 1.5$.

SOLUTION

The mean value of y for a given value of x is

$$E(y) = \beta_0 + \beta_1 x + \beta_2 x^2$$

Therefore, the marginal increase in y for $x = 1.5$ is

$$\frac{dE(y)}{dx} = \beta_1 + 2\beta_2 x$$
$$= \beta_1 + 2(1.5)\beta_2$$

Or,

$$E(\ell) = \beta_1 + 3\beta_2 \qquad \text{when} \quad x = 1.5$$

Note from the prediction equation, $\hat{y} = 2.6 + .7x - .2x^2$, that $\hat{\beta}_1 = .7$ and $\hat{\beta}_2 = -.2$. Therefore,

$$\ell = \hat{\beta}_1 + 3\hat{\beta}_2 = .7 + 3(-.2) = .1$$

and

$$\mathbf{a} - \begin{bmatrix} a_0 \\ a_1 \\ a_2 \end{bmatrix} = \begin{bmatrix} 0 \\ 1 \\ 3 \end{bmatrix}$$

We next calculate

$$\mathbf{a}'(\mathbf{X}'\mathbf{X})^{-1}\mathbf{a} = \begin{bmatrix} 0 & 1 & 3 \end{bmatrix} \begin{bmatrix} .0025 & .0005 & -.0070 \\ .0005 & .0055 & 0 \\ -.0070 & 0 & .0050 \end{bmatrix} \begin{bmatrix} 0 \\ 1 \\ 3 \end{bmatrix} = .0505$$

Then, since s is based on $n - (k + 1) = 100 - 3 = 97$ df, $t_{.025} \approx 1.96$, and a 95% confidence interval for the marginal increase in productivity when $x = 1.5$ is

$$\ell \pm t_{.025} s \sqrt{\mathbf{a}'(\mathbf{X}'\mathbf{X})^{-1}\mathbf{a}}$$

or

$$.1 \pm (1.96)(.14)\sqrt{.0505}$$
$$.1 \pm .062$$

Thus, the marginal increase in productivity, the slope of the tangent to the curve

$$E(y) = \beta_0 + \beta_1 x + \beta_2 x^2$$

is estimated to lie in the interval $.1 \pm .062$ at $x = 1.5$. A graph of $\hat{y} = 2.6 + .7x - .2x^2$ is shown in Figure A.5 (page 724).

FIGURE A.5
A Graph of
$\hat{y} = 2.6 + .7x - .2x^2$

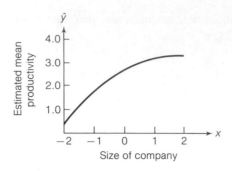

Size of company

| | | | | | | | | | | | |
SECTION A.9

A PREDICTION INTERVAL FOR SOME VALUE OF *y* TO BE OBSERVED IN THE FUTURE

We have indicated in Sections 3.9 and 4.7 that two of the most important applications of the least squares predictor $\hat{y}$ are estimating the mean value of y (the topic of the preceding section) and predicting a new value of y, yet unobserved, for specific values of $x_1, x_2, \ldots, x_k$. The difference between these two inferential problems (when each would be pertinent) was explained in Chapters 3 and 4, but we will give another example to make certain that the distinction is clear at this point.

Suppose you are the manager of a manufacturing plant and that y, the daily profit is a function of various process variables $x_1, x_2, \ldots, x_k$. Suppose you want to know how much money you would make *in the long run* if the x's are set at specific values. For this case, you would be interested in finding a confidence interval for the mean profit per day, $E(y)$. In contrast, suppose you planned to operate the plant for only one more day! Then you would be interested in predicting the value of y, the profit associated with tomorrow's production.

We have indicated that the error of prediction is always larger than the error of estimating $E(y)$. You can see this by comparing the formula for the prediction interval (shown in the box) with the formula for the confidence interval for $E(y)$ that was given in Section A.8.

EXAMPLE A.12

Refer to the sales–advertising expenditure example (Example A.10), and find a 95% prediction interval for the sales revenue next month, if it is known that next month's advertising expenditure will be $x = 4$.

SOLUTION

The 95% prediction interval for sales revenue y is

$$\hat{y} \pm t_{\alpha/2}s\sqrt{1 + a'(X'X)^{-1}a}$$

From Example A.10, when $x = 4$, $\hat{y} = -.1 + .7x = -.1 + (.7)(4) = 2.7$, $s = .61$, $t_{.025} = 3.182$, and $a'(X'X)^{-1}a = .3$. Then the 95% prediction interval for y is

$$2.7 \pm (3.182)(.61)\sqrt{1 + .3}$$

$$2.7 \pm 2.2$$

You will find that this is the same solution as obtained in Example 3.4.

A 100(1 − α)% PREDICTION INTERVAL FOR y

$$\hat{y} \pm t_{\alpha/2}s\sqrt{1 + \mathbf{a}'(\mathbf{X}'\mathbf{X})^{-1}\mathbf{a}}$$

where

$$\hat{y} = \hat{\beta}_0 + \hat{\beta}_1 x_1 + \cdots + \hat{\beta}_k x_k$$

s and $(\mathbf{X}'\mathbf{X})^{-1}$ are obtained from the least squares analysis,

$$\mathbf{a} = \begin{bmatrix} 1 \\ x_1 \\ x_2 \\ \cdot \\ \cdot \\ \cdot \\ x_k \end{bmatrix}$$

contains the numerical values of $x_1, x_2, \ldots, x_k$ and $t_{\alpha/2}$ is based on the number of degrees of freedom associated with s, namely $n - (k + 1)$.

EXERCISES A.20–A.26

A.20 Refer to Exercise A.14. Find a 90% confidence interval for $E(y)$ when $x = 1$. Interpret the interval.

A.21 Refer to Exercise A.14. Suppose you plan to observe y for $x = 1$. Find a 90% prediction interval for that value of y. Interpret the interval.

A.22 Refer to Exercise A.15. Find a 90% confidence interval for $E(y)$ when $x = 2$. Interpret the interval.

A.23 Refer to Exercise A.15. Find a 90% prediction interval for a value of y to be observed in the future when $x = 2$. Interpret the interval.

A.24 Refer to Exercise A.16. Find a 90% confidence interval for the mean value of y when $x = 1$. Interpret the interval.

A.25 Refer to Exercise A.16. Find a 90% prediction interval for a value of y to be observed in the future when $x = 1$.

A.26 The productivity (items produced per hour) per worker on a manufacturing assembly line is expected to increase as piecework pay rate (in dollars) increases and then is expected to stabilize after a certain pay rate has been reached. The productivity of five different workers was recorded for each of five piecework pay rates, $.80, $.90, $1.00, $1.10, $1.20, thus giving $n = 25$ data points. A multiple regression analysis using a second-order model,

$$E(y) = \beta_0 + \beta_1 x + \beta_2 x^2$$

gave

$$\hat{y} = 2.08 + 8.42x - 1.65x^2$$

$$\text{SSE} = 26.62, \text{ SS}_{yy} = 784.11, \text{ and}$$

$$(\mathbf{X'X})^{-1} = \begin{bmatrix} .020 & -.010 & .015 \\ -.010 & .040 & -.006 \\ .015 & -.006 & .028 \end{bmatrix}$$

a. Find s^2.

b. Find a 95% confidence interval for the mean productivity when the pay rate is $1.10. Interpret this interval.

c. Find a 95% prediction interval for the production of an individual worker who is paid at a rate of $1.10 per piece. Interpret the interval.

d. Find R^2 and interpret the value.

SECTION A.10

SUMMARY

Except for the tedious process of inverting a matrix (given in Appendix B), we have covered the major steps performed by a computer in fitting a linear statistical model to a set of data using the method of least squares. We have also explained how to find the confidence intervals, prediction intervals, and values of test statistics that would be pertinent in a regression analysis.

In addition to providing a better understanding of a multiple regression analysis, the most important contributions of this Appendix are contained in Sections A.8 and A.9. If you want to make a specific inference concerning the mean value of y or any linear function of the β parameters and if you are unable to obtain the results from the computer package you are using, you will find the contents of Sections A.8 and A.9 very useful. Since you will almost always be able to find a computer program package to find $(\mathbf{X'X})^{-1}$, you will be able to calculate the desired confidence interval(s) and so forth on your own.

SUPPLEMENTARY EXERCISES A.27–A.31

A.27 Use the method of least squares to fit a straight line to the six data points:

x	−5	−3	−1	1	3	5
y	1.1	1.9	3.0	3.8	5.1	6.0

a. Construct $\mathbf{Y}$ and $\mathbf{X}$ matrices for the data.

b. Find $\mathbf{X'X}$ and $\mathbf{X'Y}$.

c. Find the least squares estimates,

$$\hat{\beta} = (\mathbf{X'X})^{-1}\mathbf{X'Y}$$

[*Note*: See Theorem A.1 for information on finding $(\mathbf{X'X})^{-1}$.]

d. Give the prediction equation.

e. Find SSE and s^2.

f. Does the model contribute information for the prediction of y? Test $H_0: \beta_1 = 0$. Use $\alpha = .05$.

g. Find r^2 and interpret its value.

h. Find a 90% confidence interval for $E(y)$ when $x = .5$. Interpret the interval.

A.28 An experiment was conducted to investigate the effect of extrusion pressure P and temperature at extrusion T on the strength y of a new type of plastic. Two plastic specimens were prepared for each of five combinations of pressure and temperature. The specimens were then tested in random order, and the breaking strength for each specimen was recorded. The independent variables were coded to simplify computations, i.e.,

$$x_1 = \frac{P - 200}{10} \qquad x_2 = \frac{T - 400}{25}$$

The $n = 10$ data points are listed in the table.

y	x_1	x_2
5.2; 5.0	-2	2
.3; $-.1$	-1	-1
$-1.2; -1.1$	0	-2
2.2; 2.0	1	-1
6.2; 6.1	2	2

a. Give the **Y** and **X** matrices needed to fit the model $y = \beta_0 + \beta_1 x_1 + \beta_2 x_2 + \varepsilon$.

b. Find the least squares prediction equation.

c. Find SSE and s^2.

d. Does the model contribute information for the prediction of y? Test using $\alpha = .05$.

e. Find R^2 and interpret its value.

f. Test the null hypothesis that $\beta_1 = 0$. Use $\alpha = .05$. What is the practical implication of the test?

g. Find a 90% confidence interval for the mean strength of the plastic for $x_1 = -2$ and $x_2 = 2$.

h. Suppose a single specimen of the plastic is to be installed in the engine mount of a Douglas DC-10 aircraft. Find a 90% prediction interval for the strength of this specimen if $x_1 = -2$ and $x_2 = 2$.

A.29 Suppose we obtained two replications of the experiment described in Exercise A.15, i.e., two values of y were observed for each of the six values of x. The data are shown below:

x	1		2		3		4		5		6	
y	1.1	.5	1.8	2.0	2.0	2.9	3.8	3.4	4.1	5.0	5.0	5.8

a. Suppose (as in Exercise A.15) you wish to fit the model $E(y) = \beta_0 + \beta_1 x$. Construct **Y** and **X** matrices for the data. [*Hint:* Remember, the **Y** matrix must be of dimension 12×1.]

b. Find **X'X** and **X'Y**.

c. Compare the **X'X** matrix for two replications of the experiment with the **X'X** matrix obtained for a single replication (part **b** of Exercise A.15). What is the relationship between the elements in the two matrices?

d. Observe the $(\mathbf{X'X})^{-1}$ matrix for a single replication (see part **c** of Exercise A.15). Verify that the $(\mathbf{X'X})^{-1}$ matrix for two replications contains elements that are equal to $\frac{1}{2}$ of the values of the corresponding elements in the $(\mathbf{X'X})^{-1}$ matrix for a single replication of the experiment. [*Hint*:   Show that the product of the $(X'X)^{-1}$ matrix (for two replications) and the $X'X$ matrix from part c equals the identity matrix I.]

e. Find the prediction equation.

f. Find SSE and s^2.

g. Do the data provide sufficient information to indicate that x contributes information for the prediction of y? Test using $\alpha = .05$.

h. Find r^2 and interpret its value.

A.30 Refer to Exercise A.29.

   a. Find a 90% confidence interval for $E(y)$ when $x = 4.5$. Interpret the interval.

   b. Suppose we wish to predict the value of y if, in the future, $x = 4.5$. Find a 90% prediction interval for y and interpret the interval.

A.31 Refer to Exercise A.29. Suppose you replicated the experiment described in Exercise A.15 three times, i.e., you collected three observations on y for each value of x. Then $n = 18$.

   a. What would be the dimensions of the Y matrix?

   b. Write the X matrix for three replications. Compare with the X matrices for one and for two replications. Note the pattern.

   c. Examine the $X'X$ matrices obtained for one and two replications of the experiment (obtained in Exercises A.15 and A.29, respectively). Deduce the values of the elements of the $X'X$ matrix for three replications.

   d. See your answer to Exercise A.29, part **d**. Deduce the values of the elements in the $(X'X)^{-1}$ matrix for three replications.

   e. Suppose you wanted to find a 90% confidence interval for $E(y)$ when $x = 4.5$ based on three replications of the experiment. Find the value of $a'(X'X)^{-1}a$ that appears in the confidence interval and compare with the value of $a'(X'X)^{-1}a$ that would be obtained for a single replication of the experiment.

   f. Approximately how much of a reduction in the width of the confidence interval is obtained by using three in comparison to two replications? [*Note*:   The values of s computed from the two sets of data will almost certainly be different.]

REFERENCES

Draper, N. and Smith, H. *Applied Regression Analysis*. New York: Wiley, 1966.

Graybill, F. A. *Theory and Application of the Linear Model*. North Scituate, Mass.: Duxbury, 1976.

Kleinbaum, D. and Kupper, L. *Applied Regression Analysis and Other Multivariable Methods*. North Scituate, Mass.: Duxbury, 1978.

Mendenhall, W. *Introduction to Linear Models and the Design and Analysis of Experiments*. Belmont, Calif.: Wadsworth, 1968.

Neter, J., Wasserman, W., and Kutner, M. H. *Applied Linear Statistical Models*, 2nd ed. Homewood, Ill.: Richard D. Irwin, 1985.

Younger, M. S. *A First Course in Linear Regression*, 2nd ed. Boston: Duxbury, 1985.

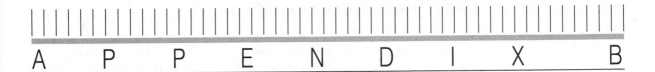

A PROCEDURE FOR INVERTING A MATRIX

There are several different methods for inverting matrices. All are tedious and time-consuming. Consequently, in practice, you will invert almost all matrices using an electronic computer. The purpose of this section is to present one method so that you will be able to invert small (2×2 or 3×3) matrices manually and so that you will appreciate the enormous computing problem involved in inverting large matrices (and, consequently, in fitting linear models containing many terms to a set of data). In particular, you will be able to understand why rounding errors creep into the inversion process and, consequently, why two different computer programs might invert the same matrix and produce inverse matrices with slightly different corresponding elements.

The procedure we will demonstrate to invert a matrix $\mathbf{A}$ requires us to perform a series of operations on the rows of the $\mathbf{A}$ matrix. For example, suppose

$$\mathbf{A} = \begin{bmatrix} 1 & -2 \\ -2 & 6 \end{bmatrix}$$

We will identify two different ways to operate on a row of a matrix:*

1. We can multiply every element in one particular row by a constant, c. For example, we could operate on the first row of the $\mathbf{A}$ matrix by multiplying every element in the row by a constant, say 2. Then the resulting row would be $[2 \quad -4]$.
2. We can operate on a row by multiplying another row of the matrix by a constant and then adding (or subtracting) the elements of that row to elements in corresponding positions in the row operated upon. For example, we could operate on the first row of the $\mathbf{A}$ matrix by multiplying the second row by a constant, say 2:

    $$2[-2 \quad 6] = [-4 \quad 12]$$

Then we add this row to row 1:

$$[(1 - 4) \quad (-2 + 12)] = [-3 \quad 10]$$

Note one important point. We operated on the *first* row of the $\mathbf{A}$ matrix. Although we used the second row of the matrix to perform the operation, *the second row would remain unchanged*. Therefore, the row operation on the $\mathbf{A}$ matrix that we have just described would produce the new matrix,

$$\begin{bmatrix} -3 & 10 \\ -2 & 6 \end{bmatrix}$$

*We omit a third row operation, because it would add little and could be confusing.

Matrix inversion using row operations is based on an elementary result from matrix algebra. It can be shown (proof omitted) that performing a series of row operations on a matrix **A** is equivalent to multiplying **A** by a matrix **B**, i.e., row operations produce a new matrix, **BA**. This result is used as follows: Place the **A** matrix and an identity matrix **I** of the same dimensions, side by side. Then perform the same series of row operations on both **A** and **I** until the **A** matrix has been changed into the identity matrix **I**. This means that you have multiplied both **A** and **I** by some matrix **B** such that:

$$
\mathbf{A} = \begin{bmatrix} & & \\ & & \\ & & \\ & & \end{bmatrix} \qquad \mathbf{I} = \begin{bmatrix} 1 & 0 & 0 & \cdots & 0 \\ 0 & 1 & 0 & \cdots & 0 \\ 0 & 0 & 1 & \cdots & 0 \\ \cdot & \cdot & \cdot & & \cdot \\ \cdot & \cdot & \cdot & & \cdot \\ \cdot & \cdot & \cdot & & \cdot \\ 0 & 0 & 0 & \cdots & 1 \end{bmatrix}
$$

$$\downarrow \qquad \leftarrow \text{Row operations change A to I} \rightarrow \qquad \downarrow$$

$$
\mathbf{I} = \begin{bmatrix} & & \\ & & \\ & & \\ & & \end{bmatrix} \qquad \mathbf{B} = \begin{bmatrix} & & \\ & & \\ & & \\ & & \end{bmatrix}
$$

$$\mathbf{BA} = \mathbf{I} \quad \text{and} \quad \mathbf{BI} = \mathbf{B}$$

Since $\mathbf{BA} = \mathbf{I}$, it follows that $\mathbf{B} = \mathbf{A}^{-1}$. Therefore, as the **A** matrix is transformed by row operations into the identity matrix **I**, the identity matrix **I** is transformed into $\mathbf{A}^{-1}$, i.e.,

$$\mathbf{BI} = \mathbf{B} = \mathbf{A}^{-1}$$

We will show you how this procedure works with two examples.

EXAMPLE B.1

Find the inverse of the matrix

$$\mathbf{A} = \begin{bmatrix} 1 & -2 \\ -2 & 6 \end{bmatrix}$$

SOLUTION

Place the **A** matrix and a 2×2 identity matrix side by side and then perform the following series of row operations (we will indicate by arrow the row operated upon in each operation):

$$A = \begin{bmatrix} 1 & -2 \\ -2 & 6 \end{bmatrix} \qquad I = \begin{bmatrix} 1 & 0 \\ 0 & 1 \end{bmatrix}$$

OPERATION 1   Multiply the first row by 2 and add to the second row:

$$\rightarrow \begin{bmatrix} 1 & -2 \\ 0 & 2 \end{bmatrix} \qquad \begin{bmatrix} 1 & 0 \\ 2 & 1 \end{bmatrix}$$

OPERATION 2   Multiply the second row by $\frac{1}{2}$:

$$\rightarrow \begin{bmatrix} 1 & -2 \\ 0 & 1 \end{bmatrix} \qquad \begin{bmatrix} 1 & 0 \\ 1 & \frac{1}{2} \end{bmatrix}$$

OPERATION 3   Multiply the second row by 2 and add it to the first row:

$$\rightarrow \begin{bmatrix} 1 & 0 \\ 0 & 1 \end{bmatrix} \qquad \begin{bmatrix} 3 & 1 \\ 1 & \frac{1}{2} \end{bmatrix}$$

Thus,

$$A^{-1} = \begin{bmatrix} 3 & 1 \\ 1 & \frac{1}{2} \end{bmatrix}$$

The final step in finding an inverse is to check your solution by finding the product $A^{-1}A$ to see if it equals the identity matrix I. To check:

$$A^{-1}A = \begin{bmatrix} 3 & 1 \\ 1 & \frac{1}{2} \end{bmatrix} \begin{bmatrix} 1 & -2 \\ -2 & 6 \end{bmatrix} = \begin{bmatrix} 1 & 0 \\ 0 & 1 \end{bmatrix}$$

Since this product is equal to the identity matrix, it follows that our solution for A^{-1} is correct.   ∎

EXAMPLE B.2

Find the inverse of the matrix

$$A = \begin{bmatrix} 2 & 0 & 3 \\ 0 & 4 & 1 \\ 3 & 1 & 2 \end{bmatrix}$$

SOLUTION

Place an identity matrix alongside the **A** matrix and perform the row operations:

OPERATION 1   Multiply row 1 by $\frac{1}{2}$:

$$\rightarrow \begin{bmatrix} 1 & 0 & \frac{3}{2} \\ 0 & 4 & 1 \\ 3 & 1 & 2 \end{bmatrix} \qquad \begin{bmatrix} \frac{1}{2} & 0 & 0 \\ 0 & 1 & 0 \\ 0 & 0 & 1 \end{bmatrix}$$

OPERATION 2   Multiply row 1 by 3 and subtract from row 3:

$$\rightarrow \begin{bmatrix} 1 & 0 & \frac{3}{2} \\ 0 & 4 & 1 \\ 0 & 1 & -\frac{5}{2} \end{bmatrix} \qquad \begin{bmatrix} \frac{1}{2} & 0 & 0 \\ 0 & 1 & 0 \\ -\frac{3}{2} & 0 & 1 \end{bmatrix}$$

OPERATION 3   Multiply row 2 by $\frac{1}{4}$:

$$\rightarrow \begin{bmatrix} 1 & 0 & \frac{3}{2} \\ 0 & 1 & \frac{1}{4} \\ 0 & 1 & -\frac{5}{2} \end{bmatrix} \qquad \begin{bmatrix} \frac{1}{2} & 0 & 0 \\ 0 & \frac{1}{4} & 0 \\ -\frac{3}{2} & 0 & 1 \end{bmatrix}$$

OPERATION 4   Subtract row 2 from row 3:

$$\rightarrow \begin{bmatrix} 1 & 0 & \frac{3}{2} \\ 0 & 1 & \frac{1}{4} \\ 0 & 0 & -\frac{11}{4} \end{bmatrix} \qquad \begin{bmatrix} \frac{1}{2} & 0 & 0 \\ 0 & \frac{1}{4} & 0 \\ -\frac{3}{2} & -\frac{1}{4} & 1 \end{bmatrix}$$

OPERATION 5   Multiply row 3 by $-\frac{4}{11}$:

$$\rightarrow \begin{bmatrix} 1 & 0 & \frac{3}{2} \\ 0 & 1 & \frac{1}{4} \\ 0 & 0 & 1 \end{bmatrix} \qquad \begin{bmatrix} \frac{1}{2} & 0 & 0 \\ 0 & \frac{1}{4} & 0 \\ \frac{12}{22} & \frac{1}{11} & -\frac{4}{11} \end{bmatrix}$$

OPERATION 6   Operate on row 2 by subtracting $\frac{1}{4}$ of row 3:

$$\rightarrow \begin{bmatrix} 1 & 0 & \frac{3}{2} \\ 0 & 1 & 0 \\ 0 & 0 & 1 \end{bmatrix} \qquad \begin{bmatrix} \frac{1}{2} & 0 & 0 \\ -\frac{3}{22} & \frac{5}{22} & \frac{1}{11} \\ \frac{12}{22} & \frac{1}{11} & -\frac{4}{11} \end{bmatrix}$$

OPERATION 7   Operate on row 1 by subtracting $\frac{3}{2}$ of row 3:

$$\rightarrow \begin{bmatrix} 1 & 0 & 0 \\ 0 & 1 & 0 \\ 0 & 0 & 1 \end{bmatrix} \qquad \begin{bmatrix} -\frac{7}{22} & -\frac{3}{22} & \frac{6}{11} \\ -\frac{3}{22} & \frac{5}{22} & \frac{1}{11} \\ \frac{6}{11} & \frac{1}{11} & -\frac{4}{11} \end{bmatrix} = \mathbf{A}^{-1}$$

To check the solution, we find the product:

$$\mathbf{A}^{-1}\mathbf{A} = \begin{bmatrix} -\frac{7}{22} & -\frac{3}{22} & \frac{6}{11} \\ -\frac{3}{22} & \frac{5}{22} & \frac{1}{11} \\ \frac{6}{11} & \frac{1}{11} & -\frac{4}{11} \end{bmatrix} \begin{bmatrix} 2 & 0 & 3 \\ 0 & 4 & 1 \\ 3 & 1 & 2 \end{bmatrix}$$

$$= \begin{bmatrix} 1 & 0 & 0 \\ 0 & 1 & 0 \\ 0 & 0 & 1 \end{bmatrix}$$

Since the product $\mathbf{A}^{-1}\mathbf{A}$ is equal to the identity matrix, it follows that our solution for $\mathbf{A}^{-1}$ is correct.   ■

Examples B.1 and B.2 indicate the strategy employed when performing row operations on the A matrix to change it into an identity matrix. Multiply the first row by a constant to change the element in the top left row into a 1. Then perform operations to change all elements in the first column into 0's. Then operate on the second row and change the second diagonal element into a 1. Then operate to change all elements in the second column beneath row 2 into 0's. Then operate on the diagonal element in row 3, etc. When all elements on the main diagonal are 1's and all below the main diagonal are 0's, perform row operations to change the last column to 0; then the next-to-last, etc., until you get back to the first column. The procedure for changing the off-diagonal elements to 0's is indicated diagrammatically as shown:

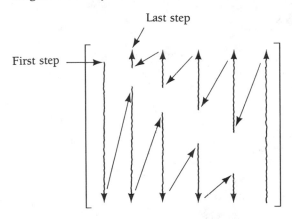

The preceding instructions on how to invert a matrix using row operations suggest that the inversion of a large matrix would involve many multiplications, subtractions, and additions and, consequently, could produce large rounding errors in the calculations unless you carry a large number of significant figures in the calculations. This explains why two different multiple regression analysis computer programs may produce different estimates of the same β parameters, and it emphasizes the importance of carrying a large number of significant figures in all computations when inverting a matrix.

Other methods for inverting matrices are discussed in the references. All work—exactly—in theory. It is only in the actual process of performing the calculations that the rounding errors occur.

EXERCISE B.1

B.1 Invert the following matrices and check your answers to make certain that $A^{-1}A = AA^{-1} = I$:

a. $A = \begin{bmatrix} 3 & 2 \\ 4 & 5 \end{bmatrix}$ **b.** $A = \begin{bmatrix} 3 & 0 & -2 \\ 1 & 4 & 2 \\ 5 & 1 & 1 \end{bmatrix}$

c. $A = \begin{bmatrix} 1 & 0 & 1 \\ 0 & 2 & 1 \\ 1 & 1 & 3 \end{bmatrix}$ **d.** $A = \begin{bmatrix} 4 & 0 & 10 \\ 0 & 10 & 0 \\ 10 & 0 & 5 \end{bmatrix}$

[*Note*: No answers are given to these exercises. You will know if your answer is correct if $A^{-1}A = I$.]

| | | | | | | | | | | | | | | | | | | | | | | | | | | | | | | | | | | | | | | | | | | | | | | | | | | | | | | | | | | | | | | | | | | | | | | | | | | | | | | | | | | | | | | | | | |

A   P   P   E   N   D   I   X   C

HOW TO PERFORM REGRESSION ANALYSIS USING FOUR COMPUTER PROGRAM PACKAGES: SAS, SPSS^x, MINITAB, AND BMDP

CONTENTS

SECTION C.1

INTRODUCTION

This Appendix explains how to perform a regression analysis using each of four popular statistical computer program packages: BMDP, Minitab, SPSS[x], and SAS. At least one of these programs is available at most college and university computing centers. The packages are also compatible for use with the computers used by many private corporations, foundations, and government agencies; moreover, all four packages have software available for use on a personal computer (PC).

A statistical computer program (or statistical software) package provides "canned" programs for a variety of statistical methods, including regression analysis. However, before you attempt to use such packages, you must learn how to enter data into your computer (either mainframe, minicomputer, or personal computer) so that the program package will be able to read and analyze it.

There are several ways in which data may be entered into a computer: disk, magnetic tape, punched cards, or a computer terminal for mainframe (or mini) computers; and floppy diskette, hard disk, or computer keyboard for PCs. The easiest and most common way to enter a small amount of data into a computer is through a computer terminal (or keyboard). The computer monitor (or screen) permits you to enter up to 80 characters (or numbers) on a single line of type. Large data sets are typically stored on disk (or floppy diskettes).

The commands given in this Appendix are appropriate for entering your data through a computer terminal or keyboard. If your data are stored on disk, diskette, or tape, consult the references given at the end of this Appendix.

The instructions to the computer and the manner in which data are entered vary, depending on the statistical software package that you use. However, all the computer packages discussed in this Appendix (SAS, SPSS[x], Minitab, and BMDP) utilize the following three basic types of instructions:*

1. Data entry instructions
2. Statistical analysis instructions
3. Input data values

Instructions to the computer on how to enter your data are called **data entry instructions**. These commands are usually followed by **statistical analysis instructions**, which command the computer to perform various graphical and numerical analyses of the input data values.

Before we give the commands for conducting a regression analysis, we will explain how to use the data entry instructions of each of the four packages, SAS, SPSS[x], Minitab, and BMDP.

---

*When using these packages on a mainframe computer, additional instructions are required. The basic function of these statements, called **job control language (JCL) instructions**, is to inform the computer which statistical program package to execute. The appropriate JCL may vary from one computing center to another. Your instructor will inform you of the appropriate JCL commands to use at your institution.

S E C T I O N C.2

ENTERING DATA INTO THE COMPUTER: SAS

The **SAS System** is probably the most versatile of the four statistical computer program packages. All SAS computer programs consist of at least two steps: (1) DATA steps—used to create data sets (data entry instructions), and (2) PROC (procedure) steps—used to analyze these data sets (statistical analysis instructions).*

Let us now create a SAS data set consisting of the sale prices for a sample of five residential properties listed in Table C.1.

TABLE C.1

Sale Prices of Five Residential Properties

$81,300
54,100
96,900
72,900
51,200

Each of the eight statements in Program C.1 is entered as a line of type on a computer terminal or keyboard, beginning in column 1.

PROGRAM C.1

SAS Statements for Creating a Data Set Consisting of Five Sale Prices

Command line	Column 1 ↓	
1	DATA SALES;	⎫
2	INPUT SALEPRIC;	⎬ Data entry instructions
3	CARDS;	⎭
4	81300	⎫
5	54100	⎪ Data values
6	96900	⎬ (one observation per line)
7	72900	⎪
8	51200	⎭

Command line 1   The first command in any SAS program is usually the DATA statement. In Program C.1, the DATA statement asks SAS to create a data set named SALES. In a DATA statement, the word DATA must be followed by a blank and the name of the data set that you want to create. The name SALES was chosen arbitrarily for our example. SAS, as with most statistical program packages, will accept any data set or variable name with a maximum of eight characters (letters or numbers), as long as the characters are *not* separated by blanks and the name does *not* begin with a number. For example, MYHOUSE is a legitimate SAS name, but MY HOUSE, 7ELEVEN, and APPRAISERS are not. The name MY HOUSE includes a blank, 7ELEVEN begins with a number, and APPRAISERS includes more than eight characters.

---

*When SAS is used on a mainframe computer, the necessary JCL statements precede all the SAS data entry and statistical analysis instructions. For example, the JCL instruction

  // EXEC SAS

commands the computer to execute the SAS program package. Your instructor will give you the JCL instructions required at your computing center.

Command line 2 The INPUT statement follows the DATA statement and describes the variables that were measured to form the data set. The variable names must follow the word INPUT and must be separated by blanks. In this example, the data set SALES contains measurements on the single variable called SALEPRIC, which again was arbitrarily named.

Command line 3 The third command in a typical SAS program contains only a single word, CARDS. This statement signals the computer that the input data values follow immediately. The computer reads the value of the variable SALE-PRIC on each of the five data lines, thereby creating a data set with a total of five observations.

Command lines 4–8 The data lines contain the actual raw data—in this case, the five sale prices listed in Table C.1. Each data line includes only a single observation—that is, a single value of SALEPRIC.

Note that all SAS statements, with the exception of the data values, end with a semicolon. In general, each SAS statement *must* end in a semicolon. The *only* exceptions to this rule are the input data lines (in this example, lines 4–8).

EXAMPLE C.1

a. Write the appropriate SAS statements to create a data set that contains information on the location (neighborhood), sale price, appraised land value, and appraised value of improvements for the three residential properties described in Table C.2.

TABLE C.2

PROPERTY	NBRHOOD	SALEPRIC	LANDVAL	IMPROVAL
1	B	72,500	12,000	56,000
2	B	97,000	15,000	79,500
3	A	121,300	22,100	85,100

b. Write the appropriate SAS statements that will compute the sum of the appraised land and improvement values and add this new variable to the data set.

SOLUTION

a. The six statements shown in Program C.2 will create the desired data set. The data set, which we have called SALES, has three observations (properties), with four variables—NBRHOOD, SALEPRIC, LANDVAL, and IMPROVAL—per observation. Notice that a dollar sign ($) follows the variable named NBRHOOD on the INPUT statement (line 2). This is because NBRHOOD is a **character** variable—that is, a qualitative variable—whose values (A or B) are characters, not numbers. In contrast, SALEPRIC, LANDVAL, and IMPROVAL are read as **numeric** variables, since their values are quantitative.

Whenever you want to include a character (nonnumeric) variable in a SAS data set, a dollar sign must follow the variable name in the INPUT statement. This informs the computer that the values of this variable to be read on the data lines are observations on a qualitative variable.

PROGRAM C.2

SAS Statements for Creating the Property Data Set

```
Command Column 1
  line      ↓
   1    DATA SALES;                                          Data entry
   2    INPUT NBRHOOD   $   SALEPRIC LANDVAL IMPROVAL;       instructions
   3    CARDS;
   4    B    72500   12000   56000
   5    B    97000   15000   79500      Data values
   6    A   121300   22100   85100      (one observation per line)
```

Note that the order in which the names of the variables appear in the INPUT statement must be the same as the order in which the values of the variables appear on the data lines (lines 4–6). Thus, the value of NBRHOOD appears first on the data lines, followed by the values of SALEPRIC, LANDVAL, and IMPROVAL, respectively. As long as these values are separated by at least one blank and appear in the proper order, they may begin in any of the 80 columns of the screen. For example, lines 4–6 could have been typed as follows:

```
Command Column 1
  line      ↓
   4         B   72500        12000       56000
   5    B              97000       15000        79500
   6                    A       121300 22100 85100
```

b. The SAS statements shown in Program C.3 expand the data set of part **a** by creating an additional variable, named TOTVAL, which represents the sum of the appraised land and improvement values. The only additional statement needed is that shown on line 3:

```
TOTVAL = LANDVAL + IMPROVAL;
```

SAS will now automatically compute TOTVAL for each of the three observations (residential properties) and include this variable on the data set.

PROGRAM C.3

SAS Statements for Creating the Property Sales Data Set with Total Appraised Value Added

```
Command Column 1
  line      ↓
   1    DATA SALES;                                          Data entry
   2    INPUT NBRHOOD   $   SALEPRIC LANDVAL IMPROVAL;       instructions
   3    TOTVAL = LANDVAL + IMPROVAL;
   4    CARDS;
   5    B    72500   12000   56000
   6    B    97000   15000   79500      Data lines
   7    A   121300   22100   85100
```

The SAS data set, which we have named SALES, is stored in the computer as shown in Table C.3.*

TABLE C.3
SAS Data Set SALES
Stored in the Computer

OBSERVATION	NBRHOOD	SALEPRIC	LANDVAL	IMPROVAL	TOTVAL
1	B	72500	12000	56000	68000
2	B	97000	15000	79500	94500
3	A	121300	22100	85100	107200

Any SAS statements that are used to create variables in addition to those on the INPUT statement should follow the INPUT statement, but precede the CARDS statement. When creating or transforming variables in SAS or in any of the other computer packages we discuss in this Appendix, you should use the standard arithmetic operation symbols, $+$, $-$, $*$, and $/$, for addition, subtraction, multiplication, and division, respectively. Two other important transformations in regression are the log and square root transformations. The SAS functions for the log and square root transformations are shown below.

```
LOGPRIC  = LOG(SALEPRIC);
SQRTPRIC = SQRT(SALEPRIC);
```

Once a SAS data set has been created, you may begin analyzing it using PROC statements. These statistical analysis instructions will follow the input data values in the SAS program. Thus, the three basic types of commands (discussed in Section C.1) appear in SAS in the following order:

1. Data entry instructions
2. Input data values
3. Statistical analysis instructions

We will illustrate the use of PROC statements in the following sections.

We end this section with a word of caution: Unlike some computer programming languages, a SAS statement may begin in *any* of the 80 columns of the computer terminal screen. Also, it is not necessary to begin each SAS statement on a new line. SAS statements can run continuously on a line, as long as each of the separate commands ends in a semicolon. However, we strongly recommend that beginning users of SAS follow the approach of Programs C.1–C.3: (1) Start typing in column 1, and (2) Begin each new SAS command on a new line. Our experience has indicated that those who follow this simple approach will make fewer programming errors and have an easier time "debugging" their programs.

---

*The storage is temporary, not permanent. That is, SAS will store the data set on the computer temporarily until the program run is complete. Once the program is finished, the data set is no longer available for use and must be recreated in future programs. Consult the SAS references if you want to store a permanent SAS data set.

SECTION C.3

ENTERING DATA INTO
THE COMPUTER:
SPSS^x

The **Statistical Package for the Social Sciences (SPSS^x)** is similar to the SAS package in the sense that its commands or control statements may be classified into one of two groups: **data creation commands** and **statistical procedure commands**. Data creation commands are data entry instructions used to create a data set, and statistical procedure commands define the statistical procedure to be used to analyze the data set.* Unlike SAS instructions, the instructions on an SPSS^x command must begin in column 1 and all continuation lines must be indented at least one column.

Consider the SPSS^x statements of Program C.4, which create a data set containing the neighborhood, sale price, appraised land value, appraised improvements, and total appraised value of the three properties of Example C.1. Each SPSS^x command begins with a **command keyword** (which may contain more than one word) in column 1. Although a few commands (such as BEGIN DATA and END DATA) are complete in themselves, most require **specifications**. Specifications are made up of variable names, subcommands, numbers, and other keywords required to complete an SPSS^x command. Unlike SAS statements, SPSS^x statements do not end with a semicolon.

PROGRAM C.4 SPSS^x Statements for Creating the Property Sales Data Set

Command   Column 1
   line   ↓

```
1    DATA LIST    FREE/SALEPRIC LANDVAL IMPROVAL * NBRHOOD (A1) ⎫ Data entry
2    COMPUTE      TOTVAL = LANDVAL + IMPROVAL                    ⎬ instructions
3    BEGIN DATA                                                  ⎭
4     72500  12000  56000  B ⎫
5     97000  15000  79500  B ⎬ Data lines
6    121300  22100  85100  A ⎭
7    END DATA
```

Command line 1   The first SPSS^x statement, the DATA LIST command, identifies the variables to be read from the input data lines. The command DATA LIST must appear as shown in columns 1–9. Specifications for this command include formats and names for the variables in the data set.

SPSS^x offers various options for arranging your data on the data lines. We have entered the SPSS^x keyword FREE after the DATA LIST command. The FREE option enables you to enter the data anywhere on the data line, as long as the variable values are separated by at least one blank and the order in which the values appear is consistent with the order given in the variable list (described in the following paragraph). Since the freefield-format mode of data entry is the simplest to describe and use, we will illustrate this method of entry in all our

*The appropriate JCL instructions for a mainframe computer must precede all the SPSS^x control statements. For example, the JCL statement

    // EXEC SPSSX

commands the computer to begin executing the SPSS^x program package. See your instructor for the JCL instructions required at your institution.

SPSS[x] examples. If you want to use another method of data entry, you will need to consult the SPSS[x] references given at the end of this Appendix.

The names of the variables in the data set are listed following the slash (/) in the DATA LIST statement. These variable names (each up to eight characters long) must be separated by blanks. As in Example C.1, we have named the variables NBRHOOD, SALEPRIC, LANDVAL, and IMPROVAL. Note that an asterisk (*) separates the numeric variables (SALEPRIC, LANDVAL, IMPROVAL) from the character variable NBRHOOD and that the numeric variables are listed first. This is necessary whenever you use the freefield format for data entry. In addition, an alphanumeric format for the character variable must be specified, in parentheses, after the variable name. In Program C.4, the format A1 implies that values of the variable NBRHOOD will occupy a single column on the data lines.*

Command line 2   The COMPUTE command is used to create a new variable, arbitrarily called TOTVAL, by summing the values of LANDVAL and IMPROVAL. The algebraic expression specified in the COMPUTE statement may begin in any column following the COMPUTE command.[†]

As with SAS, an SPSS[x] COMPUTE statement uses the standard arithmetic operation symbols, $+$, $-$, $*$, and $/$, for addition, subtraction, multiplication, and division, respectively. The log and square root transformations in SPSS[x] are illustrated below.

```
COMPUTE    LOGPRIC   =   LN(SALEPRIC)
COMPUTE    SQRTPRIC  =   SQRT(SALEPRIC)
```

Command line 3   After compiling the necessary SPSS[x] data entry instructions, the computer is ready to begin reading the input data. The BEGIN DATA command instructs SPSS[x] to start this process and signals the computer that the input data values (lines 4–6 of Program C.4) follow immediately.

Command lines 4–6   The input data values follow the BEGIN DATA command. Each data line represents a single observation (or **case**, as it is called in SPSS[x]). The only variables whose values must appear on the data lines are those named on the DATA LIST command. In our example, you do *not* need to enter the value of TOTVAL—the sum of the appraised values—on the data line. The COMPUTE statement guarantees that SPSS[x] will automatically calculate TOTVAL for each observation and include it in the data set.

Since the freefield format was specified in line 1, the values of the variables can appear anywhere on the data lines, as long as they are separated by at least one blank and the order of appearance is consistent with the order specified in

*Similarly, the format A5 would be used for a character variable with a maximum length of five columns.

[†]The COMPUTE command is an optional SPSS[x] statement. It is necessary only when you want to create additional variables or transform existing variables.

the DATA LIST command. Thus, the first value on the data line is the value of SALEPRIC, the second is the value of LANDVAL, and so forth.

Command line 7   To alert SPSS^x that all the input data values have been read, insert an END DATA statement after the last data line.

The (temporary) SPSS^x data file that results from Program C.4 is stored in the computer as indicated in Table C.4.* Now that we have created the SPSS^x data file, we are ready to begin analyzing it using SPSS^x statistical procedure commands. These analysis instructions will be inserted *after* the data entry instructions (lines 1–2 of Program C.4) but *before* the BEGIN DATA command.

TABLE C.4
SPSS^x Data File Stored in the Computer

CASE	NBRHOOD	SALEPRIC	LANDVAL	IMPROVAL	TOTVAL
1	B	72500	12000	56000	68000
2	B	97000	15000	79500	94500
3	A	121300	22100	85100	107200

To review, the three basic types of instructions (discussed in Section C.1) appear in SPSS^x programs in the following order:

1. Data entry instructions
2. Statistical analysis instructions
3. Input data values

SECTION C.4

ENTERING DATA INTO THE COMPUTER: MINITAB

Minitab is another easy-to-use statistical program package that has been designed especially for those who have no previous experience with computers. Minitab commands can be given in English; that is, the commands can be expressed in nearly the same language you would use to instruct someone to do the computations, tasks, or analysis by hand. Also, Minitab commands may begin and end in any column of the command line of your terminal. We illustrate with Example C.2.

EXAMPLE C.2

Refer to Example C.1. Write the Minitab commands that will create the property sales data set, which is reproduced in Table C.5. Include the total of the appraised land and improvement values on the data set.

TABLE C.5

PROPERTY	NBRHOOD	SALEPRIC	LANDVAL	IMPROVAL
1	B	72500	12000	56000
2	B	97000	15000	79500
3	A	121300	22100	85100

*Consult the SPSS references for instructions on how to save a permanent SPSS data file.

SOLUTION

The appropriate Minitab statements are given in Program C.5.

PROGRAM C.5
Minitab Statements for
Creating Property Sales
Data Set

Command Column 1
  line   ↓

```
1   READ  C1 C2 C3 C4} Data entry instruction
2   1      72500   12000   56000 ⎫
3   1      97000   15000   79500 ⎬ Data values
4   0     121300   22100   85100 ⎭ (one observation per line)
5   ADD    C3 C4 PUT INTO C5} Data entry instruction
6   STOP
```

Command line 1   Minitab stores data in a "worksheet" that it maintains inside the computer. The initial Minitab statement instructs the computer to read the data on lines 2–4 and to put the values of the variables into specific "columns" of the worksheet. (In contrast to the columns of a terminal screen, the columns of a Minitab worksheet may contain values with more than one digit–for example, 90, 128, 10772.) The columns are designated by the letter C and a corresponding number. Instead of referring to a variable by its name, we refer to it in Minitab by the column into which it is placed (for example, C1, C2, . . .). In this example, the computer reads the value of the first variable (NBRHOOD) and places it into column 1 (C1) of the worksheet. Likewise, the value of the second variable (SALEPRIC) is read and placed into column 2 (C2); the value of the third variable (LANDVAL) goes into column 3 (C3); and the value of the fourth variable (IMPROVAL) goes into column 4 (C4). In Minitab, the input data values always follow the READ command.

Command lines 2–4   The input data lines contain the actual values of the variables read in the worksheet columns. As is the case with SAS and SPSS^x, these values must be separated by at least one blank. Notice, however, that the values of the qualitative variable NBRHOOD in C1 are entered as numbers (0 or 1) rather than as letters (A or B). Minitab requires that all data used in statistical analyses be numerical. Thus, if we want to use the qualitative variable NBRHOOD in a statistical procedure, we need to convert its possible values to numbers. In this example, we arbitrarily let 1 represent neighborhood B and 0 represent neighborhood A.

Command line 5   In contrast to the SAS and SPSS^x computer packages, Minitab does not, in general, recognize variable names (for example, NBRHOOD or SALEPRIC) when requested to analyze or manipulate data. Thus, if you want to add the appraised values, you must refer to the variables by their column numbers—C3 and C4—rather than by their names. The Minitab command ADD in line 5 of Program C.5 instructs the computer to add the values of the variables stored in columns 3 and 4 and to place the sum in column 5 (a new column in the worksheet that represents the variable called TOTVAL in the previous sections).

Minitab commands for the other usual arithmetic operations (subtraction, multiplication, and division) and for log and square root transformations are illustrated below.

```
SUBTRACT       C2 FROM C3 PUT INTO C6
MULTIPLY       C2 BY 2 PUT INTO C7
DIVIDE         C4 BY 10 PUT INTO C8
LOGE OF C2 PUT INTO C9
SQRT OF C2 PUT INTO C10
```

Note that the words FROM, BY, OF, PUT, and INTO may be omitted from the Minitab commands for arithmetic operations.

Command line 6   All Minitab programs terminate with the STOP command.

The Minitab worksheet, stored (temporarily) in the computer as shown in Table C.6,* is now ready for data analysis.

TABLE C.6
Minitab Worksheet Stored in the Computer

OBSERVATION	C1	C2	C3	C4	C5
1	1	72500	12000	56000	68000
2	1	97000	15000	79500	94500
3	0	121300	22100	85100	107200

[*Note:*   The commands of Program C.5 were written using the minimal amount of coded text required by Minitab. However, you may insert key words within each command to help you follow the logic of the program. For example, line 1 of Program C.5 could be written as follows:

```
READ NBRHOOD IN C1, SALE PRICE IN C2, APPRAISED VALUES IN C3
AND C4
```

Minitab ignores the extraneous words (NBRHOOD, SALE PRICE, and so forth), and reads only the command name (READ) and its arguments (C1, C2, C3, C4).]

■

The statistical analysis commands of Minitab follow the input data values. Thus, the three types of instructions (discussed in Section C.1) generally appear in Minitab programs in the following order:

1. Data entry instructions
2. Input data values
3. Statistical analysis instructions

*Consult the Minitab references for instructions on how to save a permanent Minitab data file.

SECTION C.5

ENTERING DATA INTO THE COMPUTER: BMDP

The last of the statistical computer packages to be discussed in this appendix was originally developed for use in biomedical analyses and is called **BMDP Statistical Software**. BMDP employs three different types of commands: program identification (analysis) commands, data entry instructions, and input data values.

BMDP **program identification statements** are commands that identify the BMDP statistical analysis program you want to run. For example, the program identification statement

    BMDP 1D

commands a personal computer to execute the BMDP program called P1D (which generates simple descriptive statistics).* To enter data into the computer using the BMDP package, you must precede the data entry instructions and data values (to be explained in the following paragraphs) by the appropriate program identification commands (statistical analysis instructions). The appropriate BMDP program identification statements will be presented in subsequent sections.

EXAMPLE C.3

Refer to Examples C.1 and C.2. Write the BMDP commands that will create the property sales data set. Include the total appraised value on the data set.

SOLUTION

The correct BMDP data entry instructions and data lines are shown in Program C.6. BMDP data entry instructions are written in sentences (commands) that are grouped into paragraphs. Paragraphs must be separated by a slash (/). For convenience, we begin each paragraph with a (/) in column 1, followed by a blank space and an identifying paragraph name. (The paragraph name must be the first word in a paragraph.) Several paragraphs are common to all BMDP programs. In Program C.6, the BMDP instructions consist of three paragraphs: the INPUT paragraph, the VARIABLE paragraph, and the TRANSFORM paragraph. Note that each sentence within a paragraph ends with a period. Although we have written each sentence on a new line, beginning in a fixed (but arbitrary) column (card

PROGRAM C.6 BMDP Statements for Creating the Property Sales Data Set

```
Command  Column 1          Column 16
line        ↓                 ↓
  1     / INPUT       VARIABLES ARE 4,                                        ⎫
  2                   FORMAT IS FREE,                                         ⎪  Data
  3     / VARIABLE    NAMES ARE NBRHOOD, SALEPRIC, LANDVAL, IMPROVAL,         ⎬  entry
  4                   ADD = NEW,                                              ⎪  instructions
  5     / TRANSFORM   TOTVAL = LANDVAL + IMPROVAL,                            ⎪
  6     / END                                                                ⎭
  7     1    72500    12000    56000  ⎫  Data values
  8     1    97000    15000    79500  ⎬  (one observation per line)
  9     0   121300    22100    85100  ⎭
 10     /END
```

*On the mainframe computer, BMDP program ID commands are given as JCL statements, e.g.,

    // EXEC BIMED,PROG=BMDP1D

See your instructor for the appropriate JCL statements required at your computing facility.

column 16), the instructions can be typed anywhere in columns 1–80. Sentences and paragraphs may be typed continuously as long as paragraphs are separated by a slash.

Command lines 1–2   The INPUT paragraph is the first required paragraph of any BMDP program. When the data are entered from a terminal, as will most likely be the case, the INPUT paragraph must specify the number of VARIABLES to be read and their FORMAT. In our example, the sentence

```
VARIABLES ARE 4.
```

informs the computer that four variables are to be read onto a data set. The FORMAT sentence instructs the computer to read the variables in a FREE format. The FREE format method of reading data allows you to enter the data anywhere on the input data lines, as long as the data values on each line are separated by blanks.

Command lines 3–4   The VARIABLE paragraph describes the variables in the data set. Generally, the VARIABLE paragraph is used to name the variables and to indicate whether new variables are to be created by data transformations. In our example, the sentence

```
NAMES ARE NBRHOOD, SALEPRIC, LANDVAL, IMPROVAL.
```

specifies the names of the four variables to be read from the data lines. NBRHOOD is the name of the first variable to be read, SALEPRIC is the name of the second variable to be read, and so forth. As with SAS and SPSS^x, BMDP variable names are restricted to a maximum of eight characters. Note, however, that the values for NBRHOOD are numbers. BMDP, like Minitab, requires that all data be numerical. If additional variables are to be created by data transformations (such as addition, subtraction, multiplication, or division), the ADD sentence is also required. Since we will create a new variable (TOTVAL) in the next paragraph by summing the three property sales, we include the sentence

```
ADD = NEW.
```

Command line 5   The TRANSFORM paragraph provides options for computing new variables from existing ones, correcting miscoded data values, selecting only certain observations for analysis, and generating random samples. The most frequent use of the TRANSFORM paragraph is to create new variables. In this example, the sentence

```
TOTVAL = LANDVAL + IMPROVAL.
```

creates an additional variable, TOTVAL, which is the sum of the appraised land and improvement values of each property.

The TRANSFORM paragraph uses the standard arithmetic operation symbols, $+, -, *$, and $/$, for addition, subtraction, multiplication, and division, respectively.

Log and square root transformations in BMDP are illustrated in the TRANSFORM paragraph below.

```
/ TRANSFORM    LOGPRIC = LN(SALEPRIC),
              SQRTPRIC = SQRT(SALEPRIC),
```

Command line 6   BMDP data entry instructions are terminated with an END paragraph.

Command lines 7–9   The input data values follow the END paragraph. Note that each line represents an observation (property) and includes the values of the first four variables named (in their order of appearance) in the VARIABLE paragraph. These values can appear anywhere on the data lines as long as they are separated by at least one blank.

Command line 10   The last data record must be followed by /END.

The BMDP data entry instructions and input data values have created and stored in the computer a temporary data set that appears as shown in Table C.7.* The data are now in the appropriate form to be analyzed using program (analysis) identification values.

TABLE C.7
BMDP Data Set Stored in the Computer

CASE	NBRHOOD	SALEPRIC	LANDVAL	IMPROVAL	TOTVAL
1	1	72500	12000	56000	68000
2	1	97000	15000	79500	94500
3	0	121300	22100	85100	107200

To review, the three types of instructions (described in Section C.1) appear in a BMDP program in the following order:

1. Statistical analysis instructions
2. Data entry instructions
3. Input data values

SECTION C.6

RELATIVE FREQUENCY DISTRIBUTIONS, DESCRIPTIVE STATISTICS, CORRELATIONS, AND PLOTS

Prior to actually running the regression analysis, it is often helpful to conduct a preliminary analysis of the data. This may include examining relative frequency distributions and descriptive statistics (such as the mean, median, and standard deviation) for both the dependent and independent variables, correlations between pairs of variables, and plots of the dependent variable against each of the independent variables.

*Consult the BMDP references for instructions on how to save a permanent BMDP data set.

Suppose we randomly select 50 residential properties from Appendix F and record:

Dependent variable:

y = Sale price of the property (SALEPRIC)

Independent variables:

x_1 = Appraised value of the land (LANDVAL)
x_2 = Appraised value of the improvements (IMPROVAL)
x_3 = Home size in square feet (AREA)

The necessary program statements for conducting a preliminary analysis on these data are given for each of the four computer packages, SAS, SPSS[x], Minitab, and BMDP, in Programs C.7–C.10 (pages 750–753), respectively. Each program will produce:*

1. A relative frequency distribution for the sale price (y) of a residential property
2. Descriptive (univariate) statistics for sale price (y), appraised land value (x_1), appraised improvements (x_2), and home size (x_3)
3. Pairwise correlations between sale price (y), appraised land value (x_1), appraised improvements (x_2), and home size (x_3)
4. Plots of sale price (y) against appraised land value (x_1), appraised improvements (x_2), and home size (x_3)

Note that each program allows multiple observations per input data line.[†] (Our choice of the number of observations to be read per line is arbitrary. You may read in as many observations as will fit in the available 80 columns of the computer terminal.) Also, some of the statements shown in Programs C.7–C.10 are optional statements included solely for the purpose of labeling the printouts, and thus may be omitted. Consult the references at the end of this Appendix for details on optional commands and their purpose.

---

*BMDP requires that four separate programs be run, one for each of the four types of analysis. The other packages, SAS, SPSS[x], and Minitab, allow the user to run all four analyses in a single computer program.

[†]The @@ symbol in SAS, the FREE format in SPSS[x], the SET command in Minitab, and the STREAM and SLASH formats in BMDP permit multiple observations to be read per line. Consult the references for further information on these commands.

PROGRAM C.7 SAS Statements for Conducting a Preliminary Analysis on the Sample Data from Appendix F

Command
line

```
1    DATA SALES;
2    INPUT SALEPRIC LANDVAL IMPROVAL AREA @@ ;  ⎫ Data entry instructions
3    CARDS;                                      ⎬
4    49000    7200    32590   1152   50000   4000    34522   1419 ⎫
.      •       •        •       •       •      •        •       •   ⎬ Input data values
.      •       •        •       •       •      •        •       •   ⎬ (2 observations per line)
.      •       •        •       •       •      •        •       •   ⎭
28   76000   15525    67196   1943   43000   3360    30133   1079 ⎭
29   PROC CHART;                                                    ⎫ Relative frequency dis-
30   VBAR SALEPRIC/TYPE=PERCENT MIDPOINTS = 30050 TO 130050 BY 10000; ⎬ tribution for SALEPRIC
31   PROC UNIVARIATE;                                ⎫ Descriptive (univariate)
32   VAR SALEPRIC LANDVAL IMPROVAL AREA;             ⎬ statistics
33   PROC CORR;                                      ⎫ Matrix of pairwise correlations
34   VAR SALEPRIC LANDVAL IMPROVAL AREA;             ⎬ between variables
35   PROC PLOT;                                      ⎫ Plots of SALEPRIC against
36   PLOT SALEPRIC * (LANDVAL IMPROVAL AREA);        ⎬ independent variables
```

Command line 30 TYPE=PERCENT produces a percentage bar chart (that is, a relative frequency distribution). If TYPE is omitted, a frequency distribution is constructed. MIDPOINTS specifies the midpoints of the class intervals. If MIDPOINTS is omitted, SAS automatically selects classes with nicely rounded midpoints.

PROGRAM C.8 SPSS^x Statements for Conducting a Preliminary Analysis on the Sample Data from Appendix F

Command
line

```
1    DATA LIST      FREE/SALEPRIC LANDVAL IMPROVAL AREA   } Data entry instructions
2    FREQUENCIES    VARIABLES = SALEPRIC/                 ⎫ Relative frequency distribution
3                   HISTOGRAM = PERCENT INCREMENT (10000)/ ⎬ for SALEPRIC
4    CONDESCRIPTIVE SALEPRIC LANDVAL IMPROVAL AREA/  ⎫ Descriptive (univariate)
5                   STATISTICS = ALL/                ⎬ statistics
6    PEARSON CORR SALEPRIC LANDVAL IMPROVAL AREA   } Pairwise correlation matrix
7    SCATTERGRAM  SALEPRIC WITH LANDVAL IMPROVAL AREA }  Plots of SALEPRIC against independent
8    BEGIN DATA                                            variables
9    49000    7200    32590   1152   50000   4000    34522   1419 ⎫
.      •       •        •       •       •      •        •       •   ⎬ Input data values
.      •       •        •       •       •      •        •       •   ⎬ (2 observations per line)
.      •       •        •       •       •      •        •       •   ⎭
33   76000   15525    67196   1943   43000   3360    30133   1079 ⎭
34   END DATA
```

Command line 3 HISTOGRAM=PERCENT produces a percentage histogram (i.e., relative frequency distribution). If PERCENT is omitted, a frequency distribution is constructed. INCREMENT specifies the class interval width.

PROGRAM C.9   Minitab Statements for Conducting a Preliminary Analysis on the Sample Data from Appendix F

Command
line

```
 1    SET SALEPRIC IN C1   } Data entry instructions
 2    49000   50000   60900    7000   41500   75000   46000   51000   24900   40000 ⎫  Input data values
 ·       ·       ·       ·       ·       ·       ·       ·       ·       ·       ·  ⎬  (10 observations
 ·       ·       ·       ·       ·       ·       ·       ·       ·       ·       ·  ⎭  per line)
 6    35000   36000   45000   29800   13500   56600   50000   57500   76000   43000
 7    SET LANDVAL IN C2   } Data entry instructions
 8    7200    4000    2911     320    3000    5175    2000    7800     804    1242 ⎫  Input data values
 ·      ·       ·       ·       ·       ·       ·       ·       ·       ·       ·  ⎬  (10 observations
 ·      ·       ·       ·       ·       ·       ·       ·       ·       ·       ·  ⎭  per line)
12    3300    3200    1050    1400    1536    2100    2100    2300   15525    3360
13    SET IMPROVAL IN C3   } Data entry instructions
14    32590   34522   54890    3151   33686    5175   34326   30561    8319    9234 ⎫  Input data values
 ·      ·       ·       ·       ·       ·       ·       ·       ·       ·       ·  ⎬  (10 observations
 ·      ·       ·       ·       ·       ·       ·       ·       ·       ·       ·  ⎭  per line)
18    32390   31252   16933   15863   17386   45246   45873   46994   67196   30133
19    SET AREA IN C4   } Data entry instructions
20    1152    1419    1536     520    1056       0    1746    1200    1495    2200 ⎫  Input data values
 ·      ·       ·       ·       ·       ·       ·       ·       ·       ·       ·  ⎬  (10 observations
 ·      ·       ·       ·       ·       ·       ·       ·       ·       ·       ·  ⎭  per line)
24    1080    1056    1530    1008     800    1510    1620    1620    1943    1079
25    NAME C1 IS 'SALEPRICE', C2 IS 'LANDVAL', C3 IS 'IMPROVAL', C4 IS 'AREA' } Frequency
26    HISTOGRAM OF C1, FIRST MIDPOINT AT 30050, INTERVAL WIDTH 10000             } distribution
27    DESCRIBE DATA IN C1-C4   } Descriptive (univariate) statistics                for SALEPRIC
28    CORRELATION BETWEEN DATA IN C1-C4   } Pairwise correlation matrix             (C1)
29    PLOT C1 VS. C2 ⎫
30    PLOT C1 VS. C3 ⎬ Plots of SALEPRIC (C1) against independent variables
31    PLOT C1 VS. C4 ⎭
32    STOP
```

Command line 25   The optional NAME command is used to name columns for printed output.

PROGRAM C.10(a)   BMDP Statements That Produce a Relative Frequency Distribution for the Sample Sale Prices from Appendix F

Command
line

```
 1    BMDP 5D   } Program ID command: Generate a histogram
 2    /   PROBLEM      TITLE IS 'HISTOGRAM OF SALE PRICES'. ⎫  Data
 3    /   INPUT        VARIABLE IS 1.                        ⎬  entry
 4                     FORMAT IS STREAM.                     ⎭  instructions
 5    /   VARIABLE     NAME IS SALEPRIC.
 6    /   GROUP        CUTPOINTS(SALEPRIC) = 35050 TO 135050 BY 10000, ⎫
 7                     NAMES(SALEPRIC) = '30050',    '40050',  '50050', ⎬  Histogram
 8                                       '60050',    '70050',  '80050', ⎬  of SALEPRIC
 9                                       '90050',  '100050', '110050',  ⎪
10                                      '120050',  '130050'.            ⎭
11    /   PLOT         VARIABLE IS SALEPRIC.
12    /   END
13    49000   50000   60900    7000   41500   75000   46000   51000   24900    4000 ⎫  Input data values
 ·      ·       ·       ·       ·       ·       ·       ·       ·       ·       ·  ⎬  (10 observations
 ·      ·       ·       ·       ·       ·       ·       ·       ·       ·       ·  ⎭  per line)
17    35000   36000   45000   29800   13500   56600   50000   57500   76000   43000
18    /END
```

General note   Each of the four preliminary analyses must be conducted in a separate BMDP program.

Command line 6   CUTPOINTS identify upper boundaries of class intervals.

Command lines 7–10   NAMES specify midpoints of class intervals for labeling on the printout.

PROGRAM C.10(b)   BMDP Statements That Produce Descriptive Statistics for the Sample Data from Appendix F

Command
line

```
  1    BMDP  1D   } Program ID command: Descriptive statistics
  2    /   PROBLEM     TITLE IS 'DESCRIPTIVE STATISTICS FOR SAMPLE DATA'.     ⎫ Data
  3    /   INPUT       VARIABLES ARE 4.                                        ⎬ entry
  4                    FORMAT IS SLASH.                                        ⎭ instructions
  5    /   VARIABLE    NAMES ARE SALEPRIC LANDVAL IMPROVAL AREA.
  6    /   END
  7    49000    7200   32590   1152  /  50000   4000   34522   1419  /   ⎫
  .       •       •       •      •        •       •       •       •     ⎬ Input data values
  .       •       •       •      •        •       •       •       •     ⎬ (2 observations per line)
  .       •       •       •      •        •       •       •       •     ⎭
 31    76000   15525   67196   1943  /  43000   3360   30133   1079  /
 32    /END
```

Command line 4   SLASH format requires that the observations (each containing the values of four variables) be separated by a slash (/) on the data lines.

PROGRAM C.10(c)   BMDP Statements That Compute a Matrix of Correlations for the Sample Data from Appendix F

Command
line

```
  1    BMDP  8D   } Program ID command: Correlation matrix
  2    /   PROBLEM     TITLE IS 'CORRELATION MATRIX'.                          ⎫ Data
  3    /   INPUT       VARIABLES ARE 4.                                        ⎬ entry
  4                    FORMAT IS SLASH.                                        ⎭ instructions
  5    /   VARIABLE    NAMES ARE SALEPRIC LANDVAL IMPROVAL AREA.
  6    /   END
  7    49000    7200   32590   1152  /  50000   4000   34522   1419  /   ⎫
  .       •       •       •      •        •       •       •       •     ⎬ Input data values
  .       •       •       •      •        •       •       •       •     ⎬ (2 observations per line)
  .       •       •       •      •        •       •       •       •     ⎭
 31    76000   15525   67196   1943  /  43000   3360   30133   1079  /
 32    /END
```

PROGRAM C.10(d)   BMDP Statements That Produce Plots of Sale Price Against the Independent Variables for the Sample Data from Appendix F

```
Command
  line
    1    BMDP 6D  } Program ID command: Two-dimensional scattergrams
    2    /  PROBLEM     TITLE IS 'SCATTERGRAMS'.          ⎫ Data
    3    /  INPUT       VARIABLES ARE 4.                  ⎬ entry
    4                   FORMAT IS SLASH.                  ⎭ instructions
    5    /  VARIABLE    NAMES ARE SALEPRIC LANDVAL IMPROVAL AREA.
    6    /  PLOT        YVAR IS SALEPRIC.                 ⎫ Plots of SALEPRIC
    7                   XVAR ARE LANDVAL, IMPROVAL, AREA. ⎬ against
    8                   CROSS.                            ⎭ independent variables
    9    /  END
   10    49000    7200   32590  1152 /  50000  4000  34522  1419 /  ⎫ Input data values
    .        .       .       .     .       .      .      .     .  . ⎬ (2 observations per line)
    .        .       .       .     .       .      .      .     .  . ⎭
   34    76000   15525   67196  1943 /  43000  3360  30133  1079 /
   35    /END
```

Command line 8   CROSS requests that all possible pairings of the YVAR (vertical axis) and XVAR (horizontal axis) variables are to be plotted.

SECTION C.7

MULTIPLE REGRESSION

For all four of the statistical computer program packages discussed in this Appendix, simple and multiple regression analyses are performed using the same computer programs.

Suppose we want to fit the multiple regression model

$$y = \beta_0 + \beta_1 x_1 + \beta_2 x_2 + \beta_3 x_1 x_2 + \beta_4 x_1^2 + \beta_5 x_1^2 x_2 + \varepsilon$$

to data collected for $n = 50$ residential properties, where

y = Sale price of a residential property

x_1 = Total appraised value of the property

$$x_2 = \begin{cases} 1 & \text{if property located in neighborhood B} \\ 0 & \text{if property located in neighborhood A} \end{cases}$$

The SAS, SPSS^x, Minitab, and BMDP commands for conducting the multiple regression analysis are given in Programs C.11–C.14 (pages 754–756), respectively. Note that in all four programs, the values of the qualitative variable, neighborhood, are specified as 0 and 1 on the input data lines, and the higher-order terms in the model ($x_1 x_2$, x_1^2, and $x_1^2 x_2$) have been created through data transformation statements. The SPSS^x, Minitab, and BMDP regression procedures *require* that the user create the necessary qualitative dummy variables and higher-order terms for the independent variable. SAS is capable of creating dummy variables and higher-order terms internally, but this requires the use of a more general model-fitting procedure.* For the purposes of this Appendix, we give only the basic SAS regression commands.

*The GLM procedure in SAS will create dummy variables and higher-order terms internally. Consult the SAS references for details on how to use GLM in regression.

PROGRAM C.11 SAS Statements for Fitting the Multiple Regression Model

$$y = \beta_0 + \beta_1 x_1 + \beta_2 x_2 + \beta_3 x_1 x_2 + \beta_4 x_1^2 + \beta_5 x_1^2 x_2 + \varepsilon$$

Command
line

```
 1    DATA SALES;
 2    INPUT Y X1 X2 @@;      Data
 3    X1X2=X1*X2;            entry
 4    X1SQ=X1*X1;            instructions
 5    X1SQX2=X1*X1*X2;
 6    CARDS;
 7     72500    68000   1    97000   94500   0
 .       .        .     .      .       .     .     Input data values
 .       .        .     .      .       .     .     (2 observations per line)
 .       .        .     .      .       .     .
31    121300   107200   0   106900   88400   1
32    PROC REG;                                    Regression
33    MODEL Y = X1 X2 X1X2 X1SQ X1SQX2             analysis
34              /TOL VIF CLI;                      instructions
```

Command line 34 The following are options of the REG procedure:

TOL: Prints tolerance limits for β estimates

VIF: Prints variance inflation factors for β estimates

CLI: Prints predicted values, residuals, and corresponding lower and upper 95% prediction limits. Specify CLM to get 95% confidence intervals for $E(y)$.

PROGRAM C.12 SPSS[x] Statements for Fitting the Multiple Regression Model

$$y = \beta_0 + \beta_1 x_1 + \beta_2 x_2 + \beta_3 x_1 x_2 + \beta_4 x_1^2 + \beta_5 x_1^2 x_2 + \varepsilon$$

Command
line

```
 1    DATA LIST      FREE/ Y X1 X2           Data
 2    COMPUTE        X1X2=X1*X2              entry
 3    COMPUTE        X1SQ=X1*X1             instructions
 4    COMPUTE        X1SQX2=X1*X1*X2
 5    REGRESSION     VARIABLES = Y, X1, X2, X1X2, X1SQ, X1SQX2/
 6                   DESCRIPTIVES/
 7                   CRITERIA=TOLERANCE(.00001)/            Regression
 8                   STATISTICS=ALL/                        analysis
 9                   DEPENDENT=Y/ENTER X1, X2, X1X2, X1SQ, X1SQX2/   instructions
10                   CASEWISE=ALL/
11    BEGIN DATA
12     72500    68000   1    97000   94500   0
 .       .        .     .      .       .     .     Input data values
 .       .        .     .      .       .     .     (2 observations per line)
 .       .        .     .      .       .     .
36    121300   107200   0   106900   88400   1
37    END DATA
```

Command line 6   The DESCRIPTIVES option prints the means, standard deviations, and a correlation matrix for the variables (both dependent and independent) in the regression.

Command line 7   A low tolerance level (in this case, .00001) is specified in order to guarantee that all the independent variables will be entered into the regression equation. (SPSS^x will omit an independent variable from the regression if its tolerance is less than the level specified.)

Command line 8   The STATISTICS=ALL option prints a number of statistics for the regression, including tolerance limits and 95% confidence intervals for the β estimates.

Command line 10   The CASEWISE=ALL option prints predicted values and residuals for all observations (cases) in the analysis.

PROGRAM C.13   Minitab Statements for Fitting the Multiple Regression Model
$$y = \beta_0 + \beta_1 x_1 + \beta_2 x_2 + \beta_3 x_1 x_2 + \beta_4 x_1^2 + \beta_5 x_1^2 x_2 + \varepsilon$$

```
Command
 line
   1    READ Y IN C1, X1 IN C2, X2 IN C3   } Data entry instructions
   2    72500    68000    1  ⎫
   .        •        •     • ⎪  Input data values
   .        •        •     • ⎬  (1 observation
   .        •        •     • ⎪  per line)
  51    106900   88400    1  ⎭
  52    MULTIPLY C2 BY C3 PUT IN C4 ⎫
  53    MULTIPLY C2 BY C2 PUT IN C5 ⎬ Data entry instructions
  54    MULTIPLY C3 BY C5 PUT IN C6 ⎭
  55    REGRESS C1 ON 5 PREDICTORS, C2-C6   } Regression analysis instructions
```

Command line 55   Predicted values and residuals for all observations in the analysis are automatically printed. Minitab will automatically drop an independent variable from the regression equation if it is highly correlated with one or more other independent variables.

PROGRAM C.14   BMDP Statements for Fitting the Multiple Regression Model

$$y = \beta_0 + \beta_1 x_1 + \beta_2 x_2 + \beta_3 x_1 x_2 + \beta_4 x_1^2 + \beta_5 x_1^2 x_2 + \varepsilon$$

Command
line
```
 1   BMDP 1R   } Program ID command: Multiple linear regression
 2   /   PROBLEM      TITLE IS 'MULTIPLE REGRESSION.'
 3   /   INPUT        VARIABLES=3.
 4                    FORMAT IS SLASH.
 5   /   VARIABLE     NAMES=Y, X1, X2.
 6                    ADD=NEW.
 7   /   TRANSFORM    X1X2=X1*X2.
 8                    X1X1=X1*X1.
 9                    X1X1X2=X1*X1*X2.
10   /   REGRESS      DEPENDENT=Y.
11                    INDEPENDENT=X1, X2, X1X2, X1X1, X1X1X2.
12                    TOL=.01.
13   /   PRINT        DATA.
14   /   END
15       72500    68000   1 /   97000   94500   0 /
 .         .        .     . .     .       .     . .
 .         .        .     . .     .       .     . .
 .         .        .     . .     .       .     . .
39      121300   107200   0 /  106900   88400   1 /
40   /END
```

Data entry instructions (lines 2–9)
Regression analysis instructions (lines 10–13)
Input data values (2 observations per line)

Command line 12   BMDP will not allow an independent variable to enter into the regression equation if its tolerance level falls below the specified limit. TOL = .01 is the minimum tolerance limit permitted in BMDP.

Command line 13   The predicted values and residuals are printed for each observation used in the analysis when DATA is specified in the PRINT paragraph.

SECTION C.8

STEPWISE REGRESSION

Stepwise regression routines are available in each of the four computer packages discussed in this text. Although their methodologies and outputs may differ slightly, each provides an objective statistical screening procedure for eliminating unimportant variables from a large set of potential predictor variables.

The commands necessary to call the stepwise regression routines of the SAS, SPSS[x], Minitab, and BMDP packages are given in Programs C.15–C18, respectively. In these examples, stepwise regression is applied to data collected for a sample of 100 executives of a management consultant firm in order to determine which of the following ten independent variables should be included in the construction of the final model for executive salary (y):

x_1 = Years of experience

x_2 = Years of education

$x_3 = \begin{cases} 1 & \text{if male} \\ 0 & \text{if female} \end{cases}$

x_4 = Number of employees supervised

x_5 = Corporate assets (millions of dollars)

$x_6 = \begin{cases} 1 & \text{if board member} \\ 0 & \text{if not} \end{cases}$

x_7 = Years of age

x_8 = Company profits of past 12 months (millions of dollars)

$x_9 = \begin{cases} 1 & \text{if international responsibility} \\ 0 & \text{if not} \end{cases}$

x_{10} = Company total sales of past 12 months (millions of dollars)

All four programs specify the conventional "stepwise" variable selection technique. That is, after a variable has been added to the model, the program will recheck the previously entered variables and remove those that are not statistically significant at some prespecified α (see Section 4.12).

PROGRAM C.15   SAS Stepwise Regression Commands for the Executive Salary Data

Command
line

```
 1    DATA EXECS;          ⎫
 2    INPUT Y X1-X10;      ⎬ Data entry instructions
 3    CARDS;               ⎭
 4    50000   10   6   1   19   12.7   0   46   2.2   1   5.1   ⎫ Input data values
 .      .     .    .   .    .     .    .    .    .    .    .    ⎬ (1 observation
 .      .     .    .   .    .     .    .    .    .    .    .    ⎭ per line)
 .      .     .    .   .    .     .    .    .    .    .    .
103   68000   15   4   0   121  40.1   1   55   9.6   1   6.3
104   PROC STEPWISE;                         ⎫ Stepwise regression
105   MODEL Y=X1-X10/STEPWISE SLE=.15  SLS=.15; ⎬ instructions
```

Command line 104   The printed output includes R^2, the C_p-statistic, and the F-statistic for testing adequacy of the model at each step.

Command line 105   The SLE option specifies the significance level α required for a variable to enter into the model. (The default SLE is .15.) The SLS option specifies the significance level α required for a variable to stay in the model. (The default SLS is .15.) As an option, variable selection techniques other than STEP-WISE are available. These include FORWARD (forward selection), BACKWARD (backward elimination), MAXR, and MINR. Consult the SAS references for more details on the use of these techniques. (The default selection technique is STEPWISE.)

PROGRAM C.16   SPSS^x Stepwise Regression Commands for the Executive Salary Data

```
Command
line
  1    DATA LIST       FREE/Y, X1 TO X10   } Data entry instruction
  2    REGRESSION      VARIABLES=Y, X1 TO X10/         ⎫ Stepwise
  3                    CRITERIA=PIN(.05) POUT(.10)/    ⎬ regression
  4                    DEPENDENT=Y/STEPWISE/           ⎭ instructions
  5    BEGIN DATA
  6    50000   10   6   1    19    2.7   0   46    2.2   1   5.1  ⎫
  .       .     .   .   .     .     .    .    .     .    .    .   ⎪ Input data values
  .       .     .   .   .     .     .    .    .     .    .    .   ⎬ (1 observation
  .       .     .   .   .     .     .    .    .     .    .    .   ⎪ per line)
105    68000   15   4   0   121   40.1   1   55    9.6   1   6.3  ⎭
106    END DATA
```

Command line 3   The PIN option of the CRITERIA subcommand specifies the significance level α required for a variable to enter into the model. (The default PIN is .05.) The POUT option specifies the significance level α required for a variable to stay in the equation. (The default POUT is .10.)

Command line 4   As an option, variable selection techniques other than STEP-WISE are available. These include FORWARD (forward selection), BACKWARD (backward elimination), and TEST. Consult the SPSS^x references for more details on the use of these techniques.

PROGRAM C.17   Minitab Stepwise Regression Commands for the Executive Salary Data

```
Command
line
  1    READ Y IN C1, INDEPENDENTS IN C2-C11   } Data entry instruction
  2    50000   10   6   1    19    2.7   0   4.6   2.2   1   6.3  ⎫
  .       .     .   .   .     .     .    .    .     .    .    .   ⎪ Input data values
  .       .     .   .   .     .     .    .    .     .    .    .   ⎬ (1 observation
  .       .     .   .   .     .     .    .    .     .    .    .   ⎪ per line)
101    68000   15   4   0   121   40.1   1   55    9.6   1   6.3  ⎭
102    STEPWISE ON C1, PREDICTORS IN C2-C11  ⎫ Stepwise regression
103         FENTER=4   FREMOVE=4             ⎭ instructions
```

Command line 103   The FENTER subcommand specifies a value of the F-statistic required for a variable to enter into the model. If $F >$ FENTER, the variable is entered. (The default FENTER is 4.) The FREMOVE subcommand specifies a value of the F-statistic required for a variable to stay in the model. If $F <$ FREMOVE, the variable is removed. (The default FREMOVE is 4.) Forward selection and backward elimination techniques are also available. Specify FREMOVE=0 for forward selection and FREMOVE=100000 ENTER=C2–C11 for backward elimination.

PROGRAM C.18   BMDP Stepwise Regression Commands for the Executive Salary Data

Command
line

```
1    BMDP  2R   } Program ID command: Stepwise regression
2    /  PROBLEM     TITLE IS 'STEPWISE REGRESSION'.        ⎫ Data
3    /  INPUT       VARIABLES=11.                          ⎬ entry
4                   FORMAT IS FREE.                        ⎭ instructions
5    /  VARIABLE    NAMES=Y,X1,X2,X3,X4,X5,X6,X7,X8,X9,X10.
6    /  REGRESS     DEPENDENT=Y.                           ⎫ Stepwise
7                   INDEPENDENTS=X1,X2,X3,X4,X5,X6,X7,X8,X9,X10. ⎬ regression
8                   METHOD=F.                              ⎭ instructions
9                   ENTER=4. REMOVE=3.996.
10   /  END
11   50000  10  6  1   19   2.7  0  4.6  2.2  1  5.1  ⎫
 .      .    .   .  .    .    .   .   .    .   .   .   ⎪ Input data values
 .      .    .   .  .    .    .   .   .    .   .   .   ⎬ (1 observation
 .      .    .   .  .    .    .   .   .    .   .   .   ⎪ per line)
110  68000  15  4  0   121  40.1 1  55   9.6  1  6.3  ⎭
111  /END
```

Command line 7   The INDEPENDENTS sentence may be omitted if all variables are to be used in the stepwise regression.

Command line 8   METHOD=F specifies that the conventional stepwise variable selection technique which relies on F-statistics to determine the importance of each variable is to be used. Other available methods are FSWAP, R, and RSWAP. Consult the BMDP references for details on how to use these alternative techniques. (The default method is F.)

Command line 9   The ENTER sentence specifies a value of the F-statistic required for a variable to enter into the equation. If F is larger than the ENTER limit, the variable is added. (The default is ENTER = 4.0.) The REMOVE sentence specifies a value of the F-statistic required for a variable to stay in the equation. If F is less than the REMOVE limit, the variable is removed. (The default is REMOVE = 3.996.) Forward selection and backward elimination can be invoked by manipulating the ENTER and REMOVE limits. Consult the BMDP references for instructions on how to do this.

General note   An "all-possible-subsets" regression can be performed using the BMDP 9R program. The available methods are METHOD=RSQ (R^2 criterion), METHOD=ADJ (R^2-adjusted criterion) and METHOD=CP (C_p criterion). Consult the BMDP references for details.

SECTION C.9

RESIDUAL ANALYSIS

The regression routines of the four computer program packages have options for conducting a basic residual analysis—residual plots, histograms of residuals, and normal probability plots. In addition, SAS, SPSS^x, and BMDP are capable of producing influence diagnostics, including deleted residuals, leverage, and Cook's distance, for the observations used in the analysis.

Programs C.19–C.22 contain the SAS, SPSS^x, Minitab, and BMDP commands, respectively, required to conduct a residual analysis of the interaction model

$$y = \beta_0 + \beta_1 x_1 + \beta_2 x_2 + \beta_3 x_1 x_2 + \varepsilon$$

where

$y = $ Sale price of a residential property

$x_1 = $ Total appraised value of the property

$x_2 = \begin{cases} 1 & \text{if property located in neighborhood B} \\ 0 & \text{if property located in neighborhood A} \end{cases}$

PROGRAM C.19   SAS Statements for Conducting a Residual Analysis of the Model
$y = \beta_0 + \beta_1 x_1 + \beta_2 x_2 + \beta_3 x_1 x_2 + \varepsilon$

Command line

```
 1   DATA SALES;                                  ⎫
 2   INPUT Y X1 X2 @@;                            ⎬ Data entry instructions
 3   X1X2=X1*X2;                                  ⎭
 4   CARDS;
 5   72500    68000  1    97000   94500   0       ⎫
 .     .        .    .      .       .     .       ⎪ Input data values
 .     .        .    .      .       .     .       ⎬ (2 observations
 .     .        .    .      .       .     .       ⎪ per line)
29   121300  107200  0   106900   88400   1       ⎭
30   PROC REG;
31   MODEL Y = X1 X2 X1X2                         ⎫
32           / R INFLUENCE;                       ⎬ Regression analysis, listing of residuals, and
33   OUTPUT OUT=RESIDS P=YHAT R=RESID;            ⎭ influence diagnostics
34   PROC PLOT;                                   ⎫
35   PLOT RESID*(YHAT X1);                        ⎬ Residual plots
36   PROC CHART;                                  ⎫
37   VBAR RESID/TYPE=PERCENT;                     ⎬ Histogram of residuals
38   PROC UNIVARIATE PLOT NORMAL;                 ⎫ Descriptive statistics for residuals, stem
39   VAR RESID;                                   ⎭ and leaf plot, and normal probability plot
```

Command line 32   The R option produces a list of residuals, predicted values, Studentized residuals, and Cook's D-statistic for all observations in the analysis. The INFLUENCE option requests a detailed analysis of the influence of each observation on the β estimates. This includes leverage values and Studentized deleted residuals.

Command line 38   The PLOT option produces a stem and leaf plot for the variable specified (for example, RESID). The NORMAL option produces a normal probability plot for the variable.

PROGRAM C.20   SPSS^x Statements for Conducting a Residual Analysis of the Model
$$y = \beta_0 + \beta_1 x_1 + \beta_2 x_2 + \beta_3 x_1 x_2 + \varepsilon$$

Command
line

```
 2   DATA LIST        FREE/Y X1 X2 ⎫ Data entry instructions
 3   COMPUTE          X1X2=X1*X2   ⎭
 3   REGRESSION       VARIABLES=Y, X1, X2, X1X2/        ⎫
 4                    CRITERIA=TOLERANCE(.00001)/       ⎬ Regression analysis
 5                    DEPENDENT=Y/ENTER X1, X2, X1X2/   ⎭
 6                    RESIDUALS/                                        ⎫
 7                    CASEWISE=DEFAULT ALL SRESID SDRESID COOK LEVER/   ⎬ Residual analysis
 8                    SCATTERPLOT=(*RESID,*PRED)(*RESID,X1)/            ⎭
 9   BEGIN DATA
10   72500     68000   1    97000   94500   0 ⎫
 .      .        .     .      .       .     . ⎪ Input data values
 .      .        .     .      .       .     . ⎬ (2 observations
 .      .        .     .      .       .     . ⎪ per line)
34   121300  107200    0   106900   88400   1 ⎭
35   END DATA
```

Command line 6   The RESIDUALS subcommand produces a histogram and normal probability plot of the standardized residuals, and a list of the ten "worst" outliers.

Command line 7   The CASEWISE subcommand produces a list of the residuals and predicted values (DEFAULT), Studentized residuals (SRESID), Studentized deleted residuals (SDRESID), Cook's distances (COOK), and leverage values (LEVER) for all observations (ALL) used in the analysis.

Command line 8   The SCATTERPLOT subcommand produces two scatterplots— the residuals against the predicted values and the residuals against x_1.

PROGRAM C.21   Minitab Statements for Conducting a Residual Analysis of the Model
$$y = \beta_0 + \beta_1 x_1 + \beta_2 x_2 + \beta_3 x_1 x_2 + \varepsilon$$

Command
line

```
 1   READ Y IN C1, PREDICTORS IN C2 C3  } Data entry instructions
 2   72500     68000   1 ⎫
 3   97000     94500   0 ⎪
 .      .        .     . ⎬ Input data values
 .      .        .     . ⎪ (1 observation
 .      .        .     . ⎪ per line)
51   106900    88400   1 ⎭
52   MULTIPLY C2 BY C3, PUT IN C4  } Data entry instructions
53   REGRESS Y IN C1, USING 3 PREDICTORS IN C2-C4      ⎫ Regression analysis (storing
54           RESIDUALS IN C5, PREDICTED VALUES IN C6   ⎭ residuals and predicted values)
55   PLOT C5 VS. C6  ⎫
56   PLOT C5 VS. C2  ⎬ Residual analysis (plots and histogram)
57   HISTOGRAM OF C5 ⎭
58   STOP
```

Command line 54   Residuals and predicted values are stored in columns C5 and C6, respectively, for the purposes of plotting.

PROGRAM C.22   BMDP Statements for Conducting a Residual Analysis of the Model
$$y = \beta_0 + \beta_1 x_1 + \beta_2 x_2 + \beta_3 x_1 x_2 + \varepsilon$$

Command
line

```
 1    BMDP 1R   } Program ID command: Multiple regression
 2    /   PROBLEM   TITLE IS 'RESIDUAL ANALYSIS',  ⎫
 3    /   INPUT     VARIABLES=3,                    ⎪
 4                  FORMAT IS SLASH,                ⎬ Data entry instructions
 5    /   VARIABLE  NAMES=Y, X1, X2,               ⎪
 6                  ADD=NEW,                        ⎭
 7    /   TRANSFORM X1X2=X1*X2,                    ⎫
 8    /   REGRESS   DEPENDENT=Y,                    ⎪
 9                  INDEPENDENT=X1, X2, X1X2,      ⎬ Regression analysis
10                  TOL=.01,                        ⎭
11    /   PRINT     DATA,    } Listing of residuals and predicted values
12    /   PLOT      RESIDUALS, VARIABLES=X1,       ⎫ Residual plots and
13                  NORMAL.                         ⎭ normal probability plot
14    /   END
15     72500    68000   1 /    97000   94500  0 / ⎫
 ·       ·        ·      · ·      ·        ·    · · ⎪ Input data values
 ·       ·        ·      · ·      ·        ·    · · ⎬ (2 observations
 ·       ·        ·      · ·      ·        ·    · · ⎪ per line)
39    121300   107200   0 /   106900   88400  1 / ⎭
40    /END
```

Command line 12   The RESIDUALS sentence in the PLOT paragraph produces a plot of the residuals against $\hat{y}$. Specifying X1 in the VARIABLES sentence produces a plot of the residuals against X1. The NORMAL sentence produces a normal probability plot of the residuals.

General note   Influence diagnostics (Cook's distance, deleted residuals, leverage) can be obtained by using the BMDP 9R program (all-possible-subsets regression) and specifying METHOD=NONE in the REGRESS paragraph and RESI in the PRINT paragraph. Consult the BMDP references for more details.

SECTION C.10

ANALYSIS OF VARIANCE

All four computer software packages described in this text contain analysis of variance routines for designed experiments. The SAS, SPSS^x, and BMDP packages are capable of analyzing experimental designs ranging from simple one-way classifications of data (completely randomized designs) to the more sophisticated k-way classifications (k-factor factorial experiments). Minitab, however, is limited to one-way and two-way classifications of data.

As noted in Section 8.10 each experimental design possesses an underlying probabilistic regression model for the response y. These models can become quite complex, especially for large factorial experiments. Fortunately, the four software packages we discuss do not require the exact model to be specified. Rather, only the sources of variation for the analysis of variance need to be given in the appropriate command lines. For example, for the 3×3 factorial of Example 8.12, we need to specify three sources of variation: (1) ratio of raw material allocation (main effect); (2) supply of the raw material (main effect); and (3) ratio–supply interaction. The computer package will automatically fit the complete factorial model that corresponds to these sources of variation and produce an ANOVA summary table.

The SAS, SPSS^x, Minitab, and BMDP commands for conducting the analysis of variance for the factorial experiment of Example 8.12 are given in Programs C.23–C.26, respectively. The ANOVA commands for the completely randomized and randomized block designs are similar in nature.

PROGRAM C.23   SAS ANOVA Commands for a 3×3 Factorial Experiment

Command
line
```
 1    DATA MANUFACT;                                  ⎫
 2    INPUT RATIO SUPPLY PROFIT @@;                   ⎬ Data entry instructions
 3    CARDS;                                          ⎭
 4    .5   15   23   .5   15   20   .5   15   21  ⎫
 .     .    .    .    .    .    .    .    .    .   ⎪   Input data values
 .     .    .    .    .    .    .    .    .    .   ⎬   (3 observations
 .     .    .    .    .    .    .    .    .    .   ⎪   per line)
12    2    21   20   2    21   22   2    21   24  ⎭
13    PROC ANOVA;                                     ⎫
14    CLASSES RATIO SUPPLY;                           ⎬ ANOVA Instructions
15    MODEL PROFIT=RATIO SUPPLY RATIO*SUPPLY;         ⎭
```

Command line 14   The CLASSES statement identifies the main effects (factors) for the experiment.

Command line 15   The sources of variation are specified to the right of the equals sign (=) in the MODEL statement, the dependent (response) variable to the left. Interactions are specified by placing an asterisk (*) between the factors (e.g., RATIO*SUPPLY).

PROGRAM C.24   SPSS^x ANOVA Commands for a 3×3 Factorial Experiment

Command
line
```
 1    DATA LIST FREE/RATIO SUPPLY PROFIT } Data entry instruction
 2    MANOVA     PROFIT BY RATIO(0,2) SUPPLY (0,2)/       ⎫ ANOVA
 3               DESIGN=RATIO, SUPPLY, RATIO BY SUPPLY/   ⎬ instructions
 4    BEGIN DATA
 5    0   0   23   0   0   20   0   0   21 ⎫
 6    0   1   22   0   1   19   0   1   20 ⎪
 .     .   .    .    .   .    .    .   .    .  ⎬ Input data values
 .     .   .    .    .   .    .    .   .    .  ⎪ (3 observations
 .     .   .    .    .   .    .    .   .    .  ⎪ per line)
13    2   2   20   2   2   22   2   2   24 ⎭
14    END DATA
```

Command line 2 The dependent (response) variable is listed to the left of BY in the MANOVA command, while the main effects (factors) are listed to the right. The range of the coded values of the factors on the data lines must be specified in parentheses after each factor. (Note that the three values of RATIO and SUPPLY are coded as 0, 1, and 2 on the input data lines.)

Command line 3 The sources of variation are specified in the DESIGN subcommand. Interactions are indicated by placing the keyword BY between the factors (e.g., RATIO BY SUPPLY). If the DESIGN subcommand is omitted, the SPSS^x default is to fit a full factorial model, which includes sources for main effects and all interactions.

PROGRAM C.25 Minitab ANOVA Commands for a 3 × 3 Factorial Experiment

Command line

```
1    READ RATIO IN C1, SUPPLY IN C2, PROFIT IN C3   } Data entry instructions
2    .5   15   23 ⎫
.    .    .    .  ⎬ Input data lines
.    .    .    .  ⎭ (1 observation per line)
.    .    .    .
29   2    21   24
30   TWOWAY AOV ON PROFIT IN C3, FACTORS IN C1, C2   } ANOVA instructions
```

Command line 30 Minitab automatically includes interaction between the factors as a source of variation. When conducting an ANOVA for a randomized block design, you will need to pool the interaction sum of squares with the error sum of squares. This can be accomplished by including the subcommand ADDITIVE on line 31.

General For a completely randomized design, use the Minitab procedure ONEWAY.

PROGRAM C.26 BMDP ANOVA Commands for a 3 × 3 Factorial Experiment

Command line

```
1    BMDP 2V   } Program ID command: Two-way analysis of variance
2    /   PROBLEM TITLE IS '3X3 FACTORIAL'.
3    /   INPUT       VARIABLES=3,              ⎫ Data entry
4                    FORMAT=STREAM,            ⎬ instructions
5    /   VARIABLE    NAMES=RATIO, SUPPLY, PROFIT. ⎭
6    /   DESIGN      DEPENDENT=PROFIT,         ⎫
7                    GROUP=RATIO, SUPPLY,      ⎬ ANOVA instructions
8                    INCLUDE=1, 2, 12,         ⎭
9    /   END
10   .5  15  23  .5  15  20  .5  15  21 ⎫
.    .   .   .   .   .   .   .   .   .  ⎬ Input data values
.    .   .   .   .   .   .   .   .   .  ⎭ (3 observations per line)
.    .   .   .   .   .   .   .   .   .
18   2   21  20  2   21  22  2   21  24
19   /END
```

Command line 7   The main effects (factors) are specified in the GROUP sentence. Each factor is assigned a number corresponding to the order in which they appear in the GROUP sentence.

Command line 8   The sources of variation are specified in the INCLUDE sentence using the GROUP numbers. For example, "1" represents the RATIO main effect and "12" represents interaction between RATIO and SUPPLY. If the INCLUDE sentence is omitted, the BMDP default is to fit the full factorial model, which includes sources for main effects and all interactions.

General note   The ANOVA for a completely randomized design is conducted using the BMDP one-way analysis of variance program, BMDP 1V.

REFERENCES

BMDP User's Digest, 2nd ed. Mary Ann Hill (ed.). Los Angeles: BMDP Statistical Software, 1982.

BMDPC: User's Guide to BMDP on the IBM PC. Los Angeles: BMDP Statistical Software.

Dixon, W. J., Brown, M. B., Engelman, L., Frane, J. W., Hill, M. A., Jennrich, R. I., and Toporek, J. D. *BMDP Statistical Software*, 1985 ed. Berkeley: University of California Press.

Norusis, M. J. *The SPSS Guide to Data Analysis*, 1986 ed. SPSS, Inc., Suite 3000, 444 N. Michigan Avenue, Chicago, Ill. 60611.

Norusis, M. J. *SPSS^x Advanced Statistics Guide*, 1985 ed. SPSS, Inc., Suite 3000, 444 N. Michigan Avenue, Chicago, Ill. 60611.

Ryan, T. A., Joiner, B. L., and Ryan, B. F. *Minitab Reference Manual*. University Park, Pa.: Minitab Project, 1985.

Ryan, T. A., Joiner, B. L., and Ryan, B. F. *Minitab Student Handbook*. Boston, Mass.: Duxbury, 1985.

SAS/ETS User's Guide Version 5 Edition (1985). SAS Institute, Inc., Box 8000, Cary, N.C. 27511.

SAS Procedures Guide for Personal Computers, Version 6 Edition, 1986. SAS Institute, Inc., Box 8000, Cary, N.C. 27511.

SAS Statistics Guide for Personal Computers, Version 6 Edition, 1986. SAS Institute, Inc., Box 8000, Cary, N.C. 27511.

SAS Supplemental Library User's Guide, 1983 ed. SAS Institute, Inc., Box 8000, Cary, N.C. 27511.

SAS User's Guide: Basics, Version 5 Edition (1985). SAS Institute, Inc., Box 8000, Cary, N.C. 27511.

SAS User's Guide: Statistics, Version 5 Edition (1985). SAS Institute, Inc., Box 8000, Cary, N.C. 27511.

SPSS^x User's Guide. 1983 ed. SPSS, Inc., Suite 3000, 444 N. Michigan Avenue, Chicago, Ill. 60611.

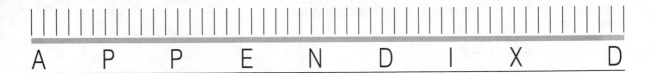

USEFUL STATISTICAL TABLES

CONTENTS

TABLE 1 Normal Curve Areas

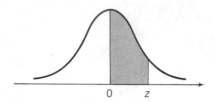

z	.00	.01	.02	.03	.04	.05	.06	.07	.08	.09
.0	.0000	.0040	.0080	.0120	.0160	.0199	.0239	.0279	.0319	.0359
.1	.0398	.0438	.0478	.0517	.0557	.0596	.0636	.0675	.0714	.0753
.2	.0793	.0832	.0871	.0910	.0948	.0987	.1026	.1064	.1103	.1141
.3	.1179	.1217	.1255	.1293	.1331	.1368	.1406	.1443	.1480	.1517
.4	.1554	.1591	.1628	.1664	.1700	.1736	.1772	.1808	.1844	.1879
.5	.1915	.1950	.1985	.2019	.2054	.2088	.2123	.2157	.2190	.2224
.6	.2257	.2291	.2324	.2357	.2389	.2422	.2454	.2486	.2517	.2549
.7	.2580	.2611	.2642	.2673	.2704	.2734	.2764	.2794	.2823	.2852
.8	.2881	.2910	.2939	.2967	.2995	.3023	.3051	.3078	.3106	.3133
.9	.3159	.3186	.3212	.3238	.3264	.3289	.3315	.3340	.3365	.3389
1.0	.3413	.3438	.3461	.3485	.3508	.3531	.3554	.3577	.3599	.3621
1.1	.3643	.3665	.3686	.3708	.3729	.3749	.3770	.3790	.3810	.3830
1.2	.3849	.3869	.3888	.3907	.3925	.3944	.3962	.3980	.3997	.4015
1.3	.4032	.4049	.4066	.4082	.4099	.4115	.4131	.4147	.4162	.4177
1.4	.4192	.4207	.4222	.4236	.4251	.4265	.4279	.4292	.4306	.4319
1.5	.4332	.4345	.4357	.4370	.4382	.4394	.4406	.4418	.4429	.4441
1.6	.4452	.4463	.4474	.4484	.4495	.4505	.4515	.4525	.4535	.4545
1.7	.4554	.4564	.4573	.4582	.4591	.4599	.4608	.4616	.4625	.4633
1.8	.4641	.4649	.4656	.4664	.4671	.4678	.4686	.4693	.4699	.4706
1.9	.4713	.4719	.4726	.4732	.4738	.4744	.4750	.4756	.4761	.4767
2.0	.4772	.4778	.4783	.4788	.4793	.4798	.4803	.4808	.4812	.4817
2.1	.4821	.4826	.4830	.4834	.4838	.4842	.4846	.4850	.4854	.4857
2.2	.4861	.4864	.4868	.4871	.4875	.4878	.4881	.4884	.4887	.4890
2.3	.4893	.4896	.4898	.4901	.4904	.4906	.4909	.4911	.4913	.4916
2.4	.4918	.4920	.4922	.4925	.4927	.4929	.4931	.4932	.4934	.4936
2.5	.4938	.4940	.4941	.4943	.4945	.4946	.4948	.4949	.4951	.4952
2.6	.4953	.4955	.4956	.4957	.4959	.4960	.4961	.4962	.4963	.4964
2.7	.4965	.4966	.4967	.4968	.4969	.4970	.4971	.4972	.4973	.4974
2.8	.4974	.4975	.4976	.4977	.4977	.4978	.4979	.4979	.4980	.4981
2.9	.4981	.4982	.4982	.4983	.4984	.4984	.4985	.4985	.4986	.4986
3.0	.4987	.4987	.4987	.4988	.4988	.4989	.4989	.4989	.4990	.4990

Source: Abridged from Table I of A. Hald, *Statistical Tables and Formulas* (New York: John Wiley & Sons, Inc.), 1952. Reproduced by permission of the publisher.

TABLE 2   Critical Values for Student's t

DEGREES OF FREEDOM	$t_{.100}$	$t_{.050}$	$t_{.025}$	$t_{.02}$	$t_{.015}$	$t_{.010}$	$t_{.0075}$	$t_{.005}$	$t_{.0025}$	$t_{.001}$	$t_{.0005}$
1	3.078	6.314	12.706	15.985	21.205	31.821	42.434	63.657	127.322	318.310	636.590
2	1.886	2.920	4.303	4.849	5.643	6.965	8.073	9.925	14.089	22.326	31.598
3	1.638	2.353	3.182	3.482	3.896	4.541	5.047	5.841	7.453	10.213	12.924
4	1.533	2.132	2.776	2.999	3.298	3.747	4.088	4.604	5.598	7.173	8.610
5	1.476	2.015	2.571	2.757	3.003	3.365	3.634	4.032	4.773	5.893	6.869
6	1.440	1.943	2.447	2.612	2.829	3.143	3.372	3.707	4.317	5.208	5.959
7	1.415	1.895	2.365	2.517	2.715	2.998	3.203	3.499	4.029	4.785	5.408
8	1.397	1.860	2.306	2.449	2.634	2.896	3.085	3.355	3.833	4.501	5.041
9	1.383	1.833	2.262	2.398	2.574	2.821	2.998	3.250	3.690	4.297	4.781
10	1.372	1.812	2.228	2.359	2.527	2.764	2.932	3.169	3.581	4.144	4.587
11	1.363	1.796	2.201	2.328	2.491	2.718	2.879	3.106	3.497	4.025	4.437
12	1.356	1.782	2.179	2.303	2.461	2.681	2.836	3.055	3.428	3.930	4.318
13	1.350	1.771	2.160	2.282	2.436	2.650	2.801	3.012	3.372	3.852	4.221
14	1.345	1.761	2.145	2.264	2.415	2.624	2.771	2.977	3.326	3.787	4.140
15	1.341	1.753	2.131	2.249	2.397	2.602	2.746	2.947	3.286	3.733	4.073
16	1.337	1.746	2.120	2.235	2.382	2.583	2.724	2.921	3.252	3.686	4.015
17	1.333	1.740	2.110	2.224	2.368	2.567	2.706	2.898	3.222	3.646	3.965
18	1.330	1.734	2.101	2.214	2.356	2.552	2.689	2.878	3.197	3.610	3.922
19	1.328	1.729	2.093	2.205	2.346	2.539	2.674	2.861	3.174	3.579	3.883
20	1.325	1.725	2.086	2.197	2.336	2.528	2.661	2.845	3.153	3.552	3.849
21	1.323	1.721	2.080	2.189	2.328	2.518	2.649	2.831	3.135	3.527	3.819
22	1.321	1.717	2.074	2.183	2.320	2.508	2.639	2.819	3.119	3.505	3.792
23	1.319	1.714	2.069	2.177	2.313	2.500	2.629	2.807	3.104	3.485	3.768
24	1.318	1.711	2.064	2.172	2.307	2.492	2.620	2.797	3.091	3.467	3.745
25	1.315	1.708	2.060	2.167	2.301	2.485	2.612	2.787	3.078	3.450	3.725
26	1.315	1.706	2.056	2.162	2.296	2.479	2.605	2.779	3.067	3.435	3.707
27	1.314	1.703	2.052	2.158	2.291	2.473	2.598	2.771	3.057	3.421	3.690
28	1.313	1.701	2.048	2.154	2.286	2.467	2.592	2.763	3.047	3.408	3.674
29	1.311	1.699	2.045	2.150	2.282	2.462	2.586	2.756	3.038	3.396	3.659
∞	1.282	1.645	1.960	2.054	2.170	2.326	2.432	2.576	2.807	3.090	3.291

TABLE 3 Critical Values for the F-Statistic: $F_{.10}$

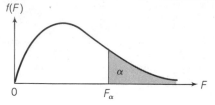

ν_1	NUMERATOR DEGREES OF FREEDOM								
ν_2	1	2	3	4	5	6	7	8	9
1	39.86	49.50	53.59	55.83	57.24	58.20	58.91	59.44	59.86
2	8.53	9.00	9.16	9.24	9.29	9.33	9.35	9.37	9.38
3	5.54	5.46	5.39	5.34	5.31	5.28	5.27	5.25	5.24
4	4.54	4.32	4.19	4.11	4.05	4.01	3.98	3.95	3.94
5	4.06	3.78	3.62	3.52	3.45	3.40	3.37	3.34	3.32
6	3.78	3.46	3.29	3.18	3.11	3.05	3.01	2.98	2.96
7	3.59	3.26	3.07	2.96	2.88	2.83	2.78	2.75	2.72
8	3.46	3.11	2.92	2.81	2.73	2.67	2.62	2.59	2.56
9	3.36	3.01	2.81	2.69	2.61	2.55	2.51	2.47	2.44
10	3.29	2.92	2.73	2.61	2.52	2.46	2.41	2.38	2.35
11	3.23	2.86	2.66	2.54	2.45	2.39	2.34	2.30	2.27
12	3.18	2.81	2.61	2.48	2.39	2.33	2.28	2.24	2.21
13	3.14	2.76	2.56	2.43	2.35	2.28	2.23	2.20	2.16
14	3.10	2.73	2.52	2.39	2.31	2.24	2.19	2.15	2.12
15	3.07	2.70	2.49	2.36	2.27	2.21	2.16	2.12	2.09
16	3.05	2.67	2.46	2.33	2.24	2.18	2.13	2.09	2.06
17	3.03	2.64	2.44	2.31	2.22	2.15	2.10	2.06	2.03
18	3.01	2.62	2.42	2.29	2.20	2.13	2.08	2.04	2.00
19	2.99	2.61	2.40	2.27	2.18	2.11	2.06	2.02	1.98
20	2.97	2.59	2.38	2.25	2.16	2.09	2.04	2.00	1.96
21	2.96	2.57	2.36	2.23	2.14	2.08	2.02	1.98	1.95
22	2.95	2.56	2.35	2.22	2.13	2.06	2.01	1.97	1.93
23	2.94	2.55	2.34	2.21	2.11	2.05	1.99	1.95	1.92
24	2.93	2.54	2.33	2.19	2.10	2.04	1.98	1.94	1.91
25	2.92	2.53	2.32	2.18	2.09	2.02	1.97	1.93	1.89
26	2.91	2.52	2.31	2.17	2.08	2.01	1.96	1.92	1.88
27	2.90	2.51	2.30	2.17	2.07	2.00	1.95	1.91	1.87
28	2.89	2.50	2.29	2.16	2.06	2.00	1.94	1.90	1.87
29	2.89	2.50	2.28	2.15	2.06	1.99	1.93	1.89	1.86
30	2.88	2.49	2.28	2.14	2.05	1.98	1.93	1.88	1.85
40	2.84	2.44	2.23	2.09	2.00	1.93	1.87	1.83	1.79
60	2.79	2.39	2.18	2.04	1.95	1.87	1.82	1.77	1.74
120	2.75	2.35	2.13	1.99	1.90	1.82	1.77	1.72	1.68
∞	2.71	2.30	2.08	1.94	1.85	1.77	1.72	1.67	1.63

Source: From M. Merrington and C. M. Thompson, "Tables of percentage points of the inverted beta (*F*)-distribution," *Biometrika*, 1943, 33, 73–88. Reproduced by permission of the *Biometrika* Trustees.

ν_1	NUMERATOR DEGREES OF FREEDOM									
ν_2	10	12	15	20	24	30	40	60	120	∞
1	60.19	60.71	61.22	61.74	62.00	62.26	62.53	62.79	63.06	63.33
2	9.39	9.41	9.42	9.44	9.45	9.46	9.47	9.47	9.48	9.49
3	5.23	5.22	5.20	5.18	5.18	5.17	5.16	5.15	5.14	5.13
4	3.92	3.90	3.87	3.84	3.83	3.82	3.80	3.79	3.78	3.76
5	3.30	3.27	3.24	3.21	3.19	3.17	3.16	3.14	3.12	3.10
6	2.94	2.90	2.87	2.84	2.82	2.80	2.78	2.76	2.74	2.72
7	2.70	2.67	2.63	2.59	2.58	2.56	2.54	2.51	2.49	2.47
8	2.54	2.50	2.46	2.42	2.40	2.38	2.36	2.34	2.32	2.29
9	2.42	2.38	2.34	2.30	2.28	2.25	2.23	2.21	2.18	2.16
10	2.32	2.28	2.24	2.20	2.18	2.16	2.13	2.11	2.08	2.06
11	2.25	2.21	2.17	2.12	2.10	2.08	2.05	2.03	2.00	1.97
12	2.19	2.15	2.10	2.06	2.04	2.01	1.99	1.96	1.93	1.90
13	2.14	2.10	2.05	2.01	1.98	1.96	1.93	1.90	1.88	1.85
14	2.10	2.05	2.01	1.96	1.94	1.91	1.89	1.86	1.83	1.80
15	2.06	2.02	1.97	1.92	1.90	1.87	1.85	1.82	1.79	1.76
16	2.03	1.99	1.94	1.89	1.87	1.84	1.81	1.78	1.75	1.72
17	2.00	1.96	1.91	1.86	1.84	1.81	1.78	1.75	1.72	1.69
18	1.98	1.93	1.89	1.84	1.81	1.78	1.75	1.72	1.69	1.66
19	1.96	1.91	1.86	1.81	1.79	1.76	1.73	1.70	1.67	1.63
20	1.94	1.89	1.84	1.79	1.77	1.74	1.71	1.68	1.64	1.61
21	1.92	1.87	1.83	1.78	1.75	1.72	1.69	1.66	1.62	1.59
22	1.90	1.86	1.81	1.76	1.73	1.70	1.67	1.64	1.60	1.57
23	1.89	1.84	1.80	1.74	1.72	1.69	1.66	1.62	1.59	1.55
24	1.88	1.83	1.78	1.73	1.70	1.67	1.64	1.61	1.57	1.53
25	1.87	1.82	1.77	1.72	1.69	1.66	1.63	1.59	1.56	1.52
26	1.86	1.81	1.76	1.71	1.68	1.65	1.61	1.58	1.54	1.50
27	1.85	1.80	1.75	1.70	1.67	1.64	1.60	1.57	1.53	1.49
28	1.84	1.79	1.74	1.69	1.66	1.63	1.59	1.56	1.52	1.48
29	1.83	1.78	1.73	1.68	1.65	1.62	1.58	1.55	1.51	1.47
30	1.82	1.77	1.72	1.67	1.64	1.61	1.57	1.54	1.50	1.46
40	1.76	1.71	1.66	1.61	1.57	1.54	1.51	1.47	1.42	1.38
60	1.71	1.66	1.60	1.54	1.51	1.48	1.44	1.40	1.35	1.29
120	1.65	1.60	1.55	1.48	1.45	1.41	1.37	1.32	1.26	1.19
∞	1.60	1.55	1.49	1.42	1.38	1.34	1.30	1.24	1.17	1.00

DENOMINATOR DEGREES OF FREEDOM

TABLE 4   Critical Values for the *F*-Statistic: $F_{.05}$

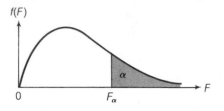

ν_1	NUMERATOR DEGREES OF FREEDOM								
ν_2	1	2	3	4	5	6	7	8	9
1	161.4	199.5	215.7	224.6	230.2	234.0	236.8	238.9	240.5
2	18.51	19.00	19.16	19.25	19.30	19.33	19.35	19.37	19.38
3	10.13	9.55	9.28	9.12	9.01	8.94	8.89	8.85	8.81
4	7.71	6.94	6.59	6.39	6.26	6.16	6.09	6.04	6.00
5	6.61	5.79	5.41	5.19	5.05	4.95	4.88	4.82	4.77
6	5.99	5.14	4.76	4.53	4.39	4.28	4.21	4.15	4.10
7	5.59	4.74	4.35	4.12	3.97	3.87	3.79	3.73	3.68
8	5.32	4.46	4.07	3.84	3.69	3.58	3.50	3.44	3.39
9	5.12	4.26	3.86	3.63	3.48	3.37	3.29	3.23	3.18
10	4.96	4.10	3.71	3.48	3.33	3.22	3.14	3.07	3.02
11	4.84	3.98	3.59	3.36	3.20	3.09	3.01	2.95	2.90
12	4.75	3.89	3.49	3.26	3.11	3.00	2.91	2.85	2.80
13	4.67	3.81	3.41	3.18	3.03	2.92	2.83	2.77	2.71
14	4.60	3.74	3.34	3.11	2.96	2.85	2.76	2.70	2.65
15	4.54	3.68	3.29	3.06	2.90	2.79	2.71	2.64	2.59
16	4.49	3.63	3.24	3.01	2.85	2.74	2.66	2.59	2.54
17	4.45	3.59	3.20	2.96	2.81	2.70	2.61	2.55	2.49
18	4.41	3.55	3.16	2.93	2.77	2.66	2.58	2.51	2.46
19	4.38	3.52	3.13	2.90	2.74	2.63	2.54	2.48	2.42
20	4.35	3.49	3.10	2.87	2.71	2.60	2.51	2.45	2.39
21	4.32	3.47	3.07	2.84	2.68	2.57	2.49	2.42	2.37
22	4.30	3.44	3.05	2.82	2.66	2.55	2.46	2.40	2.34
23	4.28	3.42	3.03	2.80	2.64	2.53	2.44	2.37	2.32
24	4.26	3.40	3.01	2.78	2.62	2.51	2.42	2.36	2.30
25	4.24	3.39	2.99	2.76	2.60	2.49	2.40	2.34	2.28
26	4.23	3.37	2.98	2.74	2.59	2.47	2.39	2.32	2.27
27	4.21	3.35	2.96	2.73	2.57	2.46	2.37	2.31	2.25
28	4.20	3.34	2.95	2.71	2.56	2.45	2.36	2.29	2.24
29	4.18	3.33	2.93	2.70	2.55	2.43	2.35	2.28	2.22
30	4.17	3.32	2.92	2.69	2.53	2.42	2.33	2.27	2.21
40	4.08	3.23	2.84	2.61	2.45	2.34	2.25	2.18	2.12
60	4.00	3.15	2.76	2.53	2.37	2.25	2.17	2.10	2.04
120	3.92	3.07	2.68	2.45	2.29	2.17	2.09	2.02	1.96
∞	3.84	3.00	2.60	2.37	2.21	2.10	2.01	1.94	1.88

DENOMINATOR DEGREES OF FREEDOM

Source: From M. Merrington and C. M. Thompson, "Tables of percentage points of the inverted beta (*F*)-distribution," *Biometrika*, 1943, 33, 73–88. Reproduced by permission of the *Biometrika* Trustees.

	ν_1	NUMERATOR DEGREES OF FREEDOM									
	ν_2	10	12	15	20	24	30	40	60	120	∞
	1	241.9	243.9	245.9	248.0	249.1	250.1	251.1	252.2	253.3	254.3
	2	19.40	19.41	19.43	19.45	19.45	19.46	19.47	19.48	19.49	19.50
	3	8.79	8.74	8.70	8.66	8.64	8.62	8.59	8.57	8.55	8.53
	4	5.96	5.91	5.86	5.80	5.77	5.75	5.72	5.69	5.66	5.63
	5	4.74	4.68	4.62	4.56	4.53	4.50	4.46	4.43	4.40	4.36
	6	4.06	4.00	3.94	3.87	3.84	3.81	3.77	3.74	3.70	3.67
	7	3.64	3.57	3.51	3.44	3.41	3.38	3.34	3.30	3.27	3.23
	8	3.35	3.28	3.22	3.15	3.12	3.08	3.04	3.01	2.97	2.93
	9	3.14	3.07	3.01	2.94	2.90	2.86	2.83	2.79	2.75	2.71
DENOMINATOR DEGREES OF FREEDOM	10	2.98	2.91	2.85	2.77	2.74	2.70	2.66	2.62	2.58	2.54
	11	2.85	2.79	2.72	2.65	2.61	2.57	2.53	2.49	2.45	2.40
	12	2.75	2.69	2.62	2.54	2.51	2.47	2.43	2.38	2.34	2.30
	13	2.67	2.60	2.53	2.46	2.42	2.38	2.34	2.30	2.25	2.21
	14	2.60	2.53	2.46	2.39	2.35	2.31	2.27	2.22	2.18	2.13
	15	2.54	2.48	2.40	2.33	2.29	2.25	2.20	2.16	2.11	2.07
	16	2.49	2.42	2.35	2.28	2.24	2.19	2.15	2.11	2.06	2.01
	17	2.45	2.38	2.31	2.23	2.19	2.15	2.10	2.06	2.01	1.96
	18	2.41	2.34	2.27	2.19	2.15	2.11	2.06	2.02	1.97	1.92
	19	2.38	2.31	2.23	2.16	2.11	2.07	2.03	1.98	1.93	1.88
	20	2.35	2.28	2.20	2.12	2.08	2.04	1.99	1.95	1.90	1.84
	21	2.32	2.25	2.18	2.10	2.05	2.01	1.96	1.92	1.87	1.81
	22	2.30	2.23	2.15	2.07	2.03	1.98	1.94	1.89	1.84	1.78
	23	2.27	2.20	2.13	2.05	2.01	1.96	1.91	1.86	1.81	1.76
	24	2.25	2.18	2.11	2.03	1.98	1.94	1.89	1.84	1.79	1.73
	25	2.24	2.16	2.09	2.01	1.96	1.92	1.87	1.82	1.77	1.71
	26	2.22	2.15	2.07	1.99	1.95	1.90	1.85	1.80	1.75	1.69
	27	2.20	2.13	2.06	1.97	1.93	1.88	1.84	1.79	1.73	1.67
	28	2.19	2.12	2.04	1.96	1.91	1.87	1.82	1.77	1.71	1.65
	29	2.18	2.10	2.03	1.94	1.90	1.85	1.81	1.75	1.70	1.64
	30	2.16	2.09	2.01	1.93	1.89	1.84	1.79	1.74	1.68	1.62
	40	2.08	2.00	1.92	1.84	1.79	1.74	1.69	1.64	1.58	1.51
	60	1.99	1.92	1.84	1.75	1.70	1.65	1.59	1.53	1.47	1.39
	120	1.91	1.83	1.75	1.66	1.61	1.55	1.50	1.43	1.35	1.25
	∞	1.83	1.75	1.67	1.57	1.52	1.46	1.39	1.32	1.22	1.00

TABLE 5 Critical Values for the F-Statistic: $F_{.025}$

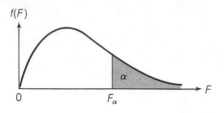

ν_2 \ ν_1	NUMERATOR DEGREES OF FREEDOM								
	1	2	3	4	5	6	7	8	9
1	647.8	799.5	864.2	899.6	921.8	937.1	948.2	956.7	963.3
2	38.51	39.00	39.17	39.25	39.30	39.33	39.36	39.37	39.39
3	17.44	16.04	15.44	15.10	14.88	14.73	14.62	14.54	14.47
4	12.22	10.65	9.98	9.60	9.36	9.20	9.07	8.98	8.90
5	10.01	8.43	7.76	7.39	7.15	6.98	6.85	6.76	6.68
6	8.81	7.26	6.60	6.23	5.99	5.82	5.70	5.60	5.52
7	8.07	6.54	5.89	5.52	5.29	5.12	4.99	4.90	4.82
8	7.57	6.06	5.42	5.05	4.82	4.65	4.53	4.43	4.36
9	7.21	5.71	5.08	4.72	4.48	4.32	4.20	4.10	4.03
10	6.94	5.46	4.83	4.47	4.24	4.07	3.95	3.85	3.78
11	6.72	5.26	4.63	4.28	4.04	3.88	3.76	3.66	3.59
12	6.55	5.10	4.47	4.12	3.89	3.73	3.61	3.51	3.44
13	6.41	4.97	4.35	4.00	3.77	3.60	3.48	3.39	3.31
14	6.30	4.86	4.24	3.89	3.66	3.50	3.38	3.29	3.21
15	6.20	4.77	4.15	3.80	3.58	3.41	3.29	3.20	3.12
16	6.12	4.69	4.08	3.73	3.50	3.34	3.22	3.12	3.05
17	6.04	4.62	4.01	3.66	3.44	3.28	3.16	3.06	2.98
18	5.98	4.56	3.95	3.61	3.38	3.22	3.10	3.01	2.93
19	5.92	4.51	3.90	3.56	3.33	3.17	3.05	2.96	2.88
20	5.87	4.46	3.86	3.51	3.29	3.13	3.01	2.91	2.84
21	5.83	4.42	3.82	3.48	3.25	3.09	2.97	2.87	2.80
22	5.79	4.38	3.78	3.44	3.22	3.05	2.93	2.84	2.76
23	5.75	4.35	3.75	3.41	3.18	3.02	2.90	2.81	2.73
24	5.72	4.32	3.72	3.38	3.15	2.99	2.87	2.78	2.70
25	5.69	4.29	3.69	3.35	3.13	2.97	2.85	2.75	2.68
26	5.66	4.27	3.67	3.33	3.10	2.94	2.82	2.73	2.65
27	5.63	4.24	3.65	3.31	3.08	2.92	2.80	2.71	2.63
28	5.61	4.22	3.63	3.29	3.06	2.90	2.78	2.69	2.61
29	5.59	4.20	3.61	3.27	3.04	2.88	2.76	2.67	2.59
30	5.57	4.18	3.59	3.25	3.03	2.87	2.75	2.65	2.57
40	5.42	4.05	3.46	3.13	2.90	2.74	2.62	2.53	2.45
60	5.29	3.93	3.34	3.01	2.79	2.63	2.51	2.41	2.33
120	5.15	3.80	3.23	2.89	2.67	2.52	2.39	2.30	2.22
∞	5.02	3.69	3.12	2.79	2.57	2.41	2.29	2.19	2.11

DENOMINATOR DEGREES OF FREEDOM

Source: From M. Merrington and C. M. Thompson, "Tables of percentage points of the inverted beta (F)-distribution," *Biometrika*, 1943, 33, 73–88. Reproduced by permission of the *Biometrika* Trustees.

ν_1 ν_2	NUMERATOR DEGREES OF FREEDOM									
	10	12	15	20	24	30	40	60	120	∞
1	968.6	976.7	984.9	993.1	997.2	1001	1006	1010	1014	1018
2	39.40	39.41	39.43	39.45	39.46	39.46	39.47	39.48	39.49	39.50
3	14.42	14.34	14.25	14.17	14.12	14.08	14.04	13.99	13.95	13.90
4	8.84	8.75	8.66	8.56	8.51	8.46	8.41	8.36	8.31	8.26
5	6.62	6.52	6.43	6.33	6.28	6.23	6.18	6.12	6.07	6.02
6	5.46	5.37	5.27	5.17	5.12	5.07	5.01	4.96	4.90	4.85
7	4.76	4.67	4.57	4.47	4.42	4.36	4.31	4.25	4.20	4.14
8	4.30	4.20	4.10	4.00	3.95	3.89	3.84	3.78	3.73	3.67
9	3.96	3.87	3.77	3.67	3.61	3.56	3.51	3.45	3.39	3.33
10	3.72	3.62	3.52	3.42	3.37	3.31	3.26	3.20	3.14	3.08
11	3.53	3.43	3.33	3.23	3.17	3.12	3.06	3.00	2.94	2.88
12	3.37	3.28	3.18	3.07	3.02	2.96	2.91	2.85	2.79	2.72
13	3.25	3.15	3.05	2.95	2.89	2.84	2.78	2.72	2.66	2.60
14	3.15	3.05	2.95	2.84	2.79	2.73	2.67	2.61	2.55	2.49
15	3.06	2.96	2.86	2.76	2.70	2.64	2.59	2.52	2.46	2.40
16	2.99	2.89	2.79	2.68	2.63	2.57	2.51	2.45	2.38	2.32
17	2.92	2.82	2.72	2.62	2.56	2.50	2.44	2.38	2.32	2.25
18	2.87	2.77	2.67	2.56	2.50	2.44	2.38	2.32	2.26	2.19
19	2.82	2.72	2.62	2.51	2.45	2.39	2.33	2.27	2.20	2.13
20	2.77	2.68	2.57	2.46	2.41	2.35	2.29	2.22	2.16	2.09
21	2.73	2.64	2.53	2.42	2.37	2.31	2.25	2.18	2.11	2.04
22	2.70	2.60	2.50	2.39	2.33	2.27	2.21	2.14	2.08	2.00
23	2.67	2.57	2.47	2.36	2.30	2.24	2.18	2.11	2.04	1.97
24	2.64	2.54	2.44	2.33	2.27	2.21	2.15	2.08	2.01	1.94
25	2.61	2.51	2.41	2.30	2.24	2.18	2.12	2.05	1.98	1.91
26	2.59	2.49	2.39	2.28	2.22	2.16	2.09	2.03	1.95	1.88
27	2.57	2.47	2.36	2.25	2.19	2.13	2.07	2.00	1.93	1.85
28	2.55	2.45	2.34	2.23	2.17	2.11	2.05	1.98	1.91	1.83
29	2.53	2.43	2.32	2.21	2.15	2.09	2.03	1.96	1.89	1.81
30	2.51	2.41	2.31	2.20	2.14	2.07	2.01	1.94	1.87	1.79
40	2.39	2.29	2.18	2.07	2.01	1.94	1.88	1.80	1.72	1.64
60	2.27	2.17	2.06	1.94	1.88	1.82	1.74	1.67	1.58	1.48
120	2.16	2.05	1.94	1.82	1.76	1.69	1.61	1.53	1.43	1.31
∞	2.05	1.94	1.83	1.71	1.64	1.57	1.48	1.39	1.27	1.00

DENOMINATOR DEGREES OF FREEDOM

TABLE 6 Critical Values for the F-Statistic: $F_{.01}$

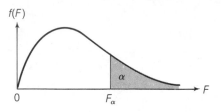

ν_2	NUMERATOR DEGREES OF FREEDOM								
ν_1	1	2	3	4	5	6	7	8	9
1	4,052	4,999.5	5,403	5,625	5,764	5,859	5,928	5,982	6,022
2	98.50	99.00	99.17	99.25	99.30	99.33	99.36	99.37	99.39
3	34.12	30.82	29.46	28.71	28.24	27.91	27.67	27.49	27.35
4	21.20	18.00	16.69	15.98	15.52	15.21	14.98	14.80	14.66
5	16.26	13.27	12.06	11.39	10.97	10.67	10.46	10.29	10.16
6	13.75	10.92	9.78	9.15	8.75	8.47	8.26	8.10	7.98
7	12.25	9.55	8.45	7.85	7.46	7.19	6.99	6.84	6.72
8	11.26	8.65	7.59	7.01	6.63	6.37	6.18	6.03	5.91
9	10.56	8.02	6.99	6.42	6.06	5.80	5.61	5.47	5.35
10	10.04	7.56	6.55	5.99	5.64	5.39	5.20	5.06	4.94
11	9.65	7.21	6.22	5.67	5.32	5.07	4.89	4.74	4.63
12	9.33	6.93	5.95	5.41	5.06	4.82	4.64	4.50	4.39
13	9.07	6.70	5.74	5.21	4.86	4.62	4.44	4.30	4.19
14	8.86	6.51	5.56	5.04	4.69	4.46	4.28	4.14	4.03
15	8.68	6.36	5.42	4.89	4.56	4.32	4.14	4.00	3.89
16	8.53	6.23	5.29	4.77	4.44	4.20	4.03	3.89	3.78
17	8.40	6.11	5.18	4.67	4.34	4.10	3.93	3.79	3.68
18	8.29	6.01	5.09	4.58	4.25	4.01	3.84	3.71	3.60
19	8.18	5.93	5.01	4.50	4.17	3.94	3.77	3.63	3.52
20	8.10	5.85	4.94	4.43	4.10	3.87	3.70	3.56	3.46
21	8.02	5.78	4.87	4.37	4.04	3.81	3.64	3.51	3.40
22	7.95	5.72	4.82	4.31	3.99	3.76	3.59	3.45	3.35
23	7.88	5.66	4.76	4.26	3.94	3.71	3.54	3.41	3.30
24	7.82	5.61	4.72	4.22	3.90	3.67	3.50	3.36	3.26
25	7.77	5.57	4.68	4.18	3.85	3.63	3.46	3.32	3.22
26	7.72	5.53	4.64	4.14	3.82	3.59	3.42	3.29	3.18
27	7.68	5.49	4.60	4.11	3.78	3.56	3.39	3.26	3.15
28	7.64	5.45	4.57	4.07	3.75	3.53	3.36	3.23	3.12
29	7.60	5.42	4.54	4.04	3.73	3.50	3.33	3.20	3.09
30	7.56	5.39	4.51	4.02	3.70	3.47	3.30	3.17	3.07
40	7.31	5.18	4.31	3.83	3.51	3.29	3.12	2.99	2.89
60	7.08	4.98	4.13	3.65	3.34	3.12	2.95	2.82	2.72
120	6.85	4.79	3.95	3.48	3.17	2.96	2.79	2.66	2.56
∞	6.63	4.61	3.78	3.32	3.02	2.80	2.64	2.51	2.41

Source: From M. Merrington and C. M. Thompson, "Tables of percentage points of the inverted beta (F)-distribution," *Biometrika*, 1943, *33*, 73–88. Reproduced by permission of the *Biometrika* Trustees.

ν_2 \ ν_1	NUMERATOR DEGREES OF FREEDOM									
	10	12	15	20	24	30	40	60	120	∞
1	6,056	6,106	6,157	6,209	6,235	6,261	6,287	6,313	6,339	6,366
2	99.40	99.42	99.43	99.45	99.46	99.47	99.47	99.48	99.49	99.50
3	27.23	27.05	26.87	26.69	26.60	26.50	26.41	26.32	26.22	26.13
4	14.55	14.37	14.20	14.02	13.93	13.84	13.75	13.65	13.56	13.46
5	10.05	9.89	9.72	9.55	9.47	9.38	9.29	9.20	9.11	9.02
6	7.87	7.72	7.56	7.40	7.31	7.23	7.14	7.06	6.97	6.88
7	6.62	6.47	6.31	6.16	6.07	5.99	5.91	5.82	5.74	5.65
8	5.81	5.67	5.52	5.36	5.28	5.20	5.12	5.03	4.95	4.86
9	5.26	5.11	4.96	4.81	4.73	4.65	4.57	4.48	4.40	4.31
10	4.85	4.71	4.56	4.41	4.33	4.25	4.17	4.08	4.00	3.91
11	4.54	4.40	4.25	4.10	4.02	3.94	3.86	3.78	3.69	3.60
12	4.30	4.16	4.01	3.86	3.78	3.70	3.62	3.54	3.45	3.36
13	4.10	3.96	3.82	3.66	3.59	3.51	3.43	3.34	3.25	3.17
14	3.94	3.80	3.66	3.51	3.43	3.35	3.27	3.18	3.09	3.00
15	3.80	3.67	3.52	3.37	3.29	3.21	3.13	3.05	2.96	2.87
16	3.69	3.55	3.41	3.26	3.18	3.10	3.02	2.93	2.84	2.75
17	3.59	3.46	3.31	3.16	3.08	3.00	2.92	2.83	2.75	2.65
18	3.51	3.37	3.23	3.08	3.00	2.92	2.84	2.75	2.66	2.57
19	3.43	3.30	3.15	3.00	2.92	2.84	2.76	2.67	2.58	2.49
20	3.37	3.23	3.09	2.94	2.86	2.78	2.69	2.61	2.52	2.42
21	3.31	3.17	3.03	2.88	2.80	2.72	2.64	2.55	2.46	2.36
22	3.26	3.12	2.98	2.83	2.75	2.67	2.58	2.50	2.40	2.31
23	3.21	3.07	2.93	2.78	2.70	2.62	2.54	2.45	2.35	2.26
24	3.17	3.03	2.89	2.74	2.66	2.58	2.49	2.40	2.31	2.21
25	3.13	2.99	2.85	2.70	2.62	2.54	2.45	2.36	2.27	2.17
26	3.09	2.96	2.81	2.66	2.58	2.50	2.42	2.33	2.23	2.13
27	3.06	2.93	2.78	2.63	2.55	2.47	2.38	2.29	2.20	2.10
28	3.03	2.90	2.75	2.60	2.52	2.44	2.35	2.26	2.17	2.06
29	3.00	2.87	2.73	2.57	2.49	2.41	2.33	2.23	2.14	2.03
30	2.98	2.84	2.70	2.55	2.47	2.39	2.30	2.21	2.11	2.01
40	2.80	2.66	2.52	2.37	2.29	2.20	2.11	2.02	1.92	1.80
60	2.63	2.50	2.35	2.20	2.12	2.03	1.94	1.84	1.73	1.60
120	2.47	2.34	2.19	2.03	1.95	1.86	1.76	1.66	1.53	1.38
∞	2.32	2.18	2.04	1.88	1.79	1.70	1.59	1.47	1.32	1.00

DENOMINATOR DEGREES OF FREEDOM

TABLE 7 Random Numbers

COLUMN ROW	1	2	3	4	5	6	7	8	9	10	11	12	13	14
1	10480	15011	01536	02011	81647	91646	69179	14194	62590	36207	20969	99570	91291	90700
2	22368	46573	25595	85393	30995	89198	27982	53402	93965	34095	52666	19174	39615	99505
3	24130	48360	22527	97265	76393	64809	15179	24830	49340	32081	30680	19655	63348	58629
4	42167	93093	06243	61680	07856	16376	39440	53537	71341	57004	00849	74917	97758	16379
5	37570	39975	81837	16656	06121	91782	60468	81305	49684	60672	14110	06927	01263	54613
6	77921	06907	11008	42751	27756	53498	18602	70659	90655	15053	21916	81825	44394	42880
7	99562	72905	56420	69994	98872	31016	71194	18738	44013	48840	63213	21069	10634	12952
8	96301	91977	05463	07972	18876	20922	94595	56869	69014	60045	18425	84903	42508	32307
9	89579	14342	63661	10281	17453	18103	57740	84378	25331	12566	58678	44947	05585	56941
10	85475	36857	53342	53988	53060	59533	38867	62300	08158	17983	16439	11458	18593	64952
11	28918	69578	88231	33276	70997	79936	56865	05859	90106	31595	01547	85590	91610	78188
12	63553	40961	48235	03427	49626	69445	18663	72695	52180	20847	12234	90511	33703	90322
13	09429	93969	52636	92737	88974	33488	36320	17617	30015	08272	84115	27156	30613	74952
14	10365	61129	87529	85689	48237	52267	67689	93394	01511	26358	85104	20285	29975	89868
15	07119	97336	71048	08178	77233	13916	47564	81056	97735	85977	29372	74461	28551	90707
16	51085	12765	51821	51259	77452	16308	60756	92144	49442	53900	70960	63990	75601	40719
17	02368	21382	52404	60268	89368	19885	55322	44819	01188	06646	64835	44919	05944	55157
18	01011	54092	33362	94904	31273	04146	18594	29852	71585	85030	51132	01915	92747	64951
19	52162	53916	46369	58586	23216	14513	83149	98736	23495	64350	94738	17752	35156	35749
20	07056	97628	33787	09998	42698	06691	76988	13602	51851	46104	88916	19509	25625	58104
21	48663	91245	85828	14346	09172	30168	90229	04734	59193	22178	30421	61666	99904	32812
22	54164	58492	22421	74103	47070	25306	76468	26384	58151	06646	21524	15227	96909	44592
23	32639	32363	05597	24200	13363	38005	94342	28728	35806	06912	17012	64161	18296	22851
24	29334	27001	87637	87308	58731	00256	45834	15398	46557	41135	10367	07684	36188	18510
25	02488	33062	28834	07351	19731	92420	60952	61280	50001	67658	32586	86679	50720	94953
26	81525	72295	04839	96423	24878	82651	66566	14778	76797	14780	13300	87074	79666	95725
27	29676	20591	68086	26432	46901	20849	89768	81536	86645	12659	92259	57102	80428	25280
28	00742	57392	39064	66432	84673	40027	32832	61362	98947	96067	64760	64584	96096	98253
29	05366	04213	25669	26422	44407	44048	37937	63904	45766	66134	75470	66520	34693	90449
30	91921	26418	64117	94305	26766	25940	39972	22209	71500	64568	91402	42416	07844	69618
31	00582	04711	87917	77341	42206	35126	74087	99547	81817	42607	43808	76655	62028	76630
32	00725	69884	62797	56170	86324	88072	76222	36086	84637	93161	76038	65855	77919	88006
33	69011	65795	95876	55293	18988	27354	26575	08625	40801	59920	29841	80150	12777	48501
34	25976	57948	29888	88604	67917	48708	18912	82271	65424	69774	33611	54262	85963	03547
35	09763	83473	73577	12908	30883	18317	28290	35797	05998	41688	34952	37888	38917	88050

36	91576	42595	27958	30134	04024	86385	29880	99730	55536	84855	29080	09250	79656	73211
37	17955	56349	90999	49127	20044	59931	06115	20542	18059	02008	73708	83517	36103	42791
38	46503	18584	18845	49618	023C4	51038	20655	58727	28168	15475	56942	53389	20562	87338
39	92157	89634	94824	78171	84610	82834	09922	25417	44137	48413	25555	21246	35509	20468
40	14577	62765	35605	81263	39667	47358	56873	56307	61607	49518	89656	20103	77490	18062
41	98427	07523	33362	64270	01638	92477	66969	98420	04880	45585	46565	04102	46880	45709
42	34914	63976	88720	82765	34476	17032	87589	40836	32427	70002	70663	88863	77775	69348
43	70060	28277	39475	46473	23219	53416	94970	25832	69975	94884	19661	72828	00102	66794
44	53976	54914	06990	67245	68350	82948	11398	42878	80287	88267	47363	46634	06541	97809
45	76072	29515	40980	07391	58745	25774	22987	80059	39911	96189	41151	14222	60697	59583
46	90725	52210	83974	29992	65831	38857	50490	83765	55657	14361	31720	57375	56228	41546
47	64364	67412	33339	31926	14883	24413	59744	92351	97473	89286	35931	04110	23726	51900
48	08962	00358	31662	25388	61642	34072	81249	35648	56891	69352	48373	45578	78547	81788
49	95012	68379	93526	70765	10592	04542	76463	54328	02349	17247	28865	14777	62730	92277
50	15664	10493	20492	38391	91132	21999	59516	81652	27195	48223	46751	22923	32261	85653
51	16408	81899	04153	53381	79401	21438	33035	92350	36693	31238	59649	91754	72772	02338
52	18629	81953	05520	91962	04739	13092	97662	24822	94730	06496	35090	04822	86774	98289
53	73115	35101	47498	87637	99016	71060	88824	71013	18735	20286	23153	72924	35165	43040
54	57491	16703	23167	49323	45021	33132	12544	41035	80780	45393	44812	12515	98931	91202
55	30405	83946	23792	14422	15059	45799	22716	19792	09983	74353	68668	30429	70735	25499
56	16631	35006	85900	98275	32388	52390	16815	69298	82732	38480	73817	32523	41961	44437
57	96773	20206	42559	78985	05300	22164	24369	54224	35083	19687	11052	91491	60383	19746
58	38935	64202	14349	82674	66523	44133	00697	35552	35970	19124	63318	29686	03387	59846
59	31624	76384	17403	53363	44167	64486	64758	75366	76554	31601	12614	33072	60332	92325
60	78919	19474	23632	27889	47914	02584	37680	20801	72152	39339	34806	08930	85001	87820
61	03931	33309	57047	74211	63445	17361	62825	39908	05607	91284	68833	25570	38818	46920
62	74426	33278	43972	10119	89917	15665	52872	73823	73144	88662	88970	74492	51805	99378
63	09066	00903	20795	95452	92648	45454	09552	88815	16553	51125	79375	97596	16296	66092
64	42238	12426	87025	14267	20979	04508	64535	31355	86064	29472	47689	05974	52468	16834
65	16153	08002	26504	41744	81959	65642	74240	56302	00033	67107	77510	70625	28725	34191
66	21457	40742	29820	96783	29400	21840	15035	34537	33310	06116	95240	15957	16572	06004
67	21581	57802	02050	89728	17937	37621	47075	42080	97403	48626	68995	43805	33386	21597
68	55612	78095	83197	33732	05310	24813	86902	60397	16489	03264	88525	42786	05269	92532
69	44657	66999	99324	51281	84463	60563	79312	93454	68876	25471	93911	25650	12682	73572

(continued)

TABLE 7 Continued

ROW \ COLUMN	1	2	3	4	5	6	7	8	9	10	11	12	13	14
70	91340	84979	46949	81973	37949	61023	43997	15263	80644	43942	89203	71795	99533	50501
71	91227	21199	31935	27022	84067	05462	35216	14486	29891	68607	41867	14951	91696	85065
72	50001	38140	66321	19924	72163	09538	12151	06878	91903	18749	34405	56087	82790	70925
73	65390	05224	72958	28609	81406	39147	25549	48542	42627	45233	57202	94617	23772	07896
74	27504	96131	83944	41575	10573	08619	64482	73923	36152	05184	94142	25299	84387	34925
75	37169	94851	39117	89632	00959	16487	65536	49071	39782	17095	02330	74301	00275	48280
76	11508	70225	51111	38351	19444	66499	71945	05422	13442	78675	84081	66938	93654	59894
77	37449	30362	06694	54690	04052	53115	62757	95348	78662	11163	81651	50245	34971	52924
78	46515	70331	85922	38329	57015	15765	97161	17869	45349	61796	66345	81073	49106	79860
79	30986	81223	42416	58353	21532	30502	32305	86482	05174	07901	54339	58861	74818	46942
80	63798	64995	46583	09785	44160	78128	83991	42865	92520	83531	80377	35909	81250	54238
81	82486	84846	99254	67632	43218	50076	21361	64816	51202	88124	41870	52689	51275	83556
82	21885	32906	92431	09060	64297	51674	64126	62570	26123	05155	59194	52799	28225	85762
83	60336	98782	07408	53458	13564	59089	26445	29789	85205	41001	12535	12133	14645	23541
84	43937	46891	24010	25560	86355	33941	25786	54990	71899	15475	95434	98227	21824	19585
85	97656	63175	89303	16275	07100	92063	21942	18611	47348	20203	18534	03862	78095	50136
86	03299	01221	05418	38982	55758	92237	26759	86367	21216	98442	08303	56613	91511	75928
87	79626	06486	03574	17668	07785	76020	79924	25651	83325	88428	85076	72811	22717	50585
88	85636	68335	47539	03129	65651	11977	02510	26113	99447	68645	34327	15152	55230	93448
89	18039	14367	61337	06177	12143	46609	32989	74014	64708	00533	35398	58408	13261	47908
90	08362	15656	60627	36478	65648	16764	53412	09013	07832	41574	17639	82163	60859	75567
91	79556	29068	04142	16268	15387	12856	66227	38358	22478	73373	88732	09443	82558	05250
92	92608	82674	27072	32534	17075	27698	98204	63863	11951	34648	88022	56148	34925	57031
93	23982	25835	40055	67006	12293	02753	14827	23235	35071	99704	37543	11601	35503	85171
94	09915	96306	05908	97901	28395	14186	00821	80703	70426	75647	76310	88717	37890	40129
95	59037	33300	26695	62247	69927	76123	50842	43834	86654	70959	79725	93872	28117	19233
96	42488	78077	69882	61657	34136	79180	97526	43092	04098	73571	80799	76536	71255	64239
97	46764	86273	63003	93017	31204	36692	40202	35275	57306	55543	53203	18098	47625	88684
98	03237	45430	55417	63282	90816	17349	88298	90183	36600	78406	06216	95787	42579	90730
99	86591	81482	52667	61582	14972	90053	89534	76036	49199	43716	97548	04379	46370	28672
100	38534	01715	94964	87288	65680	43772	39560	12918	86537	62738	19636	51132	25739	56947

Source: Abridged from W. H. Beyer (ed.). *CRC Standard Mathematical Tables*, 24th edition. (Cleveland: The Chemical Rubber Company), 1976.

TABLE 8

Critical Values of the
Sample Coefficient of
Correlation, r

SAMPLE SIZE n	$r_{.050}$	$r_{.025}$	$r_{.010}$	$r_{.005}$
3	.988	.9969	.99951	.99988
4	.900	.950	.980	.99000
5	.805	.878	.934	.959
6	.729	.811	.882	.917
7	.669	.754	.833	.875
8	.621	.707	.789	.834
9	.582	.666	.750	.798
10	.549	.632	.715	.765
11	.521	.602	.685	.735
12	.497	.576	.658	.708
13	.476	.553	.634	.684
14	.457	.532	.612	.661
15	.441	.514	.592	.641
16	.426	.497	.574	.623
17	.412	.482	.558	.606
18	.400	.468	.543	.590
19	.389	.456	.529	.575
20	.378	.444	.516	.561
21	.369	.433	.503	.549
22	.360	.423	.492	.537
27	.323	.381	.445	.487
32	.296	.349	.409	.449
37	.275	.325	.381	.418
42	.257	.304	.358	.393
47	.243	.288	.338	.372
52	.231	.273	.322	.354
62	.211	.250	.295	.325
72	.195	.232	.274	.302
82	.183	.217	.257	.283
92	.173	.205	.242	.267
102	.164	.195	.230	.254

TABLE 9 Critical Values for the Durbin–Watson d-Statistic ($\alpha = .05$)

n	k = 1		k = 2		k = 3		k = 4		k = 5	
	d_L	d_U	d_L	d_U	d_L	d_U	d_L	d_U	d_L	d_U
15	1.08	1.36	.95	1.54	.82	1.75	.69	1.97	.56	2.21
16	1.10	1.37	.98	1.54	.86	1.73	.74	1.93	.62	2.15
17	1.13	1.38	1.02	1.54	.90	1.71	.78	1.90	.67	2.10
18	1.16	1.39	1.05	1.53	.93	1.69	.82	1.87	.71	2.06
19	1.18	1.40	1.08	1.53	.97	1.68	.86	1.85	.75	2.02
20	1.20	1.41	1.10	1.54	1.00	1.68	.90	1.83	.79	1.99
21	1.22	1.42	1.13	1.54	1.03	1.67	.93	1.81	.83	1.96
22	1.24	1.43	1.15	1.54	1.05	1.66	.96	1.80	.86	1.94
23	1.26	1.44	1.17	1.54	1.08	1.66	.99	1.79	.90	1.92
24	1.27	1.45	1.19	1.55	1.10	1.66	1.01	1.78	.93	1.90
25	1.29	1.45	1.21	1.55	1.12	1.66	1.04	1.77	.95	1.89
26	1.30	1.46	1.22	1.55	1.14	1.65	1.06	1.76	.98	1.88
27	1.32	1.47	1.24	1.56	1.16	1.65	1.08	1.76	1.01	1.86
28	1.33	1.48	1.26	1.56	1.18	1.65	1.10	1.75	1.03	1.85
29	1.34	1.48	1.27	1.56	1.20	1.65	1.12	1.74.	1.05	1.84
30	1.35	1.49	1.28	1.57	1.21	1.65	1.14	1.74	1.07	1.83
31	1.36	1.50	1.30	1.57	1.23	1.65	1.16	1.74	1.09	1.83
32	1.37	1.50	1.31	1.57	1.24	1.65	1.18	1.73	1.11	1.82
33	1.38	1.51	1.32	1.58	1.26	1.65	1.19	1.73	1.13	1.81
34	1.39	1.51	1.33	1.58	1.27	1.65	1.21	1.73	1.15	1.81
35	1.40	1.52	1.34	1.58	1.28	1.65	1.22	1.73	1.16	1.80
36	1.41	1.52	1.35	1.59	1.29	1.65	1.24	1.73	1.18	1.80
37	1.42	1.53	1.36	1.59	1.31	1.66	1.25	1.72	1.19	1.80
38	1.43	1.54	1.37	1.59	1.32	1.66	1.26	1.72	1.21	1.79
39	1.43	1.54	1.38	1.60	1.33	1.66	1.27	1.72	1.22	1.79
40	1.44	1.54	1.39	1.60	1.34	1.66	1.29	1.72	1.23	1.79
45	1.48	1.57	1.43	1.62	1.38	1.67	1.34	1.72	1.29	1.78
50	1.50	1.59	1.46	1.63	1.42	1.67	1.38	1.72	1.34	1.77
55	1.53	1.60	1.49	1.64	1.45	1.68	1.41	1.72	1.38	1.77
60	1.55	1.62	1.51	1.65	1.48	1.69	1.44	1.73	1.41	1.77
65	1.57	1.63	1.54	1.66	1.50	1.70	1.47	1.73	1.44	1.77
70	1.58	1.64	1.55	1.67	1.52	1.70	1.49	1.74	1.46	1.77
75	1.60	1.65	1.57	1.68	1.54	1.71	1.51	1.74	1.49	1.77
80	1.61	1.66	1.59	1.69	1.56	1.72	1.53	1.74	1.51	1.77
85	1.62	1.67	1.60	1.70	1.57	1.72	1.55	1.75	1.52	1.77
90	1.63	1.68	1.61	1.70	1.59	1.73	1.57	1.75	1.54	1.78
95	1.64	1.69	1.62	1.71	1.60	1.73	1.58	1.75	1.56	1.78
100	1.65	1.69	1.63	1.72	1.61	1.74	1.59	1.76	1.57	1.78

Source: From J. Durbin and G. S. Watson, "Testing for serial correlation in least squares regression, II," *Biometrika*, 1951, *30*, 159–178. Reproduced by permission of the *Biometrika* Trustees.

TABLE 10   Critical Values for the Durbin–Watson d-Statistic ($\alpha = .01$)

n	k = 1 d_L	k = 1 d_U	k = 2 d_L	k = 2 d_U	k = 3 d_L	k = 3 d_U	k = 4 d_L	k = 4 d_U	k = 5 d_L	k = 5 d_U
15	.81	1.07	.70	1.25	.59	1.46	.49	1.70	.39	1.96
16	.84	1.09	.74	1.25	.63	1.44	.53	1.66	.44	1.90
17	.87	1.10	.77	1.25	.67	1.43	.57	1.63	.48	1.85
18	.90	1.12	.80	1.26	.71	1.42	.61	1.60	.52	1.80
19	.93	1.13	.83	1.26	.74	1.41	.65	1.58	.56	1.77
20	.95	1.15	.86	1.27	.77	1.41	.68	1.57	.60	1.74
21	.97	1.16	.89	1.27	.80	1.41	.72	1.55	.63	1.71
22	1.00	1.17	.91	1.28	.83	1.40	.75	1.54	.66	1.69
23	1.02	1.19	.94	1.29	.86	1.40	.77	1.53	.70	1.67
24	1.04	1.20	.96	1.30	.88	1.41	.80	1.53	.72	1.66
25	1.05	1.21	.98	1.30	.90	1.41	.83	1.52	.75	1.65
26	1.07	1.22	1.00	1.31	.93	1.41	.85	1.52	.78	1.64
27	1.09	1.23	1.02	1.32	.95	1.41	.88	1.51	.81	1.63
28	1.10	1.24	1.04	1.32	.97	1.41	.90	1.51	.83	1.62
29	1.12	1.25	1.05	1.33	.99	1.42	.92	1.51	.85	1.61
30	1.13	1.26	1.07	1.34	1.01	1.42	.94	1.51	.88	1.61
31	1.15	1.27	1.08	1.34	1.02	1.42	.96	1.51	.90	1.60
32	1.16	1.28	1.10	1.35	1.04	1.43	.98	1.51	.92	1.60
33	1.17	1.29	1.11	1.36	1.05	1.43	1.00	1.51	.94	1.59
34	1.18	1.30	1.13	1.36	1.07	1.43	1.01	1.51	.95	1.59
35	1.19	1.31	1.14	1.37	1.08	1.44	1.03	1.51	.97	1.59
36	1.21	1.32	1.15	1.38	1.10	1.44	1.04	1.51	.99	1.59
37	1.22	1.32	1.16	1.38	1.11	1.45	1.06	1.51	1.00	1.59
38	1.23	1.33	1.18	1.39	1.12	1.45	1.07	1.52	1.02	1.58
39	1.24	1.34	1.19	1.39	1.14	1.45	1.09	1.52	1.03	1.58
40	1.25	1.34	1.20	1.40	1.15	1.46	1.10	1.52	1.05	1.58
45	1.29	1.38	1.24	1.42	1.20	1.48	1.16	1.53	1.11	1.58
50	1.32	1.40	1.28	1.45	1.24	1.49	1.20	1.54	1.16	1.59
55	1.36	1.43	1.32	1.47	1.28	1.51	1.25	1.55	1.21	1.59
60	1.38	1.45	1.35	1.48	1.32	1.52	1.28	1.56	1.25	1.60
65	1.41	1.47	1.38	1.50	1.35	1.53	1.31	1.57	1.28	1.61
70	1.43	1.49	1.40	1.52	1.37	1.55	1.34	1.58	1.31	1.61
75	1.45	1.50	1.42	1.53	1.39	1.56	1.37	1.59	1.34	1.62
80	1.47	1.52	1.44	1.54	1.42	1.57	1.39	1.60	1.36	1.62
85	1.48	1.53	1.46	1.55	1.43	1.58	1.41	1.60	1.39	1.63
90	1.50	1.54	1.47	1.56	1.45	1.59	1.43	1.61	1.41	1.64
95	1.51	1.55	1.49	1.57	1.47	1.60	1.45	1.62	1.42	1.64
100	1.52	1.56	1.50	1.58	1.48	1.60	1.46	1.63	1.44	1.65

Source: From J. Durbin and G. S. Watson, "Testing for serial correlation in least squares regression, II," *Biometrika*, 1951, 30, 159–178. Reproduced by permission of the *Biometrika* Trustees.

TABLE 11   Percentage Points of the Studentized Range $q(p, \nu)$, Upper 5%

ν \ p	2	3	4	5	6	7	8	9	10	11
1	17.97	26.98	32.82	37.08	40.41	43.12	45.40	47.36	49.07	50.59
2	6.08	8.33	9.80	10.88	11.74	12.44	13.03	13.54	13.99	14.39
3	4.50	5.91	6.82	7.50	8.04	8.48	8.85	9.18	9.46	9.72
4	3.93	5.04	5.76	6.29	6.71	7.05	7.35	7.60	7.83	8.03
5	3.64	4.60	5.22	5.67	6.03	6.33	6.58	6.80	6.99	7.17
6	3.46	4.34	4.90	5.30	5.63	5.90	6.12	6.32	6.49	6.65
7	3.34	4.16	4.68	5.06	5.36	5.61	5.82	6.00	6.16	6.30
8	3.26	4.04	4.53	4.89	5.17	5.40	5.60	5.77	5.92	6.05
9	3.20	3.95	4.41	4.76	5.02	5.24	5.43	5.59	5.74	5.87
10	3.15	3.88	4.33	4.65	4.91	5.12	5.30	5.46	5.60	5.72
11	3.11	3.82	4.26	4.57	4.82	5.03	5.20	5.35	5.49	5.61
12	3.08	3.77	4.20	4.51	4.75	4.95	5.12	5.27	5.39	5.51
13	3.06	3.73	4.15	4.45	4.69	4.88	5.05	5.19	5.32	5.43
14	3.03	3.70	4.11	4.41	4.64	4.83	4.99	5.13	5.25	5.36
15	3.01	3.67	4.08	4.37	4.60	4.78	4.94	5.08	5.20	5.31
16	3.00	3.65	4.05	4.33	4.56	4.74	4.90	5.03	5.15	5.26
17	2.98	3.63	4.02	4.30	4.52	4.70	4.86	4.99	5.11	5.21
18	2.97	3.61	4.00	4.28	4.49	4.67	4.82	4.96	5.07	5.17
19	2.96	3.59	3.98	4.25	4.47	4.65	4.79	4.92	5.04	5.14
20	2.95	3.58	3.96	4.23	4.45	4.62	4.77	4.90	5.01	5.11
24	2.92	3.53	3.90	4.17	4.37	4.54	4.68	4.81	4.92	5.01
30	2.89	3.49	3.85	4.10	4.30	4.46	4.60	4.72	4.82	4.92
40	2.86	3.44	3.79	4.04	4.23	4.39	4.52	4.63	4.73	4.82
60	2.83	3.40	3.74	3.98	4.16	4.31	4.44	4.55	4.65	4.73
120	2.80	3.36	3.68	3.92	4.10	4.24	4.36	4.47	4.56	4.64
∞	2.77	3.31	3.63	3.86	4.03	4.17	4.29	4.39	4.47	4.55

ν \ p	12	13	14	15	16	17	18	19	20
1	51.96	53.20	54.33	55.36	56.32	57.22	58.04	58.83	59.56
2	14.75	15.08	15.38	15.65	15.91	16.14	16.37	16.57	16.77
3	9.95	10.15	10.35	10.52	10.69	10.84	10.98	11.11	11.24
4	8.21	8.37	8.52	8.66	8.79	8.91	9.03	9.13	9.23
5	7.32	7.47	7.60	7.72	7.83	7.93	8.03	8.12	8.21
6	6.79	6.92	7.03	7.14	7.24	7.34	7.43	7.51	7.59
7	6.43	6.55	6.66	6.76	6.85	6.94	7.02	7.10	7.17
8	6.18	6.29	6.39	6.48	6.57	6.65	6.73	6.80	6.87
9	5.98	6.09	6.19	6.28	6.36	6.44	6.51	6.58	6.64
10	5.83	5.93	6.03	6.11	6.19	6.27	6.34	6.40	6.47
11	5.71	5.81	5.90	5.98	6.06	6.13	6.20	6.27	6.33
12	5.61	5.71	5.80	5.88	5.95	6.02	6.09	6.15	6.21
13	5.53	5.63	5.71	5.79	5.86	5.93	5.99	6.05	6.11
14	5.46	5.55	5.64	5.71	5.79	5.85	5.91	5.97	6.03
15	5.40	5.49	5.57	5.65	5.72	5.78	5.85	5.90	5.96
16	5.35	5.44	5.52	5.59	5.66	5.73	5.79	5.84	5.90
17	5.31	5.39	5.47	5.54	5.61	5.67	5.73	5.79	5.84
18	5.27	5.35	5.43	5.50	5.57	5.63	5.69	5.74	5.79
19	5.23	5.31	5.39	5.46	5.53	5.59	5.65	5.70	5.75
20	5.20	5.28	5.36	5.43	5.49	5.55	5.61	5.66	5.71
24	5.10	5.18	5.25	5.32	5.38	5.44	5.49	5.55	5.59
30	5.00	5.08	5.15	5.21	5.27	5.33	5.38	5.43	5.47
40	4.90	4.98	5.04	5.11	5.16	5.22	5.27	5.31	5.36
60	4.81	4.88	4.94	5.00	5.06	5.11	5.15	5.20	5.24
120	4.71	4.78	4.84	4.90	4.95	5.00	5.04	5.09	5.13
∞	4.62	4.68	4.74	4.80	4.85	4.89	4.93	4.97	5.01

TABLE 12   Percentage Points of the Studentized Range $q(p, \nu)$, Upper 1%

ν \ p	2	3	4	5	6	7	8	9	10	11
1	90.03	135.0	164.3	185.6	202.2	215.8	227.2	237.0	245.6	253.2
2	14.04	19.02	22.29	24.72	26.63	28.20	29.53	30.68	31.69	32.59
3	8.26	10.62	12.17	13.33	14.24	15.00	15.64	16.20	16.69	17.13
4	6.51	8.12	9.17	9.96	10.58	11.10	11.55	11.93	12.27	12.57
5	5.70	6.98	7.80	8.42	8.91	9.32	9.67	9.97	10.24	10.48
6	5.24	6.33	7.03	7.56	7.97	8.32	8.61	8.87	9.10	9.30
7	4.95	5.92	6.54	7.01	7.37	7.68	7.94	8.17	8.37	8.55
8	4.75	5.64	6.20	6.62	6.96	7.24	7.47	7.68	7.86	8.03
9	4.60	5.43	5.96	6.35	6.66	6.91	7.13	7.33	7.49	7.65
10	4.48	5.27	5.77	6.14	6.43	6.67	6.87	7.05	7.21	7.36
11	4.39	5.15	5.62	5.97	6.25	6.48	6.67	6.84	6.99	7.13
12	4.32	5.05	5.50	5.84	6.10	6.32	6.51	6.67	6.81	6.94
13	4.26	4.96	5.40	5.73	5.98	6.19	6.37	6.53	6.67	6.79
14	4.21	4.89	5.32	5.63	5.88	6.08	6.26	6.41	6.54	6.66
15	4.17	4.84	5.25	5.56	5.80	5.99	6.16	6.31	6.44	6.55
16	4.13	4.79	5.19	5.49	5.72	5.92	6.08	6.22	6.35	6.46
17	4.10	4.74	5.14	5.43	5.66	5.85	6.01	6.15	6.27	6.38
18	4.07	4.70	5.09	5.38	5.60	5.79	5.94	6.08	6.20	6.31
19	4.05	4.67	5.05	5.33	5.55	5.73	5.89	6.02	6.14	6.25
20	4.02	4.64	5.02	5.29	5.51	5.69	5.84	5.97	6.09	6.19
24	3.96	4.55	4.91	5.17	5.37	5.54	5.69	5.81	5.92	6.02
30	3.89	4.45	4.80	5.05	5.24	5.40	5.54	5.65	5.76	5.85
40	3.82	4.37	4.70	4.93	5.11	5.26	5.39	5.50	5.60	5.69
60	3.76	4.28	4.59	4.82	4.99	5.13	5.25	5.36	5.45	5.53
120	3.70	4.20	4.50	4.71	4.87	5.01	5.12	5.21	5.30	5.37
∞	3.64	4.12	4.40	4.60	4.76	4.88	4.99	5.08	5.16	5.23

p ν	12	13	14	15	16	17	18	19	20
1	260.0	266.2	271.8	277.0	281.8	286.3	290.0	294.3	298.0
2	33.40	34.13	34.81	35.43	36.00	36.53	37.03	37.50	37.95
3	17.53	17.89	18.22	18.52	18.81	19.07	19.32	19.55	19.77
4	12.84	13.09	13.32	13.53	13.73	13.91	14.08	14.24	14.40
5	10.70	10.89	11.08	11.24	11.40	11.55	11.68	11.81	11.93
6	9.48	9.65	9.81	9.95	10.08	10.21	10.32	10.43	10.54
7	8.71	8.86	9.00	9.12	9.24	9.35	9.46	9.55	9.65
8	8.18	8.31	8.44	8.55	8.66	8.76	8.85	8.94	9.03
9	7.78	7.91	8.03	8.13	8.23	8.33	8.41	8.49	8.57
10	7.49	7.60	7.71	7.81	7.91	7.99	8.08	8.15	8.23
11	7.25	7.36	7.46	7.56	7.65	7.73	7.81	7.88	7.95
12	7.06	7.17	7.26	7.36	7.44	7.52	7.59	7.66	7.73
13	6.90	7.01	7.10	7.19	7.27	7.35	7.42	7.48	7.55
14	6.77	6.87	6.96	7.05	7.13	7.20	7.27	7.33	7.39
15	6.66	6.76	6.84	6.93	7.00	7.07	7.14	7.20	7.26
16	6.56	6.66	6.74	6.82	6.90	6.97	7.03	7.09	7.15
17	6.48	6.57	6.66	6.73	6.81	6.87	6.94	7.00	7.05
18	6.41	6.50	6.58	6.65	6.72	6.79	6.85	6.91	6.97
19	6.34	6.43	6.51	6.58	6.65	6.72	6.78	6.84	6.89
20	6.28	6.37	6.45	6.52	6.59	6.65	6.71	6.77	6.82
24	6.11	6.19	6.26	6.33	6.39	6.45	6.51	6.56	6.61
30	5.93	6.01	6.08	6.14	6.20	6.26	6.31	6.36	6.41
40	5.76	5.83	5.90	5.96	6.02	6.07	6.12	6.16	6.21
60	5.60	5.67	5.73	5.78	5.84	5.89	5.93	5.97	6.01
120	5.44	5.50	5.56	5.61	5.66	5.71	5.75	5.79	5.83
∞	5.29	5.35	5.40	5.45	5.49	5.54	5.57	5.61	5.65

APPENDIX E

DATA SET: REAL ESTATE APPRAISALS AND SALES DATA FOR SIX NEIGHBORHOODS IN A MID-SIZED FLORIDA CITY (SEE CASE STUDY 11)

1986 APPRAISED VALUES AND SALE PRICES FOR PROPERTIES IN SIX NEIGHBORHOODS

NEIGHBORHOOD A

OBS	LANDVAL	IMPROVAL	SALEPRIC	OBS	LANDVAL	IMPROVAL	SALEPRIC	OBS	LANDVAL	IMPROVAL	SALEPRIC
1	$20,000	$66,285	$140,000	2	$20,000	$56,755	$95,000	3	$20,000	$57,734	$95,000
4	$20,000	$47,424	$79,000	5	$30,000	$77,653	$134,000	6	$30,000	$95,031	$147,000
7	$30,000	$86,909	$165,300	8	$30,000	$74,627	$141,000	9	$25,000	$116,103	$140,000
10	$25,000	$101,358	$174,500	11	$82,080	$150,391	$400,000	12	$23,000	$51,388	$94,000
13	$23,000	$54,742	$103,000	14	$23,000	$53,766	$97,500	15	$23,000	$47,752	$83,000
16	$20,000	$42,209	$80,800	17	$23,000	$73,797	$107,500	18	$23,500	$64,005	$115,000
19	$23,500	$69,412	$113,100	20	$23,000	$52,288	$84,300	21	$17,000	$38,747	$59,900
22	$17,000	$39,009	$64,600	23	$17,000	$30,572	$57,000	24	$17,000	$37,785	$87,500
25	$32,000	$114,170	$210,000	26	$32,000	$104,532	$180,000	27	$32,000	$100,572	$200,000
28	$36,000	$94,722	$166,000	29	$32,000	$112,144	$190,000	30	$33,000	$106,110	$208,000
31	$33,000	$110,188	$200,000	32	$33,000	$90,822	$178,000	33	$49,000	$120,112	$287,500
34	$49,000	$123,417	$267,000	35	$15,000	$39,394	$71,000	36	$15,000	$43,243	$62,100
37	$22,000	$56,557	$95,000	38	$22,000	$52,435	$98,800	39	$22,000	$51,490	$84,900
40	$22,000	$50,310	$86,000	41	$22,000	$48,974	$79,900	42	$22,000	$46,067	$86,000
43	$22,000	$52,845	$84,000	44	$22,000	$63,047	$96,500	45	$22,000	$48,811	$98,000
46	$22,000	$49,355	$97,000	47	$23,000	$71,043	$112,000	48	$22,000	$47,299	$77,000
49	$20,000	$49,091	$89,900	50	$20,000	$60,960	$105,000	51	$17,000	$46,142	$89,000
52	$17,000	$61,339	$93,000	53	$17,000	$46,142	$87,500	54	$17,000	$46,142	$78,500
55	$23,000	$42,922	$72,000	56	$22,000	$61,832	$106,500	57	$22,000	$66,861	$107,500
58	$22,000	$54,131	$100,000	59	$22,000	$46,871	$89,900	60	$22,000	$48,798	$71,500
61	$22,000	$76,682	$111,000	62	$22,000	$57,800	$106,500	63	$22,000	$69,785	$105,000
64	$22,000	$46,397	$67,000	65	$22,000	$49,396	$79,900	66	$22,000	$67,375	$100,000
67	$23,000	$71,770	$110,000	68	$23,000	$50,093	$84,500	69	$23,000	$60,040	$90,000
70	$23,000	$51,662	$91,000	71	$20,000	$42,514	$74,000	72	$20,000	$32,816	$65,000
73	$32,000	$39,423	$79,900	74	$50,000	$31,238	$85,000	75	$24,000	$50,057	$112,000
76	$20,000	$44,817	$80,000	77	$20,000	$50,084	$117,000	78	$20,000	$39,224	$72,500
79	$45,000	$56,334	$102,000	80	$35,000	$61,138	$93,500	81	$38,000	$77,594	$140,000
82	$18,000	$53,859	$70,000	83	$20,000	$41,634	$75,000	84	$30,000	$69,341	$122,000
85	$24,000	$41,578	$102,000	86	$30,000	$68,669	$120,000	87	$23,000	$55,868	$86,000
88	$20,000	$58,510	$97,000	89	$20,000	$106,247	$145,000	90	$24,000	$41,185	$78,900

NEIGHBORHOOD B

OBS	LANDVAL	IMPROVAL	SALEPRIC	OBS	LANDVAL	IMPROVAL	SALEPRIC	OBS	LANDVAL	IMPROVAL	SALEPRIC
91	$37,100	$51,338	$172,500	92	$13,500	$57,576	$69,500	93	$13,500	$77,353	$102,000
94	$65,625	$152,076	$365,000	95	$12,000	$59,882	$87,500	96	$12,000	$42,191	$60,000
97	$15,000	$69,110	$95,000	98	$15,000	$61,589	$85,000	99	$15,000	$56,834	$82,000
100	$15,000	$55,853	$90,000	101	$15,000	$97,469	$145,000	102	$16,000	$61,498	$82,000
103	$15,000	$69,853	$98,500	104	$16,000	$86,506	$135,000	105	$15,000	$77,456	$111,000
106	$15,000	$55,687	$85,000	107	$65,520	$170,108	$296,000	108	$16,000	$61,275	$90,000
109	$16,000	$61,083	$90,000	110	$16,000	$65,507	$92,000	111	$15,000	$53,466	$96,500
112	$15,000	$63,025	$84,900	113	$16,000	$71,386	$109,000	114	$15,000	$47,815	$70,000
115	$15,000	$50,008	$94,500	116	$18,000	$87,698	$170,000	117	$18,000	$82,736	$120,000
118	$18,000	$60,301	$88,000	119	$18,000	$54,862	$87,500	120	$18,000	$68,243	$134,800
121	$18,000	$60,900	$95,700	122	$18,000	$78,418	$133,000	123	$18,000	$68,701	$126,000
124	$18,000	$66,156	$110,000	125	$18,000	$64,112	$88,000	126	$18,000	$74,522	$136,500
127	$18,000	$62,004	$125,000	128	$12,500	$44,454	$68,000	129	$12,000	$42,363	$60,000
130	$12,000	$41,277	$67,500	131	$12,000	$37,957	$61,500	132	$12,000	$44,868	$71,500
133	$12,000	$41,207	$67,000	134	$12,000	$47,793	$72,500	135	$12,000	$36,034	$59,000
136	$12,000	$43,243	$66,500	137	$12,000	$45,783	$71,900	138	$12,000	$43,670	$62,500
139	$12,500	$42,076	$65,900	140	$12,500	$39,488	$68,500	141	$12,500	$44,859	$64,400
142	$12,500	$37,223	$55,000	143	$8,000	$39,020	$57,500	144	$8,000	$44,028	$60,000
145	$8,000	$34,576	$54,000	146	$8,000	$48,935	$61,000	147	$8,000	$35,044	$44,200
148	$8,000	$37,193	$61,500	149	$8,000	$41,391	$54,000	150	$8,500	$41,331	$51,500
151	$11,500	$50,731	$68,000	152	$11,500	$43,963	$69,000	153	$11,500	$45,739	$68,500
154	$11,500	$36,243	$62,500	155	$11,500	$47,113	$74,000	156	$12,000	$34,897	$56,500
157	$11,500	$43,207	$67,000	158	$12,500	$39,939	$71,400	159	$11,500	$54,889	$79,500
160	$11,500	$50,162	$73,200	161	$25,000	$32,877	$67,500	162	$13,000	$69,815	$89,900
163	$13,000	$59,051	$77,000	164	$13,000	$62,580	$86,000	165	$13,000	$59,381	$83,500
166	$13,000	$61,426	$80,000	167	$13,000	$62,729	$85,000	168	$13,000	$65,068	$109,900
169	$13,000	$59,534	$90,000	170	$13,000	$52,946	$78,000				

1986 APPRAISED VALUES AND SALE PRICES FOR PROPERTIES IN SIX NEIGHBORHOODS

NEIGHBORHOOD   C

OBS	LANDVAL	IMPROVAL	SALEPRIC	OBS	LANDVAL	IMPROVAL	SALEPRIC	OBS	LANDVAL	IMPROVAL	SALEPRIC
								171	$12,000	$47,584	$65,000
172	$11,040	$44,081	$70,100	173	$10,920	$35,533	$51,500	174	$11,400	$52,649	$61,000
175	$12,000	$35,737	$57,500	176	$11,520	$40,228	$64,900	177	$11,520	$40,733	$59,900
178	$11,520	$45,824	$66,500	179	$11,400	$36,002	$56,900	180	$11,400	$42,900	$63,000
181	$11,400	$38,200	$59,900	182	$10,920	$39,120	$53,000	183	$12,000	$43,536	$69,000
184	$12,960	$41,933	$62,500	185	$12,960	$47,601	$63,500	186	$10,920	$45,656	$62,900
187	$11,400	$42,584	$66,500	188	$12,600	$42,657	$66,000	189	$11,040	$41,077	$57,000
190	$10,920	$39,083	$59,900	191	$11,040	$44,327	$72,000	192	$10,920	$47,906	$70,000
193	$10,920	$43,076	$69,500	194	$10,920	$44,833	$69,000	195	$10,920	$42,475	$66,500
196	$10,920	$40,765	$62,500	197	$11,040	$43,225	$65,700	198	$11,040	$45,657	$70,500
199	$11,040	$40,768	$65,800	200	$11,640	$49,400	$73,900	201	$11,040	$41,024	$68,500
202	$22,000	$46,135	$79,500	203	$22,000	$61,395	$114,800	204	$22,000	$53,047	$85,000
205	$22,000	$48,318	$81,000	206	$22,000	$48,264	$82,000	207	$22,000	$47,953	$86,000
208	$22,000	$51,775	$89,000	209	$22,000	$56,470	$103,000	210	$22,000	$57,097	$103,000
211	$22,000	$55,711	$99,400	212	$22,000	$40,675	$74,500	213	$23,000	$79,438	$128,200
214	$23,000	$97,182	$175,000	215	$23,000	$73,419	$110,000	216	$23,000	$72,326	$120,000
217	$12,500	$35,558	$72,000	218	$12,500	$47,773	$73,000	219	$12,500	$50,329	$68,900
220	$12,500	$35,709	$59,000	221	$12,500	$35,864	$51,500	222	$12,500	$43,135	$61,500
223	$12,500	$34,606	$59,500	224	$9,400	$41,966	$64,500	225	$9,400	$41,680	$64,000
226	$9,400	$36,176	$57,900	227	$9,400	$53,563	$73,800	228	$10,000	$53,212	$78,400
229	$9,400	$46,399	$70,900	230	$9,400	$39,925	$63,500	231	$9,400	$45,680	$70,100
232	$9,400	$55,114	$73,900	233	$9,400	$50,016	$73,000	234	$9,400	$46,361	$64,900
235	$9,400	$52,083	$72,100	236	$9,400	$40,140	$59,900	237	$9,400	$49,333	$69,500
238	$9,400	$53,578	$74,700	239	$9,400	$46,361	$65,200	240	$9,400	$40,523	$63,100
241	$9,400	$46,409	$68,300	242	$9,400	$47,322	$69,000	243	$9,400	$39,893	$61,800
244	$9,400	$46,461	$66,900	245	$10,000	$50,603	$76,500	246	$9,400	$47,481	$63,900
247	$9,400	$47,501	$72,200	248	$9,400	$53,748	$72,900	249	$9,400	$46,326	$70,400
250	$9,400	$52,083	$69,900	251	$9,400	$40,155	$64,500	252	$9,400	$46,416	$71,400
253	$9,400	$44,098	$67,200	254	$9,400	$53,148	$75,600	255	$10,000	$53,018	$72,800
256	$10,000	$48,445	$66,400	257	$14,000	$46,973	$70,400	258	$14,000	$42,272	$72,000
259	$14,000	$44,182	$75,300								

NEIGHBORHOOD   D

OBS	LANDVAL	IMPROVAL	SALEPRIC	OBS	LANDVAL	IMPROVAL	SALEPRIC	OBS	LANDVAL	IMPROVAL	SALEPRIC
				260	$8,000	$23,127	$43,500	261	$8,000	$23,042	$44,900
262	$8,000	$24,351	$43,900	263	$8,000	$24,586	$39,900	264	$8,000	$26,613	$36,000
265	$8,000	$31,864	$45,000	266	$8,000	$20,790	$39,500	267	$8,000	$31,679	$40,000
268	$8,000	$32,859	$38,500	269	$8,000	$27,564	$45,000	270	$8,000	$19,065	$43,300
271	$8,000	$22,696	$33,000	272	$8,000	$30,500	$40,000	273	$8,000	$35,258	$42,200
274	$8,000	$19,528	$35,500	275	$8,000	$28,055	$37,000	276	$8,000	$24,575	$41,000
277	$8,000	$23,776	$42,000	278	$8,000	$22,422	$37,600	279	$8,000	$40,275	$51,500
280	$8,000	$31,214	$40,900	281	$8,000	$27,672	$48,000	282	$8,000	$30,188	$46,500
283	$8,000	$32,286	$42,500	284	$8,000	$24,331	$39,900	285	$8,000	$25,604	$45,000
286	$8,000	$30,845	$50,000	287	$8,500	$43,986	$59,600	288	$8,000	$21,750	$38,000
289	$8,000	$33,831	$44,900	290	$8,000	$31,859	$45,000	291	$8,000	$23,805	$31,500
292	$8,000	$22,721	$29,000	293	$8,500	$26,242	$41,000	294	$8,000	$25,170	$40,000
295	$8,000	$24,402	$41,500	296	$8,500	$24,530	$37,500	297	$8,500	$25,031	$39,900
298	$8,000	$25,540	$45,000	299	$8,000	$24,530	$35,000	300	$6,800	$32,001	$45,300
301	$2,000	$32,201	$56,000	302	$15,500	$43,523	$105,000	303	$15,500	$52,659	$74,000
304	$10,000	$37,517	$49,500	305	$10,000	$31,899	$53,500	306	$10,000	$19,208	$36,000
307	$3,000	$31,481	$55,000	308	$9,000	$17,547	$37,500	309	$8,400	$26,200	$48,000
310	$4,560	$20,990	$40,000	311	$5,760	$23,008	$49,800	312	$4,800	$8,793	$24,000
313	$8,600	$13,258	$20,800	314	$6,400	$19,298	$30,000	315	$3,200	$13,327	$28,000
316	$8,000	$16,843	$33,000	317	$18,000	$50,556	$87,200	318	$10,000	$51,878	$72,900
319	$10,000	$42,133	$55,400	320	$5,000	$27,859	$39,000	321	$6,000	$24,999	$37,000
322	$20,000	$16,066	$42,000	323	$4,000	$13,911	$26,100	324	$4,000	$22,763	$40,000
325	$4,000	$9,256	$32,000	326	$4,800	$12,334	$28,500	327	$9,600	$15,844	$25,900
328	$9,600	$19,933	$32,700	329	$4,800	$16,504	$35,000	330	$5,000	$12,174	$26,000
331	$8,000	$22,808	$40,000	332	$8,000	$24,717	$42,500	333	$8,000	$30,653	$45,800
334	$8,000	$23,710	$32,000	335	$8,000	$22,680	$45,000	336	$8,500	$30,040	$52,000
337	$8,000	$16,700	$33,700	338	$8,000	$10,373	$13,800	339	$8,000	$24,438	$44,000
340	$8,000	$23,454	$45,500	341	$8,000	$20,177	$43,500	342	$8,000	$36,059	$48,000
343	$8,000	$31,550	$36,900	344	$8,000	$29,120	$50,000	345	$8,000	$22,991	$45,000
346	$8,000	$28,902	$46,700	347	$8,000	$29,305	$46,500	348	$8,000	$25,751	$40,500
349	$8,000	$25,717	$42,000	350	$8,000	$21,396	$40,000	351	$8,000	$23,095	$44,400
352	$16,000	$27,734	$79,900	353	$18,000	$37,624	$60,000	354	$18,000	$44,592	$72,000
355	$15,000	$35,148	$71,100	356	$13,000	$35,286	$57,700	357	$15,000	$18,698	$54,000
358	$12,000	$29,684	$40,000	359	$7,000	$12,136	$33,000	360	$8,000	$15,323	$34,400
361	$8,000	$14,740	$34,900	362	$8,000	$18,812	$37,000	363	$8,000	$15,364	$38,500
364	$8,000	$20,079	$37,500	365	$8,000	$26,843	$43,000	366	$8,000	$15,583	$31,500
367	$8,000	$16,583	$36,000	368	$8,500	$17,052	$38,500	369	$8,000	$29,693	$45,000
370	$8,000	$29,366	$36,900	371	$8,000	$39,569	$48,500	372	$8,000	$8,591	$27,000

(continued)

1986 APPRAISED VALUES AND SALE PRICES FOR PROPERTIES IN SIX NEIGHBORHOODS

NEIGHBORHOOD D

OBS	LANDVAL	IMPROVAL	SALEPRIC	OBS	LANDVAL	IMPROVAL	SALEPRIC	OBS	LANDVAL	IMPROVAL	SALEPRIC
373	$8,000	$24,535	$41,900	374	$8,000	$28,252	$45,200	375	$9,270	$16,303	$45,000
376	$9,720	$16,452	$42,500	377	$9,720	$14,155	$32,400	378	$9,270	$10,166	$35,500
379	$9,720	$12,978	$34,900	380	$4,528	$28,672	$39,000	381	$8,640	$30,596	$38,600
382	$8,000	$35,373	$49,900	383	$8,500	$19,055	$43,200	384	$8,000	$22,818	$38,500
385	$8,000	$32,886	$43,900	386	$8,000	$26,669	$43,400	387	$8,000	$26,698	$45,900
388	$8,000	$19,800	$30,000	389	$8,000	$22,775	$41,900	390	$8,000	$23,047	$43,000
391	$8,000	$22,146	$35,200	392	$8,000	$23,126	$45,000	393	$8,000	$33,769	$52,000
394	$8,000	$27,391	$41,000	395	$8,000	$23,304	$39,000	396	$8,000	$20,897	$40,900
397	$20,000	$66,388	$119,000	398	$15,600	$35,544	$58,500	399	$22,000	$48,505	$118,000

NEIGHBORHOOD E

OBS	LANDVAL	IMPROVAL	SALEPRIC	OBS	LANDVAL	IMPROVAL	SALEPRIC	OBS	LANDVAL	IMPROVAL	SALEPRIC
400	$9,000	$27,860	$48,500	401	$9,000	$34,268	$61,500	402	$9,000	$36,528	$57,500
403	$9,500	$31,439	$55,500	404	$10,000	$49,850	$67,000	405	$10,000	$50,227	$71,900
406	$10,000	$47,480	$71,700	407	$10,000	$45,347	$62,200	408	$10,500	$46,165	$74,900
409	$10,000	$42,908	$66,000	410	$11,000	$48,356	$69,700	411	$10,500	$48,348	$72,000
412	$10,000	$48,274	$68,500	413	$10,000	$45,250	$59,500	414	$10,000	$49,408	$71,000
415	$10,000	$50,592	$76,500	416	$9,500	$63,263	$73,600	417	$9,500	$45,778	$74,900
418	$16,500	$76,735	$97,500	419	$9,000	$20,739	$63,000	420	$11,500	$42,730	$55,200
421	$12,000	$45,446	$68,000	422	$12,000	$49,918	$71,700	423	$13,000	$46,185	$69,500
424	$13,000	$49,525	$78,500	425	$12,000	$45,480	$80,000	426	$11,500	$51,961	$69,000
427	$11,000	$61,227	$75,000	428	$11,000	$45,912	$68,000	429	$12,000	$41,084	$70,000
430	$10,500	$46,351	$66,900	431	$11,000	$41,099	$63,000	432	$11,000	$42,137	$68,500
433	$11,000	$44,758	$72,000	434	$11,000	$44,779	$69,900	435	$11,000	$51,886	$74,000
436	$9,500	$42,406	$60,000	437	$11,000	$42,335	$69,000	438	$10,000	$44,573	$68,500
439	$10,000	$52,387	$84,100	440	$10,000	$42,528	$63,500	441	$10,500	$50,415	$62,000
442	$10,500	$53,463	$76,000	443	$10,000	$53,133	$76,000	444	$9,500	$44,234	$68,900
445	$9,000	$34,071	$57,000	446	$9,500	$32,329	$49,500	447	$9,000	$32,606	$42,500
448	$9,000	$40,248	$58,000	449	$9,000	$40,771	$59,900	450	$9,000	$31,309	$44,900
451	$9,000	$30,399	$49,200	452	$9,000	$37,894	$55,900	453	$10,500	$39,285	$62,000
454	$10,000	$41,512	$55,000	455	$10,000	$43,851	$59,800	456	$10,000	$38,310	$58,500
457	$10,000	$35,445	$57,400	458	$10,000	$44,157	$60,000	459	$10,000	$35,920	$58,900
460	$10,000	$35,288	$56,000	461	$10,000	$43,315	$61,000	462	$10,500	$38,626	$66,900
463	$10,000	$39,745	$49,500	464	$10,000	$39,797	$58,000	465	$11,000	$47,766	$67,000
466	$9,500	$36,282	$57,900	467	$9,500	$35,257	$47,500	468	$9,500	$35,055	$51,000
469	$9,500	$33,884	$52,000	470	$9,500	$26,852	$43,500	471	$9,500	$40,006	$57,000
472	$9,500	$26,848	$48,000	473	$9,500	$36,684	$59,000	474	$9,500	$35,899	$49,700
475	$9,500	$34,991	$55,000	476	$9,500	$31,727	$45,500	477	$10,000	$35,908	$61,000
478	$9,500	$31,431	$48,900	479	$9,500	$29,294	$48,500	480	$10,000	$50,480	$66,500
481	$9,500	$41,208	$54,900	482	$9,500	$39,240	$60,000	483	$9,500	$39,664	$61,900
484	$9,500	$33,128	$55,600	485	$9,500	$32,625	$45,900	486	$9,500	$45,582	$62,000
487	$9,500	$39,128	$54,000	488	$9,500	$30,584	$46,000	489	$9,500	$38,140	$57,700
490	$9,500	$39,694	$55,000	491	$9,500	$40,883	$59,800	492	$9,500	$37,683	$48,000
493	$9,500	$38,939	$48,800	494	$9,500	$34,150	$51,000	495	$9,000	$29,767	$45,000
496	$9,000	$39,102	$55,000	497	$9,000	$38,497	$55,700	498	$9,000	$33,017	$47,500
499	$9,000	$33,920	$52,500	500	$9,000	$34,743	$51,900	501	$9,000	$33,502	$57,800
502	$9,000	$29,866	$47,000	503	$9,000	$32,625	$51,500	504	$9,000	$37,867	$49,900
505	$9,500	$36,248	$55,000	506	$9,000	$33,541	$51,000	507	$2,500	$13,499	$23,200
508	$10,000	$34,721	$56,000	509	$10,000	$34,890	$59,500	510	$10,000	$32,949	$57,900
511	$10,000	$36,088	$59,900	512	$10,000	$52,244	$70,000	513	$10,000	$34,947	$61,000
514	$10,000	$41,139	$54,000	515	$10,000	$40,141	$60,300	516	$10,000	$38,680	$60,100
517	$10,000	$43,181	$63,000	518	$10,000	$36,896	$52,000	519	$10,000	$51,098	$66,000
520	$10,500	$43,424	$62,500	521	$5,960	$9,321	$111,500	522	$9,500	$43,158	$68,000
523	$9,000	$38,113	$54,900	524	$9,500	$43,819	$59,900	525	$10,000	$42,432	$68,000
526	$10,000	$49,414	$74,900	527	$9,500	$43,214	$63,000	528	$20,000	$59,249	$89,000
529	$20,000	$62,085	$94,000	530	$18,000	$69,937	$96,600	531	$11,000	$28,526	$41,000
532	$11,000	$34,440	$59,000	533	$11,000	$28,767	$55,000	534	$11,000	$24,723	$47,500
535	$9,500	$29,527	$43,500								

1986 APPRAISED VALUES AND SALE PRICES FOR PROPERTIES IN SIX NEIGHBORHOODS

NEIGHBORHOOD   F

OBS	LANDVAL	IMPROVAL	SALEPRIC	OBS	LANDVAL	IMPROVAL	SALEPRIC	OBS	LANDVAL	IMPROVAL	SALEPRIC
				536	$11,500	$15,891	$39,900	537	$11,500	$14,713	$36,000
538	$11,500	$17,641	$37,800	539	$11,500	$19,817	$33,000	540	$11,500	$16,980	$33,900
541	$11,500	$24,752	$40,000	542	$11,500	$20,655	$37,700	543	$12,000	$32,585	$54,900
544	$12,000	$46,825	$64,500	545	$12,000	$48,511	$70,000	546	$13,500	$50,515	$73,800
547	$12,000	$45,084	$65,000	548	$12,500	$60,412	$79,900	549	$12,000	$43,301	$63,000
550	$12,500	$25,134	$47,500	551	$12,500	$44,737	$66,900	552	$12,000	$45,893	$62,000
553	$22,000	$57,543	$88,500	554	$22,000	$65,264	$100,000	555	$22,000	$62,552	$100,000
556	$22,000	$69,025	$101,200	557	$22,000	$48,691	$87,500	558	$22,000	$64,455	$106,000
559	$22,000	$61,333	$91,500	560	$22,000	$73,571	$98,000	561	$22,000	$74,829	$113,500
562	$9,000	$23,544	$43,500	563	$32,000	$99,051	$160,000	564	$34,000	$107,174	$131,200
565	$32,000	$127,940	$221,200	566	$22,000	$65,349	$79,900	567	$22,000	$63,905	$95,000
568	$22,000	$60,393	$85,000	569	$22,000	$65,644	$90,000	570	$22,000	$59,116	$96,600
571	$22,000	$81,622	$132,000	572	$22,000	$71,694	$100,000	573	$22,000	$66,280	$115,000
574	$22,000	$58,501	$95,000	575	$22,000	$69,090	$110,900	576	$22,000	$77,990	$120,000
577	$22,000	$66,313	$119,500	578	$22,000	$77,279	$130,000	579	$22,000	$64,548	$112,000
580	$23,000	$86,591	$137,000	581	$14,000	$44,112	$63,000	582	$15,500	$42,862	$83,900
583	$18,500	$48,124	$95,500	584	$18,500	$53,324	$95,900	585	$18,500	$50,662	$96,900
586	$18,500	$52,394	$111,500	587	$14,000	$47,521	$68,000	588	$14,000	$47,637	$72,000
589	$14,500	$52,195	$77,000	590	$13,000	$33,651	$50,000	591	$12,000	$35,401	$67,500
592	$12,000	$28,050	$52,000	593	$12,000	$30,125	$55,500	594	$12,000	$32,178	$48,000
595	$13,000	$31,729	$57,500	596	$13,000	$30,362	$58,000	597	$13,000	$39,430	$63,300
598	$13,000	$41,116	$67,000	599	$13,000	$45,139	$70,000	600	$13,000	$45,902	$71,300
601	$13,000	$38,291	$64,900	602	$14,000	$51,728	$80,000	603	$14,500	$41,842	$66,600
604	$14,000	$39,505	$64,900	605	$12,000	$36,523	$46,000	606	$12,000	$37,607	$49,500
607	$13,000	$28,352	$44,800	608	$13,000	$32,114	$58,700	609	$13,000	$25,970	$49,500
610	$13,000	$32,231	$55,100	611	$11,000	$21,959	$48,500	612	$11,000	$27,554	$50,000
613	$11,000	$21,216	$33,000	614	$15,000	$44,643	$80,000	615	$15,000	$46,878	$76,000
616	$15,000	$34,531	$69,000	617	$15,000	$36,478	$68,100	618	$15,000	$64,809	$107,000
619	$15,000	$30,053	$57,000	620	$13,000	$43,050	$69,000	621	$13,000	$30,962	$55,000
622	$13,000	$43,760	$64,500	623	$13,000	$31,450	$77,500	624	$13,000	$45,368	$67,500
625	$14,000	$41,723	$55,000	626	$12,000	$48,678	$62,000	627	$12,000	$37,984	$57,900
628	$12,000	$29,516	$46,900	629	$12,000	$33,323	$45,900	630	$12,000	$35,082	$57,000
631	$12,000	$44,647	$54,900	632	$12,000	$28,871	$47,900	633	$12,000	$41,273	$57,500
634	$12,000	$43,634	$61,500	635	$12,000	$33,023	$55,500	636	$12,000	$38,194	$56,000
637	$12,500	$47,547	$57,500	638	$12,000	$42,441	$70,500	639	$12,000	$45,614	$64,900
640	$12,000	$33,321	$53,000	641	$12,000	$40,726	$66,900	642	$12,500	$37,634	$50,000
643	$12,500	$30,907	$50,500	644	$25,000	$81,175	$139,000	645	$25,000	$90,076	$132,600
646	$19,000	$79,224	$113,900	647	$19,000	$81,858	$117,600	648	$19,000	$91,924	$29,700
649	$25,000	$87,410	$143,200	650	$12,000	$69,232	$86,500	651	$12,000	$45,953	$61,900

DATA SET: APPRAISALS AND SALES DATA FOR ALL 1986 RESIDENTIAL PROPERTY SALES IN A MID-SIZED FLORIDA CITY (SEE CASE STUDY 11)

REAL ESTATE SALES FOR 1986
IN A MID-SIZED FLORIDA CITY

OBS	TOWNSHIP	RANGE	SECTION	LANDVAL	IMPROVAL	SALEPRIC	SALTOAPR	AREA
1	7	17	26	$2,500	$4,883	$13,000	1.76080	576
2	7	17	33	$10,800	$44,351	$68,000	1.23298	1,170
3	7	17	33	$3,845	$49,640	$80,000	1.49575	1,822
4	7	17	34	$2,000	$26,756	$52,000	1.80832	1,800
5	8	17	2	$3,000	$34,170	$44,800	1.20527	1,107
6	8	17	2	$1,600	$2,507	$10,000	2.43487	720
7	8	17	2	$1,200	$30,329	$43,000	1.36382	960
8	8	17	2	$3,500	$30,920	$65,000	1.88844	1,544
9	8	17	2	$3,000	$22,988	$31,000	1.19286	856
10	7	17	34	$19,200	$9,995	$36,000	1.23309	836
11	7	17	34	$3,000	$10,156	$20,300	1.54302	1,108
12	8	17	3	$2,800	$22,366	$37,800	1.50203	1,292
13	8	17	3	$4,200	$25,764	$59,200	1.97570	2,052
14	8	17	3	$3,500	$37,746	$41,000	0.99404	2,200
15	8	17	3	$7,500	$40,332	$70,000	1.46346	1,392
16	8	17	3	$5,000	$32,600	$36,400	0.96809	6,250
17	8	17	3	$3,000	$15,911	$25,000	1.32198	972
18	7	17	34	$4,200	$39,553	$40,100	0.91651	1,512
19	7	17	34	$4,200	$20,001	$34,000	1.40490	1,029
20	8	17	3	$2,500	$26,731	$30,000	1.02631	960
21	8	17	3	$4,000	$23,172	$35,000	1.28809	2,400
22	8	17	3	$2,100	$18,956	$34,700	1.64799	1,152
23	8	17	3	$3,000	$7,861	$11,500	1.05883	871
24	8	17	3	$2,000	$15,147	$22,000	1.28302	1,252
25	7	17	34	$4,200	$19,312	$37,900	1.61194	1,013
26	7	17	34	$1,500	$8,985	$19,700	1.87887	585
27	7	17	34	$2,800	$16,572	$30,000	1.54863	1,038
28	7	17	34	$3,600	$31,317	$38,000	1.08830	1,056
29	7	17	34	$3,600	$29,303	$38,000	1.15491	1,040
30	7	17	34	$1,500	$3,623	$7,000	1.36639	1,150
31	8	17	3	$3,500	$10,985	$20,000	1.38074	1,519
32	8	17	3	$5,000	$19,977	$27,800	1.11302	3,381
33	8	17	3	$3,800	$11,740	$20,000	1.28700	1,014
34	8	17	3	$1,000	$35,232	$39,500	1.09020	1,056
35	8	17	3	$1,000	$10,715	$30,000	2.56082	704
36	8	17	3	$1,500	$6,506	$12,000	1.49888	1,508
37	8	17	3	$1,800	$10,924	$11,000	0.86451	693
38	8	17	3	$3,000	$27,607	$34,000	1.11086	1,056
39	8	17	3	$3,500	$12,928	$24,500	1.49136	880
40	8	17	3	$3,200	$30,188	$34,500	1.03331	1,056
41	8	17	3	$3,200	$21,665	$26,500	1.06576	1,008
42	8	17	10	$3,200	$29,608	$34,000	1.03633	1,070
43	8	17	10	$3,000	$31,977	$32,000	0.91489	1,241
44	8	17	10	$4,200	$43,685	$54,500	1.13814	1,315
45	8	17	10	$3,510	$26,606	$40,400	1.34148	1,025
46	8	17	17	$1,500	$26,987	$35,500	1.24618	1,050
47	8	17	34	$29,420	$77,837	$161,400	1.50480	2,111
48	8	17	34	$11,790	$4,985	$70,000	4.17288	1,016
49	9	17	4	$5,000	$39,539	$85,000	1.90844	2,221
50	9	17	12	$7,750	$53,083	$69,000	1.13425	1,897
51	9	17	20	$2,500	$42,514	$62,000	1.37735	1,494
52	10	17	3	$45,238	$67,862	$280,000	2.47569	2,904
53	10	17	3	$2,250	$9,100	$18,000	1.58590	572
54	10	17	4	$1,500	$45,345	$58,500	1.24880	1,326
55	10	17	4	$2,400	$27,831	$33,500	1.10813	1,056
56	10	17	4	$2,500	$34,240	$48,000	1.30648	1,118
57	10	17	4	$2,800	$31,262	$42,400	1.24479	1,272
58	10	17	4	$3,300	$34,563	$40,000	1.05644	1,080
59	10	17	4	$3,200	$28,228	$38,900	1.23775	1,008
60	10	17	4	$3,150	$26,823	$41,400	1.38124	1,008
61	10	17	4	$1,500	$5,032	$6,000	0.91855	441
62	10	17	4	$1,500	$1,150	$3,500	1.32075	300
63	10	17	4	$3,600	$36,000	$43,000	1.08586	1,226
64	10	17	4	$3,500	$10,526	$15,000	1.06944	1,930
65	10	17	4	$5,265	$29,262	$44,000	1.27436	273
66	10	17	4	$9,000	$49,887	$109,000	1.85100	1,920
67	10	17	4	$3,600	$20,600	$41,800	1.72727	1,280
68	10	17	4	$3,600	$19,471	$50,000	2.16722	1,560
69	10	17	4	$2,000	$6,093	$15,000	1.85345	1,056
70	10	17	4	$1,800	$7,187	$12,600	1.40203	960
71	10	17	4	$7,500	$12,618	$22,500	1.11840	840
72	10	17	5	$9,000	$10,288	$22,000	1.14061	960

REAL ESTATE SALES FOR 1986
IN A MID-SIZED FLORIDA CITY

OBS	TOWNSHIP	RANGE	SECTION	LANDVAL	IMPROVAL	SALEPRIC	SALTOAPR	AREA
73	10	17	9	$4,000	$35,707	$32,000	0.80590	1,190
74	10	17	9	$4,000	$37,386	$49,000	1.18398	1,240
75	10	17	9	$4,000	$44,249	$63,000	1.30573	1,360
76	10	17	9	$4,000	$42,624	$51,900	1.11316	1,266
77	10	17	24	$23,916	$44,345	$100,000	1.46497	1,860
78	6	18	33	$18,216	$19,941	$42,500	1.11382	732
79	7	18	13	$9,960	$48,221	$115,000	1.97659	2,464
80	7	18	14	$4,560	$52,846	$64,500	1.12358	1,706
81	7	18	29	$10,020	$64,285	$80,000	1.07664	1,989
82	8	18	8	$2,500	$19,772	$35,000	1.57148	1,027
83	8	18	9	$15,525	$53,287	$82,000	1.19165	1,943
84	8	18	11	$1,500	$15,022	$23,500	1.42235	744
85	8	18	14	$4,350	$33,078	$52,500	1.40269	1,178
86	8	18	13	$16,730	$76,842	$116,400	1.24396	2,552
87	8	18	14	$3,500	$31,007	$38,000	1.10123	1,080
88	8	18	14	$2,500	$30,576	$43,000	1.30004	1,092
89	8	18	14	$2,640	$23,244	$34,000	1.31355	1,125
90	8	18	14	$4,000	$8,947	$22,100	1.70696	520
91	8	18	14	$2,000	$14,147	$19,900	1.23243	884
92	8	18	14	$1,500	$20,433	$20,000	0.91187	957
93	8	18	14	$2,000	$9,690	$18,000	1.53978	672
94	8	18	15	$3,126	$13,898	$19,500	1.14544	1,410
95	8	18	15	$1,456	$13,872	$19,000	1.23956	1,560
96	8	18	15	$6,650	$52,606	$80,000	1.35007	1,428
97	8	18	15	$3,000	$27,556	$33,600	1.09962	1,008
98	8	18	15	$2,000	$33,849	$45,000	1.25527	1,293
99	8	18	15	$3,879	$19,874	$25,000	1.05250	1,960
100	8	18	15	$1,274	$14,328	$35,000	2.24330	1,375
101	8	18	15	$2,100	$14,151	$26,500	1.63067	985
102	8	18	15	$1,500	$30,813	$38,200	1.18219	1,125
103	8	18	15	$5,292	$19,371	$49,500	2.00706	1,695
104	8	18	15	$3,000	$23,074	$37,000	1.41904	1,548
105	8	18	15	$3,500	$14,681	$26,400	1.45207	775
106	8	18	15	$3,500	$31,703	$40,000	1.13627	1,125
107	8	18	15	$2,500	$14,631	$20,000	1.16747	1,578
108	8	18	18	$17,500	$26,465	$17,500	0.39804	1,440
109	8	18	19	$13,650	$15,095	$55,000	1.91338	1,148
110	8	18	23	$22,350	$117,055	$110,000	0.78907	4,320
111	8	18	13	$3,500	$26,697	$36,900	1.22198	1,032
112	8	18	25	$17,760	$53,886	$86,000	1.20035	1,622
113	8	18	27	$2,500	$27,532	$34,500	1.14877	1,104
114	8	18	28	$4,800	$30,940	$44,000	1.23111	1,071
115	8	18	30	$16,300	$8,304	$25,480	1.07948	960
116	9	18	5	$11,250	$54,399	$115,000	1.75174	1,766
117	9	18	5	$12,148	$55,100	$92,500	1.37551	1,819
118	9	18	9	$10,620	$12,092	$50,000	2.20148	792
119	9	18	9	$11,040	$13,802	$42,000	1.69069	1,464
120	9	18	14	$22,568	$51,306	$88,500	1.19799	1,832
121	9	18	14	$22,568	$53,113	$94,000	1.24206	1,686
122	9	18	14	$22,525	$71,138	$143,000	1.52675	2,480
123	9	18	14	$21,760	$50,288	$75,000	1.04097	1,456
124	9	18	14	$22,250	$60,143	$121,500	1.47464	2,016
125	9	18	15	$25,000	$53,606	$114,800	1.46045	1,659
126	9	18	17	$28,000	$65,709	$118,500	1.26455	2,209
127	9	18	22	$12,500	$90,305	$51,700	0.50289	3,504
128	9	18	25	$12,500	$44,871	$67,300	1.17307	1,362
129	9	18	25	$12,500	$42,575	$68,900	1.25102	1,280
130	9	18	25	$12,500	$59,387	$94,700	1.31735	1,906
131	9	18	25	$12,500	$46,364	$71,100	1.20787	1,362
132	9	18	25	$12,500	$45,689	$73,000	1.25453	1,434
133	9	18	25	$12,500	$49,640	$84,000	1.35179	1,566
134	9	18	25	$12,500	$44,717	$73,300	1.28109	1,229
135	9	18	25	$13,000	$49,174	$76,700	1.23363	1,542
136	9	18	25	$12,500	$40,781	$68,000	1.27625	1,224
137	9	18	25	$13,500	$53,734	$86,600	1.28804	1,616
138	9	18	25	$12,500	$41,776	$66,500	1.22522	1,212
139	9	18	25	$12,500	$50,533	$84,600	1.34215	1,554
140	9	18	25	$12,500	$41,076	$66,900	1.24869	1,224
141	9	18	25	$13,000	$54,765	$96,500	1.42640	1,694
142	9	18	25	$12,500	$45,631	$73,000	1.25578	1,436
143	9	18	25	$12,500	$57,271	$105,500	1.51209	1,835
144	9	18	25	$12,500	$40,869	$66,900	1.25354	1,224

REAL ESTATE SALES FOR 1986
IN A MID-SIZED FLORIDA CITY

OBS	TOWNSHIP	RANGE	SECTION	LANDVAL	IMPROVAL	SALEPRIC	SALTOAPR	AREA
145	9	18	25	$12,500	$47,850	$73,000	1.20961	1,380
146	9	18	25	$12,500	$53,311	$88,000	1.33716	1,715
147	9	18	25	$12,500	$49,550	$80,400	1.29573	1,623
148	9	18	25	$12,500	$44,533	$74,400	1.30451	1,344
149	9	18	25	$13,500	$41,492	$68,900	1.25291	1,247
150	9	18	25	$12,500	$46,614	$67,700	1.14524	1,496
151	9	18	25	$13,500	$45,934	$73,900	1.24340	1,343
152	9	18	25	$12,500	$49,535	$81,000	1.30571	1,534
153	9	18	25	$12,500	$51,749	$77,400	1.20469	1,548
154	9	18	25	$12,500	$57,993	$94,500	1.34056	1,273
155	9	18	25	$12,500	$45,614	$81,000	1.39381	922
156	9	18	25	$12,500	$55,265	$82,500	1.21744	1,643
157	9	18	25	$12,500	$50,658	$87,500	1.38541	1,617
158	9	18	25	$12,500	$49,662	$83,900	1.34970	1,538
159	9	18	25	$13,500	$48,895	$82,500	1.32222	1,512
160	9	18	25	$13,500	$46,386	$79,500	1.32752	1,428
161	9	18	25	$13,500	$45,539	$81,700	1.38383	1,357
162	9	18	29	$10,000	$58,260	$128,000	1.87518	2,224
163	9	18	31	$25,000	$52,130	$125,000	1.62064	1,596
164	9	18	32	$4,986	$45,714	$70,000	1.38067	1,644
165	9	18	32	$2,500	$27,458	$50,500	1.68569	1,020
166	9	18	34	$176,000	$10,213	$216,900	1.16480	912
167	9	18	34	$2,000	$12,516	$27,000	1.86002	1,066
168	9	18	35	$5,000	$40,468	$58,000	1.27562	1,624
169	9	18	35	$2,360	$9,458	$13,500	1.14233	732
170	10	18	2	$6,000	$46,779	$73,800	1.39828	1,257
171	10	18	2	$6,000	$42,704	$65,700	1.34897	1,092
172	10	18	2	$6,000	$51,195	$76,000	1.32879	1,403
173	10	18	2	$6,000	$46,767	$74,000	1.40239	1,257
174	10	18	2	$6,000	$51,137	$82,800	1.44915	1,416
175	10	18	2	$6,000	$46,779	$71,500	1.35471	1,257
176	10	18	2	$6,000	$46,039	$75,000	1.44123	1,257
177	10	18	2	$6,000	$51,917	$74,500	1.28632	1,416
178	10	18	2	$6,000	$50,420	$77,600	1.37540	1,403
179	10	18	2	$6,000	$44,641	$71,800	1.41782	1,403
180	10	18	2	$6,000	$45,679	$72,000	1.39322	1,257
181	10	18	2	$3,000	$46,369	$69,100	1.39966	1,257
182	10	18	2	$3,000	$35,578	$57,500	1.49049	1,092
183	10	18	2	$3,000	$35,575	$47,900	1.24174	1,092
184	10	18	1	$18,480	$109,835	$140,000	1.09106	2,707
185	10	18	1	$10,000	$86,464	$156,000	1.61718	2,661
186	10	18	2	$7,200	$42,375	$57,000	1.14977	1,336
187	10	18	2	$7,700	$43,256	$59,500	1.16767	1,222
188	10	18	2	$7,200	$38,582	$61,800	1.34988	1,245
189	10	18	2	$7,500	$25,741	$63,000	1.89525	1,247
190	10	18	2	$7,500	$40,202	$57,200	1.19911	1,412
191	10	18	2	$7,500	$42,028	$62,000	1.25182	1,144
192	10	18	2	$7,500	$40,076	$55,800	1.17286	1,269
193	10	18	3	$11,523	$14,336	$53,500	2.06891	1,104
194	10	18	4	$11,134	$56,165	$116,000	1.72365	1,841
195	10	18	4	$12,672	$84,954	$162,600	1.66554	2,614
196	10	18	6	$3,400	$39,296	$55,000	1.28818	1,556
197	10	18	10	$15,000	$51,796	$110,700	1.65728	1,366
198	10	18	11	$5,960	$44,967	$68,900	1.35292	1,873
199	10	18	12	$12,914	$92,288	$165,000	1.56841	2,723
200	10	18	13	$4,500	$69,212	$112,000	1.51943	2,747
201	10	18	17	$14,700	$9,996	$30,000	1.21477	720
202	10	18	19	$12,500	$14,402	$38,000	1.41253	784
203	10	18	20	$10,950	$42,767	$91,000	1.69406	1,325
204	10	18	31	$7,920	$4,725	$14,500	1.14670	1,264
205	10	18	35	$10,300	$62,910	$115,000	1.57082	1,647
206	10	18	36	$7,000	$35,827	$43,100	1.00637	1,453
207	10	18	36	$7,000	$39,310	$58,000	1.25243	1,226
208	10	18	36	$7,000	$35,684	$47,000	1.10112	1,470
209	10	18	36	$7,750	$34,847	$50,000	1.17379	1,101
210	11	18	1	$7,000	$41,196	$63,500	1.31754	1,232
211	10	18	36	$11,500	$31,508	$55,500	1.29046	1,008
212	11	18	1	$8,500	$36,865	$58,900	1.29836	1,229
213	11	18	1	$8,500	$42,152	$62,500	1.23391	1,208
214	11	18	1	$1,848	$17,740	$40,000	2.04207	552
215	11	18	1	$6,000	$55,671	$79,900	1.29558	1,881
216	11	18	2	$6,000	$29,967	$50,300	1.39850	1,185

REAL ESTATE SALES FOR 1986
IN A MID-SIZED FLORIDA CITY

OBS	TOWNSHIP	RANGE	SECTION	LANDVAL	IMPROVAL	SALEPRIC	SALTOAPR	AREA
217	10	18	35	$21,870	$65,999	$89,000	1.01287	2,300
218	11	18	2	$16,350	$57,109	$155,000	2.11002	1,934
219	11	18	2	$18,510	$58,520	$90,000	1.16838	1,913
220	11	18	2	$6,000	$31,997	$47,900	1.26063	1,025
221	11	18	2	$6,000	$31,570	$53,000	1.41070	1,325
222	11	18	3	$9,540	$17,568	$54,900	2.02523	1,334
223	11	18	9	$4,000	$31,319	$38,000	1.07591	1,086
224	11	18	10	$11,242	$19,180	$45,000	1.47919	1,152
225	11	18	10	$27,600	$12,825	$55,000	1.36054	704
226	11	18	16	$3,000	$16,382	$28,000	1.44464	1,008
227	11	18	16	$2,500	$14,520	$15,000	0.88132	1,120
228	11	18	16	$8,000	$20,885	$45,000	1.55790	1,000
229	11	18	16	$10,000	$19,326	$4,100	0.13981	960
230	11	18	17	$2,800	$9,183	$22,000	1.83593	776
231	11	18	17	$1,800	$22,798	$27,500	1.11798	1,008
232	7	19	5	$19,150	$43,548	$78,000	1.24406	1,501
233	7	19	14	$10,000	$16,454	$19,500	0.73713	1,056
234	7	19	14	$14,700	$30,411	$67,000	1.48523	960
235	8	19	8	$9,036	$12,146	$44,000	2.07724	732
236	8	19	10	$1,836	$46,874	$49,900	1.02443	1,560
237	8	19	13	$4,200	$43,515	$60,500	1.26795	1,618
238	8	19	17	$5,665	$11,267	$39,000	2.30333	1,344
239	8	19	24	$12,240	$11,231	$25,000	1.06514	672
240	8	19	25	$9,350	$23,868	$16,000	0.48167	1,456
241	8	19	28	$13,814	$41,793	$77,500	1.39371	5,000
242	8	19	28	$19,260	$69,738	$90,000	1.01126	2,366
243	8	19	28	$10,000	$61,229	$100,000	1.40392	1,558
244	8	19	28	$8,750	$59,313	$68,000	0.99907	1,800
245	8	19	28	$8,750	$40,003	$47,300	0.97020	1,536
246	8	19	21	$6,500	$36,722	$52,000	1.20309	1,259
247	8	19	21	$6,500	$41,427	$55,000	1.14758	1,355
248	8	19	21	$1,000	$5,614	$5,000	0.75597	1,014
249	8	19	21	$2,000	$23,859	$5,000	0.19336	1,440
250	8	19	28	$7,600	$49,599	$67,900	1.18708	1,530
251	8	19	28	$7,600	$55,651	$61,500	0.97232	1,461
252	8	19	28	$8,800	$45,898	$62,500	1.14264	1,483
253	8	19	28	$6,400	$42,964	$55,000	1.11417	1,344
254	8	19	34	$5,950	$46,328	$59,500	1.13815	1,300
255	8	19	0	$7,530	$96,624	$180,000	1.72821	1,933
256	9	19	2	$12,500	$15,892	$38,000	1.33841	1,200
257	9	19	6	$12,000	$16,298	$49,900	1.76338	1,152
258	9	19	2	$12,000	$16,182	$39,800	1.41225	984
259	9	19	2	$12,000	$41,696	$72,000	1.34088	1,242
260	9	19	2	$12,000	$37,629	$67,000	1.35002	1,200
261	9	19	2	$12,000	$35,250	$58,900	1.24656	1,200
262	9	19	2	$12,000	$52,361	$77,300	1.20104	1,736
263	9	19	2	$12,000	$37,092	$54,400	1.10812	1,152
264	9	19	2	$12,000	$32,194	$54,900	1.24225	1,032
265	9	19	2	$12,000	$34,462	$51,700	1.11274	1,056
266	9	19	2	$12,500	$43,594	$68,600	1.22295	1,368
267	9	19	2	$12,000	$31,301	$46,400	1.07157	912
268	9	19	2	$12,000	$32,838	$47,500	1.05937	1,056
269	9	19	2	$12,000	$32,759	$48,000	1.07241	1,056
270	9	19	2	$12,000	$35,107	$48,100	1.02108	1,152
271	9	19	2	$12,000	$28,728	$46,200	1.13435	912
272	9	19	2	$12,000	$27,317	$38,900	0.98939	864
273	9	19	2	$12,000	$34,986	$47,500	1.01094	1,152
274	9	19	4	$12,000	$47,208	$79,000	1.33428	1,148
275	9	19	10	$12,500	$54,482	$87,000	1.29886	1,586
276	9	19	10	$12,500	$52,645	$90,000	1.38153	1,611
277	9	19	10	$13,500	$58,310	$81,600	1.13633	1,692
278	9	19	10	$12,500	$47,144	$78,000	1.30776	1,464
279	9	19	10	$13,500	$48,302	$84,000	1.35918	1,542
280	9	19	12	$19,565	$91,120	$110,000	0.99381	5,000
281	9	19	13	$9,000	$27,860	$48,500	1.31579	928
282	9	19	13	$9,000	$34,268	$61,500	1.42137	1,040
283	9	19	13	$9,000	$36,528	$57,500	1.26296	1,124
284	9	19	13	$9,500	$31,439	$55,500	1.35568	1,126
285	9	19	13	$10,000	$49,850	$67,000	1.11947	1,676
286	9	19	13	$10,000	$50,227	$71,900	1.19382	1,392
287	9	19	13	$10,000	$47,480	$71,700	1.24739	1,389
288	9	19	13	$10,000	$45,347	$62,200	1.12382	1,332

REAL ESTATE SALES FOR 1986
IN A MID-SIZED FLORIDA CITY

OBS	TOWNSHIP	RANGE	SECTION	LANDVAL	IMPROVAL	SALEPRIC	SALTOAPR	AREA
289	9	19	13	$10,500	$46,165	$74,900	1.32180	1,417
290	9	19	13	$10,000	$42,908	$66,000	1.24745	1,470
291	9	19	13	$11,000	$48,356	$69,700	1.17427	1,519
292	9	19	13	$10,500	$48,348	$72,000	1.22349	1,519
293	9	19	13	$10,000	$48,274	$68,500	1.17548	1,556
294	9	19	13	$10,000	$45,250	$59,500	1.07692	1,398
295	9	19	13	$10,000	$49,408	$71,000	1.19513	1,499
296	9	19	13	$10,000	$50,592	$76,500	1.26254	1,554
297	9	19	13	$9,500	$63,263	$73,600	1.01150	1,938
298	9	19	13	$9,500	$45,778	$74,900	1.35497	1,493
299	9	19	13	$16,500	$76,735	$97,500	1.04574	3,600
300	9	19	13	$9,000	$20,739	$63,000	2.11843	900
301	9	19	13	$11,500	$42,730	$55,200	1.01789	1,438
302	9	19	13	$12,000	$45,446	$68,000	1.18372	1,588
303	9	19	13	$12,000	$49,918	$71,700	1.15798	1,782
304	9	19	13	$13,000	$46,185	$69,500	1.17428	1,584
305	9	19	13	$13,000	$49,525	$78,500	1.25550	1,735
306	9	19	13	$12,000	$45,480	$80,000	1.39179	1,603
307	9	19	13	$11,500	$51,961	$69,000	1.08728	1,716
308	9	19	13	$11,000	$61,227	$75,000	1.03839	1,682
309	9	19	13	$11,000	$45,912	$68,000	1.19483	1,589
310	9	19	13	$12,000	$41,084	$70,000	1.31866	1,357
311	9	19	13	$10,500	$46,351	$66,900	1.17676	1,459
312	9	19	13	$11,000	$41,099	$63,000	1.20924	1,309
313	9	19	13	$11,000	$42,137	$68,500	1.28912	1,324
314	9	19	13	$11,000	$44,758	$72,000	1.29129	1,667
315	9	19	13	$11,000	$44,779	$69,900	1.25316	1,512
316	9	19	13	$11,000	$51,886	$74,000	1.17673	1,534
317	9	19	13	$9,500	$42,406	$60,000	1.15594	1,378
318	9	19	13	$11,000	$42,335	$69,000	1.29371	1,416
319	9	19	13	$10,000	$44,573	$68,500	1.25520	1,557
320	9	19	13	$10,000	$52,387	$84,100	1.34804	1,564
321	9	19	13	$10,000	$42,528	$63,500	1.20888	1,351
322	9	19	13	$10,500	$50,415	$62,000	1.01781	1,539
323	9	19	13	$10,500	$53,463	$76,000	1.18819	1,888
324	9	19	13	$10,000	$53,113	$76,000	1.20381	1,614
325	9	19	13	$9,500	$44,234	$68,900	1.28224	1,435
326	9	19	13	$9,000	$34,071	$57,000	1.32340	1,100
327	9	19	13	$9,500	$32,329	$49,500	1.18339	1,036
328	9	19	13	$9,000	$32,606	$42,500	1.02149	1,070
329	9	19	13	$9,000	$40,248	$58,000	1.17771	1,170
330	9	19	13	$9,000	$40,771	$59,900	1.20351	1,196
331	9	19	13	$9,000	$31,309	$44,900	1.11390	1,064
332	9	19	13	$9,000	$30,399	$49,200	1.24876	1,002
333	9	19	13	$9,000	$37,894	$55,900	1.19205	1,194
334	9	19	13	$10,500	$39,285	$62,000	1.24536	1,247
335	9	19	13	$10,000	$41,512	$55,000	1.06771	1,324
336	9	19	13	$10,000	$43,851	$59,800	1.11047	1,392
337	9	19	13	$10,000	$38,310	$58,500	1.21093	1,230
338	9	19	13	$10,000	$35,445	$57,400	1.26307	1,319
339	9	19	13	$10,000	$44,157	$60,000	1.10789	1,170
340	9	19	13	$10,000	$35,920	$58,900	1.28267	1,334
341	9	19	13	$10,000	$35,288	$56,000	1.23653	1,196
342	9	19	13	$10,000	$43,315	$61,000	1.14414	1,552
343	9	19	13	$10,500	$38,626	$66,900	1.36180	1,397
344	9	19	13	$10,000	$39,745	$49,500	0.99507	1,302
345	9	19	13	$10,000	$39,797	$58,000	1.16473	1,324
346	9	19	13	$11,000	$47,746	$67,000	1.14012	1,380
347	9	19	13	$9,500	$36,282	$57,900	1.26469	1,267
348	9	19	13	$9,500	$35,257	$47,500	1.06129	1,062
349	9	19	13	$9,500	$35,055	$51,000	1.14465	1,148
350	9	19	13	$9,500	$33,884	$52,000	1.19860	1,344
351	9	19	13	$9,500	$26,852	$43,500	1.19663	1,008
352	9	19	13	$9,500	$40,006	$57,000	1.15138	1,344
353	9	19	13	$9,500	$26,848	$48,000	1.32057	1,008
354	9	19	13	$9,500	$36,684	$59,000	1.27750	1,267
355	9	19	13	$9,500	$35,899	$49,700	1.09474	1,308
356	9	19	13	$9,500	$34,991	$55,000	1.23621	1,050
357	9	19	13	$9,500	$31,727	$45,500	1.10365	1,075
358	9	19	13	$10,000	$35,908	$61,000	1.32874	1,326
359	9	19	13	$9,500	$31,431	$48,900	1.19469	1,075
360	9	19	13	$9,500	$29,294	$48,500	1.25019	1,075

REAL ESTATE SALES FOR 1986
IN A MID-SIZED FLORIDA CITY

OBS	TOWNSHIP	RANGE	SECTION	LANDVAL	IMPROVAL	SALEPRIC	SALTOAPR	AREA
361	9	19	13	$10,000	$50,480	$66,500	1.09954	1,288
362	9	19	13	$9,500	$41,208	$54,900	1.08267	1,462
363	9	19	13	$9,500	$39,240	$60,000	1.23102	1,267
364	9	19	13	$9,500	$39,664	$61,900	1.25905	1,105
365	9	19	13	$9,500	$33,128	$55,600	1.30431	1,182
366	9	19	13	$9,500	$32,625	$45,900	1.08961	1,025
367	9	19	13	$9,500	$45,582	$62,000	1.12559	1,750
368	9	19	13	$9,500	$39,128	$54,000	1.11047	1,170
369	9	19	13	$9,500	$30,584	$46,000	1.14759	1,000
370	9	19	13	$9,500	$38,140	$57,700	1.21117	1,208
371	9	19	13	$9,500	$39,694	$55,000	1.11802	1,222
372	9	19	13	$9,500	$40,883	$59,800	1.18691	1,181
373	9	19	13	$9,500	$37,683	$48,000	1.01732	1,403
374	9	19	13	$9,500	$38,939	$48,800	1.00745	1,184
375	9	19	13	$9,500	$34,150	$51,000	1.16838	1,104
376	9	19	13	$9,000	$29,767	$45,000	1.16078	965
377	9	19	13	$9,000	$39,102	$55,000	1.14340	1,158
378	9	19	13	$9,000	$38,497	$55,700	1.17271	1,170
379	9	19	13	$9,000	$33,017	$47,500	1.13049	1,061
380	9	19	13	$9,000	$33,920	$52,500	1.22321	1,065
381	9	19	13	$9,000	$34,743	$51,900	1.18648	1,075
382	9	19	13	$9,000	$33,502	$57,800	1.35994	1,061
383	9	19	13	$9,000	$29,866	$47,000	1.20928	1,014
384	9	19	13	$9,000	$32,625	$51,500	1.23724	1,008
385	9	19	13	$9,000	$37,867	$49,900	1.06472	1,193
386	9	19	13	$9,500	$36,248	$55,000	1.20224	1,108
387	9	19	13	$9,000	$33,541	$51,000	1.19884	1,202
388	9	19	13	$2,500	$13,499	$23,200	1.45009	440
389	9	19	14	$10,000	$36,560	$54,900	1.17912	1,060
390	9	19	14	$10,000	$38,742	$50,300	1.03196	1,150
391	9	19	14	$10,000	$39,592	$62,000	1.25020	1,265
392	9	19	14	$10,000	$41,878	$63,700	1.22788	1,329
393	9	19	14	$10,000	$50,774	$65,600	1.07941	800
394	9	19	14	$10,000	$36,634	$59,000	1.26517	1,132
395	9	19	14	$10,000	$38,403	$56,900	1.17555	1,190
396	9	19	14	$10,000	$40,552	$62,100	1.22892	1,243
597	9	19	15	$13,300	$34,531	$68,000	1.42167	1,309
398	9	19	15	$12,500	$52,051	$74,900	1.16032	1,612
399	9	19	15	$12,500	$43,760	$79,000	1.40419	1,798
400	9	19	15	$12,500	$48,699	$78,000	1.27453	1,682
401	9	19	15	$12,500	$44,612	$71,100	1.24492	1,487
402	9	19	15	$12,500	$38,251	$67,000	1.32017	1,260
403	9	19	15	$12,500	$38,622	$68,800	1.34580	1,344
404	9	19	15	$12,500	$48,698	$89,000	1.45430	1,655
405	9	19	15	$12,500	$53,009	$87,000	1.32806	1,399
406	9	19	15	$12,500	$57,496	$84,000	1.20007	1,752
407	9	19	15	$12,500	$42,627	$67,900	1.23170	1,240
408	9	19	15	$12,500	$39,983	$68,500	1.30518	1,286
409	9	19	15	$12,500	$46,208	$61,500	1.04756	1,411
410	9	19	15	$12,500	$49,420	$77,000	1.24354	1,422
411	9	19	15	$12,500	$37,624	$64,500	1.28681	1,226
412	9	19	15	$13,300	$50,315	$67,000	1.05321	1,466
413	9	19	15	$12,500	$39,370	$56,800	1.09505	1,384
414	9	19	15	$12,500	$42,891	$68,000	1.22764	1,340
415	9	19	15	$12,500	$40,614	$72,000	1.35557	1,383
416	9	19	15	$12,500	$45,988	$80,000	1.36780	1,493
417	9	19	15	$12,500	$45,505	$74,000	1.27575	1,554
418	9	19	15	$12,500	$52,623	$75,000	1.15167	1,482
419	9	19	15	$12,500	$50,437	$82,000	1.30289	1,540
420	9	19	15	$13,800	$50,569	$70,000	1.08748	1,388
421	9	19	15	$13,800	$44,464	$72,500	1.24434	1,436
422	9	19	15	$13,776	$47,146	$79,600	1.30659	1,498
423	9	19	15	$12,485	$55,593	$85,000	1.24857	1,384
424	9	19	15	$12,485	$66,519	$115,900	1.46701	1,988
425	9	19	15	$12,915	$53,357	$78,000	1.17697	1,564
426	9	19	15	$12,485	$43,862	$70,000	1.24230	1,408
427	9	19	15	$12,485	$52,384	$77,500	1.19472	1,264
428	9	19	15	$12,485	$60,845	$98,000	1.33642	2,050
429	9	19	15	$12,485	$49,890	$91,500	1.46693	1,256
430	9	19	15	$12,485	$52,543	$81,000	1.24562	1,621
431	9	19	15	$12,485	$50,671	$89,800	1.42188	1,123
432	9	19	15	$12,485	$48,428	$66,000	1.08351	1,278

REAL ESTATE SALES FOR 1986
IN A MID-SIZED FLORIDA CITY

OBS	TOWNSHIP	RANGE	SECTION	LANDVAL	IMPROVAL	SALEPRIC	SALTOAPR	AREA
433	9	19	15	$12,485	$54,285	$82,000	1.22810	1,674
434	9	19	15	$12,485	$52,578	$88,000	1.35254	1,240
435	9	19	15	$12,485	$52,344	$80,000	1.23402	1,531
436	9	19	15	$12,915	$54,440	$87,500	1.29909	1,568
437	9	19	15	$20,000	$65,611	$129,000	1.50682	2,450
438	9	19	15	$20,000	$76,723	$128,500	1.32854	2,165
439	9	19	15	$18,000	$68,097	$125,000	1.45185	2,278
440	9	19	19	$20,000	$70,554	$110,000	1.21474	2,292
441	9	19	15	$20,000	$73,620	$114,000	1.21769	2,277
442	9	19	15	$18,000	$79,793	$125,400	1.28230	2,290
443	9	19	16	$18,480	$78,142	$136,100	1.40858	2,360
444	9	19	17	$17,000	$76,013	$98,000	1.05362	2,007
445	9	19	18	$33,300	$103,263	$156,500	1.14599	2,770
446	9	19	18	$26,700	$82,415	$130,300	1.19415	2,289
447	9	19	18	$23,850	$82,639	$139,900	1.31375	2,812
448	9	19	20	$18,000	$73,297	$129,900	1.42283	1,704
449	9	19	20	$18,000	$63,445	$98,000	1.20327	2,091
450	9	19	20	$17,000	$61,918	$106,000	1.34317	2,168
451	9	19	20	$17,000	$61,181	$90,000	1.15117	1,814
452	9	19	20	$17,000	$75,131	$124,900	1.35568	2,588
453	9	19	20	$17,000	$83,495	$134,700	1.34037	1,537
454	9	19	20	$17,000	$57,601	$100,000	1.34046	1,278
455	9	19	20	$17,000	$70,932	$114,000	1.29646	1,205
456	9	19	20	$17,000	$68,809	$115,000	1.34019	1,601
457	9	19	21	$17,250	$57,135	$87,500	1.17631	1,669
458	9	19	21	$17,250	$59,242	$94,500	1.23542	1,836
459	9	19	21	$17,250	$64,999	$90,000	1.09424	2,320
460	9	19	21	$17,250	$54,874	$83,500	1.15773	1,738
461	9	19	21	$17,250	$51,738	$88,000	1.27558	1,767
462	9	19	21	$17,250	$58,494	$88,000	1.16181	1,787
463	9	19	21	$18,000	$64,640	$93,000	1.12536	1,941
464	9	19	21	$18,000	$52,834	$71,500	1.00940	1,716
465	9	19	21	$18,000	$51,196	$77,900	1.12579	1,582
466	9	19	21	$18,000	$58,383	$110,000	1.44011	2,034
467	9	19	21	$17,213	$75,713	$144,900	1.55931	1,404
468	9	19	21	$17,213	$73,014	$109,900	1.21804	1,314
469	9	19	21	$19,500	$69,618	$103,000	1.15577	2,064
470	9	19	21	$17,250	$59,819	$86,000	1.11588	2,400
471	9	19	21	$17,250	$61,047	$95,900	1.22482	1,848
472	9	19	21	$33,700	$127,765	$227,000	1.40588	3,450
473	9	19	21	$15,300	$76,880	$137,500	1.49165	1,404
474	9	19	21	$15,300	$71,245	$147,900	1.70894	2,394
475	9	19	21	$18,000	$74,214	$125,000	1.35554	2,200
476	9	19	21	$18,000	$72,827	$116,500	1.28266	2,214
477	9	19	21	$18,000	$62,595	$87,000	1.07947	1,856
478	9	19	21	$18,000	$79,118	$97,000	0.99878	2,262
479	9	19	21	$19,050	$73,850	$105,000	1.13025	2,224
480	9	19	21	$18,000	$79,951	$124,000	1.26594	2,513
481	9	19	21	$19,500	$79,534	$129,800	1.31066	1,870
482	9	19	21	$16,500	$74,745	$123,000	1.34802	2,416
483	9	19	21	$16,500	$58,245	$84,500	1.13051	1,916
484	9	19	21	$18,000	$69,446	$124,000	1.41802	2,117
485	9	19	21	$18,000	$58,105	$90,000	1.18258	1,768
486	9	19	21	$18,000	$80,353	$125,100	1.27195	2,168
487	9	19	21	$18,000	$73,006	$115,000	1.26365	2,260
488	9	19	22	$22,910	$223,920	$249,200	1.00960	1,644
489	9	19	22	$12,000	$48,741	$62,000	1.02073	1,554
490	9	19	22	$10,000	$52,082	$69,000	1.11143	1,798
491	9	19	23	$10,000	$34,721	$56,000	1.25221	1,497
492	9	19	23	$10,000	$34,890	$59,500	1.32546	1,497
493	9	19	23	$10,000	$32,949	$57,900	1.34811	1,274
494	9	19	23	$10,000	$36,088	$59,900	1.29969	1,367
495	9	19	23	$10,000	$52,244	$70,000	1.12461	1,674
496	9	19	23	$10,000	$34,947	$61,000	1.35715	1,300
497	9	19	23	$10,000	$41,139	$54,000	1.05595	1,324
498	9	19	23	$10,000	$40,141	$60,300	1.20261	1,418
499	9	19	23	$10,000	$38,680	$60,100	1.23459	1,358
500	9	19	23	$10,000	$43,181	$63,000	1.18463	1,582
501	9	19	23	$10,000	$36,896	$52,000	1.10884	1,456
502	9	19	23	$10,000	$51,098	$66,000	1.08023	1,737
503	9	19	23	$10,500	$43,424	$62,500	1.15904	1,651
504	9	19	23	$5,960	$9,321	$111,500	7.29664	192

REAL ESTATE SALES FOR 1986
IN A MID-SIZED FLORIDA CITY

OBS	TOWNSHIP	RANGE	SECTION	LANDVAL	IMPROVAL	SALEPRIC	SALTOAPR	AREA
505	9	19	23	$9,500	$43,158	$68,000	1.29135	1,614
506	9	19	23	$9,000	$38,113	$54,900	1.16528	1,248
507	9	19	23	$9,500	$43,819	$59,900	1.12343	1,500
508	9	19	23	$10,000	$42,432	$68,000	1.29692	1,626
509	9	19	23	$10,000	$49,414	$74,900	1.26065	1,671
510	9	19	23	$9,500	$43,214	$63,000	1.19513	1,404
511	9	19	23	$20,000	$59,249	$89,000	1.12304	2,153
512	9	19	23	$20,000	$62,085	$94,000	1.14515	2,040
513	9	19	23	$18,000	$69,937	$96,600	1.09851	2,237
514	9	19	23	$11,000	$28,526	$41,000	1.03729	1,351
515	9	19	23	$11,000	$34,440	$59,000	1.29842	1,662
516	9	19	23	$11,000	$28,767	$55,000	1.38306	1,160
517	9	19	23	$11,000	$24,723	$47,500	1.32968	1,160
518	9	19	24	$2,900	$44,786	$75,000	1.57279	1,369
519	9	19	24	$2,400	$34,679	$45,000	1.21362	1,206
520	9	19	24	$2,300	$44,949	$49,500	1.04764	1,579
521	9	19	24	$2,300	$44,949	$54,900	1.16193	1,579
522	9	19	24	$2,300	$44,949	$63,500	1.34394	1,579
523	9	19	24	$6,400	$39,316	$63,000	1.37807	1,259
524	9	19	24	$6,500	$58,479	$89,000	1.36967	2,094
525	9	19	24	$6,500	$46,241	$74,800	1.41825	1,540
526	9	19	24	$8,500	$25,065	$42,200	1.25726	912
527	9	19	24	$9,000	$31,178	$50,500	1.25691	1,026
528	9	19	24	$8,500	$27,133	$46,000	1.29094	1,020
529	9	19	24	$8,500	$29,546	$44,500	1.16964	972
530	9	19	24	$8,500	$45,166	$59,000	1.09939	1,269
531	9	19	24	$8,500	$30,752	$45,000	1.14644	1,026
532	9	19	24	$8,500	$44,245	$56,000	1.06171	1,341
533	9	19	24	$8,500	$47,702	$55,500	0.98751	1,474
534	9	19	24	$8,500	$37,515	$49,000	1.06487	1,100
535	9	19	24	$8,500	$35,395	$56,500	1.28716	1,286
536	9	19	24	$8,500	$36,267	$55,200	1.23305	1,070
537	9	19	24	$8,500	$29,601	$48,000	1.25981	1,044
538	9	19	24	$8,500	$34,977	$50,500	1.16153	1,131
539	9	19	24	$8,500	$32,970	$55,500	1.33832	954
540	9	19	24	$8,500	$26,420	$46,000	1.31730	912
541	9	19	24	$9,000	$30,948	$50,500	1.26732	1,020
542	9	19	24	$8,500	$27,588	$44,000	1.21924	1,008
543	9	19	24	$8,500	$27,852	$52,500	1.44421	960
544	9	19	24	$8,500	$27,282	$43,700	1.22128	912
545	9	19	24	$8,500	$25,555	$41,900	1.23036	912
546	9	19	24	$8,500	$27,317	$45,000	1.25639	912
547	9	19	24	$8,500	$28,298	$47,000	1.27724	972
548	9	19	24	$8,500	$28,360	$49,900	1.35377	1,152
549	9	19	24	$8,500	$25,356	$42,000	1.24055	912
550	9	19	24	$8,000	$26,727	$44,700	1.28718	972
551	9	19	24	$9,000	$28,072	$49,900	1.34603	1,020
552	9	19	24	$8,500	$26,757	$46,900	1.33023	972
553	9	19	24	$8,500	$29,981	$51,700	1.34352	1,020
554	9	19	24	$8,500	$32,473	$52,000	1.26913	1,146
555	9	19	24	$8,500	$38,906	$53,500	1.12855	1,260
556	9	19	24	$8,500	$31,141	$45,000	1.13519	1,027
557	9	19	24	$9,500	$30,001	$49,900	1.26326	1,008
558	9	19	24	$8,500	$32,931	$47,500	1.14648	990
559	9	19	24	$8,500	$36,665	$56,000	1.23990	1,178
560	9	19	24	$8,500	$39,052	$59,900	1.25967	1,296
561	9	19	24	$8,500	$35,716	$59,800	1.35245	1,146
562	9	19	24	$10,500	$35,438	$59,300	1.29087	1,008
563	9	19	24	$10,500	$38,828	$58,900	1.19405	1,204
564	9	19	24	$9,100	$31,983	$56,000	1.36309	1,248
565	9	19	24	$9,700	$32,055	$48,000	1.14956	1,321
566	9	19	24	$10,300	$35,374	$56,000	1.22608	1,290
567	9	19	24	$9,700	$22,330	$42,000	1.31127	900
568	9	19	24	$9,600	$25,010	$48,000	1.38688	1,056
569	9	19	24	$9,600	$26,704	$46,000	1.26708	1,050
570	9	19	23	$9,500	$29,527	$43,500	1.11461	1,103
571	9	19	24	$9,500	$35,493	$58,900	1.30909	1,327
572	9	19	24	$9,500	$34,292	$50,000	1.14176	1,260
573	9	19	24	$9,900	$37,863	$63,900	1.33786	1,210
574	9	19	24	$10,700	$26,821	$45,600	1.21532	1,008
575	9	19	24	$11,100	$32,771	$48,900	1.11463	1,226
576	9	19	24	$9,600	$33,522	$52,800	1.22443	1,226

REAL ESTATE SALES FOR 1986
IN A MID-SIZED FLORIDA CITY

OBS	TOWNSHIP	RANGE	SECTION	LANDVAL	IMPROVAL	SALEPRIC	SALTOAPR	AREA
577	9	19	24	$10,100	$27,047	$50,500	1.35946	1,026
578	9	19	24	$9,500	$28,140	$44,500	1.18225	1,008
579	9	19	24	$9,500	$26,040	$49,900	1.40405	960
580	9	19	24	$10,700	$39,468	$63,000	1.25578	1,353
581	9	19	24	$10,400	$41,732	$60,000	1.15092	1,358
582	9	19	24	$9,500	$25,930	$46,500	1.31245	912
583	9	19	24	$9,500	$26,200	$46,900	1.31373	960
584	9	19	24	$9,500	$32,592	$43,300	1.02870	1,226
585	9	19	24	$9,500	$31,238	$56,700	1.39182	1,144
586	9	19	24	$9,600	$25,213	$50,300	1.44486	972
587	9	19	24	$9,600	$24,100	$45,000	1.33531	912
588	9	19	24	$11,200	$41,140	$56,800	1.08521	1,242
589	9	19	24	$10,900	$30,247	$46,500	1.13009	1,032
590	9	19	24	$9,500	$35,019	$50,000	1.12312	1,226
591	9	19	24	$10,000	$34,496	$66,500	1.49452	1,226
592	9	19	24	$9,600	$32,336	$53,000	1.26383	1,176
593	9	19	24	$11,000	$31,646	$51,000	1.19589	1,226
594	9	19	24	$9,500	$25,552	$44,000	1.25528	931
595	9	19	24	$10,400	$31,700	$49,000	1.16390	1,074
596	9	19	24	$9,300	$35,217	$54,000	1.21302	1,240
597	9	19	24	$9,300	$27,839	$49,500	1.33283	1,050
598	9	19	24	$8,500	$28,560	$51,000	1.37615	1,225
599	9	19	24	$8,500	$24,879	$47,500	1.42305	1,008
600	9	19	24	$8,700	$30,465	$44,000	1.12345	1,120
601	9	19	24	$10,300	$31,425	$49,500	1.18634	1,118
602	9	19	24	$9,200	$37,274	$50,000	1.07587	1,504
603	9	19	24	$9,300	$37,267	$63,200	1.35718	1,320
604	9	19	24	$9,700	$40,259	$53,500	1.07088	1,280
605	9	19	24	$9,500	$38,114	$63,000	1.32314	1,350
606	9	19	24	$9,700	$33,085	$53,900	1.25979	1,248
607	9	19	24	$9,700	$27,439	$47,000	1.26552	1,025
608	9	19	24	$9,500	$36,228	$54,000	1.18090	1,430
609	9	19	24	$9,500	$37,564	$61,500	1.30673	1,400
610	9	19	24	$9,700	$37,366	$57,000	1.21107	1,356
611	9	19	24	$9,900	$39,458	$59,000	1.19535	1,384
612	9	19	24	$9,900	$45,572	$62,000	1.11768	1,498
613	9	19	24	$9,500	$34,951	$56,300	1.26656	1,269
614	9	19	24	$9,500	$37,211	$57,000	1.22027	1,360
615	9	19	24	$9,500	$35,818	$55,000	1.21365	1,170
616	9	19	24	$9,500	$38,407	$54,000	1.12718	1,346
617	9	19	24	$9,500	$35,831	$45,500	1.00373	1,170
618	9	19	24	$9,900	$39,794	$65,000	1.30800	1,440
619	9	19	24	$9,800	$45,186	$63,500	1.15484	1,389
620	9	19	24	$9,500	$40,652	$55,900	1.11461	1,496
621	9	19	25	$16,000	$70,227	$104,000	1.20612	1,407
622	9	19	25	$17,000	$62,583	$100,000	1.25655	1,827
623	9	19	25	$17,000	$62,980	$105,000	1.31283	1,821
624	9	19	25	$38,000	$137,512	$235,000	1.33894	3,194
625	9	19	25	$14,000	$50,965	$76,400	1.17602	1,646
626	9	19	25	$15,000	$49,243	$84,300	1.31221	1,396
627	9	19	25	$15,000	$47,940	$55,500	0.88179	1,897
628	9	19	25	$15,000	$49,236	$74,000	1.15200	1,498
629	9	19	25	$15,000	$51,891	$83,500	1.24830	2,183
630	9	19	25	$16,000	$49,301	$97,000	1.48543	1,972
631	9	19	25	$16,000	$69,554	$95,000	1.11041	2,329
632	9	19	25	$14,000	$53,798	$78,000	1.15048	1,799
633	9	19	25	$11,000	$56,319	$71,800	1.06656	1,829
634	9	19	25	$11,000	$51,645	$78,300	1.24990	1,763
635	9	19	25	$16,000	$53,993	$82,500	1.17869	1,917
636	9	19	25	$16,000	$49,262	$68,500	1.04962	1,759
637	9	19	25	$23,500	$99,763	$151,500	1.22908	2,506
638	9	19	25	$23,500	$107,505	$160,000	1.22133	2,661
639	9	19	25	$16,000	$13,885	$40,000	1.33846	834
640	9	19	25	$12,000	$28,549	$53,700	1.32432	1,056
641	9	19	25	$12,000	$37,776	$68,900	1.38420	1,517
642	9	19	25	$12,000	$48,014	$61,900	1.03143	1,972
643	9	19	25	$10,200	$53,732	$87,900	1.37490	2,146
644	9	19	25	$12,000	$38,711	$58,900	1.16148	1,566
645	9	19	25	$12,000	$45,509	$67,500	1.17373	1,828
646	9	19	25	$13,000	$48,006	$72,200	1.18349	1,801
647	9	19	25	$12,000	$50,198	$77,700	1.24924	1,846
648	9	19	25	$12,000	$48,038	$70,000	1.16593	1,728

REAL ESTATE SALES FOR 1986
IN A MID-SIZED FLORIDA CITY

OBS	TOWNSHIP	RANGE	SECTION	LANDVAL	IMPROVAL	SALEPRIC	SALTOAPR	AREA
649	9	19	25	$16,000	$52,156	$75,000	1.10042	2,126
650	9	19	25	$16,000	$56,536	$91,000	1.25455	2,024
651	9	19	25	$16,000	$59,697	$123,900	1.63679	1,901
652	9	19	25	$16,000	$45,488	$84,000	1.36612	1,560
653	9	19	25	$16,000	$70,940	$103,000	1.18473	2,170
654	9	19	25	$17,500	$59,349	$110,000	1.43138	1,756
655	9	19	25	$16,000	$72,615	$121,000	1.36546	2,292
656	9	19	25	$16,000	$82,760	$110,000	1.11381	2,671
657	9	19	25	$13,000	$56,153	$78,000	1.12793	1,928
658	9	19	25	$13,000	$48,658	$69,000	1.11908	1,681
659	9	19	25	$13,000	$65,354	$78,000	0.99548	2,028
660	9	19	25	$13,000	$50,523	$80,000	1.25939	1,677
661	9	19	25	$13,000	$59,730	$90,000	1.23745	2,358
662	9	19	25	$13,000	$78,767	$104,000	1.13331	2,996
663	9	19	25	$13,000	$63,142	$83,000	1.09007	2,377
664	9	19	25	$13,000	$48,412	$66,500	1.08285	1,664
665	9	19	26	$10,000	$29,496	$53,000	1.34191	1,427
666	9	19	26	$7,140	$55,326	$85,000	1.36074	1,646
667	9	19	26	$16,200	$47,283	$72,000	1.13416	1,571
668	9	19	26	$21,000	$55,873	$96,000	1.24881	1,800
669	9	19	26	$21,000	$51,911	$77,100	1.05745	1,792
670	9	19	26	$22,000	$54,088	$84,500	1.11056	1,912
671	9	19	26	$28,000	$66,002	$121,000	1.28721	2,445
672	9	19	26	$28,000	$58,132	$105,900	1.22951	2,242
673	9	19	26	$21,000	$62,144	$90,000	1.08246	2,234
674	9	19	26	$21,000	$53,743	$82,500	1.10378	1,664
675	9	19	26	$21,000	$60,841	$110,000	1.34407	2,044
676	9	19	26	$21,000	$54,619	$103,000	1.36209	1,949
677	9	19	26	$21,000	$61,018	$104,500	1.27411	2,118
678	9	19	26	$21,000	$52,388	$87,000	1.18348	1,725
679	9	19	26	$21,000	$67,664	$120,000	1.35342	2,292
680	9	19	26	$15,500	$38,289	$59,000	1.09688	1,696
681	9	19	26	$15,500	$44,272	$55,000	0.92016	1,702
682	9	19	26	$32,000	$51,598	$106,000	1.26797	1,725
683	9	19	26	$25,000	$49,323	$85,800	1.15442	1,608
684	9	19	26	$25,000	$50,395	$89,900	1.19239	1,688
685	9	19	26	$25,000	$47,871	$86,900	1.19252	1,631
686	9	19	26	$25,000	$51,215	$80,000	1.04966	1,671
687	9	19	26	$25,000	$52,154	$83,000	1.07577	1,810
688	9	19	26	$25,000	$64,200	$110,000	1.23318	1,977
689	9	19	26	$25,000	$53,365	$80,000	1.02086	1,827
690	9	19	26	$23,000	$59,176	$88,000	1.07087	1,018
691	9	19	26	$30,000	$65,401	$105,000	1.10062	2,266
692	9	19	26	$25,000	$66,672	$103,000	1.12357	2,023
693	9	19	26	$25,000	$58,287	$97,500	1.17065	2,138
694	9	19	26	$25,000	$63,104	$105,000	1.19177	2,154
695	9	19	26	$22,500	$53,642	$79,900	1.04936	1,889
696	9	19	26	$25,000	$70,628	$125,400	1.31133	2,418
697	9	19	26	$23,000	$44,863	$70,000	1.03149	1,744
698	9	19	26	$23,000	$37,001	$66,200	1.10331	1,320
699	9	19	26	$30,000	$61,548	$110,000	1.20156	1,975
700	9	19	26	$30,000	$68,498	$125,000	1.26906	2,142
701	9	19	26	$10,800	$25,643	$40,000	1.09760	1,192
702	9	19	26	$19,500	$50,563	$85,000	1.21319	1,454
703	9	19	26	$2,000	$46,947	$71,390	1.45852	1,059
704	9	19	26	$3,500	$46,947	$63,800	1.26469	1,059
705	9	19	26	$4,000	$45,143	$66,800	1.35930	1,223
706	9	19	26	$4,000	$55,308	$87,700	1.47872	1,542
707	9	19	26	$4,000	$44,696	$75,200	1.54427	1,224
708	9	19	26	$4,000	$44,696	$68,500	1.40669	1,224
709	9	19	26	$4,000	$38,517	$63,100	1.48411	1,050
710	9	19	26	$4,000	$44,696	$51,600	1.05964	1,224
711	9	19	26	$39,900	$130,000	$220,500	1.29782	2,600
712	9	19	26	$16,000	$48,965	$70,000	1.07750	2,079
713	9	19	26	$16,000	$59,041	$86,000	1.14604	2,104
714	9	19	26	$16,000	$55,673	$93,000	1.29756	2,104
715	9	19	26	$16,000	$50,122	$85,000	1.28550	2,107
716	9	19	27	$19,300	$57,706	$92,300	1.19861	1,467
717	9	19	27	$15,000	$31,656	$45,000	0.96451	1,643
718	9	19	27	$13,000	$40,527	$63,500	1.18632	1,254
719	9	19	27	$13,000	$39,645	$63,800	1.21189	1,208
720	9	19	27	$13,500	$42,916	$68,600	1.21597	1,238

REAL ESTATE SALES FOR 1986
IN A MID-SIZED FLORIDA CITY

OBS	TOWNSHIP	RANGE	SECTION	LANDVAL	IMPROVAL	SALEPRIC	SALTOAPR	AREA
721	9	19	27	$14,000	$42,695	$66,000	1.16412	1,220
722	9	19	27	$13,000	$40,968	$62,700	1.16180	1,162
723	9	19	27	$13,000	$43,913	$66,500	1.16845	1,360
724	9	19	27	$13,000	$42,591	$67,100	1.20703	1,242
725	9	19	27	$13,000	$44,168	$60,000	1.04954	1,174
726	9	19	27	$13,000	$39,534	$61,400	1.16877	1,208
727	9	19	27	$13,000	$35,323	$34,600	0.71602	1,012
728	9	19	27	$13,000	$40,137	$63,500	1.19502	1,238
729	9	19	27	$13,000	$42,388	$62,000	1.11938	1,220
730	9	19	27	$13,500	$45,736	$63,600	1.07367	1,387
731	9	19	27	$13,500	$40,216	$65,000	1.21007	1,158
732	9	19	27	$13,000	$41,242	$61,500	1.13381	1,212
733	9	19	27	$13,000	$44,561	$61,500	1.06843	1,200
734	9	19	27	$13,500	$41,366	$59,750	1.08902	1,200
735	9	19	27	$13,500	$45,026	$62,000	1.05936	1,162
736	9	19	27	$13,500	$44,964	$68,500	1.17166	1,180
737	9	19	26	$18,000	$52,940	$76,000	1.07133	2,169
738	9	19	26	$18,000	$45,462	$74,000	1.16605	1,672
739	9	19	26	$18,000	$48,691	$76,000	1.13958	1,662
740	9	19	26	$18,000	$50,512	$79,000	1.15308	2,031
741	9	19	26	$18,000	$52,900	$90,000	1.26939	2,016
742	9	19	26	$18,000	$45,612	$72,000	1.13186	1,716
743	9	19	26	$18,000	$55,886	$92,100	1.24651	1,922
744	9	19	26	$18,000	$47,195	$75,700	1.16113	1,796
745	9	19	26	$18,000	$59,289	$90,500	1.17093	1,965
746	9	19	26	$25,000	$52,804	$82,000	1.05393	1,822
747	9	19	26	$18,000	$45,223	$54,500	0.86203	1,695
748	9	19	27	$10,000	$43,789	$72,200	1.34228	1,699
749	9	19	27	$13,000	$45,609	$71,700	1.22336	1,490
750	9	19	27	$13,000	$42,157	$64,500	1.16939	1,298
751	9	19	27	$13,000	$36,967	$63,500	1.27084	1,250
752	9	19	27	$13,000	$42,208	$63,500	1.15020	1,252
753	9	19	27	$13,000	$44,308	$68,900	1.20228	1,387
754	9	19	27	$13,000	$38,311	$60,700	1.18298	1,219
755	9	19	27	$13,000	$46,699	$68,000	1.13905	1,431
756	9	19	27	$13,000	$41,905	$58,500	1.06548	1,264
757	9	19	27	$8,000	$44,910	$49,000	0.92610	926
758	9	19	27	$8,000	$44,910	$49,500	0.93555	926
759	9	19	27	$8,000	$27,492	$36,900	1.03967	818
760	9	19	27	$8,000	$29,856	$40,000	1.05664	899
761	9	19	27	$8,000	$29,856	$31,000	0.81889	899
762	9	19	27	$8,000	$29,856	$38,000	1.00380	899
763	9	19	27	$8,000	$31,914	$43,500	1.08984	929
764	9	19	27	$13,000	$43,933	$68,500	1.20317	1,438
765	9	19	27	$13,000	$43,713	$68,000	1.19902	1,397
766	9	19	27	$13,000	$44,390	$68,500	1.19359	1,400
767	9	19	27	$13,000	$39,577	$65,900	1.25340	1,230
768	9	19	27	$13,000	$58,176	$79,900	1.12257	1,432
769	9	19	27	$12,500	$38,730	$61,500	1.20047	1,282
770	9	19	27	$12,500	$47,058	$70,000	1.17532	1,496
771	9	19	27	$12,500	$45,039	$58,400	1.01496	1,536
772	9	19	27	$12,000	$36,422	$54,000	1.11520	1,202
773	9	19	27	$12,000	$37,983	$60,000	1.20041	1,197
774	9	19	27	$12,000	$36,905	$61,000	1.24732	1,232
775	9	19	27	$12,000	$46,185	$61,000	1.04838	1,400
776	9	19	27	$12,000	$36,927	$50,100	1.02397	1,200
777	9	19	27	$12,000	$40,548	$63,800	1.21413	1,415
778	9	19	27	$12,000	$40,514	$61,500	1.17112	1,312
779	9	19	27	$12,000	$38,463	$56,900	1.12756	1,170
780	9	19	27	$12,000	$37,576	$55,000	1.10941	1,246
781	9	19	27	$11,000	$36,880	$57,500	1.20092	1,144
782	9	19	27	$11,000	$36,095	$49,900	1.05956	1,160
783	9	19	27	$11,000	$37,898	$42,000	0.85893	1,204
784	9	19	27	$11,000	$36,047	$49,900	1.06064	1,368
785	9	19	27	$11,000	$35,358	$49,900	1.07641	1,032
786	9	19	27	$11,000	$39,635	$60,000	1.18495	1,214
787	9	19	27	$11,000	$38,276	$55,000	1.11616	1,303
788	9	19	27	$11,000	$43,202	$61,000	1.12542	1,419
789	9	19	27	$11,000	$39,961	$55,000	1.07926	1,284
790	9	19	27	$11,000	$39,062	$57,800	1.15457	1,288
791	9	19	27	$11,000	$45,566	$65,000	1.14910	1,248
792	9	19	27	$11,000	$41,232	$59,700	1.14298	1,456

REAL ESTATE SALES FOR 1986
IN A MID-SIZED FLORIDA CITY

OBS	TOWNSHIP	RANGE	SECTION	LANDVAL	IMPROVAL	SALEPRIC	SALTOAPR	AREA
793	9	19	27	$11,000	$42,713	$66,000	1.22875	1,428
794	9	19	27	$12,000	$45,186	$61,500	1.07544	1,400
795	9	19	27	$11,500	$41,067	$56,900	1.08243	1,238
796	9	19	27	$11,000	$38,544	$58,500	1.18077	1,222
797	9	19	27	$11,000	$35,415	$62,500	1.34655	1,148
798	9	19	28	$13,500	$44,419	$78,000	1.34671	1,161
799	9	19	28	$13,000	$45,245	$72,800	1.24989	1,398
800	9	19	28	$13,000	$47,632	$71,900	1.18584	1,530
801	9	19	28	$13,000	$49,988	$78,500	1.24627	1,603
802	9	19	28	$13,500	$42,013	$70,000	1.26097	1,528
803	9	19	28	$14,000	$55,829	$84,000	1.20294	1,583
804	9	19	28	$13,000	$58,393	$89,000	1.24662	2,134
805	9	19	28	$13,000	$61,588	$106,000	1.42114	2,147
806	9	19	28	$13,000	$57,676	$83,000	1.17437	1,790
807	9	19	28	$13,000	$60,112	$82,800	1.13251	2,106
808	9	19	28	$13,000	$45,999	$91,100	1.54409	1,632
809	9	19	28	$13,000	$52,834	$75,900	1.15290	832
810	9	19	28	$24,000	$68,730	$139,900	1.50868	2,129
811	9	19	28	$13,000	$49,690	$66,300	1.05758	1,696
812	9	19	28	$13,000	$53,719	$78,000	1.16908	1,912
813	9	19	28	$13,000	$57,942	$82,000	1.15587	1,738
814	9	19	28	$13,000	$52,286	$79,600	1.21925	1,766
815	9	19	29	$8,500	$50,115	$81,300	1.38702	1,616
816	9	19	29	$8,500	$49,487	$79,000	1.36237	888
817	9	19	29	$8,500	$56,275	$84,000	1.29680	1,838
818	9	19	29	$11,500	$42,053	$65,700	1.22682	1,432
819	9	19	29	$11,500	$37,935	$56,600	1.14494	1,300
820	9	19	30	$10,000	$39,426	$76,600	1.54979	1,214
821	9	19	30	$21,000	$47,917	$103,400	1.50036	1,559
822	9	19	30	$21,000	$47,937	$100,800	1.46220	1,559
823	9	19	30	$21,000	$46,822	$100,300	1.47887	1,403
824	9	19	30	$10,000	$43,268	$90,900	1.70647	1,403
825	9	19	30	$21,000	$49,221	$119,300	1.69892	1,571
826	9	19	30	$4,000	$36,882	$61,400	1.50188	1,131
827	9	19	30	$9,500	$47,590	$84,000	1.47136	1,030
828	9	19	30	$9,500	$54,795	$15,000	0.23330	1,620
829	9	19	30	$10,500	$51,788	$85,500	1.37266	1,678
830	9	19	30	$10,500	$35,495	$59,500	1.29362	1,293
831	9	19	31	$15,000	$58,712	$100,000	1.35663	1,986
832	9	19	31	$15,000	$51,380	$82,000	1.23531	1,496
833	9	19	31	$15,000	$57,195	$91,000	1.26048	1,780
834	9	19	31	$15,000	$58,072	$95,000	1.30009	1,569
835	9	19	31	$29,000	$143,439	$275,000	1.59477	3,064
836	9	19	31	$29,000	$83,646	$155,000	1.37599	2,890
837	9	19	31	$14,000	$72,248	$117,000	1.35655	2,714
838	9	19	31	$14,000	$60,650	$83,500	1.11855	1,893
839	9	19	31	$14,000	$62,647	$115,500	1.50691	1,929
840	9	19	31	$14,000	$58,311	$95,000	1.31377	2,016
841	9	19	31	$14,000	$55,513	$87,000	1.25156	1,318
842	9	19	31	$14,000	$77,662	$119,000	1.29825	2,505
843	9	19	31	$14,000	$54,703	$92,000	1.33910	1,616
844	9	19	31	$14,000	$68,328	$118,500	1.43936	1,285
845	9	19	31	$14,000	$54,016	$94,000	1.38203	1,669
846	9	19	31	$14,000	$56,392	$94,500	1.34248	1,825
847	9	19	31	$14,000	$58,266	$96,500	1.33534	1,896
848	9	19	31	$14,000	$59,949	$91,500	1.23734	1,825
849	9	19	31	$14,000	$52,033	$76,100	1.15245	1,560
850	9	19	33	$743,713	$41,802	1500000	1.90958	841
851	9	19	33	$40,500	$140,715	$158,000	0.87189	3,544
852	9	19	33	$30,000	$77,708	$127,900	1.18747	2,516
853	9	19	33	$33,750	$98,789	$147,500	1.11288	3,359
854	9	19	33	$40,500	$117,044	$217,000	1.37739	3,676
855	9	19	33	$27,500	$94,753	$165,000	1.34966	3,510
856	9	19	33	$27,500	$63,012	$98,500	1.08825	2,229
857	9	19	33	$27,500	$94,110	$150,000	1.23345	3,200
858	9	19	33	$39,000	$67,132	$127,000	1.19662	2,488
859	9	19	33	$39,000	$77,482	$166,000	1.42511	3,156
860	9	19	33	$43,680	$50,163	$112,000	1.19348	1,874
861	9	19	33	$30,000	$78,211	$140,000	1.29377	2,590
862	9	19	33	$27,500	$95,135	$157,300	1.28267	4,116
863	9	19	34	$11,500	$15,891	$39,900	1.45668	848
864	9	19	34	$11,500	$14,713	$36,000	1.37336	775

REAL ESTATE SALES FOR 1986
IN A MID-SIZED FLORIDA CITY

OBS	TOWNSHIP	RANGE	SECTION	LANDVAL	IMPROVAL	SALEPRIC	SALTOAPR	AREA
865	9	19	34	$11,500	$17,641	$37,800	1.29714	720
866	9	19	34	$11,500	$19,817	$33,000	1.05374	936
867	9	19	34	$11,500	$16,980	$33,900	1.19031	864
868	9	19	34	$11,500	$24,752	$40,000	1.10339	1,016
869	9	19	34	$11,500	$20,655	$37,700	1.17245	960
870	9	19	34	$12,000	$32,585	$54,900	1.23136	1,274
871	9	19	34	$12,000	$46,825	$64,500	1.09647	1,766
872	9	19	34	$12,000	$48,511	$70,000	1.15681	1,766
873	9	19	34	$13,500	$50,515	$73,800	1.15285	1,857
874	9	19	34	$12,000	$45,084	$65,000	1.13867	1,612
875	9	19	34	$12,500	$60,412	$79,900	1.09584	2,147
876	9	19	34	$12,000	$43,301	$63,000	1.13922	1,555
877	9	19	34	$12,500	$25,134	$47,500	1.26216	1,056
878	9	19	34	$12,500	$44,737	$66,900	1.16882	1,677
879	9	19	34	$12,000	$45,893	$62,000	1.07094	1,686
880	9	19	34	$22,000	$57,543	$88,500	1.11261	1,926
881	9	19	34	$22,000	$65,264	$100,000	1.14595	2,520
882	9	19	34	$22,000	$62,552	$100,000	1.18270	2,257
883	9	19	34	$22,000	$69,025	$101,200	1.11178	2,276
884	9	19	34	$22,000	$48,691	$87,500	1.23778	1,551
885	9	19	34	$22,000	$64,455	$106,000	1.22607	2,194
886	9	19	34	$22,000	$61,333	$91,500	1.09800	2,204
887	9	19	34	$22,000	$73,571	$98,000	1.02542	2,231
888	9	19	34	$22,000	$74,829	$113,500	1.17217	2,342
889	9	19	34	$9,000	$23,544	$43,500	1.33665	1,024
890	9	19	34	$32,000	$99,051	$160,000	1.22090	2,258
891	9	19	34	$34,000	$107,174	$131,200	0.92935	2,763
892	9	19	34	$32,000	$127,940	$221,200	1.38302	3,137
893	9	19	34	$22,000	$65,349	$79,900	0.91472	2,286
894	9	19	34	$22,000	$63,905	$95,000	1.10587	2,651
895	9	19	34	$22,000	$60,393	$85,000	1.03164	2,052
896	9	19	34	$22,000	$65,644	$90,000	1.02688	2,320
897	9	19	34	$22,000	$59,116	$96,600	1.19089	2,181
898	9	19	34	$22,000	$81,622	$132,000	1.27386	2,676
899	9	19	34	$22,000	$71,694	$100,000	1.06730	2,903
900	9	19	34	$22,000	$66,280	$115,000	1.30267	2,469
901	9	19	34	$22,000	$58,501	$95,000	1.18011	1,056
902	9	19	34	$22,000	$69,090	$110,900	1.21748	1,120
903	9	19	34	$22,000	$77,990	$120,000	1.20012	3,528
904	9	19	34	$22,000	$66,313	$119,500	1.35314	2,478
905	9	19	34	$22,000	$77,279	$130,000	1.30944	3,044
906	9	19	34	$22,000	$64,548	$112,000	1.29408	2,371
907	9	19	34	$23,000	$86,591	$137,000	1.25010	3,276
908	9	19	35	$14,000	$44,112	$63,000	1.08411	1,779
909	9	19	35	$15,500	$42,862	$83,900	1.43758	1,406
910	9	19	35	$18,500	$48,124	$95,500	1.43342	1,044
911	9	19	35	$18,500	$53,324	$95,900	1.33521	902
912	9	19	35	$18,500	$50,662	$96,900	1.40106	1,136
913	9	19	35	$18,500	$52,394	$111,500	1.57277	1,136
914	9	19	35	$14,000	$47,521	$68,000	1.10531	1,488
915	9	19	35	$14,000	$47,637	$72,000	1.16813	1,504
916	9	19	35	$14,500	$52,195	$77,000	1.15451	1,645
917	9	19	35	$13,000	$33,651	$50,000	1.07179	1,225
918	9	19	35	$12,000	$35,401	$67,500	1.42402	1,500
919	9	19	35	$12,000	$28,050	$52,000	1.29838	1,197
920	9	19	35	$12,000	$30,125	$55,500	1.31751	1,284
921	9	19	35	$12,000	$32,178	$48,000	1.08651	1,272
922	9	19	35	$13,000	$31,729	$57,500	1.28552	1,201
923	9	19	35	$13,000	$30,362	$58,000	1.33758	1,202
924	9	19	35	$13,000	$39,430	$63,300	1.20732	1,424
925	9	19	35	$13,000	$41,116	$67,000	1.23808	1,453
926	9	19	35	$13,000	$45,139	$70,000	1.20401	1,647
927	9	19	35	$13,000	$45,902	$71,300	1.21049	1,726
928	9	19	35	$13,000	$38,291	$64,900	1.26533	1,388
929	9	19	35	$14,000	$51,728	$80,000	1.21714	1,684
930	9	19	35	$14,500	$41,842	$66,600	1.18207	1,711
931	9	19	35	$14,000	$39,505	$64,900	1.21297	1,666
932	9	19	35	$12,000	$36,523	$46,000	0.94800	1,419
933	9	19	35	$12,000	$37,607	$49,500	0.99784	1,326
934	9	19	35	$13,000	$28,352	$44,800	1.08338	1,161
935	9	19	35	$13,000	$32,114	$58,700	1.30115	1,305
936	9	19	35	$13,000	$25,970	$49,500	1.27021	1,203

REAL ESTATE SALES FOR 1986
IN A MID-SIZED FLORIDA CITY

OBS	TOWNSHIP	RANGE	SECTION	LANDVAL	IMPROVAL	SALEPRIC	SALTOAPR	AREA
937	9	19	35	$13,000	$32,231	$55,100	1.21819	1,209
938	9	19	35	$11,000	$21,959	$48,500	1.47153	969
939	9	19	35	$11,000	$27,554	$50,000	1.29688	1,081
940	9	19	35	$11,000	$21,216	$33,000	1.02434	912
941	9	19	35	$15,000	$44,643	$80,000	1.34131	1,635
942	9	19	35	$15,000	$46,878	$76,000	1.22822	2,132
943	9	19	35	$15,000	$34,531	$69,000	1.39307	1,558
944	9	19	35	$15,000	$36,478	$68,100	1.32290	1,539
945	9	19	35	$15,000	$64,809	$107,000	1.34070	2,566
946	9	19	35	$15,000	$30,053	$57,000	1.26518	1,624
947	9	19	35	$13,000	$43,050	$69,000	1.23104	1,683
948	9	19	35	$13,000	$30,962	$55,000	1.25108	1,517
949	9	19	35	$13,000	$43,760	$64,500	1.13636	1,764
950	9	19	35	$13,000	$31,450	$77,500	1.74353	1,405
951	9	19	35	$13,000	$45,368	$67,500	1.15646	1,936
952	9	19	35	$14,000	$41,723	$55,000	0.98703	1,514
953	9	19	35	$12,000	$48,678	$62,000	1.02179	1,758
954	9	19	35	$12,000	$37,984	$57,900	1.15837	1,656
955	9	19	35	$12,000	$29,516	$46,900	1.12968	1,251
956	9	19	35	$12,000	$33,323	$45,900	1.01273	1,313
957	9	19	35	$12,000	$35,082	$57,000	1.21065	1,460
958	9	19	35	$12,000	$44,647	$54,900	0.96916	1,952
959	9	19	35	$12,000	$28,871	$47,900	1.17198	1,200
960	9	19	35	$12,000	$41,273	$57,500	1.07935	1,736
961	9	19	35	$12,000	$43,634	$61,500	1.10544	1,727
962	9	19	35	$12,000	$33,023	$55,500	1.23270	1,431
963	9	19	35	$12,000	$38,194	$56,000	1.11567	1,612
964	9	19	35	$12,500	$47,547	$57,500	0.95758	1,792
965	9	19	35	$12,000	$42,441	$70,500	1.29498	1,792
966	9	19	35	$12,000	$45,614	$64,900	1.12646	2,077
967	9	19	35	$12,000	$33,321	$53,000	1.16944	1,344
968	9	19	35	$12,000	$40,726	$66,900	1.26882	1,726
969	9	19	35	$12,500	$37,634	$50,000	0.99733	1,672
970	9	19	35	$12,500	$30,907	$50,500	1.16341	1,224
971	9	19	35	$25,000	$81,175	$139,000	1.30916	2,072
972	9	19	35	$25,000	$90,076	$132,600	1.15228	2,413
973	9	19	35	$19,000	$79,224	$113,900	1.15959	2,064
974	9	19	35	$19,000	$81,858	$117,600	1.16600	1,987
975	9	19	35	$19,000	$91,924	$29,700	0.26775	2,434
976	9	19	35	$25,000	$87,410	$143,200	1.27391	2,338
977	9	19	35	$12,000	$69,232	$86,500	1.06485	1,468
978	9	19	35	$12,000	$45,953	$61,900	1.06811	1,484
979	9	19	36	$20,000	$66,285	$140,000	1.62253	2,128
980	9	19	36	$20,000	$56,755	$95,000	1.23770	2,030
981	9	19	36	$20,000	$57,734	$95,000	1.22212	2,168
982	9	19	36	$20,000	$47,424	$79,000	1.17169	1,788
983	9	19	36	$30,000	$77,653	$134,000	1.24474	2,870
984	9	19	36	$30,000	$95,031	$147,000	1.17571	3,468
985	9	19	36	$30,000	$86,909	$165,300	1.41392	3,519
986	9	19	36	$30,000	$74,627	$141,000	1.34764	2,443
987	9	19	36	$25,000	$116,103	$140,000	0.99218	3,771
988	9	19	36	$25,000	$101,358	$174,500	1.38100	2,496
989	9	19	36	$82,080	$150,391	$400,000	1.72064	4,579
990	9	19	36	$23,000	$51,388	$94,000	1.26364	2,132
991	9	19	36	$23,000	$54,742	$103,000	1.32490	2,051
992	9	19	36	$23,000	$53,766	$97,500	1.27009	2,060
993	9	19	36	$23,000	$47,752	$83,000	1.17311	1,803
994	9	19	36	$20,000	$42,209	$80,800	1.29885	1,512
995	9	19	36	$23,000	$73,797	$107,500	1.11057	1,771
996	9	19	36	$23,500	$64,005	$115,000	1.31421	2,096
997	9	19	36	$23,500	$69,412	$113,100	1.21728	2,144
998	9	19	36	$23,000	$52,288	$84,300	1.11970	1,738
999	9	19	36	$17,000	$38,747	$59,900	1.07450	1,826
1000	9	19	36	$17,000	$39,009	$64,600	1.15339	1,624
1001	9	19	36	$17,000	$30,572	$57,000	1.19818	1,304
1002	9	19	36	$17,000	$37,785	$87,500	1.59715	1,587
1003	9	19	36	$32,000	$114,170	$210,000	1.43668	2,830
1004	9	19	36	$32,000	$104,532	$180,000	1.31837	2,885
1005	9	19	36	$32,000	$100,572	$200,000	1.50861	2,479
1006	9	19	36	$36,000	$94,722	$166,000	1.26987	2,704
1007	9	19	36	$32,000	$112,144	$190,000	1.31813	2,696
1008	9	19	36	$33,000	$106,110	$208,000	1.49522	2,752

REAL ESTATE SALES FOR 1986
IN A MID-SIZED FLORIDA CITY

OBS	TOWNSHIP	RANGE	SECTION	LANDVAL	IMPROVAL	SALEPRIC	SALTOAPR	AREA
1009	9	19	36	$33,000	$110,188	$200,000	1.39677	2,608
1010	9	19	36	$33,000	$90,822	$178,000	1.43755	2,307
1011	10	19	1	$45,000	$99,739	$245,000	1.69270	4,048
1012	10	19	1	$15,000	$23,607	$58,500	1.51527	1,051
1013	10	19	1	$15,000	$19,366	$54,500	1.58587	1,006
1014	10	19	1	$15,000	$17,539	$51,500	1.58272	968
1015	10	19	1	$14,000	$26,330	$48,700	1.20754	1,098
1016	10	19	1	$14,000	$24,461	$52,500	1.36502	1,314
1017	10	19	1	$14,000	$37,690	$57,000	1.10273	2,016
1018	10	19	1	$14,000	$30,465	$52,000	1.16946	1,353
1019	10	19	1	$8,500	$24,136	$46,500	1.42481	1,134
1020	10	19	1	$12,750	$25,730	$68,900	1.79054	1,540
1021	10	19	1	$21,000	$51,514	$94,500	1.30320	2,417
1022	10	19	1	$17,000	$31,650	$60,000	1.23330	1,258
1023	10	19	1	$12,800	$21,145	$60,000	1.76757	998
1024	10	19	1	$21,250	$23,263	$125,000	2.80817	1,382
1025	10	19	1	$8,500	$22,171	$56,500	1.84213	1,170
1026	10	19	1	$13,000	$40,254	$66,000	1.23934	2,027
1027	10	19	1	$15,000	$29,942	$68,500	1.52419	1,286
1028	10	19	1	$18,000	$41,500	$84,000	1.41176	1,890
1029	10	19	1	$18,000	$56,301	$101,000	1.35934	2,518
1030	10	19	1	$17,500	$41,464	$82,500	1.39916	2,129
1031	10	19	1	$17,500	$45,348	$65,000	1.03424	1,858
1032	10	19	1	$17,500	$45,976	$79,000	1.24456	2,109
1033	10	19	1	$17,500	$37,148	$74,000	1.35412	1,714
1034	10	19	1	$17,500	$45,720	$70,000	1.10724	1,876
1035	10	19	1	$17,500	$40,097	$63,000	1.09381	1,410
1036	10	19	1	$17,500	$42,447	$60,000	1.00088	1,650
1037	10	19	1	$17,500	$43,026	$65,400	1.08053	2,170
1038	10	19	1	$17,500	$53,126	$58,200	0.82406	2,053
1039	10	19	1	$17,500	$41,247	$65,000	1.10644	1,679
1040	10	19	1	$17,500	$72,586	$83,500	0.92689	2,281
1041	10	19	1	$14,000	$56,633	$81,000	1.14677	2,115
1042	10	19	1	$17,500	$38,637	$69,000	1.22914	1,260
1043	10	19	1	$24,000	$96,838	$150,000	1.24133	3,588
1044	10	19	1	$17,500	$56,270	$81,500	1.10479	2,358
1045	10	19	1	$35,000	$85,274	$115,500	0.96031	4,136
1046	10	19	1	$43,000	$86,561	$152,400	1.17628	1,312
1047	10	19	1	$43,000	$61,857	$137,000	1.30654	1,768
1048	10	19	2	$13,000	$26,921	$50,000	1.25247	1,236
1049	10	19	2	$13,000	$19,100	$43,000	1.33956	925
1050	10	19	2	$13,000	$23,324	$42,500	1.17003	925
1051	10	19	2	$13,000	$21,707	$45,000	1.29657	1,032
1052	10	19	2	$13,000	$29,271	$53,900	1.27511	1,410
1053	10	19	2	$13,000	$24,608	$47,900	1.27367	1,156
1054	10	19	2	$13,000	$24,490	$48,000	1.28034	1,100
1055	10	19	2	$13,000	$28,291	$49,250	1.19275	1,100
1056	10	19	1	$7,200	$45,397	$59,900	1.13885	1,692
1057	10	19	2	$14,500	$30,163	$47,000	1.05233	1,556
1058	10	19	2	$14,500	$26,035	$46,000	1.13482	1,080
1059	10	19	2	$14,500	$21,798	$45,000	1.23974	962
1060	10	19	2	$14,500	$24,548	$42,000	1.07560	999
1061	10	19	2	$14,500	$23,997	$40,400	1.04943	1,326
1062	10	19	2	$14,500	$22,524	$34,000	0.91832	999
1063	10	19	2	$14,500	$34,170	$51,700	1.06226	1,426
1064	10	19	2	$14,500	$28,289	$43,000	1.00493	1,131
1065	10	19	2	$15,000	$27,922	$49,900	1.16257	1,334
1066	10	19	2	$15,000	$31,216	$60,500	1.30907	1,920
1067	10	19	2	$12,500	$44,294	$62,000	1.09166	1,441
1068	10	19	2	$12,500	$41,149	$60,000	1.11838	1,404
1069	10	19	2	$13,000	$39,488	$73,900	1.40794	1,301
1070	10	19	2	$52,480	$130,679	$270,000	1.47413	6,220
1071	10	19	2	$10,000	$46,962	$74,000	1.29911	1,918
1072	10	19	2	$14,000	$49,760	$87,500	1.37233	1,872
1073	10	19	2	$12,000	$53,877	$73,700	1.11875	2,340
1074	10	19	2	$10,000	$41,425	$66,200	1.28731	1,320
1075	10	19	2	$10,000	$40,430	$60,000	1.18977	1,302
1076	10	19	2	$10,000	$53,963	$66,000	1.03185	2,018
1077	10	19	2	$10,000	$30,165	$62,000	1.54363	1,248
1078	10	19	2	$10,000	$33,472	$59,000	1.35720	1,453
1079	10	19	2	$10,000	$41,833	$59,500	1.14792	1,586
1080	10	19	2	$10,000	$54,646	$68,500	1.05962	2,220

REAL ESTATE SALES FOR 1986
IN A MID-SIZED FLORIDA CITY

OBS	TOWNSHIP	RANGE	SECTION	LANDVAL	IMPROVAL	SALEPRIC	SALTOAPR	AREA
1081	10	19	2	$10,000	$44,541	$80,000	1.46679	1,659
1082	10	19	2	$13,000	$58,842	$75,900	1.05649	1,601
1083	10	19	3	$10,000	$35,740	$60,400	1.32051	756
1084	10	19	3	$10,000	$42,144	$62,400	1.19669	756
1085	10	19	3	$10,000	$42,144	$62,400	1.19669	756
1086	10	19	3	$10,000	$42,648	$63,000	1.19663	756
1087	10	19	3	$10,000	$42,344	$61,900	1.18256	756
1088	10	19	3	$13,000	$47,571	$65,900	1.08798	1,633
1089	10	19	3	$20,000	$9,379	$52,000	1.76997	1,161
1090	10	19	3	$12,000	$45,615	$55,000	0.95461	1,820
1091	10	19	3	$10,000	$46,011	$58,000	1.03551	1,656
1092	10	19	3	$10,000	$50,168	$67,500	1.12186	2,061
1093	10	19	3	$10,000	$49,832	$65,900	1.10142	1,552
1094	10	19	3	$10,000	$45,023	$62,500	1.13589	1,536
1095	10	19	3	$11,000	$39,129	$63,000	1.25676	1,428
1096	10	19	4	$3,500	$28,343	$42,800	1.34409	498
1097	10	19	4	$3,500	$28,343	$42,500	1.33467	498
1098	10	19	4	$3,500	$28,343	$22,000	0.69089	498
1099	10	19	4	$37,100	$51,338	$172,500	1.95052	1,440
1100	10	19	4	$13,500	$57,576	$69,500	0.97783	2,852
1101	10	19	4	$13,500	$77,353	$102,000	1.12269	3,860
1102	10	19	5	$65,625	$152,076	$365,000	1.67661	4,626
1103	10	19	5	$12,000	$59,882	$87,500	1.21727	1,668
1104	10	19	5	$12,000	$42,191	$60,000	1.10719	1,804
1105	10	19	5	$15,000	$69,110	$95,000	1.12947	1,982
1106	10	19	5	$15,000	$61,589	$85,000	1.10982	1,904
1107	10	19	5	$15,000	$56,834	$82,000	1.14152	2,099
1108	10	19	5	$15,000	$55,853	$90,000	1.27024	1,714
1109	10	19	5	$15,000	$97,469	$145,000	1.28924	2,340
1110	10	19	5	$16,000	$61,498	$82,000	1.05809	2,189
1111	10	19	5	$15,000	$69,853	$98,500	1.16083	2,057
1112	10	19	5	$16,000	$86,506	$135,000	1.31700	2,602
1113	10	19	5	$15,000	$77,456	$111,000	1.20057	2,312
1114	10	19	5	$15,000	$55,687	$85,000	1.20248	1,944
1115	10	19	5	$65,520	$170,108	$296,000	1.25622	4,674
1116	10	19	5	$16,000	$61,275	$90,000	1.16467	1,960
1117	10	19	5	$16,000	$61,083	$90,000	1.16757	2,252
1118	10	19	5	$16,000	$65,507	$92,000	1.12874	2,219
1119	10	19	5	$15,000	$53,466	$96,500	1.40946	2,074
1120	10	19	5	$15,000	$63,025	$84,900	1.08811	1,744
1121	10	19	5	$16,000	$71,386	$109,000	1.24734	2,319
1122	10	19	5	$15,000	$47,815	$70,000	1.11438	1,692
1123	10	19	5	$15,000	$50,008	$94,500	1.45367	1,596
1124	10	19	5	$18,000	$87,698	$170,000	1.60836	2,637
1125	10	19	5	$18,000	$82,736	$120,000	1.19123	2,509
1126	10	19	5	$18,000	$60,301	$88,000	1.12387	2,148
1127	10	19	5	$18,000	$54,862	$87,500	1.20090	1,870
1128	10	19	5	$18,000	$68,243	$134,800	1.56303	1,963
1129	10	19	5	$18,000	$60,900	$95,700	1.21293	1,936
1130	10	19	5	$18,000	$78,418	$133,000	1.37941	2,904
1131	10	19	5	$18,000	$68,701	$126,000	1.45327	2,241
1132	10	19	5	$18,000	$66,156	$110,000	1.30710	2,855
1133	10	19	5	$18,000	$64,112	$88,000	1.07171	2,092
1134	10	19	5	$18,000	$74,522	$136,500	1.47532	1,777
1135	10	19	5	$18,000	$62,004	$125,000	1.56242	2,337
1136	10	19	5	$12,500	$44,454	$68,000	1.19395	1,439
1137	10	19	5	$12,000	$42,363	$60,000	1.10369	1,248
1138	10	19	5	$12,000	$41,277	$67,500	1.26696	1,312
1139	10	19	5	$12,000	$37,957	$61,500	1.23106	1,138
1140	10	19	5	$12,000	$44,868	$71,500	1.25730	1,398
1141	10	19	5	$12,000	$41,207	$67,000	1.25923	1,302
1142	10	19	5	$12,000	$47,793	$72,500	1.21252	1,615
1143	10	19	5	$12,000	$36,034	$59,000	1.22830	1,171
1144	10	19	5	$12,000	$43,243	$66,500	1.20377	1,420
1145	10	19	5	$12,000	$45,783	$71,900	1.24431	1,416
1146	10	19	5	$12,000	$43,670	$62,500	1.12269	1,365
1147	10	19	5	$12,500	$42,076	$65,900	1.20749	1,232
1148	10	19	5	$12,500	$39,488	$68,500	1.31761	1,173
1149	10	19	5	$12,500	$44,859	$64,400	1.12275	1,283
1150	10	19	5	$12,500	$37,223	$55,000	1.10613	1,144
1151	10	19	6	$11,000	$38,027	$66,500	1.35640	1,429
1152	10	19	6	$9,900	$49,633	$85,000	1.42778	1,756

REAL ESTATE SALES FOR 1986
IN A MID-SIZED FLORIDA CITY

OBS	TOWNSHIP	RANGE	SECTION	LANDVAL	IMPROVAL	SALEPRIC	SALTOAPR	AREA
1153	10	19	6	$11,000	$48,014	$69,000	1.16921	1,670
1154	10	19	6	$11,000	$49,693	$66,000	1.08744	1,780
1155	10	19	6	$11,000	$60,727	$97,900	1.36490	1,999
1156	10	19	6	$9,500	$42,191	$77,600	1.50123	1,416
1157	10	19	6	$11,000	$45,843	$62,000	1.09072	1,756
1158	10	19	6	$11,000	$54,948	$85,000	1.28889	1,937
1159	10	19	6	$9,350	$39,843	$67,900	1.38028	1,704
1160	10	19	6	$11,000	$46,225	$68,900	1.20402	1,517
1161	10	19	6	$11,000	$42,199	$66,500	1.25002	1,538
1162	10	19	6	$11,000	$48,455	$68,000	1.14372	1,308
1163	10	19	6	$11,000	$45,591	$73,000	1.28996	1,626
1164	10	19	6	$16,000	$63,273	$95,900	1.20974	1,902
1165	10	19	6	$16,000	$81,293	$156,000	1.60340	2,393
1166	10	19	6	$16,000	$71,542	$148,000	1.69062	2,331
1167	10	19	6	$16,000	$64,654	$112,000	1.38865	1,914
1168	10	19	7	$8,000	$33,663	$70,000	1.68015	1,287
1169	10	19	7	$14,000	$63,734	$96,000	1.23498	1,953
1170	10	19	7	$15,000	$47,592	$68,000	1.08640	1,736
1171	10	19	7	$15,800	$59,992	$105,000	1.38537	2,277
1172	10	19	8	$8,000	$39,020	$57,500	1.22288	1,206
1173	10	19	8	$8,000	$44,028	$60,000	1.15323	1,395
1174	10	19	8	$8,000	$34,576	$54,000	1.26832	1,185
1175	10	19	8	$8,000	$48,935	$61,000	1.07140	1,359
1176	10	19	8	$8,000	$35,044	$44,200	1.02686	1,140
1177	10	19	8	$8,000	$37,193	$61,500	1.36083	1,140
1178	10	19	8	$8,000	$41,391	$54,000	1.09332	1,178
1179	10	19	8	$8,500	$41,331	$51,500	1.03349	1,185
1180	10	19	8	$11,500	$50,731	$68,000	1.09270	1,397
1181	10	19	8	$11,500	$43,963	$69,000	1.24407	1,336
1182	10	19	8	$11,500	$45,739	$68,500	1.19674	1,452
1183	10	19	8	$11,500	$36,243	$62,500	1.30909	1,130
1184	10	19	8	$11,500	$47,113	$74,000	1.26252	1,444
1185	10	19	8	$12,000	$34,897	$56,500	1.20477	1,120
1186	10	19	8	$11,500	$43,207	$67,000	1.22471	1,379
1187	10	19	8	$12,500	$39,939	$71,400	1.36158	1,233
1188	10	19	8	$11,500	$54,889	$79,500	1.19749	1,749
1189	10	19	8	$11,500	$50,162	$73,200	1.18712	1,510
1190	10	19	8	$25,000	$32,877	$67,500	1.16627	1,558
1191	10	19	9	$9,600	$39,734	$61,500	1.24660	1,848
1192	10	19	9	$11,000	$42,466	$44,600	0.83417	1,323
1193	10	19	9	$10,500	$45,435	$61,500	1.09949	1,313
1194	10	19	9	$8,500	$34,517	$51,000	1.18558	1,078
1195	10	19	9	$8,500	$34,589	$54,900	1.27411	1,050
1196	10	19	9	$8,500	$35,960	$57,900	1.30229	1,108
1197	10	19	9	$8,500	$33,785	$58,900	1.39293	991
1198	10	19	9	$8,500	$35,990	$58,900	1.32389	1,108
1199	10	19	9	$8,500	$35,479	$61,800	1.40522	1,050
1200	10	19	9	$8,500	$34,625	$57,900	1.34261	991
1201	10	19	9	$8,500	$36,112	$57,900	1.29786	1,118
1202	10	19	9	$8,500	$35,459	$56,900	1.29439	1,050
1203	10	19	9	$8,500	$34,946	$56,900	1.30967	1,055
1204	10	19	9	$8,500	$34,674	$57,900	1.34108	1,055
1205	10	19	9	$8,500	$34,482	$56,900	1.32381	1,050
1206	10	19	9	$8,500	$31,026	$49,000	1.23969	900
1207	10	19	9	$8,500	$33,979	$50,900	1.19824	976
1208	10	19	9	$8,500	$33,941	$52,400	1.23466	976
1209	10	19	9	$8,500	$35,443	$55,900	1.27210	1,074
1210	10	19	9	$8,500	$34,993	$10,000	0.22992	1,071
1211	10	19	9	$8,500	$34,474	$57,900	1.34733	1,044
1212	10	19	9	$8,500	$34,532	$57,900	1.34551	1,050
1213	10	19	9	$8,500	$31,165	$49,900	1.25804	900
1214	10	19	9	$5,100	$22,984	$31,700	1.12876	788
1215	10	19	9	$5,100	$26,834	$29,500	0.92378	790
1216	10	19	9	$19,600	$73,430	$96,000	1.03193	3,240
1217	10	19	9	$5,100	$25,035	$36,500	1.21122	878
1218	10	19	9	$5,100	$26,948	$38,500	1.20132	923
1219	10	19	10	$6,000	$43,152	$67,900	1.38143	1,220
1220	10	19	10	$4,000	$43,152	$65,900	1.39761	1,220
1221	10	19	10	$6,000	$58,585	$94,200	1.45854	1,202
1222	10	19	10	$4,000	$57,605	$75,300	1.22230	1,202
1223	10	19	10	$4,000	$58,401	$85,700	1.37338	1,202
1224	10	19	10	$4,000	$57,605	$74,600	1.21094	1,202

REAL ESTATE SALES FOR 1986
IN A MID-SIZED FLORIDA CITY

OBS	TOWNSHIP	RANGE	SECTION	LANDVAL	IMPROVAL	SALEPRIC	SALTOAPR	AREA
1225	10	19	10	$4,000	$57,605	$74,300	1.20607	1,202
1226	10	19	10	$6,000	$58,605	$88,000	1.36212	1,202
1227	10	19	10	$4,000	$52,890	$73,700	1.29548	1,130
1228	10	19	10	$4,000	$36,355	$63,800	1.58097	1,056
1229	10	19	10	$4,000	$46,117	$75,300	1.50248	1,260
1230	10	19	10	$4,000	$44,349	$70,800	1.46435	1,260
1231	10	19	10	$6,000	$44,349	$70,000	1.39030	1,260
1232	10	19	10	$4,000	$56,181	$81,300	1.35092	1,130
1233	10	19	8	$13,000	$69,815	$89,900	1.08555	1,952
1234	10	19	8	$13,000	$59,051	$77,000	1.06869	1,746
1235	10	19	8	$13,000	$62,580	$86,000	1.13787	1,862
1236	10	19	8	$13,000	$59,381	$83,500	1.15362	1,733
1237	10	19	8	$13,000	$61,426	$80,000	1.07489	1,931
1238	10	19	8	$13,000	$62,729	$85,000	1.12242	1,748
1239	10	19	8	$13,000	$65,068	$109,900	1.40775	2,095
1240	10	19	8	$13,000	$59,534	$90,000	1.24080	1,830
1241	10	19	8	$13,000	$52,946	$78,000	1.18279	1,532
1242	10	19	10	$4,000	$44,154	$51,800	1.07572	830
1243	10	19	10	$4,000	$44,154	$52,300	1.08610	830
1244	10	19	10	$4,000	$45,054	$52,000	1.06006	830
1245	10	19	10	$4,000	$38,986	$47,900	1.11432	830
1246	10	19	10	$4,000	$44,154	$52,500	1.09025	830
1247	10	19	10	$4,000	$44,154	$50,200	1.04249	830
1248	10	19	10	$4,000	$38,986	$48,500	1.12827	830
1249	10	19	10	$4,000	$44,154	$51,400	1.06741	830
1250	10	19	10	$4,000	$44,154	$51,800	1.07572	830
1251	10	19	10	$4,000	$44,154	$52,300	1.08610	830
1252	10	19	10	$4,000	$44,154	$51,800	1.07572	830
1253	10	19	10	$4,000	$39,886	$49,100	1.11881	830
1254	10	19	10	$4,000	$44,154	$51,400	1.06741	830
1255	10	19	10	$4,000	$44,154	$52,000	1.07987	830
1256	10	19	10	$4,000	$45,054	$53,300	1.08656	830
1257	10	19	10	$4,000	$44,154	$51,100	1.06118	830
1258	10	19	10	$4,000	$44,154	$51,800	1.07572	830
1259	10	19	10	$4,000	$44,154	$48,500	1.00719	830
1260	10	19	10	$4,000	$44,154	$51,100	1.06118	830
1261	10	19	10	$4,000	$44,154	$51,400	1.06741	830
1262	10	19	10	$4,000	$44,154	$52,300	1.08610	830
1263	10	19	10	$4,000	$44,154	$51,800	1.07572	830
1264	10	19	10	$4,000	$44,154	$51,800	1.07572	830
1265	10	19	10	$4,000	$38,986	$47,800	1.11199	830
1266	10	19	10	$4,000	$44,154	$51,400	1.06741	830
1267	10	19	10	$4,000	$44,154	$53,000	1.10064	830
1268	10	19	10	$4,000	$44,154	$51,900	1.07779	830
1269	10	19	10	$4,000	$38,986	$45,400	1.05616	830
1270	10	19	10	$4,000	$44,154	$50,800	1.05495	830
1271	10	19	10	$4,000	$44,154	$50,000	1.03834	830
1272	10	19	11	$13,253	$109,645	$200,000	1.62737	2,986
1273	10	19	11	$19,440	$42,132	$420,000	6.82128	2,240
1274	10	19	11	$16,000	$28,021	$150,000	3.40746	1,440
1275	10	19	11	$6,000	$28,738	$45,900	1.32132	522
1276	10	19	11	$6,000	$28,945	$42,500	1.21620	522
1277	10	19	11	$6,000	$28,738	$43,500	1.25223	522
1278	10	19	11	$4,000	$28,517	$27,000	0.83033	522
1279	10	19	11	$4,000	$26,358	$42,000	1.38349	492
1280	10	19	11	$69,000	$40,473	1075000	9.81977	2,457
1281	10	19	13	$1,500	$12,283	$15,000	1.08830	787
1282	10	19	17	$22,400	$54,961	$135,000	1.74507	2,001
1283	10	19	17	$22,400	$69,631	$95,000	1.03226	2,288
1284	10	19	17	$22,400	$79,580	$119,000	1.16690	2,878
1285	10	19	17	$21,700	$85,462	$108,500	1.01249	3,296
1286	10	19	17	$22,400	$59,106	$105,100	1.28948	2,184
1287	10	19	17	$23,100	$72,675	$135,000	1.40955	2,373
1288	10	19	17	$23,100	$68,795	$140,600	1.53001	2,258
1289	10	19	17	$20,775	$15,504	$48,500	1.33686	960
1290	10	19	17	$42,075	$63,542	$106,000	1.00363	2,020
1291	10	19	18	$16,000	$66,919	$87,500	1.05525	1,862
1292	10	19	18	$13,500	$62,114	$79,900	1.05668	1,645
1293	10	19	18	$12,800	$48,335	$83,000	1.35765	1,526
1294	10	19	18	$13,000	$54,763	$84,900	1.25290	1,814
1295	10	19	18	$12,800	$58,128	$97,500	1.37463	1,671
1296	10	19	18	$12,800	$49,713	$81,400	1.30213	1,554

REAL ESTATE SALES FOR 1986
IN A MID-SIZED FLORIDA CITY

OBS	TOWNSHIP	RANGE	SECTION	LANDVAL	IMPROVAL	SALEPRIC	SALTOAPR	AREA
1297	10	19	18	$13,000	$59,082	$86,500	1.20002	1,816
1298	10	19	18	$12,800	$56,451	$89,000	1.28518	1,868
1299	10	19	20	$12,000	$47,584	$65,000	1.09090	1,224
1300	10	19	20	$11,040	$44,081	$70,100	1.27175	1,410
1301	10	19	20	$10,920	$35,533	$51,500	1.10865	1,140
1302	10	19	20	$11,400	$52,649	$61,000	0.95240	1,422
1303	10	19	20	$12,000	$35,737	$57,500	1.20452	1,184
1304	10	19	20	$11,520	$40,228	$64,900	1.25415	1,312
1305	10	19	20	$11,520	$40,733	$59,900	1.14635	1,335
1306	10	19	20	$11,520	$45,824	$66,500	1.15967	1,434
1307	10	19	20	$11,400	$36,002	$56,900	1.20037	1,168
1308	10	19	20	$11,400	$42,900	$63,000	1.16022	1,377
1309	10	19	20	$11,400	$38,200	$59,900	1.20766	1,312
1310	10	19	20	$10,920	$39,120	$53,000	1.05915	1,255
1311	10	19	20	$12,000	$43,536	$69,000	1.24244	1,360
1312	10	19	20	$12,960	$41,933	$62,500	1.13858	1,196
1313	10	19	20	$12,960	$47,601	$63,500	1.04853	1,413
1314	10	19	20	$10,920	$45,656	$62,900	1.11178	1,244
1315	10	19	20	$11,400	$42,584	$66,500	1.23185	1,192
1316	10	19	20	$12,600	$42,657	$66,000	1.19442	1,359
1317	10	19	20	$11,040	$41,077	$57,000	1.09369	1,122
1318	10	19	20	$10,920	$39,083	$59,900	1.19793	1,301
1319	10	19	20	$11,040	$44,327	$72,000	1.30041	1,440
1320	10	19	20	$10,920	$47,906	$70,000	1.18995	1,485
1321	10	19	20	$10,920	$43,076	$69,500	1.28713	1,298
1322	10	19	20	$10,920	$44,833	$69,000	1.23760	1,356
1323	10	19	20	$10,920	$42,475	$66,500	1.24543	1,276
1324	10	19	20	$10,920	$40,765	$62,500	1.20925	1,170
1325	10	19	20	$11,040	$43,225	$65,700	1.21073	1,362
1326	10	19	20	$11,040	$45,657	$70,500	1.24345	1,398
1327	10	19	20	$11,040	$40,768	$65,800	1.27007	1,176
1328	10	19	20	$11,640	$49,400	$73,900	1.21068	1,541
1329	10	19	20	$11,040	$41,024	$68,500	1.31569	1,287
1330	10	19	20	$22,000	$46,135	$79,500	1.16680	1,476
1331	10	19	20	$22,000	$61,395	$114,800	1.37658	2,016
1332	10	19	20	$22,000	$53,047	$85,000	1.13262	1,592
1333	10	19	20	$22,000	$48,318	$81,000	1.15191	1,476
1334	10	19	20	$22,000	$48,264	$82,000	1.16703	1,494
1335	10	19	20	$22,000	$47,953	$86,000	1.22940	1,434
1336	10	19	20	$22,000	$51,775	$89,000	1.20637	1,519
1337	10	19	20	$22,000	$56,470	$103,000	1.31260	1,724
1338	10	19	20	$22,000	$57,097	$103,000	1.30220	1,728
1339	10	19	20	$22,000	$55,711	$99,400	1.27910	1,663
1340	10	19	20	$22,000	$40,675	$74,500	1.18867	1,215
1341	10	19	20	$23,000	$79,438	$128,200	1.25149	2,426
1342	10	19	20	$23,000	$97,182	$175,000	1.45612	2,632
1343	10	19	20	$23,000	$73,419	$110,000	1.14085	1,979
1344	10	19	20	$23,000	$72,326	$120,000	1.25884	2,184
1345	10	19	20	$12,500	$35,558	$72,000	1.49819	730
1346	10	19	20	$12,500	$47,773	$73,000	1.21116	1,205
1347	10	19	20	$12,500	$50,329	$68,900	1.09663	1,353
1348	10	19	20	$12,500	$35,709	$59,000	1.22384	1,008
1349	10	19	20	$12,500	$35,864	$51,500	1.06484	1,025
1350	10	19	20	$12,500	$43,135	$61,500	1.10542	1,071
1351	10	19	20	$12,500	$34,606	$59,500	1.26311	986
1352	10	19	20	$9,400	$41,966	$64,500	1.25569	1,149
1353	10	19	20	$9,400	$41,680	$64,000	1.25294	1,144
1354	10	19	20	$9,400	$36,176	$57,900	1.27041	1,040
1355	10	19	20	$9,400	$53,563	$73,800	1.17212	1,348
1356	10	19	20	$10,000	$53,212	$78,400	1.24027	1,332
1357	10	19	20	$9,400	$46,399	$70,900	1.27063	1,175
1358	10	19	20	$9,400	$39,925	$63,500	1.28738	1,001
1359	10	19	20	$9,400	$45,680	$70,100	1.27269	1,131
1360	10	19	20	$9,400	$55,114	$73,900	1.14549	914
1361	10	19	20	$9,400	$50,016	$73,000	1.22863	1,294
1362	10	19	20	$9,400	$46,361	$64,900	1.16390	1,171
1363	10	19	20	$9,400	$52,083	$72,100	1.17268	1,348
1364	10	19	20	$9,400	$40,140	$59,900	1.20912	1,001
1365	10	19	20	$9,400	$49,333	$69,500	1.18332	1,276
1366	10	19	20	$9,400	$53,578	$74,700	1.18613	1,348
1367	10	19	20	$9,400	$46,361	$65,200	1.16928	1,171
1368	10	19	20	$9,400	$40,523	$63,100	1.26395	983

REAL ESTATE SALES FOR 1986
IN A MID-SIZED FLORIDA CITY

OBS	TOWNSHIP	RANGE	SECTION	LANDVAL	IMPROVAL	SALEPRIC	SALTOAPR	AREA
1369	10	19	20	$9,400	$46,409	$68,300	1.22382	1,175
1370	10	19	20	$9,400	$47,322	$69,000	1.21646	1,275
1371	10	19	20	$9,400	$39,893	$61,800	1.25373	1,000
1372	10	19	20	$9,400	$46,461	$66,900	1.19762	1,171
1373	10	19	20	$10,000	$50,603	$76,500	1.26231	1,294
1374	10	19	20	$9,400	$47,481	$63,900	1.12340	1,171
1375	10	19	20	$9,400	$47,501	$72,200	1.26887	1,171
1376	10	19	20	$9,400	$53,748	$72,900	1.15443	1,348
1377	10	19	20	$9,400	$46,326	$70,400	1.26332	1,155
1378	10	19	20	$9,400	$52,083	$69,900	1.13690	1,348
1379	10	19	20	$9,400	$40,155	$64,500	1.30158	1,001
1380	10	19	20	$9,400	$46,416	$71,400	1.27920	1,171
1381	10	19	20	$9,400	$44,098	$67,200	1.25612	969
1382	10	19	20	$9,400	$53,148	$75,600	1.20867	1,348
1383	10	19	20	$10,000	$53,018	$72,800	1.15523	1,348
1384	10	19	20	$10,000	$48,445	$66,400	1.13611	1,234
1385	10	19	19	$9,500	$31,715	$50,000	1.21315	952
1386	10	19	19	$9,500	$36,267	$54,600	1.19300	576
1387	10	19	19	$9,500	$31,715	$53,700	1.30292	952
1388	10	19	20	$14,000	$46,973	$70,400	1.15461	1,365
1389	10	19	20	$14,000	$42,272	$72,000	1.27950	1,359
1390	10	19	20	$14,000	$44,182	$75,300	1.29421	1,405
1391	10	19	21	$6,000	$3,598	$10,500	1.09398	612
1392	10	19	21	$6,000	$3,414	$14,000	1.48715	720
1393	10	19	21	$6,000	$3,622	$14,500	1.50696	624
1394	10	19	21	$6,000	$3,375	$12,000	1.28000	432
1395	10	19	21	$6,000	$21,833	$39,900	1.43355	1,040
1396	10	19	21	$5,100	$22,609	$14,000	0.50525	788
1397	10	19	21	$5,100	$22,609	$32,500	1.17290	788
1398	10	19	21	$5,100	$22,562	$34,900	1.26166	788
1399	10	19	21	$6,100	$22,739	$33,700	1.16856	810
1400	10	19	21	$6,100	$32,960	$47,700	1.22120	704
1401	10	19	21	$5,100	$32,374	$47,800	1.27555	704
1402	10	19	21	$5,100	$23,597	$32,500	1.13949	814
1403	10	19	21	$5,100	$23,539	$32,000	1.11736	814
1404	10	19	21	$5,100	$32,439	$47,800	1.27334	704
1405	10	19	21	$6,000	$21,073	$32,000	1.18199	972
1406	10	19	21	$6,000	$23,488	$36,900	1.25136	1,169
1407	10	19	21	$5,500	$2,376	$17,000	2.15846	510
1408	10	19	21	$4,200	$2,700	$20,000	2.89855	510
1409	10	19	21	$17,230	$67,133	$103,000	1.22091	3,536
1410	10	19	21	$17,230	$67,314	$105,700	1.25024	3,348
1411	10	19	22	$8,500	$31,259	$51,500	1.29530	994
1412	10	19	22	$8,500	$35,699	$55,000	1.24437	1,206
1413	10	19	22	$56,000	$17,682	$61,900	0.84010	897
1414	10	19	22	$3,500	$7,941	$30,000	2.62215	705
1415	10	19	22	$3,000	$31,828	$57,000	1.63661	1,325
1416	10	19	22	$2,000	$7,867	$13,500	1.36820	520
1417	10	19	22	$2,500	$8,804	$17,000	1.50389	732
1418	10	19	22	$2,800	$21,686	$49,000	2.00114	1,170
1419	10	19	22	$6,000	$29,598	$35,900	1.00848	470
1420	10	19	22	$6,000	$29,598	$35,900	1.00848	470
1421	10	19	22	$6,000	$32,540	$39,900	1.03529	575
1422	10	19	22	$6,000	$26,282	$39,900	1.23598	890
1423	10	19	22	$6,000	$25,793	$39,900	1.25499	892
1424	10	19	22	$6,000	$25,793	$40,200	1.26443	892
1425	10	19	29	$8,100	$33,105	$49,900	1.21102	1,116
1426	10	19	29	$8,100	$47,132	$67,500	1.22212	1,464
1427	10	19	29	$8,100	$35,491	$57,500	1.31908	1,138
1428	10	19	29	$8,100	$39,117	$59,000	1.24955	1,204
1429	10	19	29	$8,100	$36,175	$62,000	1.40034	1,202
1430	10	19	29	$8,100	$30,212	$46,000	1.20067	1,014
1431	10	19	29	$8,100	$39,857	$59,900	1.24904	1,245
1432	10	19	29	$8,100	$40,969	$59,500	1.21258	1,232
1433	10	19	30	$8,000	$50,175	$89,500	1.53846	1,771
1434	10	19	0	$8,500	$50,932	$72,000	1.21147	1,773
1435	10	19	0	$8,500	$57,343	$71,000	1.07832	1,770
1436	10	19	0	$8,500	$60,010	$77,000	1.12392	2,248
1437	10	19	0	$25,375	$68,301	$126,000	1.34506	2,256
1438	10	19	0	$25,375	$62,499	$110,000	1.25179	2,177
1439	10	19	0	$40,000	$66,379	$100,000	0.94004	1,916
1440	10	19	0	$40,000	$73,379	$129,000	1.13778	1,991

REAL ESTATE SALES FOR 1986
IN A MID-SIZED FLORIDA CITY

OBS	TOWNSHIP	RANGE	SECTION	LANDVAL	IMPROVAL	SALEPRIC	SALTOAPR	AREA
1441	10	19	0	$40,000	$88,638	$150,000	1.16606	3,085
1442	10	19	0	$40,000	$126,098	$165,000	0.99339	2,057
1443	10	19	0	$24,000	$52,307	$90,000	1.17945	1,529
1444	10	19	0	$22,800	$45,226	$83,000	1.22012	1,488
1445	10	19	0	$38,500	$40,533	$110,000	1.39182	2,122
1446	10	19	0	$3,000	$22,319	$59,900	2.36581	1,496
1447	10	19	0	$10,800	$40,259	$62,000	1.21428	1,479
1448	10	19	0	$10,800	$44,507	$65,000	1.17526	1,632
1449	10	19	0	$12,240	$51,961	$76,900	1.19780	1,736
1450	10	19	0	$12,000	$60,349	$70,000	0.96753	2,280
1451	10	19	0	$3,030	$21,822	$59,800	2.40624	1,008
1452	10	19	0	$10,800	$43,926	$61,500	1.12378	1,600
1453	10	19	0	$10,800	$39,658	$55,000	1.09002	1,316
1454	10	19	0	$13,000	$34,744	$59,000	1.23576	1,441
1455	10	19	0	$12,000	$66,449	$81,000	1.03252	1,998
1456	10	19	0	$23,800	$48,463	$220,000	3.04443	2,032
1457	10	19	0	$42,000	$132,976	$235,000	1.34304	3,363
1458	10	19	0	$8,000	$27,827	$50,500	1.40955	1,248
1459	10	19	0	$8,500	$29,161	$50,000	1.32763	1,488
1460	10	19	0	$9,000	$26,642	$63,500	1.78161	1,284
1461	10	19	0	$18,000	$30,212	$75,800	1.57222	1,522
1462	10	19	0	$10,000	$43,770	$65,000	1.20885	1,660
1463	10	19	0	$9,000	$65,663	$70,900	0.94960	2,081
1464	10	19	0	$7,500	$40,862	$69,500	1.43708	1,342
1465	10	19	0	$7,500	$36,179	$61,900	1.41716	1,851
1466	10	19	0	$18,500	$76,235	$70,500	0.74418	3,970
1467	10	19	0	$16,000	$66,772	$94,300	1.13927	3,348
1468	10	19	0	$4,000	$29,900	$36,900	1.08850	908
1469	10	19	0	$4,000	$26,051	$32,000	1.06486	820
1470	10	19	0	$4,000	$29,574	$35,900	1.06928	908
1471	10	19	0	$4,000	$25,737	$29,900	1.00548	820
1472	10	19	0	$4,000	$25,737	$29,900	1.00548	820
1473	10	19	0	$4,000	$29,574	$36,600	1.09013	908
1474	10	19	0	$4,000	$26,365	$29,900	0.98469	840
1475	10	19	0	$4,000	$26,365	$29,900	0.98469	840
1476	10	19	0	$4,000	$36,722	$43,900	1.07804	766
1477	10	19	0	$4,000	$36,722	$43,700	1.07313	766
1478	10	19	0	$4,000	$29,900	$36,900	1.08850	908
1479	10	19	0	$4,000	$29,900	$36,900	1.08850	908
1480	10	19	0	$4,000	$29,900	$36,900	1.08850	908
1481	10	19	0	$16,100	$60,639	$100,000	1.30312	3,348
1482	10	19	0	$10,000	$37,848	$50,000	1.04498	1,860
1483	10	19	0	$10,000	$37,443	$33,300	0.70189	1,800
1484	10	19	0	$6,500	$35,752	$125,000	2.95844	2,093
1485	11	19	5	$10,720	$33,674	$54,100	1.21863	1,084
1486	11	19	5	$10,560	$42,667	$63,000	1.18361	1,368
1487	11	19	5	$10,500	$50,785	$71,900	1.17321	1,613
1488	11	19	6	$7,410	$21,408	$29,900	1.03755	1,248
1489	11	19	6	$7,410	$12,684	$22,250	1.10730	784
1490	11	19	16	$20,000	$35,043	$73,000	1.32624	1,248
1491	11	19	16	$20,000	$31,884	$70,100	1.35109	1,068
1492	11	19	20	$15,655	$91,858	$150,000	1.39518	2,374
1493	11	19	20	$16,306	$59,539	$105,000	1.38440	1,616
1494	11	19	29	$13,280	$45,999	$65,000	1.09651	1,368
1495	8	20	5	$15,000	$81,820	$130,000	1.34270	1,546
1496	8	20	9	$1,800	$6,701	$5,000	0.58817	520
1497	8	20	18	$15,824	$74,919	$140,000	1.54282	2,480
1498	9	20	7	$10,000	$44,437	$59,800	1.09852	1,366
1499	9	20	7	$10,000	$41,176	$58,600	1.14507	1,198
1500	9	20	7	$10,000	$35,973	$57,500	1.25073	1,198
1501	9	20	7	$10,000	$49,178	$64,600	1.09162	1,332
1502	9	20	7	$10,000	$41,007	$66,200	1.29786	1,314
1503	9	20	7	$11,000	$39,185	$67,000	1.33506	1,359
1504	9	20	7	$11,000	$48,084	$71,800	1.21522	1,422
1505	9	20	12	$2,500	$47,593	$59,000	1.17781	1,093
1506	9	20	12	$2,500	$8,573	$35,100	3.16987	1,080
1507	9	20	12	$2,000	$11,460	$21,500	1.59733	995
1508	9	20	18	$6,800	$37,615	$56,000	1.26084	1,860
1509	9	20	19	$126,000	$10,404	$173,400	1.27122	552
1510	9	20	19	$9,000	$40,448	$59,900	1.21137	1,196
1511	9	20	19	$6,500	$36,397	$50,000	1.16558	1,318
1512	9	20	19	$6,500	$36,397	$53,000	1.23552	1,318

REAL ESTATE SALES FOR 1986
IN A MID-SIZED FLORIDA CITY

OBS	TOWNSHIP	RANGE	SECTION	LANDVAL	IMPROVAL	SALEPRIC	SALTOAPR	AREA
1513	9	20	19	$6,500	$34,212	$53,000	1.30183	1,216
1514	9	20	19	$4,500	$132,954	$50,000	0.36376	5,110
1515	9	20	19	$8,500	$34,560	$56,000	1.30051	1,227
1516	9	20	19	$8,500	$26,233	$52,000	1.49714	1,046
1517	9	20	19	$9,000	$34,211	$51,000	1.18026	1,290
1518	9	20	19	$8,500	$35,110	$53,000	1.21532	1,075
1519	9	20	19	$8,500	$39,265	$50,800	1.06354	1,248
1520	9	20	19	$8,500	$36,220	$52,400	1.17174	1,195
1521	9	20	19	$8,500	$30,852	$49,000	1.24517	1,000
1522	9	20	19	$8,500	$37,618	$51,400	1.11453	1,265
1523	9	20	19	$8,500	$33,108	$48,400	1.16324	1,050
1524	9	20	19	$8,500	$39,572	$53,900	1.12123	1,198
1525	9	20	19	$8,500	$37,006	$52,900	1.16248	1,131
1526	9	20	19	$8,500	$36,989	$55,900	1.22887	1,131
1527	9	20	19	$8,500	$34,404	$45,500	1.06051	1,104
1528	9	20	19	$8,500	$30,956	$51,000	1.29258	1,014
1529	9	20	19	$8,500	$31,526	$40,000	0.99935	988
1530	9	20	19	$8,500	$31,839	$48,000	1.18992	988
1531	9	20	19	$8,500	$45,560	$66,000	1.22087	1,368
1532	9	20	19	$8,500	$45,392	$65,000	1.20612	1,216
1533	9	20	19	$8,500	$34,367	$51,000	1.18973	1,253
1534	9	20	19	$17,000	$37,806	$60,500	1.10389	1,582
1535	9	20	19	$8,500	$34,519	$48,000	1.11579	1,280
1536	9	20	19	$9,000	$29,713	$55,000	1.42071	1,118
1537	9	20	19	$8,500	$38,751	$56,900	1.20421	1,316
1538	9	20	19	$8,550	$38,756	$58,500	1.23663	1,304
1539	9	20	19	$9,500	$40,413	$57,500	1.15200	1,392
1540	9	20	19	$10,200	$40,734	$62,800	1.23297	1,409
1541	9	20	19	$9,000	$34,797	$64,900	1.48184	1,118
1542	9	20	19	$9,000	$40,483	$57,500	1.16202	1,309
1543	9	20	19	$1,500	$13,034	$23,000	1.58250	888
1544	9	20	19	$8,700	$36,309	$51,500	1.14422	1,143
1545	9	20	19	$8,700	$36,309	$53,500	1.18865	1,143
1546	9	20	19	$10,500	$40,864	$65,000	1.26548	918
1547	9	20	19	$9,000	$26,196	$42,500	1.20752	1,075
1548	9	20	19	$8,500	$24,594	$37,600	1.13616	1,055
1549	9	20	19	$8,500	$39,182	$53,500	1.12202	1,456
1550	9	20	19	$8,500	$35,768	$59,900	1.35312	1,306
1551	9	20	19	$8,500	$24,648	$41,000	1.23688	998
1552	9	20	19	$8,500	$31,387	$51,500	1.29115	1,210
1553	9	20	19	$8,500	$24,496	$50,000	1.51534	987
1554	9	20	19	$9,500	$45,785	$78,000	1.41087	1,807
1555	9	20	20	$9,000	$29,349	$47,500	1.23862	1,725
1556	9	20	20	$9,000	$29,302	$40,000	1.04433	1,684
1557	9	20	20	$9,000	$25,865	$51,000	1.46279	1,524
1558	9	20	20	$11,000	$31,818	$68,000	1.58812	2,400
1559	9	20	20	$8,000	$14,970	$20,000	0.87070	1,036
1560	9	20	20	$8,000	$24,470	$39,400	1.21343	1,394
1561	9	20	20	$4,000	$30,121	$36,600	1.07265	1,253
1562	9	20	20	$9,000	$20,163	$45,000	1.54305	990
1563	9	20	20	$9,000	$25,512	$42,000	1.21697	1,548
1564	9	20	20	$9,000	$23,412	$46,000	1.41923	939
1565	9	20	20	$9,000	$13,914	$36,500	1.59291	1,468
1566	9	20	21	$43,152	$117,247	$202,000	1.25936	4,212
1567	9	20	22	$9,000	$47,578	$55,500	0.98095	1,548
1568	9	20	22	$7,300	$31,631	$43,500	1.11736	1,200
1569	9	20	22	$7,300	$40,166	$40,500	0.85324	1,080
1570	9	20	22	$7,300	$39,866	$56,000	1.18730	1,276
1571	9	20	25	$1,500	$10,077	$33,400	2.88503	672
1572	9	20	27	$8,000	$29,984	$46,000	1.21104	1,242
1573	9	20	27	$8,000	$25,125	$42,400	1.28000	1,197
1574	9	20	27	$8,000	$26,954	$46,000	1.31602	1,166
1575	9	20	27	$8,000	$40,075	$44,000	0.91524	1,450
1576	9	20	27	$8,000	$26,163	$46,000	1.34649	1,075
1577	9	20	27	$8,000	$23,639	$45,500	1.43810	1,056
1578	9	20	27	$8,000	$22,329	$42,000	1.38481	936
1579	9	20	27	$8,000	$22,053	$39,000	1.29771	945
1580	9	20	27	$8,000	$24,235	$38,000	1.17884	972
1581	9	20	27	$8,000	$23,932	$39,900	1.24953	972
1582	9	20	27	$8,000	$22,137	$43,400	1.44009	1,035
1583	9	20	27	$8,500	$22,073	$45,000	1.47189	945
1584	9	20	27	$8,000	$22,053	$40,000	1.33098	945

REAL ESTATE SALES FOR 1986
IN A MID-SIZED FLORIDA CITY

OBS	TOWNSHIP	RANGE	SECTION	LANDVAL	IMPROVAL	SALEPRIC	SALTOAPR	AREA
1585	9	20	27	$8,000	$23,376	$51,500	1.64138	1,032
1586	9	20	27	$8,000	$25,696	$45,000	1.33547	1,035
1587	9	20	27	$8,000	$22,323	$38,300	1.26307	892
1588	9	20	27	$8,500	$33,452	$48,000	1.14416	1,515
1589	9	20	27	$8,000	$26,348	$43,700	1.27227	1,008
1590	9	20	27	$8,500	$26,806	$38,800	1.09896	980
1591	9	20	27	$8,000	$24,055	$37,000	1.15427	936
1592	9	20	27	$16,200	$64,404	$78,000	0.96769	2,500
1593	9	20	27	$16,200	$35,966	$94,000	1.80194	5,000
1594	9	20	27	$10,524	$28,863	$75,000	1.90418	1,470
1595	9	20	28	$8,000	$23,127	$43,500	1.39750	1,010
1596	9	20	28	$8,000	$23,042	$44,900	1.44643	962
1597	9	20	28	$8,000	$24,351	$43,900	1.35699	1,162
1598	9	20	28	$8,000	$24,586	$39,900	1.22445	1,110
1599	9	20	28	$8,000	$26,613	$36,000	1.04007	1,182
1600	9	20	28	$8,000	$31,864	$45,000	1.12884	1,388
1601	9	20	28	$8,000	$20,790	$39,500	1.37200	988
1602	9	20	28	$8,000	$31,679	$40,000	1.00809	1,529
1603	9	20	28	$8,000	$32,859	$38,500	0.94226	1,125
1604	9	20	28	$8,000	$27,564	$45,000	1.26532	1,308
1605	9	20	28	$8,000	$19,065	$43,300	1.59985	900
1606	9	20	28	$8,000	$22,696	$33,000	1.07506	962
1607	9	20	28	$8,000	$30,500	$40,000	1.03896	1,200
1608	9	20	28	$8,000	$35,258	$42,200	0.97554	1,616
1609	9	20	28	$8,000	$19,528	$35,500	1.28960	900
1610	9	20	28	$8,000	$28,055	$37,000	1.02621	1,367
1611	9	20	28	$8,000	$24,575	$41,000	1.25863	1,125
1612	9	20	28	$8,000	$23,776	$42,000	1.32175	1,025
1613	9	20	28	$8,000	$22,422	$37,600	1.23595	900
1614	9	20	28	$8,000	$40,275	$51,500	1.06680	1,162
1615	9	20	28	$8,000	$31,214	$40,900	1.04299	1,383
1616	9	20	28	$8,000	$27,672	$48,000	1.34559	1,389
1617	9	20	28	$8,000	$30,188	$46,500	1.21766	1,547
1618	9	20	28	$8,000	$32,286	$42,500	1.05496	1,604
1619	9	20	28	$8,000	$24,331	$39,900	1.23411	1,117
1620	9	20	28	$8,000	$25,604	$45,000	1.33913	1,152
1621	9	20	28	$8,000	$30,845	$50,000	1.28717	1,445
1622	9	20	28	$8,500	$43,986	$59,600	1.13554	1,913
1623	9	20	29	$30,375	$48,091	$260,000	3.31354	5,000
1624	9	20	28	$42,300	$132,008	$221,000	1.26787	3,141
1625	9	20	28	$42,300	$20,108	$120,000	1.92283	1,240
1626	9	20	28	$8,000	$21,750	$38,000	1.27731	972
1627	9	20	28	$8,000	$33,831	$44,900	1.07337	1,170
1628	9	20	28	$8,000	$31,859	$45,000	1.12898	1,402
1629	9	20	28	$8,000	$23,805	$31,500	0.99041	891
1630	9	20	28	$8,000	$22,721	$29,000	0.94398	989
1631	9	20	28	$8,500	$26,242	$41,000	1.18013	936
1632	9	20	28	$8,000	$25,170	$40,000	1.20591	975
1633	9	20	28	$8,000	$24,402	$41,500	1.28079	933
1634	9	20	28	$8,500	$24,530	$37,500	1.13533	936
1635	9	20	28	$8,500	$25,031	$39,900	1.18994	975
1636	9	20	28	$8,000	$25,540	$45,000	1.34168	936
1637	9	20	28	$8,000	$24,530	$35,000	1.07593	936
1638	9	20	28	$6,800	$32,001	$45,300	1.16750	1,032
1639	9	20	29	$2,000	$32,201	$56,000	1.63738	1,737
1640	9	20	29	$15,500	$43,523	$105,000	1.77897	1,550
1641	9	20	29	$15,500	$52,659	$74,000	1.08570	1,780
1642	9	20	29	$10,000	$37,517	$49,500	1.04173	1,251
1643	9	20	29	$10,000	$31,899	$53,500	1.27688	1,479
1644	9	20	29	$10,000	$19,208	$36,000	1.23254	984
1645	9	20	29	$3,000	$31,481	$55,000	1.59508	1,700
1646	9	20	29	$9,000	$17,547	$37,500	1.41259	841
1647	9	20	29	$8,400	$26,200	$48,000	1.38728	1,172
1648	9	20	29	$4,560	$20,990	$40,000	1.56556	1,320
1649	9	20	29	$5,760	$23,008	$49,800	1.73109	1,320
1650	9	20	29	$4,800	$8,793	$24,000	1.76561	880
1651	9	20	29	$8,600	$13,258	$20,800	0.95160	984
1652	9	20	29	$6,400	$19,298	$30,000	1.16741	912
1653	9	20	29	$3,200	$13,327	$28,000	1.69420	648
1654	9	20	29	$8,000	$16,843	$33,000	1.32834	806
1655	9	20	29	$18,000	$50,556	$87,200	1.27195	2,150
1656	9	20	29	$10,000	$51,878	$72,900	1.17812	1,796

REAL ESTATE SALES FOR 1986
IN A MID-SIZED FLORIDA CITY

OBS	TOWNSHIP	RANGE	SECTION	LANDVAL	IMPROVAL	SALEPRIC	SALTOAPR	AREA
1657	9	20	29	$10,000	$42,133	$55,400	1.06267	1,651
1658	9	20	29	$5,000	$27,859	$39,000	1.18689	1,104
1659	9	20	29	$6,000	$24,999	$37,000	1.19359	1,235
1660	9	20	29	$20,000	$16,066	$42,000	1.16453	881
1661	9	20	29	$4,000	$13,911	$26,100	1.45721	780
1662	9	20	29	$4,000	$22,763	$40,000	1.49460	936
1663	9	20	32	$7,800	$22,183	$38,000	1.26738	1,196
1664	9	20	32	$3,900	$10,145	$33,000	2.34959	806
1665	9	20	29	$4,000	$9,256	$32,000	2.41400	768
1666	9	20	29	$4,800	$12,334	$28,500	1.66336	1,152
1667	9	20	29	$9,600	$15,844	$25,900	1.01792	1,055
1668	9	20	29	$9,600	$19,933	$32,700	1.10724	960
1669	9	20	29	$4,800	$16,504	$35,000	1.64288	817
1670	9	20	29	$5,000	$12,174	$26,000	1.51392	754
1671	9	20	30	$7,000	$8,649	$26,000	1.66145	928
1672	9	20	30	$14,000	$15,138	$31,700	1.08793	1,190
1673	9	20	30	$7,000	$8,459	$18,900	1.22259	768
1674	9	20	30	$7,000	$15,896	$28,500	1.24476	720
1675	9	20	30	$7,000	$7,339	$27,500	1.91785	520
1676	9	20	30	$8,000	$20,327	$38,400	1.35560	998
1677	9	20	30	$10,000	$29,350	$46,500	1.18170	1,053
1678	9	20	30	$10,000	$31,625	$60,100	1.44384	1,196
1679	9	20	30	$10,000	$37,884	$52,900	1.10475	1,196
1680	9	20	30	$10,000	$36,217	$52,000	1.12513	1,198
1681	9	20	30	$11,000	$34,080	$50,000	1.10914	1,075
1682	9	20	30	$10,000	$28,401	$43,000	1.11976	952
1683	9	20	30	$23,914	$73,399	$180,000	1.84970	4,836
1684	9	20	30	$10,000	$25,402	$46,000	1.29936	925
1685	9	20	30	$10,000	$30,077	$45,000	1.12284	1,040
1686	9	20	30	$10,000	$33,284	$48,200	1.11358	1,147
1687	9	20	30	$10,000	$34,380	$49,900	1.12438	1,258
1688	9	20	30	$10,000	$19,074	$38,500	1.32421	912
1689	9	20	30	$10,000	$20,485	$34,000	1.11530	1,012
1690	9	20	30	$10,000	$19,529	$45,000	1.52393	1,012
1691	9	20	30	$12,000	$22,422	$19,800	1.15024	1,056
1692	9	20	30	$11,500	$28,241	$49,500	1.24557	920
1693	9	20	30	$11,500	$37,167	$52,000	1.06849	1,278
1694	9	20	30	$11,000	$37,068	$49,000	1.01939	1,188
1695	9	20	30	$11,000	$35,505	$46,900	1.00849	1,204
1696	9	20	30	$11,000	$35,889	$52,000	1.10900	1,180
1697	9	20	30	$11,000	$33,097	$47,900	1.08624	1,046
1698	9	20	30	$11,000	$27,571	$45,300	1.17446	996
1699	9	20	30	$11,000	$32,104	$57,900	1.34326	1,136
1700	9	20	30	$11,000	$25,305	$40,000	1.11003	920
1701	9	20	30	$11,000	$43,162	$53,400	0.98593	1,518
1702	9	20	30	$11,000	$26,307	$46,000	1.23301	1,024
1703	9	20	30	$11,000	$32,033	$46,000	1.06895	1,000
1704	9	20	30	$11,000	$58,592	$70,000	1.00586	1,986
1705	9	20	30	$11,000	$35,832	$51,000	1.08900	1,307
1706	9	20	30	$11,000	$37,690	$57,000	1.17067	1,223
1707	9	20	30	$11,000	$30,394	$48,000	1.15959	984
1708	9	20	30	$11,000	$39,471	$62,100	1.23041	1,302
1709	9	20	30	$11,000	$36,466	$45,400	0.95647	1,149
1710	9	20	30	$49,000	$120,112	$287,500	1.70006	3,345
1711	9	20	30	$49,000	$123,417	$267,000	1.54857	2,551
1712	9	20	30	$13,000	$39,543	$56,000	1.06579	1,428
1713	9	20	30	$13,000	$44,556	$61,000	1.05984	1,728
1714	9	20	31	$15,000	$39,394	$71,000	1.30529	2,004
1715	9	20	31	$15,000	$43,243	$62,100	1.06622	1,923
1716	9	20	31	$22,000	$56,557	$95,000	1.20931	1,908
1717	9	20	31	$22,000	$52,435	$98,800	1.32733	2,258
1718	9	20	31	$22,000	$51,490	$84,900	1.15526	2,016
1719	9	20	31	$22,000	$50,310	$86,000	1.18932	2,157
1720	9	20	31	$22,000	$48,974	$79,900	1.12576	1,834
1721	9	20	31	$22,000	$46,067	$86,000	1.26346	1,789
1722	9	20	31	$22,000	$52,845	$84,000	1.12232	2,012
1723	9	20	31	$22,000	$63,047	$96,500	1.13467	2,172
1724	9	20	31	$22,000	$48,811	$98,000	1.38397	1,896
1725	9	20	31	$22,000	$49,355	$97,000	1.35940	2,026
1726	9	20	31	$23,000	$71,043	$112,000	1.19094	2,792
1727	9	20	31	$22,000	$47,299	$77,000	1.11113	1,756
1728	9	20	31	$20,000	$49,091	$89,900	1.30118	1,707

APPENDIX F   DATA SET: APPRAISALS AND SALES DATA

REAL ESTATE SALES FOR 1986
IN A MID-SIZED FLORIDA CITY

OBS	TOWNSHIP	RANGE	SECTION	LANDVAL	IMPROVAL	SALEPRIC	SALTOAPR	AREA
1729	9	20	31	$20,000	$60,960	$105,000	1.29694	2,048
1730	9	20	31	$17,000	$46,142	$89,000	1.40952	1,527
1731	9	20	31	$17,000	$61,339	$93,000	1.18715	2,120
1732	9	20	31	$17,000	$46,142	$87,500	1.38577	1,527
1733	9	20	31	$17,000	$46,142	$78,500	1.24323	1,527
1734	9	20	31	$9,000	$34,364	$55,900	1.28909	1,639
1735	9	20	31	$23,000	$42,922	$72,000	1.09220	1,725
1736	9	20	31	$22,000	$61,832	$106,500	1.27040	2,707
1737	9	20	31	$22,000	$66,861	$107,500	1.20975	2,712
1738	9	20	31	$22,000	$54,131	$100,000	1.31353	2,121
1739	9	20	31	$22,000	$46,871	$89,900	1.30534	1,846
1740	9	20	31	$22,000	$48,798	$71,500	1.00992	1,404
1741	9	20	31	$22,000	$76,682	$111,000	1.12483	2,819
1742	9	20	31	$22,000	$57,800	$106,500	1.33459	2,000
1743	9	20	31	$22,000	$69,785	$105,000	1.14398	2,129
1744	9	20	31	$22,000	$46,397	$67,000	0.97958	1,772
1745	9	20	31	$22,000	$49,396	$79,900	1.11911	1,793
1746	9	20	31	$22,000	$67,375	$100,000	1.11888	2,587
1747	9	20	31	$23,000	$71,770	$110,000	1.16070	2,522
1748	9	20	31	$23,000	$50,093	$84,500	1.15606	1,917
1749	9	20	31	$23,000	$60,040	$90,000	1.08382	2,872
1750	9	20	31	$23,000	$51,662	$91,000	1.21883	2,016
1751	9	20	31	$20,000	$42,514	$74,000	1.18373	2,068
1752	9	20	31	$20,000	$32,816	$65,000	1.23069	1,391
1753	9	20	31	$32,000	$39,423	$79,900	1.11869	1,890
1754	9	20	31	$50,000	$31,238	$85,000	1.04631	1,456
1755	9	20	31	$24,000	$50,057	$112,000	1.51235	1,412
1756	9	20	31	$20,000	$44,817	$80,000	1.23424	1,543
1757	9	20	31	$20,000	$50,084	$117,000	1.66943	1,847
1758	9	20	31	$20,000	$39,224	$72,500	1.22417	1,842
1759	9	20	31	$45,000	$56,334	$102,000	1.00657	2,543
1760	9	20	31	$35,000	$61,138	$93,500	0.97256	2,907
1761	9	20	31	$38,000	$77,594	$140,000	1.21114	3,845
1762	9	20	31	$18,000	$53,859	$70,000	0.97413	1,752
1763	9	20	31	$20,000	$41,634	$75,000	1.21686	1,750
1764	9	20	31	$30,000	$69,341	$122,000	1.22809	2,444
1765	9	20	31	$24,000	$41,578	$102,000	1.55540	1,648
1766	9	20	31	$20,000	$25,343	$48,500	1.06962	1,425
1767	9	20	31	$20,000	$35,755	$60,000	1.07614	1,428
1768	9	20	31	$20,000	$34,980	$54,500	0.99127	1,834
1769	9	20	31	$30,000	$68,669	$120,000	1.21619	3,083
1770	9	20	31	$23,000	$55,868	$86,000	1.09043	2,220
1771	9	20	31	$20,000	$58,510	$97,000	1.23551	2,455
1772	9	20	31	$20,000	$106,247	$145,000	1.14854	3,800
1773	9	20	31	$24,000	$41,185	$78,900	1.21040	1,558
1774	9	20	32	$4,000	$11,216	$25,500	1.67587	600
1775	9	20	32	$4,000	$21,983	$30,900	1.18924	1,152
1776	9	20	32	$4,000	$15,160	$16,000	0.83507	780
1777	9	20	32	$3,700	$13,478	$24,000	1.39714	672
1778	9	20	32	$6,700	$22,164	$32,700	1.13290	1,128
1779	9	20	32	$6,600	$24,212	$45,000	1.46047	960
1780	9	20	32	$3,700	$9,935	$22,000	1.61349	976
1781	9	20	32	$3,700	$18,181	$32,200	1.47160	1,090
1782	9	20	32	$4,890	$25,578	$59,000	1.93646	1,594
1783	9	20	32	$8,200	$23,161	$40,000	1.27547	1,088
1784	9	20	32	$8,293	$63,313	$85,000	1.18705	822
1785	9	20	32	$8,293	$63,313	$85,000	1.18705	822
1786	9	20	32	$13,108	$63,876	$85,000	1.10413	821
1787	9	20	32	$120,000	$193,170	$450,000	1.43692	6,128
1788	9	20	32	$10,000	$18,879	$41,000	1.41972	1,009
1789	9	20	32	$10,000	$20,244	$47,000	1.55403	908
1790	9	20	32	$10,000	$23,788	$48,000	1.42062	1,093
1791	9	20	32	$18,522	$12,924	$31,500	1.00172	840
1792	9	20	32	$16,000	$24,683	$50,000	1.22901	1,235
1793	9	20	32	$16,000	$33,628	$56,400	1.13646	1,491
1794	9	20	32	$16,000	$37,537	$55,000	1.02733	1,657
1795	9	20	32	$14,000	$39,348	$71,000	1.33088	810
1796	9	20	32	$14,000	$28,204	$54,900	1.30082	855
1797	9	20	32	$14,000	$33,574	$55,000	1.15609	1,110
1798	9	20	32	$14,000	$35,336	$55,000	1.11480	1,092
1799	9	20	32	$14,000	$35,336	$59,900	1.21412	1,092
1800	9	20	32	$25,300	$47,351	$80,000	1.10115	1,742

REAL ESTATE SALES FOR 1986
IN A MID-SIZED FLORIDA CITY

OBS	TOWNSHIP	RANGE	SECTION	LANDVAL	IMPROVAL	SALEPRIC	SALTOAPR	AREA
1801	9	20	32	$23,920	$10,477	$67,000	1.94784	920
1802	9	20	32	$13,800	$125,031	$230,000	1.65669	5,673
1803	9	20	32	$22,800	$17,869	$59,500	1.46303	1,188
1804	9	20	32	$12,000	$18,765	$75,000	2.43784	1,676
1805	9	20	32	$14,000	$27,753	$49,500	1.18554	1,597
1806	9	20	32	$14,000	$21,562	$36,000	1.01232	1,032
1807	9	20	32	$12,000	$16,057	$35,000	1.24746	1,182
1808	9	20	32	$7,688	$77,880	$118,000	1.37902	2,700
1809	9	20	32	$5,100	$33,814	$38,000	0.97651	1,080
1810	9	20	32	$25,000	$30,298	$51,800	0.93674	1,152
1811	9	20	32	$5,800	$7,770	$40,100	2.95505	1,056
1812	9	20	32	$4,900	$7,496	$28,500	2.29913	864
1813	9	20	33	$8,000	$22,808	$40,000	1.29836	1,165
1814	9	20	33	$8,000	$24,717	$42,500	1.29902	1,165
1815	9	20	33	$8,000	$30,653	$45,800	1.18490	1,355
1816	9	20	33	$8,000	$23,710	$32,000	1.00915	1,130
1817	9	20	33	$8,000	$22,680	$45,000	1.46675	1,165
1818	9	20	33	$8,500	$30,040	$52,000	1.34925	1,682
1819	9	20	33	$8,000	$16,700	$33,700	1.36437	930
1820	9	20	33	$8,000	$10,373	$13,800	0.75110	675
1821	9	20	33	$8,000	$24,438	$44,000	1.35643	1,151
1822	9	20	33	$8,000	$23,454	$45,500	1.44656	1,151
1823	9	20	33	$8,000	$20,177	$43,500	1.54381	1,231
1824	9	20	33	$8,000	$36,059	$48,000	1.08945	2,172
1825	9	20	33	$8,000	$31,550	$36,900	0.93300	1,815
1826	9	20	33	$8,000	$29,120	$50,000	1.34698	1,730
1827	9	20	33	$8,000	$22,991	$45,000	1.45203	1,573
1828	9	20	33	$8,000	$28,902	$46,700	1.26551	1,130
1829	9	20	33	$8,000	$29,305	$46,500	1.24648	1,130
1830	9	20	33	$8,000	$25,751	$40,500	1.19996	1,256
1831	9	20	33	$8,000	$25,717	$42,000	1.24566	1,380
1832	9	20	33	$8,000	$21,396	$40,000	1.36073	1,147
1833	9	20	33	$8,000	$23,095	$44,400	1.42788	1,048
1834	9	20	33	$16,000	$27,734	$79,900	1.82695	1,431
1835	9	20	33	$18,000	$37,624	$60,000	1.07867	1,866
1836	9	20	33	$18,000	$44,592	$72,000	1.15031	1,888
1837	9	20	33	$15,000	$35,148	$71,100	1.41780	1,679
1838	9	20	33	$13,000	$35,286	$57,700	1.19496	1,171
1839	9	20	33	$15,000	$18,698	$54,000	1.60247	1,305
1840	9	20	33	$12,000	$29,684	$40,000	0.95960	1,256
1841	9	20	33	$7,000	$12,136	$33,000	1.72450	759
1842	9	20	33	$8,000	$15,323	$34,400	1.47494	892
1843	9	20	33	$8,000	$14,740	$34,900	1.53474	784
1844	9	20	33	$8,000	$18,812	$37,000	1.37998	904
1845	9	20	33	$8,000	$15,364	$38,500	1.64785	953
1846	9	20	33	$8,000	$20,079	$37,500	1.33552	1,188
1847	9	20	33	$8,000	$26,843	$43,000	1.23411	1,276
1848	9	20	33	$8,000	$15,583	$31,500	1.33571	1,014
1849	9	20	33	$8,000	$16,583	$36,000	1.46443	861
1850	9	20	33	$8,500	$17,052	$38,500	1.50673	861
1851	9	20	33	$8,000	$29,693	$45,000	1.19386	1,395
1852	9	20	33	$8,000	$29,366	$36,900	0.98753	1,470
1853	9	20	33	$8,000	$39,569	$48,500	1.01957	1,758
1854	9	20	33	$8,000	$8,591	$27,000	1.62739	884
1855	9	20	33	$8,000	$24,535	$41,900	1.28784	1,222
1856	9	20	33	$8,000	$28,252	$45,200	1.24683	1,232
1857	9	20	33	$9,270	$16,303	$45,000	1.75967	1,066
1858	9	20	33	$9,720	$16,452	$42,500	1.62387	920
1859	9	20	33	$9,720	$14,155	$32,400	1.35707	1,085
1860	9	20	33	$9,270	$10,166	$35,500	1.82651	1,284
1861	9	20	33	$9,720	$12,978	$34,900	1.53758	726
1862	9	20	33	$4,528	$28,672	$39,000	1.17470	928
1863	9	20	33	$8,640	$30,596	$38,600	0.98379	1,986
1864	9	20	33	$8,000	$35,373	$49,900	1.15049	1,642
1865	9	20	33	$8,500	$19,055	$43,200	1.56777	1,094
1866	9	20	33	$8,000	$22,818	$38,500	1.24927	1,104
1867	9	20	33	$8,000	$32,886	$43,900	1.07372	1,664
1868	9	20	33	$8,000	$26,669	$43,400	1.25184	1,170
1869	9	20	33	$8,000	$26,698	$45,900	1.32284	1,170
1870	9	20	33	$8,000	$19,800	$30,000	1.07914	925
1871	9	20	33	$8,000	$22,775	$41,900	1.36149	1,053
1872	9	20	33	$8,000	$23,047	$43,000	1.38500	1,175

REAL ESTATE SALES FOR 1986
IN A MID-SIZED FLORIDA CITY

OBS	TOWNSHIP	RANGE	SECTION	LANDVAL	IMPROVAL	SALEPRIC	SALTOAPR	AREA
1873	9	20	33	$8,000	$22,146	$35,200	1.16765	1,236
1874	9	20	33	$8,000	$23,126	$45,000	1.44574	1,170
1875	9	20	33	$8,000	$33,769	$52,000	1.24494	1,432
1876	9	20	33	$8,000	$27,391	$41,000	1.15849	1,375
1877	9	20	33	$8,000	$23,304	$39,000	1.24585	1,075
1878	9	20	33	$8,000	$20,897	$40,900	1.41537	1,173
1879	9	20	34	$8,000	$26,490	$47,500	1.37721	1,210
1880	9	20	34	$8,000	$39,111	$44,900	0.95307	1,544
1881	9	20	34	$8,000	$23,549	$40,000	1.26787	1,081
1882	9	20	34	$8,000	$23,286	$43,000	1.37442	1,081
1883	9	20	34	$8,000	$32,790	$45,000	1.10321	1,515
1884	9	20	34	$8,000	$23,314	$60,000	1.91608	1,008
1885	9	20	34	$8,000	$23,343	$39,500	1.26025	1,025
1886	9	20	34	$8,000	$23,346	$47,000	1.49939	1,118
1887	9	20	34	$8,000	$31,178	$48,300	1.23283	1,345
1888	9	20	34	$8,000	$19,089	$27,500	1.01517	920
1889	9	20	34	$8,000	$18,953	$25,000	0.92754	1,025
1890	9	20	34	$8,000	$24,611	$40,000	1.22658	920
1891	9	20	34	$4,000	$4,989	$38,000	4.22739	1,490
1892	9	20	34	$8,000	$33,613	$47,000	1.12945	1,728
1893	9	20	34	$4,000	$13,294	$33,900	1.96022	840
1894	9	20	34	$8,000	$28,622	$48,500	1.32434	1,161
1895	9	20	34	$8,000	$27,237	$42,500	1.20612	1,150
1896	9	20	34	$8,000	$26,760	$47,000	1.35213	1,105
1897	9	20	34	$1,000	$5,788	$27,000	3.97761	746
1898	9	20	34	$1,155	$19,419	$29,000	1.40955	825
1899	10	20	1	$5,500	$37,913	$48,400	1.11487	1,638
1900	10	20	1	$5,000	$29,038	$46,300	1.36024	1,204
1901	10	20	1	$5,500	$31,986	$43,500	1.16043	1,179
1902	10	20	1	$5,000	$27,910	$40,500	1.23063	1,144
1903	10	20	1	$5,000	$37,400	$45,000	1.06132	1,557
1904	10	20	1	$5,000	$31,449	$43,500	1.19345	1,320
1905	10	20	1	$5,000	$31,043	$42,000	1.16527	1,188
1906	10	20	1	$7,000	$27,572	$28,000	0.80990	1,025
1907	10	20	1	$7,000	$28,276	$33,900	0.96099	1,050
1908	10	20	1	$6,000	$26,110	$39,900	1.24260	1,260
1909	10	20	1	$6,000	$30,282	$42,500	1.17138	1,430
1910	10	20	1	$7,500	$17,286	$35,500	1.43226	936
1911	10	20	1	$4,500	$18,237	$37,300	1.64050	1,163
1912	10	20	1	$5,000	$15,182	$29,000	1.43692	801
1913	10	20	1	$4,500	$16,140	$28,000	1.35659	728
1914	10	20	1	$4,500	$17,507	$35,400	1.60858	980
1915	10	20	1	$4,000	$16,213	$35,700	1.76619	936
1916	10	20	2	$20,300	$5,923	$30,000	1.14403	744
1917	10	20	3	$1,000	$20,217	$32,000	1.50822	1,000
1918	10	20	3	$1,000	$8,444	$22,000	2.32952	760
1919	10	20	3	$1,560	$8,120	$24,600	2.54132	624
1920	10	20	3	$780	$6,855	$15,000	1.96464	832
1921	10	20	3	$1,056	$18,768	$30,800	1.55367	990
1922	10	20	3	$1,500	$27,016	$37,000	1.29752	1,527
1923	10	20	3	$1,050	$25,580	$38,500	1.44574	1,092
1924	10	20	3	$4,000	$53,934	$74,000	1.27732	1,110
1925	10	20	3	$4,400	$49,789	$67,500	1.24564	2,091
1926	10	20	3	$1,200	$23,530	$30,500	1.23332	1,205
1927	10	20	3	$2,000	$9,289	$40,000	3.54327	1,118
1928	10	20	3	$4,263	$17,796	$35,000	1.58665	840
1929	10	20	4	$600	$16,146	$25,000	1.49289	944
1930	10	20	4	$562	$11,447	$28,500	2.37322	728
1931	10	20	4	$1,800	$6,917	$23,000	2.63852	966
1932	10	20	3	$1,650	$3,415	$10,000	1.97433	600
1933	10	20	4	$600	$17,052	$13,000	0.73646	804
1934	10	20	3	$1,200	$7,616	$39,600	4.49183	1,204
1935	10	20	4	$624	$3,383	$3,000	0.74869	682
1936	10	20	4	$16,000	$49,727	$95,000	1.44537	3,112
1937	10	20	4	$40,000	$44,034	$175,000	2.08249	3,286
1938	10	20	4	$8,500	$27,075	$28,000	0.78707	2,035
1939	10	20	4	$10,000	$11,889	$59,500	2.71826	1,085
1940	10	20	4	$35,000	$61,570	$280,000	2.89945	4,284
1941	10	20	4	$4,200	$22,068	$32,000	1.21821	1,194
1942	10	20	4	$14,385	$21,206	$115,000	3.23115	1,100
1943	10	20	4	$4,000	$8,469	$56,000	4.49114	1,090
1944	10	20	4	$7,500	$10,460	$25,000	1.39198	708

REAL ESTATE SALES FOR 1986
IN A MID-SIZED FLORIDA CITY

OBS	TOWNSHIP	RANGE	SECTION	LANDVAL	IMPROVAL	SALEPRIC	SALTOAPR	AREA
1945	10	20	4	$7,500	$10,737	$31,500	1.72726	720
1946	10	20	4	$10,000	$9,954	$25,000	1.25288	690
1947	10	20	4	$10,000	$17,004	$26,000	0.96282	921
1948	10	20	4	$10,000	$18,407	$34,800	1.22505	892
1949	10	20	4	$14,000	$35,830	$63,500	1.27433	2,466
1950	9	20	33	$20,000	$66,388	$119,000	1.37751	2,189
1951	9	20	33	$15,600	$35,544	$58,500	1.14383	1,648
1952	10	20	4	$15,000	$16,076	$48,000	1.54460	918
1953	10	20	4	$15,000	$7,528	$30,000	1.33168	360
1954	10	20	4	$15,000	$27,385	$62,000	1.46278	1,218
1955	9	20	33	$22,000	$48,505	$118,000	1.67364	2,420
1956	10	20	4	$15,000	$28,794	$52,000	1.18738	1,470
1957	10	20	4	$15,000	$20,692	$40,000	1.12070	1,123
1958	10	20	4	$16,000	$49,611	$130,000	1.98138	2,211
1959	10	20	4	$8,800	$8,126	$25,100	1.48293	1,045
1960	10	20	4	$18,000	$26,055	$60,000	1.36193	1,569
1961	10	20	4	$8,500	$14,053	$31,000	1.37454	782
1962	10	20	4	$8,500	$15,626	$38,500	1.59579	845
1963	10	20	4	$8,500	$21,707	$33,800	1.11895	1,091
1964	10	20	4	$9,000	$16,101	$37,500	1.49396	850
1965	10	20	4	$12,000	$28,420	$67,000	1.65760	1,382
1966	10	20	4	$10,000	$23,729	$43,900	1.30155	1,061
1967	10	20	4	$10,000	$20,722	$30,000	0.97650	1,011
1968	10	20	4	$10,000	$22,549	$37,000	1.13675	1,004
1969	10	20	4	$10,000	$19,345	$35,000	1.19271	912
1970	10	20	4	$10,000	$16,052	$35,000	1.34347	823
1971	10	20	4	$10,000	$18,398	$35,000	1.23248	912
1972	10	20	4	$10,000	$25,093	$27,500	0.78363	1,140
1973	10	20	4	$18,000	$24,624	$53,000	1.24343	1,554
1974	10	20	4	$9,500	$35,186	$56,000	1.25319	1,484
1975	10	20	4	$8,500	$34,476	$64,100	1.49153	1,474
1976	10	20	4	$8,500	$34,826	$60,000	1.38485	1,477
1977	10	20	4	$15,000	$55,670	$105,000	1.48578	2,127
1978	10	20	4	$8,000	$9,338	$19,000	1.09586	1,260
1979	10	20	4	$15,000	$31,178	$79,000	1.71077	1,812
1980	10	20	4	$21,000	$25,781	$71,000	1.51771	2,912
1981	10	20	4	$26,000	$18,702	$25,000	0.55926	1,728
1982	10	20	4	$600	$3,835	$21,200	4.78016	880
1983	10	20	5	$16,000	$9,098	$45,000	1.79297	1,276
1984	10	20	4	$9,800	$12,724	$50,900	2.25981	1,392
1985	10	20	5	$14,837	$15,480	$70,000	2.30894	2,120
1986	10	20	5	$23,706	$18,572	$100,000	2.36530	6,700
1987	10	20	5	$10,675	$14,471	$55,000	2.18723	1,724
1988	10	20	5	$7,500	$8,161	$16,100	1.02803	826
1989	10	20	5	$8,000	$11,227	$24,000	1.24824	1,042
1990	10	20	5	$10,500	$56,248	$80,000	1.19854	1,960
1991	10	20	5	$5,600	$12,140	$25,000	1.40924	1,022
1992	10	20	5	$26,820	$37,995	$110,000	1.69714	3,272
1993	10	20	5	$35,400	$15,406	$69,500	1.36795	700
1994	10	20	5	$13,944	$38,683	$58,000	1.10210	2,348
1995	10	20	5	$17,424	$40,958	$72,900	1.24867	2,980
1996	10	20	5	$6,000	$12,215	$26,500	1.45484	975
1997	10	20	5	$16,269	$64,922	$119,800	1.47553	3,520
1998	10	20	5	$5,000	$6,411	$6,000	0.52581	768
1999	10	20	5	$12,000	$15,672	$26,000	0.93958	1,122
2000	10	20	5	$12,150	$6,644	$24,000	1.27700	725
2001	10	20	5	$51,240	$177,724	$226,000	0.98705	5,565
2002	10	20	5	$6,500	$21,672	$40,000	1.41985	1,144
2003	10	20	5	$5,000	$45,373	$80,500	1.59808	2,896
2004	10	20	5	$17,600	$71,369	$125,000	1.40498	2,367
2005	10	20	5	$25,300	$74,388	$155,000	1.55485	2,668
2006	10	20	5	$3,000	$4,965	$6,000	0.75330	880
2007	10	20	5	$13,000	$6,102	$25,000	1.30876	934
2008	10	20	5	$1,500	$4,946	$4,700	0.72913	733
2009	10	20	5	$6,000	$17,220	$19,000	0.81826	1,066
2010	10	20	5	$3,000	$9,060	$14,500	1.20232	877
2011	10	20	4	$10,000	$22,423	$46,000	1.41875	1,320
2012	10	20	5	$7,000	$14,168	$20,000	0.94482	1,032
2013	10	20	6	$24,000	$56,847	$78,000	0.96479	2,481
2014	10	20	6	$20,000	$53,824	$107,000	1.44939	2,342
2015	10	20	6	$10,710	$26,826	$55,000	1.46526	1,560
2016	10	20	6	$21,353	$21,409	$96,000	2.24498	1,584

REAL ESTATE SALES FOR 1986
IN A MID-SIZED FLORIDA CITY

OBS	TOWNSHIP	RANGE	SECTION	LANDVAL	IMPROVAL	SALEPRIC	SALTOAPR	AREA
2017	10	20	6	$19,926	$34,060	$70,000	1.29663	2,380
2018	10	20	6	$18,000	$15,041	$37,000	1.11982	812
2019	10	20	6	$18,000	$31,744	$74,500	1.49767	2,035
2020	10	20	6	$20,250	$45,620	$84,900	1.28890	2,028
2021	10	20	6	$15,000	$44,994	$72,300	1.20512	2,698
2022	10	20	6	$20,000	$43,372	$82,500	1.30184	1,800
2023	10	20	6	$15,000	$36,291	$66,000	1.28678	1,510
2024	10	20	6	$15,000	$31,269	$73,500	1.58854	1,749
2025	10	20	6	$15,000	$30,957	$63,000	1.37085	1,604
2026	10	20	6	$77,220	$82,325	$270,000	1.69231	4,977
2027	10	20	6	$21,000	$27,694	$52,000	1.06789	1,389
2028	10	20	6	$16,000	$12,913	$25,000	0.86466	780
2029	10	20	6	$16,000	$10,305	$28,700	1.09105	780
2030	10	20	6	$16,000	$14,261	$36,000	1.18965	898
2031	10	20	6	$16,000	$16,855	$38,800	1.18095	898
2032	10	20	6	$16,000	$14,221	$34,000	1.12505	898
2033	10	20	6	$20,000	$47,359	$78,000	1.15797	2,385
2034	10	20	6	$16,000	$9,317	$20,000	0.78998	488
2035	10	20	6	$18,000	$22,670	$56,500	1.38923	1,413
2036	10	20	6	$16,200	$23,523	$52,900	1.33172	944
2037	10	20	6	$18,000	$32,077	$57,000	1.13825	1,476
2038	10	20	6	$18,000	$40,302	$65,000	1.11488	2,005
2039	10	20	6	$21,600	$47,085	$75,000	1.09194	1,851
2040	10	20	6	$16,200	$31,876	$62,000	1.28962	1,561
2041	10	20	6	$16,200	$31,896	$62,500	1.29948	1,424
2042	10	20	6	$14,400	$30,988	$66,000	1.45413	1,261
2043	10	20	6	$15,750	$43,984	$43,000	0.71986	1,965
2044	10	20	6	$31,500	$38,019	$86,500	1.24426	1,657
2045	10	20	6	$17,500	$40,256	$71,000	1.22931	1,711
2046	10	20	6	$19,250	$43,808	$80,000	1.26867	1,814
2047	10	20	6	$21,000	$74,584	$136,500	1.42806	3,728
2048	10	20	6	$17,500	$52,223	$85,000	1.21911	2,135
2049	10	20	6	$17,500	$25,583	$47,000	1.09092	1,206
2050	10	20	6	$16,500	$38,116	$67,000	1.22675	1,618
2051	10	20	8	$16,000	$44,582	$66,400	1.09604	1,953
2052	10	20	8	$18,000	$29,820	$56,000	1.17106	2,097
2053	10	20	8	$16,000	$24,189	$59,500	1.48050	1,108
2054	10	20	8	$11,761	$1,503	$30,000	2.26176	864
2055	10	20	0	$37,080	$65,640	$210,000	2.04439	2,562
2056	10	20	8	$6,200	$42,751	$70,000	1.43000	1,662
2057	10	20	0	$49,000	$55,244	$150,000	1.43893	1,139
2058	10	20	8	$5,800	$13,780	$34,600	1.76711	4,400
2059	10	20	8	$3,040	$32,202	$50,000	1.41876	7,257
2060	10	20	0	$8,184	$78,804	$110,000	1.26454	4,346
2061	10	20	0	$14,000	$47,969	$91,000	1.46848	1,736
2062	10	20	8	$10,000	$37,776	$65,800	1.37726	1,232
2063	10	20	8	$12,000	$55,910	$69,400	1.02194	1,515
2064	10	20	8	$10,000	$45,185	$62,000	1.12349	1,456
2065	10	20	8	$10,000	$46,502	$69,900	1.23712	1,532
2066	10	20	0	$45,877	$92,252	$160,000	1.15834	2,775
2067	10	20	8	$4,410	$75,393	$200,000	2.50617	8,000
2068	10	20	9	$1,200	$19,023	$22,500	1.11259	910
2069	10	20	9	$1,500	$43,417	$31,500	0.70129	1,522
2070	10	20	9	$6,500	$3,804	$39,000	3.78494	736
2071	10	20	9	$300	$2,796	$3,000	0.96899	752
2072	10	20	9	$4,000	$55,428	$76,000	1.27886	3,596
2073	10	20	9	$19,400	$106,505	$165,000	1.31051	6,500
2074	10	20	9	$4,000	$5,541	$32,500	3.40635	1,197
2075	10	20	16	$5,500	$19,797	$36,000	1.42309	1,008
2076	10	20	10	$3,500	$10,553	$12,900	0.91795	1,507
2077	10	20	10	$2,400	$3,674	$9,000	1.48173	337
2078	10	20	10	$9,000	$6,953	$20,000	1.25368	840
2079	10	20	10	$5,480	$15,924	$15,000	0.70080	759
2080	10	20	10	$5,480	$18,048	$32,000	1.36008	960
2081	10	20	10	$5,480	$27,520	$39,000	1.18182	1,068
2082	10	20	10	$5,480	$27,363	$36,400	1.10830	1,068
2083	10	20	11	$5,000	$24,369	$30,000	1.02149	1,040
2084	10	20	11	$5,000	$35,994	$48,000	1.17090	1,247
2085	10	20	11	$5,000	$38,603	$42,500	0.97470	1,642
2086	10	20	11	$5,000	$38,712	$50,000	1.14385	1,595
2087	10	20	11	$5,000	$32,296	$48,500	1.30041	1,202
2088	10	20	11	$5,000	$30,133	$36,100	1.02752	1,012

REAL ESTATE SALES FOR 1986
IN A MID-SIZED FLORIDA CITY

OBS	TOWNSHIP	RANGE	SECTION	LANDVAL	IMPROVAL	SALEPRIC	SALTOAPR	AREA
2089	10	20	11	$4,000	$33,985	$37,000	0.97407	1,176
2090	10	20	11	$3,500	$23,436	$34,500	1.28081	1,025
2091	10	20	11	$4,000	$26,972	$32,900	1.06225	975
2092	10	20	11	$4,000	$17,491	$35,000	1.62859	864
2093	10	20	11	$4,000	$29,841	$32,000	0.94560	1,920
2094	10	20	11	$4,000	$23,862	$36,000	1.29208	1,025
2095	10	20	11	$4,000	$21,879	$26,300	1.01627	1,000
2096	10	20	11	$4,000	$24,647	$31,500	1.09959	950
2097	10	20	11	$5,000	$26,332	$41,000	1.30857	1,225
2098	10	20	11	$9,000	$3,971	$14,500	1.11788	624
2099	10	20	11	$2,600	$23,693	$31,000	1.17902	1,131
2100	10	20	11	$2,600	$23,783	$33,300	1.26218	1,131
2101	10	20	11	$2,450	$24,043	$35,400	1.33620	1,131
2102	10	20	11	$2,500	$27,510	$36,500	1.21626	1,133
2103	10	20	12	$4,000	$6,439	$10,400	0.99626	1,584
2104	10	20	12	$6,000	$25,754	$38,300	1.20615	1,289
2105	10	20	12	$6,000	$13,508	$33,000	1.69161	660
2106	10	20	12	$6,000	$23,982	$36,500	1.21740	1,161
2107	10	20	12	$6,000	$23,853	$23,500	0.78719	1,040
2108	10	20	12	$5,000	$23,110	$38,000	1.35183	936
2109	10	20	12	$5,500	$19,896	$38,500	1.51599	936
2110	10	20	12	$5,500	$20,258	$36,000	1.39762	972
2111	10	20	12	$5,500	$19,547	$36,000	1.43730	936
2112	10	20	12	$5,500	$32,242	$44,900	1.18966	1,196
2113	10	20	12	$5,500	$20,423	$36,000	1.38873	936
2114	10	20	12	$5,000	$29,239	$38,000	1.10985	1,008
2115	10	20	12	$5,000	$32,852	$41,000	1.08317	1,362
2116	10	20	12	$4,000	$20,859	$57,200	2.30098	1,344
2117	10	20	12	$4,000	$20,859	$56,000	2.25271	1,344
2118	10	20	12	$2,000	$14,722	$31,200	1.86581	672
2119	10	20	12	$3,000	$19,268	$25,000	1.12269	1,008
2120	10	20	12	$3,000	$2,863	$2,300	0.39229	390
2121	10	20	12	$2,000	$13,884	$24,000	1.51095	672
2122	10	20	12	$6,500	$21,688	$41,600	1.47581	1,040
2123	10	20	12	$5,100	$6,258	$20,000	2.15721	916
2124	10	20	12	$6,000	$19,451	$33,600	1.32018	1,000
2125	10	20	12	$7,000	$23,779	$37,000	1.20212	1,134
2126	10	20	13	$2,400	$20,164	$43,500	1.92785	952
2127	10	20	14	$6,000	$32,034	$48,000	1.26203	1,520
2128	10	20	14	$6,000	$30,267	$44,000	1.21322	1,292
2129	10	20	14	$6,000	$29,140	$38,000	1.08139	1,240
2130	10	20	14	$9,160	$52,248	$66,500	1.08292	1,880
2131	10	20	14	$7,320	$49,042	$65,000	1.15326	1,524
2132	10	20	15	$5,500	$27,260	$40,000	1.22100	1,174
2133	10	20	15	$5,500	$20,276	$32,000	1.24146	1,008
2134	10	20	15	$4,500	$29,983	$42,500	1.23249	1,130
2135	10	20	16	$5,000	$38,677	$45,000	1.03029	1,649
2136	10	20	16	$5,000	$27,562	$42,500	1.30520	1,104
2137	11	20	0	$12,690	$61,310	$101,000	1.36486	1,852
2138	11	20	0	$21,165	$15,169	$47,500	1.30732	884
2139	11	20	0	$18,000	$39,812	$80,000	1.38380	1,701
2140	11	20	0	$4,025	$25,455	$78,500	2.66282	1,392
2141	11	20	21	$4,000	$22,115	$60,000	2.29753	1,597
2142	11	20	26	$4,500	$11,459	$20,000	1.25321	1,368
2143	11	20	26	$4,500	$13,277	$33,500	1.88446	610
2144	11	20	26	$4,500	$49,927	$56,800	1.04360	1,513
2145	11	20	26	$4,500	$18,294	$6,500	0.28516	1,352
2146	11	20	26	$16,000	$78,600	$175,000	1.84989	2,790
2147	11	20	26	$5,500	$14,404	$30,000	1.50723	742
2148	11	20	26	$5,000	$34,891	$50,000	1.25342	1,330
2149	11	20	34	$34,350	$37,581	$220,000	3.05849	1,448
2150	11	20	34	$2,500	$12,395	$13,000	0.87278	572
2151	11	20	34	$3,855	$8,642	$33,000	2.64063	1,104
2152	11	20	36	$24,000	$64,900	$125,000	1.40607	2,079
2153	8	21	11	$3,400	$14,830	$15,500	0.85025	768
2154	8	21	23	$1,000	$4,098	$40,000	7.84621	603
2155	8	21	23	$1,800	$13,369	$34,200	2.25460	1,417
2156	8	21	23	$2,040	$29,891	$26,000	0.81426	1,040
2157	8	21	23	$3,100	$4,289	$9,000	1.21803	564
2158	8	21	23	$3,100	$4,825	$10,000	1.26183	672
2159	8	21	23	$2,000	$9,906	$15,000	1.25987	680
2160	8	21	23	$3,500	$6,845	$8,500	0.82165	732

REAL ESTATE SALES FOR 1986
IN A MID-SIZED FLORIDA CITY

OBS	TOWNSHIP	RANGE	SECTION	LANDVAL	IMPROVAL	SALEPRIC	SALTOAPR	AREA
2161	8	21	23	$1,500	$6,956	$16,000	1.89215	644
2162	8	21	23	$7,410	$17,409	$39,900	1.60764	2,079
2163	8	21	31	$7,938	$11,267	$31,000	1.61416	1,344
2164	8	21	31	$8,064	$33,935	$61,000	1.45242	1,255
2165	8	21	32	$7,308	$12,465	$31,000	1.56779	1,344
2166	9	21	6	$1,000	$1,454	$6,500	2.64874	288
2167	9	21	6	$1,600	$13,233	$15,000	1.01126	1,344
2168	9	21	6	$6,902	$10,340	$20,000	1.15996	624
2169	9	21	6	$4,500	$22,610	$37,000	1.36481	988
2170	9	21	7	$6,700	$11,061	$6,040	0.34007	728
2171	9	21	7	$6,700	$11,581	$13,900	0.76035	784
2172	9	21	18	$7,395	$12,693	$43,900	2.18538	784
2173	9	21	22	$2,124	$3,600	$4,000	0.69881	611
2174	9	21	31	$1,500	$17,407	$36,000	1.90406	1,008
2175	9	21	31	$2,000	$54,625	$35,000	0.61810	1,904
2176	10	21	2	$2,925	$13,768	$23,000	1.37782	544
2177	10	21	3	$2,700	$27,136	$34,900	1.16973	1,008
2178	10	21	10	$19,032	$30,703	$34,800	0.69971	1,399
2179	10	21	11	$12,000	$54,663	$70,000	1.05006	1,885
2180	10	21	6	$4,000	$24,708	$42,000	1.46301	1,386
2181	10	21	6	$4,000	$31,509	$39,800	1.12084	1,393
2182	10	21	6	$4,000	$35,318	$40,000	1.01735	1,188
2183	10	21	6	$4,000	$27,062	$46,000	1.48091	1,118
2184	10	21	6	$2,500	$33,640	$47,000	1.30050	1,440
2185	10	21	6	$5,000	$7,718	$18,500	1.45463	1,008
2186	10	21	6	$5,000	$15,327	$18,500	0.91012	798
2187	10	21	6	$5,000	$19,671	$35,000	1.41867	1,030
2188	10	21	25	$3,980	$10,731	$21,990	1.49480	1,438
2189	10	21	28	$1,200	$19,776	$29,900	1.42544	1,089
2190	11	21	29	$5,520	$38,348	$64,100	1.46120	1,040
2191	11	21	36	$5,000	$22,834	$50,000	1.79636	1,108
2192	11	21	36	$3,400	$35,948	$50,000	1.27071	1,076
2193	11	21	36	$3,600	$5,512	$20,000	2.19491	672
2194	11	21	36	$3,500	$4,022	$15,500	2.06062	612
2195	11	21	36	$3,500	$2,831	$14,000	2.21134	560
2196	11	21	36	$4,000	$41,756	$68,000	1.48614	1,600
2197	12	21	1	$3,500	$14,843	$35,000	1.90808	840
2198	8	22	6	$5,000	$40,720	$59,000	1.29046	1,284
2199	8	22	19	$3,500	$5,268	$13,900	1.58531	732
2200	8	22	19	$8,300	$22,537	$45,000	1.45929	1,248
2201	8	22	19	$8,300	$23,544	$37,500	1.17762	1,344
2202	8	22	20	$5,000	$16,022	$8,800	0.41861	952
2203	8	22	20	$4,000	$13,770	$32,000	1.80079	1,260
2204	8	22	28	$28,875	$12,603	$75,000	1.80819	720
2205	8	22	28	$27,000	$47,514	$88,000	1.18099	1,672
2206	8	22	28	$27,500	$7,702	$50,000	1.42037	576
2207	8	22	30	$3,000	$8,277	$39,000	3.45837	1,618
2208	8	22	30	$50,200	$19,481	$95,000	1.36336	512
2209	8	22	32	$5,260	$17,163	$31,900	1.42265	1,344
2210	8	22	33	$50,210	$8,058	$71,000	1.21851	475
2211	8	22	33	$27,500	$41,700	$130,000	1.87861	1,672
2212	9	22	7	$7,140	$3,648	$40,000	3.70782	564
2213	9	22	7	$1,500	$12,884	$27,000	1.87709	1,166
2214	9	22	10	$3,000	$21,148	$41,600	1.72271	896
2215	9	22	10	$12,000	$7,536	$26,000	1.33088	660
2216	9	22	10	$19,200	$25,708	$57,000	1.26926	1,188
2217	9	22	11	$23,760	$33,623	$132,000	2.30033	1,936
2218	9	22	13	$1,670	$25,333	$38,000	1.40725	1,560
2219	9	22	13	$13,000	$45,426	$80,000	1.36925	1,349
2220	9	22	13	$13,000	$34,095	$66,000	1.40142	1,120
2221	9	22	13	$9,320	$34,092	$60,000	1.38211	1,072
2222	9	22	13	$10,320	$19,758	$38,500	1.28001	680
2223	10	22	1	$1,000	$26,548	$45,000	1.63351	950
2224	10	22	2	$2,625	$13,340	$20,000	1.25274	1,356
2225	10	22	13	$12,144	$12,825	$39,900	1.59798	784
2226	10	22	15	$7,500	$6,573	$15,000	1.06587	732
2227	10	22	17	$13,500	$50,004	$79,900	1.25819	2,184
2228	10	22	22	$3,000	$2,155	$6,500	1.26091	248
2229	10	22	26	$5,875	$24,787	$40,000	1.30455	1,680
2230	10	22	35	$700	$23,366	$23,000	0.95571	1,000
2231	10	22	26	$1,500	$5,779	$22,400	3.07735	962
2232	10	22	26	$5,500	$27,829	$58,000	1.74023	1,218

REAL ESTATE SALES FOR 1986
IN A MID-SIZED FLORIDA CITY

OBS	TOWNSHIP	RANGE	SECTION	LANDVAL	IMPROVAL	SALEPRIC	SALTOAPR	AREA
2233	10	22	26	$3,000	$31,995	$45,000	1.28590	1,404
2234	10	22	26	$3,000	$24,739	$36,500	1.31584	1,223
2235	10	22	26	$3,000	$35,189	$40,500	1.06051	1,116
2236	10	22	26	$3,000	$32,169	$40,000	1.13737	1,113
2237	10	22	26	$2,000	$10,927	$30,000	2.32072	1,148
2238	10	22	26	$2,500	$9,648	$35,000	2.88113	350
2239	10	22	26	$2,500	$27,031	$59,000	1.99790	1,180
2240	10	22	26	$1,400	$8,454	$21,000	2.13111	1,296
2241	10	22	29	$9,000	$26,708	$52,000	1.45626	1,848
2242	10	22	29	$3,960	$15,817	$26,000	1.31466	1,148
2243	11	22	1	$5,100	$34,358	$51,500	1.30519	1,065
2244	11	22	1	$12,000	$47,619	$75,000	1.25799	1,704
2245	11	22	2	$2,500	$56,439	$65,000	1.10284	2,144
2246	11	22	22	$8,040	$17,141	$49,000	1.94591	840
2247	11	22	22	$6,400	$10,504	$41,000	2.42546	732
2248	11	22	27	$801	$15,275	$29,000	1.80393	720
2249	11	22	27	$2,660	$13,978	$24,000	1.44248	784
2250	11	22	27	$7,000	$12,811	$32,000	1.61526	1,148
2251	11	22	27	$21,000	$9,543	$50,000	1.63704	525

APPENDIX G

DATA SET: 1970–1975 LOW BID PRICES FOR BREAD SUPPLIED TO FLORIDA PUBLIC SCHOOLS (SEE CASE STUDY 13)

1970–1975 LOW BID PRICES (DOLLARS) FOR BREAD
SUPPLIED TO FLORIDA PUBLIC SCHOOLS

OBS	MARKET	YEAR	LBPRICE
1	3	70	.150267
2	3	71	.166667
3	3	72	.183333
4	3	73	.213333
5	3	74	.306667
6	3	75	.280000
7	3	70	.200000
8	3	71	.210000
9	3	72	.179867
10	3	73	.153333
11	3	74	.200000
12	3	75	.280000
13	1	70	.206800
14	1	71	.220900
15	1	72	.220900
16	1	73	.255000
17	1	75	.364000
18	3	75	.280000
19	5	75	.248000
20	8	70	.209000
21	8	71	.228000
22	8	72	.230000
23	8	73	.230733
24	8	74	.276667
25	8	75	.242533
26	1	70	.202909
27	1	71	.228000
28	1	72	.225455
29	1	73	.240000
30	1	74	.334545
31	1	75	.340000
32	7	71	.203333
33	7	72	.256000
34	7	73	.256000
35	7	74	.360000
36	7	75	.266667
37	3	70	.191333
38	3	71	.206667
39	3	72	.213333
40	3	73	.226667
41	3	74	.304000
42	3	75	.266667
43	3	70	.190000
44	3	71	.215000
45	3	72	.220000
46	3	74	.353440
47	3	75	.260000
48	7	71	.200000
49	7	72	.213333
50	7	73	.233333
51	7	74	.270000
52	7	75	.266667
53	3	70	.160000
54	3	71	.173333
55	3	72	.186667
56	3	73	.240667
57	3	74	.297667
58	3	75	.276000
59	8	70	.195556
60	8	72	.222222
61	8	73	.235556
62	8	74	.320000
63	8	75	.311111
64	4	73	.230000
65	4	74	.313333
66	4	75	.266667
67	3	70	.175667
68	3	71	.216667
69	3	72	.170000
70	3	73	.153333
71	3	74	.313333
72	3	75	.266000

1970-1975 LOW BID PRICES (DOLLARS) FOR BREAD SUPPLIED TO FLORIDA PUBLIC SCHOOLS

OBS	MARKET	YEAR	LBPRICE
73	1	70	.200000
74	1	71	.200000
75	1	72	.200000
76	1	73	.216000
77	1	74	.328889
78	1	75	.328889
79	3	72	.205000
80	3	73	.297920
81	3	74	.313333
82	3	75	.253333
83	2	71	.208800
84	2	72	.208800
85	2	73	.255000
86	2	74	.320000
87	2	75	.256000
88	2	70	.184800
89	2	71	.206800
90	2	72	.206800
91	2	73	.255000
92	2	74	.330000
93	2	75	.320000
94	1	70	.220000
95	1	71	.211500
96	1	72	.211500
97	1	74	.300000
98	1	75	.326667
99	3	70	.173333
100	3	71	.185000
101	3	72	.185000
102	3	73	.225000
103	3	74	.289750
104	3	75	.291000
105	4	70	.216000
106	4	71	.190000
107	4	72	.220000
108	4	73	.234667
109	4	74	.313333
110	4	75	.265667
111	7	70	.192000
112	4	70	.206667
113	4	71	.213333
114	4	72	.213333
115	4	73	.193333
116	4	74	.306667
117	4	75	.266667
118	4	72	.228000
119	4	73	.228000
120	4	74	.286667
121	4	75	.260000
122	1	70	.213400
123	1	71	.227950
124	1	72	.223250
125	1	73	.255000
126	1	74	.306667
127	1	75	.340000
128	6	70	.166667
129	6	71	.208000
130	6	72	.224000
131	6	73	.248000
132	6	74	.312000
133	6	75	.264000
134	1	70	.192200
135	1	71	.218550
136	1	72	.218550
137	1	73	.240000
138	1	74	.306667
139	2	70	.205200
140	2	71	.220000
141	2	72	.220000
142	2	73	.260000
143	2	74	.264000
144	2	75	.306667

1970-1975 LOW BID PRICES (DOLLARS) FOR BREAD
SUPPLIED TO FLORIDA PUBLIC SCHOOLS

OBS	MARKET	YEAR	LBPRICE
145	3	70	.145000
146	3	71	.165000
147	3	72	.175000
148	5	70	.196923
149	5	71	.200000
150	5	72	.200000
151	5	73	.240000
152	5	74	.300000
153	5	75	.266667
154	7	71	.190000
155	7	72	.190000
156	7	73	.195000
157	7	74	.290000
158	7	75	.266667
159	2	70	.184800
160	2	71	.211500
161	2	72	.206800
162	2	73	.252450
163	2	74	.330000
164	3	70	.186667
165	3	71	.200000
166	3	72	.200000
167	3	73	.240667
168	3	74	.297667
169	3	75	.276000
170	2	70	.186667
171	2	71	.206667
172	2	72	.206667
173	2	73	.204000
174	2	74	.306667
175	2	75	.256000
176	4	70	.200000
177	4	71	.206667
178	4	72	.200000
179	4	73	.226667
180	4	74	.300000
181	4	75	.266667
182	3	71	.184167
183	3	72	.185867
184	3	73	.220800
185	3	74	.304000
186	3	75	.294400
187	6	70	.208000
188	6	71	.216000
189	6	72	.224000
190	6	73	.248000
191	6	74	.312000
192	6	75	.280000
193	8	70	.153333
194	8	71	.186667
195	8	72	.186667
196	8	73	.238133
197	8	74	.297667
198	8	75	.297667
199	3	71	.172500
200	3	72	.188750
201	3	73	.210000
202	3	74	.290000
203	3	75	.270000
204	1	70	.200000
205	1	71	.205800
206	1	72	.205800
207	1	73	.235000
208	1	74	.279300
209	1	75	.311111
210	6	70	.208000
211	6	71	.224000
212	6	72	.208000
213	6	73	.248000
214	6	74	.312000
215	6	75	.264000
216	5	72	.206667

1970-1975 LOW BID PRICES (DOLLARS) FOR BREAD
SUPPLIED TO FLORIDA PUBLIC SCHOOLS

OBS	MARKET	YEAR	LBPRICE
217	5	73	.233333
218	5	74	.304000
219	5	75	.250000
220	5	70	.245000
221	5	71	.255000
222	5	72	.210000
223	5	73	.240000
224	5	74	.410000
225	6	71	.216000
226	6	72	.212000
227	6	74	.312000
228	6	75	.264000
229	4	70	.284444
230	4	71	.297778
231	4	72	.304000
232	4	73	.304000
233	4	74	.405333
234	4	75	.346667
235	4	74	.304000
236	4	75	.253333
237	3	70	.160000
238	3	71	.170000
239	3	72	.180000
240	3	73	.253333
241	3	74	.283650
242	3	75	.274500
243	1	70	.200000
244	1	71	.200000
245	1	72	.200000
246	1	73	.240000
247	1	74	.328889
248	1	75	.328889
249	4	70	.193333
250	4	71	.216667
251	4	72	.213333
252	4	73	.226667
253	4	74	.288000
254	4	75	.266667
255	5	74	.303333
256	5	75	.253333
257	3	70	.208000
258	3	71	.206667
259	3	73	.253333
260	3	74	.297667
261	3	75	.270000
262	6	70	.208000
263	6	71	.216000
264	6	72	.212000
265	6	73	.248000
266	6	74	.312000
267	6	75	.264000
268	4	71	.193333
269	4	72	.216667
270	4	73	.230000
271	4	74	.273333
272	4	75	.266667
273	2	70	.232000
274	2	71	.248000
275	2	72	.248000
276	2	73	.268000
277	2	74	.328000
278	2	75	.352000
279	3	70	.180000
280	3	71	.200000
281	3	72	.200000
282	3	73	.153333
283	3	74	.226667
284	3	75	.270000
285	5	70	.184500
286	5	71	.193500
287	5	72	.201250
288	5	73	.153333

1970-1975 LOW BID PRICES (DOLLARS) FOR BREAD
SUPPLIED TO FLORIDA PUBLIC SCHOOLS

OBS	MARKET	YEAR	LBPRICE
289	5	74	.303333
290	5	75	.253333
291	2	70	.198000
292	2	71	.207000
293	2	72	.198000
294	2	73	.255000
295	2	74	.440000
296	2	75	.320000
297	1	73	.240000
298	1	74	.285000
299	1	75	.328889
300	1	72	.223250
301	1	73	.240000
302	1	74	.306667
303	1	75	.375000

DATA SET: CONDOMINIUM SALES DATA (SEE CASE STUDY 15)

OBS	PRICPAID	FLOORHGT	DISTELEV	VIEWOCEN	ENDUNT	FURN
1	$19,900	1	2	0	0	0
2	$20,400	1	13	0	0	0
3	$27,500	1	15	1	0	0
4	$20,900	2	14	0	0	0
5	$25,900	1	2	1	0	0
6	$26,400	1	9	1	0	0
7	$26,400	1	12	1	0	0
8	$26,400	1	13	1	0	0
9	$25,400	3	9	1	0	0
10	$26,400	3	15	1	0	0
11	$27,900	8	15	1	0	0
12	$20,400	1	12	0	0	0
13	$19,900	2	13	0	0	0
14	$24,900	2	15	1	0	0
15	$26,400	3	11	1	0	0
16	$20,900	3	14	0	0	0
17	$25,900	4	15	1	0	0
18	$20,400	1	14	0	0	0
19	$27,400	7	15	1	0	0
20	$19,900	1	3	0	0	0
21	$19,900	1	11	0	0	0
22	$27,400	2	14	1	0	0
23	$18,000	3	11	0	0	0
24	$25,400	3	12	1	0	0
25	$28,400	4	14	1	0	0
26	$26,900	3	5	1	0	0
27	$15,900	2	3	0	0	0
28	$19,900	1	6	0	0	0
29	$20,400	1	9	0	0	0
30	$19,000	2	11	0	0	0
31	$26,100	2	12	1	0	0
32	$15,900	2	12	0	0	0
33	$26,100	2	13	1	0	0
34	$16,900	3	6	0	0	0
35	$19,900	4	14	0	0	0
36	$18,900	5	14	0	0	0
37	$29,400	6	15	1	0	0
38	$19,900	6	14	0	0	0
39	$20,400	2	6	0	0	0
40	$27,900	3	14	1	0	0
41	$15,900	6	9	0	0	0
42	$15,100	2	4	0	0	0
43	$15,100	2	5	0	0	0
44	$15,900	2	10	0	0	0
45	$19,000	8	12	0	0	0
46	$19,000	8	13	0	0	0
47	$17,500	1	4	0	0	0
48	$17,500	1	5	0	0	0
49	$17,500	1	10	0	0	0
50	$15,900	2	2	0	0	0
51	$16,300	3	13	0	0	0
52	$21,900	8	14	0	0	0
53	$16,100	3	5	0	0	0
54	$16,900	3	3	0	0	0
55	$16,900	3	4	0	0	0
56	$16,500	3	10	0	0	0
57	$16,900	3	12	0	0	0
58	$26,900	3	13	1	0	0
59	$17,600	4	11	0	0	0
60	$18,600	5	11	0	0	0
61	$19,500	8	11	0	0	0
62	$25,900	1	3	1	0	0
63	$23,500	1	4	1	0	0
64	$24,000	1	11	1	0	0
65	$15,900	2	1	0	0	0
66	$16,900	3	1	0	0	0
67	$26,000	4	1	1	0	0

OBS	PRICPAID	FLOORHGT	DISTELEV	VIEWOCEN	ENDUNT	FURN
68	$20,900	4	1	0	0	0
69	$17,600	4	6	0	0	0
70	$17,900	4	9	0	0	0
71	$17,900	4	12	0	0	0
72	$27,400	4	13	1	0	0
73	$17,900	5	13	0	0	0
74	$18,000	6	2	0	0	0
75	$19,900	6	6	0	0	0
76	$19,000	6	12	0	0	0
77	$21,000	1	1	1	0	0
78	$22,500	1	5	1	0	1
79	$20,000	1	6	1	1	1
80	$22,000	1	10	1	0	0
81	$21,000	2	1	1	0	0
82	$22,500	2	3	1	0	1
83	$22,500	2	4	1	0	1
84	$22,000	2	5	1	0	1
85	$19,500	2	6	1	1	0
86	$19,000	2	9	1	0	0
87	$19,500	2	10	1	0	0
88	$20,000	3	1	1	0	0
89	$21,000	3	2	1	0	1
90	$17,000	3	2	0	0	1
91	$20,000	3	3	1	0	0
92	$18,000	3	6	1	1	0
93	$20,500	3	10	1	0	1
94	$21,500	4	2	1	0	0
95	$17,500	4	2	0	0	1
96	$20,000	4	3	1	0	0
97	$17,500	4	3	0	0	1
98	$19,500	4	4	1	0	0
99	$16,500	4	4	0	0	1
100	$19,500	4	5	1	0	0
101	$17,500	4	6	1	1	0
102	$20,000	4	9	1	0	1
103	$19,000	4	10	1	0	0
104	$16,000	4	10	0	0	0
105	$20,500	4	11	1	0	1
106	$19,000	4	12	1	0	0
107	$20,500	5	1	1	0	0
108	$17,500	5	1	0	0	1
109	$21,500	5	2	1	0	1
110	$22,000	5	3	1	0	1
111	$26,500	5	4	1	0	1
112	$17,500	5	4	0	0	1
113	$20,000	5	5	1	0	1
114	$19,500	5	6	1	1	1
115	$18,500	5	6	0	0	1
116	$19,500	5	9	1	0	1
117	$16,000	5	9	0	0	1
118	$20,500	5	10	1	0	1
119	$17,500	5	10	0	0	1
120	$20,500	5	11	1	0	1
121	$20,500	5	12	1	0	1
122	$16,600	5	12	0	0	1
123	$19,500	5	13	1	0	0
124	$20,500	5	15	1	0	0
125	$21,000	6	1	0	0	1
126	$16,500	6	1	0	0	0
127	$22,000	6	2	1	0	0
128	$18,000	6	2	0	0	1
129	$21,500	6	3	1	0	0
130	$17,500	6	3	0	0	0
131	$20,000	6	4	1	0	1
132	$16,500	6	4	0	0	0
133	$17,500	6	5	0	0	1
134	$18,500	6	6	1	0	0
135	$20,000	6	9	1	0	1
136	$16,500	6	9	0	0	0
137	$17,500	6	10	1	0	0
138	$17,000	6	10	0	0	1

OBS	PRICPAID	FLOORHGT	DISTELEV	VIEWOCEN	ENDUNT	FURN
139	$19,000	6	11	1	0	0
140	$17,000	6	11	0	0	0
141	$20,000	6	12	1	0	1
142	$19,500	6	13	1	0	0
143	$19,500	6	14	1	0	0
144	$20,500	7	1	1	0	0
145	$18,000	7	1	0	0	1
146	$19,500	7	2	1	0	0
147	$17,000	7	2	0	0	0
148	$19,500	7	3	1	0	0
149	$18,060	7	3	0	0	1
150	$20,000	7	4	1	0	0
151	$17,500	7	4	0	0	1
152	$19,000	7	5	1	0	0
153	$19,000	7	6	1	1	1
154	$19,500	7	9	1	0	0
155	$17,000	7	9	0	0	0
156	$20,000	7	10	1	0	0
157	$18,000	7	10	0	0	1
158	$19,000	7	11	1	0	0
159	$18,000	7	11	0	0	1
160	$19,000	7	12	1	0	0
161	$17,500	7	12	0	0	0
162	$19,500	7	14	1	0	0
163	$19,000	8	1	1	0	0
164	$18,000	8	1	0	0	1
165	$20,500	8	2	1	0	0
166	$17,500	8	2	0	0	0
167	$21,500	8	3	1	0	1
168	$17,500	8	3	0	0	0
169	$19,500	8	4	1	0	0
170	$17,000	8	4	0	0	0
171	$18,500	8	5	1	0	0
172	$17,500	8	5	0	0	0
173	$17,500	8	6	1	1	0
174	$17,500	8	6	0	0	0
175	$20,300	8	7	1	0	1
176	$16,500	8	9	0	0	0
177	$20,000	8	10	1	0	0
178	$17,500	8	10	0	0	0
179	$20,500	8	11	1	0	0
180	$19,000	8	12	1	0	0
181	$20,500	8	13	1	0	1
182	$19,500	8	14	1	0	0
183	$15,000	3	13	0	0	1
184	$14,000	5	1	1	0	1
185	$13,000	5	2	0	0	1
186	$18,200	6	5	1	0	1
187	$17,500	5	1	0	0	1
188	$16,400	5	14	1	0	1
189	$24,900	4	1	1	0	1
190	$21,500	7	13	1	0	1
191	$19,900	2	12	0	0	1
192	$19,000	8	5	0	0	1
193	$29,500	1	13	1	0	1
194	$21,700	2	13	0	0	1
195	$26,500	3	10	1	0	1
196	$20,200	4	1	0	0	1
197	$17,700	4	4	0	0	1
198	$18,900	8	11	0	0	1
199	$19,700	3	12	0	0	1
200	$20,500	3	1	0	0	1
201	$20,600	2	13	0	0	1
202	$23,100	7	4	1	0	1
203	$25,500	4	10	1	0	1
204	$25,000	7	14	1	0	1
205	$23,000	1	14	0	0	1
206	$27,900	4	12	1	0	1
207	$21,500	6	1	0	0	1
208	$30,600	3	15	1	0	1
209	$19,000	3	6	0	0	1

ANSWERS TO SELECTED EXERCISES

Note: *Your answers may vary slightly from those shown below due to rounding.*

CHAPTER 2

2.1a. 3, 7, 2.646   **b.** 2, 4.4, 2.098   **c.** 8, 18.67, 4.320   **d.** 4.75, 12.92, 3.594
2.2a. 10, 91.5, 9.566   **b.** 52.0, 3,336, 57.76   **c.** −2, 1.6, 1.265   **d.** .333, .0587, .242
2.3a. 15.13, 8.671, 2.945   **b.** At least 75% (if mound shape is not assumed); approx. 95% (if mound shape is assumed)
c. 96.7%   **d.** At least 10
2.4a. 41.56, 448.74, 21.28   **b.** At least 75% (if mound shape is not assumed); approx. 95% (if mound shape is assumed)
c. 100%
2.5a. 4,695.7; 27,240,870, 5,219.3   **b.** At least 75% (if mound shape is not assumed); approx. 95% (if mound shape is assumed)
c. 95.7%
2.6a. 12.22, 32.10, 5.67   **b.** At least 75% (if mound shape is not assumed); approx. 95% (if mound shape is assumed)
c. 96.0%   **d.** $\bar{y} = .876$, $s^2 = .125$, $s = .354$; 92.0%   **e.** $\bar{y} = 12.53$, $s^2 = 22.46$, $s = 4.74$; 92.0%
2.7 $R = 900 - 50 = 850$; $s \approx \frac{850}{4} = 212.5$ **2.8b.** .95   **c.** Between 7.06 and 27.70   **d.** Between .80 and 28.32
2.9a. .4772   **b.** .4332   **c.** .4987   **d.** .1915
2.10a. .7745   **b.** .4649   **c.** .1359   **d.** .9319   **e.** Approx. .3085   **f.** .6687
2.11a. .6826   **b.** .9500   **c.** .9000   **d.** .9974   **2.12a.** 1.645   **b.** 1.96   **c.** −1.96   **d.** 2.0
2.13a. −3   **b.** 1.96   **c.** 1.645   **d.** 1.0   **e.** −.15
2.14a. .9544   **b.** .1587   **c.** .1587   **d.** .8185   **e.** .1498   **f.** .9974
2.15a. .95   **b.** .90   **c.** .9974   **d.** .9759   **e.** .1574   **f.** .9319   **2.16a.** .1351   **b.** .0307
2.17a. .2843   **b.** .0228   **2.18** .0571   **2.19a.** .1423   **b.** .0526   **c.** −50.50
2.20a. Yes   **c.** $P(D \geq BE) = .8413$; decision is to market the fans   **d.** 3580   **2.21b.** 4.68   **c.** 7.931
2.22a. Approx. normal; $\mu_{\bar{y}} = 400$; $\sigma_{\bar{y}} = 11.86$ **b.** .0174; .5
2.23a. Approx. normal; $\mu_{\bar{y}} = 24.7$; $\sigma_{\bar{y}} = 1.93$ **b.** .0057
c. Waiting time distribution under the new operating procedure has changed (either the mean has decreased or the standard deviation has increased).
2.24a. Approx. normal; $\mu_{\bar{y}} = 71$; $\sigma_{\bar{y}} = 1.70$ **b.** .6034   **c.** .0094   **d.** No
2.25 In repeated sampling, 95% of all such confidence intervals contain μ.
2.26 20.3 ± .548   **2.27** 75 ± 5.152   **2.28** 779,030 ± 375,296.49   **2.30a.** 2.262   **b.** 3.747   **c.** −2.861   **d.** −1.796
2.31a. 53,000 ± 35,760.46   **b.** Relative frequency distribution of the sampled population is approx. normal.
c. Increase n or decrease confidence coefficient; increase n **2.32a.** .604 ± .117   **2.33** .0817 ± .0071
2.40a. .025   **b.** .05   **c.** .005   **d.** .0985   **e.** 10   **f.** .01   **2.41** $z = 5.69$; yes   **2.42** $t = 2.50$; reject H_0
2.43 $z = 3.54$; yes   **2.44a.** Yes; $z = 6.67$ **b.** Approx. 0
2.45 $t = -3.0$; yes; assume the population from which the sample is drawn is normal   **2.46** No; $z = -1.82$
2.47 $t = -12.60$; reject H_0 **2.49a.** $z = -4.22$; yes   **b.** −1.1 ± .429   **2.50a.** Yes; $z = -14.14$ **b.** −14.08 ± 1.95
2.51 $t = 1.30$; no   **2.52** −5 ± 1.36
2.53a. $z = 2.77$; yes   **b.** The two samples were randomly and independently selected.   **c.** $z = 2.17$; yes
2.54 Populations have equal variances and approx. normal distributions; samples are random and independent.
2.55 No; $t = -1.76$ **2.56a.** $s_1^2 = 88.62$; $s_2^2 = 68.49$ **c.** $t = .605$; no   **2.57a.** $z = 1.29$; no   **b.** 2.92 ± 3.715
2.58b. $t = 1.535$; no
2.59 The two sampled populations are normally distributed; the samples are randomly and independently selected from their respective populations.   **2.60a.** 3.73   **b.** 3.09   **c.** 6.52   **d.** 3.85   **e.** 2.52   **f.** 2.94   **2.61** $F = 3.41$; reject H_0
2.62 $F = 1.25$; no   **2.63** $F = 2.626$; reject H_0; assembly line 1   **2.64** $F = 10$; yes
2.65a. 5, 4.637, 21.5   **b.** 16.75, 6.021, 36.25   **c.** 4.857, 5.460, 29.81   **d.** 4, 0, 0
2.66a. At least $\frac{3}{4} = 75\%$ **b.** At least $\frac{8}{9} = 88.9\%$ **c.** At least $\frac{1.25}{2.25} = 55.6\%$
2.67a. .6826   **b.** .9544   **c.** .3830   **d.** .9974
2.68a. .0228   **b.** .0228   **c.** .0250   **d.** .5   **e.** .3085   **f.** .0250

2.69a. 4 below the mean   **b.** .5 above the mean   **c.** 0   **d.** 6 above
2.70a. -1.56 **b.** -2.23 **c.** Zero-coupon bond   **2.71a.** 0   **b.** .05   **c.** .44   **d.** -1.09
2.72a. .3085   **b.** .1587   **c.** .1359   **d.** .6915   **e.** 0   **f.** .9938   **2.73a.** .095   **b.** .9987   **c.** .3085
2.74a. 3.333, 2.186, 1.479   **b.** At least 75% (if mound shape not assumed); approx. 95% (if mound shape assumed)
c. 100%
2.75a. 7,666.5   **b.** .4562   **2.76a.** .0062   **b.** $412,600 \pm 600,000$ **2.77a.** .0020   **b.** .4522
2.78a. $t = -4.96$; reject H_0 **b.** Relative frequency distribution of sampled population is approx. normal
2.79a. 6.38 ± 3.74 **b.** Relative frequency distribution of sampled population is approx. normal
2.80a. $\mu_{\bar{y}} = 6$; $\sigma_{\bar{y}} = .3536$ **b.** .5222   **c.** .0793   **d.** Less spread
2.81a. $\bar{y} = 4,157.04$; $s = 2,721.46$ **c.** Approx. .95   **2.82a.** -1.321 **b.** 9.925   **c.** 2.447   **d.** 2.807   **2.83** Small
2.84a. $t = 2.099$; no   **b.** 4 ± 3.304 **2.85** $F = 1.11$; do not reject H_0 **2.86a.** 12.2 ± 1.645 **b.** 166   **c.** $z = 1.3$; no
2.87 Yes; $z = 6.15$ **2.88a.** $z = 1.414$; no
2.89 $t = 4.19$; yes; assume the population from which the sample is drawn is normal   **2.90** Yes; $t = 2.89$
2.91 Yes; $z = 5.68$ **2.92** Yes; $F = 2.40$ **2.93** $z = -4.03$; reject H_0
2.94a. $F = 20.64$; yes   **b.** Normal populations   **c.** $t = -.80$; no   **d.** Normal populations; equal variances (violated)
e. $100,250 \pm 14,895.44$

CHAPTER 3

3.2 $\beta_0 = 1$; $\beta_1 = 1$ **3.3a.** $\beta_0 = 2$; $\beta_1 = 2$ **b.** $\beta_0 = 4$; $\beta_1 = 1$ **c.** $\beta_0 = -2$; $\beta_1 = 4$ **d.** $\beta_0 = -4$; $\beta_1 = -1$
3.5a. $\beta_1 = 2$; $\beta_0 = 3$ **b.** $\beta_1 = 1$; $\beta_0 = 1$ **c.** $\beta_1 = 3$; $\beta_0 = -2$ **d.** $\beta_1 = 5$; $\beta_0 = 0$ **e.** $\beta_1 = -2$; $\beta_0 = 4$
3.6a. $\beta_0 = 0$; $\beta_1 = .8571$ **3.7a.** $\beta_0 = 2$; $\beta_1 = -1.2$ **3.8b.** $\beta_0 = -.125$; $\beta_1 = 3.125$ **d.** 15.5
3.9 $\beta_0 = -1.430$; $\beta_1 = .1947$ **3.10b.** $\hat{y} = 5.977 + 74.068x$ **d.** 11.162   **3.11a.** $\hat{y} = 133.1 - .2647x$ **3.12** .0313
3.13a. 1.143; .2857   **b.** 1.60; .5333   **c.** 11.0; 3.667   **d.** .0395; .0099   **e.** 77.414; 4.301   **f.** 2,235.9; 171.99
3.14a. $\hat{y} = -45.78 + 1.562x$ **c.** SSE $= 19.60$; $s^2 = 4.90$ **d.** 2.21
3.15b. $\hat{y} = .3537 + .000004426x$ **d.** .3892   **e.** SSE $= .10218$, $s^2 = .01022$ **f.** .10108
3.16a. $\hat{y} = -.2462 + 1.315x$ **c.** 7.645; 11.59   **d.** SSE $= 12.44$; $s^2 = 2.487$ **e.** 1.577
3.17a. $t = 6.708$; yes   **b.** $t = -5.196$; yes   **c.** $t = 5.84$; yes   **d.** $t = 8.49$; yes   **e.** $t = 4.48$; yes   **f.** $t = -1.919$; no
3.18 $t = 3.233$; yes   **3.19** $t = .256$; no   **3.20** $t = 8.122$; yes   **3.21a.** $t = -.44$; do not reject H_0 **b.** $-.01748 \pm .06536$
3.22a. $\hat{y} = 1.4478 + 1.4468x$ **b.** No, do not reject H_0: $t = 1.55$ **c.** $.14468 \pm .21125$
3.23 $\hat{\beta}_1 = 7.70$, $t = 2.556$; yes ($\alpha = .05$)
3.24a. $\hat{y} = 46,426 - 24.24x$ **b.** SSE $= 159,266,379.31$; $s^2 = 19,908,297.41$ **c.** No, do not reject H_0; $t = -.583$
d. -24.14 ± 95.53 **3.25** 5.638 ± 1.448 **3.26** $r = .9805$, $r^2 = .9614$ **3.27b.** No; $-r_{.005} = -.254$
3.28 $r = .6086$, $r^2 = .3705$ **3.29a.** Yes; $-r_{.05} = -.549$ **b.** Yes; $-r_{.05} = -.549$ **c.** .8464   **d.** .5041
3.30 $r_1 = -.9674$; $r_2 = -.1105$; x_1 **3.31a.** Yes; $r_{.01} = .230$ **b.** .139
3.32a. Negative   **b.** $-.855$ **c.** Yes   **d.** .731   **3.33b.** Probably; however, we cannot conduct a test since n is not given
3.34a. SSE $= 4.055$; $s^2 = .225$ **b.** $10.6 \pm .22$ **c.** $8.90 \pm .32$ **d.** $12.3 \pm .32$ **e.** Widens   **f.** 12.3 ± 1.05
3.35 $.7115 \pm .1778$ **3.36a.** 11.162 ± 3.739 **b.** 11.162 ± 1.023 **3.39** $.3891 \pm .2423$ **3.40** 217.9 ± 10.35
3.41a. $\hat{y} = 14,291 - 148.978x$ **b.** $t = -4.36$; reject H_0 **c.** $8,977.9 \pm 1,388.6$
d. $x = 35$ is outside the range of the sample data
3.43a. $\hat{y} = -.535 + 15.526x$ **b.** SSE $= 6.974$, $s^2 = .8718$, $s = .9337$, $r^2 = .9760$ **c.** $t = 18.05$ **d.** .001
e. (3.357, 7.994)   **3.44a.** $\hat{y} = -.0292 + 2.2985x$ **b.** .0764   **c.** .1044   **d.** Yes; $t = 3.83$
3.45a. $\hat{y} = 30.117 + .316x$ **b.** SSE $= 281.487$, $s^2 = 15.638$, $s = 3.954$, $r^2 = .1716$ **c.** $t = 1.93$ **d.** .0694
e. (34.0568, 51.4619)
3.46a. $\hat{y} = 3.1583x$ **b.** SSE $= 8.9833$; $s^2 = 1.2833$; $s = 1.1328$ **c.** Yes; $t = 43.19$ **d.** $3.1583 \pm .1729$
e. 22.1083 ± 1.2106 **f.** 22.1083 ± 2.9400
3.47a. $\hat{y} = -9.2667x$ **b.** SSE $= 12.8667$; $s^2 = 3.2167$; $s = 1.7935$ **c.** Yes; $t = -28.30$ **d.** $-9.2667 \pm .9090$
e. $-9.2667 \pm .9090$ **f.** -9.2667 ± 5.0611
3.48a. $\hat{y} = .20846x$ **b.** SSE $= 22.664$; $s^2 = 2.518$; $s = 1.587$ **c.** Yes; $t = 52.28$ **d.** $.20846 \pm .00902$
e. 26.057 ± 1.127 **f.** 26.057 ± 3.762 **3.49a.** $\hat{y} = 5.364x$ **b.** Yes; $t = 25.28$ **c.** 18.774 ± 6.300
3.50a. $\hat{y} = 46.40x$ **b.** $\hat{y} = 478 + 45.2x$ **d.** $t = .91$; no
3.51a. $\hat{y} = 51.18x$ **b.** Yes; $t = 154.56$ **c.** $\hat{y} = 1,855.35 + 47.07x$; yes, $t = 93.36$
d. $y = \beta_0 + \beta_1 x + \varepsilon$ (reject H_0: $\beta_0 = 0$, $t = 8.37$)
3.52b. $\hat{y} = 67.131 + .06142x$ **d.** $t = .03$; do not reject H_0 **e.** $.06142 \pm 4.35088$ **f.** .0081   **g.** .000065
h. 67.98 ± 13.64 **i.** Model not useful   **3.53a.** $t = 24.0$; yes   **b.** $t = 9.33$; yes   **c.** .941, .774

3.54a. $r = -.805$, $r^2 = .649$ **b.** Yes; $-r_{.05} = .621$ **3.55a.** Yes; $-r_{.05} < -.164$ **b.** .04
3.56a. Yes; $t = 2.51$ **c.** Probably not, since r^2 is very small
3.57a. $\hat{y} = 22.32 + 2.650x$ **c.** $t = 6.280$; yes ($\alpha = .05$) **d.** $r = .9421$; $r^2 = .8875$ **e.** 109.8 ± 2.155
f. 109.76 ± 6.0879
3.58a. $\hat{y} = -1.329 + 1.762x$ **b.** $r^2 = .995$; $s = .744$ **c.** Yes; $t = 32.54$ **d.** $1.762 \pm .218$ **e.** Aggressive
3.59a. Aggressive: $\hat{y} = 1.4508 + .0163x$; defensive: $\hat{y} = .4594 - .0046x$; neutral: $\hat{y} = .9112 + .00873x$
b. Aggressive: reject H_0, $t = 4.38$; defensive: reject H_0, $t = -5.49$; neutral: reject H_0, $t = 5.67$
c. Aggressive: $.0163 \pm .0088$; defensive: $-.0046 \pm .0020$; neutral: $.00873 \pm .00364$; aggressive and neutral stocks increase
linearly; defensive stocks decrease linearly
3.60a. Yes; $r_{.05} = .183$ **b.** .2304 **c.** No; $r_{.025} = .482$ **d.** .1681
3.61a. Yes **b.** Yes **d.** 8.29 **e.** 21.15
3.62a. $\hat{y} = 2.084 + .6682x$ **c.** $t = 16.28$; yes ($\alpha = .05$) **d.** .9815 **e.** $35.49 \pm .9787$ **f.** 38.83 ± 2.766
3.63a. $\hat{y} = 16.28 + 2.078x$ **c.** $t = 6.672$; yes ($\alpha = .05$) **d.** 57.84 ± 6.146
3.64a. $\hat{y} = 15.11 + .9720x$ **c.** $r = .9815$; $r^2 = .9634$ **d.** $22.88 \pm .9793$ **e.** 21.91 ± 2.574
3.65a. $\hat{y} = 13.48 + .0560x$ **c.** $r = .7402$; $r^2 = .5479$ **d.** 17.40 ± 2.474
3.66a. $\hat{y} = 44.17 - .0255x$ **c.** $t = -.0294$; do not reject H_0 **3.67** Yes
3.68a. Yes; yes **b.** Yes; no, the relationship for successful companies is negative

CHAPTER 4

4.2 $t = 2.107$; reject H_0 **4.3a.** Yes; positive **b.** Yes; positive **c.** $t = 24.56$; yes **d.** $1.278 \pm .505$
4.4a. $t = 1.18$; no **b.** $.10 \pm .743$ **c.** Yes **d.** 4.70 **4.5b.** No; $t = -.15$
4.6a. $F = 31.98$; yes **b.** $t = -2.273$; yes **4.7** $F = 1.056$; do not reject H_0 **4.8c.** Yes; $F = 5.41$ **d.** Yes; $t = -2.39$
4.9a. $F = 17.8$; yes **b.** $t = -3.50$; reject H_0: $\beta_1 = 0$ **c.** -6.38 ± 4.723
4.10a. Nuclear: $\hat{y} = -17.556 + .519x_1 + 10.889x_2 + .1322x_1x_2$; nonnuclear: $\hat{y} = -44.682 + 2.880x_1 + 25.062x_2 - .959x_1x_2$
b. Nuclear: yes, $F = 13.22$ (p-value $= .0018$), nonnuclear: yes, $F = 60.85$ (p-value $= .0001$)
c. Nuclear: no, $t = .12$ (p-value $= .9093$); nonnuclear: no, $t = -1.39$ (p-value $= .1904$)
4.11b. $\hat{y} = 95.75 - .3199x$ **c.** Quadratic model
4.12a. $(44.2595, 46.5857)$ **c.** No; $x_1 = 30$ is outside the range of the sample data
4.13a. $(7.4528, 26.8903)$ **c.** No; $x_2 = 1.10$ is outside the range of the sample data
4.14a. $\hat{y} = .0456 + .00079x_1 + .2374x_2 - .0004x_1x_2$ **b.** SSE $= 2.715$; $s^2 = .1697$
4.15a. $F = 84.96$; yes **b.** $t = -2.99$; yes **d.** No; interaction is significant **4.16** 7.323 to 9.449
4.17a. $F = 3,909.25$; reject H_0 **c.** H_0: $\beta_4 = 0$; H_a: $\beta_4 < 0$ **e.** No; since β_4 is positive **4.18** $E(y) = \beta_0 + \beta_1x_1 + \beta_2x_2$
4.19 $E(y) = \beta_0 + \beta_1x_1 + \beta_2x_2 + \beta_3x_3 + \beta_4x_4$ **4.20** $E(y) = \beta_0 + \beta_1x_1 + \beta_2x_2 + \beta_3x_1x_2 + \beta_4x_1^2 + \beta_5x_2^2$
4.21 $E(y) = \beta_0 + \beta_1x_1 + \beta_2x_2 + \beta_3x_3 + \beta_4x_1x_2 + \beta_5x_1x_3 + \beta_6x_2x_3 + \beta_7x_1^2 + \beta_8x_2^2 + \beta_9x_3^2$
4.22 $E(y) = \beta_0 + \beta_1x$, where $x = \begin{cases} 1 & \text{if A} \\ 0 & \text{if B} \end{cases}$; $\beta_0 = \mu_B$, $\beta_1 = \mu_A - \mu_B$
4.23 $E(y) = \beta_0 + \beta_1x_1 + \beta_2x_2 + \beta_3x_3$, where $x_1 = \begin{cases} 1 & \text{if A} \\ 0 & \text{if not} \end{cases}$, $x_2 = \begin{cases} 1 & \text{if B} \\ 0 & \text{if not} \end{cases}$, $x_3 = \begin{cases} 1 & \text{if C} \\ 0 & \text{if not} \end{cases}$,
$\beta_0 = \mu_D$, $\beta_1 = \mu_A - \mu_D$, $\beta_2 = \mu_B - \mu_D$, $\beta_3 = \mu_C - \mu_D$
4.24b. First-order **c.** Parallel lines (same slope) **d.** Parallel lines **4.25c.** Parallel lines
4.26b. Second-order **c.** Identical shapes, different y-intercepts **d.** No
4.27b. Second-order **c.** Different shapes **d.** Yes **e.** Shift curves along the x_1-axis
4.28a. $E(y) = \beta_0 + \beta_1x_1 + \beta_2x_2 + \beta_3x_1x_2 + \beta_4x_1^2 + \beta_5x_2^2$ **b.** Possibly
c. $\hat{y} = 54.5 + .007697x_1 + .554111x_2 + .000113x_1x_2$ **d.** Not likely **e.** Reject H_0: $F = 44.67$ (p-value $= .0005$)
f. No; $t = .68$ (p-value $= .5280$)
4.29a. $E(y) = \beta_0 + \beta_1x_1 + \beta_2x_2$ **b.** $E(y) = \beta_0 + \beta_1x_1 + \beta_2x_2 + \beta_3x_1x_2$
c. $E(y) = \beta_0 + \beta_1x_1 + \beta_2x_2 + \beta_3x_1x_2 + \beta_4x_1^2 + \beta_5x_2^2$ **d.** .600393 **e.** No; $F = 1.80$ (p-value $= .2465$)
f. $(.13543132, .57739014)$; inadequate model
4.30b. Yes; $F = 10.60$ **c.** No; $t = -1.56$ **d.** $E(y) = \beta_0 + \beta_1x_1 + \beta_2x_2 + \beta_3x_1x_2$ **e.** Lines will have different slopes
4.31a. $\hat{y} = 22.0189 - .1807x_1 - .2498x_2 - 4.6910x_3 + 3.6745x_4 + 22.5201x_5$ **b.** .5996 **c.** 8.657
d. Yes; $F = 87.45$ ($F_{.05} = 2.21$) **e.** Yes; $t = -4.65$ **f.** $3.6745 \pm .7893$ **g.** 22.5201 ± 7.0480
4.32a. β_7, β_8, and β_9 **b.** H_0: $\beta_7 = \beta_8 = \beta_9 = 0$ **c.** H_0: $\beta_3 = \beta_5 = \beta_6 = \beta_9 = 0$
4.33a. $F = 2.94$; yes **b.** $F = 1.50$; no **c.** $F = 5.73$; reject H_0 **4.34a.** H_0: $\beta_2 = \beta_3 = 0$ **b.** $F = 6.99$; reject H_0
4.35 $F = 1.409$; do not reject H_0 **4.36** $F = 2.596$; do not reject H_0

4.37a. $\beta_3, \beta_4, \ldots, \beta_{11}$ **b.** $H_0: \beta_3 = \beta_4 = \cdots = \beta_{11} = 0$ **c.** $H_0: \beta_2 = \beta_9 = \beta_{10} = \beta_{11} = 0$
4.39a. **(i)** 4; **(ii)** 6; **(iii)** 6; **(iv)** 1
4.41a. $F = 106.48$; yes   **b.** -725 ± 198.7 **d.** $t = 1.72$; reject $H_0: \beta_6 = 0$ **4.42a.** $F = 1.28$; do not reject H_0
4.43a. 13.68   **4.44b.** $F = 52.21$; reject H_0 **c.** $t = -3.333$; yes ($\alpha = .05$)   **d.** $t = 4$; yes ($\alpha = .05$)
4.45 $t = -1.104$; do not reject H_0 **4.46b.** $t = 40.54$; yes   **c.** 17.79
4.47a. $\hat{y} = 2.14 - .15x_1 + .03x_2 + 2.54x_3 - .34x_4 - .26x_5 - .72x_6$ **b.** Yes; $F = 25.27$ **c.** Yes; p-value $< .05$
d. No; p-value $> .05$ **e.** $H_0: \beta_4 = \beta_5 = \beta_6 = 0$
4.48b. $F = 16.1$; yes   **c.** $t = 2.5$; yes   **d.** 945
e. The estimate of the difference in mean attendance between weekends and weekdays is $15x_3 - 700$.
4.49b. $t = 5$; yes   **c.** 825   **d.** Support   **4.50a.** $H_0: \beta_5 = \beta_6 = 0$ **b.** Do not reject H_0; $F = 1.19$
4.51a. Yes; $F = 69.17$ **b.** Yes; $F = 91.39$ **c.** Yes; $F = 205.54$ **d.** Yes; $F = 129.89$ **4.52a.** $F = 6.614$; yes
4.53a. Test $H_0: \beta_2 = \beta_3 = 0$ **c.** $t = -3.333$; yes
4.54a. .904318   **b.** 4.548548   **c.** Reject H_0; $F = 33.079$ (p-value $= .0003$)   **d.** Yes; $t = 2.13$ (p-value $= \frac{.0706}{2} = .0353$)
e. No, do not reject $H_0: \beta_3 = \beta_4 = 0$; $F = 1.21$

CHAPTER 5

5.7 Unable to test model adequacy since there are no degrees of freedom available for estimating σ^2 (i.e., df $= n - 3 = 0$)
5.8 No; appears that multicollinearity exists (large pairwise correlations between independent variables and significant F-values in presence of nonsignificant t-values)
5.9a. $\hat{y} = 5.7106 + .62597x$ **b.** 193.501 ± 26.767
c. The prediction interval should be used cautiously, or not at all, since $x = 300$ lies outside the range of the sample data.
5.10a. $\hat{y} = 2.7433 + .800976x_1$; yes, $t = 15.92$ **b.** $\hat{y} = 1.6647 + 12.3954x_2$; yes, $t = 11.76$
c. $\hat{y} = -11.7953 + 25.0682x_3$; yes (at $\alpha = .05$), $t = 2.51$
5.11 The least squares prediction equation should be used cautiously, or not at all, since the data point ($x_1 = 95.18$, $x_2 = 1.19$) lies outside the experimental region.
5.12 Possibly, since two of the three t-tests are nonsignificant while the F-test for overall model adequacy is significant; however, the low variance inflation factors indicate that multicollinearity, if it exists, is not severe.
5.13 3; n greater than 3   **5.14** 2 levels each; n greater than 4   **5.15** 3 levels each; n greater than 6
5.16a. Yes; $F = 23.18$ (p-value $= .0020$)   **b.** Multicollinearity   **c.** Inflated standard error of $\hat{\beta}_3$ due to multicollinearity
d. $r_{12} = .327$, $r_{13} = .231$, $r_{14} = .166$, $r_{23} = .790$, $r_{24} = .791$, $r_{34} = .881$; yes, correlation coefficients r_{23}, r_{24}, and r_{34} are significant at $\alpha = .05$
5.17a. Curvilinear   **b.** Straight-line   **c.** $\log y = 10.6364 - 2.16985 \log x$; yes, $t = -13.44$ **d.** 25.95
5.18a. .0025; no   **b.** .4341; no   **c.** No   **d.** $\hat{y} = -45.154 + 3.097x_1 + 1.032x_2$; $F = 39,222.34$; $R^2 = .9998$
e. $-.8998$; x_1 and x_2 are highly correlated   **f.** No; $t = 252.31$ (for $H_0: \beta_1 = 0$); $t = 280.08$ (for $H_0: \beta_2 = 0$)
5.19a. $\hat{y} = .018201 + .005927x_1$; reject $H_0: \beta_1 = 0$, $t = 2.735$
b. $\hat{y} = -.040203 + .000308x_1 + .003905x_2 + .184777x_3$; reject $H_0: \beta_1 = \beta_2 = \beta_3 = 0$; $F = 4.53$
c. Do not reject $H_0: \beta_1 = 0$, $t = .098$; yes; possible multicollinearity   **d.** VIF $= 2.295$; $R^2 = .5642$

CHAPTER 6

6.1a. $-.406$, $-.206$, $-.047$, $.053$, $.112$, $.212$, $.271$, $.471$, $.330$, $.230$, $-.411$, $-.611$ **b.** Yes; needs curvature
6.2a. 3.197, 3.215, -2.258, $.269$, -1.713, -7.186, -2.659, 3.359, -2.132, 5.904 **b.** Yes; needs curvature
6.3a. $\hat{y} = 40.0714 - .214286x$ **b.** -4.643, -1.429, -3.857, 5.000, 4.214, $.429$, -6.357 **c.** Yes; needs curvature
d. $\hat{y} = -1,199.9286 + 75.2143x - 1.142857x^2$; yes, $t = -9.37$
6.4a. 6.35, -6.15, $.35$, -3.74, 4.67, -2.06, $.35$, 3.48, -6.15, $.35$, 2.58 **d.** Yes   **e.** Needs curvature
6.5a. $.796$, $-.219$, -1.004, $.077$, -1.071, 1.116, $.868$, $.827$, $.223$, -1.612
d. 43.83, 20.36, 36.42, 31.88, 26.99, 40.41, 45.77, 25.15, 35.77, 45.16
e. 7.19, 2.34, 10.50, 3.91, 9.15, 6.23, 9.81, 8.49, 9.17, $.94$
6.6a. -294.5, -36.4, 263.1, 334.9, 56.0, -116.7, -137.0, -49.0, 35.3, 63.8, 47.7, -278.1, -218.6, 234.2, 95.3
6.7 Yes; needs curvature   **6.8** Yes; needs curvature   **6.9a.** 2.714   **b.** (3.8612, 4.5156)   **c.** No
6.10a. Residuals: $.75668$, $.00802$, 1.38235, $.13102$, $-.18182$, $-.93048$, $-.30481$, $-.24332$, $-.61765$
b. Assumption may be violated; yes
c. $\sin^{-1}\sqrt{y}$; residuals using transformation $\sin^{-1}\sqrt{y}$: $.01124$, $-.00055$, $.01899$, $.00195$, $-.00340$, $-.01273$, $-.00425$, $-.00296$, $-.00830$

6.11a. $\hat{y} = .94 - .214x$ **b.** 0, .02, $-.026$, .034, .088, $-.112$, $-.058$, .002, .036, .016
c. Football shape; unequal variances   **d.** Use the transformation $y^* = \sin^{-1}\sqrt{y}$ and fit the model $y^* = \beta_0 + \beta_1 x + \varepsilon$
e. $\hat{y}^* = 1.307 - .2496x$; possibly
6.12a. Residuals; -29.29, -8.05, -51.55, .16, 23.69, -1.09, -29.68, -11.83, 41.37, -3.57, 38.26, 18.08, 21.91, .46, 31.14, -13.92, 6.07, -9.14, -9.08, -13.92 **b.** Yes; evidence of heteroscedasticity
c. $MSE_1 = 186.30$, $MSE_2 = 1,305.03$; $F = 7.00$, reject H_0 **d.** Use the variance-stabilizing transformation $y^* = \log y$
6.16a. $F = 83.54$; reject H_0 **b.** .182, .114, .007, $-.154$, $-.436$ **6.17** No outliers   **6.18** No outliers   **6.19** No outliers
6.20a. -161.40, -77.72, -36.48, 176.0, -118.0, 53.77, 142.1, 50.39, 116.5, 71.20, 17.64, -73.38, 52.34, -34.81, -117.0, 209.8, 101.2, -117.8, 213.5, 176.1, -145.4, 22.33, -185.2, -33.63, 175.3, 76.61, -189.2, -138.5, 107.3, 15.35, -207.2, -141.7 **c.** 0   **d.** None of the observations appear to be influential   **6.21** No outliers
6.22b. Residuals; -50.76, -47.44, -15.96, -33.72, -54.64, -78.24, 16.84, -46.52, 43.24, -64.96, -55.96, -50.52, 436.52, 37.88, -36.00; almost all residuals are negative, one large positive residual   **c.** Omit this employee from the analysis
d. Cook's $D = .425$ **e.** $\hat{y} = 191.26 - 9.58x$; $r^2 = .483$; model is adequate ($t = -3.35$)
6.23a. $d_L = 1.21$; $d_U = 1.65$ **b.** $d_L = 1.25$; $d_U = 1.34$ **c.** $d_L = 1.16$; $d_U = 1.80$
6.24a. H_0: No residual correlation; H_a: Positive residual correlation; test statistic: d; rejection region: $d < 1.39$ **b.** Reject H_0
6.25a. $F = 164.07$; yes   **b.** $d = 1.01$; yes   **c.** Result may be overoptimistic.
6.26a. Residuals: .0135, .0176, .0327, .0288, $-.0062$, $-.0251$, $-.0190$, $-.0149$, $-.0138$, $-.0257$, $-.0406$, $-.0385$, $-.0194$, .0097, .0358, .0649; yes, positively correlated residuals   **b.** $d = .368$; reject H_0 **6.27** $d = 1.23$; test is inconclusive
6.28a. $\hat{y} = 687.79 - 1.1693t$
b. -103.2, -30.5, 71.9, 38.0, 121.4, 46.6, 4.5, -74.9, -110.6, 13.5, -74.2, -113.1, -44.6, -26.2, 98.6, 119.1, 139.7, 10.0, -32.1, 80.4, 176.2, -63.7, -145.1, -101.7 **c.** $d = .998$ **d.** Yes; reject H_0
e. No; assumption of independent errors violated
6.29a. No, do not reject H_0; $d_U = 1.60$ **b.** Yes; reject H_0; $d_L = 1.39$
6.30a. $\hat{y} = 200.4 - .78x + .000987x^2$; residuals: -2.27, 11.97, 10.73, 14.39, -7.15, -8.08, $-.64$, -3.87, -5.23, -8.31, 9.41, -8.03 **b.** No   **c.** No   **d.** Residuals are skewed to the left
e. Model appears to be adequate; assumption of normality appears to be violated
6.31a. Residuals: 15.40854, 9.64800, -2.07231, -16.51283, -18.67385, -7.79455, 11.12499, 27.72461, -5.47462, -19.27347, -4.47270, 10.36830   **c.** Plot suggests second-order model.
d. Residuals for $\hat{y} = 322.738 - 8.051x + .061x^2$, $R^2 = .920$; $F = 51.66$ (significant at $\alpha = .0001$); -6.14601, 6.66728, 2.86898, -1.60023, -3.48557, -3.96127, 1.88268, 4.22146, -3.51475, -2.05895, 7.61832, -2.49196
6.32a. Residuals: .793, -2.004, 3.769, 1.839, -6.013, -3.957, 6.396, 3.973, -3.537, -1.259; no; no
b. Leverage value for observation 10 exceeds $2h = 6$ and thus is influential with respect to its x-value.
6.33a. Yes; $F_{.05} = 2.51$ **b.** No; $d_L = 1.03$, $d_U = 1.85$ **6.34** Reject H_0, $d = .7762$ ($d_L = .95$)
d. Yes; residual for Nov. 1987 (lies just over 2 standard deviations from 0)
e. No; x_1 and x_2 do not appear to contribute information for the prediction of y; $y = \beta_0 + \beta_3 x_3 + \beta_4 x_4 + \beta_5 x_5 + \varepsilon$
6.35a. Residuals: 1.27, 12.63, -10.27, $-.13$, 20.72, 13.76, -34.47, $-.05$, 11.95, -11.91, -9.35, 3.88, 4.21, 4.32, -6.54
b. -72.18, -62.41, -77.41, -65.69, -54.32, -67.60, -113.46, -82.98, -70.98, -94.84, -100.18, -90.90, -86.62, -82.56, -93.42 **c.** 350.2, 358.7, 335.8, 340.2, 358.2, 345.5, 300.1, 343.1, 355.1, 334.1, 339.5, 358.5, 364.5, 367.5, 359.5
6.36a. $\hat{y} = .072 + .002x_1 + .048x_2 + .21x_3$; $R^2 = .7906$
b. Data are recorded as proportions—random errors may be heteroscedastic.   **c.** $y^* = \sin^{-1}\sqrt{y}$
d. $\hat{y}^* = .2705 + .0017x_1 + .0792x_2 + .2862x_3$; $R^2 = .8073$; $F = 22.35$ (significant at $\alpha = .0001$)
6.37a. $\hat{y} = 1,604.81 - 64.1044t$
b. Residuals: -773.7, -550.6, -241.5, 314.6, 773.7, 671.8, 709.9, 322.0, 39.1, -330.8, -359.7, -282.6, -292.5; yes, runs of positive and negative residuals   **c.** $d = .337$, reject H_0
6.38c. Horse #3 has a residual outside $0 \pm 2s$
d. $\hat{y} = 5.0684 + .0623x_1$; $R^2 = .9704$; $F = 130.71$ (significant at $\alpha = .0003$); yes
6.39a. .162, $-.277$, $-.039$, .243, .237, $-.328$

CHAPTER 7

7.1 a, b, g quantitative; c–f, h qualitative   **7.2** Assumption of normality   **7.3** $E(y) = \beta_0 + \beta_1 x + \beta_2 x^2$, where $\beta_2 > 0$
7.4b. Linear; linear; quadratic
7.5a. Both are quantitative.   **b.** $E(y) = \beta_0 + \beta_1 x_1 + \beta_2 x_2$ **c.** $E(y) = \beta_0 + \beta_1 x_1 + \beta_2 x_2 + \beta_3 x_1 x_2$
d. $E(y) = \beta_0 + \beta_1 x_1 + \beta_2 x_2 + \beta_3 x_1 x_2 + \beta_4 x_1^2 + \beta_5 x_2^2$

7.7a. $E(y) = \beta_0 + \beta_1 x_1 + \beta_2 x_2 + \beta_3 x_3 + \beta_4 x_1 x_2 + \beta_5 x_1 x_3 + \beta_6 x_2 x_3$
b. $E(y) = \beta_0 + \beta_1 x_1 + \beta_2 x_2 + \beta_3 x_3 + \beta_4 x_1 x_2 + \beta_5 x_1 x_3 + \beta_6 x_2 x_3 + \beta_7 x_1^2 + \beta_8 x_2^2 + \beta_9 x_3^2$
7.8a. $E(y) = \beta_0 + \beta_1 x_1 + \beta_2 x_2 + \beta_3 x_3 + \beta_4 x_4 + \beta_5 x_5 + \beta_6 x_6$ **c.** Positive
d. $E(y) = \beta_0 + \beta_1 x_1 + \beta_2 x_2 + \beta_3 x_3 + \beta_4 x_4 + \beta_5 x_5 + \beta_6 x_6 + \beta_7 x_1 x_2 + \beta_8 x_1 x_3 + \beta_9 x_1 x_4 + \beta_{10} x_1 x_5 + \beta_{11} x_1 x_6 + \beta_{12} x_2 x_3 + \beta_{13} x_2 x_4 + \beta_{14} x_2 x_5 + \beta_{15} x_2 x_6 + \beta_{16} x_3 x_4 + \beta_{17} x_3 x_5 + \beta_{18} x_3 x_6 + \beta_{19} x_4 x_5 + \beta_{20} x_4 x_6 + \beta_{21} x_5 x_6$
e. Test $H_0: \beta_{11} = \beta_{15} = \beta_{18} = \beta_{20} = \beta_{21} = 0$ using a partial F-test
7.9a. $E(y) = \beta_0 + \beta_1 x_1 + \beta_2 x_2 + \beta_3 x_1 x_2 + \beta_4 x_1^2 + \beta_5 x_2^2$ **b.** $E(y) = \beta_0 + \beta_1 x_1 + \beta_2 x_2$
c. $E(y) = \beta_0 + \beta_1 x_1 + \beta_2 x_2 + \beta_3 x_1 x_2$ **d.** $\beta_1 + \beta_3 x_2$ **e.** $\beta_2 + \beta_3 x_1$

7.10a. $u = \dfrac{x - 85.1}{14.81}$ **b.** $-.668, .446, 1.026, -1.411, -.223, 1.695, -.527, -.338$ **c.** $.997$ **d.** $.377$
e. $\hat{y} = 110.95 + 14.376u + 7.425u^2$

7.11a. $u = \dfrac{x - 33}{2.16}$ **a.** $-1.389, -.926, -.463, 0, .463, .926, 1.389$ **c.** $.99966$ **d.** 0
e. $\hat{y} = 37.5714 - .4629u - 5.3333u^2$

7.12a. $u_1 = \dfrac{x_1 - 90}{8.32}$, $u_2 = \dfrac{x_2 - 55}{4.16}$ **b.** $r_{(x_1, x_1^2)} = .99949; r_{(u_1, u_1^2)} = 0$ **c.** $r_{(x_2, x_2^2)} = .99966; r_{(u_2, u_2^2)} = 0$
d. $\hat{y} = 94.926 - 7.6225u_1 + 3.2773u_2 - 5.0365u_1u_2 - 9.2346u_1^2 - 19.8038u_2^2$

7.13a. Qualitative   **b.** $E(y) = \beta_0 + \beta_1 x_1 + \beta_2 x_2$, where $x_1 = \begin{cases} 1 & \text{if brand } B_2 \\ 0 & \text{otherwise} \end{cases}$ and $x_2 = \begin{cases} 1 & \text{if brand } B_3 \\ 0 & \text{otherwise} \end{cases}$
d. $E(y) = \beta_0 + \beta_2$ **7.14a.** 60   **b.** 510
7.15a. $E(y) = \beta_0 + \beta_1 x_1 + \beta_2 x_2 + \beta_3 x_3$,
where $x_1 = \begin{cases} 1 & \text{if moderate concentration} \\ 0 & \text{if not} \end{cases}$, $x_2 = \begin{cases} 1 & \text{if high concentration} \\ 0 & \text{if not} \end{cases}$, $x_3 = \begin{cases} 1 & \text{if consumer type} \\ 0 & \text{if not} \end{cases}$
c. $E(y) = \beta_0 + \beta_1 x_1 + \beta_2 x_2 + \beta_3 x_3 + \beta_4 x_1 x_3 + \beta_5 x_2 x_3$ **e.** Test $H_0: \beta_4 = \beta_5 = 0$ using a partial F-test
7.16a. 3   **b.** $E(y) = \beta_0 + \beta_1 x_1 + \beta_2 x_2 + \beta_3 x_3$,
where $x_1 = \begin{cases} 1 & \text{if premium} \\ 0 & \text{if not} \end{cases}$, $x_2 = \begin{cases} 1 & \text{if lead-free} \\ 0 & \text{if not} \end{cases}$, $x_3 = \begin{cases} 1 & \text{if full-service} \\ 0 & \text{if self-service} \end{cases}$
c. $E(y) = \beta_0 + \beta_1 x_1 + \beta_2 x_2 + \beta_3 x_3 + \beta_4 x_1 x_3 + \beta_5 x_2 x_3$ **d.** Yes   **7.17a.** 5   **b.** -1 **c.** -1 versus 1
7.19a. $E(y) = \beta_0 + \beta_1 x_1 + \beta_2 x_2 + \beta_3 x_3$ **b.** Test $H_0: \beta_3 = 0$ versus $H_a: \beta_3 < 0$ using a t-test
c. $E(y) = \beta_0 + \beta_1 x_1 + \beta_2 x_2 + \beta_3 x_1 x_2 + \beta_4 x_1^2 + \beta_5 x_2^2 + \beta_6 x_3 + \beta_7 x_1 x_3 + \beta_8 x_2 x_3 + \beta_9 x_1 x_2 x_3 + \beta_{10} x_1^2 x_3 + \beta_{11} x_2^2 x_3$
d. Test $H_0: \beta_1 = \beta_3 = \beta_4 = \beta_7 = \beta_9 = \beta_{10} = 0$ using a partial F-test
7.20a. $E(y) = \beta_0 + \beta_1 x_1$

b. $E(y) = \beta_0 + \beta_1 x_1 + \beta_2 x_2 + \beta_3 x_3$, where $x_2 = \begin{cases} 1 & \text{if brand } B_2 \\ 0 & \text{otherwise} \end{cases}$ and $x_3 = \begin{cases} 1 & \text{if brand } B_3 \\ 0 & \text{otherwise} \end{cases}$
c. $E(y) = \beta_0 + \beta_1 x_1 + \beta_2 x_2 + \beta_3 x_3 + \beta_4 x_1 x_2 + \beta_5 x_1 x_3$
7.21a. $E(y) = \beta_0 + \beta_1 x_1 + \beta_2 x_2 + \beta_3 x_3$ **b.** $H_0: \beta_2 = \beta_3 = 0$ **7.22a.** $H_0: \beta_3 = 0$ **b.** $H_0: \beta_1 = \beta_3 = 0$
7.23 $E(y) = (\beta_0 + \beta_6 + \beta_8 + \beta_{10}) + (\beta_1 + \beta_{11} + \beta_{21} + \beta_{31})x_1 + (\beta_2 + \beta_{12} + \beta_{22} + \beta_{32})x_2 + (\beta_3 + \beta_{13} + \beta_{23} + \beta_{33})x_1 x_2 + (\beta_4 + \beta_{14} + \beta_{24} + \beta_{34})x_1^2 + (\beta_5 + \beta_{15} + \beta_{25} + \beta_{35})x_2^2$
7.24a. $E(y) = \beta_0 + \beta_1 x$; y-intercept $= \beta_0$, slope $= \beta_1$
b. $E(y) = (\beta_0 + \beta_2) + (\beta_1 + \beta_3)x_1$; y-intercept $= (\beta_0 + \beta_2)$; slope $= (\beta_1 + \beta_3)$ **c.** $H_0: \beta_1 = \beta_2 = \beta_3 = 0$
d. $H_0: \beta_3 = 0$ **e.** Stocks 2, 5, 6, 7, 8, 14, 18, 20, 23, 24, and 26
7.25a. 1, quantitative, 0–4 years; 2, qualitative, two levels; 3, qualitative, two levels; 4, quantitative, 6–24 inches; 5, qualitative, three levels   **b.** $E(y) = \beta_0 + \beta_1 x_1 + \beta_2 x_2 + \beta_3 x_3 + \beta_4 x_4 + \beta_5 x_5 + \beta_6 x_6$
c. No interaction terms; include $\beta_7 x_4 x_5 + \beta_8 x_4 x_6$ in model   **7.26** No estimate of σ^2
7.27a. $E(y) = \beta_0 + \beta_1 x_1 + \beta_2 x_2$, where $x_1 = $ years of education, $x_2 = \begin{cases} 1 & \text{if certified} \\ 0 & \text{if not} \end{cases}$
b. $E(y) = \beta_0 + \beta_1 x_1 + \beta_2 x_2 + \beta_3 x_1 x_2$ **c.** $E(y) = \beta_0 + \beta_1 x_1 + \beta_2 x_1^2 + \beta_3 x_2 + \beta_4 x_1 x_2 + \beta_5 x_1^2 x_2$
7.28a. $E(y) = \beta_0 + \beta_1 x + \beta_2 x^2$ **b.** No; need at least four different baking times
7.29 $E(y) = \beta_0 + \beta_1 x_1 + \beta_2 x_2 + \beta_3 x_3 + \beta_4 x_1 x_2 + \beta_5 x_1 x_3 + \beta_6 x_1^2 + \beta_7 x_1^2 x_2 + \beta_8 x_1^2 x_3$
7.30 $\hat{y} = 835.59 - 148.92u + 29.464u^2$, where $u = \dfrac{p - 3.35}{.245}$; $r_{(u, u^2)} = 0$
7.31b. $H_0: \beta_2 = \beta_5 = 0$ **c.** $H_0: \beta_3 = \beta_4 = \beta_5 = 0$ **7.32** $F = 3.45$; reject H_0 ($\alpha = .05$)
7.33a. $E(y) = \beta_0 + \beta_1 x_1 + \beta_2 x_2 + \beta_3 x_3 + \beta_4 x_4$; no interaction   **b.** Include $\beta_5 x_1 x_2 + \beta_6 x_1 x_3 + \beta_7 x_1 x_4$ in model

7.34a. $F = 2.33$; do not reject H_0

7.35a. $E(y) = \beta_0 + \beta_1 x_1 + \beta_2 x_2 + \beta_3 x_3$

b. Include $\beta_4 x_1 x_2 + \beta_5 x_1 x_3 + \beta_6 x_2 x_3 + \beta_7 x_1^2 + \beta_8 x_2^2 + \beta_9 x_1^2 x_3 + \beta_{10} x_2^2 x_3 + \beta_{11} x_1 x_2 x_3$

c. H_0: $\beta_4 = \beta_5 = \beta_6 = \beta_7 = \beta_8 = \beta_9 = \beta_{10} = \beta_{11} = 0$; use F-test

7.36a. $F = .64$; no   **b.** $F = 133$; reject H_0 ($\alpha = .01$)

7.37a. $E(y) = \beta_0 + \beta_1 x_1 + \beta_2 x_1^2 + \beta_3 x_2 + \beta_4 x_3 + \beta_5 x_1 x_2 + \beta_6 x_1 x_3 + \beta_7 x_1^2 x_2 + \beta_8 x_1^2 x_3$

b. $\hat{y} = 3.803 + .5563 u_1 - .5816 u_1^2 - .6657 x_2 - 1.2771 x_3 - .1464 u_1 x_2 - .01464 u_1 x_3 - .09184 u_1^2 x_2 + .16837 u_1^2 x_3$

c. Yes; reject H_0: $\beta_3 = \beta_4 = \beta_5 = \beta_6 = \beta_7 = \beta_8 = 0$, $F = 24.66$

CHAPTER 8

8.1a. Types of work schedule: flextime, staggered, and fixed   **b.** Randomly assign 20 workers to each type of work schedule.

8.2a. df(Error) = 30; SSE = 37.7; MST = 6.175; MSE = 1.257; $F = 4.91$ **b.** 5   **c.** Yes; reject H_0

8.3a. 7.502   **b.** 2.381   **c.** 1   **d.** 11   **e.** 3.15   **f.**

SOURCE	df	SS	MS	F
Treatments	1	7.502	7.502	3.15
Error	11	26.190	2.381	
Total	12	33.692		

g. Reject H_0 if $F > 4.84$ **h.** Do not reject H_0

8.4a. $t = -1.78$; do not reject H_0 **c.** Two-tailed   **8.5** No; do not reject H_0, $F = .10$

8.6a.

SOURCE	df	SS	MS	F
Industries	2	685,129.33	342,564.67	6.45
Error	21	1,115,232.50	53,106.31	
Total	23	1,800,361.83		

b. Yes; $F_{01} = 5.78$ **c.** 957.75 ± 140.22

8.7a. H_0: $\mu_1 = \mu_2 = \mu_3$

b. df(size) = 2, df(error) = 121; SS(size) = 15.5; SS(Total) = 205.5; MS(size) = 7.75; MSE = 1.57   **c.** Yes; $F_{.05} = 3.07$

8.8a. No; $F = 2.878$ **b.** 4.5 ± 32.00

8.9a. 457,802.42   **b.** 509.871   **c.** 12,259.959   **d.** 12,769.830

e.

SOURCE	df	SS	MS	F
Groups	2	509.871	254.94	1.87
Error	90	12,259.959	136.22	
Total	92	12,769.830		

f. No, $F_{.05} \approx 3.10$

8.10 Yes; $F = 9.50$ **8.11** Reject H_0; $F = 17.66$

8.12a.

SOURCE	df	SS	MS	F
Periods	2	11,357.825	5,678.91	.78
Error	15	109,152.675	7,276.85	
Total	17	120,510.500		

b. No; $F = .78$

8.13a. Yes; $F = 13.00$ **b.** $(\mu_X - \mu_Y)$: $-.97 \pm .702$; $(\mu_X - \mu_Z)$: $.148 \pm .702$; $(\mu_Y - \mu_Z)$: $1.118 \pm .702$ **c.** Theory Y

8.14a. Auditing methods: method 1 and method 2   **b.** Estimate market share of each beer brand using both methods.

8.15a. df(Error) = 15, df(Total) = 23, SSB = 74.5, SS(Total) = 135, MST = 9.033, MSE = 2.227, F(Treatments) = 4.06, F(Blocks) = 6.69   **b.** No; $F = 4.06$ **c.** Yes, reject H_0 at $\alpha = .05$; $F = 6.69$

8.16a.

SOURCE	df	SS	MS	F
Supermarkets	3	.3352	.11173	4.54
Items	6	5.2311	.87185	35.40
Error	18	.4433	.02463	
Total	27	6.0096		

b. Do not reject H_0; $F = 4.54$ **c.** $-.033 \pm .241$

8.17a.

SOURCE	df	SS	MS	F
Panelists	3	1,298.75	432.92	1.73
Years	3	554.75	184.92	.74
Error	9	2,252.25	250.25	
Total	15	4,105.75		

b. No; $F = 1.73$ **c.** No; $F = .74$

8.18a. Yes; $F = 8.72$ **b.** No; $F = 1.41$ **c.** 6 ± 23.41 **8.19** Yes; reject H_0, $F = 52.81$

8.20a.

SOURCE	df	SS	MS	F
Methods	2	.1911	.0956	.07
Brands	5	605.6978	121.1396	92.84
Error	10	13.0489	1.3049	
Total	17	618.9378		

b. No; $F = .07$ **c.** $-.233 \pm 1.469$

8.21a.

SOURCE	df	SS	MS	F
Raters	2	2.66667	1.3333	1.96
Candidates	6	7.11905	1.1865	1.74
Error	12	8.16666	.6805	
Total	20	17.95238		

b. No; $F = 1.96$ **c.** No; $F = 1.74$ **d.** $-.57 \pm .961$

8.22b.

SOURCE	df	SS	MS	F
Periods	2	9,726.50	4,863.25	1.04
Months	3	68,258.25	22,752.75	4.88
Error	6	27,947.50	4,657.92	
Total	11	105,932.25		

c. No; $F = 1.04$ **d.** 68.25 (compared to 85.30)   **e.** $F = 4.88$; reject H_0

8.23a. 3×3 factorial   **b.** Factors: pay rate (quantitative) and length of workday (qualitative)
c. Treatments: P_1L_1, P_1L_2, P_1L_3, P_2L_1, P_2L_2, P_2L_3, P_3L_1, P_3L_2, P_3L_3
8.24a. Factors: advertising expenditure (5 levels) and markets (2 levels)
b. advertising expenditure: quantitative; markets: qualitative
c. 10; $(M_1, 15)$, $(M_2, 15)$, $(M_1, 18)$, $(M_2, 18)$, $(M_1, 21)$, $(M_2, 21)$, $(M_1, 24)$, $(M_2, 24)$, $(M_1, 27)$, $(M_2, 27)$ **8.25a.** No
8.26a. $df(A) = 2$, $df(Error) = 18$, $df(Total) = 23$; $SS(B) = 559$, $SS(AB) = 5$, $SSE = 36$; $MS(A) = 50$, $MS(B) = 559$; $F(A) = 25$, $F(B) = 279.5$, $F(AB) = 1.25$ **b.** Do not reject H_0; $F = 1.25$ **c.** Reject H_0; $F = 25.0$ **d.** Reject H_0; $F = 279.5$

8.27a.

SOURCE	df	SS	MS	F
Machines (A)	1	.101	.101	22.95
Materials (B)	2	.812	.406	92.27
AB	2	.768	.384	87.27
Error	12	.053	.0044	
Total	17	1.734		

b. Yes; $F = 87.27$ **d.** No; interaction is present   **c.** $.413 \pm .118$

8.28a.

SOURCE	df	SS	MS	F
Ratio (R)	2	32.000	16.000	2.68
Length (L)	2	10.667	5.3335	.89
RL	4	95.333	23.8333	4.00
Error	27	161.000	5.9630	
Total	35	299.000		

b.

		RATIO		
		2:1	4:1	8:1
	6	26.50	28.25	25.75
LENGTH	9	28.00	26.25	28.25
	12	22.00	28.00	28.00

d. Yes; $F = 4.00$ **e.** No; $F = .89$ **f.** When no interaction is present   **g.** 28 ± 2.08

8.29a. $df(V) = 1$; $df(VP) = 1$; $df(Error) = 16$; $df(Total) = 19$; $SSE = 240$; $SS(V) = 69.15$; $SS(VP) = 170.85$; $MS(V) = 69.15$; $MS(VP) = 170.85$; remaining missing entries cannot be determined from the table   **c.** $F = 11.39$; reject H_0 ($F_{.05} = 4.49$)

8.30a.

SOURCE	df.	SS	MS	F
Times (T)	1	51.842	51.842	41.15
Metals (M)	1	76.832	76.832	60.99
TM	1	2.738	2.738	2.17
Error	16	20.156	1.25975	
Total	19	151.568		

c. No; $F = 2.17$

8.31a.

SOURCE	df	SS	MS	F
Union (U)	1	7,168.444	7,168.444	26.34
Plan (P)	2	5,425.167	2,712.584	9.97
UP	2	159.389	79.695	.29
Error	30	8,165.000	272.167	
Total	35	20,918.000		

c. No; $F = .29$ **d.** 354.67 ± 11.079 **e.** -25 ± 15.668

8.32a. $SS(Sex) = 1.21$, $SS(Position) = .01$, $SS(SP) = 6.25$ **b.** $SSE = 201.60$, $SS(Total) = 209.07$

c.

SOURCE	df	SS	MS	F
Sex (S)	1	1.21	1.21	.58
Position (P)	1	.01	.01	.005
(SP)	1	6.25	6.25	2.98
Error	96	201.60	2.10	
Total	99	209.07		

e. No; $F = 2.98$

8.33 Evidence of factor interaction; $F = 4.89$, reject H_0 ($F_{.05} \approx 1.32$)

8.34a.

	LIGHT	HEAVY
Female	146.40	116.00
Male	104.00	98.00

b. $6,739.605$ **c.** $SS(Sex) = 114.005$, $SS(Weight) = 41.405$, $SS(SW) = 18.605$

d.

	SAMPLE VARIANCE	SS(DEVIATIONS WITHIN)
Female, Light	46.3761	324.6327
Female, Heavy	8.5849	60.0943
Male, Light	25.4016	177.8112
Male, Heavy	32.4900	227.4300

e. 789.968 **f.** 963.983

g.

SOURCE	df	SS	MS	F
Sex (S)	1	114.005	114.005	4.04
Weight (W)	1	41.405	41.405	1.47
(SW)	1	18.605	18.605	.66
Error	28	789.968	28.213	
Total	31	963.983		

h. No; $F = .66$ **i.** -5.30 ± 5.44 **j.** -2.25 ± 5.44

8.35 Means for the following treatment pairs appear to differ: (small, medium), (small, large)

8.36 $\omega = 8.675$; means for PD-1 and IADC 5-1-7 appear to differ

8.37 $\omega = 95.76$ for all pairs involving location 4, $\omega = 90.28$ for all other pairs; means for the following treatment pairs appear to differ: (4, 5), (4, 1), (4, 3), (2, 1), and (2, 3)   **8.38** $\omega = .302$; none of the means appear to differ

8.39 $\omega = 1.175$; none of the means appear to differ

8.40 $\omega = .182$; means for the following treatment pairs appear to differ: (A_1B_1, A_1B_3), (A_1B_1, A_2B_1), (A_1B_1, A_2B_2), (A_1B_1, A_2B_3), (A_1B_1, A_1B_2), (A_1B_3, A_2B_3), (A_1B_3, A_1B_2), (A_2B_1, A_2B_3), (A_2B_1, A_1B_2), (A_2B_2, A_2B_3), (A_2B_2, A_1B_2)

8.41 $\omega = 5.49$; none of the means appear to differ under each pricing ratio   **8.42** $\omega = 7.01$; none of the means appear to differ

8.43 $\omega = .815$; means for the following CEO pairs do not appear to differ: (3, 7), (1, 11), (11, 9), (9, 6), (9, 10), (9, 4), (9, 5), (9, 8), (6, 10), (6, 4), (6, 5), (6, 8), (10, 4), (10, 5), (10, 8), (4, 5), (4, 8), (5, 8)

8.44a. $S = 9.06$; means for PD-1 and IADC 5-1-7 appear to differ

b. $B \approx 10.09$; means for PD-1 and IADC 5-1-7 appear to differ

8.45a. $S = .214$; means for the following treatment pairs appear to differ: (A_1B_1, A_1B_3), (A_1B_1, A_2B_1), (A_1B_1, A_2B_2), (A_1B_1, A_2B_3), (A_1B_1, A_1B_2), (A_1B_3, A_2B_3), (A_1B_3, A_1B_2), (A_2B_1, A_2B_3), (A_2B_1, A_1B_2), (A_2B_2, A_2B_3), (A_2B_2, A_1B_2)

b. $B = .218$; means for the following treatment pairs appear to differ: (A_1B_1, A_1B_3), (A_1B_1, A_2B_1), (A_1B_1, A_2B_2), (A_1B_1, A_2B_3), (A_1B_1, A_1B_2), (A_1B_3, A_2B_3), (A_1B_3, A_1B_2), (A_2B_1, A_2B_3), (A_2B_1, A_1B_2), (A_2B_2, A_2B_3), (A_2B_2, A_1B_2)

8.46a. $S = 7.64$; none of the means appear to differ   **b.** $B = 7.97$; none of the means appear to differ

8.47a. Complete model: $E(y) = \beta_0 + \beta_1 x_1 + \beta_2 x_2$, where

$x_1 = \begin{cases} 1 & \text{if utilities} \\ 0 & \text{if not} \end{cases}$, $x_2 = \begin{cases} 1 & \text{if office/computers} \\ 0 & \text{if not} \end{cases}$; Reduced model: $E(y) = \beta_0$

8.48a. Complete model: $E(y) = \beta_0 + \beta_1 x_1 + \beta_2 x_2 + \beta_3 x_3 + \beta_4 x_4$

where $x_1 = \begin{cases} 1 & \text{if II} \\ 0 & \text{if not} \end{cases}$, $x_2 = \begin{cases} 1 & \text{if III} \\ 0 & \text{if not} \end{cases}$, $x_3 = \begin{cases} 1 & \text{if IV} \\ 0 & \text{if not} \end{cases}$, $x_4 = \begin{cases} 1 & \text{if V} \\ 0 & \text{if not} \end{cases}$;

Reduced model: $E(y) = \beta_0$

8.49a. Complete model: $E(y) = \beta_0 + \beta_1 x_1 + \beta_2 x_2 + \beta_3 x_3 + \beta_4 x_4 + \beta_5 x_5 + \beta_6 x_6$,

where $x_1 = \begin{cases} 1 & \text{if Cappiello} \\ 0 & \text{if not} \end{cases}$, $x_2 = \begin{cases} 1 & \text{if Stovall} \\ 0 & \text{if not} \end{cases}$, $x_3 = \begin{cases} 1 & \text{if Randall} \\ 0 & \text{if not} \end{cases}$

$x_4 = \begin{cases} 1 & \text{if 1979} \\ 0 & \text{if not} \end{cases}$, $x_5 = \begin{cases} 1 & \text{if 1980} \\ 0 & \text{if not} \end{cases}$, $x_6 = \begin{cases} 1 & \text{if 1981} \\ 0 & \text{if not} \end{cases}$;

Reduced model (for testing treatments): $E(y) = \beta_0 + \beta_4 x_4 + \beta_5 x_5 + \beta_6 x_6$

8.50a. Complete model: $E(y) = \beta_0 + \beta_1 x_1 + \beta_2 x_2 + \beta_3 x_3 + \beta_4 x_4 + \beta_5 x_1 x_3 + \beta_6 x_1 x_4 + \beta_7 x_2 x_3 + \beta_8 x_2 x_4$

where $x_1 = \begin{cases} 1 & \text{if 9 hours} \\ 0 & \text{if not} \end{cases}$, $x_2 = \begin{cases} 1 & \text{if 12 hours} \\ 0 & \text{if not} \end{cases}$, $x_3 = \begin{cases} 1 & \text{if 4:1} \\ 0 & \text{if not} \end{cases}$, $x_4 = \begin{cases} 1 & \text{if 8:1} \\ 0 & \text{if not} \end{cases}$;

Reduced model (for testing interaction): $E(y) = \beta_0 + \beta_1 x_1 + \beta_2 x_2 + \beta_3 x_3 + \beta_4 x_4$

8.51a. $E(y) = \beta_0 + \beta_1 x_1 + \beta_2 x_2 + \beta_3 x_3 + \beta_4 x_4 + \beta_5 x_5 + \beta_6 x_6 + \beta_7 x_7 + \beta_8 x_8 + \beta_9 x_9 + \beta_{10} x_{10} + \beta_{11} x_{11} + \beta_{12} x_{12}$

b. $E(y) = \beta_0 + \beta_4 x_4 + \beta_5 x_5 + \beta_6 x_6 + \beta_7 x_7 + \beta_8 x_8 + \beta_9 x_9 + \beta_{10} x_{10} + \beta_{11} x_{11} + \beta_{12} x_{12}$

c. $E(y) = \beta_0 + \beta_1 x_1 + \beta_2 x_2 + \beta_3 x_3$ **d.** Yes; $F = 5.33$; reject H_0 **e.** $F = 207.80$; reject H_0

8.52a. Pay rate (quantitative); length of workday (quantitative)

b. Three pay rates: P_1, P_2, P_3; three workday lengths: L_1, L_2, L_3;

treatments: P_1L_1, P_1L_2, P_1L_3, P_2L_1, P_2L_2, P_2L_3, P_3L_1, P_3L_2, P_3L_3

c. $E(y) = \beta_0 + \beta_1 x_1 + \beta_2 x_1^2 + \beta_3 x_2 + \beta_4 x_2^2 + \beta_5 x_1 x_2 + \beta_6 x_1^2 x_2 + \beta_7 x_1 x_2^2 + \beta_8 x_1^2 x_2^2$ **d.** Fourth-order

e. $E(y) = \beta_0 + \beta_1 x_1 + \beta_2 x_1^2 + \beta_3 x_2 + \beta_4 x_2^2 + \beta_5 x_1 x_2$ **f.** 0   **g.** 3   **h.** Complete, 9 df; second-order, 12 df

i. Second-order   **j.** Test $H_0: \beta_5 = \beta_6 = \beta_7 = \beta_8 = 0$ using the reduced model $E(y) = \beta_0 + \beta_1 x_1 + \beta_2 x_1^2 + \beta_3 x_2 + \beta_4 x_2^2$

k. Test $H_0: \beta_5 = 0$

8.53a. Yes; $F = 33.86$ is significant at $\alpha = .0001$ **b.** $R^2 = .9338$; 93.38% of total variability is explained by the model
c. Yes   **d.** Yes   **e.** No; $t = .04$ **f.** $\hat{y} = -4,026.64 + 385.08x_1 - 24.67x_1^2 + 661.58x_2 - 37.17x_2^2 + .25x_1x_2$
8.54a. $\hat{y} = 7.46 + .0455x_1 + 5.44x_2 - .0148x_1x_2$ **b.** $\hat{y} = 12.9 + .0318x_1$ **c.** $\hat{y} = 7.46 + .0466x_1$
f. $t = -1.47$; do not reject H_0
8.55a. Yes; $F = 40.78$ is significant at $\alpha = .0001$ **b.** FOIL and SHIFT*OPERATOR
c. $E(y) = \beta_0 + \beta_1x_1 + \beta_2x_2 + \beta_3x_3 + \beta_4x_4 + \beta_5x_1x_2 + \beta_6x_1x_3 + \beta_7x_1x_4 + \beta_8x_2x_3 + \beta_9x_2x_4 + \beta_{10}x_3x_4 + \beta_{11}x_1x_2x_3 + \beta_{12}x_1x_2x_4 + \beta_{13}x_1x_3x_4 + \beta_{14}x_2x_3x_4 + \beta_{15}x_1x_2x_3x_4$ **d.** 16
8.56a. SST $= 7.015$, MST $= 2.338$, SSE $= 1.883$, MSE $= .094$, $F = 24.83$ **b.** Yes; reject H_0; p-value $= .0001$
8.57a. SST $= 3.74$, SSB $= 29.62$, SSE $= .885$; MSE $= .1475$, F(Treatments) $= 8.45$, F(Blocks) $= 100.41$
b. Yes; reject H_0 at $\alpha = .05$; p-value $= .0142$ **c.** $-1.3 \pm .77$
8.58a. SS(Time, T) $= 62.762$, SS(Material, M) $= 78.03$, SS(TM) $= 1.755$, SSE $= 3.89$, MSE $= .6483$, F(Time) $= 48.40$, F(Material) $= 120.35$, $F(TM) = 1.35$ **b.** No, do not reject H_0 at $\alpha = .05$; p-value $= .3272$
8.63 8 treatments: A_1B_1, A_1B_2, A_1B_3, A_1B_4, A_2B_1, A_2B_2, A_2B_3, A_2B_4
8.64 df(Error) $= 30$, SSE $= 136.8$, SS(Total) $= 337.5$, MSE $= 6.69$, $F = 2.27$
8.65 df(Blocks) $= 7$, SST $= 7.0$, SSB $= 21.0$, MST $= 3.50$, F(Treatments) $= 1.53$, F(Blocks) $= 1.31$
8.66 df(B) $= 3$, df(Error) $= 16$, SS(A) $= 354.5$, SS(AB) $= 13.5$, SSE $= 112$, MS(A) $= 354.5$, MS(B) $= 40.0$, $F(A) = 50.64$, $F(B) = 5.71$, $F(AB) = .64$

8.67a. Completely randomized   **b.**

SOURCE	df	SS	MS	F
Plans	3	154.112	51.371	10.21
Error	13	65.417	5.032	
Total	16	219.529		

c. Yes; $F = 10.21$ **d.** -4.58 ± 3.70

8.68a. Completely randomized   **b.** $F = 6.301$; yes   **c.** 25 ± 2.343
d. $E(y) = \beta_0 + \beta_1x_1 + \beta_2x_2 + \beta_3x_3$, where $x_1 = \begin{cases} 1 & \text{if B} \\ 0 & \text{if not} \end{cases}$, $x_2 = \begin{cases} 1 & \text{if C} \\ 0 & \text{if not} \end{cases}$, $x_3 = \begin{cases} 1 & \text{if D} \\ 0 & \text{if not} \end{cases}$

8.69a. Randomized block   **b.**

SOURCE	df	SS	MS	F
Displays	2	784.083	392.042	2.99
Periods	7	4,602.292	657.470	5.02
Error	14	1,832.583	130.899	
Total	23	7,218.958		

c. No; $F = 2.99$ **d.** Yes; $F = 5.02$ **f.** $\omega = 14.97$; none of the treatment means appear to differ
8.70a. Randomized block   **b.** $F = 2.77$; do not reject H_0
c. $E(y) = \beta_0 + \beta_1x_1 + \beta_2x_2 + \beta_3x_3 + \beta_4x_4 + \beta_5x_5 + \beta_6x_6 + \beta_7x_7 + \beta_8x_8 + \beta_9x_9$,
where $x_1 = \begin{cases} 1 & \text{if 13 weeks} \\ 0 & \text{if not} \end{cases}$, $x_2 = \begin{cases} 1 & \text{if 26 weeks} \\ 0 & \text{if not} \end{cases}$
and $x_3, x_4, \ldots, x_9$ are dummy variables for blocks (i.e., date of buy signal)
d. Test H_0: $\beta_1 = \beta_2 = 0$ using the reduced model $E(y) = \beta_0 + \beta_3x_3 + \cdots + \beta_9x_9$ **e.** -10.65 ± 9.79

8.71a. 3×3 factorial   **b.**

SOURCE	df	SS	MS	F
T	2	8,200.389	4,100.195	856.52
E	2	4,376.722	2,188.361	457.14
TE	4	103.278	25.820	5.39
Error	27	129.250	4.787	
Total	35	12,809.639		

c. Yes; $F = 5.39$ **d.** 50 ± 1.86 **e.** -22.25 ± 2.63
f. $\omega = 5.213$; (E_2T_2, E_1T_3) and (E_2T_1, E_1T_2) are the only pairs of treatment means that do not appear to differ

8.72a. Completely randomized   **b.**

SOURCE	df	SS	MS	F
Groups	2	273.068	136.534	79.22
Error	342	589.418	1.723	
Total	344	862.486		

c. Yes; $F = 79.22$ **d.** $-1.6 \pm .423$ **e.** $-.53 \pm .463$; yes
f. Fit the model $E(y) = \beta_0 + \beta_1 x_1 + \beta_2 x_2$,

where $x_1 = \begin{cases} 1 & \text{if no training} \\ 0 & \text{if not} \end{cases}$, $x_2 = \begin{cases} 1 & \text{if computer-assisted training} \\ 0 & \text{if not} \end{cases}$,

and test $H_0: \beta_1 = \beta_2 = 0$
8.73a. Randomized block   **b.** $F = 83.82$; yes   **c.** $-1.64 \pm .4877$
e. Fit the model $E(y) = \beta_0 + \beta_1 x_1 + \beta_2 x_2 + \beta_3 x_3 + \beta_4 x_4 + \beta_5 x_5 + \beta_6 x_6$,

where $x_1 = \begin{cases} 1 & \text{if II} \\ 0 & \text{if not} \end{cases}$, $x_2 = \begin{cases} 1 & \text{if III} \\ 0 & \text{if not} \end{cases}$,

and x_3, x_4, x_5, and x_6 are dummy variables for blocks (shipments), and test $H_0: \beta_1 = \beta_2 = 0$

8.74a. 3×4 factorial   **b.**

SOURCE	df	SS	MS	F
Regions	3	21.986489	7.32883	120.744
Salespersons	2	.1463167	.0731584	1.2053
Region $\times$ Salesperson	6	1.744061	.290677	4.7889
Error	24	1.45673	.060697	
Total	35	25.3336		

d. Yes; $F = 4.7889$
e. Fit the model $E(y) = \beta_0 + \beta_1 x_1 + \beta_2 x_2 + \beta_3 x_3 + \beta_4 x_4 + \beta_5 x_5 + \beta_6 x_1 x_3 + \beta_7 x_1 x_4 + \beta_8 x_1 x_5 + \beta_9 x_2 x_3 + \beta_{10} x_2 x_4 + \beta_{11} x_2 x_5$
where

$x_1 = \begin{cases} 1 & \text{if } S_2 \\ 0 & \text{if not} \end{cases}$, $x_2 = \begin{cases} 1 & \text{if } S_3 \\ 0 & \text{if not} \end{cases}$, $x_3 = \begin{cases} 1 & \text{if } R_2 \\ 0 & \text{if not} \end{cases}$, $x_4 = \begin{cases} 1 & \text{if } R_3 \\ 0 & \text{if not} \end{cases}$, $x_5 = \begin{cases} 1 & \text{if } R_4 \\ 0 & \text{if not} \end{cases}$,

and test $H_0: \beta_6 = \beta_7 = \beta_8 = 0$
f. $\omega = .555$; S_1: only the means for the treatment pair (R_3, R_2) do not appear to differ; S_2: all pairs of treatment means appear to differ; S_3: only the means for the treatment pair (R_4, R_2) do not appear to differ

8.75a.

SOURCE	df	SS	MS	F
Buyers (B)	1	227.8126	227.8126	.52
Sellers (S)	1	4,018.6126	4,018.6126	9.11
(BS)	1	1,757.8122	1,757.8122	3.99
Error	76	33,516.0000	441.0000	
Total	79	39,520.2374		

c. Reject $H_0: F = 3.99$

d. Complete model: $E(y) = \beta_0 + \beta_1 x_1 + \beta_2 x_2 + \beta_3 x_1 x_2$, where
$x_1 = \begin{cases} 1 & \text{if BC} \\ 0 & \text{if BNC} \end{cases}$, $x_2 = \begin{cases} 1 & \text{if SC} \\ 0 & \text{if SNC} \end{cases}$;
Reduced model: $E(y) = \beta_0 + \beta_1 x_1 + \beta_2 x_2$ **e.** -12.75 ± 10.92
8.76a. $t = .4635$ **b.** $t = .46$; reject if $|t| > 2.145$ or $t^2 > (2.145)^2 = 4.6$ **c.** $F = .2148 = (.4635)^2$; reject if $F > 4.6$

8.77a.

SOURCE	df	SS	MS	F
Rate, x_1	3	95.62	31.873	94.79
Temperature, x_2	3	38.83	12.943	38.49
$x_1 x_2$	9	15.51	1.723	5.124
Error	16	5.38	.336	
Total	31	155.34		

b. $E(y) = \beta_0 + \beta_1 x_1 + \beta_2 x_2 + \beta_3 x_1^2 + \beta_4 x_2^2 + \beta_5 x_1 x_2 + \beta_6 x_1^2 x_2 + \beta_7 x_1 x_2^2 + \beta_8 x_1^3 + \beta_9 x_2^3 + \beta_{10} x_1 x_2^3 + \beta_{11} x_1^3 x_2 + \beta_{12} x_1^2 x_2^2 + \beta_{13} x_1^3 x_2^3 + \beta_{14} x_1^3 x_2^2 + \beta_{15} x_1^3 x_2^3$ **c.** Yes; $F = 29.7318$ **d.** Yes; $F = 5.125$ **e.** $R^2 = .96537$

8.78a. $E(y) = \beta_0 + \beta_1 x_1 + \beta_2 x_2 + \beta_3 x_1 x_2 + \beta_4 x_1^2 + \beta_5 x_2^2$

b. $\hat{y} = 29.85625 + .56 x_1 - .1625 x_2 - .1135 x_1 x_2 - .275 x_1^2 - .23125 x_2^2$ **c.** The two models are different

d. $R^2 = .842068$ **e.** Yes; $F = 5.70$ **8.79a.** Yes; $F = 41.5$ **b.** Yes; $t = 7.74$ **c.** $-.62$

8.80a. $F = 3.168$; no   **b.** 473.3 ± 22.93 **c.** 37.05 ± 30.76

CHAPTER 9

9.1b.

	I	II	III	IV
1982	—	—	252.5	285.3
1983	327.7	343.5	365.2	383.0
1984	395.1	412.8	406.9	396.5
1985	400.7	413.5	428.3	436.1
1986	446.3	445.4	439.4	—

d. 86.25   **e.** 388.13 (using $M_{21} = 450$)

9.2a.

	I	II	III	IV
1982	165.4	187.1	206.6	222.5
1983	237.3	278.6	292.3	308.5
1984	320.3	354.7	367.4	363.8
1985	356.3	386.9	403.3	404.4
1986	395.0	426.0	433.9	424.0

b. 416.1

c.

QUARTER	FORECAST
I	366.39
II	522.83
III	471.62
IV	396.28

9.3b.

1973	1974	1975	1976	1977	1978	1979	1980	1981	1982	1983	1984	1985	1986
—	67.10	67.17	66.70	66.27	64.67	62.43	57.30	50.87	42.03	36.57	32.07	32.83	—

c.

1973	1974	1975	1976	1977	1978	1979	1980	1981	1982	1983	1984	1985	1986
67.80	67.83	67.16	67.41	67.14	66.29	65.36	63.69	59.25	54.65	48.24	43.52	39.58	38.39

d. Approx. 30.00   **e.** 37.55   **f.** 20.82

9.4a.

	JAN.	FEB.	MAR.	APR.	MAY	JUNE
1984	—	—	—	—	—	—
1985	125.33	127.04	127.28	127.23	127.65	129.14
1986	142.69	142.78	143.79	144.68	144.17	143.13

	JULY	AUG.	SEPT.	OCT.	NOV.	DEC.
1984	116.10	117.95	119.93	121.02	122.08	123.83
1985	131.83	133.09	134.48	136.33	138.81	140.79 ; forecast ≈ 138
1986	140.18	—	—	—	—	—

b.

	JAN.	FEB.	MAR.	APR.	MAY	JUNE
1984	114.10	112.15	113.08	113.34	110.47	118.03
1985	129.45	131.73	129.36	127.88	128.19	125.90
1986	147.96	149.33	149.21	152.71	152.50	149.45

	JULY	AUG.	SEPT.	OCT.	NOV.	DEC.
1984	109.37	116.53	120.42	122.51	122.10	122.60
1985	128.60	127.55	125.62	127.66	133.63	144.52 ; forecast = 122.17
1986	140.93	139.76	137.08	130.29	128.70	124.35

c. Forecast = 117.41

d.

METHOD	FORECAST ERROR
Moving average	−5.00
Exponential smoothing	10.83
Holt–Winters	15.59

9.5b.

1970	1971	1972	1973	1974	1975	1976	1977	1978
—	—	128.78	137.76	147.60	158.84	171.26	185.20	202.32

1979	1980	1981	1982	1983	1984	1985	1986	1987
222.70	244.22	264.82	283.56	298.64	309.84	318.12	—	— ; forecast ≈ 348

c.

1970	1971	1972	1973	1974	1975	1976	1977	1978
116.3	118.3	121.1	125.98	134.67	145.28	155.37	165.82	177.65

1979	1980	1981	1982	1983	1984	1985	1986	1987
193.55	214.85	237.87	258.36	274.38	289.07	302.32	312.75	319.85 ; forecast = 324.11

d. Forecast = 350.01

9.6a.

	I	II	III	IV
1975	—	—	90.15	95.01
1976	97.28	102.62	104.94	103.85
1977	102.90	100.72	97.63	95.33
1978	94.09	95.60	95.85	98.94
1979	100.79	102.48	105.44	106.38
1980	109.09	113.54	119.42	126.81
1981	130.65	128.55	125.48	121.01
1982	116.20	116.94	121.86	130.35
1983	144.17	155.91	162.24	164.27
1984	161.29	160.65	161.43	166.80
1985	176.46	180.46	191.47	204.39
1986	217.75	231.80	241.13	—

c.

QUARTER	FORECAST
I	253
II	259
III	265
IV	271

d.

	I	II	III	IV
1975	83.36	86.91	86.00	89.66
1976	93.59	96.80	99.33	101.77
1977	100.76	100.68	99.43	98.13
1978	95.46	95.48	97.60	97.15
1979	98.48	99.81	102.66	104.25
1980	104.58	107.32	113.27	118.72
1981	123.59	125.23	123.29	122.04
1982	120.56	117.34	118.65	124.69
1983	132.61	142.36	150.25	154.41
1984	155.84	155.04	158.36	161.02
1985	166.92	174.40	176.70	187.07
1986	200.65	214.05	221.31	229.50

e.

QUARTER	FORECAST
I	235.23
II	235.23
III	235.23
IV	235.23

f.

QUARTER	FORECAST
I	280.08
II	284.21
III	277.71
IV	295.17

9.7a.

1970	1971	1972	1973	1974	1975	1976	1977	1978
—	45.42	65.89	105.37	139.64	148.63	144.83	155.53	216.53

1979	1980	1981	1982	1983	1984	1985	1986	1987
369.10	454.81	476.94	424.61	394.50	375.54	348.49	364.69	—

b. Forecast = 355

c.

1970	1971	1972	1973	1974	1975	1976	1977	1978
36.41	40.28	54.94	89.24	145.61	158.24	131.49	144.94	183.79

1979	1980	1981	1982	1983	1984	1985	1986	1987
283.00	541.41	468.79	393.10	437.84	375.80	329.00	360.10	399.15

d. Forecast = 406.96   **e.** Forecast = 408.48

9.8b. $E(y_t) = \beta_0 + \beta_1 t$ **c.** $\hat{y}_t = 74.2 + 1.9165t$ **d.** $r^2 = .975$ **e.** 102.95 ± 3.380

9.9b. $\hat{y}_t = 39.4879 + 19.13032t - 1.31529t^2$ **d.** $(-31.2506, 48.9693)$

9.10a. $E(y_t) = \beta_0 + \beta_1 t + \beta_2 Q_1 + \beta_3 Q_2 + \beta_4 Q_3$, where Q_1, Q_2, and Q_3 are dummy variables for quarters

b. $\hat{y} = 66.057 + 2.861t - 2.202Q_1 - .859Q_2 - 2.570Q_3$

c.

QUARTER	FORECAST	95% PREDICTION INTERVAL
I	204.1	(158.1, 250.0)
II	208.3	(162.3, 254.2)
II	209.4	(163.5, 255.4)
IV	214.8	(168.9, 260.8)

9.11b. Railroads: $\hat{y}_t = 74.527 - 8.239t$; buses: $\hat{y}_t = 35.260 - 2.442t$; air carriers: $\hat{y}_t = -13.173 + 10.915t$

d. Railroads: -16.11, $(-40.82, 8.60)$; buses: 8.40, $(-8.69, 25.49)$; air carriers: 106.90, $(90.78, 123.00)$

9.12a. Yes   **b.** $\hat{y}_t = 6.609 + 1.704t$ **c.** $r^2 = .9873$; $t = 30.60$ (reject H_0: $\beta_1 = 0$)

d.

YEAR	FORECAST	95% PREDICTION INTERVAL
1985	32.17	(30.70, 34.27)
1986	33.88	(31.72, 36.03)
1987	35.58	(33.36, 37.80)
1988	37.28	(35.00, 39.57)
1989	38.99	(36.64, 41.34)

9.13a. $\hat{y}_t = 113.456 + .860t$ **b.** Yes   **c.** $t = 5.50$; reject H_0 **d.** 145.3 ± 22.8

9.14a. $E(y_t) = \beta_0 + \beta_1 t + \beta_2 M_1 + \beta_3 M_2 + \cdots + \beta_{12} M_{11}$, where $M_1, M_2, \ldots, M_{11}$ are dummy variables for months.

b. $\hat{y} = 108.758 + 1.003t + 12.134M_1 + 8.831M_2 + 6.228M_3 + 7.258M_4 + 3.655M_5 - 1.615M_6 - 3.018M_7 + .779M_8 - 2.357M_9 - 4.894M_{10} - 2.364M_{11}$ **c.** $F = .64$; do not reject H_0 **d.** $(129.4, 186.6)$

9.15a. No; $t = -1.39$ $(-t_{.05} = -1.645)$ **b.** $2,003.48$ **c.** No; $t = -1.61$ $(-t_{.05} = -1.645)$ **d.** $1,901.81$

9.16a. $\phi = .9$: $.9$, $.81$, $.729$, $.6561$, $.5905$, $.5314$, $.4783$, $.4305$, $.3874$, $.3487$;

$\phi = -.9$: $-.9$, $.81$, $-.729$, $.6561$, $-.5905$, $.5314$, $-.4783$, $.4305$, $-.3874$, $.3487$;

$\phi = .2$: $.2$, $.04$, $.008$, $.0016$, $.00032$, $.000064$, $.0000128$, $.0000026$, $.0000005$, $.0000001$;

$\phi = -.2$: $-.2$, $.04$, $-.008$, $.0016$, $-.00032$, $.000064$, $-.0000128$, $.0000026$, $-.0000005$, $.0000001$

9.17a. $0, 0, 0, .5, 0, 0, 0, .25, 0, 0, 0, .125, 0, 0, 0, .0625, 0, 0, 0, .03125$

b. Autocorrelations for first-order model: $.5, .25, .125, .0625, .03125, .0156, .0078, .0039, .0019, .0010, .0005, .0002, .0001, 0, 0,$ $0, 0, 0, 0, 0$ **9.18** $R_t = (-.6)R_{t-1} + \varepsilon_t$ **9.19** $R_t = \phi_1 R_{t-1} + \phi_2 R_{t-2} + \phi_3 R_{t-3} + \phi_4 R_{t-4} + \varepsilon_t$

9.20a. $E(y_t) = \beta_0 + \beta_1 t$ **b.** $E(y_t) = \beta_0 + \beta_1 t + \beta_2 t^2$ **c.** $R_t = \phi R_{t-1} + \varepsilon_t$

9.21a. $E(y_t) = \beta_0 + \beta_1 x_{1t} + \beta_2 x_{2t} + \beta_3 x_{3t} + \beta_4 t$

b. $E(y_t) = \beta_0 + \beta_1 x_{1t} + \beta_2 x_{2t} + \beta_3 x_{3t} + \beta_4 t + \beta_5 x_{1t} t + \beta_6 x_{2t} t + \beta_7 x_{3t} t$ **c.** $R_t = \phi R_{t-1} + \varepsilon_t$

9.22a. $E(y_t) = \beta_0 + \beta_1 S_1 + \beta_2 S_2 + \beta_3 S_3$, where $S_1 = \begin{cases} 1 & \text{if fall} \\ 0 & \text{otherwise} \end{cases}$, $S_2 = \begin{cases} 1 & \text{if winter} \\ 0 & \text{otherwise} \end{cases}$, $S_3 = \begin{cases} 1 & \text{if spring} \\ 0 & \text{otherwise} \end{cases}$

b. $R_t = \phi_1 R_{t-1} + \phi_2 R_{t-2} + \phi_3 R_{t-3} + \phi_4 R_{t-4} + \varepsilon_t$

c. $y_t = \beta_0 + \beta_1 S_1 + \beta_2 S_2 + \beta_3 S_3 + \phi_1 R_{t-1} + \phi_2 R_{t-2} + \phi_3 R_{t-3} + \phi_4 R_{t-4} + \varepsilon_t$

d. $y_t = \beta_0 + \beta_1 S_1 + \beta_2 S_2 + \beta_3 S_3 + \beta_4 t + \beta_5 S_1 t + \beta_6 S_2 t + \beta_7 S_3 t + \phi_1 R_{t-1} + \phi_2 R_{t-2} + \phi_3 R_{t-3} + \phi_4 R_{t-4} + \varepsilon_t$

9.23 $E(y_t) = \beta_0 + \beta_1\left(\cos\frac{2\pi}{365}t\right) + \beta_2\left(\sin\frac{2\pi}{365}t\right)$

c. $E(y_t) = \beta_0 + \beta_1\left(\cos\frac{2\pi}{365}t\right) + \beta_2\left(\sin\frac{2\pi}{365}t\right) + \beta_3 t + \beta_4 t\left(\cos\frac{2\pi}{365}t\right) + \beta_5\left(\sin\frac{2\pi}{365}t\right)$ **d.** No; $R_t = \phi R_{t-1} + \varepsilon_t$

9.24a. $E(y_t) = \beta_0 + \beta_1 t + \beta_2 t^2 + \beta_3 S_1 + \beta_4 S_2 + \beta_5 S_3 + \beta_6 S_4$,

where $S_1 = \begin{cases} 1 & \text{if January} \\ 0 & \text{otherwise} \end{cases}$, $S_2 = \begin{cases} 1 & \text{if February} \\ 0 & \text{otherwise} \end{cases}$, $S_3 = \begin{cases} 1 & \text{if March} \\ 0 & \text{otherwise} \end{cases}$, $S_4 = \begin{cases} 1 & \text{if April} \\ 0 & \text{otherwise} \end{cases}$

b. Include interaction between time t and S_1, S_2, S_3, S_4: $E(y_t) = \beta_0 + \beta_1 t + \beta_2 t^2 + \beta_3 S_1 + \beta_4 S_2 + \beta_5 S_3 + \beta_6 S_4 + \beta_7 tS_1 + \beta_8 tS_2 + \beta_9 tS_3 + \beta_{10} tS_4 + \beta_{11} t^2 S_1 + \beta_{12} t^2 S_2 + \beta_{13} t^2 S_3 + \beta_{14} t^2 S_4$

9.25a. $y_t = \beta_0 + \beta_1 t + \phi R_{t-1} + \varepsilon_t$ **b.** $\hat{y}_t = 1,327.29 + 61.03t + .825\hat{R}_{t-1}$ **d.** $.9976$

9.26a. $y_t = \beta_0 + \beta_1 t + \phi R_{t-1} + \varepsilon_t$ **b.** $\hat{y}_t = 74.05 + 1.9358t + .2344\hat{R}_{t-1}$

9.27a. $y_t = \beta_0 + \beta_1 t + \phi R_{t-1} + \varepsilon_t$ **b.** $\hat{y}_t = 116.083 + .636t + .7166\hat{R}_{t-1}$

9.29a. $\hat{y}_t = 820.22 + 8.257t + .1964\hat{R}_{t-1}$ **b.** $.1964$

9.30a. $\hat{y}_{31} = 85.58$; $\hat{y}_{32} = 88.77$; $\hat{y}_{33} = 91.71$ **b.** y_{31}: 85.58 ± 4.15; y_{32}: 88.77 ± 4.92; y_{33}: 91.71 ± 5.21

9.31a. $\hat{y}_{49} = 336.91$; $\hat{y}_{50} = 323.41$; $\hat{y}_{51} = 309.46$ **b.** y_{49}: 336.91 ± 6.48; y_{50}: 323.41 ± 8.34; y_{51}: 309.46 ± 9.36

9.32a. y_{45}: $4,135.6 \pm 78.9$; y_{46}: $4,185.8 \pm 102.3$; y_{47}: $4,237.8 \pm 115.6$; y_{48}: $4,291.5 \pm 123.8$; yes   **9.33** 103.29 ± 2.73

9.34 124.71 ± 13.41 **9.35a.** $2,136.2$ **b.** $(1,404.3, 3,249.7)$ **c.** $1,944.0$; $(1,301.8, 2,902.9)$

9.36

YEAR	FORECAST	95% PREDICTION INTERVAL
1984	1,047.3	$1,047.3 \pm 214.5$
1985	1,012.4	$1,012.4 \pm 218.6$
1986	1,012.2	$1,012.2 \pm 218.8$

9.37a. $R_t = \phi R_{t-1} + \varepsilon_t$ **b.** $y_t = \beta_0 + \beta_1 t + \beta_2 t^2 + \phi R_{t-1} + \varepsilon_t$ **c.** $\hat{y}_t = 1381.79 + 205.69t + 14.32t^2 + .504\hat{R}_{t-1}$

d.

YEAR	FORECAST	95% PREDICTION INTERVAL
1986	8,437.0	$8,437.0 \pm 323.8$
1987	9,067.1	$9,067.1 \pm 362.5$

9.38a. 9.00   **b.** 11.87   **c.** 13.73   **d.** 9.35 ± 4.56 **e.** 9.38 ± 3.18

9.39a.

	JAN.	FEB.	MAR.	APR.	MAY	JUNE
1982	—	—	—	—	—	—
1983	92.84	93.54	94.42	95.24	95.99	96.83
1984	103.83	104.36	105.24	105.71	106.44	107.16
1985	110.48	111.24	112.08	112.93	113.48	113.91
1986	116.95	117.41	117.68	118.58	119.19	119.41

	JULY	AUG.	SEPT.	OCT.	NOV.	DEC.
1982	89.63	89.96	90.19	90.79	91.29	91.87
1983	97.83	98.87	99.99	100.86	101.73	102.88
1984	107.69	108.18	108.33	108.81	109.53	110.24
1985	114.51	115.10	115.46	115.82	116.05	116.48
1986	120.46	—	—	—	—	—

b. 109.55 (using $M_{61} = 123$)

c.

	JAN.	FEB.	MAR.	APR.	MAY	JUNE
1982	77.34	76.66	82.61	85.82	88.81	88.91
1983	90.78	83.64	89.71	92.26	95.61	98.61
1984	102.01	96.88	101.27	103.15	108.22	110.49
1985	108.39	100.56	106.16	110.27	116.23	115.37
1986	115.40	105.95	110.90	113.78	120.75	120.54

	JULY	AUG.	SEPT.	OCT.	NOV.	DEC.
1982	90.29	89.90	88.86	90.39	92.67	104.93
1983	99.18	99.81	98.71	99.71	102.48	116.08
1984	107.80	109.60	106.00	107.86	110.94	123.28
1985	115.27	118.59	115.72	115.77	117.17	130.09
1986	120.64	122.72	123.85	123.40	121.84	139.52

d. 146.59   **e.** 106.54

f. $y_t = \beta_0 + \beta_1 t + \beta_2 M_1 + \beta_3 M_2 + \cdots + \beta_{12} M_{11} + \phi R_{t-1} + \varepsilon_t$, where $M_1, M_2, \ldots, M_{11}$ are dummy variables for months

g. $\hat{y}_t = 108.50 + .654t - 33.57M_1 - 36.74M_2 - 24.42M_3 - 23.81M_4 - 18.30M_5 - 20.77M_6 - 22.25M_7 - 20.32M_8 - 24.46M_9 - 22.78M_{10} - 21.29M_{11} + .229\hat{R}_{t-1}$ **h.** 115.89 ± 4.86

9.40b. Seasonal trend is more obvious for Phoenix data.

c. $y_t = \beta_0 + \beta_1 t + \beta_2\left(\cos\frac{2\pi}{12}t\right) + \beta_3\left(\sin\frac{2\pi}{12}t\right) + \phi R_{t-1} + \varepsilon_t$

d. Atlanta: $\hat{y}_t = 63.604 + .094t - 4.210\left(\cos\frac{2\pi}{12}t\right) + .460\left(\sin\frac{2\pi}{12}t\right) + .0214\hat{R}_{t-1}$;

Phoenix: $\hat{y}_t = 63.220 + .115t + 4.977\left(\cos\frac{2\pi}{12}t\right) + 12.433\left(\sin\frac{2\pi}{12}t\right) + .1630\hat{R}_{t-1}$ **e.** Yes

f. No; long-term forecast will not be very reliable

9.41b. $y_t = \beta_0 + \beta_1 t + \beta_2 Q_1 + \beta_3 Q_2 + \beta_4 Q_3 + \phi R_{t-1} + \varepsilon_t$, where Q_1, Q_2, and Q_3 are dummy variables for quarters; $\hat{y}_t = -173.933 + 1{,}623.684t - 859.249Q_1 - 141.867Q_2 + 826.733Q_3 + .1702\hat{R}_{t-1}$ **c.** 20,429.19; 20,429.19 ± 2,915.69

9.42b.

	3-POINT	5-POINT	7-POINT		3-POINT	5-POINT	7-POINT
1966	—	—	—	1976	5,405.7	5,473.2	5,658.6
1967	6,818.3	—	—	1977	5,937.3	6,029.0	6,210.6
1968	6,491.7	6,354.4	—	1978	6,637.0	6,784.0	6,973.1
1969	5,948.7	5,911.4	5,869.7	1979	7,539.0	7,715.6	7,998.4
1970	5,352.7	5,432.4	5,565.3	1980	8,695.3	8,937.2	9,461.9
1971	4,885.7	5,091.6	5,232.4	1981	10,252.7	10,748.2	11,099.9
1972	4,713.7	4,824.4	5,001.7	1982	12,544.3	12,754.2	—
1973	4,680.0	4,739.0	4,908.0	1983	15,034.7	—	—
1974	4,793.0	4,854.8	4,999.7	1984	—	—	—
1975	4,958.0	5,136.4	5,252.3				

9.43a. $\hat{y}_t = 150.08 + 4.165t$ **b.** 208.39
c. Residuals: $-5.65, -2.21, -1.98, -3.04, -.41, 4.03, 4.96, 10.70, 6.13, 4.47, .10, -7.07, -10.03$; yes
d. Durbin–Watson test; reject H_0, $d = .426$ **e.** $y_t = \beta_0 + \beta_1 t + \phi R_{t-1} + \varepsilon_t$; $\hat{y}_t = 149.73 + 3.986t + .627\hat{R}_{t-1}$
f. 200.92 ± 8.75

9.44b.

	JAN.	FEB.	MAR.	APR.	MAY	JUNE
1984	—	—	—	—	—	—
1985	11.25	12.05	12.83	13.57	14.18	14.60
1986	25.38	25.98	26.23	26.68	26.83	26.95

	JULY	AUG.	SEPT.	OCT.	NOV.	DEC.
1984	9.63	9.74	9.84	9.84	10.21	10.55
1985	15.07	16.15	17.26	19.44	22.38	24.53
1986	27.11	—	—	—	—	—

c.

MONTH	FORECAST
Jan.	28.08
Feb.	28.16
Mar.	37.44

d.

MONTH	FORECAST
Jan.	6.49
Feb.	.91
Mar.	-4.46

e. $y_t = \beta_0 + \beta_1 t + \beta_2 M_1 + \beta_3 M_2 + \cdots + \beta_{12} M_{11} + \phi R_{t-1} + \varepsilon_t$, where $M_1, M_2, \ldots, M_{11}$ are dummy variables for months; $\hat{y}_t = .053 + .592t + 6.099M_1 + 6.402M_2 + 12.150M_3 + 16.172M_4 + 12.086M_5 + 6.407M_6 + 5.449M_7 + 3.222M_8 + 2.837M_9 + 2.473M_{10} + .410M_{11} + .719\hat{R}_{t-1}$ **f.** 33.06 ± 12.24
9.45b. $\hat{y}_t = 3{,}817.91 - 3.514t - 331.864 \cos\left(\frac{2\pi}{12}t\right) - 531.403 \sin\left(\frac{2\pi}{12}t\right) - .215\hat{R}_{t-1}$ **c.** 3,106.38 ± 428.50
9.46b. $y_t = \beta_0 + \beta_1 t + \phi R_{t-1} + \varepsilon_t$ **d.** $\hat{y}_t = 1{,}278.01 - 39.166t + .722\hat{R}_{t-1}$ **e.** 542.26 ± 880.90

CHAPTER 10

10.1a. $E(y) = \beta_0 + \beta_1 x_1 + \beta_2 (x_1 - 15)x_2$, where $x_1 = x$ and $x_2 = \begin{cases} 1 & \text{if } x_1 > 15 \\ 0 & \text{if not} \end{cases}$

b.

	y-INTERCEPT	SLOPE
$x \le 15$	β_0	β_1
$x > 15$	$\beta_0 - 15\beta_2$	$\beta_1 + \beta_2$

c. Test H_0: $\beta_2 = 0$

10.2a. $E(y) = \beta_0 + \beta_1 x_1 + \beta_2(x_1 - 1.45)x_2 + \beta_3(x_1 - 5.20)x_3$,
where $x_1 = x$, $x_2 = \begin{cases} 1 & \text{if } x_1 > 1.45 \\ 0 & \text{if not} \end{cases}$, and $x_3 = \begin{cases} 1 & \text{if } x_1 > 5.20 \\ 1 & \text{if not} \end{cases}$

b.

	y-INTERCEPT	SLOPE
$x \le 1.45$	β_0	β_1
$1.45 < x \le 5.20$	$\beta_0 - 1.45\beta_2$	$\beta_1 + \beta_2$
$x > 5.20$	$\beta_0 - 1.45\beta_2 - 5.20\beta_3$	$\beta_1 + \beta_2 + \beta_3$

c. Test H_0: $\beta_2 = \beta_3 = 0$

10.3a. $E(y) = \beta_0 + \beta_1 x_1 + \beta_2(x_1 - 320)x_2 + \beta_3 x_2$, where $x_1 = x$ and $x_2 = \begin{cases} 1 & \text{if } x_1 > 320 \\ 0 & \text{if not} \end{cases}$

b.

	y-INTERCEPT	SLOPE
$x \le 320$	β_0	β_1
$x > 320$	$\beta_0 - 320\beta_2 + \beta_3$	$\beta_1 + \beta_2$

c. Test H_0: $\beta_2 = 0$

10.4b. $E(y) = \beta_0 + \beta_1 x_1 + \beta_2(x_1 - 1970)x_2$, where $x_1 = x$ and $x_2 = \begin{cases} 1 & \text{if } x_1 > 1970 \\ 0 & \text{if not} \end{cases}$
c. $\hat{y} = -180{,}276.25 + 93.93x_1 + 125.83(x_1 - 1970)x_2$ **d.** Yes; $F = 81.33$ (p-value $= .0001$)
e. 1950–1970: 93.93; 1971–1986: 219.76   **f.** Yes; $t = 2.63$
10.5a. $E(y) = \beta_0 + \beta_1 x_1 + \beta_2(x_1 - 1{,}000)x_2$, where $x_1 = x$ and $x_2 = \begin{cases} 1 & \text{if } x_1 > 1{,}000 \\ 0 & \text{if not} \end{cases}$
b. $\hat{y} = 4.024 - .0021x_1 - .00139(x_1 - 1{,}000)x_2$ **c.** Yes; $F = 363.44$ (p-value $= .0001$)   **d.** $-.0021 \pm .000366$
10.6 34.92 ± 85.05 [Note: The approximation may not be appropriate since $D = 4.40$ is relatively large.]
10.7a. 473.3 ± 540.2 **b.** Model is inadequate   **10.8** $1.61 \pm .895$
10.9a. $\hat{y} = -2.03 + 6.06x$ **b.** Reject H_0; $t = 10.35$ **c.** $1.985 \pm .687$
10.10a. $\dfrac{1}{x_i^2}$ **b.** $\dfrac{1}{\sqrt{x_i}}$ **c.** $\dfrac{1}{x_i}$ **d.** n_i **e.** x_i
10.11a. $\hat{y} = -1.2667 + .176x$; yes, $t = 4.09$ (p-value $= .0013$)
b. Residuals: $-1.33, 6.67, -5.33, -2.33, 1.67, -6.13, 3.87, -5.13, 9.87, -1.13, -7.93, 14.07, -6.93, 4.07, -3.93$;
appears to be violated

c.

x	VARIANCE OF RESIDUALS
100	20.7
150	44.3
200	85.2

; $w_i = \dfrac{1}{x_i^2}$ **d.** $\hat{y} = -1.5143 + .1777x$

10.14 $\hat{y} = .02664272 + .000011594x$; no, $t = 1.88$
10.15b. $\hat{y} = -.5279 + .0750x_1 + .0747x_2 + .3912x_3$ **c.** Reject H_0; $F = 21.79$ **d.** Yes; $t = 4.01$ **e.** $(-.2122, .2047)$
10.16 $\hat{y} = \dfrac{\exp(-2.01884 + .00005468x)}{1 + \exp(-2.01884 + .00005468x)}$
10.17 $\hat{y} = \dfrac{\exp(-14.24825 + 1.15488x_1 + .909848x_2 + 5.603676x_3)}{1 + \exp(-14.24825 + 1.15488x_1 + .909848x_2 + 5.603676x_3)}$

APPENDIX A

A.1a. $\begin{bmatrix} 6 & 3 \\ -2 & -5 \end{bmatrix}$ **b.** $\begin{bmatrix} 3 & 0 & 9 \\ -9 & 4 & 5 \end{bmatrix}$ **c.** $\begin{bmatrix} 5 & 4 \\ 1 & -4 \end{bmatrix}$ **A.2a.** $\begin{bmatrix} 15 \\ 14 \\ -8 \end{bmatrix}$ **b.** $[7]$
c. No; the number of elements in a row of A does not match the number of elements in a column of B.
A.3a. 3×4 **b.** No; the number of elements in a row of B does not match the number of elements in a column of A.
A.4a. 1×1 **b.** 3×3 **c.** $\begin{bmatrix} 3 & 0 & 6 \\ 0 & 0 & 0 \\ 2 & 0 & 4 \end{bmatrix}$ **A.5a.** $\begin{bmatrix} 2 & 3 \\ -9 & 0 \\ 8 & -2 \end{bmatrix}$ **b.** $[3 \quad 0 \quad 4]$ **c.** $[14 \quad 7]$

A.6a. [12]   **b.** $\begin{bmatrix} 6 & 0 & -2 & 4 \\ -3 & 0 & 1 & -2 \\ 0 & 0 & 0 & 0 \\ 9 & 0 & -3 & 6 \end{bmatrix}$ **A.7a.** $\begin{bmatrix} 1 & 0 \\ 0 & 1 \end{bmatrix}$ **c.** $\begin{bmatrix} 1 & 0 & 0 \\ 0 & 1 & 0 \\ 0 & 0 & 1 \end{bmatrix}$

A.12a. $A = \begin{bmatrix} 3 & 1 \\ 1 & -1 \end{bmatrix}; V = \begin{bmatrix} v_1 \\ v_2 \end{bmatrix}; G = \begin{bmatrix} 5 \\ 3 \end{bmatrix}$ **c.** $\begin{bmatrix} 2 \\ -1 \end{bmatrix}$

A.13a. $A = \begin{bmatrix} 10 & 0 & 20 \\ 0 & 20 & 0 \\ 20 & 0 & 68 \end{bmatrix}; V = \begin{bmatrix} v_1 \\ v_2 \\ v_3 \end{bmatrix}; G = \begin{bmatrix} 60 \\ 60 \\ 176 \end{bmatrix}$ **c.** $\begin{bmatrix} 2 \\ 3 \\ 2 \end{bmatrix}$

A.14a. $Y = \begin{bmatrix} 4 \\ .8 \\ 3 \\ 1 \\ -1 \end{bmatrix}; X = \begin{bmatrix} 1 & -2 \\ 1 & -1 \\ 1 & 0 \\ 1 & 1 \\ 1 & 2 \end{bmatrix}$ **b.** $X'X = \begin{bmatrix} 5 & 0 \\ 0 & 10 \end{bmatrix}; X'Y = \begin{bmatrix} 10 \\ -12 \end{bmatrix}$ **c.** $\hat{\beta} = \begin{bmatrix} 2 \\ -1.2 \end{bmatrix}$ **d.** $\hat{y} = 2 - 1.2x$

A.15a. $Y = \begin{bmatrix} 1 \\ 2 \\ 2 \\ 3 \\ 5 \\ 5 \end{bmatrix}; X = \begin{bmatrix} 1 & 1 \\ 1 & 2 \\ 1 & 3 \\ 1 & 4 \\ 1 & 5 \\ 1 & 6 \end{bmatrix}$ **b.** $X'X = \begin{bmatrix} 6 & 21 \\ 21 & 91 \end{bmatrix}; X'Y = \begin{bmatrix} 18 \\ 78 \end{bmatrix}$ **d.** $\hat{\beta} = \begin{bmatrix} 0 \\ .8571 \end{bmatrix}$ **e.** $\hat{y} = .8571x$

A.16a. Y is 10×1; X is 10×3 **d.** $\hat{\beta} = \begin{bmatrix} 2.789 \\ .715 \\ -.039 \end{bmatrix}; \hat{y} = 2.789 + .715x - .039x^2$

A.17 $t = -5.196$; reject H_0 **A.18** $(-1.7434, -.6566)$ **A.19** $t = -2.222$; reject H_0 **A.20** $(-.1412, 1.7412)$
A.21 $(-1.1593, 2.7593)$ **A.22** $(1.1057, 2.3227)$ **A.23** $(.4153, 3.0131)$ **A.24** $(3.3071, 3.6229)$
A.25 $(3.2276, 3.7024)$ **A.26a.** 1.21   **b.** $(8.5967, 10.0943)$ **c.** $(6.9444, 11.7466)$ **d.** .966051

A.27a. $X = \begin{bmatrix} 1 & -5 \\ 1 & -3 \\ 1 & -1 \\ 1 & 1 \\ 1 & 3 \\ 1 & 5 \end{bmatrix}; Y = \begin{bmatrix} 1.1 \\ 1.9 \\ 3.0 \\ 3.8 \\ 5.1 \\ 6.0 \end{bmatrix}$ **b.** $X'X = \begin{bmatrix} 6 & 0 \\ 0 & 70 \end{bmatrix}; X'Y = \begin{bmatrix} 20.9 \\ 34.9 \end{bmatrix}$ **c.** $\hat{\beta} = \begin{bmatrix} 3.4833 \\ .4986 \end{bmatrix}$ **d.** $\hat{y} = 3.4833 + .4986x$

e. SSE = .0682; s^2 = .0170   **f.** t = 31.9477; yes   **g.** r^2 = .99609   **h.** $(3.61776, 3.84747)$

A.28a. $X = \begin{bmatrix} 1 & -2 & 2 \\ 1 & -2 & 2 \\ 1 & -1 & -1 \\ 1 & -1 & -1 \\ 1 & 0 & -2 \\ 1 & 0 & -2 \\ 1 & 1 & -1 \\ 1 & 1 & -1 \\ 1 & 2 & 2 \\ 1 & 2 & 2 \end{bmatrix}; Y = \begin{bmatrix} 5.2 \\ 5.0 \\ .3 \\ -.1 \\ -1.2 \\ -1.1 \\ 2.2 \\ 2.0 \\ 6.2 \\ 6.1 \end{bmatrix}$ **b.** $\hat{y} = 2.46 + .41x_1 + 1.6143x_2$ **c.** SSE = 2.43629; s^2 = .34804

d. $F = 109.7$; yes   **e.** $R^2 = .96907$ **f.** $t = 3.108$; reject H_0 **g.** $(4.1246, 5.6125)$ **h.** $(3.5257, 6.2114)$

A.29a. $X = \begin{bmatrix} 1 & 1 \\ 1 & 1 \\ 1 & 2 \\ 1 & 2 \\ 1 & 3 \\ 1 & 3 \\ 1 & 4 \\ 1 & 4 \\ 1 & 5 \\ 1 & 5 \\ 1 & 6 \\ 1 & 6 \end{bmatrix}$; $Y = \begin{bmatrix} 1.1 \\ .5 \\ 1.8 \\ 2.0 \\ 2.0 \\ 2.9 \\ 3.8 \\ 3.4 \\ 4.1 \\ 5.0 \\ 5.0 \\ 5.8 \end{bmatrix}$ **b.** $X'X = \begin{bmatrix} 12 & 42 \\ 42 & 182 \end{bmatrix}$, $X'Y = \begin{bmatrix} 37.4 \\ 163.0 \end{bmatrix}$

c. The elements of $X'X$ are increased by a factor of 2.   **d.** $(X'X)^{-1} = \begin{bmatrix} .43333 & -.1 \\ -.1 & .02857 \end{bmatrix}$

e. $\hat{\beta} = \begin{bmatrix} -.09333 \\ .91714 \end{bmatrix}$; $\hat{y} = -.09333 + .91714x$ **f.** $SSE = 1.5568$; $s^2 = .15568$ **g.** $t = 13.7515$; yes   **h.** $r^2 = .94977$

A.30a. $(3.79463, 4.27297)$ **b.** $(3.27990, 4.78770)$

A.31a. 18×1 **b.** $X = \begin{bmatrix} 1 & 1 \\ 1 & 1 \\ 1 & 1 \\ 1 & 2 \\ 1 & 2 \\ 1 & 2 \\ 1 & 3 \\ 1 & 3 \\ 1 & 3 \\ 1 & 4 \\ 1 & 4 \\ 1 & 4 \\ 1 & 5 \\ 1 & 5 \\ 1 & 5 \\ 1 & 6 \\ 1 & 6 \\ 1 & 6 \end{bmatrix}$ **c.** $\begin{bmatrix} 18 & 63 \\ 63 & 273 \end{bmatrix}$ **d.** $(X'X)^{-1} = \begin{bmatrix} .28889 & -.06667 \\ -.06667 & .019048 \end{bmatrix}$ **e.** $a'(X'X)^{-1}a = .0746$

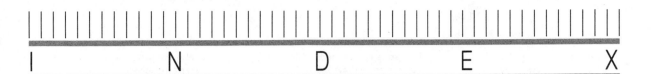

INDEX

Text designer: Janet Bollow
Cover designer: John Williams
Technical artists: Art by AYXA and Folium
Proofreader: Ellen Z. Curtin
Editor and production coordinator: Susan Reiland
Typesetter: Typeset in 10/12 Berkeley Old Style
              by Jonathan Peck Typographers, Ltd.